Basic Heat-Transfer Relations

Fourier's law of heat conduction:

$$q_x = -kA\frac{\partial T}{\partial x}$$

Characteristic thermal resistance for conduction $= \Delta x/kA$

Characteristic thermal resistance for convection $= 1/hA$

Overall heat transfer $= \Delta T_{\text{overall}}/\Sigma R_{\text{thermal}}$

Convection heat transfer from a surface:

$$q = hA(T_{\text{surface}} - T_{\text{free stream}}) \qquad \text{for exterior flows}$$

$$q = hA(T_{\text{surface}} - T_{\text{fluid bulk}}) \qquad \text{for flow in channels}$$

Forced convection: $\text{Nu} = f(\text{Re}, \text{Pr})$ (Chapters 5 and 6, Tables 5-2 and 6-8)

Free convection: $\text{Nu} = f(\text{Gr}, \text{Pr})$ (Chapter 7, Table 7-5)

$$\text{Re} = \frac{\rho u x}{\mu} \qquad \text{Gr} = \frac{\rho^2 g \beta \, \Delta T x^3}{\mu^2} \qquad \text{Pr} = \frac{c_p \mu}{k}$$

$x =$ characteristic dimension

General procedure for analysis of convection problems: Section 7-14, Figure 7-15, Inside back cover.

Radiation heat transfer (Chapter 8)

Blackbody emissive power, $\dfrac{\text{energy emitted by blackbody}}{\text{area} \cdot \text{time}} = \sigma T^4$

$$\text{Radiosity} = \frac{\text{energy leaving surface}}{\text{area} \cdot \text{time}}$$

$$\text{Irradiation} = \frac{\text{energy incident on surface}}{\text{area} \cdot \text{time}}$$

Radiation shape factor $F_{mn} =$ fraction of energy leaving surface m and arriving at surface n

Reciprocity relation: $A_m F_{mn} = A_n F_{nm}$

Radiation heat transfer from surface with area A_1, emissivity ϵ_1, and temperature $T_1(K)$ to large enclosure at temperature $T_2(K)$:

$$q = \sigma A_1 \epsilon_1 (T_1^4 - T_2^4)$$

LMTD method for heat exchangers (Section 10-5):

$$q = UAF \, \Delta T_m$$

where $F =$ factor for specific heat exchanger; $\Delta T_m =$ LMTD for counterflow double-pipe heat exchanger with same inlet and exit temperatures

Effectiveness-NTU method for heat exchangers (Section 10-6, Table 10-3):

$$\epsilon = \frac{\text{Temperaure difference for fluid with minimum value of } mc}{\text{Largest temperature difference in heat exchanger}}$$

$$\text{NTU} = \frac{UA}{C_{\min}} \qquad \epsilon = f(\text{NTU}, C_{\min}/C_{\max})$$

See List of Symbols on page xvii for definitions of terms.

Heat Transfer

McGraw-Hill Series in Mechanical Engineering

Anderson
Computational Fluid Dynamics

Anderson
Modern Compressible Flow:
With Historical Perspective

Barber
Intermediate Mechanics of Materials

Baruh
Analytical Dynamics

Beer and Johnston
Vector Mechanics for Engineers:
Statics and Dynamics

Beer, Johnston and DeWolf
Mechanics of Materials

Borman and Ragland
Combustion Engineering

Budynas
Advanced Strength and Applied Stress

Çengel and Boles
Thermodynamics: An Engineering
Approach

Çengel and Turner
Fundamentals of Thermal-Fluid Sciences

Çengel
Heat Transfer: A Practical Approach

Çengel
Introduction to Thermodynamics and
Heat Transfer

Chapra and Canale
Numerical Methods for Engineers

Condoor
Mechanical Design Modeling
with ProEngineer

Courtney
Mechanical Behavior of Materials

Dieter
Engineering Design: A Materials and
Processing Approach

Doebelin
Measurement Systems: Application
and Design

Hamrock
Fundamentals of Machine Elements

Mattingly
Elements of Gas Turbine Propulsion

Meirovitch
Fundamentals of Vibrations

Modest
Radiative Heat Transfer

Norton
Design of Machinery

Oosthuizen and Carscallen
Compressible Fluid Flow

Oosthuizen and Naylor
Introduction to Convective Heat
Transfer Analysis

Palm
Introduction to MATLAB 6 for Engineers

Palm
MATLAB for Engineering Applications

Reddy
Introduction to Finite Element Method

Ribando
Heat Transfer Tools

Rizzoni
Principles and Applications for
Electrical Engineering

Schey
Introduction to Manufacturing Processes

Schlichting
Boundary Layer Theory

SDRC, Inc.
I-DEAS Student Edition

SDRC, Inc.
I-DEAS Student Guide

Shames
Mechanics of Fluids

Shigley and Mischke
Mechanical Engineering Design

Stoecker
Design of Thermal Systems

Turns
An Introduction to Combustion:
Concepts and Applications

Heywood
Internal Combustion Engine Fundamentals

Histand and Alciatore
Introduction to Mechatronics and
Measurement Systems

Hsu
MEMS and Microsystems: Design
and Manufacturing

Holman
Experimental Methods for Engineers

Kays and Crawford
Convective Heat and Mass Transfer

Kelly
Fundamentals of Mechanical Vibrations

Kreider, Rabl and Curtiss
Heating and Cooling of Buildings

Ullman
The Mechanical Design Process

Ugural
Stresses in Plates and Shells

Vu and Esfandiari
Dynamic Systems: Modeling and Analysis

Wark
Advanced Thermodynamics for Engineers

Wark and Richards
Thermodynamics

White
Fluid Mechanics

White
Viscous Fluid Flow

Zeid
CAD/CAM Theory and Practice

Heat Transfer

Tenth Edition

J. P. Holman
Department of Mechanical Engineering
Southern Methodist University

Boston   Burr Ridge, IL   Dubuque, IA   New York   San Francisco   St. Louis
Bangkok   Bogotá   Caracas   Kuala Lumpur   Lisbon   London   Madrid   Mexico City
Milan   Montreal   New Delhi   Santiago   Seoul   Singapore   Sydney   Taipei   Toronto

 **Higher Education**

HEAT TRANSFER, TENTH EDITION

Published by McGraw-Hill, a business unit of The McGraw-Hill Companies, Inc., 1221
Avenue of the Americas, New York, NY 10020.

2 3 4 5 6 7 8 9 0 VNH/VNH 0 9

ISBN 978–0–07–352936–3
MHID 0–07–352936–2

Global Publisher: *Raghothaman Srinivasan*
Senior Sponsoring Editor: *Bill Stenquist*
Director of Development: *Kristine Tibbetts*
Developmental Editor: *Lora Neyens*
Senior Marketing Manager: *Curt Reynolds*
Senior Project Manager: *Kay J. Brimeyer*
Lead Production Supervisor: *Sandy Ludovissy*
Senior Media Project Manager: *Tammy Juran*
Associate Design Coordinator: *Brenda A. Rolwes*
Cover Designer: *Studio Montage, St. Louis, Missouri*
Cover Image: *Interferometer photo of air flow across a heated cylinder, digitally enhanced by the author.*
Compositor: *S4Carlisle Publishing Services*
Typeface: *10.5/12 Times Roman*
Printer: *R. R. Donnelley, Jefferson City, MO*

Library of Congress Cataloging-in-Publication Data

Holman, J. P. (Jack Philip)
    Heat transfer / Jack P. Holman.—10th ed.
      p. cm.—(Mcgraw-Hill series in mechanical engineering)
    Includes index.
    ISBN 978–0–07–352936–3—ISBN 0–07–352936–2 (hard copy : alk. paper)
    1. Heat-Transmission.   I. Title.

QC320.H64 2010
621.402′2—dc22
                                          2008033196

www.mhhe.com

CONTENTS

GUIDE TO WORKED EXAMPLES

PREFACE

This book presents an elementary treatment of the principles of heat transfer. As a text it contains more than enough material for a one-semester course that may be presented at the junior level, or higher, depending on individual course objectives. The course is normally required in chemical and mechanical engineering curricula but is recommended for electrical engineering students as well, because of the significance of cooling problems in various electronics applications. In the author's experience, electrical engineering students do quite well in a heat-transfer course, even with no formal coursework background in thermodynamics or fluid mechanics. A background in ordinary differential equations is helpful for proper understanding of the material.

Presentation of the subject follows classical lines of separate discussions for conduction, convection, and radiation, although it is emphasized that the physical mechanism of convection heat transfer is one of conduction through the stationary fluid layer near the heat-transfer surface. Throughout the book emphasis has been placed on physical understanding while, at the same time, relying on meaningful experimental data in those circumstances that do not permit a simple analytical solution.

Conduction is treated from both the analytical and the numerical viewpoint, so that the reader is afforded the insight that is gained from analytical solutions as well as the important tools of numerical analysis that must often be used in practice. A liberal number of numerical examples are given that include heat sources and radiation boundary conditions, non-uniform mesh size, and one example of a three-dimensional nodal system. A similar procedure is followed in the presentation of convection heat transfer. An integral analysis of both free- and forced-convection boundary layers is used to present a physical picture of the convection process. From this physical description, inferences may be drawn that naturally lead to the presentation of empirical and practical relations for calculating convection heat-transfer coefficients. Because it provides an easier instruction vehicle than other methods, the radiation-network method is used extensively in the introduction of analysis of radiation systems, while a more generalized formulation is given later. Systems of nonlinear equations requiring iterative solutions are also discussed in the conduction and radiation chapters but the details of solution are relegated to cited software references. The assumption is made that the well-disposed reader should select his or her own preferred vehicle for solution of systems of nonlinear equations.

The log-mean-temperature-difference and effectiveness approaches are presented in heat-exchanger analysis since both are in wide use and each offers its own advantages to the designer. A brief introduction to diffusion and mass transfer is presented in order to acquaint the reader with these processes and to establish more firmly the important analogies between heat, mass, and momentum transfer. A new Chapter 12 has been added on summary and design information. Numerous calculation charts are offered in this chapter as an aid in preliminary design work where speed and utility may be more important than the accuracy that may be required in final design stages. Eleven new examples are presented in this chapter illustrating use of the charts.

Problems are included at the end of each chapter. Some of these problems are of a routine nature to familiarize the student with the numerical manipulations and orders of magnitude of various parameters that occur in the subject of heat transfer. Other problems extend the subject matter by requiring students to apply the basic principles to new situations and develop their own equations. Both types of problems are important.

There is also a section at the end of each problem set designated as "Design-Oriented Problems." The problems in these sections typically are open-ended and do not result in a unique answer. In some cases they are rather extended in length and require judgment decisions during the solution process. Over 100 such problems are included in the text.

The subject of heat transfer is not static. New developments occur quite regularly, and better analytical solutions and empirical data are continuously made available to the professional in the field. Because of the huge amount of information that is available in the research literature, the beginning student could easily be overwhelmed if too many of the nuances of the subject were displayed and expanded. The book is designed to serve as an elementary text, so the author has assumed a role of interpreter of the literature with those findings and equations being presented that can be of immediate utility to the reader. It is hoped that the student's attention is called to more extensive works in a sufficient number of instances to emphasize the depth that is available on most of the subjects of heat transfer. For the serious student, then, the end-of-chapter references offer an open door to the literature of heat transfer that can pyramid upon further investigation. In several chapters the number of references offered is much larger than necessary, and older citations of historical interest have been retained freely. The author feels this is a luxury that will not be intrusive on the reader or detract from the utility of the text.

A book in its tenth edition obviously reflects many compromises and evolutionary processes over the years. While the basic physical mechanisms of heat transfer have not changed, analytical techniques and experimental data have been revised and improved. In this edition some trimming of out-of-date material has been effected, new problems added, and old problems refreshed. Sixteen new worked examples have been added. All worked examples are now referenced by page number at the front of the book, just following the Table of Contents. The listing of such examples is still retained at the end of each chapter.

A feature is the use of Microsoft Excel for solution of both steady-state and transient conduction heat-transfer problems. Excel is given a rather full discussion in a new Appendix D, which includes treatment of heat source and radiation boundary conditions, steady-state and transient conditions, and interfaces between composite materials. A special template is provided that automatically writes nodal equations for most common boundary conditions. Ten examples of the use of Excel for solution of problems are provided, including some modifications and expansions of examples that appear in Chapters 3 and 4. One example illustrates the progression of transient solution to yield the steady-state solution for sufficiently long-time duration.

In addition to the summary tables of convection formulas provided at the conclusion of each of the main convection chapters (Chapters 5, 6, 7), an overall procedure is now offered for analysis of all convection problems, and is included in the inside book cover as well as in the body of the text. While one might interpret this as a cookbook approach, the true intent is to help heat-transfer practitioners avoid common and disarmingly simple pitfalls in the analysis and solution of convection problems.

The SI (metric) system of units is the primary one for the text. Because the Btu-ft-pound system is still in wide use, answers and intermediate steps to examples are occasionally stated in these units. A few examples and problems are in English units.

It is not possible to cover all the topics in this book in either a quarter- or semester-term course, but it is hoped that the variety of topics and problems will provide the necessary flexibility for many applications.

ACKNOWLEDGMENTS

With a book at this stage of revision, the list of persons who have been generous with their comments and suggestions has grown very long indeed. The author hopes that a blanket note of thanks for all these individuals contributions will suffice. As in the past, all comments from users will be appreciated and acknowledged. The author and McGraw-Hill editorial staff would like to acknowledge the following people for their helpful comments and suggestions while developing the plan for the new edition:

Neil L. Book, *University of Missouri–Rolla*

Rodney D.W. Bowersox, *Texas A & M University*

Kyle V. Camarda, *University of Kansas*

Richard Davis, *University of Minnesota–Duluth*

Roy W. Knight, *Auburn University*

Frank A. Kulacki, *University of Minnesota*

Ian H. Leslie, *New Mexico State University*

Daniela S. Mainardi, *Louisiana Tech University*

Randall D. Manteufel, *University of Texas at San Antonio*

M. Pinar Menguc, *University of Kentucky*

Samuel Paolucci, *University of Notre Dame*

Paul D. Ronney, *University of Southern California*

Harris Wong, *Louisiana State University*

J. P. Holman

ACKNOWLEDGMENTS

ABOUT THE AUTHOR

J. P. Holman received the Ph.D. in mechanical engineering from Oklahoma State University. After two years as a research scientist at the Wright Aerospace Research Laboratory, he joined the faculty of Southern Methodist University, where he is presently Professor Emeritus of Mechanical Engineering. He has also held administrative positions as Director of the Thermal and Fluid Sciences Center, Head of the Civil and Mechanical Engineering Department, and Assistant Provost for Instructional Media. During his tenure at SMU he has been voted the outstanding faculty member by the student body 13 times.

Dr. Holman has published over 30 papers in several areas of heat transfer and his three widely used textbooks, *Heat Transfer* (9th edition, 2002), *Experimental Methods for Engineers* (7th edition, 2001), and *Thermodynamics* (4th edition, 1988), all published by McGraw-Hill, have been translated into Spanish, Portuguese, Japanese, Chinese, Korean, and Indonesian, and are distributed worldwide. He is also the author of the utilitarian monograph *What Every Engineer Should Know About EXCEL* (2006), published by CRC Press. Dr. Holman also consults for industry in the fields of heat transfer and energy systems.

A member of ASEE, he is past Chairman of the National Mechanical Engineering Division and past chairman of the Region X Mechanical Engineering Department Heads. Dr. Holman is a Fellow of ASME and recipient of several national awards: the George Westinghouse Award from ASEE for distinguished contributions to engineering education (1972), the James Harry Potter Gold Medal from ASME for contributions to thermodynamics (1986), the Worcester Reed Warner Gold Medal from ASME for outstanding contributions to the permanent literature of engineering (1987), and the Ralph Coats Roe Award from ASEE as the outstanding mechanical engineering educator of the year (1995). In 1993 he was the recipient of the Lohmann Medal from Oklahoma State University, awarded annually to a distinguished engineering alumnus of that institution.

LIST OF SYMBOLS

a Local velocity of sound

a Attenuation coefficient (Chap. 8)

A Area

A Albedo (Chap. 8)

A_m Fin profile area (Chap. 2)

c Specific heat, usually kJ/kg·°C

C Concentration (Chap. 11)

C_D Drag coefficient, defined by Eq. (6-13)

C_f Friction coefficient, defined by Eq. (5-52)

c_p Specific heat at constant pressure, usually kJ/kg·°C

c_v Specific heat at constant volume, usually kJ/kg·°C

d Diameter

D Depth or diameter

D Diffusion coefficient (Chap. 11)

D_H Hydraulic diameter, defined by Eq. (6-14)

e Internal energy per unit mass, usually kJ/kg

E Internal energy, usually kJ

E Emissive power, usually W/m^2 (Chap. 8)

E_{b0} Solar constant (Chap. 8)

$E_{b\lambda}$ Blackbody emissive power per unit wave-length, defined by Eq. (8-12)

f Friction factor

F Force, usually N

F_{m-n} or F_{ij} Radiation shape factor for radiation from surface i to surface j

g Acceleration of gravity

g_c Conversion factor, defined by Eq. (1-14)

G Irradiation (Chap. 8)

$G = \dfrac{\dot{m}}{A}$ Mass velocity

h Heat-transfer coefficient, usually W/m^2·°C

$\bar{h}$ Average heat-transfer coefficient

h_{fg} Enthalpy of vaporization, kJ/kg

h_r Radiation heat-transfer coefficient (Chap. 8)

i Enthalpy, usually kJ/kg

I Intensity of radiation

I Solar insolation (Chap. 8)

I_0 Solar insolation at outer edge of atmosphere

J Radiosity (Chap. 8)

k Thermal conductivity, usually W/m·°C

k_e Effective thermal conductivity of enclosed spaces (Chap. 7)

k_λ Scattering coefficient (Chap. 8)

K Mass-transfer coefficient, m/h

L	Length		x, y, z	Space coordinates in cartesian system
L_c	Corrected fin length (Chap. 2)		$\alpha = \dfrac{k}{pc}$	Thermal diffusivity, usually m²/s
m	Mass		α	Absorptivity (Chap. 8)
$\dot{m}$	Mass rate of flow		α	Accommodation coefficient (Chap. 7)
M	Molecular weight (Chap. 11)		α	Solar altitude angle, deg (Chap. 8)
n	Molecular density		β	Volume coefficient of expansion, 1/K
N	Molal diffusion rate, moles per unit time (Chap. 11)		β	Temperature coefficient of thermal conductivity, 1/°C
p	Pressure, usually N/m², Pa		$\gamma = \dfrac{c_p}{c_v}$	Isentropic exponent, dimensionless
P	Perimeter		Γ	Condensate mass flow per unit depth of plate (Chap. 9)
q	Heat-transfer rate, kJ per unit time		δ	Hydrodynamic-boundary-layer thickness
q''	Heat flux, kJ per unit time per unit area		δ_t	Thermal-boundary-layer thickness
$\dot{q}$	Heat generated per unit volume		ϵ	Heat-exchanger effectiveness
Q	Heat, kJ		ϵ	Emissivity
r	Radius or radial distance		ϵ_H, ϵ_M	Eddy diffusivity of heat and momentum (Chap. 5)
r	Recovery factor, defined by Eq. (5-120)		$\zeta = \dfrac{\delta_t}{\delta}$	Ratio of thermal-boundary-layer thickness to hydrodynamic-boundary-layer thickness
R	Fixed radius			
R	Gas constant			
R_{th}	Thermal resistance, usually °C/W			
s	A characteristic dimension (Chap. 4)		η	Similarity variable, defined by Eq. (B-6)
S	Conduction shape factor, usually m		η_f	Fin efficiency, dimensionless
t	Thickness, applied to fin problems (Chap. 2)		θ	Angle in spherical or cylindrical coordinate system
t, T	Temperature		θ	Temperature difference, $T - T_{\text{reference}}$
u	Velocity			
v	Velocity			The reference temperature is chosen differently for different systems (see Chaps. 2 to 4)
v	Specific volume, usually m³/kg			
V	Velocity			
V	Molecular volume (Chap. 11)			
W	Weight, usually N			

λ Wavelength (Chap. 8)

λ Mean free path (Chap. 7)

μ Dynamic viscosity

ν Kinematic viscosity, m^2/s

ν Frequency of radiation (Chap. 8)

ρ Density, usually kg/m^3

ρ Reflectivity (Chap. 8)

σ Stefan-Boltzmann constant

σ Surface tension of liquid-vapor interface (Chap. 9)

τ Time

τ Shear stress between fluid layers

τ Transmissivity (Chap. 8)

ϕ Angle in spherical or cylindrical coordinate system

ψ Stream function

Dimensionless Groups

$Bi = \dfrac{hs}{k}$ Biot number

$Fo = \dfrac{\alpha\tau}{s^2}$ Fourier number

$Gr = \dfrac{g\beta(T_w - T_\infty)x^3}{\nu^2}$ Grashof number

$Gr^* = Gr\,Nu$ Modified Grashof number for constant heat flux

$Gz = Re\,Pr\,\dfrac{d}{L}$ Graetz number

$Kn = \dfrac{\lambda}{L}$ Knudsen number

$Le = \dfrac{\alpha}{D}$ Lewis number (Chap. 11)

$M = \dfrac{u}{a}$ Mach number

$Nu = \dfrac{hx}{k}$ Local Nusselt number

$Nu = \dfrac{\overline{h}x}{k}$ Average Nusselt number

$Pe = Re\,Pr$ Peclet number

$Pr = \dfrac{c_p\mu}{k}$ Prandtl number

$Ra = Gr\,Pr$ Rayleigh number

$Re = \dfrac{\rho u x}{\mu}$ Reynolds number

$Sc = \dfrac{\nu}{D}$ Schmidt number (Chap. 11)

$Sh = \dfrac{Kx}{D}$ Sherwood number (Chap. 11)

$St = \dfrac{h}{\rho c_p u}$ Stanton number

$\overline{St} = \dfrac{\overline{h}}{\rho c_p u}$ Average Stanton number

Subscripts

aw Adiabatic wall conditions

b Refers to blackbody conditions (Chap. 8)

xxii List of Symbols

b	Evaluated at bulk conditions	s	Evaluated at condition of surroundings
d	Based on diameter	w	Evaluated at wall conditions
f	Evaluated at film conditions	x	Denotes some local position with respect to x coordinate
g	Saturated vapor conditions (Chap. 9)	0	Denotes stagnation flow conditions (Chap. 5) or some initial condition at time zero
i	Initial or inlet conditions		
L	Based on length of plate	$*$	(Superscript) Properties evaluated at reference temperature, given by Eq. (5-124)
m	Mean flow conditions		
m, n	Denotes nodal positions in numerical solution (see Chap. 3, 4)	∞	Evaluation at free-stream conditions
r	At specified radial position		

Introduction

Heat transfer is the science that seeks to predict the energy transfer that may take place between material bodies as a result of a temperature difference. Thermodynamics teaches that this energy transfer is defined as heat. The science of heat transfer seeks not merely to explain how heat energy may be transferred, but also to predict the rate at which the exchange will take place under certain specified conditions. The fact that a heat-transfer *rate* is the desired objective of an analysis points out the difference between heat transfer and thermodynamics. Thermodynamics deals with systems in equilibrium; it may be used to predict the amount of energy required to change a system from one equilibrium state to another; it may not be used to predict how fast a change will take place since the system is not in equilibrium during the process. Heat transfer supplements the first and second principles of thermodynamics by providing additional experimental rules that may be used to establish energy-transfer rates. As in the science of thermodynamics, the experimental rules used as a basis of the subject of heat transfer are rather simple and easily expanded to encompass a variety of practical situations.

As an example of the different kinds of problems that are treated by thermodynamics and heat transfer, consider the cooling of a hot steel bar that is placed in a pail of water. Thermodynamics may be used to predict the final equilibrium temperature of the steel bar–water combination. Thermodynamics will not tell us how long it takes to reach this equilibrium condition or what the temperature of the bar will be after a certain length of time before the equilibrium condition is attained. Heat transfer may be used to predict the temperature of both the bar and the water as a function of time.

Most readers will be familiar with the terms used to denote the three modes of heat transfer: conduction, convection, and radiation. In this chapter we seek to explain the mechanism of these modes qualitatively so that each may be considered in its proper perspective. Subsequent chapters treat the three types of heat transfer in detail.

1-1 | CONDUCTION HEAT TRANSFER

When a temperature gradient exists in a body, experience has shown that there is an energy transfer from the high-temperature region to the low-temperature region. We say that the energy is transferred by conduction and that the heat-transfer rate per unit area is proportional to the normal temperature gradient:

$$\frac{q_x}{A} \sim \frac{\partial T}{\partial x}$$

When the proportionality constant is inserted,

$$q_x = -kA \frac{\partial T}{\partial x} \qquad\qquad \textbf{[1-1]}$$

Figure 1-1 | Sketch showing direction of heat flow.

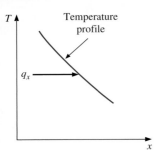

where q_x is the heat-transfer rate and $\partial T/\partial x$ is the temperature gradient in the direction of the heat flow. The positive constant k is called the *thermal conductivity* of the material, and the minus sign is inserted so that the second principle of thermodynamics will be satisfied; i.e., heat must flow downhill on the temperature scale, as indicated in the coordinate system of Figure 1-1. Equation (1-1) is called Fourier's law of heat conduction after the French mathematical physicist Joseph Fourier, who made very significant contributions to the analytical treatment of conduction heat transfer. It is important to note that Equation (1-1) is the defining equation for the thermal conductivity and that k has the units of watts per meter per Celsius degree in a typical system of units in which the heat flow is expressed in watts.

We now set ourselves the problem of determining the basic equation that governs the transfer of heat in a solid, using Equation (1-1) as a starting point.

Consider the one-dimensional system shown in Figure 1-2. If the system is in a steady state, i.e., if the temperature does not change with time, then the problem is a simple one, and we need only integrate Equation (1-1) and substitute the appropriate values to solve for the desired quantity. However, if the temperature of the solid is changing with time, or if there are heat sources or sinks within the solid, the situation is more complex. We consider the general case where the temperature may be changing with time and heat sources may be present within the body. For the element of thickness dx, the following energy balance may be made:

Energy conducted in left face + heat generated within element

= change in internal energy + energy conducted out right face

These energy quantities are given as follows:

$$\text{Energy in left face} = q_x = -kA \frac{\partial T}{\partial x}$$

$$\text{Energy generated within element} = \dot{q}A \, dx$$

Figure 1-2 | Elemental volume for one-dimensional heat-conduction analysis.

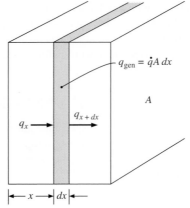

$$\text{Change in internal energy} = \rho c A \frac{\partial T}{\partial \tau} \, dx$$

$$\text{Energy out right face} = q_{x+dx} = -kA \left. \frac{\partial T}{\partial x} \right]_{x+dx}$$

$$= -A \left[k \frac{\partial T}{\partial x} + \frac{\partial}{\partial x} \left(k \frac{\partial T}{\partial x} \right) dx \right]$$

where

$\dot{q}$ = energy generated per unit volume, W/m^3

c = specific heat of material, $J/kg \cdot °C$

ρ = density, kg/m^3

Combining the relations above gives

$$-kA \frac{\partial T}{\partial x} + \dot{q} A \, dx = \rho c A \frac{\partial T}{\partial \tau} \, dx - A \left[k \frac{\partial T}{\partial x} + \frac{\partial}{\partial x} \left(k \frac{\partial T}{\partial x} \right) dx \right]$$

or

$$\frac{\partial}{\partial x} \left(k \frac{\partial T}{\partial x} \right) + \dot{q} = \rho c \frac{\partial T}{\partial \tau} \qquad\qquad \textbf{[1-2]}$$

This is the one-dimensional heat-conduction equation. To treat more than one-dimensional heat flow, we need consider only the heat conducted in and out of a unit volume in all three coordinate directions, as shown in Figure 1-3a. The energy balance yields

$$q_x + q_y + q_z + q_{\text{gen}} = q_{x+dx} + q_{y+dy} + q_{z+dz} + \frac{dE}{d\tau}$$

and the energy quantities are given by

$$q_x = -k \, dy \, dz \, \frac{\partial T}{\partial x}$$

$$q_{x+dx} = -\left[k \frac{\partial T}{\partial x} + \frac{\partial}{\partial x} \left(k \frac{\partial T}{\partial x} \right) dx \right] dy \, dz$$

$$q_y = -k \, dx \, dz \, \frac{\partial T}{\partial y}$$

$$q_{y+dy} = -\left[k \frac{\partial T}{\partial y} + \frac{\partial}{\partial y} \left(k \frac{\partial T}{\partial y} \right) dy \right] dx \, dz$$

$$q_z = -k \, dx \, dy \, \frac{\partial T}{\partial z}$$

$$q_{z+dz} = -\left[k \frac{\partial T}{\partial z} + \frac{\partial}{\partial z} \left(k \frac{\partial T}{\partial z} \right) dz \right] dx \, dy$$

$$q_{\text{gen}} = \dot{q} \, dx \, dy \, dz$$

$$\frac{dE}{d\tau} = \rho c \, dx \, dy \, dz \, \frac{\partial T}{\partial \tau}$$

Figure 1-3 | Elemental volume for three-dimensional heat-conduction analysis:
(*a*) cartesian coordinates; (*b*) cylindrical coordinates; (*c*) spherical coordinates.

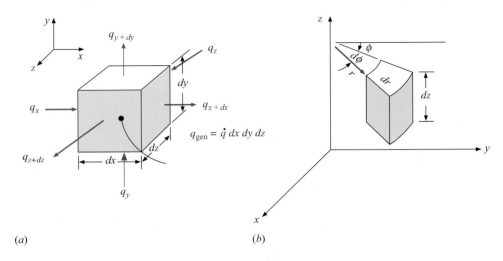

(*a*)   (*b*)

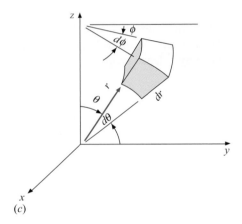

(*c*)

so that the general three-dimensional heat-conduction equation is

$$\frac{\partial}{\partial x}\left(k\frac{\partial T}{\partial x}\right) + \frac{\partial}{\partial y}\left(k\frac{\partial T}{\partial y}\right) + \frac{\partial}{\partial z}\left(k\frac{\partial T}{\partial z}\right) + \dot{q} = \rho c\frac{\partial T}{\partial \tau}$$ **[1-3]**

For constant thermal conductivity, Equation (1-3) is written

$$\frac{\partial^2 T}{\partial x^2} + \frac{\partial^2 T}{\partial y^2} + \frac{\partial^2 T}{\partial z^2} + \frac{\dot{q}}{k} = \frac{1}{\alpha}\frac{\partial T}{\partial \tau}$$ **[1-3a]**

where the quantity $\alpha = k/\rho c$ is called the *thermal diffusivity* of the material. The larger the value of α, the faster heat will diffuse through the material. This may be seen by examining the quantities that make up α. A high value of α could result either from a high value of thermal conductivity, which would indicate a rapid energy-transfer rate, or from a low value of the thermal heat capacity ρc. A low value of the heat capacity would mean that less of the energy moving through the material would be absorbed and used to raise the temperature of the material; thus more energy would be available for further transfer. Thermal diffusivity α has units of square meters per second.

In the derivations above, the expression for the derivative at $x + dx$ has been written in the form of a Taylor-series expansion with only the first two terms of the series employed for the development.

Equation (1-3a) may be transformed into either cylindrical or spherical coordinates by standard calculus techniques. The results are as follows:

Cylindrical coordinates:

$$\frac{\partial^2 T}{\partial r^2} + \frac{1}{r}\frac{\partial T}{\partial r} + \frac{1}{r^2}\frac{\partial^2 T}{\partial \phi^2} + \frac{\partial^2 T}{\partial z^2} + \frac{\dot{q}}{k} = \frac{1}{\alpha}\frac{\partial T}{\partial \tau} \qquad \textbf{[1-3}\boldsymbol{b}\textbf{]}$$

Spherical coordinates:

$$\frac{1}{r}\frac{\partial^2}{\partial r^2}(rT) + \frac{1}{r^2 \sin\theta}\frac{\partial}{\partial \theta}\left(\sin\theta\,\frac{\partial T}{\partial \theta}\right) + \frac{1}{r^2 \sin^2\theta}\frac{\partial^2 T}{\partial \phi^2} + \frac{\dot{q}}{k} = \frac{1}{\alpha}\frac{\partial T}{\partial \tau} \qquad \textbf{[1-3}\boldsymbol{c}\textbf{]}$$

The coordinate systems for use with Equations (1-3b) and (1-3c) are indicated in Figure 1-3b and c, respectively.

Many practical problems involve only special cases of the general equations listed above. As a guide to the developments in future chapters, it is worthwhile to show the reduced form of the general equations for several cases of practical interest.

Steady-state one-dimensional heat flow (no heat generation):

$$\frac{d^2 T}{dx^2} = 0 \qquad \textbf{[1-4]}$$

Note that this equation is the same as Equation (1-1) when $q = $ constant.

Steady-state one-dimensional heat flow in cylindrical coordinates (no heat generation):

$$\frac{d^2 T}{dr^2} + \frac{1}{r}\frac{dT}{dr} = 0 \qquad \textbf{[1-5]}$$

Steady-state one-dimensional heat flow with heat sources:

$$\frac{d^2 T}{dx^2} + \frac{\dot{q}}{k} = 0 \qquad \textbf{[1-6]}$$

Two-dimensional steady-state conduction without heat sources:

$$\frac{\partial^2 T}{\partial x^2} + \frac{\partial^2 T}{\partial y^2} = 0 \qquad \textbf{[1-7]}$$

1-2 | THERMAL CONDUCTIVITY

Equation (1-1) is the defining equation for thermal conductivity. On the basis of this definition, experimental measurements may be made to determine the thermal conductivity of different materials. For gases at moderately low temperatures, analytical treatments in the kinetic theory of gases may be used to predict accurately the experimentally observed values. In some cases, theories are available for the prediction of thermal conductivities in liquids and solids, but in general, many open questions and concepts still need clarification where liquids and solids are concerned.

The mechanism of thermal conduction in a gas is a simple one. We identify the kinetic energy of a molecule with its temperature; thus, in a high-temperature region, the molecules have higher velocities than in some lower-temperature region. The molecules are in continuous random motion, colliding with one another and exchanging energy and momentum. The molecules have this random motion whether or not a temperature gradient exists in the gas. If a molecule moves from a high-temperature region to a region of lower temperature, it transports kinetic energy to the lower-temperature part of the system and gives up this energy through collisions with lower-energy molecules.

Table 1-1 lists typical values of the thermal conductivities for several materials to indicate the relative orders of magnitude to be expected in practice. More complete tabular information is given in Appendix A. In general, the thermal conductivity is strongly temperature-dependent.

Table 1-1 | Thermal conductivity of various materials at 0°C.

Material	Thermal conductivity k	
	W/m · °C	Btu/h · ft · °F
Metals:		
Silver (pure)	410	237
Copper (pure)	385	223
Aluminum (pure)	202	117
Nickel (pure)	93	54
Iron (pure)	73	42
Carbon steel, 1% C	43	25
Lead (pure)	35	20.3
Chrome-nickel steel (18% Cr, 8% Ni)	16.3	9.4
Nonmetallic solids:		
Diamond	2300	1329
Quartz, parallel to axis	41.6	24
Magnesite	4.15	2.4
Marble	2.08−2.94	1.2−1.7
Sandstone	1.83	1.06
Glass, window	0.78	0.45
Maple or oak	0.17	0.096
Hard rubber	0.15	0.087
Polyvinyl chloride	0.09	0.052
Styrofoam	0.033	0.019
Sawdust	0.059	0.034
Glass wool	0.038	0.022
Ice	2.22	1.28
Liquids:		
Mercury	8.21	4.74
Water	0.556	0.327
Ammonia	0.540	0.312
Lubricating oil, SAE 50	0.147	0.085
Freon 12, CCl_2F_2	0.073	0.042
Gases:		
Hydrogen	0.175	0.101
Helium	0.141	0.081
Air	0.024	0.0139
Water vapor (saturated)	0.0206	0.0119
Carbon dioxide	0.0146	0.00844

Figure 1-4 | Thermal conductivities of some typical gases
[1 W/m · °C = 0.5779 Btu/h · ft · °F].

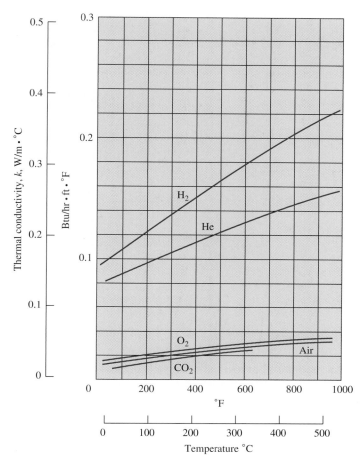

We noted that thermal conductivity has the units of watts per meter per Celsius degree when the heat flow is expressed in watts. Note that a heat *rate* is involved, and the numerical value of the thermal conductivity indicates how fast heat will flow in a given material. How is the rate of energy transfer taken into account in the molecular model discussed above? Clearly, the faster the molecules move, the faster they will transport energy. Therefore the thermal conductivity of a gas should be dependent on temperature. A simplified analytical treatment shows the thermal conductivity of a gas to vary with the square root of the absolute temperature. (It may be recalled that the velocity of sound in a gas varies with the square root of the absolute temperature; this velocity is approximately the mean speed of the molecules.) Thermal conductivities of some typical gases are shown in Figure 1-4. For most gases at moderate pressures the thermal conductivity is a function of temperature alone. This means that the gaseous data for 1 atmosphere (atm), as given in Appendix A, may be used for a rather wide range of pressures. When the pressure of the gas becomes of the order of its critical pressure or, more generally, when nonideal-gas behavior is encountered, other sources must be consulted for thermal-conductivity data.

Figure 1-5 | Thermal conductivities of some typical liquids.

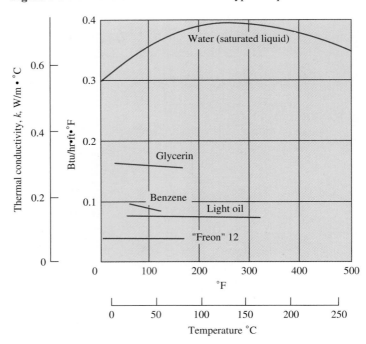

The physical mechanism of thermal-energy conduction in liquids is qualitatively the same as in gases; however, the situation is considerably more complex because the molecules are more closely spaced and molecular force fields exert a strong influence on the energy exchange in the collision process. Thermal conductivities of some typical liquids are shown in Figure 1-5.

In the English system of units, heat flow is expressed in British thermal units per hour (Btu/h), area in square feet, and temperature in degrees Fahrenheit. Thermal conductivity will then have units of Btu/h · ft · °F.

Thermal energy may be conducted in solids by two modes: lattice vibration and transport by free electrons. In good electrical conductors a rather large number of free electrons move about in the lattice structure of the material. Just as these electrons may transport electric charge, they may also carry thermal energy from a high-temperature region to a low-temperature region, as in the case of gases. In fact, these electrons are frequently referred to as the *electron gas*. Energy may also be transmitted as vibrational energy in the lattice structure of the material. In general, however, this latter mode of energy transfer is not as large as the electron transport, and for this reason good electrical conductors are almost always good heat conductors, namely, copper, aluminum, and silver, and electrical insulators are usually good heat insulators. A notable exception is diamond, which is an electrical insulator, but which can have a thermal conductivity five times as high as silver or copper. It is this fact that enables a jeweler to distinguish between genuine diamonds and fake stones. A small instrument is available that measures the response of the stones to a thermal heat pulse. A true diamond will exhibit a far more rapid response than the nongenuine stone.

Thermal conductivities of some typical solids are shown in Figure 1-6. Other data are given in Appendix A.

Figure 1-6 │ Thermal conductivities of some typical solids.

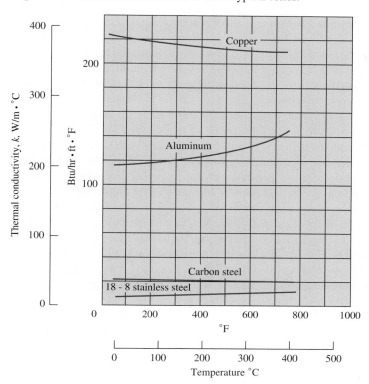

Table 1-2 │ Effective thermal conductivities of cryogenic insulating materials for use in range 15°C to −195°C. Density range 30 to 80 kg/m³.

Type of insulation	Effective k, mW/m · °C
1. Foams, powders, and fibers, unevacuated	7–36
2. Powders, evacuated	0.9–6
3. Glass fibers, evacuated	0.6–3
4. Opacified powders, evacuated	0.3–1
5. Multilayer insulations, evacuated	0.015–0.06

The thermal conductivities of various insulating materials are also given in Appendix A. Some typical values are 0.038 W/m · °C for glass wool and 0.78 W/m · °C for window glass. At high temperatures, the energy transfer through insulating materials may involve several modes: conduction through the fibrous or porous solid material; conduction through the air trapped in the void spaces; and, at sufficiently high temperatures, radiation.

An important technical problem is the storage and transport of cryogenic liquids like liquid hydrogen over extended periods of time. Such applications have led to the development of *superinsulations* for use at these very low temperatures (down to about −250°C). The most effective of these superinsulations consists of multiple layers of highly reflective materials separated by insulating spacers. The entire system is evacuated to minimize air conduction, and thermal conductivities as low as 0.3 m W/m · °C are possible. A convenient summary of the thermal conductivities of a few insulating materials at cryogenic temperatures is given in Table 1-2. Further information on multilayer insulation is given in References 2 and 3.

1-3 | CONVECTION HEAT TRANSFER

It is well known that a hot plate of metal will cool faster when placed in front of a fan than when exposed to still air. We say that the heat is convected away, and we call the process *convection heat transfer.* The term *convection* provides the reader with an intuitive notion concerning the heat-transfer process; however, this intuitive notion must be expanded to enable one to arrive at anything like an adequate analytical treatment of the problem. For example, we know that the velocity at which the air blows over the hot plate obviously influences the heat-transfer rate. But does it influence the cooling in a linear way; i.e., if the velocity is doubled, will the heat-transfer rate double? We should suspect that the heat-transfer rate might be different if we cooled the plate with water instead of air, but, again, how much difference would there be? These questions may be answered with the aid of some rather basic analyses presented in later chapters. For now, we sketch the physical mechanism of convection heat transfer and show its relation to the conduction process.

Consider the heated plate shown in Figure 1-7. The temperature of the plate is T_w, and the temperature of the fluid is T_∞. The velocity of the flow will appear as shown, being reduced to zero at the plate as a result of viscous action. Since the velocity of the fluid layer at the wall will be zero, the heat must be transferred only by conduction at that point. Thus we might compute the heat transfer, using Equation (1-1), with the thermal conductivity of the fluid and the fluid temperature gradient at the wall. Why, then, if the heat flows by conduction in this layer, do we speak of *convection* heat transfer and need to consider the velocity of the fluid? The answer is that the temperature gradient is dependent on the rate at which the fluid carries the heat away; a high velocity produces a large temperature gradient, and so on. Thus the temperature gradient at the wall depends on the flow field, and we must develop in our later analysis an expression relating the two quantities. Nevertheless, it must be remembered that the physical mechanism of heat transfer at the wall is a conduction process.

To express the overall effect of convection, we use Newton's law of cooling:

$$q = hA\,(T_w - T_\infty) \tag{1-8}$$

Here the heat-transfer rate is related to the overall temperature difference between the wall and fluid and the surface area A. The quantity h is called the *convection heat-transfer coefficient,* and Equation (1-8) is the defining equation. An analytical calculation of h may be made for some systems. For complex situations it must be determined experimentally. The heat-transfer coefficient is sometimes called the *film conductance* because of its relation to the conduction process in the thin stationary layer of fluid at the wall surface. From Equation (1-8) we note that the units of h are in watts per square meter per Celsius degree when the heat flow is in watts.

In view of the foregoing discussion, one may anticipate that convection heat transfer will have a dependence on the viscosity of the fluid in addition to its dependence on the

Figure 1-7 | Convection heat transfer from a plate.

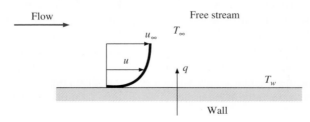

thermal properties of the fluid (thermal conductivity, specific heat, density). This is expected because viscosity influences the velocity profile and, correspondingly, the energy-transfer rate in the region near the wall.

If a heated plate were exposed to ambient room air without an external source of motion, a movement of the air would be experienced as a result of the density gradients near the plate. We call this *natural,* or *free,* convection as opposed to *forced* convection, which is experienced in the case of the fan blowing air over a plate. Boiling and condensation phenomena are also grouped under the general subject of convection heat transfer. The approximate ranges of convection heat-transfer coefficients are indicated in Table 1-3.

Convection Energy Balance on a Flow Channel

The energy transfer expressed by Equation (1-8) is used for evaluating the convection loss for flow over an external surface. Of equal importance is the convection gain or loss resulting from a fluid flowing inside a channel or tube as shown in Figure 1-8. In this case, the heated wall at T_w loses heat to the cooler fluid, which consequently rises in temperature as it flows

Table 1-3 | Approximate values of convection heat-transfer coefficients.

	h	
Mode	W/m$^2 \cdot$ °C	Btu/h $\cdot$ ft$^2 \cdot$ °F
Across 2.5-cm air gap evacuated to a pressure of 10^{-6} atm and subjected to $\Delta T = 100°C - 30°C$	0.087	0.015
Free convection, $\Delta T = 30°C$		
Vertical plate 0.3 m [1 ft] high in air	4.5	0.79
Horizontal cylinder, 5-cm diameter, in air	6.5	1.14
Horizontal cylinder, 2-cm diameter, in water	890	157
Heat transfer across 1.5-cm vertical air gap with $\Delta T = 60°C$	2.64	0.46
Fine wire in air, $d = 0.02$ mm, $\Delta T = 55°C$	490	86
Forced convection		
Airflow at 2 m/s over 0.2-m square plate	12	2.1
Airflow at 35 m/s over 0.75-m square plate	75	13.2
Airflow at Mach number $= 3$, $p = 1/20$ atm, $T_\infty = -40°C$, across 0.2-m square plate	56	9.9
Air at 2 atm flowing in 2.5-cm-diameter tube at 10 m/s	65	11.4
Water at 0.5 kg/s flowing in 2.5-cm-diameter tube	3500	616
Airflow *across* 5-cm-diameter cylinder with velocity of 50 m/s	180	32
Liquid bismuth at 4.5 kg/s and 420°C in 5.0-cm-diameter tube	3410	600
Airflow at 50 m/s across fine wire, $d = 0.04$ mm	3850	678
Boiling water		
In a pool or container	2500–35,000	440–6200
Flowing in a tube	5000–100,000	880–17,600
Condensation of water vapor, 1 atm		
Vertical surfaces	4000–11,300	700–2000
Outside horizontal tubes	9500–25,000	1700–4400
Dropwise condensation	170,000–290,000	30,000–50,000

Figure 1-8 | Convection in a channel.

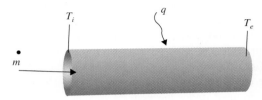

from inlet conditions at T_i to exit conditions at T_e. Using the symbol i to designate enthalpy (to avoid confusion with h, the convection coefficient), the energy balance on the fluid is

$$q = \dot{m}(i_e - i_i)$$

where $\dot{m}$ is the fluid mass flow rate. For many single-phase liquids and gases operating over reasonable temperature ranges $\triangle i = c_p \triangle T$ and we have

$$q = \dot{m}c_p(T_e - T_i)$$

which may be equated to a convection relation like Equation (1-8)

$$q = \dot{m}c_p(T_e - T_i) = hA(T_{w,\,\text{avg}} - T_{\text{fluid, avg}}) \qquad \textbf{[1-8a]}$$

In this case, the fluid temperatures T_e, T_i, and T_{fluid} are called *bulk* or *energy average* temperatures. A is the surface area of the flow channel in contact with the fluid. We shall have more to say about the notions of computing convection heat transfer for external and internal flows in Chapters 5 and 6. For now, we simply want to alert the reader to the distinction between the two types of flows.

We must be careful to distinguish between the surface area for convection that is employed in convection Equation (1-8) and the cross-sectional area that is used to calculate the flow rate from

$$\dot{m} = \rho u_{\text{mean}} A_c$$

where $A_c = \pi d^2/4$ for flow in a circular tube. The surface area for convection in this case would be πdL, where L is the tube length. The surface area for convection is always the area of the heated surface in contact with the fluid.

1-4 | RADIATION HEAT TRANSFER

In contrast to the mechanisms of conduction and convection, where energy transfer through a material medium is involved, heat may also be transferred through regions where a perfect vacuum exists. The mechanism in this case is electromagnetic radiation. We shall limit our discussion to electromagnetic radiation that is propagated as a result of a temperature difference; this is called *thermal radiation.*

Thermodynamic considerations show[*] that an ideal thermal radiator, or *blackbody*, will emit energy at a rate proportional to the fourth power of the absolute temperature of the body and directly proportional to its surface area. Thus

$$q_{\text{emitted}} = \sigma A T^4 \qquad \textbf{[1-9]}$$

---

[*]See, for example, J. P. Holman, *Thermodynamics.* 4th ed. New York: McGraw-Hill, 1988, p. 705.

where σ is the proportionality constant and is called the Stefan-Boltzmann constant with the value of 5.669×10^{-8} W/m$^2 \cdot$K^4. Equation (1-9) is called the Stefan-Boltzmann law of thermal radiation, and it applies only to blackbodies. It is important to note that this equation is valid only for thermal radiation; other types of electromagnetic radiation may not be treated so simply.

Equation (1-9) governs only radiation *emitted* by a blackbody. The net radiant *exchange* between two surfaces will be proportional to the difference in absolute temperatures to the fourth power; i.e.,

$$\frac{q_{\text{net exchange}}}{A} \propto \sigma(T_1^4 - T_2^4) \qquad \textbf{[1-10]}$$

We have mentioned that a blackbody is a body that radiates energy according to the T^4 law. We call such a body *black* because black surfaces, such as a piece of metal covered with carbon black, approximate this type of behavior. Other types of surfaces, such as a glossy painted surface or a polished metal plate, do not radiate as much energy as the blackbody; however, the total radiation emitted by these bodies still generally follows the T^4 proportionality. To take account of the "gray" nature of such surfaces we introduce another factor into Equation (1-9), called the emissivity ϵ, which relates the radiation of the "gray" surface to that of an ideal black surface. In addition, we must take into account the fact that not all the radiation leaving one surface will reach the other surface since electromagnetic radiation travels in straight lines and some will be lost to the surroundings. We therefore introduce two new factors in Equation (1-9) to take into account both situations, so that

$$q = F_\epsilon F_G \sigma A \, (T_1^4 - T_2^4) \qquad \textbf{[1-11]}$$

where F_ϵ is the emissivity function, and F_G is the geometric "view factor" function. The determination of the form of these functions for specific configurations is the subject of a subsequent chapter. It is important to alert the reader at this time, however, to the fact that these functions usually are not independent of one another as indicated in Equation (1-11).

Radiation in an Enclosure

A simple radiation problem is encountered when we have a heat-transfer surface at temperature T_1 completely enclosed by a much larger surface maintained at T_2. We will show in Chapter 8 that the net radiant exchange in this case can be calculated with

$$q = \epsilon_1 \sigma A_1 \, (T_1^4 - T_2^4) \qquad \textbf{[1-12]}$$

Values of ϵ are given in Appendix A.

Radiation heat-transfer phenomena can be exceedingly complex, and the calculations are seldom as simple as implied by Equation (1-11). For now, we wish to emphasize the difference in physical mechanism between radiation heat-transfer and conduction-convection systems. In Chapter 8 we examine radiation in detail.

1-5 | DIMENSIONS AND UNITS

In this section we outline the systems of units that are used throughout the book. One must be careful not to confuse the meaning of the terms *units* and *dimensions*. A dimension is a physical variable used to specify the behavior or nature of a particular system. For example, the length of a rod is a dimension of the rod. In like manner, the temperature of a gas may be considered one of the thermodynamic dimensions of the gas. When we say the rod is so many meters long, or the gas has a temperature of so many degrees Celsius, we have given the units with which we choose to measure the dimension. In our development of heat transfer we use the dimensions

$$L = \text{length}$$
$$M = \text{mass}$$
$$F = \text{force}$$
$$\tau = \text{time}$$
$$T = \text{temperature}$$

All the physical quantities used in heat transfer may be expressed in terms of these fundamental dimensions. The units to be used for certain dimensions are selected by somewhat arbitrary definitions that usually relate to a physical phenomenon or law. For example, Newton's second law of motion may be written

$$\text{Force} \sim \text{time rate of change of momentum}$$

$$F = k \frac{d(mv)}{d\tau}$$

where k is the proportionality constant. If the mass is constant,

$$F = kma \qquad\qquad \textbf{[1-13]}$$

where the acceleration is $a = dv/d\tau$. Equation (1-11) is usually written

$$F = \frac{1}{g_c} ma \qquad\qquad \textbf{[1-14]}$$

with $1/g_c = k$. Equation (1-14) is used to define our systems of units for mass, force, length, and time. Some typical systems of units are

1. 1-pound force will accelerate a 1-lb mass 32.17 ft/s^2.
2. 1-pound force will accelerate a 1-slug mass 1 ft/s^2.
3. 1-dyne force will accelerate a 1-g mass 1 cm/s^2.
4. 1-newton force will accelerate a 1-kg mass 1 m/s^2.
5. 1-kilogram force will accelerate a 1-kg mass 9.806 m/s^2.

The 1-kg force is sometimes called a *kilopond* (kp).

Since Equation (1-14) must be dimensionally homogeneous, we shall have a different value of the constant g_c for each of the unit systems in items 1 to 5 above. These values are

1. $g_c = 32.17 \ \text{lb}_m \cdot \text{ft}/\text{lb}_f \cdot \text{s}^2$
2. $g_c = 1 \ \text{slug} \cdot \text{ft}/\text{lb}_f \cdot \text{s}^2$
3. $g_c = 1 \ \text{g} \cdot \text{cm}/\text{dyn} \cdot \text{s}^2$
4. $g_c = 1 \ \text{kg} \cdot \text{m}/\text{N} \cdot \text{s}^2$
5. $g_c = 9.806 \ \text{kg}_m \cdot \text{m}/\text{kg}_f \cdot \text{s}^2$

It matters not which system of units is used so long as it is consistent with these definitions.

Work has the dimensions of a product of force times a distance. Energy has the same dimensions. The units for work and energy may be chosen from any of the systems used on the previous page, and would be

1. $lb_f \cdot ft$
2. $lb_f \cdot ft$
3. $dyn \cdot cm = 1$ erg
4. $N \cdot m = 1$ joule (J)
5. $kg_f \cdot m = 9.806$ J

In addition, we may use the units of energy that are based on thermal phenomena:

1 Btu will raise 1 lb_m of water 1°F at 68°F.

1 cal will raise 1 g of water 1°C at 20°C.

1 kcal will raise 1 kg of water 1°C at 20°C.

Some conversion factors for the various units of work and energy are

$$1 \text{ Btu} = 778.16 \ lb_f \cdot ft$$
$$1 \text{ Btu} = 1055 \text{ J}$$
$$1 \text{ kcal} = 4182 \text{ J}$$
$$1 \ lb_f \cdot ft = 1.356 \text{ J}$$
$$1 \text{ Btu} = 252 \text{ cal}$$

Other conversion factors are given in Appendix A.

The weight of a body is defined as the force exerted on the body as a result of the acceleration of gravity. Thus

$$W = \frac{g}{g_c} m \qquad \textbf{[1-15]}$$

where W is the weight and g is the acceleration of gravity. Note that the weight of a body has the dimensions of a force. We now see why systems 1 and 5 were devised; 1 lb_m will weigh 1 lb_f at sea level, and 1 kg_m will weigh 1 kg_f.

Temperature conversions are performed with the familiar formulas

$$°F = \tfrac{9}{5}°C + 32$$
$$°R = °F + 459.69$$
$$K = °C + 273.16$$
$$°R = \tfrac{9}{5}K$$

Unfortunately, *all* of these unit systems are used in various places throughout the world. While the foot, pound force, pound mass, second, degree Fahrenheit, Btu system is still widely used in the United States, there is increasing impetus to institute the SI (Système International d'Unités) units as a worldwide standard. In this system, the fundamental units are meter, newton, kilogram mass, second, and degree Celsius; a "thermal" energy unit is not used; i.e., the joule (newton-meter) becomes the energy unit used throughout. The watt (joules per second) is the unit of power in this system. In the SI system, the standard units for thermal conductivity would become

$$k \text{ in } W/m \cdot °C$$

Table 1-4 | Multiplier factors for SI units.

Multiplier	Prefix	Abbreviation
10^{12}	tera	T
10^{9}	giga	G
10^{6}	mega	M
10^{3}	kilo	k
10^{2}	hecto	h
10^{-2}	centi	c
10^{-3}	milli	m
10^{-6}	micro	μ
10^{-9}	nano	n
10^{-12}	pico	p
10^{-18}	atto	a

Table 1-5 | SI quantities used in heat transfer.

Quantity	Unit abbreviation
Force	N (newton)
Mass	kg (kilogram mass)
Time	s (second)
Length	m (meter)
Temperature	°C or K
Energy	J (joule)
Power	W (watt)
Thermal conductivity	W/m · °C
Heat-transfer coefficient	W/m^2 · °C
Specific heat	J/kg · °C
Heat flux	W/m^2

and the convection heat-transfer coefficient would be expressed as

$$h \text{ in } W/m^2 \cdot {}^\circ C$$

Because SI units are so straightforward we shall use them as the standard in this text, with intermediate steps and answers in examples also given parenthetically in the Btu–pound mass system. A worker in heat transfer must obtain a feel for the order of magnitudes in both systems. In the SI system the concept of g_c is not normally used, and the newton is *defined* as

$$1 \text{ N} = 1 \text{ kg} \cdot m/s^2 \qquad \qquad \textbf{[1-16]}$$

Even so, one should keep in mind the physical relation between force and mass as expressed by Newton's second law of motion.

The SI system also specifies standard multiples to be used to conserve space when numerical values are expressed. They are summarized in Table 1-4. Standard symbols for quantities normally encountered in heat transfer are summarized in Table 1-5. Conversion factors are given in Appendix A.

EXAMPLE 1-1	Conduction Through Copper Plate

One face of a copper plate 3 cm thick is maintained at 400°C, and the other face is maintained at 100°C. How much heat is transferred through the plate?

■ **Solution**

From Appendix A, the thermal conductivity for copper is 370 W/m · °C at 250°C. From Fourier's law

$$\frac{q}{A} = -k\frac{dT}{dx}$$

Integrating gives

$$\frac{q}{A} = -k\frac{\Delta T}{\Delta x} = \frac{-(370)(100-400)}{3\times10^{-2}} = 3.7\text{ MW/m}^2 \quad [1.173\times10^6\text{ Btu/h}\cdot\text{ft}^2]$$

Convection Calculation — EXAMPLE 1-2

Air at 20°C blows over a hot plate 50 by 75 cm maintained at 250°C. The convection heat-transfer coefficient is 25 W/m² · °C. Calculate the heat transfer.

■ **Solution**

From Newton's law of cooling

$$q = hA\,(T_w - T_\infty)$$
$$= (25)(0.50)(0.75)(250-20)$$
$$= 2.156\text{ kW} \quad [7356\text{ Btu/h}]$$

Multimode Heat Transfer — EXAMPLE 1-3

Assuming that the plate in Example 1-2 is made of carbon steel (1%) 2 cm thick and that 300 W is lost from the plate surface by radiation, calculate the inside plate temperature.

■ **Solution**

The heat conducted through the plate must be equal to the sum of convection and radiation heat losses:

$$q_{\text{cond}} = q_{\text{conv}} + q_{\text{rad}}$$

$$-kA\frac{\Delta T}{\Delta x} = 2.156 + 0.3 = 2.456\text{ kW}$$

$$\Delta T = \frac{(-2456)(0.02)}{(0.5)(0.75)(43)} = -3.05°\text{C} \quad [-5.49°\text{F}]$$

where the value of k is taken from Table 1-1. The inside plate temperature is therefore

$$T_i = 250 + 3.05 = 253.05°\text{C}$$

Heat Source and Convection — EXAMPLE 1-4

An electric current is passed through a wire 1 mm in diameter and 10 cm long. The wire is submerged in liquid water at atmospheric pressure, and the current is increased until the water boils. For this situation $h = 5000$ W/m$^2 \cdot$°C, and the water temperature will be 100°C. How much electric power must be supplied to the wire to maintain the wire surface at 114°C?

■ **Solution**

The total convection loss is given by Equation (1-8):

$$q = hA\,(T_w - T_\infty)$$

For this problem the surface area of the wire is

$$A = \pi dL = \pi(1 \times 10^{-3})(10 \times 10^{-2}) = 3.142 \times 10^{-4}\ \text{m}^2$$

The heat transfer is therefore

$$q = (5000\ \text{W/m}^2 \cdot °\text{C})(3.142 \times 10^{-4}\ \text{m}^2)(114 - 100) = 21.99\ \text{W}\quad[75.03\ \text{Btu/h}]$$

and this is equal to the electric power that must be applied.

EXAMPLE 1-5	Radiation Heat Transfer

Two infinite black plates at 800°C and 300°C exchange heat by radiation. Calculate the heat transfer per unit area.

■ **Solution**

Equation (1-10) may be employed for this problem, so we find immediately

$$q/A = \sigma(T_1^4 - T_2^4)$$
$$= (5.669 \times 10^{-8})(1073^4 - 573^4)$$
$$= 69.03\ \text{kW/m}^2\quad[21{,}884\ \text{Btu/h} \cdot \text{ft}^2]$$

EXAMPLE 1-6	Total Heat Loss by Convection and Radiation

A horizontal steel pipe having a diameter of 5 cm is maintained at a temperature of 50°C in a large room where the air and wall temperature are at 20°C. The surface emissivity of the steel may be taken as 0.8. Using the data of Table 1-3, calculate the total heat lost by the pipe per unit length.

■ **Solution**

The total heat loss is the sum of convection and radiation. From Table 1-3 we see that an estimate for the heat-transfer coefficient for *free* convection with this geometry and air is $h = 6.5$ W/m$^2 \cdot$°C. The surface area is πdL, so the convection loss per unit length is

$$q/L]_{\text{conv}} = h(\pi d)(T_w - T_\infty)$$
$$= (6.5)(\pi)(0.05)(50 - 20) = 30.63\ \text{W/m}$$

The pipe is a body surrounded by a large enclosure so the radiation heat transfer can be calculated from Equation (1-12). With $T_1 = 50°\text{C} = 323°\text{K}$ and $T_2 = 20°\text{C} = 293°\text{K}$, we have

$$q/L]_{\text{rad}} = \epsilon_1(\pi d_1)\sigma(T_1^4 - T_2^4)$$
$$= (0.8)(\pi)(0.05)(5.669 \times 10^{-8})(323^4 - 293^4)$$
$$= 25.04\ \text{W/m}$$

The total heat loss is therefore

$$q/L]_{tot} = q/L]_{conv} + q/L]_{rad}$$
$$= 30.63 + 25.04 = 55.67 \text{ W/m}$$

In this example we see that the convection and radiation are about the same. To neglect either would be a serious mistake.

1-6 | SUMMARY

We may summarize our introductory remarks very simply. Heat transfer may take place by one or more of three modes: conduction, convection, and radiation. It has been noted that the physical mechanism of convection is related to the heat conduction through the thin layer of fluid adjacent to the heat-transfer surface. In both conduction and convection Fourier's law is applicable, although fluid mechanics must be brought into play in the convection problem in order to establish the temperature gradient.

Radiation heat transfer involves a different physical mechanism—that of propagation of electromagnetic energy. To study this type of energy transfer we introduce the concept of an ideal radiator, or blackbody, which radiates energy at a rate proportional to its absolute temperature to the fourth power.

It is easy to envision cases in which all three modes of heat transfer are present, as in Figure 1-9. In this case the heat conducted through the plate is removed from the plate surface by a combination of convection and radiation. An energy balance would give

$$-kA \left. \frac{dT}{dy} \right]_{wall} = hA\,(T_w - T_\infty) + F_\epsilon F_G \sigma A\,(T_w^4 - T_s^4)$$

where

T_s = temperature of surroundings

T_w = surface temperature

T_∞ = fluid temperature

To apply the science of heat transfer to practical situations, a thorough knowledge of all three modes of heat transfer must be obtained.

Figure 1-9 | Combination of conduction, convection, and radiation heat transfer.

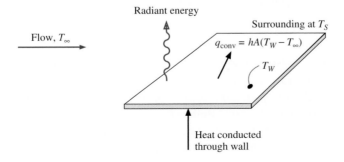

Radiant energy

Surrounding at T_S

Flow, T_∞

$q_{conv} = hA(T_W - T_\infty)$

T_W

Heat conducted through wall

Table 1-6 | Listing of equation summary tables in text.

Table	Topic
1-3	Approximate values of convection heat-transfer coefficients
3-1	Conduction shape factors
3-2	Summary of steady-state nodal equations for $\triangle x = \triangle y$
4-1	Examples of lumped capacities
4-2	Summary of transient nodal equations for $\triangle x = \triangle y$
5-2	Forced-convection relations for flow over flat plates
6-8	Forced-convection relations for internal and external flows (nonflat plates)
7-2	Simplified relations for free convection from heated objects in room air
7-5	Summary of free-convection relations
Section 7-14 and Figure 7-15	Summary procedure for all convection calculations
8-7	Radiation formulas for diffuse, gray-body enclosures
10-3	Effectiveness relations for heat exchangers
10-4	NTU relations for heat exchangers

About Areas

The reader will note that area is an important part of the calculation for all three modes of heat transfer: The larger the area through which heat is conducted, the larger the heat transfer; the larger the surface area in contact with the fluid, the larger the potential convection heat transfer; and a larger surface will emit more thermal radiation than a small surface. For conduction, the heat transfer will almost always be directly proportional to the area. For convection, the heat transfer is a complicated function of the fluid mechanics of the problem, which in turn is a function of both the geometric configuration of the heated surface and the thermal and viscous fluid properties of the convecting medium. Radiation heat transfer also involves a complex interaction between the surface emissive properties and the geometry of the enclosure that involves the radiant transfer. Despite these remarks, the general principle is that an increased area means an increase in heat transfer.

Summary Tables Available in Text

As our discussion progresses we will present several tables which summarize equations and empirical correlations for convenience of the reader. A listing of some of these tables and/or figures along with their topical content is given in Table 1-6.

REVIEW QUESTIONS

1. Define thermal conductivity.
2. Define the convection heat-transfer coefficient.
3. Discuss the mechanism of thermal conduction in gases and solids.
4. Discuss the mechanism of heat convection.
5. What is the order of magnitude for the convection heat-transfer coefficient in free convection? Forced convection? Boiling?
6. When may one expect radiation heat transfer to be important?
7. Name some good conductors of heat; some poor conductors.

8. What is the order of magnitude of thermal conductivity for (*a*) metals, (*b*) solid insulating materials, (*c*) liquids, (*d*) gases?

9. Suppose a person stated that heat cannot be transferred in a vacuum. How do you respond?

10. Review any standard text on thermodynamics and define: (*a*) heat, (*b*) internal energy, (*c*) work, (*d*) enthalpy.

11. Define and discuss g_c.

LIST OF WORKED EXAMPLES

1-1 Conduction through copper plate
1-2 Convection calculation
1-3 Multimode heat transfer
1-4 Heat source and convection
1-5 Radiation heat transfer
1-6 Total heat loss by convection and radiation

PROBLEMS

1-1 If 3 kW is conducted through a section of insulating material 0.6 m^2 in cross section and 2.5 cm thick and the thermal conductivity may be taken as 0.2 W/m · °C, compute the temperature difference across the material.

1-2 A temperature difference of 85°C is impressed across a fiberglass layer of 13 cm thickness. The thermal conductivity of the fiberglass is 0.035 W/m · °C. Compute the heat transferred through the material per hour per unit area.

1-3 A truncated cone 30 cm high is constructed of aluminum. The diameter at the top is 7.5 cm, and the diameter at the bottom is 12.5 cm. The lower surface is maintained at 93°C; the upper surface, at 540°C. The other surface is insulated. Assuming one-dimensional heat flow, what is the rate of heat transfer in watts?

1-4 The temperatures on the faces of a plane wall 15 cm thick are 375 and 85°C. The wall is constructed of a special glass with the following properties: $k = 0.78$ W/m · °C, $\rho = 2700$ kg/m^3, $c_P = 0.84$ kJ/kg · °C. What is the heat flow through the wall at steady-state conditions?

1-5 A certain superinsulation material having a thermal conductivity of 2×10^{-4} W/m · °C is used to insulate a tank of liquid nitrogen that is maintained at −196°C; 199 kJ is required to vaporize each kilogram mass of nitrogen at this temperature. Assuming that the tank is a sphere having an inner diameter (ID) of 0.52 m, estimate the amount of nitrogen vaporized per day for an insulation thickness of 2.5 cm and an ambient temperature of 21°C. Assume that the outer temperature of the insulation is 21°C.

1-6 Rank the following materials in order of (*a*) transient response and (*b*) steady-state conduction. Taking the material with the highest rank, give the other materials as a percentage of the maximum: aluminum, copper, silver, iron, lead, chrome steel (18% Cr, 8% Ni), magnesium. What do you conclude from this ranking?

1-7 A 50-cm-diameter pipeline in the Arctic carries hot oil at 30°C and is exposed to a surrounding temperature of −20°C. A special powder insulation 5 cm thick surrounds the pipe and has a thermal conductivity of 7 mW/m · °C. The convection heat-transfer coefficient on the outside of the pipe is 9 W/m² · °C. Estimate the energy loss from the pipe per meter of length.

1-8 Some people might recall being told to be sure to put on a hat when outside in cold weather because "you lose all the heat out the top of your head." Comment on the validity of this statement.

1-9 A 5-cm layer of loosely packed asbestos is placed between two plates at 100 and 200°C. Calculate the heat transfer across the layer.

1-10 A certain insulation has a thermal conductivity of 10 W/m · °C. What thickness is necessary to effect a temperature drop of 500°C for a heat flow of 400 W/m²?

1-11 Assuming that the heat transfer to the sphere in Problem 1-5 occurs by free convection with a heat-transfer coefficient of 2.7 W/m² · °C, calculate the temperature difference between the outer surface of the sphere and the environment.

1-12 Two perfectly black surfaces are constructed so that all the radiant energy leaving a surface at 800°C reaches the other surface. The temperature of the other surface is maintained at 250°C. Calculate the heat transfer between the surfaces per hour and per unit area of the surface maintained at 800°C.

1-13 Two very large parallel planes having surface conditions that very nearly approximate those of a blackbody are maintained at 1100 and 425°C, respectively. Calculate the heat transfer by radiation between the planes per unit time and per unit surface area.

1-14 Calculate the radiation heat exchange in 1 day between two black planes having the area of the surface of a 0.7-m-diameter sphere when the planes are maintained at 70 K and 300 K.

1-15 Two infinite black plates at 500 and 100°C exchange heat by radiation. Calculate the heat-transfer rate per unit area. If another perfectly black plate is placed between the 500 and 100°C plates, by how much is the heat transfer reduced? What is the temperature of the center plate?

1-16 Water flows at the rate of 0.5 kg/s in a 2.5-cm-diameter tube having a length of 3 m. A constant heat flux is imposed at the tube wall so that the tube wall temperature is 40°C higher than the water temperature. Calculate the heat transfer and estimate the temperature rise in the water. The water is pressurized so that boiling cannot occur.

1-17 Steam at 1 atm pressure ($T_{sat} = 100°C$) is exposed to a 30-by-30-cm vertical square plate that is cooled such that 3.78 kg/h is condensed. Calculate the plate temperature. Consult steam tables for any necessary properties.

1-18 Boiling water at 1 atm may require a surface heat flux of 3×10^4 Btu/h · ft² for a surface temperature of 232°F. What is the value of the heat-transfer coefficient?

1-19 A small radiant heater has metal strips 6 mm wide with a total length of 3 m. The surface emissivity of the strips is 0.85. To what temperature must the strips be heated if they are to dissipate 2000 W of heat to a room at 25°C?

1-20 Calculate the energy emitted by a blackbody at 1000°C.

1-21 If the radiant flux from the sun is 1350 W/m², what would be its equivalent blackbody temperature?

1-22 A 4.0-cm-diameter sphere is heated to a temperature of 200°C and is enclosed in a large room at 20°C. Calculate the radiant heat loss if the surface emissivity is 0.6.

1-23 A flat wall is exposed to an environmental temperature of 38°C. The wall is covered with a layer of insulation 2.5 cm thick whose thermal conductivity is 1.4 W/m · °C, and the temperature of the wall on the inside of the insulation is 315°C. The wall loses heat to the environment by convection. Compute the value of the convection heat-transfer coefficient that must be maintained on the outer surface of the insulation to ensure that the outer-surface temperature does not exceed 41°C.

1-24 Consider a wall heated by convection on one side and cooled by convection on the other side. Show that the heat-transfer rate through the wall is

$$q = \frac{T_1 - T_2}{1/h_1 A + \Delta x/kA + 1/h_2 A}$$

where T_1 and T_2 are the fluid temperatures on each side of the wall and h_1 and h_2 are the corresponding heat-transfer coefficients.

1-25 One side of a plane wall is maintained at 100°C, while the other side is exposed to a convection environment having $T = 10$°C and $h = 10$ W/m$^2 \cdot$ °C. The wall has $k = 1.6$ W/m $\cdot$ °C and is 40 cm thick. Calculate the heat-transfer rate through the wall.

1-26 How does the free-convection heat transfer from a vertical plate compare with pure conduction through a vertical layer of air having a thickness of 2.5 cm and a temperature difference the same at $T_w - T_\infty$? Use information from Table 1-3.

1-27 A $\frac{1}{4}$-in steel plate having a thermal conductivity of 25 Btu/h $\cdot$ ft $\cdot$ °F is exposed to a radiant heat flux of 1500 Btu/h $\cdot$ ft^2 in a vacuum space where the convection heat transfer is negligible. Assuming that the surface temperature of the steel exposed to the radiant energy is maintained at 100°F, what will be the other surface temperature if all the radiant energy striking the plate is transferred through the plate by conduction?

1-28 A solar radiant heat flux of 700 W/m^2 is absorbed in a metal plate that is perfectly insulated on the back side. The convection heat-transfer coefficient on the plate is 11 W/m$^2 \cdot$ °C, and the ambient air temperature is 30°C. Calculate the temperature of the plate under equilibrium conditions.

1-29 A 5.0-cm-diameter cylinder is heated to a temperature of 200°C, and air at 30°C is forced across it at a velocity of 50 m/s. If the surface emissivity is 0.7, calculate the total heat loss per unit length if the walls of the enclosing room are at 10°C. Comment on this calculation.

1-30 A vertical square plate, 30 cm on a side, is maintained at 50°C and exposed to room air at 20°C. The surface emissivity is 0.8. Calculate the total heat lost by both sides of the plate.

1-31 A black 20-by-20-cm plate has air forced over it at a velocity of 2 m/s and a temperature of 0°C. The plate is placed in a large room whose walls are at 30°C. The back side of the plate is perfectly insulated. Calculate the temperature of the plate resulting from the convection-radiation balance. Use information from Table 1-3. Are you surprised at the result?

1-32 Two large black plates are separated by a vacuum. On the outside of one plate is a convection environment of $T = 80$°C and $h = 100$ W/m$^2 \cdot$ °C, while the outside of the other plate is exposed to 20°C and $h = 15$ W/m$^2 \cdot$ °C. Make an energy balance on the system and determine the plate temperatures. For this problem $F_G = F_\epsilon = 1.0$.

1-33 Using the basic definitions of units and dimensions given in Section 1-5, arrive at expressions (*a*) to convert joules to British thermal units, (*b*) to convert dyne-centimeters to joules, and (*c*) to convert British thermal units to calories.

1-34 Beginning with the three-dimensional heat-conduction equation in cartesian coordinates [Equation (1-3*a*)], obtain the general heat-conduction equation in cylindrical coordinates [Equation (1-3*b*)].

1-35 Write the simplified heat-conduction equation for (*a*) steady one-dimensional heat flow in cylindrical coordinates in the *azimuth* (ϕ) direction, and (*b*) steady one-dimensional heat flow in spherical coordinates in the azimuth (ϕ) direction.

1-36 Using the approximate values of convection heat-transfer coefficients given in Table 1-3, estimate the surface temperature for which the free convection heat loss will just equal the radiation heat loss from a vertical 0.3-m-square plate or a 5-cm-diameter cylinder exposed to room air at 20°C. Assume the surfaces are blackened such that $\epsilon = 1.0$ and the radiation surrounding temperature may be taken the same as the room air temperature.

Design-Oriented Problems

1-37 A woman informs an engineer that she frequently feels cooler in the summer when standing in front of an open refrigerator. The engineer tells her that she is only "imagining things" because there is no fan in the refrigerator to blow the cool air over her. A lively argument ensues. Whose side of the argument do you take? Why?

1-38 A woman informs her engineer husband that "hot water will freeze faster than cold water." He calls this statement nonsense. She answers by saying that she has actually timed the freezing process for ice trays in the home refrigerator and found that hot water does indeed freeze faster. As a friend, you are asked to settle the argument. Is there any logical explanation for the woman's observation?

1-39 An air-conditioned classroom in Texas is maintained at 72°F in the summer. The students attend classes in shorts, sandals, and tee shirts and are quite comfortable. In the same classroom during the winter, the same students wear wool slacks, long-sleeve shirts, and sweaters, and are equally comfortable with the room temperature maintained at 75°F. Assuming that humidity is not a factor, explain this apparent anomaly in "temperature comfort."

1-40 A vertical cylinder 6 ft tall and 1 ft in diameter might be used to approximate a man for heat-transfer purposes. Suppose the surface temperature of the cylinder is 78°F, $h = 2$ Btu/h · ft² · °F, the surface emissivity is 0.9, and the cylinder is placed in a large room where the air temperature is 68°F and the wall temperature is 45°F. Calculate the heat lost from the cylinder. Repeat for a wall temperature of 80°F. What do you conclude from these calculations?

1-41 An ice-skating rink is located in an indoor shopping mall with an environmental air temperature of 22°C and radiation surrounding walls of about 25°C. The convection heat-transfer coefficient between the ice and air is about 10 W/m² · °C because of air movement and the skaters' motion. The emissivity of the ice is about 0.95. Calculate the cooling required to maintain the ice at 0°C for an ice rink having dimensions of 12 by 40 m. Obtain a value for the heat of fusion of ice and estimate how long it would take to melt 3 mm of ice from the surface of the rink if no cooling is supplied and the surface is considered insulated on the back side.

1-42 In energy conservation studies, cost is usually expressed in terms of Btu of energy, or some English unit of measure such as the gallon. Some typical examples are

Overall cost: $/10^6$ Btu

Transportation results: passenger miles per 10^6 Btu or per gallon of fuel
ton-miles of freight per 10^6 Btu or per gallon of fuel

Consult whatever sources are needed, and devise suitable measures for energy consumption and cost using the SI system of units. How would you price such items as

Energy content of various types of coal

Energy content of gasoline

Energy content of natural gas

Energy "content" of electricity

After devising the SI system of cost measures, construct a table of conversion factors like that given in the front inside cover of this book, to convert from SI to English and from English to SI.

1-43 Using information developed in Problem 1-42, investigate the energy cost saving that results from the installation of a layer of glass wool 15 cm thick on a steel building 12 by 12 m in size and 5 m high. Assume the building is subjected to a temperature difference of 30°C and the floor of the building does not participate in the heat lost. Assume that the outer surface of the building loses heat by convection to a surrounding temperature of -10°C with a convection coefficient $h = 13$ W/m$^2 \cdot$°C.

1-44 A boy-scout counselor gives the following advice to his scout troop regarding camping out in cold weather. "Be careful when setting up your cot/bunk—you may have provided for plenty of blankets to cover the top of your body, but don't forget that you can lose heat from the bottom through the thin layer of the cot/bunk. Provide a layer of insulation for your bottom side also." Investigate the validity of this statement by making suitable assumptions regarding exterior body temperature, thermal conductivity of blankets and cot/bunk materials, and the like.

REFERENCES

1. Glaser, P. E., I. A. Black, and P. Doherty. Multilayer Insulation, *Mech. Eng.,* August 1965, p. 23.
2. Barron, R. *Cryogenic Systems*. New York: McGraw-Hill, 1967.
3. Dewitt, W. D., N. C. Gibbon, and R. L. Reid. "Multifoil Type Thermal Insulation," *IEEE Trans. Aerosp. Electron. Syst.,* vol. 4, no. 5, suppl. pp. 263–71, 1968.

Steady-State Conduction— One Dimension

2-1 | INTRODUCTION

We now wish to examine the applications of Fourier's law of heat conduction to calculation of heat flow in some simple one-dimensional systems. Several different physical shapes may fall in the category of one-dimensional systems: cylindrical and spherical systems are one-dimensional when the temperature in the body is a function only of radial distance and is independent of azimuth angle or axial distance. In some two-dimensional problems the effect of a second-space coordinate may be so small as to justify its neglect, and the multidimensional heat-flow problem may be approximated with a one-dimensional analysis. In these cases the differential equations are simplified, and we are led to a much easier solution as a result of this simplification.

2-2 | THE PLANE WALL

First consider the plane wall where a direct application of Fourier's law [Equation (1-1)] may be made. Integration yields

$$q = -\frac{kA}{\Delta x}(T_2 - T_1) \qquad \text{[2-1]}$$

when the thermal conductivity is considered constant. The wall thickness is Δx, and T_1 and T_2 are the wall-face temperatures. If the thermal conductivity varies with temperature according to some linear relation $k = k_0(1 + \beta T)$, the resultant equation for the heat flow is

$$q = -\frac{k_0 A}{\Delta x}\left[(T_2 - T_1) + \frac{\beta}{2}(T_2^2 - T_1^2)\right] \qquad \text{[2-2]}$$

If more than one material is present, as in the multilayer wall shown in Figure 2-1, the analysis would proceed as follows: The temperature gradients in the three materials are shown, and the heat flow may be written

$$q = -k_A A \frac{T_2 - T_1}{\Delta x_A} = -k_B A \frac{T_3 - T_2}{\Delta x_B} = -k_C A \frac{T_4 - T_3}{\Delta x_C}$$

Note that the heat flow must be the same through all sections.

Figure 2-1 | One-dimensional heat transfer through a composite wall and electrical analog.

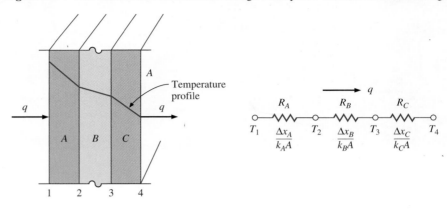

Solving these three equations simultaneously, the heat flow is written

$$q = \frac{T_1 - T_4}{\Delta x_A / k_A A + \Delta x_B / k_B A + \Delta x_C / k_C A} \qquad \text{[2-3]}$$

At this point we retrace our development slightly to introduce a different conceptual view-point for Fourier's law. The heat-transfer rate may be considered as a flow, and the combination of thermal conductivity, thickness of material, and area as a resistance to this flow. The temperature is the potential, or driving, function for the heat flow, and the Fourier equation may be written

$$\text{Heat flow} = \frac{\text{thermal potential difference}}{\text{thermal resistance}} \qquad \text{[2-4]}$$

a relation quite like Ohm's law in electric-circuit theory. In Equation (2-1) the thermal resistance is $\Delta x / kA$, and in Equation (2-3) it is the sum of the three terms in the denominator. We should expect this situation in Equation (2-3) because the three walls side by side act as three thermal resistances in series. The equivalent electric circuit is shown in Figure 2-1*b*.

The electrical analogy may be used to solve more complex problems involving both series and parallel thermal resistances. A typical problem and its analogous electric circuit are shown in Figure 2-2. The one-dimensional heat-flow equation for this type of problem may be written

$$q = \frac{\Delta T_{\text{overall}}}{\sum R_{\text{th}}} \qquad \text{[2-5]}$$

where the R_{th} are the thermal resistances of the various materials. The units for the thermal resistance are °C/W or °F·h/Btu.

It is well to mention that in some systems, like that in Figure 2-2, two-dimensional heat flow may result if the thermal conductivities of materials *B*, *C*, and *D* differ by an appreciable amount. In these cases other techniques must be employed to effect a solution.

2-3 | INSULATION AND *R* VALUES

In Chapter 1 we noted that the thermal conductivities for a number of insulating materials are given in Appendix A. In classifying the performance of insulation, it is a common practice

Figure 2-2 | Series and parallel one-dimensional heat transfer through a composite wall and electrical analog.

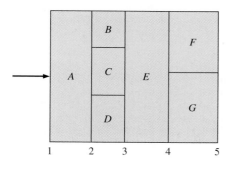

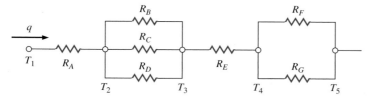

in the building industry to use a term called the *R value,* which is defined as

$$R = \frac{\Delta T}{q/A}$$ [2-6]

The units for R are $°C \cdot m^2/W$ or $°F \cdot ft^2 \cdot h/Btu$. Note that this differs from the thermal-resistance concept discussed above in that a heat flow *per unit area* is used.

At this point it is worthwhile to classify insulation materials in terms of their application and allowable temperature ranges. Table 2-1 furnishes such information and may be used as a guide for the selection of insulating materials.

2-4 | RADIAL SYSTEMS

Cylinders

Consider a long cylinder of inside radius r_i, outside radius r_o, and length L, such as the one shown in Figure 2-3. We expose this cylinder to a temperature differential $T_i - T_o$ and ask what the heat flow will be. For a cylinder with length very large compared to diameter, it may be assumed that the heat flows only in a radial direction, so that the only space coordinate needed to specify the system is r. Again, Fourier's law is used by inserting the proper area relation. The area for heat flow in the cylindrical system is

$$A_r = 2\pi r L$$

so that Fourier's law is written

$$q_r = -kA_r \frac{dT}{dr}$$ [2-7]

or

$$q_r = -2\pi k r L \frac{dT}{dr}$$

Table 2-1 | Insulation types and applications.

Type		Temperature range, °C	Thermal conductivity, mW/m · °C	Density, kg/m³	Application
1	Linde evacuated superinsulation	−240–1100	0.0015–0.72	Variable	Many
2	Urethane foam	−180–150	16–20	25–48	Hot and cold pipes
3	Urethane foam	−170–110	16–20	32	Tanks
4	Cellular glass blocks	−200–200	29–108	110–150	Tanks and pipes
5	Fiberglass blanket for wrapping	−80–290	22–78	10–50	Pipe and pipe fittings
6	Fiberglass blankets	−170–230	25–86	10–50	Tanks and equipment
7	Fiberglass preformed shapes	−50–230	32–55	10–50	Piping
8	Elastomeric sheets	−40–100	36–39	70–100	Tanks
9	Fiberglass mats	60–370	30–55	10–50	Pipe and pipe fittings
10	Elastomeric preformed shapes	−40–100	36–39	70–100	Pipe and fittings
11	Fiberglass with vapor barrier blanket	−5–70	29–45	10–32	Refrigeration lines
12	Fiberglass without vapor barrier jacket	to 250	29–45	24–48	Hot piping
13	Fiberglass boards	20–450	33–52	25–100	Boilers, tanks, heat exchangers
14	Cellular glass blocks and boards	20–500	29–108	110–150	Hot piping
15	Urethane foam blocks and boards	100–150	16–20	25–65	Piping
16	Mineral fiber preformed shapes	to 650	35–91	125–160	Hot piping
17	Mineral fiber blankets	to 750	37–81	125	Hot piping
18	Mineral wool blocks	450–1000	52–130	175–290	Hot piping
19	Calcium silicate blocks, boards	230–1000	32–85	100–160	Hot piping, boilers, chimney linings
20	Mineral fiber blocks	to 1100	52–130	210	Boilers and tanks

Figure 2-3 | One-dimensional heat flow through a hollow cylinder and electrical analog.

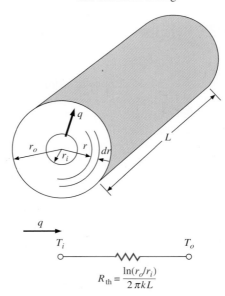

$$R_{\text{th}} = \frac{\ln(r_o/r_i)}{2\pi k L}$$

Figure 2-4 | One-dimensional heat flow through multiple cylindrical sections and electrical analog.

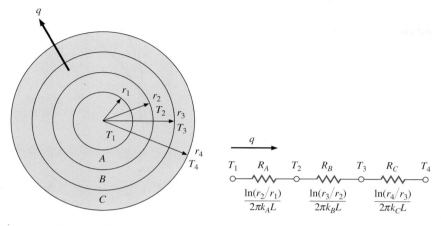

with the boundary conditions

$$T = T_i \qquad \text{at } r = r_i$$
$$T = T_o \qquad \text{at } r = r_o$$

The solution to Equation (2-7) is

$$q = \frac{2\pi k L\,(T_i - T_o)}{\ln\,(r_o/r_i)} \qquad\qquad \text{[2-8]}$$

and the thermal resistance in this case is

$$R_{\text{th}} = \frac{\ln\,(r_o/r_i)}{2\pi k L}$$

The thermal-resistance concept may be used for multiple-layer cylindrical walls just as it was used for plane walls. For the three-layer system shown in Figure 2-4 the solution is

$$q = \frac{2\pi L\,(T_1 - T_4)}{\ln\,(r_2/r_1)/k_A + \ln\,(r_3/r_2)/k_B + \ln\,(r_4/r_3)/k_C} \qquad\qquad \text{[2-9]}$$

The thermal circuit is also shown in Figure 2-4.

Spheres

Spherical systems may also be treated as one-dimensional when the temperature is a function of radius only. The heat flow is then

$$q = \frac{4\pi k\,(T_i - T_o)}{1/r_i - 1/r_o} \qquad\qquad \text{[2-10]}$$

The derivation of Equation (2-10) is left as an exercise.

Multilayer Conduction · EXAMPLE 2-1

An exterior wall of a house may be approximated by a 4-in layer of common brick [$k = 0.7$ W/m · °C] followed by a 1.5-in layer of gypsum plaster [$k = 0.48$ W/m · °C]. What thickness of loosely packed rock-wool insulation [$k = 0.065$ W/m · °C] should be added to reduce the heat loss (or gain) through the wall by 80 percent?

■ **Solution**

The overall heat loss will be given by

$$q = \frac{\Delta T}{\sum R_{th}}$$

Because the heat loss with the rock-wool insulation will be only 20 percent (80 percent reduction) of that before insulation

$$\frac{q \text{ with insulation}}{q \text{ without insulation}} = 0.2 = \frac{\sum R_{th} \text{ without insulation}}{\sum R_{th} \text{ with insulation}}$$

We have for the brick and plaster, for unit area,

$$R_b = \frac{\Delta x}{k} = \frac{(4)(0.0254)}{0.7} = 0.145 \text{ m}^2 \cdot {}^\circ C/W$$

$$R_p = \frac{\Delta x}{k} = \frac{(1.5)(0.0254)}{0.48} = 0.079 \text{ m}^2 \cdot {}^\circ C/W$$

so that the thermal resistance without insulation is

$$R = 0.145 + 0.079 = 0.224 \text{ m}^2 \cdot {}^\circ C/W$$

Then

$$R \text{ with insulation} = \frac{0.224}{0.2} = 1.122 \text{ m}^2 \cdot {}^\circ C/W$$

and this represents the sum of our previous value and the resistance for the rock wool

$$1.122 = 0.224 + R_{rw}$$

$$R_{rw} = 0.898 = \frac{\Delta x}{k} = \frac{\Delta x}{0.065}$$

so that

$$\Delta x_{rw} = 0.0584 \text{ m} = 2.30 \text{ in}$$

Figure Example 2-2

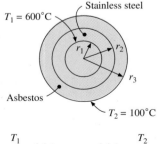

| EXAMPLE 2-2 | Multilayer Cylindrical System |

A thick-walled tube of stainless steel [18% Cr, 8% Ni, $k = 19$ W/m · °C] with 2-cm inner diameter (ID) and 4-cm outer diameter (OD) is covered with a 3-cm layer of asbestos insulation [$k = 0.2$ W/m · °C]. If the inside wall temperature of the pipe is maintained at 600°C, calculate the heat loss per meter of length. Also calculate the tube–insulation interface temperature.

■ **Solution**

Figure Example 2-2 shows the thermal network for this problem. The heat flow is given by

$$\frac{q}{L} = \frac{2\pi (T_1 - T_2)}{\ln(r_2/r_1)/k_s + \ln(r_3/r_2)/k_a} = \frac{2\pi (600 - 100)}{(\ln 2)/19 + (\ln \frac{5}{2})/0.2} = 680 \text{ W/m}$$

This heat flow may be used to calculate the interface temperature between the outside tube wall and the insulation. We have

$$\frac{q}{L} = \frac{T_a - T_2}{\ln(r_3/r_2)/2\pi k_a} = 680 \text{ W/m}$$

where T_a is the interface temperature, which may be obtained as

$$T_a = 595.8°C$$

The largest thermal resistance clearly results from the insulation, and thus the major portion of the temperature drop is through that material.

Convection Boundary Conditions

We have already seen in Chapter 1 that convection heat transfer can be calculated from

$$q_{conv} = h A (T_w - T_\infty)$$

An electric-resistance analogy can also be drawn for the convection process by rewriting the equation as

$$q_{conv} = \frac{T_w - T_\infty}{1/h A} \qquad \textbf{[2-11]}$$

where the $1/h A$ term now becomes the convection resistance.

2-5 | THE OVERALL HEAT-TRANSFER COEFFICIENT

Consider the plane wall shown in Figure 2-5 exposed to a hot fluid A on one side and a cooler fluid B on the other side. The heat transfer is expressed by

$$q = h_1 A (T_A - T_1) = \frac{kA}{\Delta x} (T_1 - T_2) = h_2 A (T_2 - T_B)$$

The heat-transfer process may be represented by the resistance network in Figure 2-5, and the overall heat transfer is calculated as the ratio of the overall temperature difference to the sum of the thermal resistances:

$$q = \frac{T_A - T_B}{1/h_1 A + \Delta x/kA + 1/h_2 A} \qquad \textbf{[2-12]}$$

Figure 2-5 | Overall heat transfer through a plane wall.

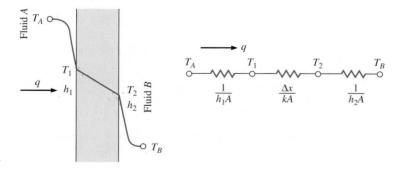

Figure 2-6 | Resistance analogy for hollow cylinder with convection boundaries.

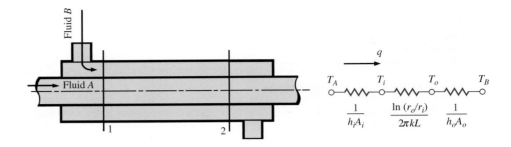

Observe that the value $1/hA$ is used to represent the convection resistance. The overall heat transfer by combined conduction and convection is frequently expressed in terms of an overall heat-transfer coefficient U, defined by the relation

$$q = UA\,\Delta T_{\text{overall}} \qquad\qquad \textbf{[2-13]}$$

where A is some suitable area for the heat flow. In accordance with Equation (2-12), the overall heat-transfer coefficient would be

$$U = \frac{1}{1/h_1 + \Delta x/k + 1/h_2}$$

The overall heat-transfer coefficient is also related to the R value of Equation (2-6) through

$$U = \frac{1}{R\text{ value}}$$

For a hollow cylinder exposed to a convection environment on its inner and outer surfaces, the electric-resistance analogy would appear as in Figure 2-6 where, again, T_A and T_B are the two fluid temperatures. Note that the area for convection is not the same for both fluids in this case, these areas depending on the inside tube diameter and wall thickness. The overall heat transfer would be expressed by

$$q = \frac{T_A - T_B}{\dfrac{1}{h_i A_i} + \dfrac{\ln(r_o/r_i)}{2\pi kL} + \dfrac{1}{h_o A_o}} \qquad\qquad \textbf{[2-14]}$$

in accordance with the thermal network shown in Figure 2-6. The terms A_i and A_o represent the inside and outside surface areas of the inner tube. The overall heat-transfer coefficient may be based on either the inside or the outside area of the tube. Accordingly,

$$U_i = \frac{1}{\dfrac{1}{h_i} + \dfrac{A_i \ln(r_o/r_i)}{2\pi kL} + \dfrac{A_i}{A_o}\dfrac{1}{h_o}} \qquad\qquad \textbf{[2-15]}$$

$$U_o = \frac{1}{\dfrac{A_o}{A_i}\dfrac{1}{h_i} + \dfrac{A_o \ln(r_o/r_i)}{2\pi kL} + \dfrac{1}{h_o}} \qquad\qquad \textbf{[2-16]}$$

The general notion, for either the plane wall or cylindrical coordinate system, is that

$$UA = 1/\Sigma\,R_{\text{th}} = 1/R_{\text{th,overall}}$$

Calculations of the convection heat-transfer coefficients for use in the overall heat-transfer coefficient are made in accordance with the methods described in later chapters. Some typical values of the overall heat-transfer coefficient for heat exchangers are given in Table 10-1. Some values of U for common types of building construction system are given in Table 2-2 and may be employed for calculations involving the heating and cooling of buildings.

Table 2-2 | Overall heat transfer coefficients for common construction systems according to James and Goss [12].

	Description of construction system	U, Btu/hr·ft^2·°F	U, W/m^2·°C
1	2 × 3 in double-wood stud wall, 406 mm OC, polyisocyanurate (0.08-mm vapor retarder, 19-mm insulation), fiberglass batts in cavity, 12.7-mm plywood	0.027	0.153
2	2 × 4 in wood stud wall, 406 mm OC, polyisocyanurate foil-faced, fiberglass batts in cavity, 15-mm plywood	0.060	0.359
3	2 × 4 in wood stud wall, 406 mm OC, 38-mm polyisocyanurate, foil-faced, cellular polyurethane in cavity, 19-mm plywood	0.039	0.221
4	2 × 4 in wood stud wall, 406 mm OC, 15-mm exterior sheathing, 0.05-mm polyethylene vapor barrier, no fill in cavity	0.326	1.85
5	Nominal 4-in concrete-block wall with brick facade and extruded polystyrene insulation	0.080	0.456
6	2 × 4 in wood stud wall, 406 mm OC, fiberglass batt insulation in cavity, 16-mm plywood	0.084	0.477
7	2 × 4 in wood stud wall, 406 mm OC, fiberglass batt insulation in cavity, 16-mm plywood, clay brick veneer	0.060	0.341
8	2 × 4 in wood stud wall, 406 mm OC, fiberglass batt in cavity, 13-mm plywood, aluminum or vinyl siding	0.074	0.417
9	2 × 4 in wood stud wall, 406 mm OC, polyurethane foam in cavity, extruded polystyrene sheathing, aluminum siding	0.040	0.228
10	2 × 4 in steel stud wall, 406 mm OC, fiberglass batts in cavity, 41-mm air space, 13-mm plaster board	0.122	0.691
11	Aluminum motor home roof with fiberglass insulation in cavity (32 mm)	0.072	0.41
12	2 × 6 in wood stud ceiling, 406 mm OC, fiberglass foil-faced insulation in cavity, reflective airspace ($\varepsilon \approx 0.05$)	0.065	0.369
13	8-in (203-mm) normal-weight structural concrete ($\rho = 2270$ kg/m^3) wall, 18-mm board insulation, painted off-white	0.144	0.817
14	10-in (254-mm) concrete-block-brick cavity wall, no insulation in cavities	0.322	1.83
15	8-in (203-mm) medium-weight concrete block wall, perlite insulation in cores	0.229	1.3
16	8-in (203-mm) normal-weight structural concrete, ($\rho = 2270$ kg/m^3) including steel reinforcement bars (Note: actual thickness of concrete is 211 mm.)	0.764	4.34
17	8-in (203-mm) lightweight structural concrete ($\rho = 1570$ kg/m^3) including steel reinforcement bars (Note: Actual thickness of concrete is 210 mm.)	0.483	2.75
18	8-in (203-mm) low-density concrete wall ($\rho = 670$ kg/m^3) including steel reinforcement bars (Note: Actual thickness of concrete is 216 mm.)	0.216	1.23
19	Corrugated sheet steel wall with 10.2-in (260-mm.) fiberglass batt in cavity	0.030	0.17
20	Corrugated sheet steel wall with (159-mm) fiberglass batt in cavity	0.054	0.31
21	Metal building roof deck, 25 mm polyisocyanurate, foil-faced ($\varepsilon \approx 0.03$), 203-mm reflective air space	0.094	0.535
22	Metal building roof deck, 25-mm foil-faced polyisocyanurate, 38-mm fiberglass batts in cavity	0.065	0.366

| EXAMPLE 2-3 | Heat Transfer Through a Composite Wall |

"Two-by-four" wood studs have actual dimensions of 4.13×9.21 cm and a thermal conductivity of 0.1 W/m · °C. A typical wall for a house is constructed as shown Figure Example 2-3. Calculate the overall heat-transfer coefficient and R value of the wall.

Figure Example 2-3 | (*a*) Construction of a dwelling wall; (*b*) thermal resistance model.

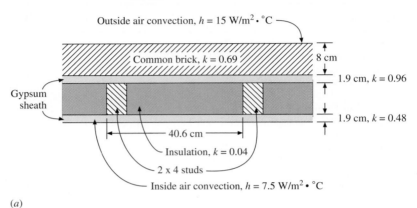

(*a*)

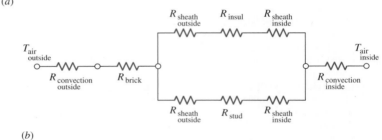

(*b*)

■ **Solution**

The wall section may be considered as having two parallel heat-flow paths: (1) through the studs, and (2) through the insulation. We will compute the thermal resistance for each, and then combine the values to obtain the overall heat-transfer coefficient.

1. *Heat transfer through studs* ($A = 0.0413$ m^2 for unit depth). This heat flow occurs through six thermal resistances:

   a. Convection resistance outside of brick

   $$R = \frac{1}{hA} = \frac{1}{(15)(0.0413)} = 1.614 \,°C/W$$

   b. Conduction resistance in brick

   $$R = \Delta x / kA = \frac{0.08}{(0.69)(0.0413)} = 2.807 \,°C/W$$

   c. Conduction resistance through outer sheet

   $$R = \frac{\Delta x}{kA} = \frac{0.019}{(0.96)(0.0413)} = 0.48 \,°C/W$$

d. Conduction resistance through wood stud

$$R = \frac{\Delta x}{kA} = \frac{0.0921}{(0.1)(0.0413)} = 22.3 \,°\text{C/W}$$

e. Conduction resistance through inner sheet

$$R = \frac{\Delta x}{kA} = \frac{0.019}{(0.48)(0.0413)} = 0.96 \,°\text{C/W}$$

f. Convection resistance on inside

$$R = \frac{1}{hA} = \frac{1}{(7.5)(0.0413)} = 3.23 \,°\text{C/W}$$

The total thermal resistance through the wood stud section is

$$R_{\text{total}} = 1.614 + 2.807 + 0.48 + 22.3 + 0.96 + 3.23 = 31.39 \,°\text{C/W} \qquad [a]$$

2. *Insulation section* ($A = 0.406 - 0.0413$ m^2 for unit depth). Through the insulation section, five of the materials are the same, but the resistances involve different area terms, i.e., $40.6 - 4.13$ cm instead of 4.13 cm, so that each of the previous resistances must be multiplied by a factor of $4.13/(40.6 - 4.13) = 0.113$. The resistance through the insulation is

$$R = \frac{\Delta x}{kA} = \frac{0.0921}{(0.04)(0.406 - 0.0413)} = 6.31$$

and the total resistance through the insulation section is

$$R_{\text{total}} = (1.614 + 2.807 + 0.48 + 0.96 + 3.23)(0.113) + 6.31 = 7.337 \,°\text{C/W} \qquad [b]$$

The overall resistance for the section is now obtained by combining the parallel resistances in Equations (*a*) and (*b*) to give

$$R_{\text{overall}} = \frac{1}{(1/31.39) + (1/7.337)} = 5.947 \,°\text{C/W} \qquad [c]$$

This value is related to the overall heat-transfer coefficient by

$$q = UA\Delta T = \frac{\Delta T}{R_{\text{overall}}} \qquad [d]$$

where A is the area of the total section = 0.406 m^2. Thus,

$$U = \frac{1}{RA} = \frac{1}{(5.947)(0.406)} = 0.414 \,\text{W/m}^2 \cdot °\text{C}$$

As we have seen, the R value is somewhat different from thermal resistance and is given by

$$R \text{ value} = \frac{1}{U} = \frac{1}{0.414} = 2.414 °\text{C} \cdot \text{m}^2/\text{W}$$

■ **Comment**
This example illustrates the relationships between the concepts of thermal resistance, the overall heat-transfer coefficient, and the R value. Note that the R value involves a unit area concept, while the thermal resistance does not.

| EXAMPLE 2-4 | Cooling Cost Savings with Extra Insulation |

A small metal building is to be constructed of corrugated steel sheet walls with a total wall surface area of about 300 m^2. The air conditioner consumes about 1 kW of electricity for every 4 kW of cooling supplied.[1] Two wall constructions are to be compared on the basis of cooling costs. Assume that electricity costs $0.15/kWh. Determine the electrical energy savings of using 260 mm of fiberglass batt insulation instead of 159 mm of fiberglass insulation in the wall. Assume an overall temperature difference across the wall of 20°C on a hot summer day in Texas.

■ **Solution**

Consulting Table 2-2 (Numbers 19 and 20) we find that overall heat transfer coefficients for the two selected wall constructions are

$$U(260\text{-mm fiberglass}) = 0.17 \text{ W/m}^2 \cdot {}^\circ\text{C}$$

$$U(159\text{-mm fiberglass}) = 0.31 \text{ W/m}^2 \cdot {}^\circ\text{C}$$

The heat gain is calculated from $q = UA\Delta T$, so for the two constructions

$$q \text{ (260-mm fiberglass)} = (0.17)(300)(20) = 1020 \text{ W}$$

$$q \text{ (159-mm fiberglass)} = (0.31)(300)(20) = 1860 \text{ W}$$

$$\text{Savings due to extra insulation} = 840 \text{ W}$$

The energy consumed to supply this extra cooling is therefore

$$\text{Extra electric power required} = (840)(1/4) = 210 \text{ W}$$

and the cost is

$$\text{Cost} = (0.210\text{kW})(0.15\$/\text{kWh}) = 0.0315 \text{ \$/hr}$$

Assuming 10-h/day operation for 23 days/month this cost becomes

$$(0.0315)(10)(23) = \$7.25/\text{month}$$

Both of these cases are rather well insulated. If one makes a comparison to a 2 × 4 wood stud wall with no insulation (Number 4 in Table 2-2) fill in the cavity ($U = 1.85 \text{ W/m}^2 \cdot {}^\circ\text{C}$), the heating load would be

$$q = (1.85)(300)(20) = 11,100 \text{ W}$$

and the savings compared with the 260-mm fiberglass insulation would be

$$11,100 - 1020 = 10,080 \text{ W}$$

producing a corresponding electric power saving of $0.378/h or $86.94/month. Clearly the insulated wall will pay for itself. It is a matter of conjecture whether the 260-mm of insulation will pay for itself in comparison to the 159-mm insulation.

---

[1]This is not getting something for nothing. Consult any standard thermodynamics text for the reason for this behavior.

Overall Heat-Transfer Coefficient for a Tube	**EXAMPLE 2-5**

Water flows at 50°C inside a 2.5-cm-inside-diameter tube such that $h_i = 3500$ W/m$^2 \cdot$ °C. The tube has a wall thickness of 0.8 mm with a thermal conductivity of 16 W/m $\cdot$ °C. The outside of the tube loses heat by free convection with $h_o = 7.6$ W/m$^2 \cdot$ °C. Calculate the overall heat-transfer coefficient and heat loss per unit length to surrounding air at 20°C.

■ **Solution**

There are three resistances in series for this problem, as illustrated in Equation (2-14). With $L = 1.0$ m, $d_i = 0.025$ m, and $d_o = 0.025 + (2)(0.0008) = 0.0266$ m, the resistances may be calculated as

$$R_i = \frac{1}{h_i A_i} = \frac{1}{(3500)\pi(0.025)(1.0)} = 0.00364 \text{ °C/W}$$

$$R_t = \frac{\ln(d_o/d_i)}{2\pi k L}$$

$$= \frac{\ln(0.0266/0.025)}{2\pi(16)(1.0)} = 0.00062 \text{ °C/W}$$

$$R_o = \frac{1}{h_o A_o} = \frac{1}{(7.6)\pi(0.0266)(1.0)} = 1.575 \text{ °C/W}$$

Clearly, the outside convection resistance is the largest, and *overwhelmingly so*. This means that it is the controlling resistance for the total heat transfer because the other resistances (in series) are negligible in comparison. We shall base the overall heat-transfer coefficient on the outside tube area and write

$$q = \frac{\Delta T}{\sum R} = U A_o \Delta T \qquad\qquad [a]$$

$$U_o = \frac{1}{A_o \sum R} = \frac{1}{[\pi(0.0266)(1.0)](0.00364 + 0.00062 + 1.575)}$$

$$= 7.577 \text{ W/m}^2 \cdot \text{°C}$$

or a value very close to the value of $h_o = 7.6$ for the outside convection coefficient. The heat transfer is obtained from Equation (a), with

$$q = U A_o \, \Delta T = (7.577)\pi(0.0266)(1.0)(50 - 20) = 19 \text{ W (for 1.0 m length)}$$

■ **Comment**

This example illustrates the important point that many practical heat-transfer problems involve multiple modes of heat transfer acting in combination; in this case, as a series of thermal resistances. It is not unusual for one mode of heat transfer to dominate the overall problem. In this example, the total heat transfer could have been computed very nearly by just calculating the free convection heat loss from the outside of the tube maintained at a temperature of 50°C. Because the inside convection and tube wall resistances are so small, there are correspondingly small temperature drops, and the outside temperature of the tube will be very nearly that of the liquid inside, or 50°C.

2-6 | CRITICAL THICKNESS OF INSULATION

Let us consider a layer of insulation which might be installed around a circular pipe, as shown in Figure 2-7. The inner temperature of the insulation is fixed at T_i, and the outer

Figure 2-7 | Critical insulation thickness.

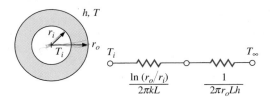

surface is exposed to a convection environment at T_∞. From the thermal network the heat transfer is

$$q = \frac{2\pi L\,(T_i - T_\infty)}{\dfrac{\ln\,(r_o/r_i)}{k} + \dfrac{1}{r_o h}}$$

[2-17]

Now let us manipulate this expression to determine the outer radius of insulation r_o, which will maximize the heat transfer. The maximization condition is

$$\frac{dq}{dr_o} = 0 = \frac{-2\pi L\,(T_i - T_\infty)\left(\dfrac{1}{kr_o} - \dfrac{1}{hr_o^2}\right)}{\left[\dfrac{\ln\,(r_o/r_i)}{k} + \dfrac{1}{r_o h}\right]^2}$$

which gives the result

$$r_o = \frac{k}{h}$$

[2-18]

Equation (2-18) expresses the critical-radius-of-insulation concept. If the outer radius is less than the value given by this equation, then the heat transfer will be *increased* by adding more insulation. For outer radii greater than the critical value an increase in insulation thickness will cause a decrease in heat transfer. The central concept is that for sufficiently small values of h the convection heat loss may actually increase with the addition of insulation because of increased surface area.

EXAMPLE 2-6	Critical Insulation Thickness

Calculate the critical radius of insulation for asbestos $[k = 0.17\ \text{W/m}\cdot{}^\circ\text{C}]$ surrounding a pipe and exposed to room air at 20°C with $h = 3.0\ \text{W/m}^2\cdot{}^\circ\text{C}$. Calculate the heat loss from a 200°C, 5.0-cm-diameter pipe when covered with the critical radius of insulation and without insulation.

■ **Solution**
From Equation (2-18) we calculate r_o as

$$r_o = \frac{k}{h} = \frac{0.17}{3.0} = 0.0567\ \text{m} = 5.67\ \text{cm}$$

The inside radius of the insulation is $5.0/2 = 2.5$ cm, so the heat transfer is calculated from Equation (2-17) as

$$\frac{q}{L} = \frac{2\pi\,(200 - 20)}{\dfrac{\ln\,(5.67/2.5)}{0.17} + \dfrac{1}{(0.0567)(3.0)}} = 105.7\ \text{W/m}$$

Without insulation the convection from the outer surface of the pipe is

$$\frac{q}{L} = h(2\pi r)(T_i - T_o) = (3.0)(2\pi)(0.025)(200 - 20) = 84.8\ \text{W/m}$$

So, the addition of 3.17 cm (5.67 − 2.5) of insulation actually *increases* the heat transfer by 25 percent.

As an alternative, fiberglass having a thermal conductivity of 0.04 W/m · °C might be employed as the insulation material. Then, the critical radius would be

$$r_o = \frac{k}{h} = \frac{0.04}{3.0} = 0.0133 \text{ m} = 1.33 \text{ cm}$$

Now, the value of the critical radius is less than the outside radius of the pipe (2.5 cm), so addition of *any* fiberglass insulation would cause a *decrease* in the heat transfer. In a practical pipe insulation problem, the total heat loss will also be influenced by radiation as well as convection from the outer surface of the insulation.

2-7 | HEAT-SOURCE SYSTEMS

A number of interesting applications of the principles of heat transfer are concerned with systems in which heat may be generated internally. Nuclear reactors are one example; electrical conductors and chemically reacting systems are others. At this point we shall confine our discussion to one-dimensional systems, or, more specifically, systems where the temperature is a function of only one space coordinate.

Plane Wall with Heat Sources

Consider the plane wall with uniformly distributed heat sources shown in Figure 2-8. The thickness of the wall in the x direction is $2L$, and it is assumed that the dimensions in the other directions are sufficiently large that the heat flow may be considered as one-dimensional. The heat generated per unit volume is $\dot{q}$, and we assume that the thermal conductivity does not vary with temperature. This situation might be produced in a practical situation by passing a current through an electrically conducting material. From Chapter 1,

Figure 2-8 | Sketch illustrating one-dimensional conduction problem with heat generation.

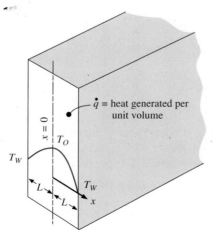

the differential equation that governs the heat flow is

$$\frac{d^2T}{dx^2} + \frac{\dot{q}}{k} = 0 \qquad \text{[2-19]}$$

For the boundary conditions we specify the temperatures on either side of the wall, i.e.,

$$T = T_w \qquad \text{at } x = \pm L \qquad \text{[2-20]}$$

The general solution to Equation (2-19) is

$$T = -\frac{\dot{q}}{2k}x^2 + C_1 x + C_2 \qquad \text{[2-21]}$$

Because the temperature must be the same on each side of the wall, C_1 must be zero. The temperature at the midplane ($x = 0$) is denoted by T_0 and from Equation (2-21)

$$T_0 = C_2$$

The temperature distribution is therefore

$$T - T_0 = -\frac{\dot{q}}{2k}x^2 \qquad \text{[2-22a]}$$

or

$$\frac{T - T_0}{T_w - T_0} = \left(\frac{x}{L}\right)^2 \qquad \text{[2-22b]}$$

a parabolic distribution. An expression for the midplane temperature T_0 may be obtained through an energy balance. At steady-state conditions the total heat generated must equal the heat lost at the faces. Thus

$$2\left(-kA\frac{dT}{dx}\bigg]_{x=L}\right) = \dot{q}A\,2L$$

where A is the cross-sectional area of the plate. The temperature gradient at the wall is obtained by differentiating Equation (2-22b):

$$\frac{dT}{dx}\bigg]_{x=L} = (T_w - T_0)\left(\frac{2x}{L^2}\right)\bigg]_{x=L} = (T_w - T_0)\frac{2}{L}$$

Then

$$-k(T_w - T_0)\frac{2}{L} = \dot{q}L$$

and

$$T_0 = \frac{\dot{q}L^2}{2k} + T_w \qquad \text{[2-23]}$$

This same result could be obtained by substituting $T = T_w$ at $x = L$ into Equation (2-22a).

The equation for the temperature distribution could also be written in the alternative form

$$\frac{T - T_w}{T_0 - T_w} = 1 - \frac{x^2}{L^2} \qquad \text{[2-22c]}$$

2-8 | CYLINDER WITH HEAT SOURCES

Consider a cylinder of radius R with uniformly distributed heat sources and constant thermal conductivity. If the cylinder is sufficiently long that the temperature may be considered a function of radius only, the appropriate differential equation may be obtained by neglecting the axial, azimuth, and time-dependent terms in Equation (1-3b),

$$\frac{d^2T}{dr^2} + \frac{1}{r}\frac{dT}{dr} + \frac{\dot{q}}{k} = 0 \qquad\qquad \textbf{[2-24]}$$

The boundary conditions are

$$T = T_w \qquad \text{at } r = R$$

and heat generated equals heat lost at the surface:

$$\dot{q}\pi R^2 L = -k2\pi R L \left.\frac{dT}{dr}\right]_{r=R}$$

Since the temperature function must be continuous at the center of the cylinder, we could specify that

$$\frac{dT}{dr} = 0 \qquad \text{at } r = 0$$

However, it will not be necessary to use this condition since it will be satisfied automatically when the two boundary conditions are satisfied.

We rewrite Equation (2-24)

$$r\frac{d^2T}{dr^2} + \frac{dT}{dr} = \frac{-\dot{q}r}{k}$$

and note that

$$r\frac{d^2T}{dr^2} + \frac{dT}{dr} = \frac{d}{dr}\left(r\frac{dT}{dr}\right)$$

Then integration yields

$$r\frac{dT}{dr} = \frac{-\dot{q}r^2}{2k} + C_1$$

and

$$T = \frac{-\dot{q}r^2}{4k} + C_1 \ln r + C_2$$

From the second boundary condition above,

$$\left.\frac{dT}{dr}\right]_{r=R} = \frac{-\dot{q}R}{2k} = \frac{-\dot{q}R}{2k} + \frac{C_1}{R}$$

Thus

$$C_1 = 0$$

We could also note that C_1 must be zero because at $r = 0$ the logarithm function becomes infinite.

From the first boundary condition,

$$T = T_w = \frac{-\dot{q}R^2}{4k} + C_2 \qquad \text{at } r = R$$

so that

$$C_2 = T_w + \frac{\dot{q}R^2}{4k}$$

The final solution for the temperature distribution is then

$$T - T_w = \frac{\dot{q}}{4k}(R^2 - r^2) \qquad \textbf{[2-25a]}$$

or, in dimensionless form,

$$\frac{T - T_w}{T_0 - T_w} = 1 - \left(\frac{r}{R}\right)^2 \qquad \textbf{[2-25b]}$$

where T_0 is the temperature at $r = 0$ and is given by

$$T_0 = \frac{\dot{q}R^2}{4k} + T_w \qquad \textbf{[2-26]}$$

It is left as an exercise to show that the temperature gradient at $r = 0$ is zero.

For a hollow cylinder with uniformly distributed heat sources the appropriate boundary conditions would be

$$T = T_i \quad \text{at } r = r_i \text{ (inside surface)}$$
$$T = T_o \quad \text{at } r = r_o \text{ (outside surface)}$$

The general solution is still

$$T = -\frac{\dot{q}r^2}{4k} + C_1 \ln r + C_2$$

Application of the new boundary conditions yields

$$T - T_o = \frac{\dot{q}}{4k}(r_o^2 - r^2) + C_1 \ln \frac{r}{r_o} \qquad \textbf{[2-27]}$$

where the constant C_1 is given by

$$C_1 = \frac{T_i - T_o + \dot{q}(r_i^2 - r_o^2)/4k}{\ln(r_i/r_o)} \qquad \textbf{[2-28]}$$

EXAMPLE 2-7	Heat Source with Convection

A current of 200 A is passed through a stainless-steel wire $[k = 19 \text{ W/m} \cdot °\text{C}]$ 3 mm in diameter. The resistivity of the steel may be taken as 70 $\mu\Omega \cdot$ cm, and the length of the wire is 1 m. The wire is submerged in a liquid at 110 °C and experiences a convection heat-transfer coefficient of 4 k W/m$^2 \cdot$ °C. Calculate the center temperature of the wire.

■ **Solution**

All the power generated in the wire must be dissipated by convection to the liquid:

$$P = I^2 R = q = hA(T_w - T_\infty) \qquad \textbf{[a]}$$

The resistance of the wire is calculated from

$$R = \rho \frac{L}{A} = \frac{(70 \times 10^{-6})(100)}{\pi(0.15)^2} = 0.099 \ \Omega$$

where ρ is the resistivity of the wire. The surface area of the wire is πdL, so from Equation (a),

$$(200)^2(0.099) = 4000\pi(3 \times 10^{-3})(1)(T_w - 110) = 3960 \text{ W}$$

and

$$T_w = 215°C \quad [419°F]$$

The heat generated per unit volume $\dot{q}$ is calculated from

$$P = \dot{q}V = \dot{q}\pi r^2 L$$

so that

$$\dot{q} = \frac{3960}{\pi(1.5 \times 10^{-3})^2(1)} = 560.2 \text{ MW/m}^3 \quad [5.41 \times 10^7 \text{ Btu/h} \cdot \text{ft}^3]$$

Finally, the center temperature of the wire is calculated from Equation (2-26):

$$T_0 = \frac{\dot{q}r_o^2}{4k} + T_w = \frac{(5.602 \times 10^8)(1.5 \times 10^{-3})^2}{(4)(19)} + 215 = 231.6°C \quad [449°F]$$

2-9 | CONDUCTION-CONVECTION SYSTEMS

The heat that is conducted through a body must frequently be removed (or delivered) by some convection process. For example, the heat lost by conduction through a furnace wall must be dissipated to the surroundings through convection. In heat-exchanger applications a finned-tube arrangement might be used to remove heat from a hot liquid. The heat transfer from the liquid to the finned tube is by convection. The heat is conducted through the material and finally dissipated to the surroundings by convection. Obviously, an analysis of combined conduction-convection systems is very important from a practical standpoint.

We shall defer part of our analysis of conduction-convection systems to Chapter 10 on heat exchangers. For the present we wish to examine some simple extended-surface problems. Consider the one-dimensional fin exposed to a surrounding fluid at a temperature T_∞ as shown in Figure 2-9. The temperature of the base of the fin is T_0. We approach the problem by making an energy balance on an element of the fin of thickness dx as shown in the figure. Thus

Energy in left face = energy out right face + energy lost by convection

The defining equation for the convection heat-transfer coefficient is recalled as

$$q = hA(T_w - T_\infty) \qquad \text{[2-29]}$$

where the area in this equation is the surface area for convection. Let the cross-sectional area of the fin be A and the perimeter be P. Then the energy quantities are

$$\text{Energy in left face} = q_x = -kA\frac{dT}{dx}$$

$$\text{Energy out right face} = q_{x+dx} = -kA\frac{dT}{dx}\bigg]_{x+dx}$$

$$= -kA\left(\frac{dT}{dx} + \frac{d^2T}{dx^2}dx\right)$$

$$\text{Energy lost by convection} = hP\,dx(T - T_\infty)$$

Figure 2-9 | Sketch illustrating one-dimensional conduction and convection through a rectangular fin.

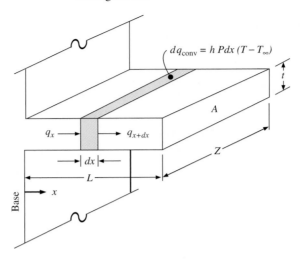

Here it is noted that the differential surface area for convection is the product of the perimeter of the fin and the differential length dx. When we combine the quantities, the energy balance yields

$$\frac{d^2T}{dx^2} - \frac{hP}{kA}(T - T_\infty) = 0 \qquad\qquad \textbf{[2-30a]}$$

Let $\theta = T - T_\infty$. Then Equation (2-30a) becomes

$$\frac{d^2\theta}{dx^2} - \frac{hP}{kA}\theta = 0 \qquad\qquad \textbf{[2-30b]}$$

One boundary condition is

$$\theta = \theta_0 = T_0 - T_\infty \qquad \text{at } x = 0$$

The other boundary condition depends on the physical situation. Several cases may be considered:

CASE 1 The fin is very long, and the temperature at the end of the fin is essentially that of the surrounding fluid.

CASE 2 The fin is of finite length and loses heat by convection from its end.

CASE 3 The end of the fin is insulated so that $dT/dx = 0$ at $x = L$.

If we let $m^2 = hP/kA$, the general solution for Equation (2-30b) may be written

$$\theta = C_1 e^{-mx} + C_2 e^{mx} \qquad\qquad \textbf{[2-31]}$$

For case 1 the boundary conditions are

$$\theta = \theta_0 \quad \text{at } x = 0$$
$$\theta = 0 \quad \text{at } x = \infty$$

and the solution becomes

$$\frac{\theta}{\theta_0} = \frac{T - T_\infty}{T_0 - T_\infty} = e^{-mx}$$

[2-32]

For case 3 the boundary conditions are

$$\theta = \theta_0 \text{ at } x = 0$$
$$\frac{d\theta}{dx} = 0 \text{ at } x = L$$

Thus

$$\theta_0 = C_1 + C_2$$
$$0 = m(-C_1 e^{-mL} + C_2 e^{mL})$$

Solving for the constants C_1 and C_2, we obtain

$$\frac{\theta}{\theta_0} = \frac{e^{-mx}}{1 + e^{-2mL}} + \frac{e^{mx}}{1 + e^{2mL}}$$

[2-33a]

$$= \frac{\cosh[m(L - x)]}{\cosh mL}$$

[2-33b]

The hyperbolic functions are defined as

$$\sinh x = \frac{e^x - e^{-x}}{2} \qquad \cosh x = \frac{e^x + e^{-x}}{2}$$

$$\tanh x = \frac{\sinh x}{\cosh x} = \frac{e^x - e^{-x}}{e^x + e^{-x}}$$

The solution for case 2 is more involved algebraically, and the result is

$$\frac{T - T_\infty}{T_o - T_\infty} = \frac{\cosh m(L - x) + (h/mk)\sinh m(L - x)}{\cosh mL + (h/mk)\sinh mL}$$

[2-34]

All of the heat lost by the fin must be conducted into the base at $x = 0$. Using the equations for the temperature distribution, we can compute the heat loss from

$$q = -kA \left. \frac{dT}{dx} \right]_{x=0}$$

An alternative method of integrating the convection heat loss could be used:

$$q = \int_0^L hP(T - T_\infty)\,dx = \int_0^L hP\theta\,dx$$

In most cases, however, the first equation is easier to apply. For case 1,

$$q = -kA(-m\theta_0 e^{-m(0)}) = \sqrt{hPkA}\,\theta_0$$

[2-35]

For case 3,

$$q = -kA\theta_0 m \left(\frac{1}{1 + e^{-2mL}} - \frac{1}{1 + e^{+2mL}} \right)$$

[2-36]

$$= \sqrt{hPkA}\,\theta_0 \tanh mL$$

The heat flow for case 2 is

$$q = \sqrt{h\,PkA}\,(T_0 - T_\infty)\,\frac{\sinh mL + (h/mk)\cosh mL}{\cosh mL + (h/mk)\sinh mL} \qquad \textbf{[2-37]}$$

In this development it has been assumed that the substantial temperature gradients occur only in the x direction. This assumption will be satisfied if the fin is sufficiently thin. For most fins of practical interest the error introduced by this assumption is less than 1 percent. The overall accuracy of practical fin calculations will usually be limited by uncertainties in values of the convection coefficient h. It is worthwhile to note that the convection coefficient is seldom uniform over the entire surface, as has been assumed above. If severe nonuniform behavior is encountered, numerical finite-difference techniques must be employed to solve the problem. Such techniques are discussed in Chapter 3.

2-10 | FINS

In the foregoing development we derived relations for the heat transfer from a rod or fin of uniform cross-sectional area protruding from a flat wall. In practical applications, fins may have varying cross-sectional areas and may be attached to circular surfaces. In either case the area must be considered as a variable in the derivation, and solution of the basic differential equation and the mathematical techniques become more tedious. We present only the results for these more complex situations. The reader is referred to References 1 and 8 for details on the mathematical methods used to obtain the solutions.

To indicate the effectiveness of a fin in transferring a given quantity of heat, a new parameter called *fin efficiency* is defined by

$$\text{Fin efficiency} = \frac{\text{actual heat transferred}}{\begin{array}{c}\text{heat that would be transferred}\\\text{if entire fin area were}\\\text{at base temperature}\end{array}} = \eta_f$$

For case 3, the fin efficiency becomes

$$\eta_f = \frac{\sqrt{h\,P\,kA}\,\theta_0 \tanh mL}{h\,PL\theta_0} = \frac{\tanh mL}{mL} \qquad \textbf{[2-38]}$$

The fins discussed were assumed to be sufficiently deep that the heat flow could be considered one-dimensional. The expression for mL may be written

$$mL = \sqrt{\frac{hP}{kA}}\,L = \sqrt{\frac{h(2z + 2t)}{kzt}}\,L$$

where z is the depth of the fin, and t is the thickness. Now, if the fin is sufficiently deep, the term $2z$ will be large compared with $2t$, and

$$mL = \sqrt{\frac{2hz}{ktz}}\,L = \sqrt{\frac{2h}{kt}}\,L$$

Multiplying numerator and denominator by $L^{1/2}$ gives

$$mL = \sqrt{\frac{2h}{kLt}}\,L^{3/2}$$

Lt is called the profile area of the fin, which we define as

$$A_m = Lt$$

so that

$$mL = \sqrt{\frac{2h}{kA_m}} L^{3/2} \qquad \text{[2-39]}$$

We may therefore use the expression in Equation (2-39) to compute the efficiency of a fin with insulated tip as given by Equation (2-38).

Harper and Brown [2] have shown that the solution in case 2 may be expressed in the same form as Equation (2-38) when the length of the fin is extended by one-half the thickness of the fin. In effect, lengthening of the fin by $t/2$ is assumed to represent the same convection heat transfer as half the fin tip area placed on top and bottom of the fin. A corrected length L_c is then used in all the equations that apply for the case of the fin with an insulated tip. Thus

$$L_c = L + \frac{t}{2} \qquad \text{[2-40]}$$

The error that results from this approximation will be less than 8 percent when

$$\left(\frac{ht}{2k}\right)^{1/2} \le \frac{1}{2} \qquad \text{[2-41]}$$

If a straight cylindrical rod extends from a wall, the corrected fin length is calculated from

$$L_c = L + \frac{\pi d^2/4}{\pi d} = L + d/4 \qquad \text{[2-42]}$$

Again, the real fin is extended a sufficient length to produce a circumferential area equal to that of the tip area.

Examples of other types of fins are shown in Figure 2-10. Figure 2-11 presents a comparison of the efficiencies of a triangular fin and a straight rectangular fin corresponding to case 2. Figure 2-12 shows the efficiencies of circumferential fins of rectangular cross-sectional area. Notice that the corrected fin lengths L_c and profile area A_m have been used in Figures 2-11 and 2-12. We may note that as $r_{2c}/r_1 \to 1.0$, the efficiency of the circumferential fin becomes identical to that of the straight fin of rectangular profile.

It is interesting to note that the fin efficiency reaches its maximum value for the trivial case of $L = 0$, or no fin at all. Therefore, we should not expect to be able to maximize fin performance with respect to fin length. It is possible, however, to maximize the efficiency with respect to the quantity of fin material (mass, volume, or cost), and such a maximization process has rather obvious economic significance. We have not discussed the subject of radiation heat transfer from fins. The radiant transfer is an important consideration in a

Figure 2-10 | Different types of finned surfaces. (*a*) Straight fin of rectangular profile on plane wall, (*b*) straight fin of rectangular profile on circular tube, (*c*) cylindrical tube with radial fin of rectangular profile, (*d*) cylindrical-spine or circular-rod fin.

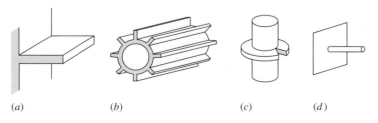

(*a*)　　　　　(*b*)　　　　　(*c*)　　　　(*d*)

Figure 2-11 | Efficiencies of straight rectangular and triangular fins.

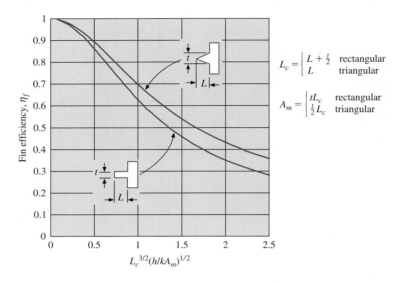

Figure 2-12 | Efficiencies of circumferential fins of rectangular profile, according to Reference 3.

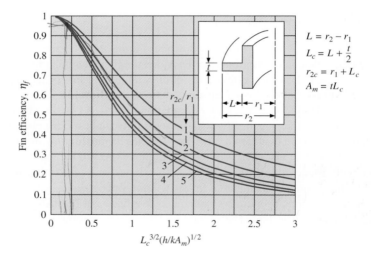

number of applications, and the interested reader should consult Siegel and Howell [9] for information on this subject.

In some cases a valid method of evaluating fin performance is to compare the heat transfer with the fin to that which would be obtained without the fin. The ratio of these quantities is

$$\frac{q \text{ with fin}}{q \text{ without fin}} = \frac{\eta_f A_f h \theta_0}{h A_b \theta_0}$$

where A_f is the total surface area of the fin and A_b is the base area. For the insulated-tip fin described by Equation (2-36),

$$A_f = PL$$
$$A_b = A$$

and the heat ratio would become

$$\frac{q \text{ with fin}}{q \text{ without fin}} = \frac{\tanh mL}{\sqrt{hA/kP}}$$

This term is sometimes called the *fin effectiveness.*

Thermal Resistance for Fin-Wall Combinations

Consider a fin attached to a wall as illustrated in either Figure 2-11 or Figure 2-12. We may calculate a thermal resistance for the wall using either $R_w = \Delta x/kA$ for a plane wall, or $R_w = \ln(r_o/r_i)/2\pi kL$ for a cylindrical wall. In the absence of the fin the convection resistance at the surface would be $1/hA$. The *combined* conduction and convection resistance R_f for the fin is related to the heat lost by the fin through

$$q_f = \eta_f A_f h \theta_o = \frac{\theta_o}{R_f} \qquad \textbf{[2-43]}$$

or, the fin resistance may be expressed as

$$R_f = \frac{1}{\eta_f A_f h} \qquad \textbf{[2-44]}$$

The overall heat transfer through the fin-wall combination is then

$$q_f = \frac{T_i - T_\infty}{R_{wf} + R_f} \qquad \textbf{[2-45]}$$

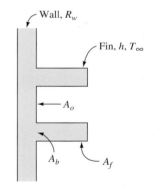

Figure 2-13 | Heat loss from fin-wall combination.

where T_i is the inside wall temperature and R_{wf} is the wall resistance at the fin position. This heat transfer is only for the fin portion of the wall. Now consider the wall section shown in Figure 2-13, having a wall area A_b for the fin and area A_o for the open section of the wall exposed directly to the convection environment. The open wall heat transfer is

$$q_o = \frac{T_i - T_\infty}{R_{wo} + R_o} \qquad \textbf{[2-46]}$$

where now

$$R_o = \frac{1}{hA_o} \qquad \textbf{[2-47]}$$

and R_{wo} is the wall resistance for the open wall section. This value is $R_{wo} = \Delta x/k_w A_o$ for a plane wall, where Δx is the wall thickness. A logarithmic form would be employed for a cylindrical wall, as noted above. The total heat lost by the wall is therefore

$$q_{\text{total}} = q_f + q_o \qquad \textbf{[2-48]}$$

which may be expressed in terms of the thermal resistances by

$$q_{total} = (T_i - T_\infty) \left[\frac{1}{R_{wf} + R_f} + \frac{1}{R_{wo} + R_o} \right]$$

$$= (T_i - T_\infty) \frac{R_{wo} + R_o + R_{wf} + R_f}{(R_{wf} + R_f)(R_{wo} + R_o)}$$ [2-49]

Conditions When Fins Do Not Help

At this point we should remark that the installation of fins on a heat-transfer surface will not necessarily increase the heat-transfer rate. If the value of h, the convection coefficient, is large, as it is with high-velocity fluids or boiling liquids, the fin may produce a *reduction* in heat transfer because the conduction resistance then represents a larger impediment to the heat flow than the convection resistance. To illustrate the point, consider a stainless-steel pin fin that has $k = 16$ W/m · °C, $L = 10$ cm, $d = 1$ cm and that is exposed to a boiling-water convection situation with $h = 5000$ W/m^2 · °C. From Equation (2-36) we can compute

$$\frac{q \text{ with fin}}{q \text{ without fin}} = \frac{\tanh mL}{\sqrt{hA/kp}}$$

$$= \frac{\tanh \left\{ \left[\dfrac{5000\pi(1 \times 10^{-2})(4)}{16\pi(1 \times 10^{-2})^2} \right]^{1/2} (10 \times 10^{-2}) \right\}}{\left[\dfrac{5000\pi(1 \times 10^{-2})^2}{(4)(16)\pi(1 \times 10^{-2})} \right]^{1/2}}$$

$$= 1.13$$

Thus, this rather large pin produces an increase of only 13 percent in the heat transfer.

Still another method of evaluating fin performance is discussed in Problem 2-68. Kern and Kraus [8] give a very complete discussion of extended-surface heat transfer. Some photographs of different fin shapes used in electronic cooling applications are shown in Figure 2-14. These fins are obviously not one-dimensional, i.e., they cannot be characterized with a single space coordinate.

Cautionary Remarks Concerning Convection Coefficients for Fins

We have already noted that the convection coefficient may vary with type of fluid, flow velocity, geometry, etc. As we shall see in Chapters 5, 6, and 7, empirical correlations for h frequently have uncertainties of the order of ±25 percent. Moreover, the correlations are based on controlled laboratory experiments that are infrequently matched in practice. What this means is that the assumption of constant h used in the derivation of fin performance may be in considerable error and the value of h may vary over the fin surface. For the heat-transfer practitioner, complex geometries like those shown in Figure 2-14 must be treated with particular care. These configurations usually must be tested under near or actual operating conditions in order to determine their performance with acceptable reliability. These remarks are not meant to discourage the reader, but rather to urge prudence when estimating the performance of complex finned surfaces for critical applications.

Figure 2-14 | Some fin arrangements used in electronic cooling applications.

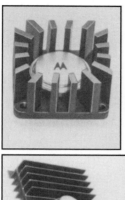

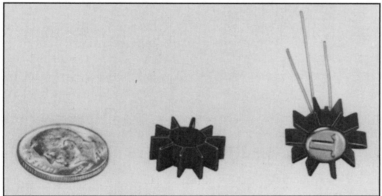

Source: Courtesy Wakefield Engineering Inc., Wakefield, Mass.

Influence of Thermal Conductivity on Fin Temperature Profiles

EXAMPLE 2-8

Compare the temperature distributions in a straight cylindrical rod having a diameter of 2 cm and a length of 10 cm and exposed to a convection environment with $h = 25$ W/m$^2 \cdot °$C, for three fin materials: copper [$k = 385$ W/m $\cdot °$C], stainless steel [$k = 17$ W/m $\cdot °$C], and glass [$k = 0.8$ W/m $\cdot °$C]. Also compare the relative heat flows and fin efficiencies.

■ **Solution**
We have

$$\frac{hP}{kA} = \frac{(25)\pi(0.02)}{k\pi(0.01)^2} = \frac{5000}{k}$$

The terms of interest are therefore

Material	$\frac{hP}{kA}$	m	mL
Copper	12.99	3.604	0.3604
Stainless steel	294.1	17.15	1.715
Glass	6250	79.06	7.906

These values may be inserted into Equation (2-33a) to calculate the temperatures at different x locations along the rod, and the results are shown in Figure Example 2-8. We notice that the glass behaves as a "very long" fin, and its behavior could be calculated from Equation (2-32). The fin efficiencies are calculated from Equation (2-38) by using the corrected length approximation of Equation (2-42). We have

$$L_c = L + \frac{d}{4} = 10 + \frac{2}{4} = 10.5 \text{ cm}$$

Figure Example 2-8

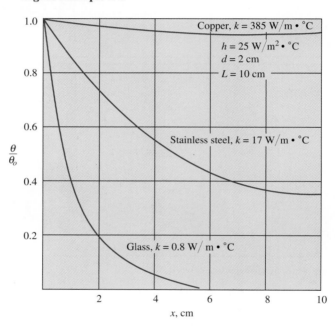

The parameters of interest for the heat-flow and efficiency comparisons are now tabulated as

Material	$hPkA$	mL_c
Copper	0.190	0.3784
Stainless steel	0.0084	1.8008
Glass	3.9×10^{-4}	8.302

To compare the heat flows we could either calculate the values from Equation (2-36) for a unit value of θ_0 or observe that the fin efficiency gives a relative heat-flow comparison because the maximum heat transfer is the same for all three cases; i.e., we are dealing with the same fin size, shape, and value of h. We thus calculate the values of η_f from Equation (2-38) and the above values of mL_c.

Material	η_f	q relative to copper, %
Copper	0.955	100
Stainless steel	0.526	53.1
Glass	0.124	12.6

The temperature profiles in the accompanying figure can be somewhat misleading. The glass has the steepest temperature gradient at the base, but its much lower value of k produces a lower heat-transfer rate.

Straight Aluminum Fin   EXAMPLE 2-9

An aluminum fin [$k = 200$ W/m · °C] 3.0 mm thick and 7.5 cm long protrudes from a wall, as in Figure 2-9. The base is maintained at 300°C, and the ambient temperature is 50°C with $h = 10$ W/m² · °C. Calculate the heat loss from the fin per unit depth of material.

■ **Solution**
We may use the approximate method of solution by extending the fin a fictitious length $t/2$ and then computing the heat transfer from a fin with insulated tip as given by Equation (2-36). We have

$$L_c = L + t/2 = 7.5 + 0.15 = 7.65 \text{ cm [3.01 in]}$$
$$m = \sqrt{\frac{hP}{kA}} = \left[\frac{h(2z+2t)}{ktz}\right]^{1/2} \approx \sqrt{\frac{2h}{kt}}$$

when the fin depth $z \gg t$. So,

$$m = \left[\frac{(2)(10)}{(200)(3 \times 10^{-3})}\right]^{1/2} = 5.774$$

From Equation (2-36), for an insulated-tip fin

$$q = (\tanh m L_c)\sqrt{hPkA}\,\theta_0$$

For a 1 m depth

$$A = (1)(3 \times 10^{-3}) = 3 \times 10^{-3} \text{ m}^2 \text{ [4.65 in}^2\text{]}$$

and

$$q = (5.774)(200)(3 \times 10^{-3})(300 - 50)\tanh[(5.774)(0.0765)]$$
$$= 359 \text{ W/m [373.5 Btu/h · ft]}$$

Circumferential Aluminum Fin   EXAMPLE 2-10

Aluminum fins 1.5 cm wide and 1.0 mm thick are placed on a 2.5-cm-diameter tube to dissipate the heat. The tube surface temperature is 170°, and the ambient-fluid temperature is 25°C. Calculate the heat loss per fin for $h = 130$ W/m² · °C. Assume $k = 200$ W/m · °C for aluminum.

■ **Solution**
For this example we can compute the heat transfer by using the fin-efficiency curves in Figure 2-12. The parameters needed are

$$L_c = L + t/2 = 1.5 + 0.05 = 1.55 \text{ cm}$$
$$r_1 = 2.5/2 = 1.25 \text{ cm}$$
$$r_{2c} = r_1 + L_c = 1.25 + 1.55 = 2.80 \text{ cm}$$

$$r_{2c}/r_1 = 2.80/1.25 = 2.24$$

$$A_m = t(r_{2c} - r_1) = (0.001)(2.8 - 1.25)(10^{-2}) = 1.55 \times 10^{-5} \ m^2$$

$$L_c^{3/2} \left(\frac{h}{kA_m} \right)^{1/2} = (0.0155)^{3/2} \left[\frac{130}{(200)(1.55 \times 10^{-5})} \right]^{1/2} = 0.396$$

From Figure 2-12, $\eta_f = 82$ percent. The heat that would be transferred if the entire fin were at the base temperature is (both sides of fin exchanging heat)

$$q_{max} = 2\pi(r_{2c}^2 - r_1^2)h(T_0 - T_\infty)$$

$$= 2\pi(2.8^2 - 1.25^2)(10^{-4})(130)(170 - 25)$$

$$= 74.35 \ W \ [253.7 \ Btu/h]$$

The actual heat transfer is then the product of the heat flow and the fin efficiency:

$$q_{act} = (0.82)(74.35) = 60.97 \ W \ [208 \ Btu/h]$$

EXAMPLE 2-11 Rod with Heat Sources

A rod containing uniform heat sources per unit volume $\dot{q}$ is connected to two temperatures as shown in Figure Example 2-11. The rod is also exposed to an environment with convection coefficient h and temperature T_∞. Obtain an expression for the temperature distribution in the rod.

Figure Example 2-11

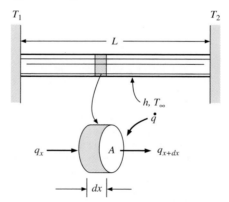

■ Solution

We first must make an energy balance on the element of the rod shown, similar to that used to derive Equation (2-30). We have

Energy in left face + heat generated in element

= energy out right face + energy lost by convection

or

$$-kA \frac{dT}{dx} + \dot{q}A \, dx = -kA \left(\frac{dT}{dx} + \frac{d^2T}{dx^2} \, dx \right) + hP \, dx \, (T - T_\infty)$$

Simplifying, we have

$$\frac{d^2T}{dx^2} - \frac{hP}{kA}(T - T_\infty) + \frac{\dot{q}}{k} = 0 \qquad\qquad [a]$$

or, with $\theta = T - T_\infty$ and $m^2 = hP/kA$

$$\frac{d^2\theta}{dx} - m^2\theta + \frac{\dot{q}}{k} = 0 \qquad [b]$$

We can make a further variable substitution as

$$\theta' = \theta - \dot{q}/km^2$$

so that our differential equation becomes

$$\frac{d^2\theta'}{dx^2} - m^2\theta' = 0 \qquad [c]$$

which has the general solution

$$\theta' = C_1 e^{-mx} + C_2 e^{mx} \qquad [d]$$

The two end temperatures are used to establish the boundary conditions:

$$\theta' = \theta'_1 = T_1 - T_\infty - \dot{q}/km^2 = C_1 + C_2$$
$$\theta' = \theta'_2 = T_2 - T_\infty - \dot{q}/km^2 = C_1 e^{-mL} + C_2 e^{mL}$$

Solving for the constants C_1 and C_2 gives

$$\theta' = \frac{(\theta'_1 e^{2mL} - \theta'_2 e^{mL})e^{-mx} + (\theta'_2 e^{mL} - \theta'_1)e^{mx}}{e^{2mL} - 1} \qquad [e]$$

For an infinitely long heat-generating fin with the left end maintained at T_1, the temperature distribution becomes

$$\theta'/\theta'_1 = e^{-mx} \qquad [f]$$

a relation similar to Equation (2-32) for a non-heat-generating fin.

■ **Comment**
Note that the above relationships assume one-dimensional behavior, i.e., temperature dependence only on the x-coordinate and temperature uniformity across the area A. For sufficiently large heat generation rates and/or cross-section areas, the assumption may no longer be valid. In these cases, the problem must be treated as multidimensional using the techniques described in Chapter 3.

2-11 | THERMAL CONTACT RESISTANCE

Imagine two solid bars brought into contact as indicated in Figure 2-15, with the sides of the bars insulated so that heat flows only in the axial direction. The materials may have different thermal conductivities, but if the sides are insulated, the heat flux must be the same through both materials under steady-state conditions. Experience shows that the actual temperature profile through the two materials varies approximately as shown in Figure 2-15b. The temperature drop at plane 2, the contact plane between the two materials, is said to be the result of a *thermal contact resistance*. Performing an energy balance on the two materials, we obtain

$$q = k_A A \frac{T_1 - T_{2A}}{\Delta x_A} = \frac{T_{2A} - T_{2B}}{1/h_c A} = k_B A \frac{T_{2B} - T_3}{\Delta x_B}$$

or

$$q = \frac{T_1 - T_3}{\Delta x_A/k_A A + 1/h_c A + \Delta x_B/k_B A} \qquad [2-50]$$

where the quantity $1/h_c A$ is called the thermal contact resistance and h_c is called the contact coefficient. This factor can be extremely important in a number of applications because of the many heat-transfer situations that involve mechanical joining of two materials.

The physical mechanism of contact resistance may be better understood by examining a joint in more detail, as shown in Figure 2-16. The actual surface roughness is exaggerated to implement the discussion. No real surface is perfectly smooth, and the actual surface roughness is believed to play a central role in determining the contact resistance. There are two principal contributions to the heat transfer at the joint:

1. The solid-to-solid conduction at the spots of contact
2. The conduction through entrapped gases in the void spaces created by the contact

The second factor is believed to represent the major resistance to heat flow, because the thermal conductivity of the gas is quite small in comparison to that of the solids.

Figure 2-15 | Illustrations of thermal-contact-resistance effect: (*a*) physical situation; (*b*) temperature profile.

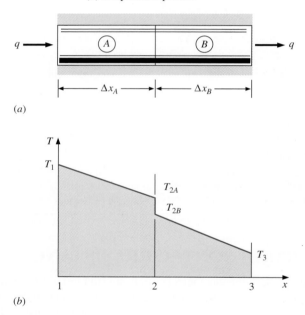

Figure 2-16 | Joint-roughness model for analysis of thermal contact resistance.

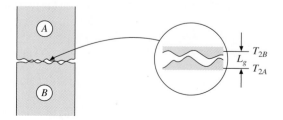

Table 2-3 | Contact conductance of typical surfaces.

Surface type	Roughness μ in	Roughness μm	Temperature, °C	Pressure, atm	1/h_c h·ft²·°F/ Btu	1/h_c m²·°C/W × 10⁴
416 Stainless, ground, air	100	2.54	90–200	3–25	0.0015	2.64
304 Stainless, ground, air	45	1.14	20	40–70	0.003	5.28
416 Stainless, ground, with 0.001-in brass shim, air	100	2.54	30–200	7	0.002	3.52
Aluminum, ground, air	100	2.54	150	12–25	0.0005	0.88
	10	0.25	150	12–25	0.0001	0.18
Aluminum, ground, with 0.001-in brass shim, air	100	2.54	150	12–200	0.0007	1.23
Copper, ground, air	50	1.27	20	12–200	0.00004	0.07
Copper, milled, air	150	3.81	20	10–50	0.0001	0.18
Copper, milled, vacuum	10	0.25	30	7–70	0.0005	0.88

Designating the contact area by A_c and the void area by A_v, we may write for the heat flow across the joint

$$q = \frac{T_{2A} - T_{2B}}{L_g/2k_A A_c + L_g/2k_B A_c} + k_f A_v \frac{T_{2A} - T_{2B}}{L_g} = \frac{T_{2A} - T_{2B}}{1/h_c A}$$

where L_g is the thickness of the void space and k_f is the thermal conductivity of the fluid which fills the void space. The *total* cross-sectional area of the bars is A. Solving for h_c, the contact coefficient, we obtain

$$h_c = \frac{1}{L_g} \left(\frac{A_c}{A} \frac{2k_A k_B}{k_A + k_B} + \frac{A_v}{A} k_f \right) \qquad \textbf{[2-51]}$$

In most instances, air is the fluid filling the void space and k_f is small compared with k_A and k_B. If the contact area is small, the major thermal resistance results from the void space. The main problem with this simple theory is that it is extremely difficult to determine effective values of A_c, A_v, and L_g for surfaces in contact.

From the physical model, we may tentatively conclude:

1. The contact resistance should increase with a decrease in the ambient gas pressure when the pressure is decreased below the value where the mean free path of the molecules is large compared with a characteristic dimension of the void space, since the effective thermal conductance of the entrapped gas will be decreased for this condition.

2. The contact resistance should be decreased for an increase in the joint pressure since this results in a deformation of the high spots of the contact surfaces, thereby creating a greater contact area between the solids.

A very complete survey of the contact-resistance problem is presented in References 4, 6, 7, 10, 11. Unfortunately, there is no satisfactory theory that will predict thermal contact resistance for all types of engineering materials, nor have experimental studies yielded completely reliable empirical correlations. This is understandable because of the many complex surface conditions that may be encountered in practice.

Radiation heat transfer across the joint can also be important when high temperatures are encountered. This energy transfer may be calculated by the methods discussed in Chapter 8.

For design purposes the contact conductance values given in Table 2-3 may be used in the absence of more specific information. Thermal contact resistance can be reduced markedly, perhaps as much as 75 percent, by the use of a "thermal grease" like Dow 340.

EXAMPLE 2-12

Influence of Contact Conductance on Heat Transfer

Two 3.0-cm-diameter 304 stainless-steel bars, 10 cm long, have ground surfaces and are exposed to air with a surface roughness of about 1 μm. If the surfaces are pressed together with a pressure of 50 atm and the two-bar combination is exposed to an overall temperature difference of 100°C, calculate the axial heat flow and temperature drop across the contact surface.

■ **Solution**

The overall heat flow is subject to three thermal resistances, one conduction resistance for each bar, and the contact resistance. For the bars

$$R_{th} = \frac{\Delta x}{kA} = \frac{(0.1)(4)}{(16.3)\pi(3 \times 10^{-2})^2} = 8.679°\text{C/W}$$

From Table 2-2 the contact resistance is

$$R_c = \frac{1}{h_c A} = \frac{(5.28 \times 10^{-4})(4)}{\pi(3 \times 10^{-2})^2} = 0.747°\text{C/W}$$

The total thermal resistance is therefore

$$\sum R_{th} = (2)(8.679) + 0.747 = 18.105$$

and the overall heat flow is

$$q = \frac{\Delta T}{\sum R_{th}} = \frac{100}{18.105} = 5.52 \text{ W} \quad [18.83 \text{ Btu/h}]$$

The temperature drop across the contact is found by taking the ratio of the contact resistance to the total thermal resistance:

$$\Delta T_c = \frac{R_c}{\sum R_{th}} \Delta T = \frac{(0.747)(100)}{18.105} = 4.13°\text{C} \quad [39.43°\text{F}]$$

In this problem the contact resistance represents about 4 percent of the total resistance.

REVIEW QUESTIONS

1. What is meant by the term *one-dimensional* when applied to conduction problems?
2. What is meant by thermal resistance?
3. Why is the one-dimensional heat-flow assumption important in the analysis of fins?
4. Define fin efficiency.
5. Why is the insulated-tip solution important for the fin problems?
6. What is meant by thermal contact resistance? Upon what parameters does this resistance depend?

LIST OF WORKED EXAMPLES

2-1 Multilayer conduction
2-2 Multilayer cylindrical system
2-3 Heat transfer through a composite wall
2-4 Cooling cost savings with extra insulation
2-5 Overall heat-transfer coefficient for a tube

PROBLEMS

2-1 A wall 2 cm thick is to be constructed from material that has an average thermal conductivity of 1.3 W/m · °C. The wall is to be insulated with material having an average thermal conductivity of 0.35 W/m · °C, so that the heat loss per square meter will not exceed 1830 W. Assuming that the inner and outer surface temperatures of the insulated wall are 1300 and 30°C, calculate the thickness of insulation required.

2-2 A certain material 2.5 cm thick, with a cross-sectional area of 0.1 m^2, has one side maintained at 35°C and the other at 95°C. The temperature at the center plane of the material is 62°C, and the heat flow through the material is 1 kW. Obtain an expression for the thermal conductivity of the material as a function of temperature.

2-3 A composite wall is formed of a 2.5-cm copper plate, a 3.2-mm layer of asbestos, and a 5-cm layer of fiberglass. The wall is subjected to an overall temperature difference of 560°C. Calculate the heat flow per unit area through the composite structure.

2-4 Find the heat transfer per unit area through the composite wall in Figure P2-4. Assume one-dimensional heat flow.

Figure P2-4

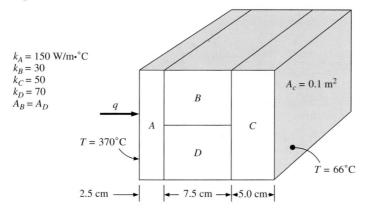

$k_A = 150$ W/m·°C
$k_B = 30$
$k_C = 50$
$k_D = 70$
$A_B = A_D$

$A_c = 0.1$ m^2

q

B

A C

$T = 370°C$

D

$T = 66°C$

2.5 cm   7.5 cm   5.0 cm

2-5 One side of a copper block 5 cm thick is maintained at 250°C. The other side is covered with a layer of fiberglass 2.5 cm thick. The outside of the fiberglass is maintained at 35°C, and the total heat flow through the copper-fiberglass combination is 52 kW. What is the area of the slab?

2-6 An outside wall for a building consists of a 10-cm layer of common brick and a 2.5-cm layer of fiberglass [$k = 0.05$ W/m · °C]. Calculate the heat flow through the wall for a 25°C temperature differential.

2-7 One side of a copper block 4 cm thick is maintained at 175°C. The other side is covered with a layer of fiberglass 1.5 cm thick. The outside of the fiberglass is maintained at 80°C, and the total heat flow through the composite slab is 300 W. What is the area of the slab?

2-8 A plane wall is constructed of a material having a thermal conductivity that varies as the square of the temperature according to the relation $k = k_0(1 + \beta T^2)$. Derive an expression for the heat transfer in such a wall.

2-9 A steel tube having $k = 46$ W/m · °C has an inside diameter of 3.0 cm and a tube wall thickness of 2 mm. A fluid flows on the inside of the tube producing a convection coefficient of 1500 W/m^2 · °C on the inside surface, while a second fluid flows across the outside of the tube producing a convection coefficient of 197 W/m^2 · °C on the outside tube surface. The inside fluid temperature is 223°C while the outside fluid temperature is 57°C. Calculate the heat lost by the tube per meter of length.

2-10 A certain material has a thickness of 30 cm and a thermal conductivity of 0.04 W/m · °C. At a particular instant in time, the temperature distribution with x, the distance from the left face, is $T = 150x^2 - 30x$, where x is in meters. Calculate the heat-flow rates at $x = 0$ and $x = 30$ cm. Is the solid heating up or cooling down?

2-11 A 0.025-mm-diameter stainless steel wire having $k = 16$ W/m · °C is connected to two electrodes. The length of the wire is 80 cm and it is exposed to a convection environment at 20°C with $h = 500$ W/m^2 · °C. A voltage is impressed on the wire that produces temperatures at each electrode of 200°C. Determine the total heat lost by the wire.

2-12 A wall is constructed of 2.0 cm of copper, 3.0 mm of asbestos sheet [$k = 0.166$ W/m · °C], and 6.0 cm of fiberglass. Calculate the heat flow per unit area for an overall temperature difference of 500°C.

2-13 A certain building wall consists of 6.0 in of concrete [$k = 1.2$ W/m · °C], 2.0 in of fiberglass insulation, and $\frac{3}{8}$ in of gypsum board [$k = 0.05$ W/m · °C]. The inside and outside convection coefficients are 2.0 and 7.0 Btu/h · ft^2 · °F, respectively. The outside air temperature is 20°F, and the inside temperature is 72°F. Calculate the overall heat-transfer coefficient for the wall, the R value, and the heat loss per unit area.

2-14 A wall is constructed of a section of stainless steel [$k = 16$ W/m · °C] 4.0 mm thick with identical layers of plastic on both sides of the steel. The overall heat-transfer coefficient, considering convection on both sides of the plastic, is 120 W/m^2 · °C. If the overall temperature difference across the arrangement is 60°C, calculate the temperature difference across the stainless steel.

2-15 An ice chest is constructed of Styrofoam [$k = 0.033$ W/m · °C] with inside dimensions of 25 by 40 by 100 cm. The wall thickness is 5.0 cm. The outside of the chest is exposed to air at 25°C with $h = 10$ W/m^2 · °C. If the chest is completely filled with ice, calculate the time for the ice to completely melt. State your assumptions. The enthalpy of fusion for water is 330 kJ/kg.

2-16 A spherical tank, 1 m in diameter, is maintained at a temperature of 120°C and exposed to a convection environment. With $h = 25$ W/m^2 · °C and $T_\infty = 15$°C, what thickness of urethane foam should be added to ensure that the outer temperature of the insulation does not exceed 40°C? What percentage reduction in heat loss results from installing this insulation?

2-17 A hollow sphere is constructed of aluminum with an inner diameter of 4 cm and an outer diameter of 8 cm. The inside temperature is 100°C and the outer temperature is 50°C. Calculate the heat transfer.

2-18 Suppose the sphere in Problem 2-16 is covered with a 1-cm layer of an insulating material having $k = 50$ m W/m · °C and the outside of the insulation is exposed to an environment with $h = 20$ W/m² · °C and $T_\infty = 10$°C. The inside of the sphere remains at 100°C. Calculate the heat transfer under these conditions.

2-19 In Appendix A, dimensions of standard steel pipe are given. Suppose a 3-in schedule 80 pipe is covered with 1 in of an insulation having $k = 60$ m W/m · °C and the outside of the insulation is exposed to an environment having $h = 10$ W/m² · °C and $T_\infty = 20$°C. The temperature of the inside of the pipe is 250°C. For unit length of the pipe calculate (*a*) overall thermal resistance and (*b*) heat loss.

2-20 A steel pipe with 5-cm OD is covered with a 6.4-mm asbestos insulation [$k = 0.096$ Btu/h · ft · °F] followed by a 2.5-cm layer of fiberglass insulation [$k = 0.028$ Btu/h · ft · °F]. The pipe-wall temperature is 315°C, and the outside insulation temperature is 38°C. Calculate the interface temperature between the asbestos and fiberglass.

2-21 Derive an expression for the thermal resistance through a hollow spherical shell of inside radius r_i and outside radius r_o having a thermal conductivity k. (See Equation 2–10.)

2-22 A 1.0-mm-diameter wire is maintained at a temperature of 400°C and exposed to a convection environment at 40°C with $h = 120$ W/m² · °C. Calculate the thermal conductivity that will just cause an insulation thickness of 0.2 mm to produce a "critical radius." How much of this insulation must be added to reduce the heat transfer by 75 percent from that which would be experienced by the bare wire?

2-23 A 2.0-in schedule 40 steel pipe (see Appendix A) has $k = 27$ Btu/h · ft · °F. The fluid inside the pipe has $h = 30$ Btu/h · ft² · °F, and the outer surface of the pipe is covered with 0.5-in fiberglass insulation with $k = 0.023$ Btu/h · ft · °F. The convection coefficient on the outer insulation surface is 2.0 Btu/h · ft² · °F. The inner fluid temperature is 320°F and the ambient temperature is 70°F. Calculate the heat loss per foot of length.

2-24 Derive a relation for the critical radius of insulation for a sphere.

2-25 A cylindrical tank 80 cm in diameter and 2.0 m high contains water at 80°C. The tank is 90 percent full, and insulation is to be added so that the water temperature will not drop more than 2°C per hour. Using the information given in this chapter, specify an insulating material and calculate the thickness required for the specified cooling rate.

2-26 A hot steam pipe having an inside surface temperature of 250°C has an inside diameter of 8 cm and a wall thickness of 5.5 mm. It is covered with a 9-cm layer of insulation having $k = 0.5$ W/m · °C, followed by a 4-cm layer of insulation having $k = 0.25$ W/m · °C. The outside temperature of the insulation is 20°C. Calculate the heat lost per meter of length. Assume $k = 47$ W/m · °C for the pipe.

2-27 A house wall may be approximated as two 1.2-cm layers of fiber insulating board, an 8.0-cm layer of loosely packed asbestos, and a 10-cm layer of common brick. Assuming convection heat-transfer coefficients of 12 W/m² · °C on both sides of the wall, calculate the overall heat-transfer coefficient for this arrangement.

2-28 Calculate the R value for the following insulations: (*a*) urethane foam, (*b*) fiberglass mats, (*c*) mineral wool blocks, (*d*) calcium silicate blocks.

2-29 An insulation system is to be selected for a furnace wall at 1000°C using first a layer of mineral wool blocks followed by fiberglass boards. The outside of the insulation is exposed to an environment with $h = 15$ W/m² · °C and $T_\infty = 40$°C. Using the data of Table 2-1, calculate the thickness of each insulating material such that the interface temperature is not greater than 400°C and the outside temperature is not greater than 55°C. Use mean values for the thermal conductivities. What is the heat loss in this wall in watts per square meter?

2-30 Derive an expression for the temperature distribution in a plane wall having uniformly distributed heat sources and one face maintained at a temperature T_1 while the other face is maintained at a temperature T_2. The thickness of the wall may be taken as $2L$.

2-31 A 5-cm-diameter steel pipe is covered with a 1-cm layer of insulating material having $k = 0.22$ W/m · °C followed by a 3-cm-thick layer of another insulating material having $k = 0.06$ W/m · °C. The entire assembly is exposed to a convection surrounding condition of $h = 60$ W/m^2 · °C and $T_\infty = 15$°C. The outside surface temperature of the steel pipe is 400°C. Calculate the heat lost by the pipe-insulation assembly for a pipe length of 20 m. Express in Watts.

2-32 Derive an expression for the temperature distribution in a plane wall in which distributed heat sources vary according to the linear relation

$$\dot{q} = \dot{q}_w[1 + \beta(T - T_w)]$$

where $\dot{q}_w$ is a constant and equal to the heat generated per unit volume at the wall temperature T_w. Both sides of the plate are maintained at T_w, and the plate thickness is $2L$.

2-33 A circumferential fin of rectangular profile is constructed of stainless steel with $k = 43$ W/m · °C and a thickness of 1.0 mm. The fin is installed on a tube having a diameter of 3.0 cm and the outer radius of the fin is 4.0 cm. The inner tube is maintained at 250°C and the assembly is exposed to a convection environment having $T_\infty = 35$°C and $h = 45$ W/m^2 · °C. Calculate the heat lost by the fin.

2-34 A plane wall 6.0 cm thick generates heat internally at the rate of 0.3 MW/m^3. One side of the wall is insulated, and the other side is exposed to an environment at 93°C. The convection heat-transfer coefficient between the wall and the environment is 570 W/m^2 · °C. The thermal conductivity of the wall is 21 W/m · °C. Calculate the maximum temperature in the wall.

2-35 Consider a shielding wall for a nuclear reactor. The wall receives a gamma-ray flux such that heat is generated within the wall according to the relation

$$\dot{q} = \dot{q}_0 e^{-ax}$$

where $\dot{q}_0$ is the heat generation at the inner face of the wall exposed to the gamma-ray flux and a is a constant. Using this relation for heat generation, derive an expression for the temperature distribution in a wall of thickness L, where the inside and outside temperatures are maintained at T_i and T_0, respectively. Also obtain an expression for the maximum temperature in the wall.

2-36 Repeat Problem 2-35, assuming that the outer surface is adiabatic while the inner surface temperature is maintained at T_i.

2-37 Rework Problem 2-32 assuming that the plate is subjected to a convection environment on both sides of temperature T_∞ with a heat-transfer coefficient h. T_w is now some reference temperature not necessarily the same as the surface temperature.

2-38 Heat is generated in a 2.5-cm-square copper rod at the rate of 35.3 MW/m^3. The rod is exposed to a convection environment at 20°C, and the heat-transfer coefficient is 4000 W/m^2 · °C. Calculate the surface temperature of the rod.

2-39 A plane wall of thickness $2L$ has an internal heat generation that varies according to $\dot{q} = \dot{q}_0 \cos ax$, where $\dot{q}_0$ is the heat generated per unit volume at the center of the wall ($x = 0$) and a is a constant. If both sides of the wall are maintained at a constant temperature of T_w, derive an expression for the total heat loss from the wall per unit surface area.

2-40 A certain semiconductor material has a conductivity of 0.0124 W/cm · °C. A rectangular bar of the material has a cross-sectional area of 1 cm^2 and a length of 3 cm. One end is maintained at 300°C and the other end at 100°C, and the bar carries a current of 50 A. Assuming the longitudinal surface is insulated, calculate the midpoint temperature in the bar. Take the resistivity as $1.5 \times 10^{-3} \Omega \cdot$ cm.

2-41 The temperature distribution in a certain plane wall is

$$\frac{T - T_1}{T_2 - T_1} = C_1 + C_2 x^2 + C_3 x^3$$

where T_1 and T_2 are the temperatures on each side of the wall. If the thermal conductivity of the wall is constant and the wall thickness is L, derive an expression for the heat generation per unit volume as a function of x, the distance from the plane where $T = T_1$. Let the heat-generation rate be $\dot{q}_0$ at $x = 0$.

2-42 Electric heater wires are installed in a solid wall having a thickness of 8 cm and $k = 2.5$ W/m · °C. The right face is exposed to an environment with $h = 50$ W/m^2 · °C and $T_\infty = 30°$C, while the left face is exposed to $h = 75$ W/m^2 · °C and $T_\infty = 50°$C. What is the maximum allowable heat-generation rate such that the maximum temperature in the solid does not exceed 300°C?

2-43 Two 5.0-cm-diameter aluminum bars, 2 cm long, have ground surfaces and are joined in compression with a 0.025-mm brass shim at a pressure exceeding 20 atm. The combination is subjected to an overall temperature difference of 200°C. Calculate the temperature drop across the contact join.

2-44 A 3.0-cm-thick plate has heat generated uniformly at the rate of 5×10^5 W/m^3. One side of the plate is maintained at 200°C and the other side at 45°C. Calculate the temperature at the center of the plate for $k = 16$ W/m · °C.

2-45 Heat is generated uniformly in a stainless steel plate having $k = 20$ W/m · °C. The thickness of the plate is 1.0 cm and the heat-generation rate is 500 MW/m^3. If the two sides of the plate are maintained at 100 and 200°C, respectively, calculate the temperature at the center of the plate.

2-46 A plate having a thickness of 4.0 mm has an internal heat generation of 200 MW/m^3 and a thermal conductivity of 25 W/m · °C. One side of the plate is insulated and the other side is maintained at 100°C. Calculate the maximum temperature in the plate.

2-47 A 3.2-mm-diameter stainless-steel wire 30 cm long has a voltage of 10 V impressed on it. The outer surface temperature of the wire is maintained at 93°C. Calculate the center temperature of the wire. Take the resistivity of the wire as 70 $\mu\Omega \cdot$ cm and the thermal conductivity as 22.5 W/m · °C.

2-48 The heater wire of Example 2-7 is submerged in a fluid maintained at 93°C. The convection heat-transfer coefficient is 5.7 kW/m^2 · °C. Calculate the center temperature of the wire.

2-49 An electric current is used to heat a tube through which a suitable cooling fluid flows. The outside of the tube is covered with insulation to minimize heat loss to the surroundings, and thermocouples are attached to the outer surface of the tube to measure the temperature. Assuming uniform heat generation in the tube, derive an expression for the convection heat-transfer coefficient on the inside of the tube in terms of the measured variables: voltage E, current I, outside tube wall temperature T_0, inside and outside radii r_i and r_o, tube length L, and fluid temperature T_f.

2-50 Derive an expression for the temperature distribution in a sphere of radius r with uniform heat generation $\dot{q}$ and constant surface temperature T_w.

2-51 A stainless-steel sphere [$k = 16$ W/m · °C] having a diameter of 4 cm is exposed to a convection environment at 20°C, $h = 15$ W/m² · °C. Heat is generated uniformly in the sphere at the rate of 1.0 MW/m³. Calculate the steady-state temperature for the center of the sphere.

2-52 An aluminum-alloy electrical cable has $k = 190$ W/m · °C, a diameter of 30 mm, and carries an electric current of 230 A. The resistivity of the cable is 2.9 $\mu\Omega$ · cm, and the outside surface temperature of the cable is 180°C. Calculate the maximum temperature in the cable if the surrounding air temperature is 15°C.

2-53 Derive an expression for the temperature distribution in a hollow cylinder with heat sources that vary according to the linear relation

$$\dot{q} = a + br$$

with $\dot{q}_i$ the generation rate per unit volume at $r = r_i$. The inside and outside temperatures are $T = T_i$ at $r = r_i$ and $T = T_o$ at $r = r_o$.

2-54 The outside of a copper wire having a diameter of 2 mm is exposed to a convection environment with $h = 5000$ W/m² · °C and $T_\infty = 100$°C. What current must be passed through the wire to produce a center temperature of 150°C? Repeat for an aluminum wire of the same diameter. The resistivity of copper is 1.67 $\mu\Omega$ · cm.

2-55 A hollow tube having an inside diameter of 2.5 cm and a wall thickness of 0.4 mm is exposed to an environment at $h = 100$ W/m² · °C and $T_\infty = 40$°C. What heat-generation rate in the tube will produce a maximum tube temperature of 250°C for $k = 24$ W/m · °C?

2-56 Water flows on the inside of a steel pipe with an ID of 2.5 cm. The wall thickness is 2 mm, and the convection coefficient on the inside is 500 W/m² · °C. The convection coefficient on the outside is 12 W/m² · °C. Calculate the overall heat-transfer coefficient. What is the main determining factor for U?

2-57 The pipe in Problem 2-56 is covered with a layer of asbestos [$k = 0.18$ W/m · °C] while still surrounded by a convection environment with $h = 12$ W/m² · °C. Calculate the critical insulation radius. Will the heat transfer be increased or decreased by adding an insulation thickness of (a) 0.5 mm, (b) 10 mm?

2-58 Calculate the overall heat-transfer coefficient for Problem 2-4.

2-59 Calculate the overall heat-transfer coefficient for Problem 2-5.

2-60 Air flows at 120°C in a thin-wall stainless-steel tube with $h = 65$ W/m² · °C. The inside diameter of the tube is 2.5 cm and the wall thickness is 0.4 mm. $k = 18$ W/m · °C for the steel. The tube is exposed to an environment with $h = 6.5$ W/m² · °C and $T_\infty = 15$°C. Calculate the overall heat-transfer coefficient and the heat loss per meter of length. What thickness of an insulation having $k = 40$ mW/m · °C should be added to reduce the heat loss by 90 percent?

2-61 An insulating glass window is constructed of two 5-mm glass plates separated by an air layer having a thickness of 4 mm. The air layer may be considered stagnant so that pure conduction is involved. The convection coefficients for the inner and outer surfaces are 12 and 50 W/m² · °C, respectively. Calculate the overall heat-transfer coefficient for this arrangement, and the R value. Repeat the calculation for a single glass plate 5 mm thick.

2-62 A wall consists of a 1-mm layer of copper, a 4-mm layer of 1 percent carbon steel, a 1-cm layer of asbestos sheet, and 10 cm of fiberglass blanket. Calculate the overall heat-transfer coefficient for this arrangement. If the two outside surfaces are at 10 and 150°C, calculate each of the interface temperatures.

2-63 A circumferential fin of rectangular profile has a thickness of 0.7 mm and is installed on a tube having a diameter of 3 cm that is maintained at a temperature of 200°C. The length of the fin is 2 cm and the fin material is copper. Calculate the heat lost by the fin to a surrounding convection environment at 100°C with a convection heat-transfer coefficient of 524 W/m^2 · °C.

2-64 A thin rod of length L has its two ends connected to two walls which are maintained at temperatures T_1 and T_2, respectively. The rod loses heat to the environment at T_∞ by convection. Derive an expression (*a*) for the temperature distribution in the rod and (*b*) for the total heat lost by the rod.

2-65 A rod of length L has one end maintained at temperature T_0 and is exposed to an environment of temperature T_∞. An electrical heating element is placed in the rod so that heat is generated uniformly along the length at a rate $\dot{q}$. Derive an expression (*a*) for the temperature distribution in the rod and (*b*) for the total heat transferred to the environment. Obtain an expression for the value of $\dot{q}$ that will make the heat transfer zero at the end that is maintained at T_0.

2-66 One end of a copper rod 30 cm long is firmly connected to a wall that is maintained at 200°C. The other end is firmly connected to a wall that is maintained at 93°C. Air is blown across the rod so that a heat-transfer coefficient of 17 W/m^2 · °C is maintained. The diameter of the rod is 12.5 mm. The temperature of the air is 38°C. What is the net heat lost to the air in watts?

2-67 Verify the temperature distribution for case 2 in Section 2-9, i.e., that

$$\frac{T - T_\infty}{T_0 - T_\infty} = \frac{\cosh m(L - x) + (h/mk) \sinh m(L - x)}{\cosh mL + (h/mk) \sinh mL}$$

Subsequently show that the heat transfer is

$$q = \sqrt{h\,PkA}\,(T_0 - T_\infty)\frac{\sinh mL + (h/mk) \cosh mL}{\cosh mL + (h/mk) \sinh mL}$$

2-68 An aluminum rod 2.0 cm in diameter and 12 cm long protrudes from a wall that is maintained at 250°C. The rod is exposed to an environment at 15°C. The convection heat-transfer coefficient is 12 W/m^2 · °C. Calculate the heat lost by the rod.

2-69 Derive Equation (2-35) by integrating the convection heat loss from the rod of case 1 in Section 2-9.

2-70 Derive Equation (2-36) by integrating the convection heat loss from the rod of case 3 in Section 2-9.

2-71 A long, thin copper rod 5 mm in diameter is exposed to an environment at 20°C. The base temperature of the rod is 120°C. The heat-transfer coefficient between the rod and the environment is 20 W/m^2 · °C. Calculate the heat given up by the rod.

2-72 A very long copper rod [$k = 372$ W/m · °C] 2.5 cm in diameter has one end maintained at 90°C. The rod is exposed to a fluid whose temperature is 40°C. The heat-transfer coefficient is 3.5 W/m^2 · °C. How much heat is lost by the rod?

2-73 An aluminum fin 1.5 mm thick is placed on a circular tube with 2.7-cm OD. The fin is 6 mm long. The tube wall is maintained at 150°C, the environment temperature is 15°C, and the convection heat-transfer coefficient is 20 W/m² · °C. Calculate the heat lost by the fin.

2-74 A straight fin of rectangular profile has a thermal conductivity of 14 W/m · °C, thickness of 2.0 mm, and length of 23 mm. The base of the fin is maintained at a temperature of 220°C while the fin is exposed to a convection environment at 23°C with $h = 25$ W/m² · °C. Calculate the heat lost per meter of fin depth.

2-75 A circumferential fin of rectangular profile is constructed of a material having $k = 55$ W/m · °C and is installed on a tube having a diameter of 3 cm. The length of fin is 3 cm and the thickness is 2 mm. If the fin is exposed to a convection environment at 20°C with a convection coefficient of 68 W/m² · °C and the tube wall temperature is 100°C, calculate the heat lost by the fin.

2-76 The total efficiency for a finned surface may be defined as the ratio of the total heat transfer of the combined area of the surface and fins to the heat that would be transferred if this total area were maintained at the base temperature T_0. Show that this efficiency can be calculated from

$$\eta_t = 1 - \frac{A_f}{A}(1 - \eta_f)$$

where

η_t = total efficiency

A_f = surface area of all fins

A = total heat-transfer area, including fins and exposed tube or other surface

η_f = fin efficiency

2-77 A triangular fin of stainless steel (18% Cr, 8% Ni) is attached to a plane wall maintained at 460°C. The fin thickness is 6.4 mm, and the length is 2.5 cm. The environment is at 93°C, and the convection heat-transfer coefficient is 28 W/m² · °C. Calculate the heat lost from the fin.

2-78 A 2.5-cm-diameter tube has circumferential fins of rectangular profile spaced at 9.5-mm increments along its length. The fins are constructed of aluminum and are 0.8 mm thick and 12.5 mm long. The tube wall temperature is maintained at 200°C, and the environment temperature is 93°C. The heat-transfer coefficient is 110 W/m² · °C. Calculate the heat loss from the tube per meter of length.

2-79 A circumferential fin of rectangular profile surrounds a 2-cm-diameter tube. The length of the fin is 5 mm, and the thickness is 2.5 mm. The fin is constructed of mild steel. If air blows over the fin so that a heat-transfer coefficient of 25 W/m² · °C is experienced and the temperatures of the base and air are 260 and 93°C, respectively, calculate the heat transfer from the fin.

2-80 A straight rectangular fin 2.0 cm thick and 14 cm long is constructed of steel and placed on the outside of a wall maintained at 200°C. The environment temperature is 15°C, and the heat-transfer coefficient for convection is 20 W/m² · °C. Calculate the heat lost from the fin per unit depth.

2-81 An aluminum fin 1.6 mm thick surrounds a tube 2.5 cm in diameter. The length of the fin is 12.5 mm. The tube-wall temperature is 200°C, and the environment temperature is 20°C. The heat-transfer coefficient is 60 W/m² · °C. What is the heat lost by the fin?

2-82 Obtain an expression for the optimum thickness of a straight rectangular fin for a given profile area. Use the simplified insulated-tip solution.

2-83 Derive a differential equation (do not solve) for the temperature distribution in a straight triangular fin. For convenience, take the coordinate axis as shown in Figure P2-83 and assume one-dimensional heat flow.

Figure P2-83

2-84 A circumferential fin of rectangular profile is installed on a 10-cm-diameter tube maintained at 120°C. The fin has a length of 15 cm and thickness of 2 mm. The fin is exposed to a convection environment at 23°C with $h = 60$ W/m$^2 \cdot$°C and the fin conductivity is 120 W/m $\cdot$ °C. Calculate the heat lost by the fin expressed in watts.

2-85 A long stainless-steel rod [$k = 16$ W/m $\cdot$ °C] has a square cross section 12.5 by 12.5 mm and has one end maintained at 250°C. The heat-transfer coefficient is 40 W/m$^2 \cdot$°C, and the environment temperature is 90°C. Calculate the heat lost by the rod.

2-86 A straight fin of rectangular profile is constructed of duralumin (94% Al, 3% Cu) with a thickness of 2.1 mm. The fin is 17 mm long, and it is subjected to a convection environment with $h = 75$ W/m$^2 \cdot$°C. If the base temperature is 100°C and the environment is at 30°C, calculate the heat transfer per unit length of fin.

2-87 A certain internal-combustion engine is air-cooled and has a cylinder constructed of cast iron [$k = 35$ Btu/h $\cdot$ ft $\cdot$ °F]. The fins on the cylinder have a length of $\frac{5}{8}$ in and thickness of $\frac{1}{8}$ in. The convection coefficient is 12 Btu/h $\cdot$ ft$^2 \cdot$ °F. The cylinder diameter is 4 in. Calculate the heat loss per fin for a base temperature of 450°F and environment temperature of 100°F.

2-88 A 1.5-mm-diameter stainless-steel rod [$k = 19$ W/m $\cdot$ °C] protrudes from a wall maintained at 45°C. The rod is 12 mm long, and the convection coefficient is 500 W/m$^2 \cdot$ °C. The environment temperature is 20°C. Calculate the temperature of the tip of the rod. Repeat the calculation for $h = 200$ and 1500 W/m$^2 \cdot$ °C.

2-89 An aluminum block is cast with an array of pin fins protruding like that shown in Figure 2-10d and subjected to room air at 20°C. The convection coefficient between the pins and the surrounding air may be assumed to be $h = 13.2$ W/m$^2 \cdot$ °C. The pin diameters are 2 mm and their length is 25 mm. The base of the aluminum block may be assumed constant at 70°C. Calculate the total heat lost by an array of 15 by 15, that is, 225 fins.

2-90 A finned tube is constructed as shown in Figure 2-10b. Eight fins are installed as shown and the construction material is aluminum. The base temperature of the fins may be assumed to be 100°C and they are subjected to a convection environment at 30°C with $h = 15$ W/m$^2 \cdot$ °C. The longitudinal length of the fins is 15 cm and the peripheral length is 2 cm. The fin thickness is 2 mm. Calculate the total heat dissipated by the finned tube. Consider only the surface area of the fins.

2-91 Circumferential fins of rectangular profile are constructed of aluminum and attached to a copper tube having a diameter of 25 mm and maintained at 100°C. The length of the fins is 2 cm and thickness is 2 mm. The arrangement is exposed to a convection environment at 30°C with $h = 15$ W/m$^2 \cdot$ °C. Assume that a number of fins is installed such that the total fin surface area equals that of the total surface fine area in Problem 2-90. Calculate the total heat lost by the fins.

2-92 A 2-cm-diameter glass rod 6 cm long [$k = 0.8$ W/m $\cdot$ °C] has a base temperature of 100°C and is exposed to an air convection environment at 20°C. The temperature at the tip of the rod is measured as 35°C. What is the convection heat-transfer coefficient? How much heat is lost by the rod?

2-93 A straight rectangular fin has a length of 2.5 cm and a thickness of 1.5 mm. The thermal conductivity is 55 W/m $\cdot$ °C, and it is exposed to a convection environment at 20°C and $h = 500$ W/m$^2 \cdot$ °C. Calculate the maximum possible heat loss for a base temperature of 200°C. What is the actual heat loss?

2-94 A straight rectangular fin has a length of 3.5 cm and a thickness of 1.4 mm. The thermal conductivity is 55 W/m · °C. The fin is exposed to a convection environment at 20°C and $h = 500$ W/m$^2 \cdot$ °C. Calculate the maximum possible heat loss for a base temperature of 150°C. What is the actual heat loss for this base temperature?

2-95 A circumferential fin of rectangular profile is constructed of 1 percent carbon steel and attached to a circular tube maintained at 150°C. The diameter of the tube is 5 cm, and the length is also 5 cm with a thickness of 2 mm. The surrounding air is maintained at 20°C and the convection heat-transfer coefficient may be taken as 100 W/m$^2 \cdot$ °C. Calculate the heat lost from the fin.

2-96 A circumferential fin of rectangular profile is constructed of aluminum and surrounds a 3-cm-diameter tube. The fin is 2 cm long and 1 mm thick. The tube wall temperature is 200°C, and the fin is exposed to a fluid at 20°C with a convection heat-transfer coefficient of 80 W/m$^2 \cdot$ °C. Calculate the heat loss from the fin.

2-97 A 1.0-cm-diameter steel rod [$k = 20$ W/m · °C] is 20 cm long. It has one end maintained at 50°C and the other at 100°C. It is exposed to a convection environment at 20°C with $h = 50$ W/m$^2 \cdot$ °C. Calculate the temperature at the center of the rod.

2-98 A circumferential fin of rectangular profile is constructed of copper and surrounds a tube having a diameter of 1.25 cm. The fin length is 6 mm and its thickness is 0.3 mm. The fin is exposed to a convection environment at 20°C with $h = 55$ W/m$^2 \cdot$ °C and the fin base temperature is 100°C. Calculate the heat lost by the fin.

2-99 A straight rectangular fin of steel (1% C) is 2 cm thick and 17 cm long. It is placed on the outside of a wall which is maintained at 230°C. The surrounding air temperature is 25°C, and the convection heat-transfer coefficient is 23 W/m$^2 \cdot$ °C. Calculate the heat lost from the fin per unit depth and the fin efficiency.

2-100 A straight fin having a triangular profile has a length of 5 cm and a thickness of 4 mm and is constructed of a material having $k = 23$ W/m · °C. The fin is exposed to surroundings with a convection coefficient of 20 W/m$^2 \cdot$ °C and a temperature of 40°C. The base of the fin is maintained at 200°C. Calculate the heat lost per unit depth of fin.

2-101 A circumferential aluminum fin is installed on a 25.4-mm-diameter tube. The length of the fin is 12.7 mm and the thickness is 1.0 mm. It is exposed to a convection environment at 30°C with a convection coefficient of 56 W/m$^2 \cdot$ °C. The base temperature is 125°C. Calculate the heat lost by the fin.

2-102 A circumferential fin of rectangular profile is constructed of stainless steel (18% Cr, 8% Ni). The thickness of the fin is 2.0 mm, the inside radius is 2.0 cm, and the length is 8.0 cm. The base temperature is maintained at 135°C and the fin is exposed to a convection environment at 15°C with $h = 20$ W/m$^2 \cdot$ °C. Calculate the heat lost by the fin.

2-103 A rectangular fin has a length of 2.5 cm and thickness of 1.1 mm. The thermal conductivity is 55 W/m · °C. The fin is exposed to a convection environment at 20°C and $h = 500$ W/m$^2 \cdot$ °C. Calculate the heat loss for a base temperature of 125°C.

2-104 A 1.0-mm-thick aluminum fin surrounds a 2.5-cm-diameter tube. The length of the fin is 1.25 cm. The fin is exposed to a convection environment at 30°C with $h = 75$ W/m$^2 \cdot$ °C. The tube surface is maintained at 100°C. Calculate the heat lost by the fin.

2-105 A glass rod having a diameter of 1 cm and length of 5 cm is exposed to a convection environment at a temperature of 20°C. One end of the rod is maintained at a temperature of 180°C. Calculate the heat lost by the rod if the convection heat-transfer coefficient is 20 W/m^2 · °C.

2-106 A stainless steel rod has a square cross section measuring 1 by 1 cm. The rod length is 8 cm, and $k = 18$ W/m · °C. The base temperature of the rod is 300°C. The rod is exposed to a convection environment at 50°C with $h = 45$ W/m^2 · °C. Calculate the heat lost by the rod and the fin efficiency.

2-107 Copper fins with a thickness of 1.0 mm are installed on a 2.5-cm-diameter tube. The length of each fin is 12 mm. The tube temperature is 275°C and the fins are exposed to air at 35°C with a convection heat-transfer coefficient of 120 W/m^2 · °C. Calculate the heat lost by each fin.

2-108 A straight fin of rectangular profile is constructed of stainless steel (18% Cr, 8% Ni) and has a length of 5 cm and a thickness of 2.5 cm. The base temperature is maintained at 100°C and the fin is exposed to a convection environment at 20°C with $h = 47$ W/m^2 · °C. Calculate the heat lost by the fin per meter of depth, and the fin efficiency.

2-109 A circumferential fin of rectangular profile is constructed of duralumin and surrounds a 3-cm-diameter tube. The fin is 3 cm long and 1 mm thick. The tube wall temperature is 200°C, and the fin is exposed to a fluid at 20°C with a convection heat-transfer coefficient of 80 W/m^2 · °C. Calculate the heat loss from the fin.

2-110 A circular fin of rectangular profile is attached to a 3.0-cm-diameter tube maintained at 100°C. The outside diameter of the fin is 9.0 cm and the fin thickness is 1.0 mm. The environment has a convection coefficient of 50 W/m^2 · °C and a temperature of 30°C. Calculate the thermal conductivity of the material for a fin efficiency of 60 percent.

2-111 A circumferential fin of rectangular profile having a thickness of 1.0 mm and a length of 2.0 cm is placed on a 2.0-cm-diameter tube. The tube temperature is 150°C, the environment temperature is 20°C, and $h = 150$ W/m^2 · °C. The fin is aluminum. Calculate the heat lost by the fin.

2-112 Two 1-in-diameter bars of stainless steel [$k = 17$ W/m · °C] are brought into end-to-end contact so that only 0.1 percent of the cross-sectional area is in contact at the joint. The bars are 7.5 cm long and subjected to an axial temperature difference of 300°C. The roughness depth in each bar ($L_g/2$) is estimated to be 1.3 μm. The surrounding fluid is air, whose thermal conductivity may be taken as 0.035 W/m · °C for this problem. Estimate the value of the contact resistance and the axial heat flow. What would the heat flow be for a continuous 15-cm stainless-steel bar?

2-113 When the *joint pressure* for two surfaces in contact is increased, the high spots of the surfaces are deformed so that the contact area A_c is increased and the roughness depth L_g is decreased. Discuss this effect in the light of the presentation of Section 2-11. (Experimental work shows that joint conductance varies almost directly with pressure.)

2-114 Two aluminum plates 5 mm thick with a ground roughness of 100 μin are bolted together with a contact pressure of 20 atm. The overall temperature difference across the plates is 80°C. Calculate the temperature drop across the contact joint.

2-115 Fins are frequently installed on tubes by a press-fit process. Consider a circumferential aluminum fin having a thickness of 1.0 mm to be installed on a 2.5-cm-diameter aluminum tube. The fin length is 1.25 cm, and the contact conductance may be taken from Table 2-2 for a 100-μin ground surface. The convection environment is at 20°C, and $h = 125$ W/m$^2 \cdot$°C. Calculate the heat transfer for each fin for a tube wall temperature of 200°C. What percentage reduction in heat transfer is caused by the contact conductance?

2-116 An aluminum fin is attached to a transistor that generates heat at the rate of 300 mW. The fin has a total surface area of 9.0 cm^2 and is exposed to surrounding air at 27°C. The contact conductance between transistor and fin is 0.9×10^{-4} m$^2 \cdot$°C/W, and the contact area is 0.5 cm^2. Estimate the temperature of the transistor, assuming the fin is uniform in temperature.

2-117 A plane wall 20 cm thick with uniform internal heat generation of 200 kW/m^3 is exposed to a convection environment on both sides at 50°C with $h = 400$ W/m$^2 \cdot$°C. Calculate the center temperature of the wall for $k = 20$ W/m $\cdot$ °C.

2-118 Suppose the wall of Problem 2-117 is only 10 cm thick and has one face insulated. Calculate the maximum temperature in the wall assuming all the other conditions are the same. Comment on the results.

2-119 A circumferential fin of rectangular profile is constructed of aluminum and placed on a 6-cm-diameter tube maintained at 120°C. The length of the fin is 3 cm and its thickness is 2 mm. The fin is exposed to a convection environment at 20°C with $h = 220$ W/m$^2 \cdot$°C. Calculate the heat lost by the fin expressed in Watts.

2-120 A straight aluminum fin of triangular profile has a base maintained at 200°C and is exposed to a convection environment at 25°C with $h = 45$ W/m$^2 \cdot$°C. The fin has a length of 8 mm and a thickness of 2.0 mm. Calculate the heat lost per unit depth of fin.

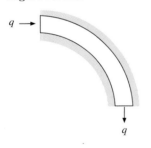

2-121 One hundred circumferential aluminum fins of rectangular profile are mounted on a 1.0-m tube having a diameter of 2.5 cm. The fins are 1 cm long and 2.0 mm thick. The base temperature is 180°C, and the convection environment is at 20°C with $h = 50$ W/m$^2 \cdot$°C. Calculate the total heat lost from the finned-tube arrangement over the 1.0-m length.

2-122 The cylindrical segment shown in Figure P2-122 has a thermal conductivity of 100 W/m $\cdot$ °C. The inner and outer radii are 1.5 and 1.7 cm, respectively, and the surfaces are insulated. Calculate the circumferential heat transfer per unit depth for an imposed temperature difference of 50°C. What is the thermal resistance?

Figure P2-123

2-123 The truncated hollow cone shown in Figure P2-123 is used in laser-cooling applications and is constructed of copper with a thickness of 0.5 mm. Calculate the thermal resistance for one-dimensional heat flow. What would be the heat transfer for a temperature difference of 300°C?

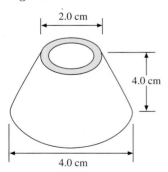

2-124 A tube assembly is constructed of copper with an inside diameter of 1.25 cm, wall thickness of 0.8 mm, and circumferential fins around the periphery. The fins have a thickness of 0.3 mm and length of 3 mm, and are spaced 6 mm apart. If the convection heat transfer coefficient from the tube and fins to the surrounding air is 50 W/m$^2 \cdot$°C, calculate the thermal resistance for a 30-cm length of the tube-fin combination. What is the fin efficiency for this arrangement? If the inside tube temperature is 100°C and the surrounding air temperature is 20°C, what is the heat loss per meter of tube length? What fraction of the loss is by the fins?

2-125 Calculate the R value for the fin-tube combination in Problem 2-116.

2-126 Repeat Problem 2-124 for aluminum fins installed on a copper tube.

2-127 Repeat Problem 2-125 for aluminum fins installed on a copper tube.

2-128   A stainless-steel rod having a length of 10 cm and diameter of 2 mm has a resistivity of 70 $\mu\Omega \cdot$ cm and thermal conductivity of 16 W/m $\cdot$ °C. The rod is exposed to a convection environment with $h = 100$ W/m$^2 \cdot$ °C and $T = 20$°C. Both ends of the rod are maintained at $T = 100$°C. What voltage must be impressed on the rod to dissipate twice as much heat to the surroundings as in a zero-voltage condition?

2-129   Suppose the rod in Problem 2-128 is very long. What would the zero-voltage heat transfer be in this case?

2-130   Suppose the cylindrical segment of Problem 2-122 has a periphery exposed to a convection environment with $h = 75$ W/m$^2 \cdot$ °C and $T_\infty = 30$°C instead of to the insulated surface. For this case, one end is at 50°C while the other end is at 100°C. What is the heat lost by the segment to the surroundings in this circumstance? What is the heat transfer at each end of the segment?

Design-Oriented Problems

2-131   Suppose you have a choice between a straight triangular or rectangular fin constructed of aluminum with a base thickness of 3.0 mm. The convection coefficient is 50 W/m$^2 \cdot$ °C. Select the fin with the least weight for a given heat flow.

2-132   Consider aluminum circumferential fins with $r_1 = 1.0$ cm, $r_2 = 2.0$ cm, and thicknesses of 1.0, 2.0, and 3.0 mm. The convection coefficient is 160 W/m$^2 \cdot$ °C. Compare the heat transfers for six 1.0-mm fins, three 2.0-mm fins, and two 3.0-mm fins. What do you conclude? Repeat for $h = 320$ W/m$^2 \cdot$ °C.

2-133   "Pin fins" of aluminum are to be compared in terms of their relative performance as a function of diameter. Three "pins" having diameters of 2, 5, and 10 mm with a length of 5 cm are exposed to a convection environment with $T_\infty = 20$°C, and $h = 40$ W/m$^2 \cdot$ °C. The base temperature is 200°C. Calculate the heat transfer for each pin. How does it vary with pin diameter?

2-134   Calculate the heat transfer per unit mass for the pin fins in Problem 2-133. How does it vary with diameter?

2-135   A straight rectangular fin has a length of 1.5 cm and a thickness of 1.0 mm. The convection coefficient is 20 W/m$^2 \cdot$ °C. Compare the heat-transfer rates for aluminum and magnesium fins.

2-136   Suppose both fins in Problem 2-129 are to dissipate the same heat. Which would be lower in weight? Assume that the thickness is the same for both fins but adjust the lengths until the heat transfers are equal.

2-137   Insulating materials are frequently installed with a reflective coating to reduce the radiation heat transfer between the surface and the surroundings. An insulating material is installed on a furnace oven wall that is maintained at 200°C. The energy cost of the fuel firing the oven is \$8.25/GJ and the insulation installation must be justified by the savings in energy costs over a three-year period. Select an appropriate insulation from Table 2-1 and/or Table A-3 and determine a suitable quantity of insulation that will pay for itself over a three-year period. For this computation assume that the outer surface of the insulation radiates like a blackbody and that the heat loss can be determined from Equation (1-12). For the calculation use Table 1-2 as a guide for selecting the convection heat-transfer coefficient. Next, consider the same type of insulating material but with a reflective coating having $\epsilon = 0.1$. The radiation transfer may still be calculated with Equation (1-12). Determine the quantity of the reflective insulating material required to be economical. How much higher cost per unit thickness or volume could be justified for the reflective material over that of the nonreflective? Comment on uncertainties which may exist in your analysis.

2-138 A thin-wall stainless-steel tube is to be used as an electric heating element that will deliver a convection coefficient of 5000 W/m$^2 \cdot$°C to water at 100°C. Devise several configurations to accomplish a total heat transfer of 10 kW. Specify the length, outside diameter, wall thickness, maximum tube temperature, and necessary voltage that must be imposed on the tube. Take the resistivity of stainless steel as 70 $\mu\Omega \cdot$ cm.

2-139 Thin cylindrical or spherical shells may be treated as a plane wall for sufficiently large diameters in relation to the thickness of the shell. Devise a scheme for quantifying the error that would result from such a treatment.

2-140 A 2.5-cm-diameter steel pipe is maintained at 100°C by condensing steam on the inside. The pipe is to be used for dissipating heat to a surrounding room at 20°C by placing circular steel fins around the outside surface of the pipe. The convection loss from the pipe and fins occurs by free convection, with $h = 8.0$ W/m$^2 \cdot$°C. Examine several cases of fin thickness, fin spacing, and fin outside diameters to determine the overall heat loss per meter of pipe length. Take $k = 43$ W/m · °C for the steel fins and assume h is uniform over all surfaces. Make appropriate conclusions about the results of your study.

2-141 A pipe having a diameter of 5.3 cm is maintained at 200°C by steam flowing inside. The pipe passes through a large factory area and loses heat by free convection from the outside with $h = 7.2$ W/m$^2 \cdot$°C. Using information from Table 2-1 and/or Table A-3, select two alternative insulating materials that could be installed to lower the outside surface temperature of the insulation to 30°C when the pipe is exposed to room air at 20°C. If the energy loss from the steam costs $8.00/10^9 J, what are the allowable costs of the insulation materials per unit volume to achieve a payback period of three years where

(energy cost saved per year) × 3
= (cost of installed insulation/unit volume) × volume

2-142 It is frequently represented that the energy savings resulting from installation of extra ceiling insulation in a home will pay for the insulation cost within a three-year period. You are asked to evaluate this claim. For the evaluation it may be assumed that 1 kW of electrical input to an air-conditioning unit will produce about 1.26×10^4 kJ/h of cooling and that electricity is priced at $0.085/kWh. Assume that an existing home has ceiling insulation with an R value of 7.0°F · ft$^2 \cdot$ h/Btu and is to be upgraded to an R value of either 15 or 30. Choose two alternative insulation materials from Table 2-1 and/or Table A-3 and calculate the allowable costs per unit volume of insulating material to accomplish the three-year payback with the two specified R values. For this calculation, (energy cost saved/year) × 3 = (insulation cost per unit volume) × volume. Make your own assumptions regarding (1) temperature difference between the interior of the house and the attic area and (2) the hours of operation for the air-conditioning system during an annual period. Comment on the results and assumptions.

2-143 A finned wall like that shown in Figure 2-10a is constructed of aluminum alloy with $k = 160$ W/m · °C. The wall thickness is 2.0 mm and the fins are straight with rectangular profile. The inside of the wall is maintained at a constant temperature of 70°C and the fins are exposed to a convection environment at 25°C with $h = 8$ W/m$^2 \cdot$°C (free convection). The assembly will be cast from the aluminum material and must dissipate 30 W of heat under the conditions noted. Assuming a square array, determine suitable combinations of numbers of fins, fin spacing, dimension of the square, and fin thickness to accomplish this cooling objective. Assume a uniform value of h for both the fin and wall surfaces.

2-144 Repeat Problem 2-143 for cooling with forced convection, which produces a convection coefficient of $h = 20$ W/m$^2 \cdot$ °C.

2-145 Consider a pin fin as shown in Figure 2-10d. Assume that the fin is exposed to an evacuated space such that convection is negligible and that the radiation loss per unit surface area is given by

$$q_{\mathrm{rad}}/A = \epsilon\sigma(T^4 - T_s^4)$$

where ϵ is a surface emissivity constant, σ is the Stefan-Boltzmann constant, and the temperatures are expressed in degrees Kelvin. Derive a differential equation for the temperature in the pin fin as a function of x, the distance from the base. Let T_0 be the base temperature, and write the appropriate boundary conditions for the differential equation.

2-146 Consider two special cases for the fin in Problem 2-145: (a) an insulated-tip fin losing heat by radiation and (b) a very long fin losing heat by radiation. Write the appropriate boundary conditions for these two cases.

2-147 Consider another special case for the fin of Problem 2-145; where the surrounding radiation boundary temperature is negligible, that is,

$$T_s^4 \ll T^4$$

Write the resulting simplified differential equation under this condition.

REFERENCES

1. Schneider, P. J. *Conduction Heat Transfer*, Reading, Mass.: Addison-Wesley Publishing Company, 1955.

2. Harper, W. B., and D. R. Brown. "Mathematical Equations for Heat Conduction in the Fins of Air-cooled Engines," *NACA Rep.* 158, 1922.

3. Gardner, K. A. "Efficiency of Extended Surfaces," *Trans. ASME*, vol. 67, pp. 621–31, 1945.

4. Moore, C. J. "Heat Transfer across Surfaces in Contact: Studies of Transients in One-dimensional Composite Systems," *Southern Methodist Univ., Thermal/Fluid Sci. Ctr. Res. Rep.* 67-2, Dallas, Tex., March 1967.

5. Ybarrondo, L. J., and J. E. Sunderland. "Heat Transfer from Extended Surfaces," *Bull. Mech. Eng. Educ.*, vol. 5, pp. 229–34, 1966.

6. Moore, C. J., Jr., H. A. Blum, and H. Atkins. "Subject Classification Bibliography for Thermal Contact Resistance Studies," *ASME Pap.* 68-WA/HT-18, December 1968.

7. Clausing, A. M. "Transfer at the Interface of Dissimilar Metals: The Influence of Thermal Strain," *Int. J. Heat Mass Transfer*, vol. 9, p. 791, 1966.

8. Kern, D. Q., and A. D. Kraus. *Extended Surface Heat Transfer*. New York: McGraw-Hill, 1972.

9. Siegel, R., and J. R. Howell. *Thermal Radiation Heat Transfer*. 2d ed., New York: McGraw-Hill, 1980.

10. Fried, E. "Thermal Conduction Contribution to Heat Transfer at Contacts," *Thermal Conductivity*, (R. P. Tye, Ed.) vol. 2, New York: Academic Press, 1969.

11. Fletcher, L. S. "Recent Developments in Contact Conductance Heat Transfer," *J. Heat Transfer*, vol. 110, no. 4(B), p. 1059, Nov. 1988.

12. James, T. B., and W. P. Goss. *Heat Transmission Coefficients for Walls, Roofs, Ceilings, and Floors*. Altanta: American Society of Heating, Refrigeration, and Air-Conditioning Engineers, 1993.

3

Steady-State Conduction—Multiple Dimensions

3-1 | INTRODUCTION

In Chapter 2 steady-state heat transfer was calculated in systems in which the temperature gradient and area could be expressed in terms of one space coordinate. We now wish to analyze the more general case of two-dimensional heat flow. For steady state with no heat generation, the Laplace equation applies.

$$\frac{\partial^2 T}{\partial x^2} + \frac{\partial^2 T}{\partial y^2} = 0 \qquad \textbf{[3-1]}$$

assuming constant thermal conductivity. The solution to this equation may be obtained by analytical, numerical, or graphical techniques.

The objective of any heat-transfer analysis is usually to predict heat flow or the temperature that results from a certain heat flow. The solution to Equation (3-1) will give the temperature in a two-dimensional body as a function of the two independent space coordinates x and y. Then the heat flow in the x and y directions may be calculated from the Fourier equations

$$q_x = -kA_x \frac{\partial T}{\partial x} \qquad \textbf{[3-2]}$$

$$q_y = -kA_y \frac{\partial T}{\partial y} \qquad \textbf{[3-3]}$$

These heat-flow quantities are directed either in the x direction or in the y direction. The total heat flow at any point in the material is the resultant of the q_x and q_y at that point. Thus the total heat-flow vector is directed so that it is perpendicular to the lines of constant temperature in the material, as shown in Figure 3-1. So if the temperature distribution in the material is known, we may easily establish the heat flow.

3-2 | MATHEMATICAL ANALYSIS OF TWO-DIMENSIONAL HEAT CONDUCTION

We first consider an analytical approach to a two-dimensional problem and then indicate the numerical and graphical methods that may be used to advantage in many other problems.

It is worthwhile to mention here that analytical solutions are not always possible to obtain; indeed, in many instances they are very cumbersome and difficult to use. In these cases numerical techniques are frequently used to advantage. For a more extensive treatment of the analytical methods used in conduction problems, the reader may consult References 1, 2, 10, and 11.

Consider the rectangular plate shown in Figure 3-2. Three sides of the plate are maintained at the constant temperature T_1, and the upper side has some temperature distribution impressed upon it. This distribution could be simply a constant temperature or something more complex, such as a sine-wave distribution. We shall consider both cases.

To solve Equation (3-1), the separation-of-variables method is used. The essential point of this method is that the solution to the differential equation is assumed to take a product form

$$T = XY \quad \text{where} \quad \begin{aligned} X &= X(x) \\ Y &= Y(y) \end{aligned} \qquad \textbf{[3-4]}$$

The boundary conditions are then applied to determine the form of the functions X and Y. The basic assumption as given by Equation (3-4) can be justified *only* if it is possible to find a solution of this form that satisfies the boundary conditions.

Figure 3-1 | Sketch showing the heat flow in two dimensions.

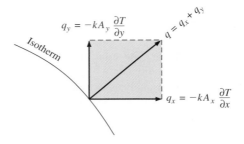

Figure 3-2 | Isotherms and heat flow lines in a rectangular plate.

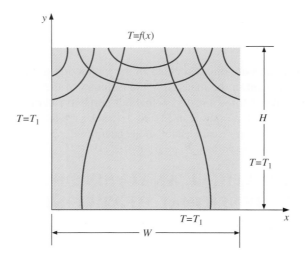

First consider the boundary conditions with a sine-wave temperature distribution impressed on the upper edge of the plate. Thus

$$T = T_1 \qquad \text{at } y = 0$$
$$T = T_1 \qquad \text{at } x = 0$$
$$T = T_1 \qquad \text{at } x = W \qquad \text{[3-5]}$$
$$T = T_m \sin\left(\frac{\pi x}{W}\right) + T_1 \quad \text{at } y = H$$

where T_m is the amplitude of the sine function. Substituting Equation (3-4) in (3-1) gives

$$-\frac{1}{X}\frac{d^2 X}{dx^2} = \frac{1}{Y}\frac{d^2 Y}{dy^2} \qquad \text{[3-6]}$$

Observe that each side of Equation (3-6) is independent of the other because x and y are independent variables. This requires that each side be equal to some constant. We may thus obtain two ordinary differential equations in terms of this constant,

$$\frac{d^2 X}{dx^2} + \lambda^2 X = 0 \qquad \text{[3-7]}$$

$$\frac{d^2 Y}{dy^2} - \lambda^2 Y = 0 \qquad \text{[3-8]}$$

where λ^2 is called the *separation constant*. Its value must be determined from the boundary conditions. Note that the form of the solution to Equations (3-7) and (3-8) will depend on the sign of λ^2; a different form would also result if λ^2 were zero. The only way that the correct form can be determined is through an application of the boundary conditions of the problem. So we shall first write down all possible solutions and then see which one fits the problem under consideration.

For $\lambda^2 = 0$:

$$X = C_1 + C_2 x$$
$$Y = C_3 + C_4 y \qquad \text{[3-9]}$$
$$T = (C_1 + C_2 x)(C_3 + C_4 y)$$

This function cannot fit the sine-function boundary condition, so the $\lambda^2 = 0$ solution may be excluded.

For $\lambda^2 < 0$:

$$X = C_5 e^{-\lambda x} + C_6 e^{\lambda x}$$
$$Y = C_7 \cos \lambda y + C_8 \sin \lambda y \qquad \text{[3-10]}$$
$$T = (C_5 e^{-\lambda x} + C_6 e^{\lambda x})(C_7 \cos \lambda y + C_8 \sin \lambda y)$$

Again, the sine-function boundary condition cannot be satisfied, so this solution is excluded also.

For $\lambda^2 > 0$:

$$X = C_9 \cos \lambda x + C_{10} \sin \lambda x$$
$$Y = C_{11} e^{-\lambda y} + C_{12} e^{\lambda y} \qquad \text{[3-11]}$$
$$T = (C_9 \cos \lambda x + C_{10} \sin \lambda x)(C_{11} e^{-\lambda y} + C_{12} e^{\lambda y})$$

Now, it is possible to satisfy the sine-function boundary condition; so we shall attempt to satisfy the other conditions. The algebra is somewhat easier to handle when the substitution

$$\theta = T - T_1$$

is made. The differential equation and the solution then retain the same form in the new variable θ, and we need only transform the boundary conditions. Thus

$$
\begin{array}{ll}
\theta = 0 & \text{at } y = 0 \\
\theta = 0 & \text{at } x = 0 \\
\theta = 0 & \text{at } x = W \\
\theta = T_m \sin \dfrac{\pi x}{W} & \text{at } y = H
\end{array}
\qquad \textbf{[3-12]}
$$

Applying these conditions, we have

$$
0 = (C_9 \cos \lambda x + C_{10} \sin \lambda x)(C_{11} + C_{12})
\qquad \textbf{[a]}
$$

$$
0 = C_9(C_{11}e^{-\lambda y} + C_{12}e^{\lambda y})
\qquad \textbf{[b]}
$$

$$
0 = (C_9 \cos \lambda W + C_{10} \sin \lambda W)(C_{11}e^{-\lambda y} + C_{12}e^{\lambda y})
\qquad \textbf{[c]}
$$

$$
T_m \sin \frac{\pi x}{W} = (C_9 \cos \lambda x + C_{10} \sin \lambda x)(C_{11}e^{-\lambda H} + C_{12}e^{\lambda H})
\qquad \textbf{[d]}
$$

Accordingly,

$$
\begin{aligned}
C_{11} &= -C_{12} \\
C_9 &= 0
\end{aligned}
$$

and from (c),

$$
0 = C_{10}C_{12} \sin \lambda W (e^{\lambda y} - e^{-\lambda y})
$$

This requires that

$$
\sin \lambda W = 0
\qquad \textbf{[3-13]}
$$

Recall that λ was an undetermined separation constant. Several values will satisfy Equation (3-13), and these may be written

$$
\lambda = \frac{n\pi}{W}
\qquad \textbf{[3-14]}
$$

where n is an integer. The solution to the differential equation may thus be written as a sum of the solutions for each value of n. This is an infinite sum, so that the final solution is the infinite series

$$
\theta = T - T_1 = \sum_{n=1}^{\infty} C_n \sin \frac{n\pi x}{W} \sinh \frac{n\pi y}{W}
\qquad \textbf{[3-15]}
$$

where the constants have been combined and the exponential terms converted to the hyperbolic function. The final boundary condition may now be applied:

$$
T_m \sin \frac{\pi x}{W} = \sum_{n=1}^{\infty} C_n \sin \frac{n\pi x}{W} \sinh \frac{n\pi H}{W}
$$

which requires that $C_n = 0$ for $n > 1$. The final solution is therefore

$$
T = T_m \frac{\sinh(\pi y/W)}{\sinh(\pi H/W)} \sin\left(\frac{\pi x}{W}\right) + T_1
\qquad \textbf{[3-16]}
$$

The temperature field for this problem is shown in Figure 3-2. Note that the heat-flow lines are perpendicular to the isotherms.

We now consider the set of boundary conditions

$$T = T_1 \quad \text{at } y = 0$$
$$T = T_1 \quad \text{at } x = 0$$
$$T = T_1 \quad \text{at } x = W$$
$$T = T_2 \quad \text{at } y = H$$

Using the first three boundary conditions, we obtain the solution in the form of Equation (3-15):

$$T - T_1 = \sum_{n=1}^{\infty} C_n \sin \frac{n\pi x}{W} \sinh \frac{n\pi y}{W} \tag{3-17}$$

Applying the fourth boundary condition gives

$$T_2 - T_1 = \sum_{n=1}^{\infty} C_n \sin \frac{n\pi x}{W} \sinh \frac{n\pi H}{W} \tag{3-18}$$

This is a Fourier sine series, and the values of the C_n may be determined by expanding the constant temperature difference $T_2 - T_1$ in a Fourier series over the interval $0 < x < W$. This series is

$$T_2 - T_1 = (T_2 - T_1) \frac{2}{\pi} \sum_{n=1}^{\infty} \frac{(-1)^{n+1} + 1}{n} \sin \frac{n\pi x}{W} \tag{3-19}$$

Upon comparison of Equation (3-18) with Equation (3-19), we find that

$$C_n = \frac{2}{\pi} (T_2 - T_1) \frac{1}{\sinh(n\pi H/W)} \frac{(-1)^{n+1} + 1}{n}$$

and the final solution is expressed as

$$\frac{T - T_1}{T_2 - T_1} = \frac{2}{\pi} \sum_{n=1}^{\infty} \frac{(-1)^{n+1} + 1}{n} \sin \frac{n\pi x}{W} \frac{\sinh(n\pi y/W)}{\sinh(n\pi H/W)} \tag{3-20}$$

An extensive study of analytical techniques used in conduction heat transfer requires a background in the theory of orthogonal functions. Fourier series are one example of orthogonal functions, as are Bessel functions and other special functions applicable to different geometries and boundary conditions. The interested reader may consult one or more of the conduction heat-transfer texts listed in the references for further information on the subject.

3-3 | GRAPHICAL ANALYSIS

Consider the two-dimensional system shown in Figure 3-3. The inside surface is maintained at some temperature T_1, and the outer surface is maintained at T_2. We wish to calculate the heat transfer. Isotherms and heat-flow lanes have been sketched to aid in this calculation. The isotherms and heat-flow lanes form groupings of curvilinear figures like that shown in Figure 3-3b. The heat flow across this curvilinear section is given by Fourier's law, assuming unit depth of material:

$$q = -k \, \Delta x(1) \frac{\Delta T}{\Delta y} \tag{3-21}$$

Figure 3-3 | Sketch showing element used for curvilinear-square analysis of two-dimensional heat flow.

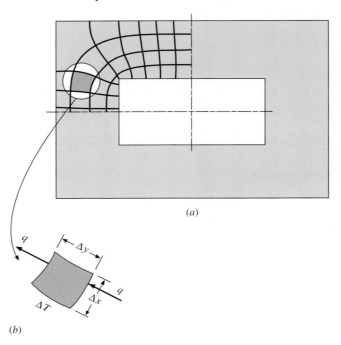

(a)

(b)

This heat flow will be the same through each section within this heat-flow lane, and the total heat flow will be the sum of the heat flows through all the lanes. If the sketch is drawn so that $\Delta x \cong \Delta y$, the heat flow is proportional to the ΔT across the element and, since this heat flow is constant, the ΔT across each element must be the same within the same heat-flow lane. Thus the ΔT across an element is given by

$$\Delta T = \frac{\Delta T_{\text{overall}}}{N}$$

where N is the number of temperature increments between the inner and outer surfaces. Furthermore, the heat flow through each lane is the same since it is independent of the dimensions Δx and Δy when they are constructed equal. Thus we write for the total heat transfer

$$q = \frac{M}{N}k\,\Delta T_{\text{overall}} = \frac{M}{N}k(T_2 - T_1) \qquad \textbf{[3-22]}$$

where M is the number of heat-flow lanes. So, to calculate the heat transfer, we need only construct these curvilinear-square plots and count the number of temperature increments and heat-flow lanes. Care must be taken to construct the plot so that $\Delta x \approx \Delta y$ and the lines are perpendicular. For the corner section shown in Figure 3-3a the number of temperature increments between the inner and outer surfaces is about $N = 4$, while the number of heat-flow lanes for the corner section may be estimated as $M = 8.2$. The total number of heat-flow lanes is four times this value, or $4 \times 8.2 = 32.8$. The ratio M/N is thus $32.8/4 = 8.2$ for the whole wall section. This ratio will be called the *conduction shape factor* in subsequent discussions.

The accuracy of this method is dependent entirely on the skill of the person sketching the curvilinear squares. Even a crude sketch, however, can frequently help to give fairly good estimates of the temperatures that will occur in a body. An electrical analogy may be employed to sketch the curvilinear squares, as discussed in Section 3-9.

The graphical method presented here is mainly of historical interest to show the relation of heat-flow lanes and isotherms. It may not be expected to be used for the solution of many practical problems.

3-4 | THE CONDUCTION SHAPE FACTOR

In a two-dimensional system where only two temperature limits are involved, we may define a conduction shape factor S such that

$$q = kS\,\Delta T_{\text{overall}} \qquad\qquad \textbf{[3-23]}$$

The values of S have been worked out for several geometries and are summarized in Table 3-1. A very comprehensive summary of shape factors for a large variety of geometries is given by Rohsenow [15] and Hahne and Grigull [17]. Note that the inverse hyperbolic cosine can be calculated from

$$\cosh^{-1} x = \ln(x \pm \sqrt{x^2 - 1})$$

For a three-dimensional wall, as in a furnace, separate shape factors are used to calculate the heat flow through the edge and corner sections, with the dimensions shown in Figure 3-4. When all the interior dimensions are greater than one-fifth of the wall thickness,

$$S_{\text{wall}} = \frac{A}{L} \qquad S_{\text{edge}} = 0.54D \qquad S_{\text{corner}} = 0.15L$$

where

A = area of wall

L = wall thickness

D = length of edge

Note that the shape factor per unit depth is given by the ratio M/N when the curvilinear-squares method is used for calculations.

Figure 3-4 | Sketch illustrating dimensions for use in calculating three-dimensional shape factors.

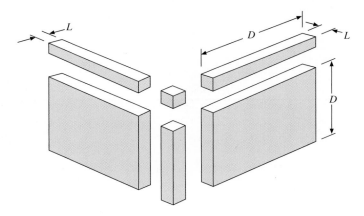

Table 3-1 | Conduction shape factors, summarized from References 6 and 7.
Note: **For buried objects the temperature difference is $\Delta T = T_{\text{object}} - T_{\text{far field}}$. The far-field temperature is taken the same as the isothermal surface temperature for semi-infinite media.**

Physical system	Schematic	Shape factor	Restrictions
Isothermal cylinder of radius r buried in semi-infinite medium having isothermal surface	Isothermal	$\dfrac{2\pi L}{\cosh^{-1}(D/r)}$	$L \gg r$
		$\dfrac{2\pi L}{\ln(D/r)}$	$L \gg r$ $D > 3r$
Isothermal sphere of radius r buried in infinite medium		$4\pi r$	
Isothermal sphere of radius r buried in semi-infinite medium having isothermal surface $\Delta T = T_{\text{surf}} - T_{\text{far field}}$	Isothermal	$\dfrac{4\pi r}{1 - r/2D}$	
Conduction between two isothermal cylinders of length L buried in infinite medium	r_1 r_2	$\dfrac{2\pi L}{\cosh^{-1}\left(\dfrac{D^2 - r_1^2 - r_2^2}{2r_1 r_2}\right)}$	$L \gg r$ $L \gg D$
Row of horizontal cylinders of length L in semi-infinite medium with isothermal surface	Isothermal	$S = \dfrac{2\pi L}{\ln\left[\left(\dfrac{1}{\pi r}\right)\sinh(2\pi D/l)\right]}$	$D > 2r$
Buried cube in infinite medium, L on a side		$8.24L$	

Table 3-1 | (*Continued*).

Physical system	Schematic	Shape factor	Restrictions
Isothermal cylinder of radius r placed in semi-infinite medium as shown	Isothermal	$\dfrac{2\pi L}{\ln(2L/r)}$	$L \gg 2r$
Isothermal rectangular parallelepiped buried in semi-infinite medium having isothermal surface	Isothermal	$1.685L\left[\log\left(1+\dfrac{b}{a}\right)\right]^{-0.59}\left(\dfrac{b}{c}\right)^{-0.078}$	See Reference 7
Plane wall		$\dfrac{A}{L}$	One-dimensional heat flow
Hollow cylinder, length L		$\dfrac{2\pi L}{\ln(r_o/r_i)}$	$L \gg r$
Hollow sphere		$\dfrac{4\pi r_o r_i}{r_o - r_i}$	
Thin horizontal disk buried in semi-infinite medium with isothermal surface	Isothermal	$4r$ $8r$ $\dfrac{4\pi r}{\pi/2 - \tan^{-1}(r/2D)}$	$D=0$ $D \gg 2r$ $D/2r > 1$ $\tan^{-1}(r/2D)$ in radians
Hemisphere buried in semi-infinite medium $\Delta T = T_{\text{sphere}} - T_{\text{far field}}$	Isothermal	$2\pi r$	
Isothermal sphere buried in semi-infinite medium with insulated surface	Insulated	$\dfrac{4\pi r}{1 + r/2D}$	
Two isothermal spheres buried in infinite medium		$\dfrac{4\pi r_2}{\dfrac{r_2}{r_1}\left[1 - \dfrac{(r_1/D)^4}{1-(r_2/D)^2}\right] - \dfrac{2r_2}{D}}$	$D > 5r_{\max}$

Table 3-1 | (*Continued*).

Physical system	Schematic	Shape factor	Restrictions
Thin rectangular plate of length L, buried in semi-infinite medium having isothermal surface	Isothermal	$\dfrac{\pi W}{\ln(4W/L)}$	$D = 0$ $W > L$
		$\dfrac{2\pi W}{\ln(4W/L)}$	$D \gg W$ $W > L$
		$\dfrac{2\pi W}{\ln(2\pi D/L)}$	$W \gg L$ $D > 2W$
Parallel disks buried in infinite medium		$\dfrac{4\pi r}{\left[\frac{\pi}{2} - \tan^{-1}(r/D)\right]}$	$D > 5r$ $\tan^{-1}(r/D)$ in radians
Eccentric cylinders of length L		$\dfrac{2\pi L}{\cosh^{-1}\left(\dfrac{r_1^2 + r_2^2 - D^2}{2 r_1 r_2}\right)}$	$L \gg r_2$
Cylinder centered in a square of length L		$\dfrac{2\pi L}{\ln(0.54 W/r)}$	$L \gg W$
Horizontal cylinder of length L centered in infinite plate	Isothermal Isothermal	$\dfrac{2\pi L}{\ln(4D/r)}$	
Thin horizontal disk buried in semi-infinite medium with adiabatic surface $\Delta T = T_{\text{disk}} - T_{\text{far field}}$	Insulated	$\dfrac{4\pi r}{\pi/2 + \tan^{-1}(r/2D)}$	$D/2r > 1$ $\tan^{-1}(r/2D)$ in radians

Buried Pipe
EXAMPLE 3-1

A horizontal pipe 15 cm in diameter and 4 m long is buried in the earth at a depth of 20 cm. The pipe-wall temperature is 75°C, and the earth surface temperature is 5°C. Assuming that the thermal conductivity of the earth is 0.8 W/m · °C, calculate the heat lost by the pipe.

■ **Solution**

We may calculate the shape factor for this situation using the equation given in Table 3-1. Since $D < 3r$,

$$S = \frac{2\pi L}{\cosh^{-1}(D/r)} = \frac{2\pi(4)}{\cosh^{-1}(20/7.5)} = 15.35 \text{ m}$$

The heat flow is calculated from

$$q = kS\Delta T = (0.8)(15.35)(75 - 5) = 859.6 \text{ W} \quad [2933 \text{ Btu/h}]$$

Cubical Furnace
EXAMPLE 3-2

A small cubical furnace 50 by 50 by 50 cm on the inside is constructed of fireclay brick $[k = 1.04 \text{ W/m} \cdot °C]$ with a wall thickness of 10 cm. The inside of the furnace is maintained at 500°C, and the outside is maintained at 50°C. Calculate the heat lost through the walls.

■ **Solution**

We compute the total shape factor by adding the shape factors for the walls, edges, and corners:

Walls:
$$S = \frac{A}{L} = \frac{(0.5)(0.5)}{0.1} = 2.5 \text{ m}$$

Edges:
$$S = 0.54D = (0.54)(0.5) = 0.27 \text{ m}$$

Corners:
$$S = 0.15L = (0.15)(0.1) = 0.015 \text{ m}$$

There are six wall sections, twelve edges, and eight corners, so that the total shape factor is

$$S = (6)(2.5) + (12)(0.27) + (8)(0.015) = 18.36 \text{ m}$$

and the heat flow is calculated as

$$q = kS\Delta T = (1.04)(18.36)(500 - 50) = 8.592 \text{ kW} \quad [29.320 \text{ Btu/h}]$$

Buried Disk
EXAMPLE 3-3

A disk having a diameter of 30 cm and maintained at a temperature of 95°C is buried at a depth of 1.0 m in a semi-infinite medium having an isothermal surface temperature of 20°C and a thermal conductivity of 2.1 W/m · °C. Calculate the heat lost by the disk.

■ **Solution**

This is an application of the conduction shape factor relation $q = kS \Delta T$. Consulting Table 3-1 we find a choice of three relations for S for the geometry of a disk buried in a semi-infinite medium with an isothermal surface. Clearly, $D \neq 0$ and D is not large compared to $2r$, so the relation we select for the shape factor is for the case $D/2r > 1.0$:

$$S = \frac{4\pi r}{[\pi/2 - \tan^{-1}(r/2D)]}$$

Note that this relation differs from the one for an insulated surface by the minus sign in the denominator. Inserting $r = 0.15$ m and $D = 1.0$ m we obtain

$$S = \frac{4\pi(0.15)}{[\pi/2 - \tan^{-1}(0.15/2)]} = \frac{4\pi(0.15)}{[\pi/2 - 0.07486]} = 1.26 \text{ m}$$

For buried objects the shape factor is based on $\Delta T = T_{object} - T_{far\ field}$. The far-field temperature is taken as the isothermal surface temperature, and the heat lost by the disk is therefore

$$q = kS\Delta T = (2.1)(1.26)(95 - 20) = 198.45 \text{ W}$$

EXAMPLE 3-4   Buried Parallel Disks

Two parallel 50-cm-diameter disks are separated by a distance of 1.5 m in an infinite medium having $k = 2.3$ W/m · °C. One disk is maintained at 80°C and the other at 20°C. Calculate the heat transfer between the disks.

■ **Solution**
This is a shape-factor problem and the heat transfer may be calculated from

$$q = kS\Delta T$$

where S is obtained from Table 3-1 as

$$S = \frac{4\pi r}{[\pi/2 - \tan^{-1}(r/D)]} \qquad \text{for } D > 5r$$

With $r = 0.25$ m and $D = 1.5$ m we obtain

$$S = \frac{4\pi(0.25)}{[\pi/2 - \tan^{-1}(0.25/1.5)]} = \frac{4\pi(0.25)}{[\pi/2 - 0.1651]} = 2.235$$

and

$$q = kS\,\Delta T = (2.3)(2.235)(80 - 20) = 308.4 \text{ W}$$

3-5 | NUMERICAL METHOD OF ANALYSIS

An immense number of analytical solutions for conduction heat-transfer problems have been accumulated in the literature over the past 150 years. Even so, in many practical situations the geometry or boundary conditions are such that an analytical solution has not been obtained at all, or if the solution has been developed, it involves such a complex series solution that numerical evaluation becomes exceedingly difficult. For such situations the most fruitful approach to the problem is one based on finite-difference techniques, the basic principles of which we shall outline in this section.

Consider a two-dimensional body that is to be divided into equal increments in both the x and y directions, as shown in Figure 3-5. The nodal points are designated as shown, the m locations indicating the x increment and the n locations indicating the y increment. We wish to establish the temperatures at any of these nodal points within the body, using Equation (3-1) as a governing condition. Finite differences are used to approximate differential increments in the temperature and space coordinates; and the smaller we choose these finite increments, the more closely the true temperature distribution will be approximated.

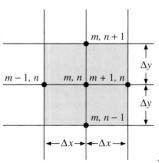

Figure 3-5 | Sketch illustrating nomenclature used in two-dimensional numerical analysis of heat conduction.

The temperature gradients may be written as follows:

$$\frac{\partial T}{\partial x}\bigg]_{m+1/2,n} \approx \frac{T_{m+1,n} - T_{m,n}}{\Delta x}$$

$$\frac{\partial T}{\partial x}\bigg]_{m-1/2,n} \approx \frac{T_{m,n} - T_{m-1,n}}{\Delta x}$$

$$\frac{\partial T}{\partial y}\bigg]_{m,n+1/2} \approx \frac{T_{m,n+1} - T_{m,n}}{\Delta y}$$

$$\frac{\partial T}{\partial y}\bigg]_{m,n-1/2} \approx \frac{T_{m,n} - T_{m,n-1}}{\Delta y}$$

$$\frac{\partial^2 T}{\partial x^2}\bigg]_{m,n} \approx \frac{\dfrac{\partial T}{\partial x}\bigg]_{m+1/2,n} - \dfrac{\partial T}{\partial x}\bigg]_{m-1/2,n}}{\Delta x} = \frac{T_{m+1,n} + T_{m-1,n} - 2T_{m,n}}{(\Delta x)^2}$$

$$\frac{\partial^2 T}{\partial y^2}\bigg]_{m,n} \approx \frac{\dfrac{\partial T}{\partial y}\bigg]_{m,n+1/2} - \dfrac{\partial T}{\partial y}\bigg]_{m,n-1/2}}{\Delta y} = \frac{T_{m,n+1} + T_{m,n-1} - 2T_{m,n}}{(\Delta y)^2}$$

Thus the finite-difference approximation for Equation (3-1) becomes

$$\frac{T_{m+1,n} + T_{m-1,n} - 2T_{m,n}}{(\Delta x)^2} + \frac{T_{m,n+1} + T_{m,n-1} - 2T_{m,n}}{(\Delta y)^2} = 0$$

If $\Delta x = \Delta y$, then

$$T_{m+1,n} + T_{m-1,n} + T_{m,n+1} + T_{m,n-1} - 4T_{m,n} = 0 \qquad \textbf{[3-24]}$$

Since we are considering the case of constant thermal conductivity, the heat flows may all be expressed in terms of temperature differentials. Equation (3-24) states very simply that the net heat flow into any node is zero at steady-state conditions. In effect, the numerical finite-difference approach replaces the continuous temperature distribution by fictitious heat-conducting rods connected between small nodal points that do not generate heat.

We can also devise a finite-difference scheme to take heat generation into account. We merely add the term $\dot{q}/k$ into the general equation and obtain

$$\frac{T_{m+1,n} + T_{m-1,n} - 2T_{m,n}}{(\Delta x)^2} + \frac{T_{m,n+1} + T_{m,n-1} - 2T_{m,n}}{(\Delta y)^2} + \frac{\dot{q}}{k} = 0$$

Then for a square grid in which $\Delta x = \Delta y$,

$$T_{m+1,n} + T_{m-1,n} + T_{m,n+1} + T_{m,n-1} + \frac{\dot{q}(\Delta x)^2}{k} - 4T_{m,n-1} = 0 \qquad \textbf{[3-24a]}$$

To utilize the numerical method, Equation (3-24) must be written for each node within the material and the resultant system of equations solved for the temperatures at the various nodes. A very simple example is shown in Figure 3-6, and the four equations for nodes 1, 2, 3, and 4 would be

$$100 + 500 + T_2 + T_3 - 4T_1 = 0$$
$$T_1 + 500 + 100 + T_4 - 4T_2 = 0$$

Figure 3-6 | Four-node problem.

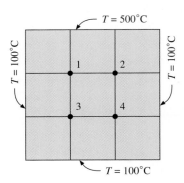

$$100 + T_1 + T_4 + 100 - 4T_3 = 0$$
$$T_3 + T_2 + 100 + 100 - 4T_4 = 0$$

These equations have the solution

$$T_1 = T_2 = 250°C \qquad T_3 = T_4 = 150°C$$

Of course, we could recognize from symmetry that $T_1 = T_2$ and $T_3 = T_4$ and would then only need two nodal equations,

$$100 + 500 + T_3 - 3T_1 = 0$$
$$100 + T_1 + 100 - 3T_3 = 0$$

Once the temperatures are determined, the heat flow may be calculated from

$$q = \sum k \, \Delta x \, \frac{\Delta T}{\Delta y}$$

where the ΔT is taken at the boundaries. In the example the heat flow may be calculated at either the 500°C face or the three 100°C faces. If a sufficiently fine grid is used, the two values should be very nearly the same. As a matter of general practice, it is usually best to take the arithmetic average of the two values for use in the calculations. In the example, the two calculations yield:

500°C face:

$$q = -k \frac{\Delta x}{\Delta y}[(250 - 500) + (250 - 500)] = 500k$$

100°C face:

$$q = -k \frac{\Delta y}{\Delta x} \, [(250 - 100) + (150 - 100) + (150 - 100) + (150 - 100)$$
$$+ (150 - 100) + (250 - 100)] = -500k$$

and the two values agree in this case. The calculation of the heat flow in cases in which curved boundaries or complicated shapes are involved is treated in References 2, 3, and 15.

When the solid is exposed to some convection boundary condition, the temperatures at the surface must be computed differently from the method given above. Consider the boundary shown in Figure 3-7. The energy balance on node (m, n) is

$$-k\,\Delta y\frac{T_{m,n} - T_{m-1,n}}{\Delta x} - k\frac{\Delta x}{2}\frac{T_{m,n} - T_{m,n+1}}{\Delta y} - k\frac{\Delta x}{2}\frac{T_{m,n} - T_{m,n-1}}{\Delta y}$$
$$= h\,\Delta y(T_{m,n} - T_\infty)$$

If $\Delta x = \Delta y$, the boundary temperature is expressed in the equation

$$T_{m,n}\left(\frac{h\,\Delta x}{k} + 2\right) - \frac{h\,\Delta x}{k}T_\infty - \frac{1}{2}(2T_{m-1,n} + T_{m,n+1} + T_{m,n-1}) = 0 \qquad \textbf{[3-25]}$$

An equation of this type must be written for each node along the surface shown in Figure 3-7. So when a convection boundary condition is present, an equation like (3-25) is used at the boundary and an equation like (3-24) is used for the interior points.

Equation (3-25) applies to a plane surface exposed to a convection boundary condition. It will not apply for other situations, such as an insulated wall or a corner exposed to a convection boundary condition. Consider the corner section shown in Figure 3-8. The energy balance for the corner section is

$$-k\frac{\Delta y}{2}\frac{T_{m,n} - T_{m-1,n}}{\Delta x} - k\frac{\Delta x}{2}\frac{T_{m,n} - T_{m,n-1}}{\Delta y} = h\frac{\Delta x}{2}(T_{m,n} - T_\infty) + h\frac{\Delta y}{2}(T_{m,n} - T_\infty)$$

If $\Delta x = \Delta y$,

$$2T_{m,n}\left(\frac{h\,\Delta x}{k} + 1\right) - 2\frac{h\,\Delta x}{k}T_\infty - (T_{m-1,n} + T_{m,n-1}) = 0 \qquad \textbf{[3-26]}$$

Other boundary conditions may be treated in a similar fashion, and a convenient summary of nodal equations is given in Table 3-2 for different geometrical and boundary situations. Situations f and g are of particular interest since they provide the calculation equations that may be employed with curved boundaries, while still using uniform increments in Δx and Δy.

Figure 3-7 | Nomenclature for nodal equation with convective boundary condition.

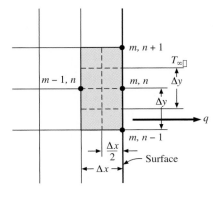

Figure 3-8 | Nomenclature for nodal equation with convection at a corner section.

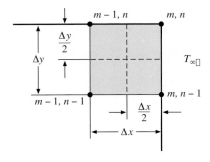

Table 3-2 | Summary of nodal formulas for finite-difference calculations. (Dashed lines indicate element volume.)[†]

Physical situation	Nodal equation for equal increments in x and y (second equation in situation is in form for Gauss-Seidel iteration)
(a) Interior node 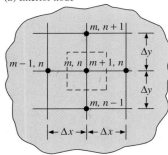	$$0 = T_{m+1,n} + T_{m,n+1} + T_{m-1,n} + T_{m,n-1} - 4T_{m,n}$$ $$T_{m,n} = (T_{m+1,n} + T_{m,n+1} + T_{m-1,n} + T_{m,n-1})/4$$
(b) Convection boundary node 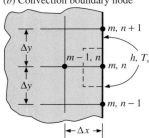	$$0 = \frac{h\Delta x}{k} T_\infty + \frac{1}{2}(2T_{m-1,n} + T_{m,n+1} + T_{m,n-1}) - \left(\frac{h\Delta x}{k} + 2\right) T_{m,n}$$ $$T_{m,n} = \frac{T_{m-1,n} + (T_{m,n+1} + T_{m,n-1})/2 + \text{Bi } T_\infty}{2 + \text{Bi}}$$ $$\text{Bi} = \frac{h\Delta x}{k}$$
(c) Exterior corner with convection boundary 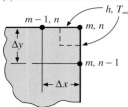	$$0 = 2\frac{h\Delta x}{k} T_\infty + (T_{m-1,n} + T_{m,n-1}) - 2\left(\frac{h\Delta x}{k} + 1\right) T_{m,n}$$ $$T_{m,n} = \frac{(T_{m-1,n} + T_{m,n-1})/2 + \text{Bi } T_\infty}{1 + \text{Bi}}$$ $$\text{Bi} = \frac{h\Delta x}{k}$$
(d) Interior corner with convection boundary 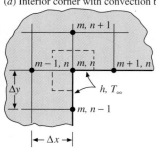	$$0 = 2\frac{h\Delta x}{k} T_\infty + 2T_{m-1,n} + T_{m,n+1} + T_{m+1,n} + T_{m,n-1} - 2\left(3 + \frac{h\Delta x}{k}\right) T_{m,n}$$ $$T_{m,n} = \frac{\text{Bi } T_\infty + T_{m,n+1} + T_{m-1,n} + (T_{m+1,n} + T_{m,n-1})/2}{3 + \text{Bi}}$$ $$\text{Bi} = \frac{h\Delta x}{k}$$

Table 3-2 | (*Continued*).

Physical situation	Nodal equation for equal increments in x and y (second equation in situation is in form for Gauss-Seidel iteration)
(*e*) Insulated boundary 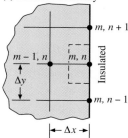	$0 = T_{m,n+1} + T_{m,n-1} + 2T_{m-1,n} - 4T_{m,n}$ $T_{m,n} = (T_{m,n+1} + T_{m,n-1} + 2T_{m-1,n})/4$
(*f*) Interior node near curved boundary‡ 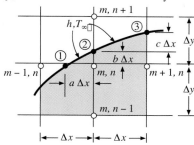	$0 = \frac{2}{b(b+1)}T_2 + \frac{2}{a+1}T_{m+1,n} + \frac{2}{b+1}T_{m,n-1} + \frac{2}{a(a+1)}T_1 - 2\left(\frac{1}{a}+\frac{1}{b}\right)T_{m,n}$
(*g*) Boundary node with convection along curved boundary—node 2 for (*f*) above§	$0 = \frac{b}{\sqrt{a^2+b^2}}T_1 + \frac{b}{\sqrt{c^2+1}}T_3 + \frac{a+1}{b}T_{m,n} + \frac{h\Delta x}{k}(\sqrt{c^2+1}+\sqrt{a^2+b^2})T_\infty$ $\quad - \left[\frac{b}{\sqrt{a^2+b^2}} + \frac{b}{\sqrt{c^2+1}} + \frac{a+1}{b} + (\sqrt{c^2+1}+\sqrt{a^2+b^2})\frac{h\Delta x}{k}\right]T_2$

† Convection boundary may be converted to insulated surface by setting $h = 0$ (Bi = 0).
‡ This equation is obtained by multiplying the resistance by $4/(a+1)(b+1)$
§ This relation is obtained by dividing the resistance formulation by 2.

Nine-Node Problem   EXAMPLE 3-5

Consider the square of Figure Example 3-5. The left face is maintained at 100°C and the top face at 500°C, while the other two faces are exposed to an environment at 100°C:

$$h = 10 \text{ W/m}^2 \cdot {}^\circ\text{C} \quad \text{and} \quad k = 10 \text{ W/m} \cdot {}^\circ\text{C}$$

The block is 1 m square. Compute the temperature of the various nodes as indicated in Figure Example 3-5 and the heat flows at the boundaries.

■ **Solution**

The nodal equation for nodes 1, 2, 4, and 5 is

$$T_{m+1,n} + T_{m-1,n} + T_{m,n+1} + T_{m,n-1} - 4T_{m,n} = 0$$

The equation for nodes 3, 6, 7, and 8 is given by Equation (3-25), and the equation for 9 is given by Equation (3-26):

$$\frac{h\Delta x}{k} = \frac{(10)(1)}{(3)(10)} = \frac{1}{3}$$

Figure Example 3-5 | Nomenclature for Example 3-5.

The equations for nodes 3 and 6 are thus written

$$2T_2 + T_6 + 567 - 4.67T_3 = 0$$
$$2T_5 + T_3 + T_9 + 67 - 4.67T_6 = 0$$

The equations for nodes 7 and 8 are given by

$$2T_4 + T_8 + 167 - 4.67T_7 = 0$$
$$2T_5 + T_7 + T_9 + 67 - 4.67T_8 = 0$$

and the equation for node 9 is

$$T_6 + T_8 + 67 - 2.67T_9 = 0$$

We thus have nine equations and nine unknown nodal temperatures. We shall discuss solution techniques shortly, but for now we just list the answers:

Node	Temperature, °C
1	280.67
2	330.30
3	309.38
4	192.38
5	231.15
6	217.19
7	157.70
8	184.71
9	175.62

The heat flows at the boundaries are computed in two ways: as conduction flows for the 100 and 500°C faces and as convection flows for the other two faces. For the 500°C face, the heat flow *into* the face is

$$q = \sum k\Delta x \frac{\Delta T}{\Delta y} = (10)\left[500 - 280.67 + 500 - 330.30 + (500 - 309.38)\left(\tfrac{1}{2}\right)\right]$$

$$= 4843.4 \text{ W/m}$$

The heat flow *out* of the 100°C face is

$$q = \sum k\Delta y \frac{\Delta T}{\Delta x} = (10)\left[280.67 - 100 + 192.38 - 100 + (157.70 - 100)\left(\tfrac{1}{2}\right)\right]$$

$$= 3019 \text{ W/m}$$

The convection heat flow *out* the right face is given by the convection relation

$$q = \sum h \Delta y (T - T_\infty)$$

$$= (10)\left(\tfrac{1}{3}\right)\left[309.38 - 100 + 217.19 - 100 + (175.62 - 100)\left(\tfrac{1}{2}\right)\right]$$

$$= 1214.6 \text{ W/m}$$

Finally, the convection heat flow *out* the bottom face is

$$q = \sum h \Delta x (T - T_\infty)$$

$$= (10)\left(\tfrac{1}{3}\right)\left[(100 - 100)\left(\tfrac{1}{2}\right) + 157.70 - 100 + 184.71 - 100 + (175.62 - 100)\left(\tfrac{1}{2}\right)\right]$$

$$= 600.7 \text{ W/m}$$

The total heat flow out is

$$q_{\text{out}} = 3019 + 1214.6 + 600.7 = 4834.3 \text{ W/m}$$

This compares favorably with the 4843.4 W/m conducted into the top face. A solution of this example using the Excel spreadsheet format is given in Appendix D.

Solution Techniques

From the foregoing discussion we have seen that the numerical method is simply a means of approximating a continuous temperature distribution with the finite nodal elements. The more nodes taken, the closer the approximation; but, of course, more equations mean more cumbersome solutions. Fortunately, computers and even programmable calculators have the capability to obtain these solutions very quickly.

In practical problems the selection of a large number of nodes may be unnecessary because of uncertainties in boundary conditions. For example, it is not uncommon to have uncertainties in h, the convection coefficient, of ± 15 to 20 percent.

The nodal equations may be written as

$$
\begin{aligned}
a_{11} T_1 + a_{12} T_2 + &\cdots + a_{1n} T_n = C_1 \\
a_{21} T_1 + a_{22} T_2 + &\cdots \qquad\qquad = C_2 \\
a_{31} T_1 + &\cdots \qquad\qquad\quad = C_3 \\
&\cdots\cdots\cdots\cdots\cdots\cdots\cdots\cdots \\
a_{n1} T_1 + a_{n2} T_2 + &\cdots + a_{nn} T_n = C_n
\end{aligned}
\qquad \textbf{[3-27]}
$$

where $T_1, T_2, \ldots, T_n$ are the unknown nodal temperatures. By using the matrix notation

$$
[A] = \begin{bmatrix}
a_{11} & a_{12} & \cdots & a_{1n} \\
a_{21} & a_{22} & \cdots & \\
a_{31} & & \cdots & \\
\multicolumn{4}{c}{\cdots\cdots\cdots\cdots\cdots} \\
a_{n1} & a_{n2} & \cdots & a_{nn}
\end{bmatrix}
\qquad
[C] = \begin{bmatrix} C_1 \\ C_2 \\ . \\ . \\ . \\ C_n \end{bmatrix}
\qquad
[T] = \begin{bmatrix} T_1 \\ T_2 \\ . \\ . \\ . \\ T_n \end{bmatrix}
$$

Equation (3-27) can be expressed as

$$[A][T] = [C] \qquad \textbf{[3-28]}$$

and the problem is to find the inverse of $[A]$ such that

$$[T] = [A]^{-1}[C] \qquad\qquad \textbf{[3-29]}$$

Designating $[A]^{-1}$ by

$$[A]^{-1} = \begin{bmatrix} b_{11} & b_{12} & \cdots & b_{1n} \\ b_{21} & b_{22} & \cdots & \\ \multicolumn{4}{c}{\dotfill} \\ b_{n1} & b_{n2} & \cdots & b_{nn} \end{bmatrix}$$

the final solutions for the unknown temperatures are written in expanded form as

$$\begin{aligned} T_1 &= b_{11}C_1 + b_{12}C_2 + \cdots + b_{1n}C_n \\ T_2 &= b_{21}C_1 + \quad\cdots \\ &\ \dotfill \\ T_n &= b_{n1}C_1 + b_{n2}C_2 + \cdots + b_{nn}C_n \end{aligned} \qquad\qquad \textbf{[3-30]}$$

Clearly, the larger the number of nodes, the more complex and time-consuming the solution, even with a high-speed computer. For most conduction problems the matrix contains a large number of zero elements so that some simplification in the procedure is afforded. For example, the matrix notation for the system of Example 3-5 would be

$$\begin{bmatrix} -4 & 1 & 0 & 1 & 0 & 0 & 0 & 0 & 0 \\ 1 & -4 & 1 & 0 & 1 & 0 & 0 & 0 & 0 \\ 0 & 2 & -4.67 & 0 & 0 & 1 & 0 & 0 & 0 \\ 1 & 0 & 0 & -4 & 1 & 0 & 1 & 0 & 0 \\ 0 & 1 & 0 & 1 & -4 & 1 & 0 & 1 & 0 \\ 0 & 0 & 1 & 0 & 2 & -4.67 & 0 & 0 & 1 \\ 0 & 0 & 0 & 2 & 0 & 0 & -4.67 & 1 & 0 \\ 0 & 0 & 0 & 0 & 2 & 0 & 1 & -4.67 & 1 \\ 0 & 0 & 0 & 0 & 0 & 1 & 0 & 1 & -2.67 \end{bmatrix} \begin{bmatrix} T_1 \\ T_2 \\ T_3 \\ T_4 \\ T_5 \\ T_6 \\ T_7 \\ T_8 \\ T_9 \end{bmatrix} = \begin{bmatrix} -600 \\ -500 \\ -567 \\ -100 \\ 0 \\ -67 \\ -167 \\ -67 \\ -67 \end{bmatrix}$$

We see that because of the structure of the equations the coefficient matrix is very sparse. For this reason iterative methods of solution may be very efficient. The Gauss-Seidel iteration method is probably the most widely used for solution of these equations in heat transfer problems, and we shall discuss that method in Section 3-7.

Software Packages for Solution of Equations

Several software packages are available for solution of simultaneous equations, including MathCAD (22), TK Solver (23), Matlab (24), and Microsoft Excel (25, 26, 27). The spreadsheet grid of Excel is particularly adaptable to formulation, solution, and graphical displays associated with the nodal equations. Details of the use of Excel as a tool for such problems are presented in Appendix D for both steady-state and transient conditions.

Figure 3-9 | General conduction node.

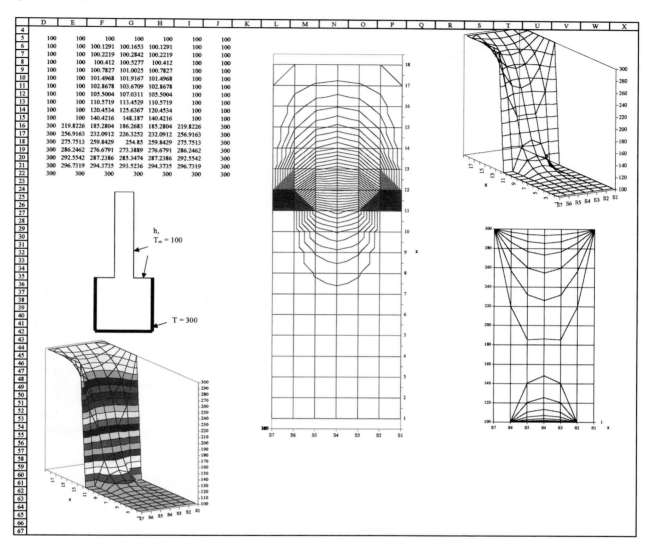

An example of an Excel worksheet is shown in Figure 3-9, which includes a schematic of a protruding fin cooled by a convection environment, numerical solution displayed to match the geometric configuration, and four graphical presentations of the results. While one might regard this presentation as graphical overkill, it does illustrate a variety of options available in the Excel format. Several examples are discussed in detail in Appendix D, including the effects of heat sources and radiation boundary conditions.

Other methods of solution include a transient analysis carried through to steady state (see Chapter 4), direct elimination (Gauss elimination [9]), or more sophisticated iterative techniques [12]. An Excel spreadsheet solution treating Example 3-5 as a transient problem carried through to steady state is given Appendix D.

3-6 | NUMERICAL FORMULATION IN TERMS OF RESISTANCE ELEMENTS

Up to this point we have shown how conduction problems can be solved by finite-difference approximations to the differential equations. An equation is formulated for each node and the set of equations solved for the temperatures throughout the body. In formulating the equations we could just as well have used a resistance concept for writing the heat transfer between nodes. Designating our node of interest with the subscript i and the adjoining nodes with subscript j, we have the general-conduction-node situation shown in Figure 3-10. At steady state the net heat input to node i must be zero or

$$q_i + \sum_j \frac{T_j - T_i}{R_{ij}} = 0 \qquad \text{[3-31]}$$

where q_i is the heat delivered to node i by heat generation, radiation, etc. The R_{ij} can take the form of convection boundaries, internal conduction, etc., and Equation (3-31) can be set equal to some residual for a relaxation solution or to zero for treatment with matrix and iterative methods.

No new information is conveyed by using a resistance formulation, but some workers may find it convenient to think in these terms. When a numerical solution is to be performed that takes into account property variations, the resistance formulation is particularly useful. In addition, there are many heat-transfer problems where it is convenient to think of convection and radiation boundary conditions in terms of the thermal resistance they impose on the system. In such cases the relative magnitudes of convection, radiation, and conduction resistances may have an important influence on the behavior of the thermal model. We shall examine different boundary resistances in the examples. It will be clear that one will want to increase thermal resistances when desiring to impede the heat flow and decrease the thermal resistance when an increase in heat transfer is sought. In some cases the term *thermal impedance* is employed as a synonym for thermal resistance, following this line of thinking.

For convenience of the reader Table 3-3 lists the resistance elements that correspond to the nodes in Table 3-2. Note that all resistance elements are for unit depth of material and $\Delta x = \Delta y$. The nomenclature for the table is that R_{m+} refers to the resistance on the positive x side of node (m, n), R_{n-} refers to the resistance on the negative y side of node (m, n), and so on.

Figure 3-10 | General conduction node.

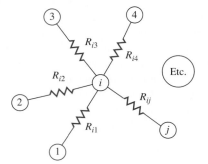

Table 3-3 | Resistances for nodes of Table 3-2 $\Delta x = \Delta y$, $\Delta z = 1$.

Physical situation	R_{m+}	R_{m-}	R_{n+}	R_{n-}	ΔV
(a) Interior node	$\frac{1}{k}$	$\frac{1}{k}$	$\frac{1}{k}$	$\frac{1}{k}$	$(\Delta x)^2$
(b) Convection boundary	$\frac{1}{h\,\Delta x}$	$\frac{1}{k}$	$\frac{2}{k}$	$\frac{2}{k}$	$\frac{(\Delta x)^2}{2}$
(c) Exterior corner, convection	$\frac{2}{h\,\Delta x}$	$\frac{2}{k}$	$\frac{2}{h\,\Delta x}$	$\frac{2}{k}$	$\frac{(\Delta x)^2}{4}$
(d) Interior corner, convection[†]	$\frac{2}{k}$	$\frac{1}{k}$	$\frac{1}{k}$	$\frac{2}{k}$	$\frac{3(\Delta x)^2}{4}$
(e) Insulated boundary	∞	$\frac{1}{k}$	$\frac{2}{k}$	$\frac{2}{k}$	$\frac{(\Delta x)^2}{2}$
(f) Interior node near curved boundary	$\frac{2}{(b+1)k}$ to node $(m+1,n)$	$\frac{2a}{(b+1)k}$ to node 1	$\frac{2b}{(a+1)k}$ to node 2	$\frac{2}{(a+1)k}$ to node $(m,n-1)$	$0.25(1+a)(1+b)(\Delta x)^2$

(g) Boundary node with curved boundary node 2 for (f) above

$$R_{23} = \frac{2\sqrt{c^2+1}}{bk}$$

$$R_{21} = \frac{2\sqrt{a^2+b^2}}{bk}$$

$$\Delta V = 0.125[(2+a)+c](\Delta x)^2$$

$$R_{2-\infty} = \frac{2}{h\,\Delta x\left(\sqrt{c^2+1}+\sqrt{a^2+b^2}\right)}$$

$$R_{n-} = \frac{2b}{k(a+1)} \text{ to node } (m,n)$$

[†] Also $R_\infty = 1/h\,\Delta x$ for convection to T_∞.

The resistance formulation is also useful for numerical solution of complicated three-dimensional shapes. The volume elements for the three common coordinate systems are shown in Figure 3-11, and internal nodal resistances for each system are given in Table 3-4. The nomenclature for the (m, n, k) subscripts is given in Table 3-3, and the plus or minus sign on the resistance subscripts designates the resistance in a positive or negative direction from the central node (m, n, k). The elemental volume ΔV is also indicated for each coordinate system. We note, of course, that in a practical problem the coordinate increments are frequently chosen so that $\Delta x = \Delta y = \Delta z$, etc., and the resistances are simplified.

3-7 | GAUSS-SEIDEL ITERATION

When the number of nodes is very large, an iterative technique may frequently yield a more efficient solution to the nodal equations than a direct matrix inversion. One such method is called the *Gauss-Seidel iteration* and is applied in the following way. From Equation (3-31) we may solve for the temperature T_i in terms of the resistances and temperatures of the adjoining nodes T_j as

$$T_i = \frac{q_i + \sum\limits_{j}(T_j/R_{ij})}{\sum\limits_{j}(1/R_{ij})} \qquad \textbf{[3-32]}$$

The Gauss-Seidel iteration makes use of the difference equations expressed in the form of Equation (3-32) through the following procedure.

1. An initial set of values for the T_i is assumed. This initial assumption can be obtained through any expedient method. For a large number of nodes to be solved on a computer the T_i's are usually assigned a zero value to start the calculation.

Figure 3-11 | Volume of resistance elements: (*a*) cartesian, (*b*) cylindrical, and (*c*) spherical coordinate systems.

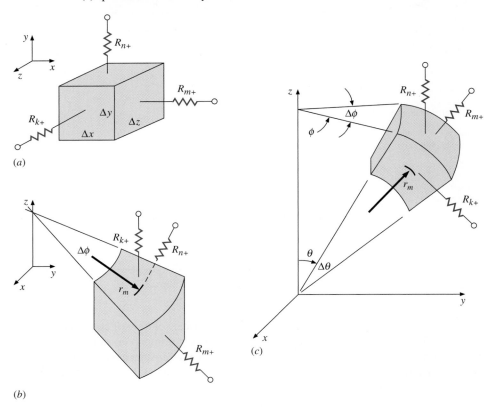

Table 3-4 | Internal nodal resistances for different coordinate systems.

	Cartesian	**Cylindrical**	**Spherical**
Nomenclature for increments	x, m	r, m	r, m
	y, n	ϕ, n	ϕ, n
	z, k	z, k	θ, k
Volume element ΔV	$\Delta x\, \Delta y\, \Delta z$	$r_m\, \Delta r\, \Delta\phi\, \Delta z$	$r_m^2\, \sin\theta\, \Delta r\, \Delta\phi\, \Delta\theta$
R_{m+}	$\dfrac{\Delta x}{\Delta y\, \Delta z\, k}$	$\dfrac{\Delta r}{(r_m+\Delta r/2)\, \Delta\phi\, \Delta z\, k}$	$\dfrac{\Delta r}{(r_m+\Delta r/2)^2\, \sin\theta\, \Delta\phi\, \Delta\theta\, k}$
R_{m-}	$\dfrac{\Delta x}{\Delta y\, \Delta z\, k}$	$\dfrac{\Delta r}{(r_m-\Delta r/2)\, \Delta\phi\, \Delta z\, k}$	$\dfrac{\Delta r}{(r_m-\Delta r/2)^2\, \sin\theta\, \Delta\phi\, \Delta\theta\, k}$
R_{n+}	$\dfrac{\Delta y}{\Delta x\, \Delta z\, k}$	$\dfrac{r_m\, \Delta\phi}{\Delta r\, \Delta z\, k}$	$\dfrac{\Delta\phi\, \sin\theta}{\Delta r\, \Delta\theta\, k}$
R_{n-}	$\dfrac{\Delta y}{\Delta x\, \Delta z\, k}$	$\dfrac{r_m\, \Delta\phi}{\Delta r\, \Delta z\, k}$	$\dfrac{\Delta\phi\, \sin\theta}{\Delta r\, \Delta\theta\, k}$
R_{k+}	$\dfrac{\Delta z}{\Delta x\, \Delta y\, k}$	$\dfrac{\Delta z}{r_m\, \Delta\phi\, \Delta r\, k}$	$\dfrac{\Delta\theta}{\sin(\theta+\Delta\theta/2)\, \Delta r\, \Delta\phi\, k}$
R_{k-}	$\dfrac{\Delta z}{\Delta x\, \Delta y\, k}$	$\dfrac{\Delta z}{r_m\, \Delta\phi\, \Delta r\, k}$	$\dfrac{\Delta\theta}{\sin(\theta-\Delta\theta/2)\, \Delta r\, \Delta\phi\, k}$

2. Next, the new values of the nodal temperatures T_i are calculated according to Equation (3-32), *always using the most recent values of the T_j.*
3. The process is repeated until successive calculations differ by a sufficiently small amount. In terms of a computer program, this means that a test may be inserted to stop the calculations when

$$|T_{i_{n+1}} - T_{i_n}| \leq \delta \qquad \text{for all } T_i$$

where δ is some selected constant and n is the number of iterations. Alternatively, a nondimensional test may be selected such that

$$\epsilon \geq \left| \frac{T_{i_{n+1}} - T_{i_n}}{T_{i_n}} \right|$$

Obviously, the smaller the value of δ, the greater the calculation time required to obtain the desired result. The reader should note, however, that the *accuracy* of the solution to the physical problem is not dependent on the value of δ alone. This constant governs the accuracy of the solution to the set of difference equations. The solution to the physical problem also depends on the selection of the increment Δx.

As we noted in the discussion of solution techniques, the matrices encountered in the numerical formulations are very sparse; they contain a large number of zeros. In solving a problem with a large number of nodes it may be quite time-consuming to enter all these zeros, and the simple form of the Gauss-Seidel equation may be preferable.

Nodal Equations for $\Delta x = \Delta y$

For nodes with $\Delta x = \Delta y$ and no heat generation, the form of Equation (3-32) has been listed as the second equation in segments of Table 3-2. The nondimensional group

$$\frac{h \Delta x}{k} = \text{Bi}$$

is called the *Biot number*. Note that equations for convection boundaries may be converted to insulated boundaries by simply setting $\text{Bi} = 0$ in the respective formula.

Heat Sources and Boundary Radiation Exchange

To include heat generation or radiation heat transfer in the nodal equations for $\Delta x = \Delta y$, one need only add a term

$$q_i / k$$

to the numerator of each of the equations. For heat sources

$$q_i = \dot{q} \Delta V$$

where $\dot{q}$ is the heat generated per unit volume and ΔV is the volume of the respective node. Note that the volume elements are indicated in Table 3-2 by dashed lines. For an interior node $\Delta V = \Delta x \Delta y$, for a plane convection boundary $\Delta V = (\Delta x / 2) \Delta y$, for an exterior corner $\Delta V = (\Delta x / 2)(\Delta y / 2)$, etc.

For radiation exchange at boundary note,

$$q_i = q''_{\text{rad},i} \times \Delta A$$

where ΔA is the surface area of the node exposed to radiation, and $q''_{\text{rad},i}$ is the *net radiation transferred to node i per unit area* as determined by the methods of Chapter 8.

For the common case of a surface exposed to a large enclosure at radiation temperature of T_r, the net radiation *to the surface* per unit area is given by Equation (1-12),

$$q''_{\text{rad},i} = \sigma \varepsilon_i \left(T_r^4 - T_i^4 \right)$$

where ε_i is the emissivity of note i and all temperatures must by expressed in degrees absolute.

3-8 | ACCURACY CONSIDERATIONS

We have already noted that the finite-difference approximation to a physical problem improves as smaller and smaller and smaller increments of Δx and Δy are used. But we have not said how to estimate the accuracy of this approximation. Two basic approaches are available.

1. Compare the numerical solution with an analytical solution for the problem, if available, or an analytical solution for a similar problem.
2. Choose progressively smaller values of Δx and observe the behavior of the solution. If the problem has been correctly formulated and solved, the nodal temperatures should converge as Δx becomes smaller. It should be noted that computational round-off errors increase with an increase in the number of nodes because of the increased number of machine calculations. This is why one needs to observe the convergence of the solution.

It can be shown that the error of the finite-difference approximation to $\partial T/\partial x$ is of the order of $(\Delta x/L)^2$ where L is some characteristic body dimension.

Analytical solutions are of limited utility in checking the accuracy of a numerical model because most problems that will need to be solved by numerical methods either do not have an analytical solution at all or, if one is available, it may be too cumbersome to compute.

Energy Balance as Check on Solution Accuracy

In discussing solution techniques for nodal equations, we stated that an accurate solution of these equations does not ensure an accurate solution to the physical problem. In many cases the final solution is in serious error simply because the problem was not formulated correctly at the start. No computer or convergence criterion can correct this kind of error. One way to check for formulation errors is to perform some sort of energy balance using the final solution. The nature of the balance varies from problem to problem but for steady state it always takes the form of energy in equals energy out. If the energy balance does not check within reasonable limits, there is a likelihood that the problem has not been formulated correctly. Perhaps a constant is wrong here or there, or an input data point is incorrect, a faulty computer statement employed, or one or more nodal equations incorrectly written. If the energy balance does check, one may then address the issue of using smaller values of Δx to improve accuracy.

In the examples we present energy balances as a check on problem formulation.

Accuracy of Properties and Boundary Conditions

From time to time we have mentioned that thermal conductivities of materials vary with temperature; however, over a temperature range of 100 to 200°C the variation is not great (on the order of 5 to 10 percent) and we are justified in assuming constant values to simplify problem solutions. Convection and radiation boundary conditions are particularly notorious for their nonconstant behavior. Even worse is the fact that for many practical problems the basic uncertainty in our knowledge of convection heat-transfer coefficients may not be better than ±25 percent. Uncertainties of surface-radiation properties of ±10 percent are not unusual at all. For example, a highly polished aluminum plate, if allowed to oxidize heavily, will absorb as much as *300 percent* more radiation than when it was polished.

These remarks are not made to alarm the reader, but rather to show that selection of a large number of nodes for a numerical formulation does not necessarily produce an accurate solution to the physical problem; we must also examine uncertainties in the boundary conditions. At this point the reader is ill-equipped to estimate these uncertainties. Later chapters on convection and radiation will clarify the matter.

Gauss-Seidel Calculation

| | EXAMPLE 3-6 |

Apply the Gauss-Seidel technique to obtain the nodal temperatures for the four nodes in Figure 3-6.

■ **Solution**

It is useful to think in terms of a resistance formulation for this problem because all the connecting resistances between the nodes in Figure 3-6 are equal; that is,

$$R = \frac{\Delta y}{k\Delta y} = \frac{\Delta x}{k\Delta y} = \frac{1}{k} \qquad [a]$$

Therefore, when we apply Equation (3-32) to each node, we obtain ($q_i = 0$)

$$T_i = \frac{\sum_j k_j T_j}{\sum_j k_j} \qquad [b]$$

Because each node has four resistances connected to it and k is assumed constant,

$$\sum_j k_j = 4k$$

and

$$T_i = \frac{1}{4} \sum_j T_j \qquad [c]$$

We now set up an iteration table as shown and use initial temperature assumptions of 300 and 200°C. Equation (c) is then applied repeatedly until satisfactory convergence is achieved. In the table, five iterations produce convergence with 0.13 degree. To illustrate the calculation, we can note the two specific cases below:

$$(T_2)_{n=1} = \tfrac{1}{4}(500 + 100 + T_4 + T_1) = \tfrac{1}{4}(500 + 100 + 200 + 275) = 268.75$$
$$(T_3)_{n=4} = \tfrac{1}{4}(100 + T_1 + T_4 + 100) = \tfrac{1}{4}(100 + 250.52 + 150.52 + 100) = 150.26$$

Number of iterations n	T_1	T_2	T_3	T_4
0	300	300	200	200
1	275	268.75	168.75	159.38
2	259.38	254.69	154.69	152.35
3	251.76	251.03	151.03	150.52
4	250.52	250.26	150.26	150.13
5	250.13	250.07	150.07	150.03

Note that in computing $(T_3)_{n=4}$ we have used the most recent information available to us for T_1 and T_4.

	Numerical Formulation with Heat Generation
EXAMPLE 3-7	

We illustrate the resistance formulation in cylindrical coordinates by considering a 4.0-mm-diameter wire with uniform heat generation of 500 MW/m^3. The outside surface temperature of the wire is 200°C, and the thermal conductivity is 19 W/m · °C. We wish to calculate the temperature distribution in the wire. For this purpose we select four nodes as shown in Figure Example 3-7a. We shall make the calculations per unit length, so we let $\Delta z = 1.0$. Because the system is one-dimensional, we take $\Delta \phi = 2\pi$. For all the elements Δr is chosen as 0.5 mm. We then compute the resistances and volume elements using the relations from Table 3-4, and the values are given below. The computation of R_{m+} for node 4 is different from the others because the heat-flow path is shorter. For node 4, r_m is 1.75 mm, so the positive resistance extending to the known surface temperature is

$$R_{m+} = \frac{\Delta r/2}{(r_m + \Delta r/4)\,\Delta\phi\,\Delta z\,k} = \frac{1}{15\pi k}$$

The temperature equation for node 4 is written as

$$T_4 = \frac{2749 + 6\pi k T_3 + 15\pi k(200)}{21\pi k}$$

where the 200 is the known outer surface temperature.

Node	r_m, mm	R_{m+}, °C/W	R_{m-}, °C/W	$\Delta V = r_m\,\Delta r\,\Delta\phi\,\Delta z$, μm^3	$q_i = \dot{q}\,\Delta V$, W
1	0.25	$\frac{1}{2\pi k}$	∞	0.785	392.5
2	0.75	$\frac{1}{4\pi k}$	$\frac{1}{2\pi k}$	2.356	1178
3	1.25	$\frac{1}{6\pi k}$	$\frac{1}{4\pi k}$	3.927	1964
4	1.75	$\frac{1}{15\pi k}$	$\frac{1}{6\pi k}$	5.498	2749

Figure Example 3-7a |
Example Schematic.

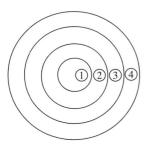

Figure Example 3-7b | Comparison of analytical and numerical solution.

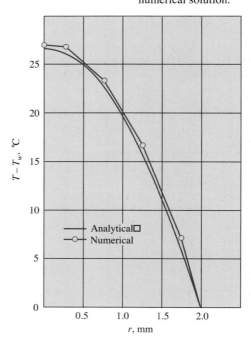

A summary of the values of $\sum(1/R_{ij})$ and T_i according to Equation (3-32) is now given to be used in a Gauss-Seidel iteration scheme.

Node	$\sum \dfrac{1}{R_{ij}}$, W/°C	$T_i = \dfrac{q_i + \sum(T_j/R_{ij})}{\sum(1/R_{ij})}$
1	$2\pi k = 119.38$	$T_1 = 3.288 + T_2$
2	$6\pi k = 358.14$	$T_2 = 3.289 + \frac{1}{3}T_1 + \frac{2}{3}T_3$
3	$10\pi k = 596.90$	$T_3 = 3.290 + 0.4T_2 + 0.6T_4$
4	$21\pi k = 1253.50$	$T_4 = 2.193 + \frac{2}{7}T_3 + 142.857$

Thirteen iterations are now tabulated:

	Node temperature, °C			
Iteration n	T_1	T_2	T_3	T_4
0	240	230	220	210
1	233.29	227.72	220.38	208.02
2	231.01	227.21	218.99	207.62
3	230.50	226.12	218.31	207.42
4	229.41	225.30	217.86	207.30
5	228.59	224.73	217.56	207.21
6	228.02	224.34	217.35	207.15
7	227.63	224.07	217.21	207.11
8	227.36	223.88	217.11	207.08
9	227.17	223.75	217.04	207.06
10	227.04	223.66	216.99	207.04
11	226.95	223.60	216.95	207.04
12	226.89	223.55	216.93	207.03
13	226.84	223.52	216.92	207.03
Analytical	225.904	222.615	216.036	206.168
Gauss-Seidel check	225.903	222.614	216.037	206.775
Exact solution of nodal equations	226.75	223.462	216.884	207.017

We may compare the iterative solution with an exact calculation which makes use of Equation (2-25a):

$$T - T_w = \frac{\dot{q}}{4k}(R^2 - r^2)$$

where T_w is the 200°C surface temperature, $R = 2.0$ mm, and r is the value of r_m for each node. The analytical values are shown following iteration 13, and then a Gauss-Seidel check is made on the analytical values. There is excellent agreement on the first three nodes and somewhat less on node 4. Finally, the exact solutions to the nodal equations are shown for comparison. These are the values the iterative scheme would converge to if carried far enough. In this limit the analytical and numerical calculations differ by a constant factor of about 0.85°C, and this difference results mainly from the way in which the surface resistance and boundary condition are handled. A smaller value of Δr near the surface would produce better agreement. A graphical comparison of the analytical and numerical solutions is shown in Figure Example 3-7b.

The total heat loss from the wire may be calculated as the conduction through R_{m+} at node 4. Then

$$q = \frac{T_4 - T_w}{R_{m+}} = 15\pi k(207.03 - 200) = 6.294 \text{ kW/m} \quad [6548 \text{ Btu/h} \cdot \text{ft}]$$

This must equal the total heat generated in the wire, or

$$q = \dot{q}V = (500 \times 10^6)\pi(2 \times 10^{-3})^2 = 6.283 \text{ kW/m} \quad [6536 \text{ Btu/h} \cdot \text{ft}]$$

The difference between the two values results from the inaccuracy in determination of T_4. Using the exact solution value of 207.017°C would give a heat loss of 6.2827 kW. For this problem the exact value of heat flow is 6.283 kW because the heat-generation calculation is independent of the finite-difference formulation.

EXAMPLE 3-8	Heat Generation with Nonuniform Nodal Elements

A layer of glass [$k = 0.8$ W/m · °C] 3 mm thick has thin 1-mm electric conducting strips attached to the upper surface, as shown in Figure Example 3-8. The bottom surface of the glass is insulated, and the top surface is exposed to a convection environment at 30°C with $h = 100$ W/m² · °C. The strips generate heat at the rate of 40 or 20 W per meter of length. Determine the steady-state temperature distribution in a typical glass section, using the numerical method for both heat-generation rates.

Figure Example 3-8 | (*a*) Physical system, (*b*) nodal boundaries.

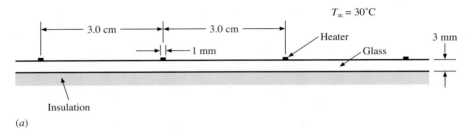

(*a*)

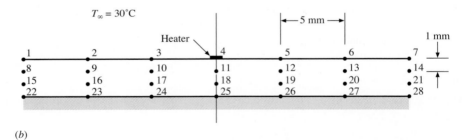

(*b*)

■ Solution
The nodal network for a typical section of the glass is shown in the figure. In this example we have *not* chosen $\Delta x = \Delta y$. Because of symmetry, $T_1 = T_7$, $T_2 = T_6$, etc., and we only need to solve for the temperatures of 16 nodes. We employ the resistance formulation. As shown, we have chosen $\Delta x = 5$ mm and $\Delta y = 1$ mm. The various resistances may now be calculated:
Nodes 1, 2, 3, 4:

$$\frac{1}{R_{m+}} = \frac{1}{R_{m-}} = \frac{k(\Delta y/2)}{\Delta x} = \frac{(0.8)(0.001/2)}{0.005} = 0.08$$

$$\frac{1}{R_{n+}} = hA = (100)(0.005) = 0.5$$

$$\frac{1}{R_{n-}} = \frac{k\Delta x}{\Delta y} = \frac{(0.8)(0.005)}{0.001} = 4.0$$

Nodes 8, 9, 10, 11, 15, 16, 17, 18:

$$\frac{1}{R_{m+}} = \frac{1}{R_{m-}} = \frac{k\Delta y}{\Delta x} = \frac{(0.8)(0.001)}{0.005} = 0.16$$

$$\frac{1}{R_{n+}} = \frac{1}{R_{n-}} = \frac{k\Delta x}{\Delta y} = 4.0$$

Nodes 22, 23, 24, 25:

$$\frac{1}{R_{m+}} = \frac{1}{R_{m-}} = \frac{k(\Delta y/2)}{\Delta x} = 0.08$$

$$\frac{1}{R_{n+}} = \frac{k\Delta x}{\Delta y} = 4.0$$

$$\frac{1}{R_{n-}} = 0 \qquad \text{(insulated surface)}$$

The nodal equations are obtained from Equation (3-31) in the general form

$$\sum (T_j/R_{ij}) + q_i - T_i \sum (1/R_{ij}) = 0$$

Only node 4 has a heat-generation term, and $q_i = 0$ for all other nodes. From the above resistances we may calculate the $\sum (1/R_{ij})$ as

Node	$\sum (1/R_{ij})$
1, 2, 3, 4	4.66
8, ..., 18	8.32
22, 23, 24, 25	4.16

For node 4 the equation is

$$(2)(0.08)T_3 + 4.0T_5 + (0.5)(30) + q_4 - 4.66T_4 = 0$$

The factor of 2 on T_3 occurs because $T_3 = T_5$ from symmetry. When all equations are evaluated and the solution obtained, the following temperatures result:

Node	q/L, W/m	
temperature, °C	20	40
1	31.90309	33.80617
2	32.78716	35.57433
3	36.35496	42.70993
4	49.81266	69.62532
8	32.10561	34.21122
9	33.08189	36.16377
10	36.95154	43.90307
11	47.82755	65.65510
15	32.23003	34.46006
16	33.26087	36.52174
17	37.26785	44.53571
18	46.71252	63.42504
22	32.27198	34.54397
23	33.32081	36.64162
24	37.36667	44.73333
25	46.35306	62.70613

The results of the model and calculations may be checked by calculating the convection heat lost by the top surface. Because all the energy generated in the small heater strip must eventually be lost by convection (the bottom surface of the glass is insulated and thus loses no heat), we know the numerical value that the convection should have. The convection loss at the top surface is given by

$$q_c = \sum h_i A_i (T_i - T_\infty)$$

$$= (2)(100) \left[\frac{\Delta x}{2} (T_1 - T_\infty) + \Delta x (T_2 + T_3 - 2T_\infty) + \frac{\Delta x}{2} (T_4 - T_\infty) \right]$$

The factor of 2 accounts for both sides of the section. With $T_\infty = 30°C$ this calculation yields

$$q_c = 19.999995 \qquad \text{for } q/L = 20 \text{ W/m}$$
$$q_c = 40.000005 \qquad \text{for } q/L = 40 \text{ W/m}$$

Obviously, the agreement is excellent.

Composite Material with Nonuniform Nodal Elements

EXAMPLE 3-9

A composite material is embedded in a high-thermal-conductivity material maintained at 400°C as shown in Figure Example 3-9a. The upper surface is exposed to a convection environment at 30°C with $h = 25$ W/m$^2 \cdot$°C. Determine the temperature distribution and heat loss from the upper surface for steady state.

■ **Solution**
For this example we choose nonsquare nodes as shown in Figure Example 3-9b. Note also that nodes 1, 4, 7, 10, 13, 14, and 15 consist of *two* materials. We again employ the resistance formulation.

For node 1:

$$\frac{1}{R_{m+}} = \frac{kA}{\Delta x} = \frac{(2.0)(0.005)}{0.015} = 0.6667$$

$$\frac{1}{R_{m-}} = \frac{kA}{\Delta x} = \frac{(0.3)(0.005)}{0.01} = 0.15$$

$$\frac{1}{R_{n+}} = hA = (25)(0.005 + 0.0075) = 0.3125$$

$$\frac{1}{R_{n-}} = \left(\frac{kA}{\Delta y} \right)_L + \left(\frac{kA}{\Delta y} \right)_R = \frac{(0.3)(0.005) + (2.0)(0.0075)}{0.01} = 1.65$$

For nodes 4, 7, 10:

$$\frac{1}{R_{m+}} = \frac{(2.0)(0.01)}{0.015} = 1.3333$$

$$\frac{1}{R_{m-}} = \frac{(0.3)(0.01)}{0.01} = 0.3$$

$$\frac{1}{R_{n+}} = \frac{1}{R_{n-}} = 1.65$$

Figure Example 3-9 | (*a*) Physical system, (*b*) nodal boundaries.

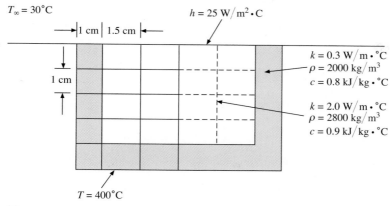

(*a*)

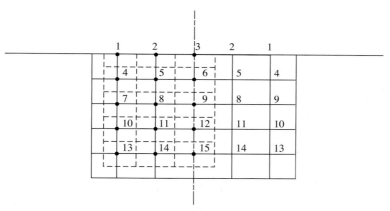

(*b*)

For node 13:

$$\frac{1}{R_{m+}} = \frac{(2.0)(0.005) + (0.3)(0.005)}{0.015} = 0.76667$$

$$\frac{1}{R_{m-}} = \frac{(0.3)(0.01)}{0.01} = 0.3$$

$$\frac{1}{R_{n+}} = 1.65$$

$$\frac{1}{R_{n-}} = \frac{(0.3)(0.0075) + (0.3)(0.005)}{0.01} = 0.375$$

For nodes, 5, 6, 8, 9, 11, 12:

$$\frac{1}{R_{m+}} = \frac{1}{R_{m-}} = \frac{(2.0)(0.01)}{0.015} = 1.3333$$

$$\frac{1}{R_{n+}} = \frac{1}{R_{n-}} = \frac{(2.0)(0.015)}{0.01} = 3.0$$

For nodes 2, 3:

$$\frac{1}{R_{m+}} = \frac{1}{R_{m-}} = \frac{(2.0)(0.005)}{0.015} = 0.6667$$

$$\frac{1}{R_{n+}} = hA = (2.5)(0.015) = 0.375$$

$$\frac{1}{R_{n-}} = 3.0$$

For nodes 14, 15:

$$\frac{1}{R_{m+}} = \frac{1}{R_{m-}} = \frac{(2.0)(0.005) + (0.3)(0.005)}{0.015} = 0.76667$$

$$\frac{1}{R_{n+}} = 3.0$$

$$\frac{1}{R_{n-}} = \frac{(0.3)(0.015)}{0.01} = 0.45$$

We shall use Equation (3-32) for formulating the nodal equations. For node 1, $\sum(1/R_{ij}) = 2.7792$, and we obtain

$$T_1 = \frac{1}{2.7792}[(400)(0.15) + (30)(0.3125) + T_2(0.6667) + 1.65T_4]$$

For node 3, $\sum(1/R_{ij}) = 4.7083$, and the nodal equation is

$$T_3 = \frac{1}{4.7083}[T_2(0.6667)(2) + 3.0T_6 + (0.375)(30)]$$

The factor of 2 on T_2 occurs because of the mirror image of T_2 to the right of T_3.

A similar procedure is followed for the other nodes to obtain 15 nodal equations with the 15 unknown temperatures. These equations may then be solved by whatever computation method is most convenient. The resulting temperatures are:

$$
\begin{array}{lll}
T_1 = 254.956 & T_2 = 247.637 & T_3 = 244.454 \\
T_4 = 287.334 & T_5 = 273.921 & T_6 = 269.844 \\
T_7 = 310.067 & T_8 = 296.057 & T_9 = 291.610 \\
T_{10} = 327.770 & T_{11} = 313.941 & T_{12} = 309.423 \\
T_{13} = 343.516 & T_{14} = 327.688 & T_{15} = 323.220
\end{array}
$$

The heat flow out the top face is obtained by summing the convection loss from the nodes:

$$
\begin{aligned}
q_{\text{conv}} &= \sum hA_i(T_i - T_\infty) \\
&= (2)(25)[(0.0125)(254.96 - 30) + (0.015)(247.64 - 30) \\
&\quad + (0.0075)(244.45 - 30)] \\
&= 382.24 \text{ W per meter of depth}
\end{aligned}
$$

As a check on this value, we can calculate the heat conducted in from the 400°C surface to nodes 1, 4, 7, 10, 13, 14, and 15:

$$q_{\text{cond}} = \sum kA_i \frac{\Delta T}{\Delta x}$$

$$
\begin{aligned}
q_{\text{cond}} &= 2\frac{0.3}{0.01}[(0.005)(400 - 254.96) + (0.01)(400 - 287.33) + (0.01)(400 - 310.07) \\
&\quad + (0.01)(400 - 327.77) + (0.0225)(400 - 343.52) + (0.015)(400 - 327.69) \\
&\quad + (0.0075)(400 - 323.22)] \\
&= 384.29 \text{ W per meter of depth}
\end{aligned}
$$

The agreement is excellent.

Radiation Boundary Condition

EXAMPLE 3-10

A 1-by-2-cm ceramic strip [$k = 3.0$ W/m · °C, $\rho = 1600$ kg/m^3, and $c = 0.8$ kJ/kg · °C] is embedded in a high-thermal-conductivity material, as shown in Figure Example 3-10, so that the sides are maintained at a constant temperature of 900°C. The bottom surface of the ceramic is insulated, and the top surface is exposed to a convection and radiation environment at $T_\infty = 50$°C; $h = 50$ W/m^2 · °C, and the radiation heat loss is calculated from

$$q = \sigma A \epsilon (T^4 - T_\infty^4)$$

where

A = surface area

$\sigma = 5.669 \times 10^{-8}$ W/m^2 · °K^4

$\epsilon = 0.7$

Solve for the steady-state temperature distribution of the nodes shown and the rate of heat loss. The radiation temperatures are in degrees Kelvin.

Figure Example 3-10

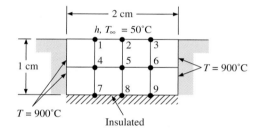

■ Solution

We shall employ the resistance formulation and note that the radiation can be written as

$$q = \sigma \epsilon A (T^4 - T_\infty^4) = \frac{T - T_\infty}{R_{\text{rad}}} \qquad [a]$$

$$\frac{1}{R_{\text{rad}}} = \sigma \epsilon A (T^2 + T_\infty^2)(T + T_\infty) \qquad [b]$$

From symmetry $T_1 = T_3$, $T_4 = T_6$, and $T_7 = T_9$, so we have only six unknown nodes. The resistances are now computed:

Nodes 1, 2:

$$\frac{1}{R_{m+}} = \frac{1}{R_{m-}} = \frac{kA}{\Delta x} = \frac{(3.0)(0.0025)}{0.005} = 1.5 \qquad \frac{1}{R_{n-}} = \frac{(3.0)(0.005)}{0.005} = 3.0$$

$$\frac{1}{R_{n+,\text{conv}}} = hA = (50)(0.005) = 0.25 \qquad [c]$$

$$\frac{1}{R_{n+,\text{rad}}} = \sigma \epsilon A (T^2 + T_\infty^2)(T + T_\infty)$$

The radiation term introduces nonlinearities and will force us to employ an iterative solution.

Nodes 4, 5:

$$\text{All } \frac{1}{R} = \frac{kA}{\Delta x} = \frac{(3.0)(0.005)}{0.005} = 3.0$$

Nodes 7, 8:

$$\frac{1}{R_{m+}} = \frac{1}{R_{m-}} = 1.5 \qquad \frac{1}{R_{n+}} = 3.0$$

Because the bottom surface is insulated, $1/R_{n-} = 0$. We now use Equation (3-32)

$$T_i = \frac{\sum(T_j/R_{ij})}{\sum(1/R_{ij})} \qquad\qquad \textbf{[3-33]}$$

and tabulate:

Node	$\sum(1/R_{ij})$
1	$6.25 + 1/R_{rad}$
2	$6.25 + 1/R_{rad}$
4	12
5	12
7	6
8	6

Our nodal equations are thus expressed in degrees Kelvin because of the radiation terms and become

$$T_1 = \frac{1}{\sum(1/R_{ij})}[1.5T_2 + 3T_4 + (1.5)(1173) + (323)(0.25)$$
$$+ \sigma\epsilon(0.005)(T_1^2 + 323^2)(T_1 + 323)(323)]$$

$$T_2 = \frac{1}{\sum(1/R_{ij})}[1.5T_1(2) + 3T_5 + (323)(0.25)$$
$$+ \sigma\epsilon(0.005)(T_2^2 + 323^2)(T_2 + 323)(323)]$$

$$T_4 = \tfrac{1}{12}[(1173)(3.0) + 3T_1 + 3T_7 + 3T_5] \qquad T_5 = \tfrac{1}{12}[2T_4(3.0) + 3T_2 + 3T_8]$$
$$T_7 = \tfrac{1}{6}[(1173)(1.5) + 3T_4 + 1.5T_8] \qquad T_8 = \tfrac{1}{6}[2T_7(1.5) + 3T_5]$$

The radiation terms create a very nonlinear set of equations. The computational algorithm we shall use is outlined as follows:

1. Assume $T_1 = T_2 = 1173$ K.
2. Compute $1/R_{rad}$ and $\sum(1/R_{ij})$ for nodes 1 and 2 on the basis of this assumption.
3. Solve the set of equations for T_1 through T_8.
4. Using new values of T_1 and T_2, recalculate $1/R_{rad}$ values.
5. Solve equations again, using new values.
6. Repeat the procedure until answers are sufficiently convergent.

The results of six iterations are shown in the table. As can be seen, the convergence is quite rapid. The temperatures are in kelvins.

Iteration	T_1	T_2	T_4	T_5	T_7	T_8
1	990.840	944.929	1076.181	1041.934	1098.951	1070.442
2	1026.263	991.446	1095.279	1068.233	1113.622	1090.927
3	1019.879	982.979	1091.827	1063.462	1110.967	1087.215
4	1021.056	984.548	1092.464	1064.344	1111.457	1087.901
5	1020.840	984.260	1092.347	1064.182	1111.367	1087.775
6	1020.879	984.313	1092.369	1064.212	1111.384	1087.798

As a practical matter, the iterations would be carried out using a commercial software package (such as those mentioned in References 22–27) and only the final set of values would be displayed on the computer.

At this point we may note that in a practical problem the value of ϵ will only be known within a tolerance of several percent, and thus there is nothing to be gained by carrying the solution to unreasonable limits of accuracy.

The heat loss is determined by calculating the radiation and convection from the top surface (nodes 1, 2, 3):

$$q_{rad} = \sum \sigma \epsilon A_i (T_i^4 - 323^4)$$
$$= (5.669 \times 10^{-8})(0.7)(0.005)[(2)(1020.88^4 - 323^4) + 984.313^4 - 323^4]$$
$$= 610.8 \text{ W/m depth}$$
$$q_{conv} = \sum h A_i (T_i - 323)$$
$$= (50)(0.005)[(2)(1020.88 - 323) + 984.313 - 323] = 514.27 \text{ W}$$
$$q_{total} = 610.8 + 514.27 = 1125.07 \text{ W/m depth}$$

This can be checked by calculating the conduction input from the 900°C surfaces:

$$q_{cond} = \sum k A_i \frac{\Delta T}{\Delta x}$$
$$= \frac{(2)(3.0)}{0.005}[(0.0025)(1173 - 1020.879) + (0.005)(1173 - 1092.369)$$
$$+ (0.0025)(1173 - 1111.384)]$$
$$= 1124.99 \text{ W/m depth}$$

The agreement is excellent.

Use of Variable Mesh Size   EXAMPLE 3-11

One may use a variable mesh size in a problem with a finer mesh to help in regions of large temperature gradients. This is illustrated in Figure Example 3-11, in which Figure 3-6 is redrawn with a fine mesh in the corner. The boundary temperatures are the same as in Figure 3-6. We wish to calculate the nodal temperatures and compare with the previous solution. Note the symmetry of the problem: $T_1 = T_2$, $T_3 = T_4$, etc.

Figure Example 3-11

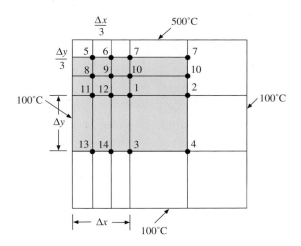

■ **Solution**

Nodes 5, 6, 8, and 9 are internal nodes with $\Delta x = \Delta y$ and have nodal equations in the form of Equation (3-24). Thus,

$$600 + T_6 + T_8 - 4T_5 = 0$$
$$500 + T_5 + T_7 + T_9 - 4T_6 = 0$$
$$100 + T_5 + T_9 + T_{11} - 4T_8 = 0$$
$$T_8 + T_6 + T_{10} + T_{12} - 4T_9 = 0$$

For node 7 we can use a resistance formulation and obtain

$$1/R_{7-6} = k$$
$$1/R_{7-500°} = \frac{k(\Delta x/6 + \Delta x/2)}{\Delta y/3} = 2k$$
$$1/R_{7-10} = 2k$$

and we find

$$1000 + T_6 + 2T_{10} - 5T_7 = 0$$

Similar resistances are obtained for node 10.

$$1/R_{10-9} = k$$
$$1/R_{10-7} = 2k = 1/R_{10-1}$$

so that

$$2T_7 + T_9 + 2T_1 - 5T_{10} = 0$$

For node 1,

$$1/R_{1-12} = \frac{k(\Delta y/6 + \Delta y/2)}{\Delta x/3} = 2k$$
$$1/R_{1-3} = \frac{k(\Delta x/6 + \Delta x/2)}{\Delta y} = 2k/3$$
$$1/R_{1-10} = 2k$$

and the nodal equation becomes

$$3T_{12} + 3T_{10} + T_3 - 7T_1 = 0$$

For node 11,

$$1/R_{11-100°} = 1/R_{11-12} = \frac{k(\Delta y/6 + \Delta y/2)}{\Delta x/3} = 2k$$
$$1/R_{11-8} = k$$
$$1/R_{11-13} = \frac{k(\Delta x/3)}{\Delta y} = k/3$$

and the nodal equation becomes

$$600 + 6T_{12} + 3T_8 + T_{13} - 16T_{11} = 0$$

Similarly, the equation for node 12 is

$$3T_9 + 6T_{11} + 6T_1 + T_{14} - 16T_{12} = 0$$

For node 13,

$$\frac{1}{R_{13-100°}} = \frac{k\,\Delta y}{\Delta x/3} = 3k = 1/R_{13-14}$$

$$1/R_{13-11} = 1/R_{13-100} = k/3$$

and we obtain

$$1000 + 9T_{14} + T_{11} - 20T_{13} = 0$$

Similarly for node 14,

$$100 + 9T_{13} + 9T_3 + T_{12} - 20T_{14} = 0$$

Finally, from resistances already found, the nodal equation for node 3 is

$$200 + 9T_{14} + 2T_1 - 13T_3 = 0$$

We choose to solve the set of equations by the Gauss-Seidel iteration technique and thus write them in the form $T_i = f(T_j)$. The solution was set up on a computer with all initial values for the T_i's taken as zero. The results of the computations are shown in the following table.

	Number of iterations				
Node	**2**	**10**	**20**	**30**	**50**
1	59.30662	232.6668	247.1479	247.7605	247.7875
2	59.30662	232.6668	247.1479	247.7605	247.7875
3	50.11073	139.5081	147.2352	147.5629	147.5773
4	50.11073	139.5081	147.2352	147.5629	147.5773
5	206.25	288.358	293.7838	294.0129	294.023
6	248.75	359.025	366.9878	367.3243	367.3391
7	291.45	390.989	398.7243	399.0513	399.0657
8	102.9297	200.5608	208.4068	208.7384	208.753
9	121.2334	264.2423	275.7592	276.2462	276.2677
10	164.5493	302.3108	313.5007	313.974	313.9948
11	70.95459	156.9976	164.3947	164.7076	164.7215
12	73.89051	203.6437	214.5039	214.9634	214.9836
13	70.18905	115.2635	119.2079	119.3752	119.3826
14	62.82942	129.8294	135.6246	135.8703	135.8811

Again, the results of the various sets of iterations are shown merely to illustrate the rapidity of convergence. In actual practice only the final set of values would be displayed on the computer. Note that these solutions for $T_1 = T_2 = 247.79°C$ and $T_3 = T_4 = 147.58°C$ are somewhat below the values of $250°C$ and $150°C$ obtained when only four nodes were employed, but only modestly so.

Three-Dimensional Numerical Formulation   EXAMPLE 3-12

To further illustrate the numerical formulation, consider the simple three-dimensional block shown in Figure Example 3-12a. The block has dimensions of $3 \times 4 \times 4$ cm with the front surface exposed to a convection environment with $T_\infty = 10°C$ and $h = 500\ \text{W/m}^2 \cdot °C$. The four sides are maintained constant at $100°C$ and the back surface is insulated. We choose $\Delta x = \Delta y = \Delta z = 1$ cm and set up the nodes as shown. The front surface has nodes 11, 12, 13, 14, 15, 16; the next Δz nodes are 21, 22, 23, 24, 25, 26 and so on. We shall use the resistance formulation in the form of Equation (3-32) to set up the nodal equations.

Figure Example 3-12a | Schematic.

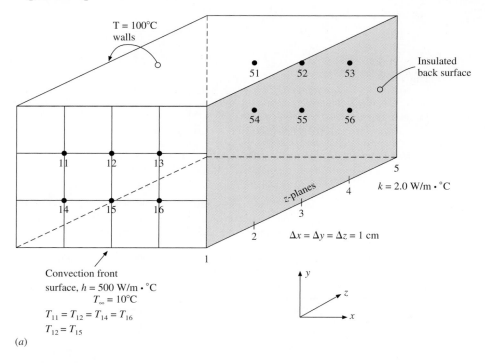

$T = 100°C$
walls

Insulated
back surface

51 52 53

54 55 56

11 12 13

14 15 16

z-planes

$k = 2.0$ W/m $\cdot$ °C

5

4

3

2    $\Delta x = \Delta y = \Delta z = 1$ cm

1

Convection front
surface, $h = 500$ W/m $\cdot$ °C
$T_\infty = 10°C$
$T_{11} = T_{12} = T_{14} = T_{16}$
$T_{12} = T_{15}$

y

z

x

(a)

Figure Example 3-12b | Results.

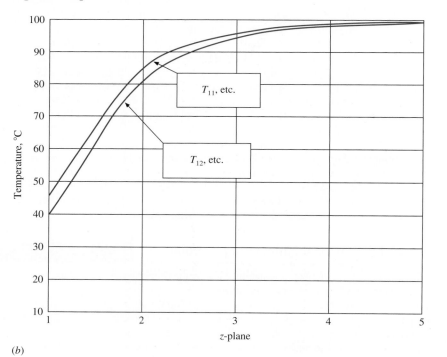

T_{11}, etc.

T_{12}, etc.

Temperature, °C

z-plane

(b)

■ **Solution**
All of the interior nodes for z-planes 2, 3, 4 have resistances of

$$1/R = k\,A/\Delta x = (2)(0.01)^2/0.01 = 0.02 = 1/R_{11-21} = 1/R_{21-22}, \text{ etc.}$$

The surface conduction resistances for surface z-plane 1 are

$$1/R_{11-12} = k\,A/\Delta x = (2)(0.01/2)/0.01 = 0.01 = 1/R_{11-14}, \text{ etc.}$$

The surface convection resistances are

$$1/R_{11-\infty} = h\,A = (500)(0.01)^2 = 0.05$$

For surface nodes like 11 the $\sum(1/R_{ij})$ term in Equation (3-32) becomes

$$\sum(1/R_{11-j}) = (4)(0.01) + 0.02 + 0.05 = 0.11$$

while, for interior nodes, we have

$$\sum(1/R_{21-j}) = (6)(0.02) = 0.12$$

For the insulated back surface nodes

$$\sum(1/R_{51-j}) = (4)(0.01) + (0.02) = 0.06$$

There are 30 nodes in total; 6 in each z-plane. We could write the equations for all of them but prefer to take advantage of the symmetry of the problem as indicated in the figure. Thus,

$$T_{11} = T_{13} = T_{14} = T_{16} \quad \text{and} \quad T_{12} = T_{15}, \text{ etc.}$$

We may then write the surface nodal equations as

$$T_{11} = [0.05\,T_\infty + 0.02\,T_{21} + (0.01)(100 + 100 + T_{14} + T_{12})]/0.11$$
$$T_{12} = [0.05\,T_\infty + 0.02\,T_{22} + (0.01)(100 + T_{11} + T_{15} + T_{13})]/0.11$$

Inserting $T_\infty = 10$ and simplifying we have

$$T_{11} = (2.5 + 0.02\,T_{21} + 0.01\,T_{12})/0.1$$
$$T_{12} = (1.5 + 0.02\,T_{22} + 0.02\,T_{11})/0.1$$

Following the same procedure for the other z-planes we obtain

$$T_{21} = (200 + T_{11} + T_{31} + T_{22})/5$$
$$T_{22} = (100 + T_{12} + T_{32} + T_{21})/5$$
$$T_{31} = (200 + T_{21} + T_{41} + T_{32})/5$$
$$T_{32} = (100 + T_{22} + T_{42} + T_{31})/5$$
$$T_{41} = (200 + T_{31} + T_{51} + T_{42})/5$$
$$T_{42} = (100 + T_{32} + T_{52} + 2\,T_{41})/5$$
$$T_{51} = (2 + 0.02\,T_{41} + 0.01\,T_{52})/0.05$$
$$T_{52} = (1 + 0.02\,T_{42} + 0.02\,T_{51})/0.05$$

Solving the 10 equations gives the following results for the temperatures in each z-plane.

z-plane	Node 1	Node 2
1	45.9	40.29
2	84.36	80.57
3	95.34	93.83
4	98.49	97.93
5	99.16	98.94

Figure Example 3-12b gives a graphical display of the results, and the behavior is as expected. The temperature drops as the cooled front surface is approached. Node 2 is cooled somewhat more than node 1 because it is in contact with only a single 100° surface.

■ **Comments**

While this is a rather simple three-dimensional example, it has illustrated the utility of the resistance formulation in solving such problems. As with two-dimensional systems, variable mesh sizes, heat generation, and variable boundary conditions can be accommodated with care and patience.

Remarks on Computer Solutions

It should be apparent by now that numerical methods and computers give the engineer powerful tools for solving very complex heat-transfer problems. Many large commercial software packages are available, and new ones appear with increasing regularity. One characteristic common to almost *all* heat-transfer software is a requirement that the user *understand* something about the subject of heat transfer. Without such an understanding it can become very easy to make gross mistakes and never detect them at all. We have shown how energy balances are one way to check the validity of a computer solution. Sometimes common sense also works well. We know, for example, that a plate will cool faster when air is blown across the plate than when exposed to still air. Later, in Chapters 5 through 7, we will see how to quantify these effects and will be able to anticipate what influence they may have on a numerical solution to a conduction problem. A similar statement can be made pertaining to radiation boundary conditions, which will be examined in Chapter 8. These developments will give the reader a "feel" for what the effects of various boundary conditions should be and insight about whether the numerical solution obtained for a problem appears realistic. Up to now, boundary conditions have been stipulated quantities, but experienced heat-transfer practitioners know that they are seldom easy to determine in the real world.

3-9 | ELECTRICAL ANALOGY FOR TWO-DIMENSIONAL CONDUCTION

Steady-state electric conduction in a homogeneous material of constant resistivity is analogous to steady-state heat conduction in a body of similar geometric shape. For two-dimensional electric conduction the Laplace equation applies:

$$\frac{\partial^2 E}{\partial x^2} + \frac{\partial^2 E}{\partial y^2} = 0$$

where E is the electric potential. A very simple way of solving a two-dimensional heat-conduction problem is to construct an electrical analog and experimentally determine the geometric shape factors for use in Equation (3-23). One way to accomplish this is to use a commercially available paper that is coated with a thin conductive film. This paper may be cut to an exact geometric model of the two-dimensional heat-conduction system. At the appropriate edges of the paper, good electrical conductors are attached to simulate the temperature boundary conditions on the problem. An electric-potential difference is then impressed on the model. It may be noted that the paper has a very high resistance in comparison with the conductors attached to the edges, so that a constant-potential condition can be maintained at the region of contact.

Once the electric potential is impressed on the paper, an ordinary voltmeter may be used to plot lines of constant electric potential. With these constant-potential lines available, the flux lines may be easily constructed since they are orthogonal to the potential lines. These equipotential and flux lines have precisely the same arrangement as the isotherms and heat-flux lines in the corresponding heat-conduction problem. The shape factor is calculated immediately using the method which was applied to the curvilinear squares.

It may be noted that the conducting-sheet analogy is not applicable to problems where heat generation is present; however, by addition of appropriate resistances, convection boundary conditions may be handled with little trouble. Schneider [2] and Ozisik [10] discuss the conducting-sheet method, as well as other analogies for treating conduction heat-transfer problems, and Kayan [4, 5] gives a detailed discussion of the conducting-sheet method. Because of the utility of numerical methods, analogue techniques for solution of heat-transfer problems are largely of historical interest.

3-10 | SUMMARY

There is a myriad of analytical solutions for steady-state conduction heat-transfer problems available in the literature. In this day of computers most of these solutions are of small utility, despite their exercise in mathematical facilities. This is not to say that we cannot use the results of past experience to anticipate answers to new problems. But, most of the time, the problem a person wants to solve can be attacked directly by numerical techniques, *except* when there is an easier way to do the job. As a summary, the following suggestions are offered:

1. When tackling a two- or three-dimensional heat-transfer problem, *first* try to reduce it to a one-dimensional problem. An example is a cylinder with length much larger than its diameter.
2. If possible, select a simple shape-factor model that may either exactly or approximately represent the physical situation. See comments under items 4 and 5.
3. Seek some simple analytical solutions but, if solutions are too complicated, go directly to the numerical techniques.
4. In practical problems, recognize that convection and radiation boundary conditions are subject to large uncertainties. This means that, in most practical situations, undue concern over accuracy of solution to numerical nodal equations is unjustified.
5. In general, approach the solution in the direction of simple to complex, and make use of checkpoints along the way.

REVIEW QUESTIONS

1. What is the main assumption in the separation-of-variables method for solving Laplace's equation?
2. Define the conduction shape factor.
3. What is the basic procedure in setting up a numerical solution to a two-dimensional conduction problem?

4. Once finite-difference equations are obtained for a conduction problem, what methods are available to effect a solution? What are the advantages and disadvantages of each method, and when would each technique be applied?

5. Investigate the computer software packages that are available at your computer center for solution of conduction heat-transfer problems.

LIST OF WORKED EXAMPLES

PROBLEMS

3-1 Beginning with the separation-of-variables solutions for $\lambda^2 = 0$ and $\lambda^2 < 0$ [Equations (3-9) and (3-10)], show that it is not possible to satisfy the boundary conditions for the constant temperature at $y = H$ with either of these two forms of solution. That is, show that, in order to satisfy the boundary conditions

$$
\begin{array}{ll}
T = T_1 & \text{at } y = 0 \\
T = T_1 & \text{at } x = 0 \\
T = T_1 & \text{at } x = W \\
T = T_2 & \text{at } y = H
\end{array}
$$

either a trivial or physically unreasonable solution results when either Equation (3-9) or (3-10) is used.

3-2 Write out the first four nonzero terms of the series solutions given in Equation (3-20). What percentage error results from using only these four terms at $y = H$ and $x = W/2$?

3-3 A horizontal pipe having a surface temperature of $67°C$ and diameter of 25 cm is buried at a depth of 1.2 m in the earth at a location where $k = 1.8$ W/m · °C. The earth surface temperature is $15°C$. Calculate the heat lost by the pipe per unit length.

3-4 A 6.0-cm-diameter pipe whose surface temperature is maintained at 210°C passes through the center of a concrete slab 45 cm thick. The outer surface temperatures of the slab are maintained at 15°C. Using the flux plot, estimate the heat loss from the pipe per unit length. Also work using Table 3-1.

3-5 A 2.5-cm-diameter pipe carrying condensing steam at 101 kPa passes through the center of an infinite plate having a thickness of 5 cm. The plate is exposed to room air at 27°C with a convection coefficient of 5.1 W/m² · °C on both sides. The plate is composed of an insulation material having $k = 0.1$ W/m · °C. Calculate the heat lost by the steam pipe per meter of length.

3-6 A heavy-wall tube of Monel, 2.5-cm ID and 5-cm OD, is covered with a 2.5-cm layer of glass wool. The inside tube temperature is 300°C, and the temperature at the outside of the insulation is 40°C. How much heat is lost per foot of length? Take $k = 11$ Btu/h · ft · °F for Monel.

3-7 A symmetrical furnace wall has the dimensions shown in Figure P3-7. Using the flux plot, obtain the shape factor for this wall.

Figure P3-7

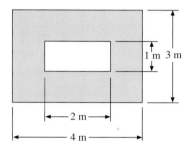

3-8 A furnace of 70- by 60- by 90-cm inside dimensions is constructed of a material having a thermal conductivity of 0.5 Btu/h · ft · °F. The wall thickness is 6 in. The inner and outer surface temperatures are 500 and 100°F, respectively. Calculate the heat loss through the furnace wall.

3-9 A cube 35 cm on each external side is constructed of fireclay brick. The wall thickness is 5.0 cm. The inner surface temperature is 500°C, and the outer surface temperature is 80°C. Compute the heat flow in watts.

3-10 Two long cylinders 8.0 and 3.0 cm in diameter are completely surrounded by a medium with $k = 1.4$ W/m · °C. The distance between centers is 10 cm, and the cylinders are maintained at 200 and 35°C. Calculate the heat-transfer rate per unit length.

3-11 A 10-cm-diameter sphere maintained at 30°C is buried in the earth at a place where $k = 1.2$ W/m · °C. The depth to the centerline is 24 cm, and the earth surface temperature is 0°C. Calculate the heat lost by the sphere.

3-12 A 20-cm-diameter sphere is totally enclosed by a large mass of glass wool. A heater inside the sphere maintains its outer surface temperature at 170°C while the temperature at the outer edge of the glass wool is 20°C. How much power must be supplied to the heater to maintain equilibrium conditions?

3-13 A large spherical storage tank, 2 m in diameter, is buried in the earth at a location where the thermal conductivity is 1.5 W/m · °C. The tank is used for the storage of an ice mixture at 0°C, and the ambient temperature of the earth is 20°C. Calculate the heat loss from the tank.

3-14 A 2.5-cm-diameter pipe carrying condensing steam at 101 kPa passes through the center of a square block of insulating material having $k = 0.04$ W/m · °C. The block is 5 cm on side and 2 m long. The outside of the block is exposed to room air at 27°C and a convection coefficient of $h = 5.1$ W/m² · °C. Calculate the heat lost by the steam pipe.

3-15 The solid shown in Figure P3-15 has the upper surface, including the half-cylinder cutout, maintained at 100°C. At a large depth in the solid the temperature is 300 K; $k = 1$ W/m · °C. What is the heat transfer at the surface for the region where $L = 30$ cm and $D = 10$ cm?

Figure P3-15

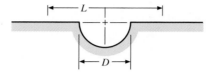

3-16 In certain locales, power transmission is made by means of underground cables. In one example an 8.0-cm-diameter cable is buried at a depth of 1.3 m, and the resistance of the cable is 1.1×10^{-4} Ω/m. The surface temperature of the ground is 25°C, and $k = 1.2$ W/m · °C for earth. Calculate the maximum allowable current if the outside temperature of the cable cannot exceed 110°C.

3-17 A copper sphere 4.0 cm in diameter is maintained at 70°C and submerged in a large earth region where $k = 1.3$ W/m · °C. The temperature at a large distance from the sphere is 12°C. Calculate the heat lost by the sphere.

3-18 Two long, eccentric cylinders having diameters of 20 and 5 cm, respectively, are maintained at 100 and 20°C and separated by a material with $k = 2.5$ W/m · °C. The distance between centers is 5.5 cm. Calculate the heat transfer per unit length between the cylinders.

3-19 Two pipes are buried in the earth and maintained at temperatures of 200 and 100°C. The diameters are 9 and 18 cm, and the distance between centers is 40 cm. Calculate the heat-transfer rate per unit length if the thermal conductivity of earth at this location is 1.1 W/m · °C.

3-20 A hot sphere having a diameter of 1.5 m is maintained at 300°C and buried in a material with $k = 1.2$ W/m · °C and outside surface temperature of 30°C. The depth of the centerline of the sphere is 3.75 m. Calculate the heat loss.

3-21 A scheme is devised to measure the thermal conductivity of soil by immersing a long electrically heated rod in the ground in a vertical position. For design purposes, the rod is taken as 2.5 cm in diameter with a length of 1 m. To avoid improper alteration of the soil, the maximum surface temperature of the rod is 55°C while the soil temperature is 10°C. Assuming a soil conductivity of 1.7 W/m · °C, what are the power requirements of the electric heater in watts?

3-22 Two pipes are buried in an insulating material having $k = 0.8$ W/m · °C. One pipe is 10 cm in diameter and carries a hot fluid at 300°C while the other pipe is 2.8 cm in diameter and carries a cool fluid at 15°C. The pipes are parallel and separated by a distance of 12 cm on centers. Calculate the heat-transfer rate between the pipes per meter of length.

3-23 At a certain location the thermal conductivity of the earth is 1.5 W/m · °C. At this location an isothermal sphere having a temperature of 5°C and a diameter of 2.0 m is buried at a centerline depth of 5.0 m. The earth temperature is 25°C. Calculate the heat gained by the sphere.

3-24 Two parallel pipes are buried very deep in the earth at a location where they are in contact with a rock formation having $k = 3.5$ W/m · °C. One pipe has a diameter of 20 cm and carries a hot fluid at 120°C while the other pipe has a diameter of 40 cm and carries a cooler fluid at 20°C. The distance between centers of the pipes is 1.0 m and both pipes are very long in respect to their diameters and spacing. Calculate the conduction heat transfer between the two pipes per unit length of pipe. Express as W/m length.

3-25 Steam pipes are sometimes carelessly buried in the earth without insulation. Consider a 10-cm pipe carrying steam at 150°C buried at a depth of 23 cm to centerline. The buried length is 100 m. Assuming that the earth thermal conductivity is 1.2 W/m · °C and the surface temperature is 15°C, estimate the heat lost from the pipe.

3-26 A hot steam pipe, 5 cm in diameter and carrying steam at 150°C, is placed in the center of a 15-cm-thick slab of lightweight structural concrete. The outside of the concrete slab is exposed to a convection environment that maintains the top and bottom of the sheet at 20°C. Calculate the heat lost per unit length of pipe.

3-27 Seven 1.0-cm-diameter tubes carrying steam at 100°C are buried in a semi-infinite medium having a thermal conductivity of 1.2 W/m · °C and surface temperature of 25°C. The depth to the centerline of the tubes is 5 cm and the spacing between centers is 3 cm. Calculate the heat lost per unit length for each tube.

3-28 Two parallel pipes 5 cm and 10 cm in diameter are totally surrounded by loosely packed asbestos. The distance between centers for the pipes is 20 cm. One pipe carries steam at 110°C while the other carries chilled water at 3°C. Calculate the heat lost by the hot pipe per unit length.

3-29 A long cylinder has its surface maintained at 135°C and is buried in a material having a thermal conductivity of 15.5 W/m · °C. The diameter of the cylinder is 3 cm and the depth to its centerline is 5 cm. The surface temperature of the material is 46°C. Calculate the heat lost by the cylinder per meter of length.

3-30 A 2.5-m-diameter sphere contains a mixture of ice and water at 0°C and is buried in a semi-infinite medium having a thermal conductivity of 0.2 W/m · °C. The top surface of the medium is isothermal at 30°C and the sphere centerline is at a depth of 8.5 m. Calculate the heat lost by the sphere.

3-31 An electric heater in the form of a 50- by-100-cm plate is laid on top of a semi-infinite insulating material having a thermal conductivity of 0.74 W/m · °C. The heater plate is maintained at a constant temperature of 120°C over all its surface, and the temperature of the insulating material a large distance from the heater is 15°C. Calculate the heat conducted into the insulating material.

3-32 A thin isothermal disk, having a diameter of 1.8 cm, is maintained at 40°C and buried in a semi-infinite medium at a depth of 2 cm. The medium has a thermal conductivity of 0.8 W/m · °C and its surface is maintained at 15°C. Calculate the heat lost by the disk.

3-33 Two parallel pipes, each having a diameter of 5 cm, carry steam at 120°C and chilled water at 5°C, respectively, and are buried in an infinite medium of fiberglass blanket ($k = 0.04$ W/m · °C). Plot the heat transfer between the pipes per unit length as a function of the centerline spacing between the pipes.

3-34 A small furnace has inside dimensions of 60 by 70 by 80 cm with a wall thickness of 5 cm. Calculate the overall shape factor for this geometry.

3-35 A 15-cm-diameter steam pipe at 150°C is buried in the earth near a 5-cm pipe carrying chilled water at 5°C. The distance between centers is 15 cm and the thermal conductivity of the earth at this location may be taken as 0.7 W/m · °C. Calculate the heat lost by the steam pipe per unit length.

3-36 Derive an equation equivalent to Equation (3-24) for an interior node in a three-dimensional heat-flow problem.

3-37 Derive an equation equivalent to Equation (3-24) for an interior node in a one-dimensional heat-flow problem.

3-38 Derive an equation equivalent to Equation (3-25) for a one-dimensional convection boundary condition.

3-39 Considering the one-dimensional fin problems of Chapter 2, show that a nodal equation for nodes along the fin in the Figure P3-39 may be expressed as

$$T_m \left[\frac{hP(\Delta x)^2}{kA} + 2 \right] - \frac{hP(\Delta x)^2}{kA} T_\infty - (T_{m-1} + T_{m+1}) = 0$$

Figure P3-39

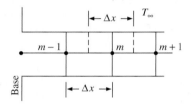

3-40 Show that the nodal equation corresponding to an insulated wall shown in Figure P3-40 is

$$T_{m,n+1} + T_{m,n-1} + 2T_{m-1,n} - 4T_{m,n} = 0$$

Figure P3-40

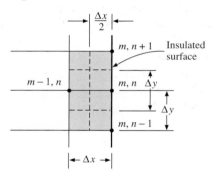

Figure P3-41

Insulated surfaces

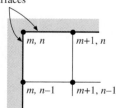

3-41 For the insulated corner section shown in Figure P3-41, derive an expression for the nodal equation of node (m, n) under steady-state conditions.

3-42 Derive the equation in Table 3-2*f*.

3-43 Derive an expression for the equation of a boundary node subjected to a constant heat flux from the environment. Use the nomenclature of Figure 3-7.

3-44 Set up the nodal equations for a modification of Example 3-7 in which the left half of the wire is insulated and the right half is exposed to a connection environment with $h = 200$ W/m$^2 \cdot$ °C and $T = 20$°C.

3-45 In a proposed solar-energy application, the solar flux is concentrated on a 5-cm-OD stainless-steel tube [$k = 16$ W/m $\cdot$ °C] 2 m long. The energy flux on the tube surface is 20,000 W/m^2, and the tube wall thickness is 2 mm. Boiling water flows inside the tube with a convection coefficient of 5000 W/m$^2 \cdot$ °C and a temperature of 250°C.

Both ends of the tube are mounted in an appropriate supporting bracket, which main-tains them at 100°C. For thermal-stress considerations the temperature gradient near the supports is important. Assuming a one-dimensional system, set up a numerical solution to obtain the temperature gradient near the supports.

3-46 An aluminum rod 2.5 cm in diameter and 15 cm long protrudes from a wall main-tained at 300°C. The environment temperature is 38°C. The heat-transfer coefficient is 17 W/m² · °C. Using a numerical technique in accordance with the result of Prob-lem 3-39, obtain values for the temperature along the rod. Subsequently obtain the heat flow from the wall at $x = 0$. *Hint:* The boundary condition at the end of the rod may be expressed by

$$T_m \left[\frac{h\,\Delta x}{k} + \frac{hP(\Delta x)^2}{2kA} + 1 \right] - T_\infty \left[\frac{h\,\Delta x}{k} + \frac{hP(\Delta x)^2}{2kA} \right] - T_{m-1} = 0$$

where m denotes the node at the tip of the fin. The heat flow at the base is

$$q_{x=0} = \frac{-kA}{\Delta x}(T_{m+1} - T_m)$$

where T_m is the base temperature and T_{m+1} is the temperature of the first increment.

3-47 Repeat Problem 3-46, using a linear variation of heat-transfer coefficient between base temperature and the tip of the fin. Assume $h = 28$ W/m² · °C at the base and $h = 11$ W/m² · °C at the tip.

3-48 For the wall in Problem 3-6 a material with $k = 1.4$ W/m · °C is used. The inner and outer wall temperatures are 650 and 150°C, respectively. Using a numerical technique, calculate the heat flow through the wall.

3-49 Repeat Problem 3-48, assuming that the outer wall is exposed to an environment at 38°C and that the convection heat-transfer coefficient is 17 W/m² · °C. Assume that the inner surface temperature is maintained at 650°C.

3-50 Repeat Problem 3-4, using the numerical technique.

3-51 In the section illustrated in Figure P3-51 the surface 1-4-7 is insulated. The convection heat transfer coefficient at surface 1-2-3 is 28 W/m² · °C. The thermal conductivity of the solid material is 5.2 W/m · °C. Using the numerical technique, compute the temperatures at nodes 1, 2, 4, and 5.

Figure P3-51

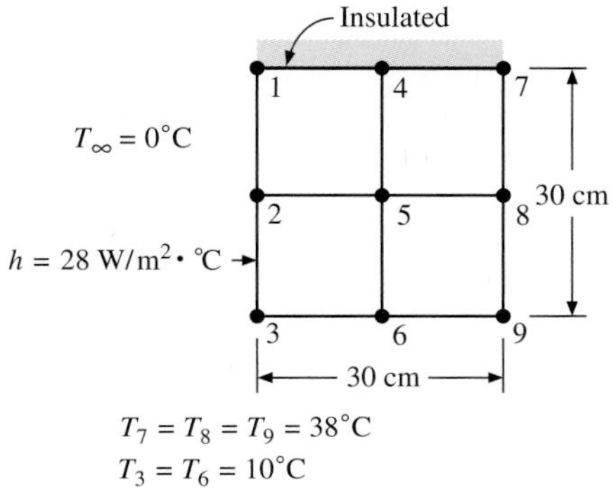

$T_7 = T_8 = T_9 = 38°C$
$T_3 = T_6 = 10°C$

3-52 A glass plate 3 by 12 by 12 in [$k = 0.7$ W/m · °C] is oriented with the 12 by 12 face in a vertical position. One face loses heat by convection to the surroundings at 70°F. The other vertical face is placed in contact with a constant-temperature block at 400°F.

The other four faces are insulated. The convection heat-transfer coefficient varies approximately as

$$h_x = 0.22(T_s - T_\infty)^{1/4} x^{-1/4} \qquad \text{Btu/h} \cdot \text{ft}^2 \cdot °F$$

where T_s and T_∞ are in degrees Fahrenheit, T_s is the local surface temperature, and x is the vertical distance from the bottom of the plate, measured in feet. Determine the convection heat loss from the plate, using an appropriate numerical analysis.

3-53 In Figure P3-53, calculate the temperatures at points 1, 2, 3, and 4 using the numerical method.

3-54 For the block shown in Figure P3-54, calculate the steady-state temperature distribution at appropriate nodal locations using the numerical method. $k = 3.2$ W/m · °C.

Figure P3-53

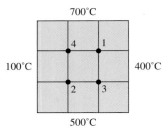

700°C

100°C   400°C

500°C

Figure P3-54

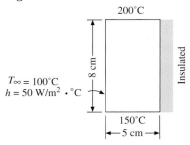

200°C

$T_\infty = 100°C$
$h = 50$ W/m² · °C

8 cm

Insulated

150°C

5 cm

3-55 The composite strip in Figure P3-55 is exposed to the convection environment at 300°C and $h = 40$ W/m² · °C. The material properties are $k_A = 20$ W/m · °C, $k_B = 1.2$ W/m · °C, and $k_C = 0.5$ W/m · °C. The strip is mounted on a plate maintained at the constant temperature of 50°C. Calculate the heat transfer from the strip to plate per unit length of strip. Assume two-dimensional heat flow.

Figure P3-55

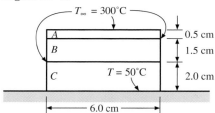

$T_\infty = 300°C$

A   0.5 cm
B   1.5 cm
C $T = 50°C$ 2.0 cm

6.0 cm

3-56 The fin shown in Figure P3-56 has a base maintained at 300°C and is exposed to the convection environment indicated. Calculate the steady-state temperatures of the nodes shown and the heat loss if $k = 1.0$ W/m · °C.

Figure P3-56

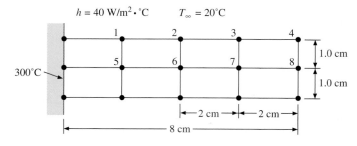

$h = 40$ W/m² · °C $T_\infty = 20°C$

1   2   3   4

5   6   7   8

300°C

1.0 cm

1.0 cm

2 cm   2 cm

8 cm

3-57 Calculate the steady-state temperatures for nodes 1 to 16 in Figure P3-57. Assume symmetry.

Figure P3-57

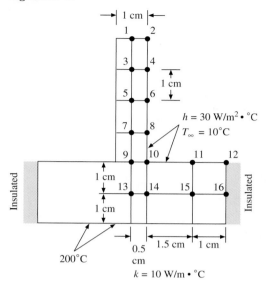

3-58 Calculate the steady-state temperatures for nodes 1 to 9 in Figure P3-58.

Figure P3-58

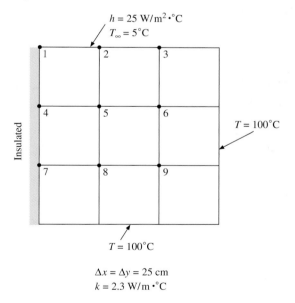

3-59 Calculate the steady-state temperatures for nodes 1 to 6 in Figure P3-59.

Figure P3-59

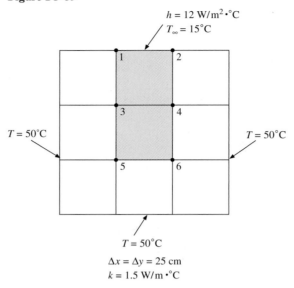

$\Delta x = \Delta y = 25$ cm
$k = 1.5$ W/m·°C

3-60 Calculate the temperatures for the nodes indicated in Figure P3-60. The entire outer surface is exposed to the convection environment and the entire inner surface is at a constant temperature of 300°C. Properties for materials A and B are given in the figure.

Figure P3-60

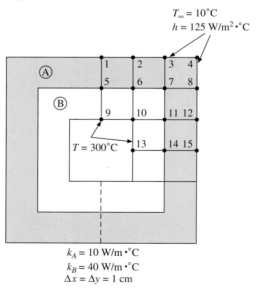

$k_A = 10$ W/m·°C
$k_B = 40$ W/m·°C
$\Delta x = \Delta y = 1$ cm

3-61 A rod having a diameter of 2 cm and a length of 10 cm has one end maintained at 200°C and is exposed to a convection environment at 25°C with $h = 40$ W/m$^2 \cdot$°C. The rod generates heat internally at the rate of 50 MW/m^3 and the thermal conductivity is 35 W/m $\cdot$°C. Calculate the temperatures of the nodes shown in the Figure P3-61 assuming one-dimensional heat flow.

Figure P3-61

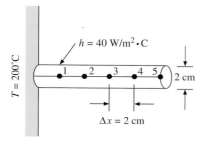

3-62 Calculate the steady-state temperatures of the nodes in Figure P3-62. The entire outer surface is exposed to the convection environment at 20°C and the entire inner surface is constant at 500°C. Assume $k = 0.2$ W/m $\cdot$°C.

Figure P3-62

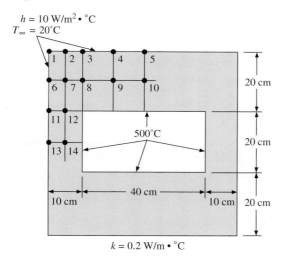

3-63 Calculate the steady-state temperatures for the nodes indicated in Figure P3-63.
Figure P3-63

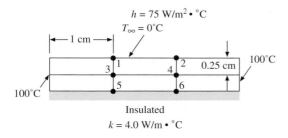

3-64 The two-dimensional solid shown in Figure P3-64 generates heat internally at the rate of 90 MW/m^3. Using the numerical method calculate the steady-state nodal temperatures for $k = 20$ W/m $\cdot$ °C.

Figure P3-64

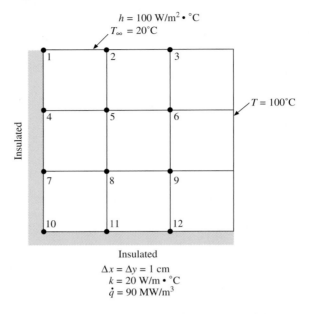

$$\Delta x = \Delta y = 1 \text{ cm}$$
$$k = 20 \text{ W/m} \bullet \text{°C}$$
$$\dot{q} = 90 \text{ MW/m}^3$$

3-65 Two parallel disks having equal diameters of 30 cm are maintained at 120°C and 34°C. The disks are spaced a distance of 80 cm apart, on centers, and immersed in a conducting medium having $k = 3.4$ W/m $\cdot$ °C. Assuming that the disks exchange heat only on the sides facing each other, calculate the heat lost by the hotter disk, expressed in watts.

3-66 The half-cylinder has $k = 20$ W/m $\cdot$ °C and is exposed to the convection environment at 20°C. The lower surface is maintained at 300°C. Compute the temperatures for the nodes shown in Figure P3-66 and the heat loss for steady state.

Figure P3-66

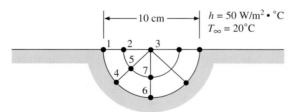

3-67 A tube has diameters of 4 mm and 5 mm and a thermal conductivity 20 W/m $\cdot$ °C. Heat is generated uniformly in the tube at a rate of 500 MW/m^3 and the outside surface temperature is maintained at 100°C. The inside surface may be assumed to be insulated. Divide the tube wall into four nodes and calculate the temperature at each using the numerical method. Check with an analytical solution.

3-68 Repeat Problem 3-67 with the inside of the tube exposed to a convection condition with $h = 40$ W/m^2 $\cdot$ °C. Check with an analytical calculation.

3-69 Rework Problem 3-57 with the surface absorbing a constant heat flux of 300 W/m^2 instead of the convection boundary condition. The bottom surface still remains at 200°C.

3-70 Rework Problem 3-60 with the inner surface absorbing a constant heat flux of 300 W/m^2 instead of being maintained at a constant temperature of 300°C.

3-71 Rework Problem 3-64 with the surface marked at a constant 100°C now absorbing a constant heat flux of 500 W/m^2. Add nodes as necessary.

3-72 The tapered aluminum pin fin shown in Figure P3-72 is circular in cross section with a base diameter of 1 cm and a tip diameter of 0.5 cm. The base is maintained at 200°C and loses heat by convection to the surroundings at $T_\infty = 10°C$, $h = 200 \text{ W/m}^2 \cdot °C$. The tip is insulated. Assume one-dimensional heat flow and use the finite-difference method to obtain the nodal equations for nodes 1 through 4 and the heat lost by the fin. The length of the fin is 6 cm.

Figure P3-72

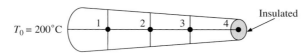

3-73 Write the nodal equations 1 through 7 for the symmetrical solid shown in Figure P3-73. $\Delta x = \Delta y = 1$ cm.

Figure P3-73

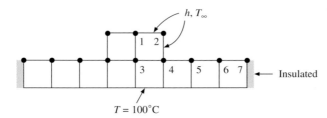

3-74 Obtain the temperature for nodes 1 through 6 shown in Figure P3-74. $\Delta x = \Delta y = 1$ cm.

Figure P3-74

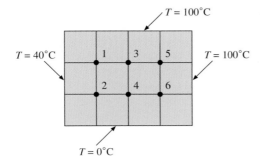

3-75   Write the nodal equations for nodes 1 through 9 shown in Figure P3-75. $\Delta x = \Delta y = 1$ cm.

Figure P3-75

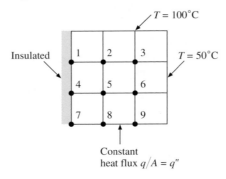

3-76   Write the nodal equation for nodes 1 through 12 shown in Figure P3-76. Express the equations in a format for Gauss-Seidel iteration.

Figure P3-76

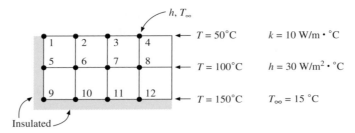

3-77   Sometimes a square grid is desired even for a circular system. Consider the quadrant of a circle shown in Figure P3-77 with $r = 10$ cm. $\Delta x = \Delta y = 3$ cm and $k = 10$ W/m · °C. Write the steady-state nodal equations for nodes 3 and 4. Make use of Tables 3-2 and 3-4.

Figure P3-77

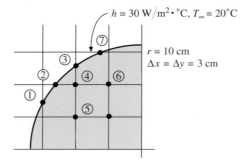

3-78 Taking Figure P3-78 as a special case of Table 3-2(*f*), write the nodal equations for nodes (m, n) and 2 for the case of $\Delta x = \Delta y$.

Figure P3-78

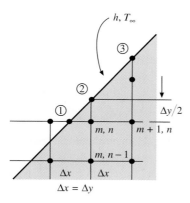

3-79 Repeat Problem 3-78 for a slanted surface that is insulated; i.e., $h = 0$.

3-80 If the slanted surface of Problem 3-78 is isothermal at T_∞, what is the nodal equation for node (m, n)?

3-81 The slanted intersection shown in Figure P3-81 involves materials A and B. Write steady-state nodal equations for nodes 3, 4, 5, and 6 using Table 3-2(*f* and *g*) as a guide.

Figure P3-81

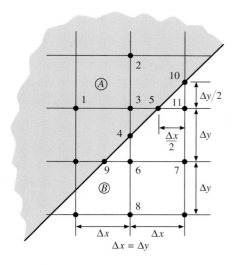

3-82 A horizontal plate, 25 by 50 cm, is maintained at a constant temperature of 78°C and buried in a semi-infinite medium at a depth of 5 m. The medium has an isothermal surface maintained at 15°C and a thermal conductivity of 2.8 W/m · °C. Calculate the heat lost by the plate.

3-83 A cube 20 cm on a side is maintained at 80°C and buried in a large medium at 10°C with a thermal conductivity of 2.3 W/m · °C. Calculate the heat lost by the cube.

How does this compare with the heat that would be lost by a 20-cm-diameter sphere? Compare these heat transfers on a unit-volume basis.

3-84 A long horizontal cylinder having a diameter of 10 cm is maintained at a temperature of 100°C and centered in a 30-cm-thick slab of material for which $k = $ W/m · °C. The outside of the slab is at 20°C. Calculate the heat lost by the cylinder per unit length.

3-85 Work Problem 3-84 using the flux plot.

3-86 A horizontal plate 20 by 150 cm is buried in a large medium at a depth of 2.0 m and maintained at 50°C. The surface of the medium is at 10°C and has $k = 1.5$ W/m · °C. Calculate the heat lost by the plate.

3-87 A thin disk 10 cm in diameter is maintained at 75°C and placed on the surface of a large medium at 15°C with $k = 3$ W/m · °C. Calculate the heat conducted into the medium.

3-88 Repeat Problem 3-87 for a square 10 cm on a side. Compare the heat transfers on a per unit area basis.

3-89 A hot steam pipe 10 cm in diameter is maintained at 200°C and centered in a square mineral-fiber insulation 20 cm on a side. The outside surface temperature of the insulation is 35°C. Calculate the heat lost by a 20-m length of pipe if the thermal conductivity of the insulation can be taken as 50 mW/m · °C.

3-90 A pipe having a diameter of 10 cm passes through the center of a concrete slab having a thickness of 70 cm. The surface temperature of the pipe is maintained at 100°C by condensing steam while the outer surfaces of the concrete are at 24°C. Calculate the heat lost by the pipe per meter of length.

3-91 Consider a circumferential fin of rectangular profile as shown in Figure 2-12. Set up nodal equations for a fin of thickness t, heat transfer coefficient h, thermal conductivity k, and heat generation rate q as a function of radial coordinate r, taking increments of Δr. Write the nodal equations for the node adjacent to the base temperature T_0, a node in the middle of the fin, and the node at the end of the fin.

3-92 Set up a nodal equation for the geometry of Problem 2-123, using increments in the height of the truncated cone as the one-dimensional variable. Then work the problem with the numerical method and compare with the one-dimensional analytical solution.

3-93 Set up nodal equations for the geometry of Problem 2-122, using increments in an angle θ as the one-dimensional variable. Then work the problem using the numerical method and compare with the one-dimensional analytical solution.

3-94 A cube 30 cm on a side is buried in an infinite medium with a thermal conductivity of 1.8 W/m · °C. The surface temperature of the cube is 30°C while the temperature of the medium is 10°C. Calculate the heat lost by the cube.

3-95 A thin horizontal disk having a diameter of 15 cm is maintained at a constant surface temperature of 87°C and buried at a depth of 20 cm in a semi-infinite medium with an adiabatic surface. The thermal conductivity of the medium is 2.7 W/m · °C and the temperature of the medium a large distance away from the disk (*not* the adiabatic the surface temperature) is 13°C. Calculate the heat lost by the disk in watts.

3-96 A copper rod has an internal heater that maintains its surface temperature at 50°C while it is buried vertically in a semi-infinite medium. The rod is 2 cm in diamter and 40 cm long and the isothermal surface of the medium is at 20°C. Calculate the heat lost by the rod if the thermal conductivity of the medium is 3.4 W/m · °C.

3-97 Rework Problem 2-122, using a numerical approach with five nodes operating in increments of the radial angle θ, and compare with the analytical results of Problem 2-122.

Design-Oriented Problems

3-98 A liner of stainless steel ($k = 20$ W/m · °C), having a thickness of 3 mm, is placed on the inside surface of the solid in Problem 3-62. Assuming now that the inside surface of the stainless steel is at 500°C, calculate new values for the nodal temperatures in the low-conductivity material. Set up your nodes in the stainless steel as necessary.

3-99 A basement for a certain home is 4×5 m with a ceiling height of 3 m. The walls are concrete having a thickness of 10 cm. In the winter the convection coefficient on the inside is 10 W/m² · °C and the soil on the outside has $k = 1.7$ W/m · °C. Analyze this problem and determine an overall heat transfer coefficient U defined by $q_{loss} = UA_{inside}(T_{inside} - T_{soil})$. Determine the heat loss when $T_{inside} = 26$°C and $T_{soil} = 15$°C.

3-100 A groundwater heat pump is a refrigeration device that rejects heat to the ground through buried pipes instead of to the local atmosphere. The heat rejection rate for such a machine at an Oklahoma location is to be 22 kW in a location where the ground temperature at depth is 17°C. The thermal conductivity of the soil at this location may be taken as 1.6 W/m · °C. Water is to be circulated through a length of horizontal buried pipe or tube with the water entering at 29°C and leaving at 23.5°C. The convection coefficient on the inside of the pipe is sufficiently high that the inner pipe wall temperature may be assumed to be the same as the water temperature. Select an appropriate pipe/tube material, size, and length to accomplish the required cooling. You may choose standard steel pipe sizes from Table A-11. Standard tubing or plastic pipe sizes are obtained from other sources. Examine several choices before making your final selection and give reasons for that selection.

3-101 Professional chefs claim that gas stove burners are superior to electric burners because of the more uniform heating afforded by the gas flame and combustion products around the bottom of a cooking pan. Advocates of electric stoves note the lack of combustion products to pollute the air in the cooking area, but acknowledge that gas heat may be more uniform. Manufacturers of thick-bottomed cookware claim that their products can achieve uniformity of cooking as good as gas heat because of the "spreading" of heat through an 8-mm-thick aluminum layer on the bottom of the pan. You are asked to verify this claim. For the evaluation assume a 200-mm-diameter pan with an 8-mm-thick aluminum bottom and the interior exposed to boiling water, which produces $h = 1500$ W/m² · °C at 1 atm (100°C). Observe the approximate spacing for the circular element in an electric burner and devise an appropriate numerical model to investigate the uniformity-of-heating claim. Consider such factors as contact resistance between the burner element and the bottom of the pan, and radiation transfer that might be present. Consider different heating rates (different burner element temperatures) and their effect. When the study is complete, make recommendations as to what the cookware manufacturers might prudently claim for their thick-bottomed product. Discuss uncertainties in your analysis.

3-102 The fin analyses of Section 2-10 assumed one-dimensional heat flows in the fins. Devise a numerical model similar to that shown in Problem 3-57 to examine the validity of this assumption. Restrict the analysis to aluminum with $k = 200$ W/m · °C. Examine several different combinations of fin thickness, fin length, and convection coefficient to determine the relative effects on temperature variation across the fin thickness. State conclusions as you think appropriate.

3-103 A small building 5 m wide by 7 m long by 3 m high (inside dimensions) is mounted on a flat concrete slab having a thickness of 15 cm. The walls of the building are constructed of concrete also, with a thickness of 7 cm. The inside of the building is used for cold storage at $-20°$C and the outside of the building is exposed to ambient air at 30°C, with a convection coefficient of 15 W/m$^2 \cdot$ °C. The inside convection coefficient for the building is estimated at 10 W/m$^2 \cdot$ °C and the floor slab is in contact with earth having $k = 1.8$ W/m $\cdot$ °C. The earth temperature may be assumed to be 15°C. Calculate the heat gained by the building in the absence of any insulating material on the outside. Next, select two alternative insulation materials for the outside of the building from Table 2-1 and/or Table A-3. The insulation objective is to raise the outside surface temperature of the insulation to 26°C for the ambient temperature of 30°C. The refrigeration system operates in such a manner that 1 kW will produce 4000 kJ/hr of cooling, and electricity costs \$0.085/kWh. Economics dictates that the insulation should pay for itself in a three-year period. What is the allowable cost per unit volume of insulation to accomplish this payback objective, for the two insulating materials selected? Suppose an outside surface temperature of 24°C is chosen as the allowable value for the insulation. What would the allowable costs be for a three-year payback in this case? Make your own assumptions as to the annual hours of operation for the cooling system.

REFERENCES

1. Carslaw, H. S., and J. C. Jaeger. *Conduction of Heat in Solids*, 2d ed. Fair Lawn, NJ: Oxford University Press, 1959.

2. Schneider, P. J. *Conduction Heat Transfer*. Reading, MA: Addison-Wesley, 1955.

3. Dusinberre, G. M. *Heat Transfer Calculations by Finite Differences*, Scranton, PA: International Textbook, 1961.

4. Kayan, C. F. "Heat Transfer Temperature Patterns of a Multicomponent Structure by Comparative Methods," *Trans ASME,* vol. 71, p. 9, 1949.

5. Rudenberg, R. Die Ausbreitung der Luft-und Erdfelder und Hochspannungsleitungen, besonders bei Erd-und Kurzschlussen, *Elektrotech. Z.,* vol. 46, p. 1342, 1925.

6. Andrews, R. V. "Solving Conductive Heat Transfer Problems with Electrical-Analogue Shape Factors," *Chem. Eng. Prog.,* vol. 51, no. 2, p. 67, 1955.

7. Sunderland, J. E., and K. R. Johnson. "Shape Factors for Heat Conduction through Bodies with Isothermal or Convective Boundary Conditions," *Trans. ASHAE,* vol. 70, pp. 237–41, 1964.

8. Richtmeyer, R. D. *Difference Methods for Initial Value Problems.* New York: Interscience Publishers, 1957.

9. Crank, J., and P. Nicolson. "A Practical Method for Numerical Evaluation of Solutions of P. D. E. of Heat Conduction Type," *Proc. Camb. Phil. Soc.,* vol. 43, p. 50, 1947.

10. Ozisik, M. N. *Boundary Value Problems of Heat Conduction.* Scranton, PA: International Textbook, 1968.

11. Arpaci, V. S. *Conduction Heat Transfer.* Reading, MA: Addison-Wesley, 1966.

12. Ames, W. F. *Nonlinear Partial Differential Equations in Engineering.* New York: Academic Press, 1965.

13. Myers, R. F. *Conduction Heat Transfer.* New York: McGraw-Hill, 1972.

14. Adams, J. A., and D. F. Rogers. *Computer Aided Analysis in Heat Transfer.* New York: McGraw-Hill, 1973.

15. Rohsenow, W. M., and J. P. Hartnett, eds. *Handbook of Heat Transfer.* 2nd ed. New York: McGraw-Hill, 1988.

16. Kern, D. Q., and A. D. Kraus. *Extended Surface Heat Transfer.* New York: McGraw-Hill, 1972.

17. Hahne, E., and U. Grigull. "Formfaktor und Formwiderstand der stationaren mehr-dimensionalen Warmeleitung," *Int. J. Heat Mass Transfer,* vol. 18, p. 751, 1975.

18. Chapra, S. C., and R. P. Canale. *Numerical Methods for Engineers.* 3rd ed. McGraw-Hill, 1996.

19. Constantinides, A. *Applied Numerical Methods with Personal Computers.* McGraw-Hill, 1987.

20. Patankar, S. V. *Numerical Heat Transfer and Fluid Flow.* Hemisphere Publishing, 1980.

21. Minkowycz, W. J., E. M. Sparrow, G. E. Schneider, and R. H. Pletcher. *Handbook of Numerical Heat Transfer.* New York: Wiley, 1988.

22. ——. *Mathcad 8,* Cambridge, MA: Mathsoft, Inc., 1999.

23. ——. *TK Solver,* Rockford, Ill.: Universal Technical Systems, 1999.

24. Palm, W. *MATLAB for Engineering Applications.* New York: McGraw-Hill, 1999.

25. Gottfried, B. *Spreadsheet Tools for Engineers—Excel 97 Version.* New York: McGraw-Hill, 1998.

26. Holman, J. P. *What Every Engineer Should Know About EXCEL,* Chap. 5. Boca Raton, FL: CRC Press, 2006.

27. Orvis, W. J., *Excel for Scientists and Engineers*, 2nd ed. San Francisco: SYBEX, 1996.

Unsteady-State Conduction

4-1 | INTRODUCTION

If a solid body is suddenly subjected to a change in environment, some time must elapse before an equilibrium temperature condition will prevail in the body. We refer to the equilibrium condition as the steady state and calculate the temperature distribution and heat transfer by methods described in Chapters 2 and 3. In the transient heating or cooling process that takes place in the interim period before equilibrium is established, the analysis must be modified to take into account the change in internal energy of the body with time, and the boundary conditions must be adjusted to match the physical situation that is apparent in the unsteady-state heat-transfer problem. Unsteady-state heat-transfer analysis is obviously of significant practical interest because of the large number of heating and cooling processes that must be calculated in industrial applications.

To analyze a transient heat-transfer problem, we could proceed by solving the general heat-conduction equation by the separation-of-variables method, similar to the analytical treatment used for the two-dimensional steady-state problem discussed in Section 3-2. We give one illustration of this method of solution for a case of simple geometry and then refer the reader to the references for analysis of more complicated cases. Consider the infinite plate of thickness $2L$ shown in Figure 4-1. Initially the plate is at a uniform temperature T_i, and at time zero the surfaces are suddenly lowered to $T = T_1$. The differential equation is

$$\frac{\partial^2 T}{\partial x^2} = \frac{1}{\alpha} \frac{\partial T}{\partial \tau} \qquad \textbf{[4-1]}$$

The equation may be arranged in a more convenient form by introduction of the variable $\theta = T - T_1$. Then

$$\frac{\partial^2 \theta}{\partial x^2} = \frac{1}{\alpha} \frac{\partial \theta}{\partial \tau} \qquad \textbf{[4-2]}$$

with the initial and boundary conditions

$$\theta = \theta_i = T_i - T_1 \quad \text{at } \tau = 0, 0 \leq x \leq 2L \qquad \textbf{[a]}$$

$$\theta = 0 \qquad \qquad \text{at } x = 0, \tau > 0 \qquad \textbf{[b]}$$

$$\theta = 0 \qquad \qquad \text{at } x = 2L, \tau > 0 \qquad \textbf{[c]}$$

Figure 4-1 | Infinite plate subjected to sudden cooling of surfaces.

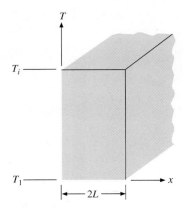

Assuming a product solution $\theta(x, \tau) = X(x)\mathcal{H}(\tau)$ produces the two ordinary differential equations

$$\frac{d^2 X}{dx^2} + \lambda^2 X = 0$$

$$\frac{d\mathcal{H}}{d\tau} + \alpha\lambda^2\mathcal{H} = 0$$

where λ^2 is the separation constant. In order to satisfy the boundary conditions it is necessary that $\lambda^2 > 0$ so that the form of the solution becomes

$$\theta = (C_1\cos\lambda x + C_2\sin\lambda x)e^{-\lambda^2\alpha\tau}$$

From boundary condition (b), $C_1 = 0$ for $\tau > 0$. Because C_2 cannot also be zero, we find from boundary condition (c) that $\sin 2L\lambda = 0$, or

$$\lambda = \frac{n\pi}{2L} \qquad n = 1, 2, 3, \ldots$$

The final series form of the solution is therefore

$$\theta = \sum_{n=1}^{\infty} C_n e^{-[n\pi/2L]^2\alpha\tau}\sin\frac{n\pi x}{2L}$$

This equation may be recognized as a Fourier sine expansion with the constants C_n determined from the initial condition (a) and the following equation:

$$C_n = \frac{1}{L}\int_0^{2L}\theta_i\sin\frac{n\pi x}{2L}\,dx = \frac{4}{n\pi}\theta_i \qquad n = 1, 3, 5, \ldots$$

The final series solution is therefore

$$\frac{\theta}{\theta_i} = \frac{T - T_1}{T_i - T_1} = \frac{4}{\pi}\sum_{n=1}^{\infty}\frac{1}{n}e^{-[n\pi/2L]^2\alpha\tau}\sin\frac{n\pi x}{2L} \qquad n = 1, 3, 5\ldots \qquad \textbf{[4-3]}$$

We note, of course, that at time zero ($\tau = 0$) the series on the right side of Equation (4-3) must converge to unity for all values of x.

In Section 4-4, this solution will be presented in graphical form for calculation purposes. For now, our purpose has been to show how the unsteady-heat-conduction equation can be solved, for at least one case, with the separation-of-variables method. Further information on analytical methods in unsteady-state problems is given in the references.

4-2 | LUMPED-HEAT-CAPACITY SYSTEM

We continue our discussion of transient heat conduction by analyzing systems that may be considered uniform in temperature. This type of analysis is called the *lumped-heat-capacity* method. Such systems are obviously idealized because a temperature gradient must exist in a material if heat is to be conducted into or out of the material. In general, the smaller the physical size of the body, the more realistic the assumption of a uniform temperature throughout; in the limit a differential volume could be employed as in the derivation of the general heat-conduction equation.

If a hot steel ball were immersed in a cool pan of water, the lumped-heat-capacity method of analysis might be used if we could justify an assumption of uniform ball temperature during the cooling process. Clearly, the temperature distribution in the ball would depend on the thermal conductivity of the ball material and the heat-transfer conditions from the surface of the ball to the surrounding fluid (i.e., the surface-convection heat-transfer coefficient). We should obtain a reasonably uniform temperature distribution in the ball if the resistance to heat transfer by conduction were small compared with the convection resistance at the surface, so that the major temperature gradient would occur through the fluid layer at the surface. The lumped-heat-capacity analysis, then, is one that assumes that the internal resistance of the body is negligible in comparison with the external resistance.

The convection heat loss from the body is evidenced as a decrease in the internal energy of the body, as shown in Figure 4-2. Thus,

$$q = hA(T - T_\infty) = -c\rho V \frac{dT}{d\tau}$$ [4-4]

where A is the surface area for convection and V is the volume. The initial condition is written

$$T = T_0 \qquad \text{at } \tau = 0$$

so that the solution to Equation (4-4) is

$$\frac{T - T_\infty}{T_0 - T_\infty} = e^{-[hA/\rho cV]\tau}$$ [4-5]

Figure 4-2 | Nomenclature for single-lump heat-capacity analysis.

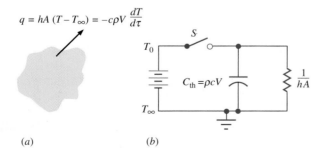

$$q = hA(T - T_\infty) = -c\rho V \frac{dT}{d\tau}$$

(a) $\qquad\qquad$ (b)

where T_∞ is the temperature of the convection environment. The thermal network for the single-capacity system is shown in Figure 4-2b. In this network we notice that the thermal capacity of the system is "charged" initially at the potential T_0 by closing the switch S. Then, when the switch is opened, the energy stored in the thermal capacitance is dissipated through the resistance $1/hA$. The analogy between this thermal system and an electric system is apparent, and we could easily construct an electric system that would behave exactly like the thermal system as long as we made the ratio

$$\frac{hA}{\rho c V} = \frac{1}{R_{\text{th}} C_{\text{th}}} \qquad R_{\text{th}} = \frac{1}{hA} \qquad C_{\text{th}} = \rho c V$$

equal to $1/R_e C_e$, where R_e and C_e are the electric resistance and capacitance, respectively. In the thermal system we store energy, while in the electric system we store electric charge. The flow of energy in the thermal system is called heat, and the flow of charge is called electric current. The quantity $c\rho V/hA$ is called the *time constant* of the system because it has the dimensions of time. When

$$\tau = \frac{c\rho V}{hA}$$

it is noted that the temperature difference $T - T_\infty$ has a value of 36.8 percent of the initial difference $T_0 - T_\infty$.

The reader should note that the lumped-capacity formulation assumes essentially uniform temperature throughout the solid at any instant of time so that the change in internal energy can be represented by $\rho c V dT/d\tau$. It does *not* require that the convection boundary condition have a constant value of h. In fact, variable values of h coupled with radiation boundary conditions are quite common. The specification of "time constant" in terms of the 36.8 percent value stated above implies a constant boundary condition.

For variable convection or radiation boundary conditions, numerical methods (see Section 4-6) are used to advantage to predict lumped capacity behavior. A rather general setup of a lumped-capacity solution using numerical methods and Microsoft Excel is given in Section D-6 of the Appendix. In some cases, multiple lumped-capacity formulations can be useful. An example involving the combined convection-radiation cooling of a box of electronic components is also given in this same section of the Appendix.

Applicability of Lumped-Capacity Analysis

We have already noted that the lumped-capacity type of analysis assumes a uniform temperature distribution throughout the solid body and that the assumption is equivalent to saying that the surface-convection resistance is large compared with the internal-conduction resistance. Such an analysis may be expected to yield reasonable estimates within about 5 percent when the following condition is met:

$$\frac{h(V/A)}{k} < 0.1 \qquad\qquad\qquad \textbf{[4-6]}$$

where k is the thermal conductivity of the solid. In sections that follow, we examine those situations for which this condition does *not* apply. We shall see that the lumped-capacity analysis has a direct relationship to the numerical methods discussed in Section 4-7. If one considers the ratio $V/A = s$ as a characteristic dimension of the solid, the dimensionless group in Equation (4-6) is called the *Biot number*:

$$\frac{hs}{k} = \text{Biot number} = \text{Bi}$$

The reader should recognize that there are many practical cases where the lumped-capacity method may yield good results. In Table 4-1 we give some examples that illustrate the relative validity of such cases.

Table 4-1 | Examples of lumped-capacity systems.

Physical situation	k, W/m·°C	Approximate value of h, W/m²·°C	$\dfrac{h(V/A)}{k}$
1. 3.0-cm steel cube cooling in room air	40	7.0	8.75×10^{-4}
2. 5.0-cm glass cylinder cooled by a 50-m/s airstream	0.8	180	2.81
3. Same as situation 2 but a copper cylinder	380	180	0.006
4. 3.0-cm hot copper cube submerged in water such that boiling occurs	380	10,000	0.132

We may point out that uncertainties in the knowledge of the convection coefficient of ±25 percent are quite common, so that the condition $\text{Bi} = h(V/A)/k < 0.1$ should allow for some leeway in application.

Do not dismiss lumped-capacity analysis because of its simplicity. Because of uncertainties in the convection coefficient, it may not be necessary to use more elaborate analysis techniques.

Steel Ball Cooling in Air | EXAMPLE 4-1

A steel ball [$c = 0.46$ kJ/kg·°C, $k = 35$ W/m·°C] 5.0 cm in diameter and initially at a uniform temperature of 450°C is suddenly placed in a controlled environment in which the temperature is maintained at 100°C. The convection heat-transfer coefficient is 10 W/m²·°C. Calculate the time required for the ball to attain a temperature of 150°C.

■ **Solution**
We anticipate that the lumped-capacity method will apply because of the low value of h and high value of k. We can check by using Equation (4-6):

$$\frac{h(V/A)}{k} = \frac{(10)[(4/3)\pi(0.025)^3]}{4\pi(0.025)^2(35)} = 0.0023 < 0.1$$

so we may use Equation (4-5). We have

$$T = 150°C \quad \rho = 7800 \text{ kg/m}^3 \quad [486 \text{ lb}_m/\text{ft}^3]$$
$$T_\infty = 100°C \quad h = 10 \text{ W/m}^2\cdot°C \quad [1.76 \text{Btu/h}\cdot\text{ft}^2\cdot°F]$$
$$T_0 = 450°C \quad c = 460 \text{ J/kg}\cdot°C \quad [0.11 \text{ Btu/lb}_m\cdot°F]$$

$$\frac{hA}{\rho cV} = \frac{(10)4\pi(0.025)^2}{(7800)(460)(4\pi/3)(0.025)^3} = 3.344 \times 10^{-4} \text{ s}^{-1}$$

$$\frac{T - T_\infty}{T_0 - T_\infty} = e^{-[hA/\rho cV]\tau}$$
$$\frac{150 - 100}{450 - 100} = e^{-3.344\times10^{-4}\tau}$$
$$\tau = 5819 \text{ s} = 1.62 \text{ h}$$

4-3 | TRANSIENT HEAT FLOW IN A SEMI-INFINITE SOLID

Consider the semi-infinite solid shown in Figure 4-3 maintained at some initial temperature T_i. The surface temperature is suddenly lowered and maintained at a temperature T_0, and we

Figure 4-3 | Nomenclature for transient heat flow in a semi-infinite solid.

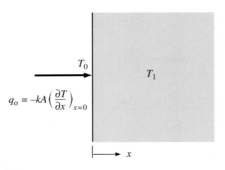

seek an expression for the temperature distribution in the solid as a function of time. This temperature distribution may subsequently be used to calculate heat flow at any x position in the solid as a function of time. For constant properties, the differential equation for the temperature distribution $T(x, \tau)$ is

$$\frac{\partial^2 T}{\partial x^2} = \frac{1}{\alpha}\frac{\partial T}{\partial \tau}$$ [4-7]

The boundary and initial conditions are

$$T(x, 0) = T_i$$
$$T(0, \tau) = T_0 \qquad \text{for } \tau > 0$$

This is a problem that may be solved by the Laplace-transform technique. The solution is given in Reference 1 as

$$\frac{T(x, \tau) - T_0}{T_i - T_0} = \text{erf}\,\frac{x}{2\sqrt{\alpha\tau}}$$ [4-8]

where the Gauss error function is defined as

$$\text{erf}\,\frac{x}{2\sqrt{\alpha\tau}} = \frac{2}{\sqrt{\pi}}\int_0^{x/2\sqrt{\alpha\tau}} e^{-\eta^2}\, d\eta$$ [4-9]

It will be noted that in this definition η is a dummy variable and the integral is a function of its upper limit. When the definition of the error function is inserted in Equation (4-8), the expression for the temperature distribution becomes

$$\frac{T(x, \tau) - T_0}{T_i - T_0} = \frac{2}{\sqrt{\pi}}\int_0^{x/2\sqrt{\alpha\tau}} e^{-\eta^2}\, d\eta$$ [4-10]

The heat flow at any x position may be obtained from

$$q_x = -kA\frac{\partial T}{\partial x}$$

Performing the partial differentiation of Equation (4-10) gives

$$\frac{\partial T}{\partial x} = (T_i - T_0)\frac{2}{\sqrt{\pi}}e^{-x^2/4\alpha\tau}\frac{\partial}{\partial x}\left(\frac{x}{2\sqrt{\alpha\tau}}\right)$$

$$= \frac{T_i - T_0}{\sqrt{\pi\alpha\tau}}e^{-x^2/4\alpha\tau}$$ [4-11]

Figure 4-4 | Response of semi-infinite solid to (*a*) sudden change in surface temperature and (*b*) instantaneous surface pulse of Q_0/A J/m^2.

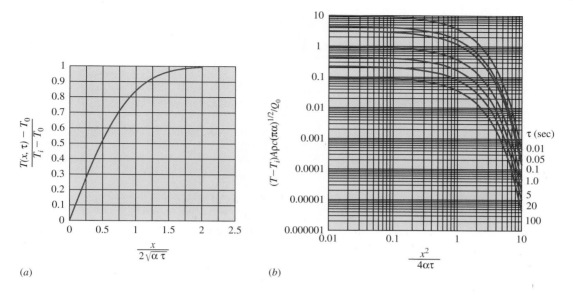

(*a*)

(*b*)

At the surface ($x = 0$) the heat flow is

$$q_0 = \frac{kA(T_0 - T_i)}{\sqrt{\pi \alpha \tau}} \qquad\qquad \textbf{[4-12]}$$

The surface heat flux is determined by evaluating the temperature gradient at $x = 0$ from Equation (4-11). A plot of the temperature distribution for the semi-infinite solid is given in Figure 4-4. Values of the error function are tabulated in Reference 3, and an abbreviated tabulation is given in Appendix A.

Constant Heat Flux on Semi-Infinite Solid

For the same uniform initial temperature distribution, we could suddenly expose the surface to a constant surface heat flux q_0/A. The initial and boundary conditions on Equation (4-7) would then become

$$T(x, 0) = T_i$$

$$\frac{q_0}{A} = -k \left. \frac{\partial T}{\partial x} \right]_{x=0} \qquad \text{for } \tau > 0$$

The solution for this case is

$$T - T_i = \frac{2q_0\sqrt{\alpha\tau/\pi}}{kA} \exp\left(\frac{-x^2}{4\alpha\tau}\right) - \frac{q_0 x}{kA}\left(1 - \operatorname{erf}\frac{x}{2\sqrt{\alpha\tau}}\right) \qquad \textbf{[4-13a]}$$

Energy Pulse at Surface

Equation (4-13*a*) presents the temperature response that results from a surface heat flux that remains constant with time. A related boundary condition is that of a short, instantaneous pulse of energy at the surface having a magnitude of Q_0/A. The resulting temperature response is given by

$$T - T_i = [Q_0/A\rho c(\pi\alpha\tau)^{1/2}] \exp(-x^2/4\alpha\tau) \qquad \textbf{[4-13b]}$$

In contrast to the constant-heat-flux case where the temperature increases indefinitely for all x and times, the temperature response to the instantaneous surface pulse will die out with time, or

$$T - T_i \to 0 \text{ for all } x \text{ as } \tau \to \infty$$

This rapid exponential decay behavior is illustrated in Figure 4-4b.

EXAMPLE 4-2   Semi-Infinite Solid with Sudden Change in Surface Conditions

A large block of steel [$k = 45$ W/m $\cdot$ °C, $\alpha = 1.4 \times 10^{-5}$ m^2/s] is initially at a uniform temperature of 35°C. The surface is exposed to a heat flux (a) by suddenly raising the surface temperature to 250°C and (b) through a constant surface heat flux of 3.2×10^5 W/m^2. Calculate the temperature at a depth of 2.5 cm after a time of 0.5 min for both these cases.

■ **Solution**
We can make use of the solutions for the semi-infinite solid given as Equations (4-8) and (4-13a). For case a,

$$\frac{x}{2\sqrt{\alpha\tau}} = \frac{0.025}{(2)[(1.4 \times 10^{-5})(30)]^{1/2}} = 0.61$$

The error function is determined from Appendix A as

$$\text{erf} \frac{x}{2\sqrt{\alpha\tau}} = \text{erf } 0.61 = 0.61164$$

We have $T_i = 35$°C and $T_0 = 250$°C, so the temperature at $x = 2.5$ cm is determined from Equation (4-8) as

$$T(x, \tau) = T_0 + (T_i - T_0) \text{erf} \frac{x}{2\sqrt{\alpha\tau}}$$
$$= 250 + (35 - 250)(0.61164) = 118.5\text{°C}$$

For the constant-heat-flux case b, we make use of Equation (4-13a). Since q_0/A is given as 3.2×10^5 W/m^2, we can insert the numerical values to give

$$T(x, \tau) = 35 + \frac{(2)(3.2 \times 10^5)[(1.4 \times 10^{-5})(30)/\pi]^{1/2}}{45} e^{-(0.61)^2}$$
$$- \frac{(0.025)(3.2 \times 10^5)}{45}(1 - 0.61164)$$
$$= 79.3\text{°C} \qquad x = 2.5 \text{ cm}, \tau = 30 \text{ s}$$

For the constant-heat-flux case the *surface* temperature after 30 s would be evaluated with $x = 0$ in Equation (4-13a). Thus,

$$T(x = 0) = 35 + \frac{(2)(3.2 \times 10^5)[(1.4 \times 10^{-5})(30)/\pi]^{1/2}}{45} = 199.4\text{°C}$$

EXAMPLE 4-3   Pulsed Energy at Surface of Semi-Infinite Solid

An instantaneous laser pulse of 10 MJ/m^2 is imposed on a slab of stainless steel having properties of $\rho = 7800$ kg/m^3, $c = 460$ J/kg $\cdot$ °C, and $\alpha = 0.44 \times 10^{-5}$ m^2/s. The slab is initially at a uniform temperature of 40 °C. Estimate the temperature at the surface and at a depth of 2.0 mm after a time of 2 s.

■ **Solution**

This problem is a direct application of Equation (4-13b). We have $Q_0/A = 10^7$ J/m^2 and at $x = 0$

$$T_0 - T_i = Q_0/A\rho c(\pi\alpha\tau)^{1.2}$$
$$= 10^7/(7800)(460)[\pi(0.44 \times 10^{-5})(2)]^{0.5} = 530°C$$

and

$$T_0 = 40 + 530 = 570°C$$
$$\text{At } x = 2.0\,\text{mm} = 0.002\,\text{m},$$

$$T - T_i = (530)\exp[-(0.002)^2/(4)(0.44 \times 10^{-5})(2)] = 473°C$$

and

$$T = 40 + 473 = 513°C$$

Heat Removal from Semi-Infinite Solid

EXAMPLE 4-4

A large slab of aluminum at a uniform temperature of 200°C suddenly has its surface temperature lowered to 70°C. What is the total heat removed from the slab per unit surface area when the temperature at a depth 4.0 cm has dropped to 120°C?

■ **Solution**

We first find the time required to attain the 120°C temperature and then integrate Equation (4-12) to find the total heat removed during this time interval. For aluminum,

$$\alpha = 8.4 \times 10^{-5}\,\text{m}^2/\text{s} \qquad k = 215\,\text{W/m} \cdot °C\,[124\,\text{Btu/h} \cdot \text{ft} \cdot °F]$$

We also have

$$T_i = 200°C \qquad T_0 = 70°C \qquad T(x, \tau) = 120°C$$

Using Equation (4-8) gives

$$\frac{120 - 70}{200 - 70} = \text{erf}\,\frac{x}{2\sqrt{\alpha\tau}} = 0.3847$$

From Figure 4-4 or Appendix A,

$$\frac{x}{2\sqrt{\alpha\tau}} = 0.3553$$

and

$$\tau = \frac{(0.04)^2}{(4)(0.3553)^2(8.4 \times 10^{-5})} = 37.72\,\text{s}$$

The total heat removed at the surface is obtained by integrating Equation (4-12):

$$\frac{Q_0}{A} = \int_0^\tau \frac{q_0}{A}\,d\tau = \int_0^\tau \frac{k(T_0 - T_i)}{\sqrt{\pi\alpha\tau}}\,d\tau = 2k(T_0 - T_i)\sqrt{\frac{\tau}{\pi\alpha}}$$

$$= (2)(215)(70 - 200)\left[\frac{37.72}{\pi(8.4 \times 10^{-5})}\right]^{1/2} = -21.13 \times 10^6\,\text{J/m}^2 \quad [-1861\,\text{Btu/ft}^2]$$

4-4 | CONVECTION BOUNDARY CONDITIONS

In most practical situations the transient heat-conduction problem is connected with a convection boundary condition at the surface of the solid. Naturally, the boundary conditions for the differential equation must be modified to take into account this convection heat transfer at the surface. For the semi-infinite-solid problem, the convection boundary condition would be expressed by

Heat convected into surface = heat conducted into surface

or

$$hA(T_\infty - T)_{x=0} = -kA \left. \frac{\partial T}{\partial x} \right]_{x=0}$$ [4-14]

The solution for this problem is rather involved and is worked out in detail by Schneider [1]. The result is

$$\frac{T - T_i}{T_\infty - T_i} = 1 - \mathrm{erf}\, X - \left[\exp\left(\frac{hx}{k} + \frac{h^2 \alpha \tau}{k^2} \right) \right] \times \left[1 - \mathrm{erf}\left(X + \frac{h\sqrt{\alpha \tau}}{k} \right) \right]$$ [4-15]

where

$X = x/(2\sqrt{\alpha \tau})$
T_i = initial temperature of solid
T_∞ = environment temperature

This solution is presented in graphical form in Figure 4-5.

Solutions have been worked out for other geometries. The most important cases are those dealing with (1) plates whose thickness is small in relation to the other dimensions, (2) cylinders where the diameter is small compared to the length, and (3) spheres. Results of analyses for these geometries have been presented in graphical form by Heisler [2], and nomenclature for the three cases is illustrated in Figure 4-6. In all cases the convection environment temperature is designated as T_∞ and the center temperature for $x = 0$ or $r = 0$ is T_0. At time zero, each solid is assumed to have a uniform initial temperature T_i. Temperatures in the solids are given in Figures 4-7 to 4-13 as functions of time and spatial position. In these charts we note the definitions

$$\theta = T(x, \tau) - T_\infty \qquad \text{or} \qquad T(r, \tau) - T_\infty$$
$$\theta_i = T_i - T_\infty$$
$$\theta_0 = T_0 - T_\infty$$

If a centerline temperature is desired, only one chart is required to obtain a value for θ_0 and then T_0. To determine an off-center temperature, two charts are required to calculate the product

$$\frac{\theta}{\theta_i} = \frac{\theta_0}{\theta_i} \frac{\theta}{\theta_0}$$

For example, Figures 4-7 and 4-10 would be employed to calculate an off-center temperature for an infinite plate.

The heat losses for the infinite plate, infinite cylinder, and sphere are given in Figures 4-14 to 4-16, where Q_0 represents the initial internal energy content of the body in reference to the environment temperature

$$Q_0 = \rho c V(T_i - T_\infty) = \rho c V \theta_i$$ [4-16]

In these figures Q is the actual heat lost by the body in time τ.

Figure 4-5 | Temperature distribution in the semi-infinite solid with convection boundary condition.

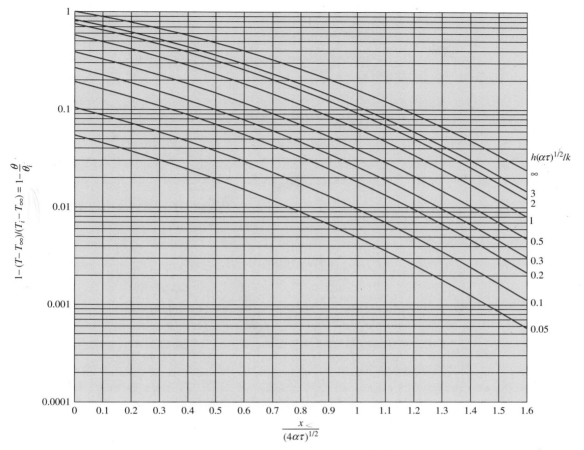

Figure 4-6 | Nomenclature for one-dimensional solids suddenly subjected to convection environment at T_∞: (a) infinite plate of thickness $2L$; (b) infinite cylinder of radius r_0; (c) sphere of radius r_0.

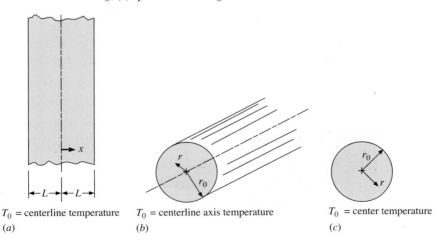

T_0 = centerline temperature
(a)

T_0 = centerline axis temperature
(b)

T_0 = center temperature
(c)

Figure 4-7. Midplane temperature for an infinite plate of thickness 2L: (a) full scale.

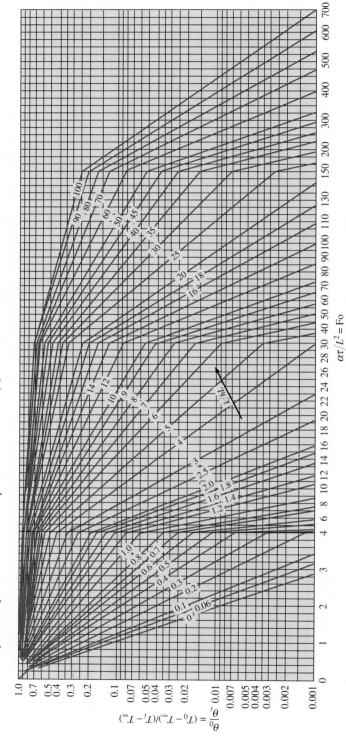

(a)

Figure 4-7 | (*Continued*). (*b*) expanded scale for 0 < Fo < 4, from Reference 2.

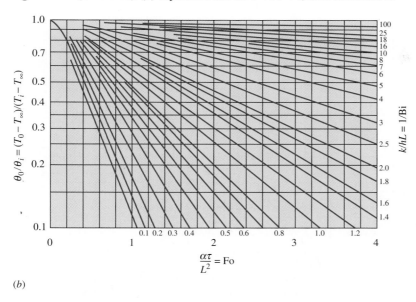

(*b*)

If one considers the solid as behaving as a lumped capacity during the cooling or heating process, that is, small internal resistance compared to surface resistance, the exponential cooling curve of Figure 4-5 may be replotted in expanded form, as shown in Figure 4-13 using the Biot-Fourier product as the abscissa. We note that the following parameters apply for the bodies considered in the Heisler charts.

$$(A/V)_{\text{inf plate}} = 1/L$$
$$(A/V)_{\text{inf cylinder}} = 2/r_0$$
$$(A/V)_{\text{sphere}} = 3/r_0$$

Obviously, there are many other practical heating and cooling problems of interest. The solutions for a large number of cases are presented in graphical form by Schneider [7], and readers interested in such calculations will find this reference to be of great utility.

The Biot and Fourier Numbers

A quick inspection of Figures 4-5 to 4-16 indicates that the dimensionless temperature profiles and heat flows may all be expressed in terms of two dimensionless parameters called the Biot and Fourier numbers:

$$\text{Biot number} = \text{Bi} = \frac{hs}{k}$$
$$\text{Fourier number} = \text{Fo} = \frac{\alpha\tau}{s^2} = \frac{k\tau}{\rho cs^2}$$

In these parameters s designates a characteristic dimension of the body; for the plate it is the half-thickness, whereas for the cylinder and sphere it is the radius. The Biot number compares the relative magnitudes of surface-convection and internal-conduction resistances

Figure 4-8 | Axis temperature for an infinite cylinder of radius r_0: (*a*) full scale.

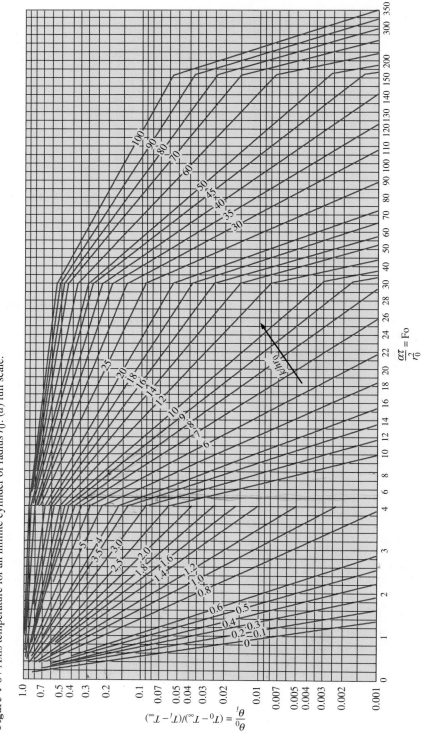

(*a*)

152

Figure 4-8 | (*Continued*). (*b*) expanded scale for $0 < \text{Fo} < 4$, from Reference 2.

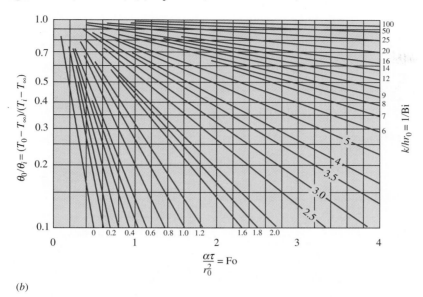

(*b*)

to heat transfer. The Fourier modulus compares a characteristic body dimension with an approximate temperature-wave penetration depth for a given time τ.

A very low value of the Biot modulus means that internal-conduction resistance is negligible in comparison with surface-convection resistance. This in turn implies that the temperature will be nearly uniform throughout the solid, and its behavior may be approximated by the lumped-capacity method of analysis. It is interesting to note that the exponent of Equation (4-5) may be expressed in terms of the Biot and Fourier numbers if one takes the ratio V/A as the characteristic dimension s. Then,

$$\frac{hA}{\rho c V}\tau = \frac{h\tau}{\rho c s} = \frac{hs}{k}\frac{k\tau}{\rho c s^2} = \text{Bi Fo}$$

Applicability of the Heisler Charts

The calculations for the Heisler charts were performed by truncating the infinite series solutions for the problems into a few terms. This restricts the applicability of the charts to values of the Fourier number greater than 0.2.

$$\text{Fo} = \frac{\alpha\tau}{s^2} > 0.2$$

For smaller values of this parameter the reader should consult the solutions and charts given in the references at the end of the chapter. Calculations using the truncated series solutions directly are discussed in Appendix C.

Figure 4-9 | Center temperature for a sphere of radius r_0: (a) full scale.

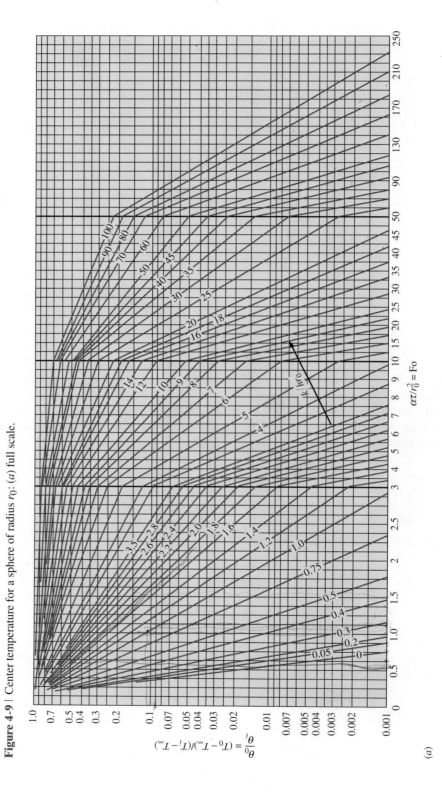

(a)

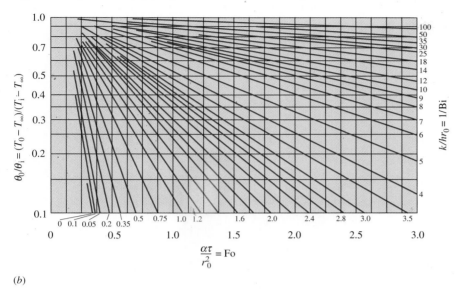

(*b*)

Figure 4-10 | Temperature as a function of center temperature in an infinite plate of thickness 2L, from Reference 2.

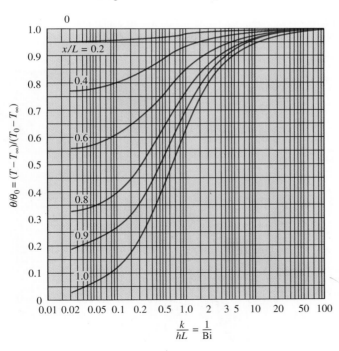

Figure 4-11 | Temperature as a function of axis temperature in an infinite cylinder of radius r_0, from Reference 2.

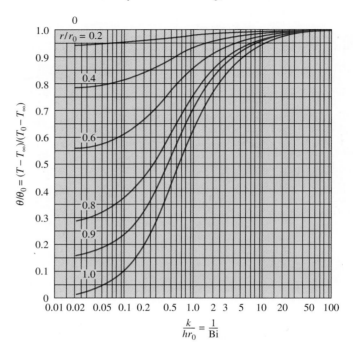

Figure 4-12 | Temperature as a function of center temperature for a sphere of radius r_0, from Reference 2.

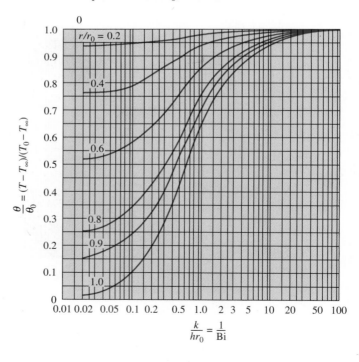

Figure 4-13 | Temperature variation with time for solids that may be treated as lumped capacities: (*a*) $0 < \text{BiFo} < 10$, (*b*) $0.1 < \text{BiFo} < 1.0$, (*c*) $0 < \text{BiFo} < 0.1$. Note: $(A/V)_{\text{inf plate}} = 1/L$, $(A/V)_{\text{inf cyl}} = 2/r_0$, $(A/V)_{\text{sphere}} = 3/r_0$. See Equations (4-5) and (4-6).

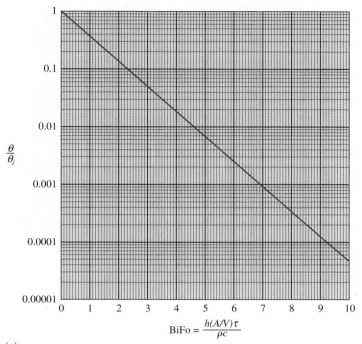

$$\text{BiFo} = \frac{h(A/V)\tau}{\rho c}$$

(*a*)

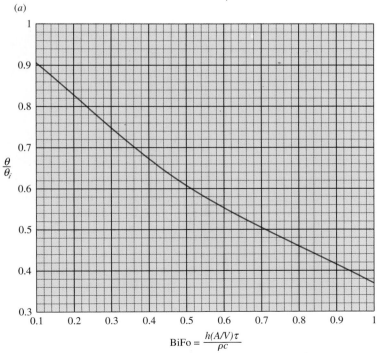

$$\text{BiFo} = \frac{h(A/V)\tau}{\rho c}$$

(*b*)

Figure 4-13 | (*Continued*).

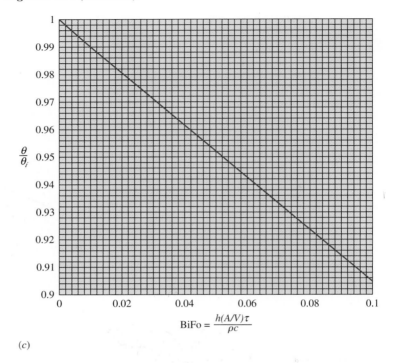

$$\text{BiFo} = \frac{h(A/V)\tau}{\rho c}$$

(*c*)

Figure 4-14 | Dimensionless heat loss Q/Q_0 of an infinite plane of thickness $2L$ with time, from Reference 6.

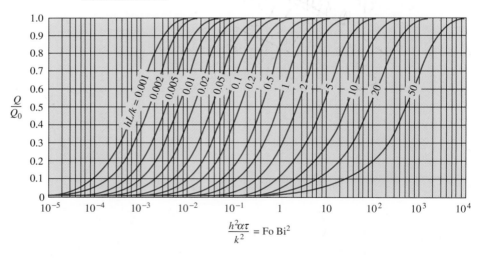

$$\frac{h^2\alpha\tau}{k^2} = \text{Fo Bi}^2$$

Figure 4-15 | Dimensionlesss heat loss Q/Q_0 of an infinite cylinder of radius r_0 with time, from Reference 6.

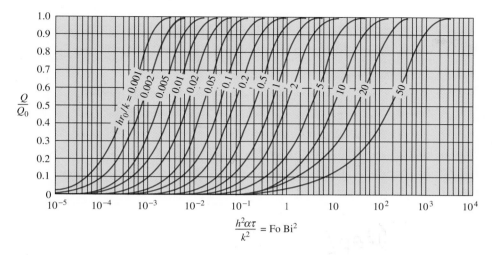

Figure 4-16 | Dimensionless heat loss Q/Q_0 of a sphere of radius r_0 with time, from Reference 6.

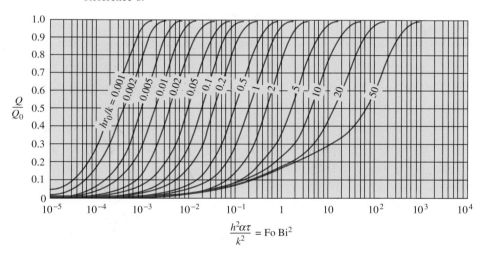

Sudden Exposure of Semi-Infinite
Slab to Convection

EXAMPLE 4-5

The slab of Example 4-4 is suddenly exposed to a convection-surface environment of 70°C with a heat-transfer coefficient of 525 W/m² · °C. Calculate the time required for the temperature to reach 120°C at the depth of 4.0 cm for this circumstance.

■ **Solution**
We may use either Equation (4-15) or Figure 4-5 for solution of this problem, but Figure 4-5 is easier to apply because the time appears in two terms. Even when the figure is used, an iterative procedure is required because the time appears in both of the variables $h\sqrt{\alpha\tau}/k$ and $x/(2\sqrt{\alpha\tau})$.

We seek the value of τ such that

$$\frac{T - T_i}{T_\infty - T_i} = \frac{120 - 200}{70 - 200} = 0.615 \qquad\qquad [a]$$

We therefore try values of τ and obtain readings of the temperature ratio from Figure 4-5 until agreement with Equation (a) is reached. The iterations are listed below. Values of k and α are obtained from Example 4-4.

τ, s	$\dfrac{h\sqrt{\alpha\tau}}{k}$	$\dfrac{x}{2\sqrt{\alpha\tau}}$	$\dfrac{T - T_i}{T_\infty - T_i}$ from Figure 4-5
1000	0.708	0.069	0.41
3000	1.226	0.040	0.61
4000	1.416	0.035	0.68

Consequently, the time required is approximately 3000 s.

EXAMPLE 4-6   Aluminum Plate Suddenly Exposed to Convection

A large plate of aluminum 5.0 cm thick and initially at 200°C is suddenly exposed to the convection environment of Example 4-5. Calculate the temperature at a depth of 1.25 cm from one of the faces 1 min after the plate has been exposed to the environment. How much energy has been removed per unit area from the plate in this time?

■ **Solution**
The Heisler charts of Figures 4-7 and 4-10 may be used for solution of this problem. We first calculate the center temperature of the plate, using Figure 4-7, and then use Figure 4-10 to calculate the temperature at the specified x position. From the conditions of the problem we have

$$\theta_i = T_i - T_\infty = 200 - 70 = 130 \qquad \alpha = 8.4 \times 10^{-5} \text{ m}^2/\text{s} \quad [3.26 \text{ ft}^2/\text{h}]$$

$$2L = 5.0 \text{ cm} \qquad L = 2.5 \text{ cm} \qquad \tau = 1 \text{ min} = 60 \text{ s}$$

$$k = 215 \text{ W/m} \cdot °\text{C} \quad [124 \text{ Btu/h} \cdot \text{ft} \cdot °\text{F}]$$

$$h = 525 \text{ W/m}^2 \cdot °\text{C} \quad [92.5 \text{ Btu/h} \cdot \text{ft}^2 \cdot °\text{F}]$$

$$x = 2.5 - 1.25 = 1.25 \text{ cm}$$

Then

$$\frac{\alpha\tau}{L^2} = \frac{(8.4 \times 10^{-5})(60)}{(0.025)^2} = 8.064 \qquad \frac{k}{hL} = \frac{215}{(525)(0.025)} = 16.38$$

$$\frac{x}{L} = \frac{1.25}{2.5} = 0.5$$

From Figure 4-7

$$\frac{\theta_0}{\theta_i} = 0.61$$

$$\theta_0 = T_0 - T_\infty = (0.61)(130) = 79.3$$

From Figure 4-10 at $x/L = 0.5$,

$$\frac{\theta}{\theta_0} = 0.98$$

and

$$\theta = T - T_\infty = (0.98)(79.3) = 77.7$$

$$T = 77.7 + 70 = 147.7°\text{C}$$

We compute the energy lost by the slab by using Figure 4-14. For this calculation we require the following properties of aluminum:

$$\rho = 2700 \text{ kg/m}^3 \qquad c = 0.9 \text{ kJ/kg} \cdot {}^\circ\text{C}$$

For Figure 4-14 we need

$$\frac{h^2 \alpha \tau}{k^2} = \frac{(525)^2 (8.4 \times 10^{-5})(60)}{(215)^2} = 0.03 \qquad \frac{hL}{k} = \frac{(525)(0.025)}{215} = 0.061$$

From Figure 4-14

$$\frac{Q}{Q_0} = 0.41$$

For unit area

$$\frac{Q_0}{A} = \frac{\rho c V \theta_i}{A} = \rho c (2L) \theta_i$$
$$= (2700)(900)(0.05)(130)$$
$$= 15.8 \times 10^6 \text{ J/m}^2$$

so that the heat removed per unit surface area is

$$\frac{Q}{A} = (15.8 \times 10^6)(0.41) = 6.48 \times 10^6 \text{ J/m}^2 \quad [571 \text{ Btu/ft}^2]$$

Long Cylinder Suddenly Exposed to Convection   EXAMPLE 4-7

A long aluminum cylinder 5.0 cm in diameter and initially at 200°C is suddenly exposed to a convection environment at 70°C and $h = 525$ W/m$^2 \cdot$ °C. Calculate the temperature at a radius of 1.25 cm and the heat lost per unit length 1 min after the cylinder is exposed to the environment.

■ Solution

This problem is like Example 4-6 except that Figures 4-8 and 4-11 are employed for the solution. We have

$$\theta_i = T_i - T_\infty = 200 - 70 = 130 \qquad \alpha = 8.4 \times 10^{-5} \text{ m}^2/\text{s}$$
$$r_0 = 2.5 \text{ cm} \qquad \tau = 1 \text{ min} = 60 \text{ s}$$
$$k = 215 \text{ W/m} \cdot {}^\circ\text{C} \qquad h = 525 \text{ W/m}^2 \cdot {}^\circ\text{C} \qquad r = 1.25 \text{ cm}$$
$$\rho = 2700 \text{ kg/m}^3 \qquad c = 0.9 \text{ kJ/kg} \cdot {}^\circ\text{C}$$

We compute

$$\frac{\alpha \tau}{r_0^2} = \frac{(8.4 \times 10^{-5})(60)}{(0.025)^2} = 8.064 \qquad \frac{k}{h r_0} = \frac{215}{(525)(0.025)} = 16.38$$
$$\frac{r}{r_0} = \frac{1.25}{2.5} = 0.5$$

From Figure 4-8

$$\frac{\theta_0}{\theta_i} = 0.38$$

and from Figures 4-11 at $r/r_0 = 0.5$

$$\frac{\theta}{\theta_0} = 0.98$$

so that

$$\frac{\theta}{\theta_i} = \frac{\theta_0}{\theta_i}\frac{\theta}{\theta_0} = (0.38)(0.98) = 0.372$$

and

$$\theta = T - T_\infty = (0.372)(130) = 48.4$$

$$T = 70 + 48.4 = 118.4°C$$

To compute the heat lost, we determine

$$\frac{h^2\alpha\tau}{k^2} = \frac{(525)^2(8.4 \times 10^{-5})(60)}{(215)^2} = 0.03 \qquad \frac{hr_0}{k} = \frac{(525)(0.025)}{215} = 0.061$$

Then from Figure 4-15

$$\frac{Q}{Q_0} = 0.65$$

For unit length

$$\frac{Q_0}{L} = \frac{\rho c V \theta_i}{L} = \rho c \pi r_0^2 \theta_i = (2700)(900)\pi(0.025)^2(130) = 6.203 \times 10^5 \text{ J/m}$$

and the actual heat lost per unit length is

$$\frac{Q}{L} = (6.203 \times 10^5)(0.65) = 4.032 \times 10^5 \text{ J/m} \quad [116.5 \text{ Btu/ft}]$$

4-5 | MULTIDIMENSIONAL SYSTEMS

The Heisler charts discussed in Section 4-4 may be used to obtain the temperature distribution in the infinite plate of thickness $2L$, in the long cylinder, or in the sphere. When a wall whose height and depth dimensions are not large compared with the thickness or a cylinder whose length is not large compared with its diameter is encountered, additional space coordinates are necessary to specify the temperature, the charts no longer apply, and we are forced to seek another method of solution. Fortunately, it is possible to combine the solutions for the one-dimensional systems in a very straightforward way to obtain solutions for the multidimensional problems.

It is clear that the infinite rectangular bar in Figure 4-17 can be formed from two infinite plates of thickness $2L_1$ and $2L_2$, respectively. The differential equation governing this situation would be

$$\frac{\partial^2 T}{\partial x^2} + \frac{\partial^2 T}{\partial z^2} = \frac{1}{\alpha}\frac{\partial T}{\partial \tau} \tag{4-17}$$

and to use the separation-of-variables method to effect a solution, we should assume a product solution of the form

$$T(x, z, \tau) = X(x)Z(z)\Theta(\tau)$$

It can be shown that the dimensionless temperature distribution may be expressed as a product of the solutions for two plate problems of thickness $2L_1$ and $2L_2$, respectively:

$$\left(\frac{T - T_\infty}{T_i - T_\infty}\right)_{\text{bar}} = \left(\frac{T - T_\infty}{T_i - T_\infty}\right)_{2L_1 \text{ plate}} \left(\frac{T - T_\infty}{T_i - T_\infty}\right)_{2L_2 \text{ plate}} \tag{4-18}$$

where T_i is the initial temperature of the bar and T_∞ is the environment temperature.

Figure 4-17 | Infinite rectangular bar.

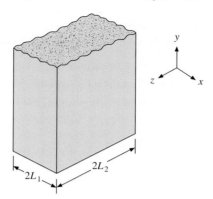

For two infinite plates the respective differential equations would be

$$\frac{\partial^2 T_1}{\partial x^2} = \frac{1}{\alpha}\frac{\partial T_1}{\partial \tau} \qquad \frac{\partial^2 T_2}{\partial z^2} = \frac{1}{\alpha}\frac{\partial T_2}{\partial \tau} \qquad\qquad \textbf{[4-19]}$$

and the product solutions assumed would be

$$T_1 = T_1(x,\tau) \qquad T_2 = T_2(z,\tau) \qquad\qquad \textbf{[4-20]}$$

We shall now show that the product solution to Equation (4-17) can be formed from a simple product of the functions (T_1, T_2), that is,

$$T(x,z,\tau) = T_1(x,\tau)T_2(z,\tau) \qquad\qquad \textbf{[4-21]}$$

The appropriate derivatives for substitution in Equation (4-17) are obtained from Equation (4-21) as

$$\frac{\partial^2 T}{\partial x^2} = T_2\frac{\partial^2 T_1}{\partial x^2} \qquad \frac{\partial^2 T}{\partial z^2} = T_1\frac{\partial^2 T_2}{\partial z^2}$$

$$\frac{\partial T}{\partial \tau} = T_1\frac{\partial T_2}{\partial \tau} + T_2\frac{\partial T_1}{\partial \tau}$$

Using Equations (4-19), we have

$$\frac{\partial T}{\partial \tau} = \alpha T_1\frac{\partial^2 T_2}{\partial z^2} + \alpha T_2\frac{\partial^2 T_1}{\partial x^2}$$

Substituting these relations in Equation (4-17) gives

$$T_2\frac{\partial^2 T_1}{\partial x_2} + T_1\frac{\partial^2 T_2}{\partial z^2} = \frac{1}{\alpha}\left(\alpha T_1\frac{\partial^2 T_2}{\partial z^2} + \alpha T_2\frac{\partial^2 T_1}{\partial x^2}\right)$$

or the assumed product solution of Equation (4-21) does indeed satisfy the original differential equation (4-17). This means that the dimensionless temperature distribution for the infinite rectangular bar may be expressed as a product of the solutions for two plate problems of thickness $2L_1$ and $2L_2$, respectively, as indicated by Equation (4-18).

In a manner similar to that described above, the solution for a three-dimensional block may be expressed as a product of three infinite-plate solutions for plates having the thickness of the three sides of the block. Similarly, a solution for a cylinder of finite length could be expressed as a product of solutions of the infinite cylinder and an infinite plate having a thickness equal to the length of the cylinder. Combinations could also be made with the infinite-cylinder and infinite-plate solutions to obtain temperature distributions in semi-infinite bars and cylinders. Some of the combinations are summarized in Figure 4-18, where

$$C(\Theta) = \text{solution for infinite cylinder}$$

$$P(X) = \text{solution for infinite plate}$$

$$S(X) = \text{solution for semi-infinite solid}$$

Figure 4-18 | Product solutions for temperatures in multidimensional systems: (*a*) semi-infinite plate; (*b*) infinite rectangular bar; (*c*) semi-infinite rectangular bar; (*d*) rectangular parallelepiped; (*e*) semi-infinite cylinder; (*f*) short cylinder.

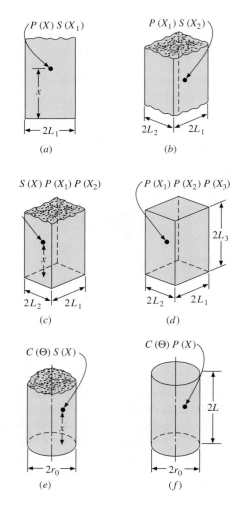

CHAPTER 4 Unsteady-State Conduction
165

The general idea is then

$$\left(\frac{\theta}{\theta_i}\right)_{\substack{\text{combined}\\\text{solid}}} = \left(\frac{\theta}{\theta_i}\right)_{\substack{\text{intersection}\\\text{solid 1}}} \left(\frac{\theta}{\theta_i}\right)_{\substack{\text{intersection}\\\text{solid 2}}} \left(\frac{\theta}{\theta_i}\right)_{\substack{\text{intersection}\\\text{solid 3}}}$$

Heat Transfer in Multidimensional Systems

Langston [16] has shown that it is possible to superimpose the heat-loss solutions for one-dimensional bodies, as shown in Figures 4-14, 4-15, and 4-16, to obtain the heat for a multidimensional body. The results of this analysis for intersection of two bodies is

$$\left(\frac{Q}{Q_0}\right)_{\text{total}} = \left(\frac{Q}{Q_0}\right)_1 + \left(\frac{Q}{Q_0}\right)_2 \left[1 - \left(\frac{Q}{Q_0}\right)_1\right]$$ [4-22]

where the subscripts refer to the two intersecting bodies. For a multidimensional body formed by intersection of three one-dimensional systems, the heat loss is given by

$$\left(\frac{Q}{Q_0}\right)_{\text{total}} = \left(\frac{Q}{Q_0}\right)_1 + \left(\frac{Q}{Q_0}\right)_2 \left[1 - \left(\frac{Q}{Q_0}\right)_1\right] + \left(\frac{Q}{Q_0}\right)_3 \left[1 - \left(\frac{Q}{Q_0}\right)_1\right]\left[1 - \left(\frac{Q}{Q_0}\right)_2\right]$$ [4-23]

If the heat loss is desired after a given time, the calculation is straightforward. On the other hand, if the *time to achieve a certain heat loss* is the desired quantity, a trial-and-error or iterative procedure must be employed. The following examples illustrate the use of the various charts for calculating temperatures and heat flows in multidimensional systems.

Semi-Infinite Cylinder Suddenly Exposed to Convection | EXAMPLE 4-8

A semi-infinite aluminum cylinder 5 cm in diameter is initially at a uniform temperature of 200°C. It is suddenly subjected to a convection boundary condition at 70°C with $h = 525$ W/m²·°C. Calculate the temperatures at the axis and surface of the cylinder 10 cm from the end 1 min after exposure to the environment.

■ **Solution**
This problem requires a combination of solutions for the infinite cylinder and semi-infinite slab in accordance with Figure 4-18e. For the slab we have

$$x = 10 \text{ cm} \qquad \alpha = 8.4 \times 10^{-5} \text{ m}^2/\text{s} \qquad k = 215 \text{ W/m·°C}$$

so that the parameters for use with Figure 4-5 are

$$\frac{h\sqrt{\alpha\tau}}{k} = \frac{(525)[(8.4 \times 10^{-5})(60)]^{1/2}}{215} = 0.173$$

$$\frac{x}{2\sqrt{\alpha\tau}} = \frac{0.1}{(2)[(8.4 \times 10^{-5})(60)]^{1/2}} = 0.704$$

From Figure 4-5

$$\left(\frac{\theta}{\theta_i}\right)_{\text{semi-infinite slab}} = 1 - 0.036 = 0.964 = S(X)$$

For the infinite cylinder we seek both the axis- and surface-temperature ratios. The parameters for use with Figure 4-8 are

$$r_0 = 2.5 \text{ cm} \qquad \frac{k}{hr_0} = 16.38 \qquad \frac{\alpha\tau}{r_0^2} = 8.064 \qquad \frac{\theta_0}{\theta_i} = 0.38$$

This is the axis-temperature ratio. To find the surface-temperature ratio, we enter Figure 4-11, using

$$\frac{r}{r_0} = 1.0 \qquad \frac{\theta}{\theta_0} = 0.97$$

Thus

$$C(\Theta) = \left(\frac{\theta}{\theta_i}\right)_{\text{inf cyl}} = \begin{cases} 0.38 & \text{at } r = 0 \\ (0.38)(0.97) = 0.369 & \text{at } r = r_0 \end{cases}$$

Combining the solutions for the semi-infinite slab and infinite cylinder, we have

$$\left(\frac{\theta}{\theta_i}\right)_{\text{semi−infinite cylinder}} = C(\Theta)S(X)$$

$$= (0.38)(0.964) = 0.366 \quad \text{at } r = 0$$

$$= (0.369)(0.964) = 0.356 \quad \text{at } r = r_0$$

The corresponding temperatures are

$$T = 70 + (0.366)(200 - 70) = 117.6 \quad \text{at } r = 0$$

$$T = 70 + (0.356)(200 - 70) = 116.3 \quad \text{at } r = r_0$$

Finite-Length Cylinder Suddenly Exposed to Convection

EXAMPLE 4-9

A short aluminum cylinder 5.0 cm in diameter and 10.0 cm long is initially at a uniform temperature of 200°C. It is suddenly subjected to a convection environment at 70°C, and $h = 525$ W/m² · °C. Calculate the temperature at a radial position of 1.25 cm and a distance of 0.625 cm from one end of the cylinder 1 min after exposure to the environment.

■ **Solution**

To solve this problem we combine the solutions from the Heisler charts for an infinite cylinder and an infinite plate in accordance with the combination shown in Figure 4-18f. For the infinite-plate problem

$$L = 5 \text{ cm}$$

The x position is measured from the center of the plate so that

$$x = 5 - 0.625 = 4.375 \text{ cm} \qquad \frac{x}{L} = \frac{4.375}{5} = 0.875$$

For aluminum

$$\alpha = 8.4 \times 10^{-5} \text{ m}^2/\text{s} \qquad k = 215 \text{ W/m} \cdot °\text{C}$$

so

$$\frac{k}{hL} = \frac{215}{(525)(0.05)} = 8.19 \qquad \frac{\alpha\tau}{L^2} = \frac{(8.4 \times 10^{-5})(60)}{(0.05)^2} = 2.016$$

From Figures 4-7 and 4-10, respectively,

$$\frac{\theta_0}{\theta_i} = 0.75 \qquad \frac{\theta}{\theta_0} = 0.95$$

so that

$$\left(\frac{\theta}{\theta_i}\right)_{\text{plate}} = (0.75)(0.95) = 0.7125$$

For the cylinder $r_0 = 2.5$ cm

$$\frac{r}{r_0} = \frac{1.25}{2.5} = 0.5 \qquad \frac{k}{hr_0} = \frac{215}{(525)(0.025)} = 16.38$$

$$\frac{\alpha\tau}{r_0^2} = \frac{(8.4 \times 10^{-5})(60)}{(0.025)^2} = 8.064$$

and from Figures 4-8 and 4-11, respectively,

$$\frac{\theta_0}{\theta_i} = 0.38 \qquad \frac{\theta}{\theta_0} = 0.98$$

so that

$$\left(\frac{\theta}{\theta_i}\right)_{\text{cyl}} = (0.38)(0.98) = 0.3724$$

Combining the solutions for the plate and cylinder gives

$$\left(\frac{\theta}{\theta_i}\right)_{\text{short cylinder}} = (0.7125)(0.3724) = 0.265$$

Thus

$$T = T_\infty + (0.265)(T_i - T_\infty) = 70 + (0.265)(200 - 70) = 104.5°\text{C}$$

Heat Loss for Finite-Length Cylinder | EXAMPLE 4-10

Calculate the heat loss for the short cylinder in Example 4-9.

■ **Solution**
We first calculate the dimensionless heat-loss ratio for the infinite plate and infinite cylinder that make up the multidimensional body. For the plate we have $L = 5\text{ cm} = 0.05\text{ m}$. Using the properties of aluminum from Example 4-9, we calculate

$$\frac{hL}{k} = \frac{(525)(0.05)}{215} = 0.122$$

$$\frac{h^2\alpha\tau}{k^2} = \frac{(525)^2(8.4 \times 10^{-5})(60)}{(215)^2} = 0.03$$

From Figure 4-14, for the plate, we read

$$\left(\frac{Q}{Q_0}\right)_p = 0.22$$

For the cylinder $r_0 = 2.5\text{ cm} = 0.025\text{ m}$, so we calculate

$$\frac{hr_0}{k} = \frac{(525)(0.025)}{215} = 0.061$$

and from Figure 4-15 we can read

$$\left(\frac{Q}{Q_0}\right)_c = 0.55$$

The two heat ratios may be inserted in Equation (4-22) to give

$$\left(\frac{Q}{Q_0}\right)_{\text{tot}} = 0.22 + (0.55)(1 - 0.22) = 0.649$$

The specific heat of aluminum is $0.896\text{ kJ/kg} \cdot °\text{C}$ and the density is 2707 kg/m^3, so we calculate Q_0 as

$$Q_0 = \rho c V\theta_i = (2707)(0.896)\pi(0.025)^2(0.1)(200 - 70)$$
$$= 61.9\text{ kJ}$$

The actual heat loss in the 1-min time is thus

$$Q = (61.9 \text{ kJ})(0.649) = 40.2 \text{ kJ}$$

4-6 | TRANSIENT NUMERICAL METHOD

The charts described in Sections 4-4 and 4-5 are very useful for calculating temperatures in certain regular-shaped solids under transient heat-flow conditions. Unfortunately, many geometric shapes of practical interest do not fall into these categories; in addition, one is frequently faced with problems in which the boundary conditions vary with time. These transient boundary conditions as well as the geometric shape of the body can be such that a mathematical solution is not possible. In these cases, the problems are best handled by a numerical technique with computers. It is the setup for such calculations that we now describe. For ease in discussion we limit the analysis to two-dimensional systems. An extension to three dimensions can then be made very easily.

Consider a two-dimensional body divided into increments as shown in Figure 4-19. The subscript m denotes the x position, and the subscript n denotes the y position. Within the solid body the differential equation that governs the heat flow is

$$k\left(\frac{\partial^2 T}{\partial x^2} + \frac{\partial^2 T}{\partial y^2}\right) = \rho c \, \frac{\partial T}{\partial \tau} \qquad \text{[4-24]}$$

assuming constant properties. We recall from Chapter 3 that the second partial derivatives may be approximated by

$$\frac{\partial^2 T}{\partial x^2} \approx \frac{1}{(\Delta x)^2}(T_{m+1,n} + T_{m-1,n} - 2T_{m,n}) \qquad \text{[4-25]}$$

$$\frac{\partial^2 T}{\partial y^2} \approx \frac{1}{(\Delta y)^2}(T_{m,n+1} + T_{m,n-1} - 2T_{m,n}) \qquad \text{[4-26]}$$

Figure 4-19 | Nomenclature for numerical solution of two-dimensional unsteady-state conduction problem.

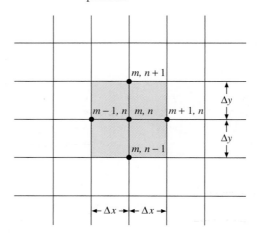

The time derivative in Equation (4-24) is approximated by

$$\frac{\partial T}{\partial \tau} \approx \frac{T_{m,n}^{p+1} - T_{m,n}^{p}}{\Delta \tau} \qquad \textbf{[4-27]}$$

In this relation the superscripts designate the time increment. Combining the relations above gives the difference equation equivalent to Equation (4-24)

$$\frac{T_{m+1,n}^{p} + T_{m-1,n}^{p} - 2T_{m,n}^{p}}{(\Delta x)^2} + \frac{T_{m,n+1}^{p} + T_{m,n-1}^{p} - 2T_{m,n}^{p}}{(\Delta y)^2} = \frac{1}{\alpha}\frac{T_{m,n}^{p+1} - T_{m,n}^{p}}{\Delta \tau} \qquad \textbf{[4-28]}$$

Thus, if the temperatures of the various nodes are known at any particular time, the temperatures after a time increment $\Delta \tau$ may be calculated by writing an equation like Equation (4-28) for each node and obtaining the values of $T_{m,n}^{p+1}$. The procedure may be repeated to obtain the distribution after any desired number of time increments. If the increments of space coordinates are chosen such that

$$\Delta x = \Delta y$$

the resulting equation for $T_{m,n}^{p+1}$ becomes

$$T_{m,n}^{p+1} = \frac{\alpha \Delta \tau}{(\Delta x)^2}\left(T_{m+1,n}^{p} + T_{m-1,n}^{p} + T_{m,n+1}^{p} + T_{m,n-1}^{p}\right) + \left[1 - \frac{4\alpha \Delta \tau}{(\Delta x)^2}\right]T_{m,n}^{p} \qquad \textbf{[4-29]}$$

If the time and distance increments are conveniently chosen so that

$$\frac{(\Delta x)^2}{\alpha \Delta \tau} = 4 \qquad \textbf{[4-30]}$$

it is seen that the temperature of node (m, n) after a time increment is simply the arithmetic average of the four surrounding nodal temperatures at the beginning of the time increment.

When a one-dimensional system is involved, the equation becomes

$$T_{m}^{p+1} = \frac{\alpha \Delta \tau}{(\Delta x)^2}\left(T_{m+1}^{p} + T_{m-1}^{p}\right) + \left[1 - \frac{2\alpha \Delta \tau}{(\Delta x)^2}\right]T_{m}^{p} \qquad \textbf{[4-31]}$$

and if the time and distance increments are chosen so that

$$\frac{(\Delta x)^2}{\alpha \Delta \tau} = 2 \qquad \textbf{[4-32]}$$

the temperature of node m after the time increment is given as the arithmetic average of the two adjacent nodal temperatures at the beginning of the time increment.

Some general remarks concerning the use of numerical methods for solution of transient conduction problems are in order at this point. We have already noted that the selection of the value of the parameter

$$M = \frac{(\Delta x)^2}{\alpha \Delta \tau}$$

governs the ease with which we may proceed to effect the numerical solution; the choice of a value of 4 for a two-dimensional system or a value of 2 for a one-dimensional system makes the calculation particularly easy.

Once the distance increments and the value of M are established, the time increment is fixed, and we may not alter it without changing the value of either Δx or M, or both. Clearly, the larger the values of Δx and $\Delta \tau$, the more rapidly our solution will proceed. On the other hand, the smaller the value of these increments in the independent variables, the more accuracy will be obtained. At first glance one might assume that small distance increments could be used for greater accuracy in combination with large time increments to speed the solution. This is not the case, however, because the finite-difference equations limit the values of $\Delta\tau$ that may be used once Δx is chosen. Note that if $M < 2$ in Equation (4-31), the coefficient of T_m^p becomes negative, and we generate a condition that will violate the second law of thermodynamics. Suppose, for example, that the adjoining nodes are equal in temperature but less than T_m^p. After the time increment $\Delta\tau$, T_m^p may not be lower than these adjoining temperatures; otherwise heat would have to flow uphill on the temperature scale, and this is impossible. A value of $M < 2$ would produce just such an effect; so we must restrict the values of M to

$$\frac{(\Delta x)^2}{\alpha\,\Delta\tau} = \begin{cases} M \geq 2 & \text{one-dimensional systems} \\ M \geq 4 & \text{two-dimensional systems} \end{cases}$$

This restriction automatically limits our choice of $\Delta\tau$, once Δx is established.

It so happens that the above restrictions, which are imposed in a physical sense, may also be derived on mathematical grounds. It may be shown that the finite-difference solutions will not converge unless these conditions are fulfilled. The problems of stability and convergence of numerical solutions are discussed in References 7, 13, and 15 in detail.

The difference equations given above are useful for determining the internal temperature in a solid as a function of space and time. At the boundary of the solid, a convection resistance to heat flow is usually involved, so that the above relations no longer apply. In general, each convection boundary condition must be handled separately, depending on the particular geometric shape under consideration. The case of the flat wall will be considered as an example.

For the one-dimensional system shown in Figure 4-20 we may make an energy balance at the convection boundary such that

$$-kA\,\frac{\partial T}{\partial x}\bigg]_{\text{wall}} = hA(T_w - T_\infty) \qquad\qquad \textbf{[4-33]}$$

The finite-difference approximation would be given by

$$-k\frac{\Delta y}{\Delta x}(T_{m+1} - T_m) = h\,\Delta y(T_{m+1} - T_\infty)$$

Figure 4-20 | Nomenclature for numerical solution of unsteady-state conduction problem with convection boundary condition.

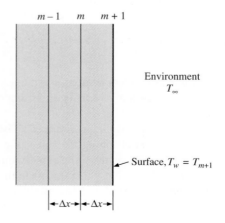

or

$$T_{m+1} = \frac{T_m + (h\,\Delta x/k)T_\infty}{1 + h\,\Delta x/k}$$

To apply this condition, we should calculate the surface temperature T_{m+1} at each time increment and then use this temperature in the nodal equations for the interior points of the solid. This is only an approximation because we have neglected the heat capacity of the element of the wall at the boundary. This approximation will work fairly well when a large number of increments in x are used because the portion of the heat capacity that is neglected is then small in comparison with the total. We may take the heat capacity into account in a general way by considering the two-dimensional wall of Figure 3-7 exposed to a convection boundary condition, which we duplicate here for convenience as Figure 4-21. We make a transient energy balance on the node (m, n) by setting the sum of the energy conducted and convected into the node equal to the increase in the internal energy of the node. Thus

$$k\,\Delta y\,\frac{T_{m-1,n}^p - T_{m,n}^p}{\Delta x} + k\,\frac{\Delta x}{2}\,\frac{T_{m,n+1}^p - T_{m,n}^p}{\Delta y} + k\,\frac{\Delta x}{2}\,\frac{T_{m,n-1}^p - T_{m,n}^p}{\Delta y}$$

$$+ h\,\Delta y(T_\infty - T_{m,n}^p) = \rho c\,\frac{\Delta x}{2}\,\Delta y\,\frac{T_{m,n}^{p+1} - T_{m,n}^p}{\Delta \tau}$$

If $\Delta x = \Delta y$, the relation for $T_{m,n}^{p+1}$ becomes

$$T_{m,n}^{p+1} = \frac{\alpha\,\Delta \tau}{(\Delta x)^2}\left\{2\frac{h\,\Delta x}{k}\,T_\infty + 2T_{m-1,n}^p + T_{m,n+1}^p + T_{m,n-1}^p\right.$$

$$\left. + \left[\frac{(\Delta x)^2}{\alpha\,\Delta \tau} - 2\frac{h\,\Delta x}{k} - 4\right]T_{m,n}^p\right\} \qquad \textbf{[4-34]}$$

The corresponding one-dimensional relation is

$$T_m^{p+1} = \frac{\alpha\,\Delta \tau}{(\Delta x)^2}\left\{2\frac{h\,\Delta x}{k}T_\infty + 2T_{m-1}^p + \left[\frac{(\Delta x)^2}{\alpha\,\Delta \tau} - 2\frac{h\,\Delta x}{k} - 2\right]T_m^p\right\} \qquad \textbf{[4-35]}$$

Notice now that the selection of the parameter $(\Delta x)^2/\alpha\,\Delta \tau$ is not as simple as it is for the interior nodal points because the heat-transfer coefficient influences the choice. It is still

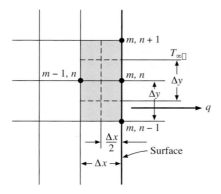

Figure 4-21 | Nomenclature for nodal
equation with convective
boundary condition.

possible to choose the value of this parameter so that the coefficient of T_m^p or $T_{m,n}^p$ will be zero. These values would then be

$$\frac{(\Delta x)^2}{\alpha \, \Delta \tau} = \begin{cases} 2\left(\dfrac{h \, \Delta x}{k} + 1\right) & \text{for the one-dimensional case} \\[2mm] 2\left(\dfrac{h \, \Delta x}{k} + 2\right) & \text{for the two-dimensional case} \end{cases}$$

To ensure convergence of the numerical solution, all selections of the parameter $(\Delta x)^2/\alpha \, \Delta \tau$ must be restricted according to

$$\frac{(\Delta x)^2}{\alpha \, \Delta \tau} \geq \begin{cases} 2\left(\dfrac{h \, \Delta x}{k} + 1\right) & \text{for the one-dimensional case} \\[2mm] 2\left(\dfrac{h \, \Delta x}{k} + 2\right) & \text{for the two-dimensional case} \end{cases}$$

Forward and Backward Differences

The equations above have been developed on the basis of a *forward-difference* technique in that the temperature of a node at a future time increment is expressed in terms of the surrounding nodal temperatures at the beginning of the time increment. The expressions are called *explicit* formulations because it is possible to write the nodal temperatures $T_{m,n}^{p+1}$ explicitly in terms of the previous nodal temperatures $T_{m,n}^p$. In this formulation, the calculation proceeds directly from one time increment to the next until the temperature distribution is calculated at the desired final state.

The difference equation may also be formulated by computing the space derivatives in terms of the temperatures at the $p+1$ time increment. Such an arrangement is called a *backward-difference* formulation because the time derivative moves backward from the times for heat conduction into the node. The equation equivalent to Equation (4-28) would then be

$$\frac{T_{m+1,n}^{p+1} + T_{m-1,n}^{p+1} - 2T_{m,n}^{p+1}}{(\Delta x)^2} + \frac{T_{m,n+1}^{p+1} + T_{m,n-1}^{p+1} - 2T_{m,n}^{p+1}}{(\Delta y)^2} = \frac{1}{\alpha} \frac{T_{m,n}^{p+1} - T_{m,n}^p}{\Delta \tau} \qquad \textbf{[4-36]}$$

The equivalence to Equation (4-29) is

$$T_{m,n}^p = \frac{-\alpha \, \Delta \tau}{(\Delta x)^2}\left(T_{m+1,n}^{p+1} + T_{m-1,n}^{p+1} + T_{m,n+1}^{p+1} + T_{m,n-1}^{p+1}\right) + \left[1 + \frac{4\alpha \, \Delta \tau}{(\Delta x)^2}\right] T_{m,n}^{p+1} \qquad \textbf{[4-37]}$$

We may now note that this backward-difference formulation does not permit the explicit calculation of the T^{p+1} in terms of T^p. Rather, a whole set of equations must be written for the entire nodal system and solved simultaneously to determine the temperatures T^{p+1}. Thus we say that the backward-difference method produces an *implicit formulation* for the future temperatures in the transient analysis. The solution to the set of equations can be performed with the methods discussed in Chapter 3.

The Biot and Fourier numbers may also be defined in the following way for problems in the numerical format:

$$\text{Bi} = \frac{h \, \Delta x}{k} \qquad \textbf{[4-38]}$$

$$\text{Fo} = \frac{\alpha \, \Delta \tau}{(\Delta x)^2} \qquad \textbf{[4-39]}$$

By using this notation, Tables 4-2 and 4-3 have been constructed to summarize some typical nodal equations in both the explicit and implicit formulations. For the cases of $\Delta x = \Delta y$

displayed in Table 4-2, the most restrictive stability requirement (smallest $\Delta\tau$) is exhibited by an exterior corner node, assuming all the convection nodes have the same value of Bi.

Table 4-2 | Explicit nodal equations. (Dashed lines indicate element volume.)[†]

Physical situation	Nodal equation for $\Delta x = \Delta y$	Stability requirement
(a) Interior node	$T_{m,n}^{p+1} = \text{Fo}\left(T_{m-1,n}^{p} + T_{m,n+1}^{p} + T_{m+1,n}^{p} + T_{m,n-1}^{p}\right)$ $+[1 - 4(\text{Fo})]T_{m,n}^{p}$ $T_{m,n}^{p+1} = \text{Fo}\left(T_{m-1,n}^{p} + T_{m,n+1}^{p} + T_{m+1,n}^{p} + T_{m,n-1}^{p} -4T_{m,n}^{p}\right) + T_{m,n}^{p}$	$\text{Fo} \leq \frac{1}{4}$
(b) Convection boundary node	$T_{m,n}^{p+1} = \text{Fo}\,[2T_{m-1,n}^{p} + T_{m,n+1}^{p} + T_{m,n-1}^{p} + 2(\text{Bi})T_{\infty}^{p}]$ $+[1 - 4(\text{Fo}) - 2(\text{Fo})(\text{Bi})]T_{m,n}^{p}$ $T_{m,n}^{p+1} = \text{Fo}\,[2\text{Bi}\,(T_{\infty}^{p} - T_{m,n}^{p}) + 2T_{m-1,n}^{p} + T_{m,n+1}^{p} +T_{m,n-1}^{p} - 4T_{m,n}^{p}] + T_{m,n}^{p}$	$\text{Fo}(2 + \text{Bi}) \leq \frac{1}{2}$
(c) Exterior corner with convection boundary	$T_{m,n}^{p+1} = 2(\text{Fo})\,[T_{m-1,n}^{p} + T_{m,n-1}^{p} + 2(\text{Bi})T_{\infty}^{p}]$ $+[1 - 4(\text{Fo}) - 4(\text{Fo})(\text{Bi})]T_{m,n}^{p}$ $T_{m,n}^{p+1} = 2\text{Fo}\,[T_{m-1,n}^{p} + T_{m,n-1}^{p} - 2T_{m,n}^{p} +2\text{Bi}(T_{\infty}^{p} - T_{m,n}^{p})] + T_{m,n}^{p}$	$\text{Fo}(1 + \text{Bi}) \leq \frac{1}{4}$
(d) Interior corner with convection boundary	$T_{m,n}^{p+1} = \frac{2}{3}(\text{Fo})\,[2T_{m,n+1}^{p} + 2T_{m+1,n}^{p}$ $+2T_{m-1,n}^{p} + T_{m,n-1}^{p} + 2(\text{Bi})T_{\infty}^{p}]$ $+[1 - 4(\text{Fo}) - \frac{4}{3}(\text{Fo})(\text{Bi})]T_{m,n}^{p}$ $T_{m,n}^{p+1} = (4/3)\text{Fo}\,[T_{m,n+1}^{p} + T_{m+1,n}^{p}$ $+T_{m-1,n}^{p} - 3T_{m,n}^{p} + \text{Bi}\,(T_{\infty}^{p} - T_{m,n}^{p})] + T_{m,n}^{p}$	$\text{Fo}(3 + \text{Bi}) \leq \frac{3}{4}$

Table 4-2 | (*Continued*).

Physical situation	Nodal equation for $\Delta x = \Delta y$	Stability requirement
(e) Insulated boundary 	$T_{m,n}^{p+1} = \text{Fo}\,[2T_{m-1,n}^{p} + T_{m,n+1}^{p} + T_{m,n-1}^{p}]$ $+ [1 - 4(\text{Fo})]T_{m,n}^{p}$	$\text{Fo} \leq \frac{1}{4}$

†Convection surfaces may be made insulated by setting $h = 0$ (Bi = 0).

Table 4-3 | Implicit nodal equations. (Dashed lines indicate volume element.)

Physical situation	Nodal equation for $\Delta x = \Delta y$
(a) Interior node 	$[1 + 4(\text{Fo})]T_{m,n}^{p+1} - \text{Fo}\left(T_{m-1,n}^{p+1} + T_{m,n+1}^{p+1} + T_{m+1,n}^{p+1}\right.$ $\left. + T_{m,n-1}^{p+1}\right) - T_{m,n}^{p} = 0$
(b) Convection boundary node 	$[1 + 2(\text{Fo})(2 + \text{Bi})]T_{m,n}^{p+1} - \text{Fo}\left[T_{m,n+1}^{p+1} + T_{m,n-1}^{p+1}\right.$ $\left. + 2T_{m-1,n}^{p+1} + 2(\text{Bi})T_{\infty}^{p+1}\right] - T_{m,n}^{p} = 0$
(c) Exterior corner with convection boundary 	$[1 + 4(\text{Fo})(1 + \text{Bi})]T_{m,n}^{p+1} - 2(\text{Fo})\left[T_{m-1,n}^{p+1} + T_{m,n-1}^{p+1}\right.$ $\left. + 2(\text{Bi})T_{\infty}^{p+1}\right] - T_{m,n}^{p} = 0$

Table 4-3 | (*Continued*).

Physical situation	Nodal equation for $\Delta x = \Delta y$
(*d*) Interior corner with convection boundary	$\left[1 + 4(\text{Fo})\left(1 + \dfrac{\text{Bi}}{3}\right)\right]T_{m,n}^{p+1} - \dfrac{2(\text{Fo})}{3} \times \left[2T_{m-1,n}^{p+1} + T_{m,n-1}^{p+1}\right.$ $\left. + 2T_{m,n+1}^{p+1} + 2T_{m+1,n}^{p+1} + 2(\text{Bi})T_\infty^{p+1}\right] - T_{m,n}^p = 0$

| (*e*) Insulated boundary | $[1 + 4(\text{Fo})]T_{m,n}^{p+1} - \text{Fo}\left(2T_{m-1,n}^{p+1} + T_{m,n+1}^{p+1} + T_{m,n-1}^{p+1}\right) - T_{m,n}^p = 0$ |

The advantage of an explicit forward-difference procedure is the direct calculation of future nodal temperatures; however, the stability of this calculation is governed by the selection of the values of Δx and $\Delta \tau$. A selection of a small value of Δx automatically forces the selection of some maximum value of $\Delta \tau$. On the other hand, no such restriction is imposed on the solution of the equations that are obtained from the implicit formulation. This means that larger time increments can be selected to speed the calculation. The obvious disadvantage of the implicit method is the larger number of calculations for each time step. For problems involving a large number of nodes, however, the implicit method may result in less total computer time expended for the final solution because very small time increments may be imposed in the explicit method from stability requirements. Much larger increments in $\Delta \tau$ can be employed with the implicit method to speed the solution.

Most problems will involve only a modest number of nodes and the explicit formulation will be quite satisfactory for a solution, particularly when considered from the standpoint of the more generalized formulation presented in the following section.

For a discussion of many applications of numerical analysis to transient heat conduction problems, the reader is referred to References 4, 8, 13, 14, and 15.

It should be obvious to the reader by now that finite-difference techniques may be applied to almost any situation with just a little patience and care. Very complicated problems then become quite easy to solve with only modest computer facilities. The use of Microsoft Excel for solution of transient heat-transfer problems is discussed in Appendix D.

Finite-element methods for use in conduction heat-transfer problems are discussed in References 9 to 13. A number of software packages are available commercially.

4-7 | THERMAL RESISTANCE AND CAPACITY FORMULATION

As in Chapter 3, we can view each volume element as a node that is connected by thermal resistances to its adjoining neighbors. For steady-state conditions the net energy transfer into the node is zero, while for the unsteady-state problems of interest in this chapter the net energy transfer into the node must be evidenced as an increase in internal energy of the element. Each volume element behaves like a small "lumped capacity," and the interaction of all the elements determines the behavior of the solid during a transient process. If the internal energy of a node i can be expressed in terms of specific heat and temperature, then its rate of change with time is approximated by

$$\frac{\Delta E}{\Delta \tau} = \rho c \Delta V \frac{T_i^{p+1} - T_i^p}{\Delta \tau}$$

where ΔV is the volume element. If we define the thermal capacity as

$$C_i = \rho_i c_i \, \Delta V_i \qquad \text{[4-40]}$$

then the general resistance-capacity formulation for the energy balance on a node is

$$q_i + \sum_j \frac{T_j^p - T_i^p}{R_{ij}} = C_i \frac{T_i^{p+1} - T_i^j}{\Delta \tau} \qquad \text{[4-41]}$$

where all the terms on the left are the same as in Equation (3-31). The resistance and volume elements for a variety of geometries and boundary conditions were given in Tables 3-3 and 3-4. Physical systems where the internal energy E involves phase changes can also be accommodated in the above formulation but are beyond the scope of our discussion.

The central point is that use of the concepts of thermal resistance and capacitance enables us to write the forward-difference equation for all nodes and boundary conditions in the single compact form of Equation (4-41). The setup for a numerical solution then becomes a much more organized process that can be adapted quickly to the computational methods at hand.

Equation (4-41) is developed by using the forward-difference concept to produce an explicit relation for each T_i^{p+1}. As in our previous discussion, we could also write the energy balance using backward differences, with the heat transfers into each ith node calculated in terms of the temperatures at the $p + 1$ time increment. Thus,

$$q_i + \sum_j \frac{T_j^{p+1} - T_i^{p+1}}{R_{ij}} = C_i \frac{T_i^{p+1} - T_i^p}{\Delta \tau} \qquad \text{[4-42]}$$

Now, as before, the set of equations produces an implicit set that must be solved simultaneously for the T_i^{p+1}, etc. The solution can be carried out by a number of methods as discussed in Chapter 3. If the solution is to be performed with a Gauss-Seidel iteration technique, then Equation (4-42) should be solved for T_i^{p+1} and expressed as

$$T_i^{p+1} = \frac{q_i + \sum_i (T_j^{p+1}/R_{ij}) + (C_i/\Delta \tau)T_i^p}{\sum_j (1/R_{ij}) + C_i/\Delta \tau} \qquad \text{[4-43]}$$

It is interesting to note that in the steady-state limit of $\Delta \tau \to \infty$ this equation becomes identical with Equation (3-32), the formulation we employed for the iterative solution in Chapter 3.

The stability requirement in the explicit formulation may be examined by solving Equation (4-41) for T_i^{p+1}:

$$T_i^{p+1} = \left(q_i + \sum_j \frac{T_j^p}{R_{ij}} \right) \frac{\Delta\tau}{C_i} + \left(1 - \frac{\Delta\tau}{C_i} \sum_j \frac{1}{R_{ij}} \right) T_i^p \qquad \textbf{[4-44]}$$

The value of q_i can influence the stability, but we can choose a safe limit by observing the behavior of the equation for $q_i = 0$. Using the same type of thermodynamic argument as with Equation (4-31), we find that the coefficient of T_i^p cannot be negative. Our stability requirement is therefore

$$1 - \frac{\Delta\tau}{C_i} \sum_j \frac{1}{R_{ij}} \geq 0 \qquad \textbf{[4-45]}$$

Suppose we have a complicated numerical problem to solve with a variety of boundary conditions, perhaps nonuniform values of the space increments, etc. Once we have all the nodal resistances and capacities formulated, we then have the task of choosing the time increment $\Delta\tau$ to use for the calculation. To ensure stability we must keep $\Delta\tau$ equal to or less than a value obtained from the most restrictive nodal relation like Equation (4-45).

Solving for $\Delta\tau$ gives

$$\Delta\tau \leq \left[\frac{C_i}{\sum_j (1/R_{ij})} \right]_{\min} \qquad \text{for stability} \qquad \textbf{[4-46]}$$

While Equation (4-44) is very useful in establishing the maximum allowable time increment, it may involve problems of round-off errors in computer solutions when small thermal resistances are employed. The difficulty may be alleviated by expressing T_i^{p+1} in the following form for calculation purposes:

$$T_i^{p+1} = \frac{\Delta\tau}{C_i} \left[q_i + \sum_j \frac{T_j^p - T_i^p}{R_{ij}} \right] + T_i^p \qquad \textbf{[4-47]}$$

In Table 4-2 the nodal equations for $\Delta x = \Delta y$ are listed in the formats of both equations (4-44) and (4-47). The equations listed in Table 4-2 in the form of Equation (4-47) do not include the heat-source term. If needed, the term may be added using

$$q_i = \dot{q}_i \, \Delta V_i$$

where $\dot{q}_i$ is the heat generation per unit volume and ΔV_i is the volume element shown by dashed lines in the table. For radiation input to the node,

$$q_i = q''_{i,\,\text{rad}} \times \Delta A_i$$

where $q''_{i,\text{rad}}$ is the *net radiant energy input to the node per unit area* and ΔA_i is the area of the node for radiant exchange, which may or may not be equal to the area for convection heat transfer.

We should remark that the resistance-capacity formulation is easily adapted to take into account thermal-property variations with temperature. One need only calculate the proper values of ρ, c, and k for inclusion in the C_i and R_{ij}. Depending on the nature of the problem and accuracy required, it may be necessary to calculate new values of C_i and R_{ij} for each time increment. Example 4-17 illustrates the effects of variable conductivity.

Steady State as a Limiting Case of Transient Solution

As we have seen, the steady-state numerical formulation results when the right side of Equation (4-41) is set equal to zero. It also results when the calculation of the unsteady case using either Equation (4-44) or (4-47) is carried out for a large number of time increments. While the latter method of obtaining a steady-state solution may appear rather cumbersome, it can proceed quite rapidly with a computer. We may recall that the Gauss-Seidel iteration method was employed for the solution of many steady-state numerical problems, which of course entailed many computer calculations. If variable thermal resistances resulting from either variable thermal conductivities or variations in convection boundary conditions are encountered, the steady-state limit of a transient solution may offer advantages over the direct steady-state solution counterpart. We will recall that when variable thermal resistances appear, the resulting steady-state nodal equations become nonlinear and their solution may be tedious. The transient solution for such cases merely requires that each resistance be recalculated at the end of each time increment $\Delta\tau$, or the resistances may be entered directly as variables in the nodal equations. The calculations are then carried out for a sufficiently large number of time increments until the values of the T_i^{p+1} no longer change by a significant amount. At this point, the steady-state solution is obtained as the resulting values of the T_i.

The formulation and solution of transient numerical problems using Microsoft Excel is described in Section D-5 of the Appendix, along with worked examples. An example is also given of a transient solution carried forward a sufficient length of time to achieve steady-state conditions.

EXAMPLE 4-11	Sudden Cooling of a Rod

A steel rod [$k = 50$ W/m · °C] 3 mm in diameter and 10 cm long is initially at a uniform temperature of 200°C. At time zero it is suddenly immersed in a fluid having $h = 50$ W/m² · °C and $T_\infty = 40$°C while one end is maintained at 200°C. Determine the temperature distribution in the rod after 100 s. The properties of steel are $\rho = 7800$ kg/m³ and $c = 0.47$ kJ/kg · °C.

Figure Example 4-11

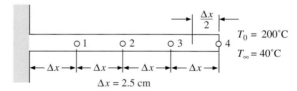

$\Delta x = 2.5$ cm

■ **Solution**

The selection of increments on the rod is as shown in the Figure Example 4-11. The cross-sectional area of the rod is $A = \pi(1.5)^2 = 7.069$ mm². The volume element for nodes 1, 2, and 3 is

$$\Delta V = A\Delta x = (7.069)(25) = 176.725 \text{ mm}^3$$

Node 4 has a ΔV of half this value, or 88.36 mm³. We can now tabulate the various resistances and capacities for use in an explicit formulation. For nodes 1, 2, and 3 we have

$$R_{m+} = R_{m-} = \frac{\Delta x}{kA} = \frac{0.025}{(50)(7.069 \times 10^{-6})} = 70.731 \text{°C/W}$$

and

$$R_\infty = \frac{1}{h(\pi d\,\Delta x)} = \frac{1}{(50)\pi(3 \times 10^{-3})(0.025)} = 84.883 \text{°C/W}$$

$$C = \rho c \, \Delta V = (7800)(470)(1.7673 \times 10^{-7}) = 0.6479 \; \text{J/°C}$$

For node 4 we have

$$R_{m+} = \frac{1}{hA} = 2829 \, \text{°C/W} \qquad R_{m-} = \frac{\Delta x}{kA} = 70.731 \, \text{°C/W}$$

$$C = \frac{\rho c \, \Delta V}{2} = 0.3240 \; \text{J/°C} \qquad R_\infty = \frac{2}{h\pi d\Delta x} = 169.77 \, \text{°C/W}$$

To determine the stability requirement we form the following table:

Node	$\sum(1/R_{ij})$	C_i	$\dfrac{C_i}{\sum(1/R_{ij})}$, s
1	0.04006	0.6479	16.173
2	0.04006	0.6479	16.173
3	0.04006	0.6479	16.173
4	0.02038	0.3240	15.897

Thus node 4 is the most restrictive, and we must select $\Delta\tau < 15.9$ s. Since we wish to find the temperature distribution at 100 s, let us use $\Delta\tau = 10$ s and make the calculation for 10 time increments using Equation (4-47) for the computation. We note, of course, that $q_i = 0$ because there is no heat generation. The calculations are shown in the following table.

Time increment	Node temperature			
	T_1	T_2	T_3	T_4
0	200	200	200	200
1	170.87	170.87	170.87	169.19
2	153.40	147.04	146.68	145.05
3	141.54	128.86	126.98	125.54
4	133.04	115.04	111.24	109.70
5	126.79	104.48	98.76	96.96
6	122.10	96.36	88.92	86.78
7	118.53	90.09	81.17	78.71
8	115.80	85.23	75.08	72.34
9	113.70	81.45	70.31	67.31
10	112.08	78.51	66.57	63.37

We can calculate the heat-transfer *rate* at the end of 100 s by summing the convection heat losses on the surface of the rod. Thus

$$q = \sum_i \frac{T_i - T_\infty}{R_{i\infty}}$$

and

$$q = \frac{200 - 40}{(2)(84.883)} + \frac{112.08 + 78.51 + 66.57 - (3)(40)}{84.883} + \left(\frac{1}{169.77} + \frac{1}{2829} \right)(63.37 - 40)$$

$$= 2.704 \; \text{W}$$

Implicit Formulation   EXAMPLE 4-12

We can illustrate the calculation scheme for the implicit formulation by reworking Example 4-11 using only two time increments, that is, $\Delta\tau = 50$ s.

For this problem we employ the formulation indicated by Equation (4-43), with $\Delta \tau = 50$ s. The following quantities are needed.

Node	$\dfrac{C_i}{\Delta \tau}$	$\sum_i \dfrac{1}{R_{ij}} + \dfrac{C_i}{\Delta \tau}$
1	0.01296	0.05302
2	0.01296	0.05302
3	0.01296	0.05302
4	0.00648	0.02686

We have already determined the R_{ij} in Example 4-11 and thus can insert them into Equation (4-43) to write the nodal equations for the end of the first time increment, taking all $T_i^p = 200°$C. We use the prime to designate temperatures at the end of the time increment. For node 1,

$$0.05302T_1' = \frac{200}{70.731} + \frac{T_2'}{70.731} + \frac{40}{84.833} + (0.01296)(200)$$

For node 2,

$$0.05302T_2' = \frac{T_1'}{70.731} + \frac{T_3'}{70.731} + \frac{40}{84.833} + (0.01296)(200)$$

For nodes 3 and 4,

$$0.05302T_3' = \frac{T_2'}{70.731} + \frac{T_4'}{70.731} + \frac{40}{84.833} + (0.01296)(200)$$

$$0.02686T_4' = \frac{T_3'}{70.731} + \frac{40}{2829} + \frac{40}{169.77} + (0.00648)(200)$$

These equations can then be reduced to

$$0.05302T_1' - 0.01414T_2' \qquad\qquad\qquad = 5.8911$$
$$-0.01414T_1' + 0.05302T_2' - 0.01414T_3' \qquad = 3.0635$$
$$-0.01414T_2' + 0.05302T_3' - 0.01414T_4' = 3.0635$$
$$-0.01414T_3' + 0.02686T_4' = 1.5457$$

which have the solution

$$T_1' = 145.81°\text{C} \quad T_2' = 130.12°\text{C}$$
$$T_3' = 125.43°\text{C} \quad T_4' = 123.56°\text{C}$$

We can now apply the backward-difference formulation a second time using the double prime to designate the temperatures at the end of the second time increment:

$$0.05302T_1'' = \frac{200}{70.731} + \frac{T_2''}{70.731} + \frac{40}{84.833} + (0.01296)(145.81)$$

$$0.05302T_2'' = \frac{T_1''}{70.731} + \frac{T_3''}{70.731} + \frac{40}{84.833} + (0.01296)(130.12)$$

$$0.05302T_3'' = \frac{T_2''}{70.731} + \frac{T_4''}{70.731} + \frac{40}{84.833} + (0.01296)(125.43)$$

$$0.02686T_4'' = \frac{T_3''}{70.731} + \frac{40}{2829} + \frac{40}{169.77} + (0.00648)(123.56)$$

and this equation set has the solution

$$T_1'' = 123.81°\text{C} \quad T_2'' = 97.27°\text{C}$$
$$T_3'' = 88.32°\text{C} \quad T_4'' = 85.59°\text{C}$$

We find this calculation in substantial disagreement with the results of Example 4-11. With a larger number of time increments, better agreement would be achieved. In a problem involving a large number of nodes, the implicit formulation might involve less computer time than the explicit method, and the purpose of this example has been to show how the calculation is performed.

Cooling of a Ceramic **EXAMPLE 4-13**

A 1 by 2 cm ceramic strip $[k = 3.0 \text{ W/m} \cdot {}^\circ\text{C}]$ is embedded in a high-thermal-conductivity material, as shown in Figure Example 4-13, so that the sides are maintained at a constant temperature of 300°C. The bottom surface of the ceramic is insulated, and the top surface is exposed to a convection environment with $h = 200 \text{ W/m}^2 \cdot {}^\circ\text{C}$ and $T_\infty = 50°\text{C}$. At time zero the ceramic is uniform in temperature at 300°C. Calculate the temperatures at nodes 1 to 9 after a time of 12 s. For the ceramic $\rho = 1600 \text{ kg/m}^3$ and $c = 0.8 \text{ kJ/kg} \cdot {}^\circ\text{C}$. Also calculate the total heat loss in this time.

Figure Example 4-13

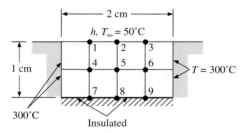

■ **Solution**

We treat this as a two-dimensional problem with $\Delta x = \Delta y = 0.5$ cm. From symmetry $T_1 = T_3$, $T_4 = T_6$, and $T_7 = T_9$, so we have six unknown nodal temperatures. We now tabulate the various nodal resistances and capacities. For nodes 4 and 5

$$R_{m+} = R_{m-} = R_{n+} = R_{n-} = \frac{\Delta x}{kA} = \frac{0.005}{(3.0)(0.005)} = 0.3333$$

For nodes 1 and 2

$$R_{m+} = R_{m-} = \frac{\Delta x}{kA} = \frac{(0.005)(2)}{(3.0)(0.005)} = 0.6667°\text{C/W} \qquad R_{n-} = 0.3333°\text{C/W}$$

$$R_{n+} = \frac{1}{h \, \Delta x} = \frac{1}{(200)(0.005)} = 1.0°\text{C/W}$$

For nodes 7 and 8

$$R_{m+} = R_{m-} = 0.6667°\text{C/W} \qquad R_{n+} = 0.3333°\text{C/W} \qquad R_{n-} = \infty$$

For nodes 1, 2, 7, and 8 the capacities are

$$C = \frac{\rho c (\Delta x)^2}{2} = \frac{(1600)(800)(0.005)^2}{2} = 16 \text{ J/°C}$$

For nodes 4 and 5

$$C = \rho c (\Delta x)^2 = 32 \text{ J/°C}$$

The stability requirement for an explicit solution is now determined by tabulating the following quantities:

Node	$\sum \dfrac{1}{R_{ij}}$	C_i	$\dfrac{C_i}{\sum(1/R_{ij})}$, s
1	7	16	2.286
2	7	16	2.286
4	12	32	2.667
5	12	32	2.667
7	6	16	2.667
8	6	16	2.667

Thus the two convection nodes control the stability requirement, and we must choose $\Delta\tau \le 2.286$ s. Let us choose $\Delta\tau = 2.0$ s and make the calculations for six time increments with Equation (4-47). We note once again the symmetry considerations when calculating the temperatures of nodes 2, 5, and 8, that is, $T_1 = T_3$, etc. The calculations are shown in the following table.

	Node temperature					
Time increment	T_1	T_2	T_4	T_5	T_7	T_8
0	300	300	300	300	300	300
1	268.75	268.75	300	300	300	300
2	258.98	253.13	294.14	294.14	300	300
3	252.64	245.31	289.75	287.55	297.80	297.80
4	284.73	239.48	285.81	282.38	295.19	293.96
5	246.67	235.35	282.63	277.79	292.34	290.08
6	243.32	231.97	279.87	273.95	289.71	286.32

The total heat loss during the 12-s time interval is calculated by summing the heat loss of each node relative to the initial temperature of 300°C. Thus

$$q = \sum C_i (300 - T_i)$$

where q is the heat *loss*. For this summation, since the constant-temperature boundary nodes experience no change in temperature, they can be left out. Recalling that $T_1 = T_3$, $T_4 = T_6$, and $T_7 = T_9$, we have

$$\sum C_i (300 - T_i) = \text{nodes } (1, 2, 3, 7, 8, 9) + \text{nodes } (4, 5, 6)$$
$$= 16[(6)(300) - (2)(243.2) - 231.97 - (2)(289.71)$$
$$- 286.32] + 32[(3)(300) - (2)(279.87) - 273.95]$$
$$= 5572.3 \text{ J/m length of strip}$$

The *average rate* of heat loss for the 12-s time interval is

$$\frac{q}{\Delta\tau} = \frac{5572.3}{12} = 464.4 \text{ W} \quad [1585 \text{ Btu/h}]$$

EXAMPLE 4-14   Cooling of a Steel Rod, Nonuniform h

A nickel-steel rod having a diameter of 2.0 cm is 10 cm long and initially at a uniform temperature of 200°C. It is suddenly exposed to atmospheric air at 30°C while one end of the rod is maintained at 200°C. The convection heat-transfer coefficient can be computed from

$$h = 9.0 \, \Delta T^{0.175} \text{ W/m}^2 \cdot °\text{C}$$

where ΔT is the temperature difference between the rod and air surroundings. The properties of nickel steel may be taken as $k = 12$ W/m·°C, $c = 0.48$ kJ/kg·°C, and $\rho = 7800$ kg/m³. Using the numerical method, (*a*) determine the temperature distribution in the rod after 250, 500, 750, 1000, 1250 s, and for steady state; (*b*) determine the steady-state temperature distribution for a constant $h = 22.11$ W/m²·°C and compare with an analytical solution.

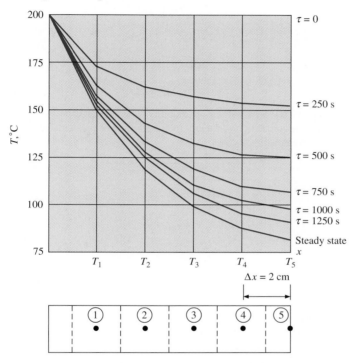

Figure Example 4-14

■ **Solution**
Five nodes are chosen as shown in Figure Example 4-14 with $\Delta x = 2.0$ cm. The capacitances are then

$$C_1 = C_2 = C_3 = C_4 = \frac{(7800)(480)\pi(0.02)^2(0.02)}{4} = 23.524 \text{ J/°C}$$

$$C_5 = \tfrac{1}{2}C_1 = 11.762 \text{ J/°C}$$

The resistances for nodes 1, 2, 3, and 4 are

$$\frac{1}{R_{m+}} = \frac{1}{R_{m-}} = \frac{kA}{\Delta x} = \frac{(12)\pi(0.02)^2}{(4)(0.02)} = 0.188496$$

$$\frac{1}{R_\infty} = hP\,\Delta x = (9.0)\pi(0.02)(0.02)(T-30)^{0.175} = (1.131 \times 10^{-2})(T-30)^{0.175}$$

For node 5

$$\frac{1}{R_{m-}} = 0.188496$$

$$\frac{1}{R_{m+}} = hA = 9.0\frac{\pi(0.02)^2}{4}(T-30)^{0.175} = (2.827 \times 10^{-3})(T-30)^{0.175}$$

$$\frac{1}{R_{5\infty}} = \frac{1}{2R_{1\infty}} = (5.655 \times 10^{-3})(T - 30)^{0.175}$$

where $T_\infty = 30°C$ for all nodes. We can compute the following table for worst-case conditions of $T = 200°C$ throughout the rod. The stability requirement so established will then work for all other temperatures.

Node	$\sum(1/R_{ij})\vert_{min}$	$\dfrac{C_i}{\sum(1/R_{ij})}$, s
1	0.4048	58.11
2	0.4048	58.11
3	0.4048	58.11
4	0.4048	58.11
5	0.2093	56.197

Thus, time steps below 56 s will ensure stability. The computational procedure is complicated by the fact that the convection-resistance elements must be recalculated for each time step. Selecting $\Delta\tau = 50$ s, we have:

Node	$\Delta\tau/C_i$
1	2.1255
2	2.1255
3	2.1255
4	2.1255
5	4.251

We then use the explicit formulation of Equation (4-47) with no heat generation. The computational algorithm is thus:

1. Compute R_∞ values for the initial condition.
2. Compute temperatures at next time increment using Equation (4-47).
3. Recalculate R_∞ values based on new temperatures.
4. Repeat temperature calculations and continue until the temperature distributions are obtained at the desired times.

Results of these calculations are shown in the accompanying figure.

To determine the steady-state distribution we could carry the unsteady method forward a large number of time increments or use the steady-state method and an iterative approach. The iterative approach is required because the equations are nonlinear as a result of the variations in the convection coefficient.

We still use a resistance formulation, which is now given as Equation (3-31):

$$\sum \frac{T_j - T_i}{R_{ij}} = 0$$

The computational procedure is:

1. Calculate R_∞ values for all nodes assuming all $T_i = 200°C$.
2. Formulate nodal equations for the T_i's.
3. Solve the equations by an appropriate method.
4. Recalculate R_∞ values based on T_i values obtained in step 3.
5. Repeat the procedure until there are only small changes in T_i's.

The results of this iteration are shown in the following table:

Iteration	T_1, °C	T_2, °C	T_3, °C	T_4, °C	T_5, °C
1	148.462	114.381	92.726	80.310	75.302
2	151.381	119.557	99.409	87.853	83.188
3	151.105	119.038	98.702	87.024	82.306
4	151.132	119.090	98.774	87.109	82.396

This steady-state temperature distribution is also plotted with the transient profiles.

The value of h for $T_i = 200$°C is 22.11 W/m$^2 \cdot$ °C, so the results of the first iteration correspond to a solution for a constant h of this value. The exact analytical solution is given in Equation (2-34) as

$$\frac{\theta}{\theta_0} = \frac{T - T_\infty}{T_0 - T_\infty} = \frac{\cosh m(L - x) + [h/km] \sinh m(L - x)}{\cosh mL + [h/km] \sinh mL}$$

The required quantities are

$$m = \left(\frac{hP}{kA}\right)^{1/2} = \left[\frac{(22.11)\pi(0.02)}{(12)\pi(0.01)^2}\right]^{1/2} = 19.1964$$

$$mL = (19.1964)(0.1) = 1.91964$$

$$h/km = \frac{22.22}{(12)(19.1964)} = 0.09598$$

The temperatures at the nodal points can then be calculated and compared with the numerical results in the following table. As can be seen, the agreement is excellent.

Node	x, m	$(\theta/\theta_0)_{num}$	$(\theta/\theta_0)_{anal}$	Percent deviation
1	0.02	0.6968	0.6949	0.27
2	0.04	0.4964	0.4935	0.59
3	0.06	0.3690	0.3657	0.9
4	0.08	0.2959	0.2925	1.16
5	0.1	0.2665	0.2630	1.33

We may also check the heat loss with that predicted by the analytical relation in Equation (2-34). When numerical values are inserted we obtain

$$q_{anal} = 11.874 \text{ W}$$

The heat loss for the numerical model is computed by summing the convection loss from the six nodes (including base node at 200°C). Using the temperatures for the first iteration corresponding to $h = 22.11$ W/m$^2 \cdot$ °C,

$$q = (22.11)\pi(0.02)(0.02)\left[(200 - 30)\left(\tfrac{1}{2}\right) + (148.462 - 30)\right.$$
$$+ (114.381 - 30) + (92.726 - 30) + (80.31 - 30)$$
$$\left. + (75.302 - 30)\left(\tfrac{1}{2}\right)\right] + (22.11)\pi(0.01)2(75.302 - 30)$$
$$= 12.082 \text{ W}$$

We may make a further check by calculating the energy conducted in the base. This must be the energy conducted to node 1 plus the convection lost by the base node or

$$q = (12)\pi(0.01)^2 \frac{(200 - 148.462)}{0.02} + (22.11)\pi(0.02)(0.01)(200 - 30)$$
$$= 12.076 \text{ W}$$

This agrees very well with the convection calculation and both are within 1.8 percent of the analytical value.

The results of this example illustrate the power of the numerical method in solving problems that could not be solved in any other way. Furthermore, only a modest number of nodes, and thus modest computation facilities, may be required to obtain a sufficiently accurate solution. For example, the accuracy with which h will be known is typically ± 10 to 15 percent. This would overshadow any inaccuracies introduced by using relatively large nodes, as was done here.

| EXAMPLE 4-15 | Radiation Heating and Cooling |

The ceramic wall shown in Figure Example 4-15a is initially uniform in temperature at 20°C and has a thickness of 3.0 cm. It is suddenly exposed to a radiation source on the right side at 1000°C. The left side is exposed to room air at 20°C with a radiation surrounding temperature of 20°C. Properties of the ceramic are $k = 3.0$ W/m · °C, $\rho = 1600$ kg/m^3, and $c = 0.8$ kJ/kg · °C. Radiation heat transfer with the surroundings at T_r may be calculated from

$$q_r = \sigma \epsilon A (T^4 - T_r{}^4) \qquad \text{W} \qquad [a]$$

where $\sigma = 5.669 \times 10^{-8}$, $\epsilon = 0.8$, and T is in degrees Kelvin. The convection heat-transfer coefficient from the left side of the plate is given by

$$h = 1.92 \Delta T^{1/4} \qquad \text{W/m}^2 \cdot \text{°C} \qquad [b]$$

Convection on the right side is negligible. Determine the temperature distribution in the plate after 15, 30, 45, 60, 90, 120, and 150 s. Also determine the steady-state temperature distribution. Calculate the total heat gained by the plate for these times.

■ Solution

We divide the wall into five nodes as shown and must express temperatures in degrees Kelvin because of the radiation boundary condition. For node 1 the transient energy equation is

$$\sigma \epsilon \left(293^4 - T_1^{p4} \right) - 1.92 \left(T_1^p - 293 \right)^{5/4} + \frac{k}{\Delta x} \left(T_2^p - T_1^p \right) = \rho c \frac{\Delta x}{2} \frac{T_1^{p+1} - T_1^p}{\Delta \tau} \qquad [c]$$

Similarly, for node 5

$$\sigma \epsilon \left(1273^4 - T_5^{p4} \right) + \frac{k}{\Delta x} \left(T_4^p - T_5^p \right) = \rho c \frac{\Delta x}{2} \frac{T_5^{p+1} - T_5^p}{\Delta \tau} \qquad [d]$$

Equations (c) and (d) may be subsequently written

$$T_1^{p+1} = \frac{\Delta \tau}{C_1} \left[\sigma \epsilon (293^2 + T_1^{p2})(293 + T_1^p)(293) - 1.92(T_1^p - 293)^{1/4}(293) + \frac{k}{\Delta x} T_2^p \right]$$

$$+ \left\{ 1 - \frac{\Delta \tau}{C_1} \left[\sigma \epsilon (293^2 + T_1^{p2})(293 + T_1^p) - 1.92(T_1^p - 293)^{1/4} + \frac{k}{\Delta x} \right] \right\} T_1^p \qquad [e]$$

$$T_5^{p+1} = \frac{\Delta \tau}{C_5} \left[\sigma \epsilon (1273^2 + T_5^{p2})(1273 + T_5^p)(1273) + \frac{k}{\Delta x} T_4^p \right]$$

$$+ \left\{ 1 - \frac{\Delta \tau}{C_5} \left[\sigma \epsilon (1273^2 + T_5^{p2})(1273 + T_5^p) + \frac{k}{\Delta x} \right] \right\} T_5^p \qquad [f]$$

where $C_1 = C_5 = \rho c \Delta x / 2$. For the other three nodes the expressions are much simpler:

$$T_2^{p+1} = \frac{\Delta \tau}{C_2} \frac{k}{\Delta x} (T_1^p + T_3^p) + \left(1 - \frac{2k \Delta \tau}{C_2 \Delta x} \right) T_2^p \qquad [g]$$

Figure Example 4-15 | (*a*) Nodal system, (*b*) transient response, (*c*) heat added.

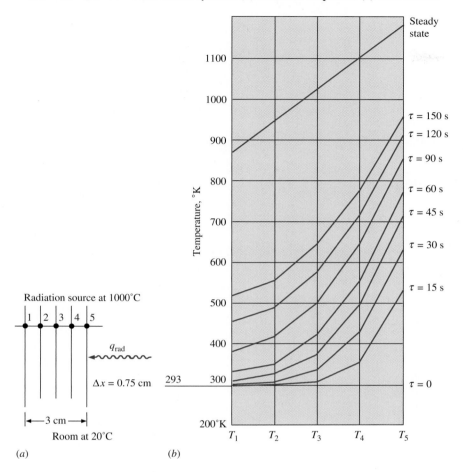

Radiation source at 1000°C

q_{rad}

$\Delta x = 0.75$ cm

293

3 cm

Room at 20°C

(*a*)

(*b*)

Steady state

$\tau = 150$ s
$\tau = 120$ s
$\tau = 90$ s
$\tau = 60$ s
$\tau = 45$ s
$\tau = 30$ s
$\tau = 15$ s
$\tau = 0$

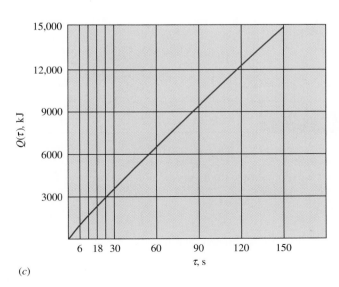

(*c*)

$$T_3^{p+1} = \frac{\Delta\tau}{C_3}\frac{k}{\Delta x}(T_2^p + T_4^p) + \left(1 - \frac{2k\,\Delta\tau}{C_3\,\Delta x}\right)T_3^p \qquad [h]$$

$$T_4^{p+1} = \frac{\Delta\tau}{C_4}\frac{k}{\Delta x}(T_3^p + T_5^p) + \left(1 - \frac{2k\,\Delta\tau}{C_4\,\Delta x}\right)T_4^p \qquad [i]$$

where $C_2 = C_3 = C_4 = \rho c\,\Delta x$. So, to determine the transient response, we simply choose a suitable value of $\Delta\tau$ and march through the calculations. The stability criterion is such that the coefficients of the last term in each equation cannot be negative. For Equations (g), (h), and (i) the maximum allowable time increment is

$$\Delta\tau_{max} = \frac{C_3\,\Delta x}{2k} = \frac{(1600)(800)(0.0075)^2}{(2)(3)} = 12\text{ s}$$

For Equation (f), the worst case is at the start when $T_5^p = 20°C = 293\text{ K}$. We have

$$C_5 = \frac{(1600)(800)(0.0075)}{2} = 4800$$

so that

$$\Delta\tau_{max} = \frac{4800}{(5.669 \times 10^{-8})(0.8)(1273^2 + 293^2)(1273 + 293) + 3.0/0.0075}$$
$$= 9.43\text{ s}$$

For node 1 [Equation (e)] the most restrictive condition occurs when $T_1^p = 293$. We have

$$C_1 = C_5 = 4800$$

so that

$$\Delta\tau_{max} = \frac{4800}{(5.669 \times 10^{-8})(0.8)(293^2 + 293^2)(293 + 293) + 3.0/0.0075}$$
$$= 11.86\text{ s}$$

So, from these calculations we see that node 5 is most restrictive and we must choose $\Delta\tau < 9.43$ s.

The calculations were performed with $\Delta\tau = 3.0$ s, and the results are shown in Figure Example 4-15b, c. Note that a straight line is obtained for the steady-state temperature distribution in the solid, which is what would be expected for a constant thermal conductivity. To compute the heat added at any instant of time we perform the sum

$$Q(\tau) = \sum C_i(T_i - 293) \qquad [j]$$

and plot the results in Figure Example 4-15c.

EXAMPLE 4-16	Transient Conduction with Heat Generation

The plane wall shown has internal heat generation of 50 MW/m^3 and thermal properties of $k = 19$ W/m · °C, $\rho = 7800$ kg/m^3, and $C = 460$ J/kg · °C. It is initially at a uniform temperature of 100°C and is suddenly subjected to the heat generation and the convective boundary conditions indicated in Figure Example 4-16A. Calculate the temperature distribution after several time increments.

■ **Solution**

We use this resistance and capacity formulation and write, for unit area,

$$1/R_{12} = kA/\Delta x = (19)(1)/0.001 = 19{,}000\text{ W/°C}$$

Figure Example 4-16a

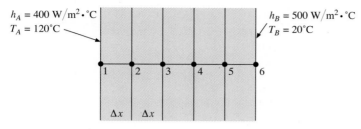

$$\Delta x = 1.0 \text{ mm}$$

All the conduction resistances have this value. Also,

$$1/R_{1A} = hA = (400)(1) = 400 \text{ W/}°\text{C}$$
$$1/R_{1B} = hA = (500)(1) = 500 \text{ W/}°\text{C}$$

The capacities are

$$C_1 = C_6 = \rho(\Delta x/2)c = (7800)(0.001/2)(460) = 1794 \text{ J/}°\text{C}$$
$$C_2 = C_3 = C_4 = C_5 = \rho(\Delta x)c = 3588 \text{ J/}°\text{C}$$

We next tabulate values.

Node	$\sum(1/R_{ij})$	C_i	$\dfrac{C_i}{\sum(1/R_{ij})}$
1	19,400	1794	0.092
2	38,000	3588	0.094
3	38,000	3588	0.094
4	38,000	3588	0.094
5	38,000	3588	0.094
6	19,500	1794	0.092

Any time increment $\Delta\tau$ less than 0.09 s will be satisfactory. The nodal equations are now written in the form of Equation (4-47) and the calculation marched forward on a computer.

The heat-generation terms are

$$q_i = \dot{q}\,\Delta V_i$$

so that

$$q_1 = q_6 = (50 \times 10^6)(1)(0.001/2) = 25,000 \text{ W}$$
$$q_2 = q_3 = q_4 = q_5 = (50 \times 106)(1)(0.001) = 50,000 \text{ W}$$

The computer results for several time increments of 0.09 s are shown in the following table. Because the solid stays nearly uniform in temperature at any instant of time it behaves almost like a lumped capacity. The temperature of node 3 is plotted versus time in Figure Example 4-16B to illustrate this behavior.

Node	Number of time increments ($\Delta\tau = 0.09$ s)			
	5	**20**	**100**	**200**
1	106.8826	123.0504	190.0725	246.3855
2	106.478	122.8867	190.9618	248.1988
3	106.1888	122.1404	190.7033	248.3325
4	105.3772	120.9763	189.3072	246.7933
5	104.4622	119.2217	186.7698	243.5786
6	102.4416	117.0056	183.0735	238.6773

Node	Number of time increments ($\Delta\tau = 0.09$ s)			
	500	800	1200	3000
1	320.5766	340.1745	346.0174	347.2085
2	323.6071	343.5267	349.4654	350.676
3	324.2577	344.3137	350.2931	351.512
4	322.5298	342.536	348.5006	349.7165
5	318.4229	338.1934	344.0877	345.2893
6	311.9341	331.2853	337.0545	338.2306

Figure Example 4-16b

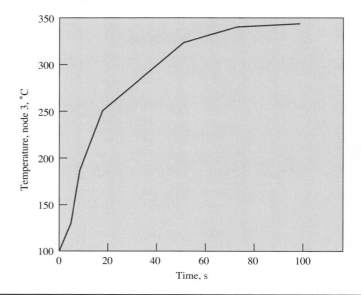

| **EXAMPLE 4-17** | Numerical Solution for Variable Conductivity |

A 4.0-cm-thick slab of stainless steel (18% Cr, 8% Ni) is initially at a uniform temperature of 0°C with the left face perfectly insulated as shown in Figure Example 4-17a. The right face is suddenly raised to a constant 1000°C by an intense radiation source. Calculate the temperature distribution after (a) 25 s, (b) 50 s, (c) 100 s, (d) an interval long enough for the slab to reach a steady state, taking into account variation in thermal conductivity. Approximate the conductivity data in Appendix A with a linear relation. Repeat the calculation for the left face maintained at 0°C.

Figure Example 4-17a

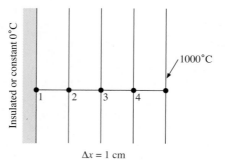

■ Solution

From Table A-2 we have $k = 16.3$ W/m·°C at 0°C and $k = 31$ W/m·°C at 1000°C. A linear relation for k is assumed so that

$$k = k_0(1 + \beta T)$$

where T is in degrees Celsius. Inserting the data gives

$$k = 16.3(1 + 9.02 \times 10^{-4} T) \text{ W/m·°C}$$

We also have $\rho = 7817$ kg/m^3 and $c = 460$ J/kg·°C, and use the thermal resistance-capacitance formula assuming that the resistances are evaluated at the arithmetic mean of their connecting nodal temperatures; i.e., R_{3-4} is evaluated at $(T_3 + T_4)/2$.

First, the thermal capacities are evaluated for unit area:

$$C_1 = \rho(\Delta x/2)c = (7817)(0.01/2)(460) = 17{,}980 \text{ J/m}^2 \cdot °\text{C}$$

$$C_2 = C_3 = C_4 = \rho(\Delta x)c = (7817)(0.01)(460) = 35{,}960 \text{ J/m}^2 \cdot °\text{C}$$

For the resistances we have the form, for unit area,

$$1/R = k/\Delta x = k_0(1 + \beta T)/\Delta x$$

Evaluating at the mean temperatures between nodes gives

$$1/R_{1-2} = (16.3)[1 + 4.51 \times 10^{-4}(T_1 + T_2)]/0.01 = 1/R_{2-1}$$

$$1/R_{2-3} = (16.3)[1 + 4.51 \times 10^{-4}(T_2 + T_3)]/0.01 = 1/R_{3-2}$$

$$1/R_{3-4} = (16.3)[1 + 4.51 \times 10^{-4}(T_3 + T_4)]/0.01 = 1/R_{4-3}$$

$$1/R_{4-1000} = (16.3)[1 + 4.51 \times 10^{-4}(T_4 + T_{1000})]/0.01 = 1/R_{1000-4}$$

The stability requirement is most severe on node 1 because it has the lowest capacity. To be on the safe side we can choose a large k of about 31 W/m·°C and calculate

$$\Delta\tau_{\text{max}} = \frac{(17{,}980)(0.01)}{31} = 5.8 \text{ s}$$

The nodal equations are now written in the form of Equation (4-47); that is to say, the equation for node 2 would be

$$T_2^{p+1} = \frac{\Delta\tau}{C_2} \left\{ 1630\,[1 + 4.51 \times 10^{-4}(T_1^p + T_2^p)](T_1^p - T_2^p) \right.$$

$$\left. + 1630\,[1 + 4.51 \times 10^{-4}(T_3^p + T_2^p)](T_3^p - T_2^p) \right\} + T_2^p$$

A computer solution has been performed with $\Delta\tau = 5$ s and the results are shown in the tables. The steady-state solution for the insulated left face is, of course, a constant 1000°C. The steady-state distribution for the left face at 0°C corresponds to Equation (2-2) of Chapter 2. Note that, because of the nonconstant thermal conductivity, the steady-state temperature profile is not a straight line.

Temperatures for left face at constant 0°C, $\Delta\tau = 5$ s

Node	25 s	50 s	100 s	Steady state
1	0	0	0	0
2	94.57888	236.9619	308.2699	317.3339
3	318.7637	486.5802	565.7786	575.9138
4	653.5105	748.1359	793.7976	799.7735

Temperatures for left face insulated, $\Delta\tau = 5\,s$				
Node	25 s	50 s	100 s	Steady state
1	30.55758	232.8187	587.021	1000
2	96.67601	310.1737	623.5018	1000
3	318.7637	505.7613	721.5908	1000
4	653.5105	752.3268	855.6965	1000

These temperatures are plotted in Figure Example 4-17b.

Figure Example 4-17b

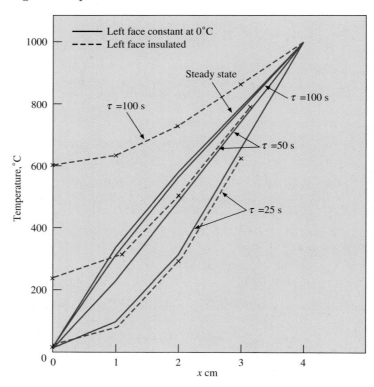

The purpose of this example has been to show how the resistance-capacity formulation can be used to take into account property variations in a rather straightforward way. These variations may or may not be important when one considers uncertainties in boundary conditions.

4-8 | SUMMARY

In progressing through this chapter the reader will have noted analysis techniques of varying complexity, ranging from simple lumped-capacity systems to numerical computer solutions. At this point some suggestions are offered for a general approach to follow in the solution of transient heat-transfer problems.

1. First, determine if a lumped-capacity analysis can apply. If so, you may be led to a much easier calculation.

2. Check to see if an analytical solution is available with such aids as the Heisler charts and approximations.

3. If analytical solutions are very complicated, *even when already available,* move directly to numerical techniques. This is particularly true where repetitive calculations must be performed.

4. When approaching a numerical solution, recognize the large uncertainties present in convection and radiation boundary conditions. Do not insist upon a large number of nodes and computer time (and chances for error) that cannot possibly improve upon the basic uncertainty in the boundary conditions.

5. Finally, recognize that it is a rare occurrence when one has a "pure" conduction problem; there is almost always a coupling with convection and radiation. The reader should keep this in mind as we progress through subsequent chapters that treat heat convection and radiation in detail.

REVIEW QUESTIONS

1. What is meant by a lumped capacity? What are the physical assumptions necessary for a lumped-capacity unsteady-state analysis to apply?

2. What is meant by a semi-infinite solid?

3. What initial conditions are imposed on the transient solutions presented in graphical form in this chapter?

4. What boundary conditions are applied to problems in this chapter?

5. Define the error function.

6. Define the Biot and Fourier numbers.

7. Describe how one-dimensional transient solutions may be used for solution of two- and three-dimensional problems.

LIST OF WORKED EXAMPLES

PROBLEMS

4-1 A copper sphere initially at a uniform temperature T_0 is immersed in a fluid. Electric heaters are placed in the fluid and controlled so that the temperature of the fluid follows a periodic variation given by

$$T_\infty - T_m = A \sin \omega\tau$$

where

T_m = time-average mean fluid temperature
A = amplitude of temperature wave
ω = frequency

Derive an expression for the temperature of the sphere as a function of time and the heat-transfer coefficient from the fluid to the sphere. Assume that the temperatures of the sphere and fluid are uniform at any instant so that the lumped-capacity method of analysis may be used.

4-2 An infinite plate having a thickness of 2.5 cm is initially at a temperature of 150°C, and the surface temperature is suddenly lowered to 30°C. The thermal diffusivity of the material is 1.8×10^{-6} m²/s. Calculate the center-plate temperature after 1 min by summing the first four nonzero terms of Equation (4-3). Check the answer using the Heisler charts.

4-3 What error would result from using the first four terms of Equation (4-3) to compute the temperature at $\tau = 0$ and $x = L$? (Note: temperature $= T_i$.)

4-4 A solid body at some initial temperature T_0 is suddenly placed in a room where the air temperature is T_∞ and the walls of the room are very large. The heat-transfer coefficient for the convection heat loss is h, and the surface of the solid may be assumed black. Assuming that the temperature in the solid is uniform at any instant, write the differential equation for the variation in temperature with time, considering both radiation and convection.

4-5 A 20 by 20 cm slab of copper 5 cm thick at a uniform temperature of 260°C suddenly has its surface temperature lowered to 35°C. Using the concepts of thermal resistance and capacitance and the lumped-capacity analysis, find the time at which the center temperature becomes 90°C; $\rho = 8900$ kg/m³, $c_p = 0.38$ kJ/kg · °C, and $k = 370$ W/m · °C.

4-6 A piece of aluminum weighing 6 kg and initially at a temperature of 300°C is suddenly immersed in a fluid at 20°C. The convection heat-transfer coefficient is 58 W/m² · °C. Taking the aluminum as a sphere having the same weight as that given, estimate the time required to cool the aluminum to 90°C, using the lumped-capacity method of analysis.

4-7 Two identical 7.5-cm cubes of copper at 425 and 90°C are brought into contact. Assuming that the blocks exchange heat only with each other and that there is no resistance to heat flow as a result of the contact of the blocks, plot the temperature of each block as a function of time, using the lumped-capacity method of analysis. That is, assume the resistance to heat transfer is the conduction resistance of the two blocks. Assume that all surfaces are insulated except those in contact.

4-8 Repeat Problem 4-7 for a 7.5-cm copper cube at 425°C in contact with a 7.5-cm steel cube at 90°C. Sketch the thermal circuit.

4-9 An infinite plate of thickness $2L$ is suddenly exposed to a constant-temperature radiation heat source or sink of temperature T_s. The plate has a uniform initial temperature of T_i. The radiation heat loss from each side of the plate is given by $q = \sigma \epsilon A (T^4 - T_s^4)$, where σ and ϵ are constants and A is the surface area. Assuming that the plate behaves as a lumped capacity, that is, $k \to \infty$, derive an expression for the temperature of the plate as a function of time.

4-10 A stainless-steel rod (18% Cr, 8% Ni) 6.4 mm in diameter is initially at a uniform temperature of 25°C and is suddenly immersed in a liquid at 150°C with $h = 120$ W/m$^2 \cdot$°C. Using the lumped-capacity method of analysis, calculate the time necessary for the rod temperature to reach 120°C.

4-11 A 5-cm-diameter copper sphere is initially at a uniform temperature of 200°C. It is suddenly exposed to an environment at 20°C having a heat-transfer coefficient $h = 28$ W/m$^2 \cdot$°C. Using the lumped-capacity method of analysis, calculate the time necessary for the sphere temperature to reach 90°C.

4-12 A stack of common building brick 1 m high, 3 m long, and 0.5 m thick leaves an oven, where it has been heated to a uniform temperature of 300°C. The stack is allowed to cool in a room at 35°C with an air-convection coefficient of 15 W/m$^2 \cdot$°C. The bottom surface of the brick is on an insulated stand. How much heat will have been lost when the bricks cool to room temperature? How long will it take to lose half this amount, and what will the temperature at the geometric center of the stack be at this time?

4-13 A copper sphere having a diameter of 3.0 cm is initially at a uniform temperature of 50°C. It is suddenly exposed to an airstream of 10°C with $h = 15$ W/m$^2 \cdot$°C. How long does it take the sphere temperature to drop to 25°C?

4-14 An aluminum sphere, 5.0 cm in diameter, is initially at a uniform temperature of 50°C. It is suddenly exposed to an outer-space radiation environment at 0 K (no convection). Assuming the surface of aluminum is blackened and lumped-capacity analysis applies, calculate the time required for the temperature of the sphere to drop to −110°C.

4-15 An aluminum can having a volume of about 350 cm^3 contains beer at 1°C. Using a lumped-capacity analysis, estimate the time required for the contents to warm to 15°C when the can is placed in a room at 20°C with a convection coefficient of 15 W/m$^2 \cdot$°C. Assume beer has the same properties as water.

4-16 A 12-mm-diameter aluminum sphere is heated to a uniform temperature of 400°C and then suddenly subjected to room air at 20°C with a convection heat-transfer coefficient of 10 W/m$^2 \cdot$°C. Calculate the time for the center temperature of the sphere to reach 200°C.

4-17 A 4-cm-diameter copper sphere is initially at a uniform temperature of 200°C. It is suddenly exposed to a convection environment at 30°C with $h = 20$ W/m$^2 \cdot$°C. Calculate the time necessary for the center of the sphere to reach a temperature of 80°C.

4-18 When a sine-wave temperature distribution is impressed on the surface of a semi-infinite solid, the temperature distribution in the solid is given by

$$T_{x,\tau} - T_m = A \exp\left(-x\sqrt{\frac{\pi n}{\alpha}}\right) \sin\left(2\pi n \tau - x\sqrt{\frac{\pi n}{\alpha}}\right)$$

where

$T_{x,\tau}$ = temperature at depth x and time τ after start of temperature wave at surface

T_m = mean surface temperature

n = frequency of wave, cycles per unit time

A = amplitude of temperature wave at surface

If a sine-wave temperature distribution is impressed on the surface of a large slab of concrete such that the temperature varies from 35 to 90°C and a complete cycle is accomplished in 15 min, find the heat flow through a plane 5 cm from the surface 2 h after the start of the initial wave.

4-19 Using the temperature distribution of Problem 4-18, show that the time lag between maximum points in the temperature wave at the surface and at a depth x is given by

$$\Delta \tau = \frac{x}{2} \sqrt{\frac{1}{\alpha \pi n}}$$

4-20 A thick concrete wall having a uniform temperature of 54°C is suddenly subjected to an airstream at 10°C. The heat-transfer coefficient is 10 W/m² · °C. Calculate the temperature in the concrete slab at a depth of 7 cm after 30 min.

4-21 A very large slab of copper is initially at a temperature of 300°C. The surface temperature is suddenly lowered to 35°C. What is the temperature at a depth of 7.5 cm 4 min after the surface temperature is changed?

4-22 On a hot summer day a concrete driveway may reach a temperature of 50°C. Suppose that a stream of water is directed on the driveway so that the surface temperature is suddenly lowered to 10°C. How long will it take to cool the concrete to 25°C at a depth of 5 cm from the surface?

4-23 A semi-infinite slab of copper is exposed to a constant heat flux at the surface of 0.5 MW/m². Assume that the slab is in a vacuum, so that there is no convection at the surface. What is the surface temperature after 5 min if the initial temperature of the slab is 20°C? What is the temperature at a distance of 15 cm from the surface after 5 min?

4-24 A semi-infinite slab of material having $k = 0.1$ W/m · °C and $\alpha = 1.1 \times 10^{-7}$ m²/s is maintained at an initially uniform temperature of 20°C. Calculate the temperature at a depth of 5 cm after 100 s if (a) the surface temperature is suddenly raised to 150°C, (b) the surface is suddenly exposed to a convection source with $h = 40$ W/m² · °C and 150°C, and (c) the surface is suddenly exposed to a constant heat flux of 350 W/m².

4-25 A brick wall having a thickness of 10 cm is initially uniform in temperature at 25°C. One side is insulated. The other side is suddenly exposed to a convection environment with $T = 0$°C and $h = 200$ W/m² · °C. Using whatever method is suitable, plot the temperature of the insulated surface as a function of time. How might this calculation be applicable to building design?

4-26 A large slab of copper is initially at a uniform temperature of 90°C. Its surface temperature is suddenly lowered to 30°C. Calculate the heat-transfer rate through a plane 7.5 cm from the surface 10 s after the surface temperature is lowered.

4-27 A large slab of aluminum at a uniform temperature of 30°C is suddenly exposed to a constant surface heat flux of 15 kW/m². What is the temperature at a depth of 2.5 cm after 2 min?

4-28 For the slab in Problem 4-27, how long would it take for the temperature to reach 150°C at the depth of 2.5 cm?

4-29 A piece of ceramic material [$k = 0.8$ W/m · °C, $\rho = 2700$ kg/m³, $c = 0.8$ kJ/kg · °C] is quite thick and initially at a uniform temperature of 30°C. The surface of the material is suddenly exposed to a constant heat flux of 650 W/m². Plot the temperature at a depth of 1 cm as a function of time.

4-30 An aluminum sphere having a diameter of 5.6 cm is initially at a uniform temperature of 355°C and is suddenly exposed to a convection environment at $T = 23°C$ with a convection heat transfer coefficient of 78 W/m$^2 \cdot$ °C. Calculate the time for the center of the sphere to cool to a temperature of 73°C. Express the answer in seconds.

4-31 A large thick layer of ice is initially at a uniform temperature of −20°C. If the surface temperature is suddenly raised to −1°C, calculate the time required for the temperature at a depth of 1.5 cm to reach −11°C. The properties of ice are $\rho = 57$ lb$_m$/ft^3, $c_p = 0.46$ Btu/lbm $\cdot$ °F, $k = 1.28$ Btu/h $\cdot$ ft $\cdot$ °F, $\alpha = 0.048$ ft^2/h.

4-32 A large slab of concrete (stone 1-2-4 mix) is suddenly exposed to a constant radiant heat flux of 900 W/m^2. The slab is initially uniform in temperature at 20°C. Calculate the temperature at a depth of 10 cm in the slab after a time of 9 h.

4-33 A very thick plate of stainless steel (18% Cr, 8% Ni) at a uniform temperature of 300°C has its surface temperature suddenly lowered to 100°C. Calculate the time required for the temperature at a depth of 3 cm to attain a value of 200°C.

4-34 A large slab has properties of common building brick and is heated to a uniform temperature of 40°C. The surface is suddenly exposed to a convection environment at 2°C with $h = 25$ W/m$^2 \cdot$ °C. Calculate the time for the temperature to reach 20°C at a depth of 8 cm.

4-35 A large block having the properties of chrome brick at 200°C is at a uniform temperature of 30°C when it is suddenly exposed to a surface heat flux of 3×10^4 W/m^2. Calculate the temperature at a depth of 3 cm after a time of 10 min. What is the surface temperature at this time?

4-36 A slab of copper having a thickness of 3.0 cm is initially at 300°C. It is suddenly exposed to a convection environment on the top surface at 80°C while the bottom surface is insulated. In 6 min the surface temperature drops to 140°C. Calculate the value of the convection heat-transfer coefficient.

4-37 A large slab of aluminum has a thickness of 10 cm and is initially uniform in temperature at 400°C. Suddenly it is exposed to a convection environment at 90°C with $h = 1400$ W/m$^2 \cdot$ °C. How long does it take the centerline temperature to drop to 180°C?

4-38 A horizontal copper plate 10 cm thick is initially uniform in temperature at 250°C. The bottom surface of the plate is insulated. The top surface is suddenly exposed to a fluid stream at 80°C. After 6 min the surface temperature has dropped to 150°C. Calculate the convection heat-transfer coefficient that causes this drop.

4-39 A large slab of aluminum has a thickness of 10 cm and is initially uniform in temperature at 400°C. It is then suddenly exposed to a convection environment at 90°C with $h = 1400$ W/m$^2 \cdot$ °C. How long does it take the center to cool to 180°C?

4-40 A plate of stainless steel (18% Cr, 8% Ni) has a thickness of 3.0 cm and is initially uniform in temperature at 500°C. The plate is suddenly exposed to a convection environment on both sides at 40°C with $h = 150$ W/m$^2 \cdot$ °C. Calculate the times for the center and face temperatures to reach 120°C.

4-41 A steel cylinder 10 cm in diameter and 10 cm long is initially at 300°C. It is suddenly immersed in an oil bath that is maintained at 40°C, with $h = 280$ W/m$^2 \cdot$ °C. Find (*a*) the temperature at the center of the solid after 2 min and (*b*) the temperature at the center of one of the regular faces after 2 min.

4-42 Derive an expression for the heat flux per unit area at depth x and time τ when a semi-infinite solid is suddenly exposed to an instantaneous energy pulse at the surface of strength Q_0/A.

4-43 Buildings of various constructions exhibit different responses to thermal changes in climate conditions. Consider a 10-cm-thick wall of normal weight structural concrete ($c = 0.9 \, \text{kJ/kg} \cdot °\text{C}$) suddenly exposed to a "blue norther" at $-10°\text{C}$ with a convection coefficient of $65 \, \text{W/m}^2 \cdot °\text{C}$. The wall is initially at $15°\text{C}$. Estimate the time required for the wall temperature to drop to $5°\text{C}$. State the assumptions.

4-44 A semi-infinite solid of aluminum is coated with a special chemical material that reacts suddenly to ultraviolet radiation and releases energy in the amount of $1.0 \, \text{MJ/m}^2$. If the solid is initially uniform in temperature at $20°\text{C}$, calculate the temperature at a depth of 2.3 cm after 1.8 s.

4-45 A semi-infinite solid of stainless steel (18% Cr, 8% Ni) is initially at a uniform temperature of $0°\text{C}$. The surface is pulsed with a laser with $10 \, \text{MJ/m}^2$ instantaneous energy. Calculate the temperature at the surface and depth of 1 cm after a time of 3 s.

4-46 What strength pulse would be necessary to produce the same temperature effect at a depth of 1.2 cm as that experienced at a depth of 1.0 cm?

4-47 Calculate the heat flux at $x = 1$ cm and $\tau = 3$ s for the conditions of Problem 4-45.

4-48 A semi-infinite solid of aluminum is to be pulsed with a laser at the surface such that a temperature of $600°\text{C}$ will be attained at a depth of 2 mm, 0.2 s after the pulse. The solid is initially at $30°\text{C}$. Calculate the strength of pulse required, expressed in MJ/m^2.

4-49 A slab of polycrystalline aluminum oxide is to be pulsed with a laser to produce a temperature of $900°\text{C}$ at a depth of 0.2 mm after a time of 0.2 s. The solid is initially at $40°\text{C}$. Calculate the strength of pulse required expressed in MJ/m^2.

4-50 Repeat Problem 4-49 for window glass.

4-51 An aluminum bar has a diameter of 11 cm and is initially uniform in temperature at $300°\text{C}$. If it is suddenly exposed to a convection environment at $50°\text{C}$ with $h = 1200 \, \text{W/m}^2 \cdot °\text{C}$, how long does it take the center temperature to cool to $80°\text{C}$? Also calculate the heat loss per unit length.

4-52 A fused-quartz sphere has a thermal diffusivity of $9.5 \times 10^{-7} \, \text{m}^2/\text{s}$, a diameter of 2.5 cm, and a thermal conductivity of $1.52 \, \text{W/m} \cdot °\text{C}$. The sphere is initially at a uniform temperature of $25°\text{C}$ and is suddenly subjected to a convection environment at $200°\text{C}$. The convection heat-transfer coefficient is $110 \, \text{W/m}^2 \cdot °\text{C}$. Calculate the temperatures at the center and at a radius of 6.4 mm after a time of 3 min.

4-53 Lead shot may be manufactured by dropping molten-lead droplets into water. Assuming that the droplets have the properties of solid lead at $300°\text{C}$, calculate the time for the center temperature to reach $120°\text{C}$ when the water is at $100°\text{C}$ with $h = 5000 \, \text{W/m}^2 \cdot °\text{C}$, $d = 1.5$ mm.

4-54 A steel sphere 10 cm in diameter is suddenly immersed in a tank of oil at $10°\text{C}$. The initial temperature of the sphere is $220°\text{C}$; $h = 5000 \, \text{W/m}^2 \cdot °\text{C}$. How long will it take the center of the sphere to cool to $120°\text{C}$?

4-55 A boy decides to place his glass marbles in an oven at $200°\text{C}$. The diameter of the marbles is 15 mm. After a while he takes them from the oven and places them in room air at $20°\text{C}$ to cool. The convection heat-transfer coefficient is approximately $14 \, \text{W/m}^2 \cdot °\text{C}$. Calculate the time the boy must wait until the center temperature of the marbles reaches $50°\text{C}$.

4-56 A lead sphere with $d = 1.5$ mm and initial temperature of $200°\text{C}$ is suddenly exposed to a convection environment at $100°\text{C}$ and $h = 5000 \, \text{W/m}^2 \cdot °\text{C}$. Calculate the time for the center temperature to reach $120°\text{C}$.

4-57 A long steel bar 5 by 10 cm is initially maintained at a uniform temperature of 250°C. It is suddenly subjected to a change such that the environment temperature is lowered to 35°C. Assuming a heat-transfer coefficient of 23 W/m² · °C, use a numerical method to estimate the time required for the center temperature to reach 90°C. Check this result with a calculation using the Heisler charts.

4-58 A steel bar 2.5 cm square and 7.5 cm long is initially at a temperature of 250°C. It is immersed in a tank of oil maintained at 30°C. The heat-transfer coefficient is 570 W/m² · °C. Calculate the temperature in the center of the bar after 3 min.

4-59 A cube of aluminum 10 cm on each side is initially at a temperature of 300°C and is immersed in a fluid at 100°C. The heat-transfer coefficient is 900 W/m² · °C. Calculate the temperature at the center of one face after 1 min.

4-60 A short concrete cylinder 15 cm in diameter and 30 cm long is initially at 25°C. It is allowed to cool in an atmospheric environment in which the temperature is 0°C. Calculate the time required for the center temperature to reach 10°C if the heat-transfer coefficient is 17 W/m² · °C.

4-61 Assume that node m in Problem 3-39 occurs along a circular rod having a diameter of 2 cm with $\Delta x = 1$ cm. The material is glass with $k = 0.8$ W/m · °C, $\rho = 2700$ kg/m³, $c = 0.84$ kJ/kg · °C. The convection surrounding condition is $h = 50$ W/m² · °C and $T_\infty = 35$°C. Write the transient nodal equation for node m and determine the corresponding maximum allowable time increment, expressed in seconds.

4-62 A 4.0-cm cube of aluminum is initially at 450°C and is suddenly exposed to a convection environment at 100°C with $h = 120$ W/m² · °C. How long does it take the cube to cool to 250°C?

4-63 A cube of aluminum 11 cm on each side is initially at a temperature of 400°C. It is suddenly immersed in a tank of oil maintained at 85°C. The convection coefficient is 1100 W/m² · °C. Calculate the temperature at the center of one face after a time of 1 min.

4-64 An aluminum cube 5 cm on a side is initially at a uniform temperature of 100°C and is suddenly exposed to room air at 25°C. The convection heat-transfer coefficient is 20 W/m² · °C. Calculate the time required for the geometric center temperature to reach 50°C.

4-65 A stainless steel cylinder (18% Cr, 8% Ni) is heated to a uniform temperature of 200°C and then allowed to cool in an environment where the air temperature is maintained constant at 30°C. The convection heat-transfer coefficient may be taken as 200 W/m² · °C. The cylinder has a diameter of 10 cm and a length of 15 cm. Calculate the temperature of the geometric center of the cylinder after a time of 10 min. Also calculate the heat loss.

4-66 A cylinder having a diameter of 15 cm and a length of 30 cm is initially uniform in temperature at 300°C. It is suddenly exposed to a convection environment at 20°C with $h = 35$ W/m² · °C. Properties of the solid are $k = 2.3$ W/m · °C, $\rho = 300$ kg/m³, and $c = 840$ J/kg · °C. Calculate the time for (a) the center and (b) the center of one face to reach a temperature of 120°C. Also calculate the heat loss for each case.

4-67 A rectangular solid is 15 by 10 by 20 cm and has the properties of fireclay brick. It is initially uniform in temperature at 300°C and then suddenly exposed to a convection environment at 80°C and $h = 110$ W/m² · °C. Calculate the time for (a) the geometric center and (b) the center of each face to reach a temperature of 190°C. Also calculate the heat loss for each of these times.

4-68 Calculate the heat loss for both cases in Problem 4-45.

Figure P4-72

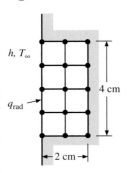

h, T_∞

q_{rad}

4 cm

←2 cm→

Figure P4-73

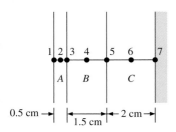

1 2 3 4   5 6   7

A   B   C

0.5 cm →| |← →|← 2 cm →|
       1.5 cm

4-69 Calculate the heat loss for the bar in Problem 4-58 per unit length.

4-70 Calculate the heat loss for the cube in Problem 4-59.

4-71 Develop a backward-difference formulation for a boundary node subjected to a convection environment. Check with Table 4-3.

4-72 The stainless-steel plate is surrounded by an insulating block as shown in Figure P4-72 and is initially at a uniform temperature of 50°C with a convection environment at 50°C. The plate is suddenly exposed to a radiant heat flux of 20 kW/m². Calculate the temperatures at the indicated nodes after 10 s, 1 min, and 10 min. Take the properties of stainless steel as $k = 16$ W/m · °C, $\rho = 7800$ kg/m³, and $c = 0.46$ kJ/kg · °C, $h = 30$ W/m² · °C. Assume all the radiation is absorbed.

4-73 The composite plate shown in Figure P4-73 has one face insulated and is initially at a uniform temperature of 100°C. At time zero the face is suddenly exposed to a convection environment at 10°C and $h = 70$ W/m² · °C. Determine the temperatures at the indicated nodes after 1 s, 10 s, 1 min, and 10 min.

Material	k, W/m · °C	ρ, kg/m³	c, kJ/kg · °C
A	20	7800	0.46
B	1.2	1600	0.85
C	0.5	2500	0.8

Figure P4-74

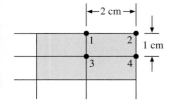

←2 cm→

1   2   1 cm

3   4

4-74 The corner shown in Figure P4-74 is initially uniform at 200°C and then suddenly exposed to convection around the edge with $h = 50$ W/m² · °C and $T = 30$°C. Assume the solid has the properties of fireclay brick. Examine nodes 1, 2, 3, and 4 and determine the maximum time increment which may be used for a transient numerical calculation.

4-75 An aluminum rod 2.5 cm in diameter and 20 cm long protrudes from a wall maintained at 200°C and is exposed to a convection environment with $h = 50$ W/m² · °C and a temperature of 20°C. Using $\Delta x = 4$ cm write a transient nodal equation for the node at the tip of the fin and determine the maximum allowable time increment for that node.

4-76 Write the nodal equation for node 3 in Figure P4-76 for use in a transient analysis. Determine the stability criterion for this node.

Figure P4-76

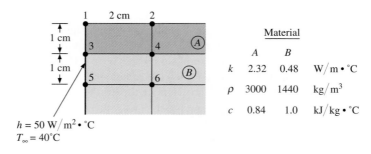

1   2 cm   2

1 cm

3   4   Ⓐ   Material

1 cm   A   B

5   6   Ⓑ   k   2.32   0.48   W/m • °C

                                ρ 3000   1440   kg/m³

$h = 50$ W/m² • °C   c   0.84   1.0   kJ/kg • °C
$T_\infty = 40$°C

Figure P4-77

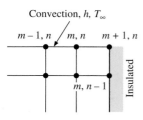

Convection, h, T_∞

$m-1, n$ m, n $m+1, n$

$m, n-1$

Insulated

4-77 Write a nodal equation for analysis of node (m, n) in Figure P4-77 to be used in a transient analysis of the solid.

4-78 Write the nodal equation and establish the stability criteria for node 1 in Figure P4-78 (transient analysis). Materials A and B have the properties given in Problem 4-73.

Figure P4-78

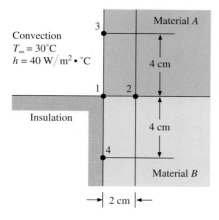

4-79 Write the transient equation for node 1 in Figure P4-79 and determine the maximum allowable time increment that may be employed in the calculation. The properties of materials A, B, and C are the same as those given in Problem 4-73.

Figure P4-79

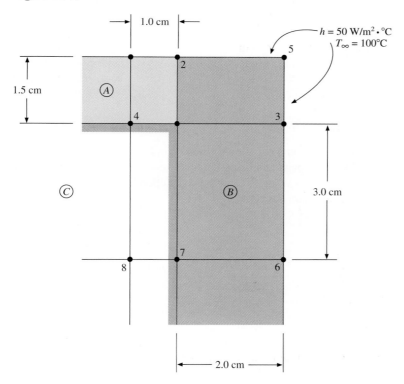

4-80 Calculate the maximum time increment that can be used for node 5 in Figure P4-80 for a transient numerical analysis. Also write the nodal equation for this node.

Figure P4-80

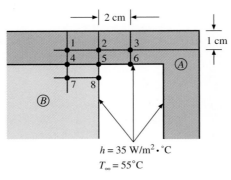

$$h = 35 \ \text{W/m}^2 \cdot ^\circ\text{C}$$
$$T_\infty = 55^\circ\text{C}$$

A = gypsum plaster
B = duralumin

4-81 The corner shown in Figure P4-81 is initially uniform at 300°C and then suddenly exposed to a convection environment at 50°C with $h = 60 \ \text{W/m}^2 \cdot ^\circ\text{C}$. Assume the solid has the properties of fireclay brick. Examine nodes 1, 2, 3, 4, and 5 and determine the maximum time increment which may be used for a transient numerical calculation.

Figure P4-81

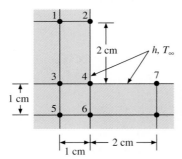

4-82 Write a steady-state nodal equation for node 3 in Figure P4-82 assuming unit depth perpendicular to the page and using the node spacing shown. The thermal conductivity of the solid is 15 W/m · °C and the convection heat-transfer coefficient on the side surface is 25 W/m² · °C.

Figure P4-82

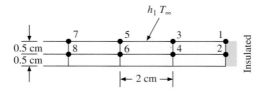

4-83 Devise a nodal equation that will take into account a change in the phase of the material. Assume that the volume remains constant during the change from the solid to liquid or liquid to solid phase, and that the energy required to liquefy or solidify a unit mass is u_{if}.

4-84 Write the transient nodal equation for node 7 in Figure P4-84 and determine the maximum allowable time increment for the node. Properties of materials A and B are given in the figure.

Figure P4-84

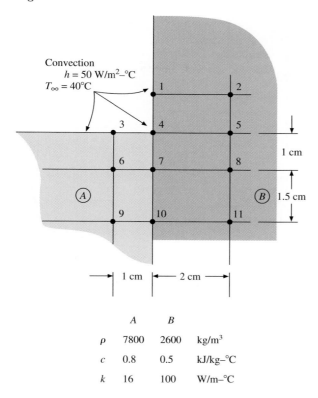

	A	B	
ρ	7800	2600	kg/m^3
c	0.8	0.5	kJ/kg–°C
k	16	100	W/m–°C

4-85 For the section shown in Figure P4-85, calculate the maximum time increment allowed for node 2 in a transient numerical analysis. Also write the entire nodal equation for this node.

Figure P4-85

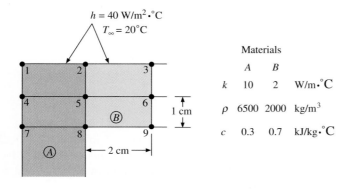

Materials

	A	B	
k	10	2	W/m·°C
ρ	6500	2000	kg/m^3
c	0.3	0.7	kJ/kg·°C

4-86 A transient numerical analysis is to be performed on the composite material section shown in Figure P4-86. Calculate the maximum time increment that can be used for node 5 to ensure convergence.

Figure P4-86

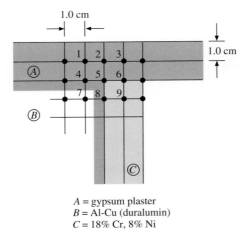

A = gypsum plaster
B = Al-Cu (duralumin)
C = 18% Cr, 8% Ni

4-87 For the section shown in Figure P4-87, calculate the maximum time increment allowed for node 4 in a transient numerical environment. Also write the complete nodal equation for node 4.

Figure P4-87

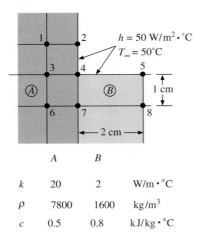

$h = 50 \ \text{W/m}^2 \cdot {}^\circ\text{C}$
$T_\infty = 50\,{}^\circ\text{C}$

		A	B	
k		20	2	W/m $\cdot$ °C
ρ		7800	1600	kg/m^3
c		0.5	0.8	kJ/kg $\cdot$ °C

4-88 Write the transient nodal temperature equation for node 1 in Figure P4-88. Also determine the maximum allowable time increment for the node. The right face is exposed to the convection condition shown. Properties for materials A and B are given in the figure.

Figure P4-88

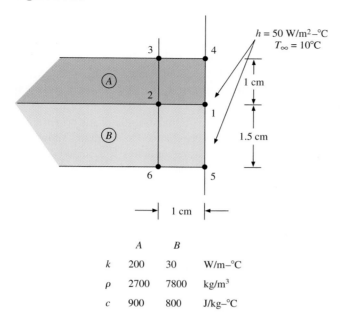

	A	B	
k	200	30	W/m–°C
ρ	2700	7800	kg/m³
c	900	800	J/kg–°C

4-89 Write the nodal equation for a transient analysis of node 2 in Figure P4-89 and determine the stability criterion for this node. The properties for materials A and B are given in the figure.

Figure P4-89

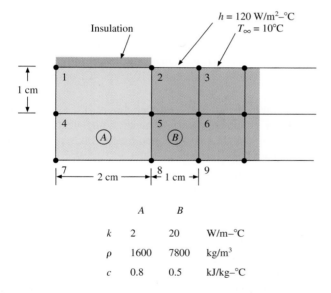

	A	B	
k	2	20	W/m–°C
ρ	1600	7800	kg/m³
c	0.8	0.5	kJ/kg–°C

4-90 A node like that shown in Table 3-2d has both x and y increments equal to 1.0 cm. The convection boundary condition is at 50°C and $h = 60$ W/m² · °C. The solid material is stainless steel (18% Cr, 8% Ni). Using the thermal resistance and capacitance formulation for a transient analysis, write the nodal equation for this node and determine the maximum allowable time increment.

4-91 The solid in Problem 3-51 is initially uniform in temperature at 10°C. At time zero the right face is suddenly changed to 38°C and the left face exposed to the convection environment. Nodes 3 and 6 remain at 10°C. Select an appropriate value for $\Delta\tau$ and calculate the temperatures of nodes 1, 2, 4, and 5 after 10 time increments. Carry the calculation forward to verify the steady-state distribution. Take $\rho = 3000 \text{ kg/m}^3$ and $c = 840 \text{ J/kg} \cdot \text{°C}$.

4-92 The solid in Problem 3-53 has $k = 11 \text{ W/m} \cdot \text{°C}$ and is initially uniform in temperature at 1000°C. At time zero the four surfaces are changed to the values shown. Select an appropriate $\Delta\tau$ and calculate the temperatures of nodes 1, 2, 3, and 4 after 10 time increments. Also obtain the limiting steady-state temperatures. Take $\rho = 2800 \text{ kg/m}^3$ and $c = 940 \text{ J/kg} \cdot \text{°C}$.

4-93 The fin in Problem 3-56 is initially uniform in temperature at 300°C and then suddenly exposed to the convection environment. Select an appropriate $\Delta\tau$ and calculate the nodal temperatures after 10 time increments. Take $\rho = 2200 \text{ kg/m}^3$ and $c = 820 \text{ J/kg} \cdot \text{°C}$.

4-94 The fin in Problem 3-57 is initially uniform in temperature at 200°C and then is suddenly exposed to the convection environment shown while maintaining the bottom face at 200°C. Select an appropriate $\Delta\tau$ and calculate the nodal temperatures after 10 time increments. Repeat for 100 $\Delta\tau$. Take $\rho = 7800 \text{ kg/m}^3$ and $c = 460 \text{ J/kg} \cdot \text{°C}$.

4-95 The solid in Problem 3-58 is initially uniform in temperature at 100°C and then suddenly exposed to the convection condition while the right and bottom faces are held constant at 100°C. Select a value for $\Delta\tau$ and calculate the nodal temperatures after 10 time increments. Take $\rho = 3000 \text{ kg/m}^3$ and $c = 800 \text{ J/kg} \cdot \text{°C}$.

4-96 The solid in Problem 3-59 is initially uniform in temperature at 50°C and suddenly is exposed to the convection condition. Select a value for $\Delta\tau$ and calculate the nodal temperatures after 10 time increments. Take $\rho = 2500 \text{ kg/m}^3$ and $c = 900 \text{ J/kg} \cdot \text{°C}$.

4-97 The solids in Problem 3-60 are initially uniform in temperature at 300°C and suddenly are exposed to the convection boundary, while the inner temperature is kept constant at 300°C. Select a value for $\Delta\tau$ and calculate the nodal temperatures after 10 time increments. Take $\rho_A = 2900 \text{ kg/m}^3$, $c_A = 810 \text{ J/kg} \cdot \text{°C}$, $\rho_B = 7800 \text{ kg/m}^3$, and $c_B = 470 \text{ J/kg} \cdot \text{°C}$.

4-98 The fin in Problem 3-61 is initially uniform in temperature at 200°C, and then suddenly exposed to the convection boundary and heat generation. Select a value for $\Delta\tau$ and calculate the nodal temperatures for 10 time increments. Take $\rho = 7600 \text{ kg/m}^3$ and $c = 450 \text{ J/kg} \cdot \text{°C}$. The base stays constant at 200°C.

4-99 The solid in Problem 3-62 is initially uniform in temperature at 500°C and suddenly exposed to the convection boundary while the inner surface is kept constant at 500°C. Select a value for $\Delta\tau$ and calculate the nodal temperatures after 10 time increments. Take $\rho = 500 \text{ kg/m}^3$ and $c = 810 \text{ J/kg} \cdot \text{°C}$.

4-100 Repeat Problem 4-99 for the steel liner of Problem 3-98. Take $\rho = 7800 \text{ kg/m}^3$ and $c = 460 \text{ J/kg} \cdot \text{°C}$ for the steel.

4-101 The plate in Problem 3-63 is initially uniform in temperature at 100°C and suddenly exposed to the convection boundary. Select a value for $\Delta\tau$ and calculate the nodal temperatures after 10 time increments. Take $\rho = 7500 \text{ kg/m}^3$ and $c = 440 \text{ J/kg} \cdot \text{°C}$.

4-102 The solid shown in Problem 3-64 is initially uniform in temperature at 100°C and suddenly exposed to the convection boundary and heat generation while the right face is kept at 100°C. Select a value for $\Delta\tau$ and calculate the nodal temperatures after 10 time increments. Take $\rho = 7600$ kg/m^3 and $c = 460$ J/kg · °C.

4-103 A steel rod 12.5 mm in diameter and 20 cm long has one end attached to a heat reservoir at 250°C. The bar is initially maintained at this temperature throughout. It is then subjected to an airstream at 30°C such that the convection heat-transfer coefficient is 35 W/m^2 · °C. Estimate the time required for the temperature midway along the length of the rod to attain a value of 190°C.

4-104 A concrete slab 15 cm thick has a thermal conductivity of 0.87 W/m · °C and has one face insulated and the other face exposed to an environment. The slab is initially uniform in temperature at 300°C, and the environment temperature is suddenly lowered to 90°C. The heat-transfer coefficient is proportional to the fourth root of the temperature difference between the surface and environment and has a value of 11 W/m^2 · °C at time zero. The environment temperature increases linearly with time and has a value of 200°C after 20 min. Using the numerical method, obtain the temperature distribution in the slab after 5, 10, 15, and 20 min.

4-105 The two-dimensional body of Figure 3-6 has the initial surface and internal temperatures as calculated. At time zero the 500°C face is suddenly lowered to 30°C. Taking $\Delta x = \Delta y = 15$ cm and $\alpha = 1.29 \times 10^{-5}$ m^2/s, calculate the temperatures at nodes 1, 2, 3, and 4 after 30 min. Perform the calculation using both a forward- and a backward-difference method. For the backward-difference method use only two time increments. Take $k = 45$ W/m · °C.

4-106 The strip of material shown in Figure P4-106 has a thermal conductivity of 20 W/m · °C and is placed firmly on the isothermal surface maintained at 50°C. At time zero the strip is suddenly exposed to an airstream with $T_\infty = 300$°C and $h = 40$ W/m^2 · °C. Using a numerical technique, calculate the temperatures at nodes 1 to 8 after 1 s, 10 s, 1 min, and steady state; $\rho = 7000$ kg/m^3 and $c = 0.5$ kJ/kg · °C.

Figure P4-106

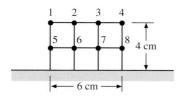

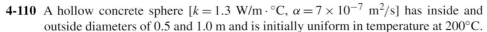

4-107 Rework Problems 4-7 and 4-8 using the numerical technique.

4-108 Rework Problem 4-103 using the numerical technique.

4-109 A blackened stainless-steel sphere of 10 cm diameter is initially uniform in temperature at 1000°K and is suddenly placed in outer space where it loses heat by radiation (no convection) according to

$$q_{rad} = \sigma A T^4 \qquad T \text{ in degrees Kelvin}$$
$$\sigma = 5.669 \times 10^{-8} \text{ W/m}^2 \cdot \text{K}^4$$

Calculate the temperatures of the nodes shown in Figure P4-109 for several increments of time and the corresponding heat losses. Use the values of k, ρ, and c from Problem 4-72.

Figure P4-109

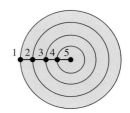

4-110 A hollow concrete sphere [$k = 1.3$ W/m · °C, $\alpha = 7 \times 10^{-7}$ m^2/s] has inside and outside diameters of 0.5 and 1.0 m and is initially uniform in temperature at 200°C.

The outside surface is suddenly lowered to 20°C. Calculate the nodal temperatures shown in Figure P4-110 for several increments of time. Assume the inside surface acts as though it were insulated.

Figure P4-110

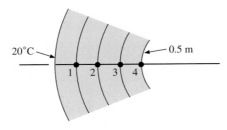

4-111 Repeat Problem 4-62 with the top surface also losing heat by radiation according to

$$q_{rad} = \sigma A \epsilon (T^4 - T_\infty^4) \qquad T \text{ in degrees Kelvin}$$
$$\sigma = 5.669 \times 10^{-8} \text{ W/m}^2 \cdot \text{K}^4$$
$$\epsilon = 0.7$$

4-112 A fireproof safe is constructed of loosely packed asbestos contained between thin sheets of stainless steel. The safe is built in the form of a cube with inside and outside dimensions of 0.5 and 1.0 m. If the safe is initially uniform in temperature at 30°C and the outside is suddenly exposed to a convection environment at 600°C, $h = 100$ W/m$^2 \cdot$ °C, calculate the time required for the inside temperature to reach 150°C. Assume the inside surface is insulated, and neglect the resistance and capacitance of the stainless steel. Take the properties of asbestos as $k = 0.16$ W/m $\cdot$ °C, $\alpha = 3.5 \times 10^{-7}$ m^2/s.

4-113 The half-cylinder in Problem 3-66 is initially uniform in temperature at 300°C and then suddenly exposed to the convection boundary while the bottom side is maintained at 300°C. Calculate the nodal temperatures for several time increments, and compute the heat loss in each period. Take $\alpha = 0.5 \times 10^{-5}$ m^2/s.

4-114 A large slab of brick [$k = 1.07$ W/m $\cdot$ °C, $\alpha = 5.4 \times 10^{-7}$m^2/s] is initially at a uniform temperature of 20°C. One surface is suddenly exposed to a uniform heat flux of 4500 W/m^2. Calculate and plot the surface temperature as a function of time. Also calculate the heat flux through the plane 2.0 cm deep when the surface temperature reaches 150°C.

4-115 A ceramic plate having a thickness of 2.0 cm is heated to a uniform temperature of 1000°K and suddenly exposed to radiation on both sides at 300°K. The properties of the solid are $k = 1.2$ W/m $\cdot$ °C, $\rho = 2500$ kg/m^3, $c = 0.9$ kJ/kg $\cdot$ °C, and $\epsilon = 0.85$. Divide the plate into eight segments ($\Delta x = 0.25$ cm) and, using a numerical technique, obtain information to plot the center and surface temperatures as a function of time.

4-116 Suppose the ceramic of Problem 4-115 is in the form of a long cylinder having a diameter of 2.0 cm. Divide the cylinder into four increments ($\Delta r = 0.25$ cm) and obtain information to plot the center and surface temperatures as a function of time.

4-117 A granite sphere having a diameter of 15 cm and initially at a uniform temperature of 120°C is suddenly exposed to a convection environment with $h = 350$ W/m$^2 \cdot °$C and $T = 30°$C. Calculate the temperature at a radius of 4.5 cm after 21 min and the energy removed from the sphere in this time. Take the properties of granite as $k = 3.2$ W/m $\cdot$ °C and $\alpha = 13 \times 10^{-7}$ m^2/s.

4-118 A 10-cm-thick brick wall having the properties of common building brick initially at a uniform temperature of 80°C is suddenly exposed to a convection environment of $T_\infty = 20°$C and $h = 100$ W/m$^2 \cdot °$C. Using $\Delta x = 2.5$ cm, calculate the time for the center temperature to reach 50°C using the numerical method. Also determine the maximum time increment for these calculations.

4-119 A chrome steel plate (1% Cr) is heated in an oven to a uniform temperature of 200°C and then subjected to a convection environment having $T_\infty = 20°$C and $h = 300$ W/m$^2 \cdot °$C on both sides. The plate thickness is 10 cm. Taking $\Delta x = 1$ cm, calculate the center temperature after 5 and 10 min using the numerical method. Also solve using the Heisler charts.

4-120 A long slab of oak 4.1 by 9.2 cm is initially at 20°C and is placed in an oven with $T_\infty = 200°$C and $h = 40$ W/m$^2 \cdot °$C. Calculate the time required for the surface to reach 120°C. Repeat for the geometric center.

4-121 Consider two solids initially at uniform temperatures of 200°C with $k = 1.4$ W/m $\cdot$ °C and $\alpha = 7 \times 10^{-7}$ m^2/s: (a) a semi-infinite solid and (b) an infinite plate 10 cm thick. Both solids are suddenly exposed to a convection environment at 25°C with $h = 40$ W/m$^2 \cdot °$C. Calculate the temperatures at the center of the plate and for $x = 5$ cm in the semi-infinite solid for 5, 10, 20, and 30 min. What do you conclude from these calculations?

4-122 Make the calculations for Problem 4-121 based on a lumped-capacity analysis and comment on the results.

4-123 For the square grid imposed on the degular quadrant shown in Figure P4-123, write the transient explicit nodal equations for nodes 3 and 4. Take $k = 10$ W/m $\cdot$ °C, $\rho = 2000$ kg/m^3, and $c = 840$ J/kg $\cdot$ °C. Use information from Tables 3-2 and 3-4. What is the maximum allowable time increment for each node?

Figure P4-123

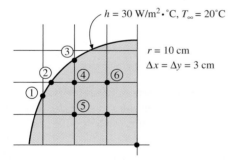

$h = 30$ W/m$^2 \cdot °$C, $T_\infty = 20°$C

$r = 10$ cm

$\Delta x = \Delta y = 3$ cm

4-124 Taking Figure P4-124 as a special case of Table 3-2(f), write an explicit formulation for nodes (m, n) and 2 using the resistance-capacity formulation and the information of Table 3-4.

Figure P4-124

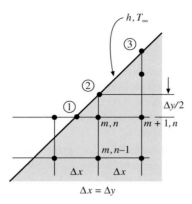

4-125 Repeat Problem 4-124 for a slanted surface that is (*a*) insulated and (*b*) isothermal at T_∞.

4-126 The slanted intersection shown in Figure P4-126 is an intersection of materials *A* and *B*. Write the transient nodal equations for nodes 3, 4, and 6 using information from Tables 3-2(*f*) and 3-3(*f* and *g*).

Figure P4-126

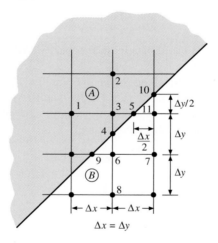

4-127 The solid of Problem 3-74 is initially uniform in temperature at 100°C, but suddenly the two surfaces are lowered to 0 and 40°C. If the solid has $k = 20$ W/m · °C and $\alpha = 5 \times 10^{-6}$ m²/s, find the steady-state temperature of each node and the nodal temperatures after 1 min.

4-128 The solid in Problem 3-76 is initially at a uniform temperature of 150°C and then suddenly exposed to the given boundary conditions with $h = 50$ W/m² · °C and $T_\infty = 20$°C. Taking the properties as $k = 61$ W/m · °C and $\alpha = 1.7 \times 10^{-5}$ m²/s, determine the steady-state values of the 12 nodes and the nodal temperature after 10 min.

4-129 The pin fin of Problem 3-72, initially uniform in temperature at 200°C, is suddenly exposed to the convection environment. Determine (*a*) the steady-state temperature distribution by a transient analysis taken to a long time and (*b*) the distribution for a time approximately equal to half the "long" time.

4-130 The solid in Problem 3-73 is initially uniform in temperature at 100°C before suddenly being exposed to $h = 100$ W/m$^2 \cdot$°C and $T_\infty = 0$°C. Take the properties as $k = 2$ W/m $\cdot$ °C and $\alpha = 7 \times 10^{-7}$ m^2/s. Determine (*a*) the steady-state temperature distribution by taking a transient analysis to a long time and (*b*) the temperature distribution at a time approximately half the "long" time.

4-131 The truncated cone shown in Problem 2-123 is insulated on the sides and initially at a uniform temperature of 20°C. While the large end is maintained at 20°C, the small end is suddenly raised to 320°C. Set up a 5-node model to predict the temperature distribution in the cone as a function of time and perform the calculations. Carry the calculation through to steady state and compare with the analytical results for Problem 2-123.

4-132 The one-dimensional solid shown in Problem 2-122 is initially at a uniform temperature of 20°C. One end is maintained at 20°C while the other end is suddenly raised to 70°C. Set up a five-node model to predict the temperature distribution in the cylindrical segment as a function of time and radial angle θ. Perform the calculations and carry through to steady state to compare with the analytical results of Problem 2-122.

4-133 The noninsulated cylindrical segment of Problem 2-130 is initially at a uniform temperature of 100°C. It is then suddenly exposed to the convection environment at 30°C, while maintaining one end of the segment at 100°C. The other end of the segment is suddenly lowered to 50°C at the same time as the exposure to the convection environment. Set up a numerical model using five nodes in the angle θ that may be used to predict the temperature behavior as a function of time. Perform the calculations and carry through to steady state to compare with the analytical results of Problem 2-130.

4-134 Apply the lumped-capacity criterion of Equation (4-6) $[h(V/A)/k < 0.1]$ to each of the geometries treated with the Heisler charts. Approximately what percent error would result for each geometry in the value of θ/θ_0 if a lumped capacity is assumed for the conditions of Equation (4-6)?

4-135 Because of symmetry, the temperature gradient $\partial T/\partial x$ at the centerline of an infinite plate will be zero when both sides are subjected to the same boundary condition in a cooling process. This may be interpreted that a half plate will act like a plate with one side insulated ($\partial T/\partial x = 0$), and the Heisler charts may be employed for the solution of problems with this boundary condition. Suppose an aluminum plate having a thickness of 5 cm is placed on an insulating material and is initially at a uniform temperature of 200°C. The exposed surface of the plate is suddenly subjected to a convection boundary with $h = 5000$ W/m$^2 \cdot$°C and $T_\infty = 25$°C. How long will it take for the back surface of the plate to reach a temperature of 90°C?

4-136 Rework Problem 4-135 for the surface temperature suddenly lowered to 25°C. This is equivalent to $h \rightarrow \infty$.

4-137 Rework Problem 4-135, assuming the plate behaves as a lumped capacity.

4-138 Rework Problem 4-135, assuming the aluminum plate behaves as a semi-infinite solid with the desired temperature occurring at $x = 5$ cm. Perform the same kind of calculation for Problem 4-136.

4-139 A concrete driveway having a thickness of 18 cm attains an essentially uniform temperature of 30°C on a warm November day in Texas. A "blue norther" arrives, which suddenly subjects the driveway to a convection boundary with $h = 23 \text{ W/m}^2 \cdot {}^\circ\text{C}$ and $T_\infty = 0{}^\circ\text{C}$. How long will it take for the surface temperature of the driveway to drop to $5{}^\circ\text{C}$? Work the problem two ways, using different assumptions.

Design-Oriented Problems

4-140 A 5-lb roast initially at $70{}^\circ\text{F}$ is placed in an oven at $350{}^\circ\text{F}$. Assuming that the heat-transfer coefficient is $2.5 \text{ Btu/h} \cdot \text{ft}^2 \cdot {}^\circ\text{F}$ and that the thermal properties of the roast may be approximated by those of water, estimate the time required for the center of the roast to attain a temperature of $200{}^\circ\text{F}$.

Figure P4-141

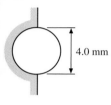

4.0 mm

4-141 The 4.0-mm-diameter stainless-steel wire shown in Figure P4-141 is initially at $20{}^\circ\text{C}$ and is exposed to a convection environment at $20{}^\circ\text{C}$ where h may be taken as $200 \text{ W/m}^2 \cdot {}^\circ\text{C}$. An electric current is applied to the wire such that there is a uniform internal heat generation of 500 MW/m^3. The left side of the wire is insulated as shown. Set up the nodal equations and stability requirement for calculating the temperature in the wire as a function of time, using increments of $\Delta r = 0.5 \text{ mm}$ and $\Delta\phi = \pi/4$. Take the properties of stainless steel as $k = 16 \text{ W/m} \cdot {}^\circ\text{C}$, $\rho = 7800 \text{ kg/m}^3$, and $c = 0.46 \text{ kJ/kg} \cdot {}^\circ\text{C}$.

4-142 Write a computer program which will solve Example 4-16 for different input properties. For nomenclature take T(N) = temperature of node N at beginning of time increment, TP(N) = temperature of node at end of time increment, X = number of nodes, W = width of plate, TA = temperature of left fluid, HA = convection coefficient of left fluid, TB = temperature of right fluid, HB = convection coefficient of right fluid, DT = time increment, C = specific heat, D = density, K = thermal conductivity, Q = heat-generation rate per unit volume, TI = total time. Write the program so that the user can easily rerun the program for new times and print out the results for each.

4-143 The stainless-steel plate shown in Figure P4-143 is initially at a uniform temperature of $150{}^\circ\text{C}$ and is suddenly exposed to a convection environment at $30{}^\circ\text{C}$ with $h = 17 \text{ W/m}^2 \cdot {}^\circ\text{C}$. Using numerical techniques, calculate the time necessary for the temperature at a depth of 6.4 mm to reach $65{}^\circ\text{C}$.

Figure P4-143

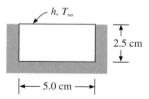

h, T_∞

2.5 cm

5.0 cm

4-144 Repeat Problem 4-143 with the top surface also losing heat by radiation according to

$$q_{\text{rad}} = \sigma A\epsilon(T^4 - T_\infty^4) \qquad T \text{ in degrees Kelvin}$$
$$\sigma = 5.669 \times 10^{-8} \text{ W/m}^2 \cdot \text{K}^4$$
$$\epsilon = 0.7$$

Repeat the calculation for 10 and 20 min.

4-145 Oranges with a diameter of about 3 in are to be cooled from room temperature of $25{}^\circ\text{C}$ to $3{}^\circ\text{C}$ using an air-convection environment with $h = 45 \text{ W/m}^2 \cdot {}^\circ\text{C}$ and $T_\infty = 0{}^\circ\text{C}$.

Assuming that the oranges have the properties of water at 10°C, calculate the time required for the cooling and the total cooling required for 100 oranges.

4-146 The rate at which cooling can be accomplished is of considerable importance in the food-processing industry. In a pizza-cooking application hot-air jets at 200°C can achieve heat-transfer coefficients of $h = 75$ W/m$^2 \cdot$ °C. Suppose the jets impinge on both sides of a pizza layer having a thickness of 1.2 cm at an initial uniform temperature of 25°C. How long does it take to reach a center temperature of 100°C? Take the properties of pizza as those of water ($k = 0.6$ W/m $\cdot$ °C, $\alpha = 1.5 \times 10^{-7}$ m^2/s).

4-147 A cold-storage building 16×35 m is built on a concrete slab having a thickness of 15 cm, which is placed on a suitable insulating material in contact with the ground. During the start-up period the interior of the building is exposed to convection air with $h = 20$ W/m$^2 \cdot$ °C and $T = -15$°C. The ground temperature may be taken as $+15$°C. The design objective is to achieve a steady-state temperature of 0°C at the inside surface of the concrete floor. Consider different insulating materials and thicknesses and recommend a selection that will achieve a cooldown within a reasonable period of time. For the design, take into account only the floor slab of the building.

4-148 A press is to be designed to heat and bond plastic layers. A transient operation is proposed whereby a 30- $\times$ 60-cm steel plate will be heated to 100°C by condensing steam in internal channels. The plate will then be brought into immediate contact with two 2.0-mm layers of plastic that bond at 50°C. Assuming the plastic has the properties of polyvinylchloride, comment on the design and estimate the time required to achieve a bonding temperature. Be sure to state all assumptions clearly.

4-149 A 2.0-mm-thick sheet of polyethylene covers a 10-cm-thick slab of high-density particle board that is perfectly insulated on the back side. The assembly is initially uniform in temperature at 20°C. If the outer surface of the plastic is suddenly exposed to a constant heat flux of 1300 W/m^2, estimate how long will it take for the insulated back surface to reach a temperature of 50°C. State your assumptions.

4-150 Free convection in air at atmospheric pressure is found to experience a convection heat-transfer coefficient that varies as $h = A(\Delta T)^n$, where ΔT is the temperature difference between the surface and the surrounding air, A is a constant, and n is some exponent. You are to devise a way to determine the constant and exponent in this equation by utilizing an experiment in combination with a lumped-capacity analysis. Consider a complex finned structure like that shown in Figure 2-13, where the mass, material of construction, and surface area can be determined. The structure is heated to a uniform initial temperature in an appropriate oven and then allowed to cool while exposed to room air at about 20°C. The initial temperature may be taken as about 200°C. The temperature of the structure is measured by a thermocouple device embedded within the structure and is displayed on a readout device. The structure is coated with a black paint so that it radiates as an ideal blackbody exchanging heat with a large enclosure according to Equation (1-12) with $\epsilon = 1.0$. Recall that the temperatures in this equation must be in degrees Kelvin. Write the finite-difference equation for cooling of the structure, taking account of both convection and radiation loss, and describe how experimental data for cooling of the body may be used to determine the values of the constant A and exponent n.

4-151 A safe is to be designed to withstand a fire at 600°C for a period of one hour while the contents remain below a temperature of 160°C during this time. Both the inner and outer shells of the safe are to be constructed of 1 percent carbon steel with an appropriate insulating material placed between. Select an interior volume for the safe and an insulating material which will withstand the temperatures. By a suitable analysis, determine the thickness of the shell and insulating materials needed to accomplish the design temperature objectives.

REFERENCES

1. Schneider, P. J. *Conduction Heat Transfer*. Reading, MA: Addison-Wesley Publishing Company, 1955.

2. Heisler, M. P. "Temperature Charts for Induction and Constant Temperature Heating," *Trans. ASME,* vol. 69, pp. 227–36, 1947.

3. Abramowitz, M., and I. Stegun (eds.). *Handbook of Mathematical Functions*, NBS AMS 55, U.S. Government Printing Office, 1964.

4. Dusinberre, G. M. *Heat Transfer Calculations by Finite Differences*. Scranton, PA: International Textbook Company, 1961.

5. Jakob, M. *Heat Transfer*, vol. 1. New York: John Wiley & Sons, 1949.

6. Gröber, H., S. Erk, and U. Grigull. *Fundamentals of Heat Transfer*. New York: McGraw-Hill, 1961.

7. Schneider, P. J. *Temperature Response Charts*. New York: John Wiley & Sons, 1963.

8. Schenck, H. *Fortran Methods in Heat Flow*. New York: The Ronald Press Company, 1963.

9. Richardson, P. D., and Y. M. Shum. "Use of Finite-Element Methods in Solution of Transient Heat Conduction Problems," *ASME Pap.* 69-WA/HT-36.

10. Emery, A. F., and W. W. Carson. "Evaluation of Use of the Finite Element Method in Computation of Temperature," *ASME Pap.* 69-WA/HT-38.

11. Wilson, E. L., and R. E. Nickell. "Application of the Finite Element Method to Heat Conduction Analysis," *Nucl. Eng. Des.*, vol. 4, pp. 276–86, 1966.

12. Zienkiewicz, O. C. *The Finite Element Method in Structural and Continuum Mechanics*. New York: McGraw-Hill, 1967.

13. Myers, G. E. *Conduction Heat Transfer*. New York: McGraw-Hill, 1972.

14. Arpaci, V. S. *Conduction Heat Transfer*. Reading, MA: Addison-Wesley Publishing Company, 1966.

15. Ozisik, M. N. *Boundary Value Problems of Heat Conduction*. Scranton, PA: International Textbook Company, 1968.

16. Langston, L. S. "Heat Transfer from Multidimensional Objects Using One-Dimensional Solutions for Heat Loss," *Int. J. Heat Mass Transfer*, vol. 25, p. 149, 1982.

17. Colakyan, M., R. Turton, and O. Levenspiel. "Unsteady State Heat Transfer to Variously Shaped Objects," *Heat Transfer Engr.*, vol. 5, p. 82, 1984.

18. Chapra, S. C., and R. P. Canale. *Numerical Methods for Engineers*, 2d ed. New York: McGraw-Hill, 1988.

19. Constantinides, A. "Applied Numerical Methods with Personal Computers," New York: McGraw-Hill, 1987.

20. Patankar, S. V. *Numerical Heat Transfer and Fluid Flow*. Hemisphere Publishing, 1980.

21. Minkowycz, W. J., E. M. Sparrow, G. E. Schneider, and R. H. Pletcher. *Handbook of Numerical Heat Transfer*. New York: Wiley, 1988.

CHAPTER 5

Principles of Convection

5-1 | INTRODUCTION

The preceding chapters have considered the mechanism and calculation of conduction heat transfer. Convection was considered only insofar as it related to the boundary conditions imposed on a conduction problem. We now wish to examine the methods of calculating convection heat transfer and, in particular, the ways of predicting the value of the convection heat-transfer coefficient h. The subject of convection heat transfer requires an energy balance along with an analysis of the fluid dynamics of the problems concerned. Our discussion in this chapter will first consider some of the simple relations of fluid dynamics and boundary-layer analysis that are important for a basic understanding of convection heat transfer. Next, we shall impose an energy balance on the flow system and determine the influence of the flow on the temperature gradients in the fluid. Finally, having obtained a knowledge of the temperature distribution, the heat-transfer rate from a heated surface to a fluid that is forced over it may be determined.

Our development in this chapter is primarily analytical in character and is concerned only with forced-convection flow systems. Subsequent chapters will present empirical relations for calculating forced-convection heat transfer and will also treat the subjects of natural convection and boiling and condensation heat transfer.

5-2 | VISCOUS FLOW

Consider the flow over a flat plate as shown in Figures 5-1 and 5-2. Beginning at the leading edge of the plate, a region develops where the influence of viscous forces is felt. These viscous forces are described in terms of a shear stress τ between the fluid layers. If this stress is assumed to be proportional to the normal velocity gradient, we have the defining equation for the viscosity,

$$\tau = \mu \frac{du}{dy} \qquad \text{[5-1]}$$

The constant of proportionality μ is called the *dynamic viscosity*. A typical set of units is newton-seconds per square meter; however, many sets of units are used for the viscosity, and care must be taken to select the proper group that will be consistent with the formulation at hand.

The region of flow that develops from the leading edge of the plate in which the effects of viscosity are observed is called the *boundary layer*. Some arbitrary point is used to

Figure 5-1 | Sketch showing different boundary-layer flow regimes on a
flat plate.

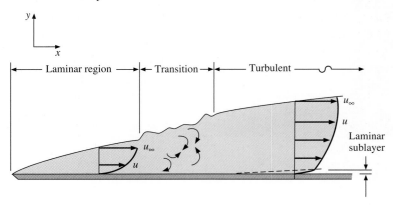

Figure 5-2 | Laminar velocity profile
on a flat plate.

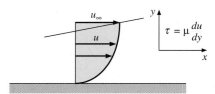

designate the y position where the boundary layer ends; this point is usually chosen as the
y coordinate where the velocity becomes 99 percent of the free-stream value.

Initially, the boundary-layer development is laminar, but at some critical distance from
the leading edge, depending on the flow field and fluid properties, small disturbances in the
flow begin to become amplified, and a transition process takes place until the flow becomes
turbulent. The turbulent-flow region may be pictured as a random churning action with
chunks of fluid moving to and fro in all directions.

The transition from laminar to turbulent flow occurs when

$$\frac{u_\infty x}{\nu} = \frac{\rho u_\infty x}{\mu} > 5 \times 10^5$$

where

u_∞ = free-stream velocity, m/s

x = distance from leading edge, m

$\nu = \mu/\rho$ = kinematic viscosity, m²/s

This particular grouping of terms is called the Reynolds number, and is dimensionless if a
consistent set of units is used for all the properties:

$$\mathrm{Re}_x = \frac{u_\infty x}{\nu} \qquad\qquad [5\text{-}2]$$

Although the critical Reynolds number for transition on a flat plate is usually taken as
5×10^5 for most analytical purposes, the critical value in a practical situation is strongly
dependent on the surface-roughness conditions and the "turbulence level" of the free stream.
The normal range for the beginning of transition is between 5×10^5 and 10^6. With very large

disturbances present in the flow, transition may begin with Reynolds numbers as low as 10^5, and for flows that are very free from fluctuations, it may not start until $\text{Re} = 2 \times 10^6$ or more. In reality, the transition process is one that covers a range of Reynolds numbers, with transition being complete and with developed turbulent flow usually observed at Reynolds numbers twice the value at which transition began.

The relative shapes for the velocity profiles in laminar and turbulent flow are indicated in Figure 5-1. The laminar profile is approximately parabolic, while the turbulent profile has a portion near the wall that is very nearly linear. This linear portion is said to be due to a laminar sublayer that hugs the surface very closely. Outside this sublayer the velocity profile is relatively flat in comparison with the laminar profile.

The physical mechanism of viscosity is one of momentum exchange. Consider the laminar-flow situation. Molecules may move from one lamina to another, carrying with them a momentum corresponding to the velocity of the flow. There is a net momentum transport from regions of high velocity to regions of low velocity, thus creating a force in the direction of the flow. This force is the viscous-shear stress, which is calculated with Equation (5-1).

The rate at which the momentum transfer takes place is dependent on the rate at which the molecules move across the fluid layers. In a gas, the molecules would move about with some average speed proportional to the square root of the absolute temperature since, in the kinetic theory of gases, we identify temperature with the mean kinetic energy of a molecule. The faster the molecules move, the more momentum they will transport. Hence we should expect the viscosity of a gas to be approximately proportional to the square root of temperature, and this expectation is corroborated fairly well by experiment. The viscosities of some typical fluids are given in Appendix A.

In the turbulent-flow region, distinct fluid layers are no longer observed, and we are forced to seek a somewhat different concept for viscous action. A qualitative picture of the turbulent-flow process may be obtained by imagining macroscopic chunks of fluid transporting energy and momentum instead of microscopic transport on the basis of individual molecules. Naturally, we should expect the larger mass of the macroscopic elements of fluid to transport more energy and momentum than the individual molecules, and we should also expect a larger viscous-shear force in turbulent flow than in laminar flow (and a larger thermal conductivity as well). This expectation is verified by experiment, and it is this larger viscous action in turbulent flow which causes the flat velocity profile indicated in Figure 5-1.

Consider the flow in a tube as shown in Figure 5-3. A boundary layer develops at the entrance, as shown. Eventually the boundary layer fills the entire tube, and the flow is said to be fully developed. If the flow is laminar, a parabolic velocity profile is experienced, as shown in Figure 5-3a. When the flow is turbulent, a somewhat blunter profile is observed, as in Figure 5-3b. In a tube, the Reynolds number is again used as a criterion for laminar and turbulent flow. For

$$\text{Re}_d = \frac{u_m d}{\nu} > 2300 \qquad \text{[5-3]}$$

the flow is usually observed to be turbulent d is the tube diameter.

Again, a range of Reynolds numbers for transition may be observed, depending on the pipe roughness and smoothness of the flow. The generally accepted range for transition is

$$2000 < \text{Re}_d < 4000$$

although laminar flow has been maintained up to Reynolds numbers of 25,000 in carefully controlled laboratory conditions.

The continuity relation for one-dimensional flow in a tube is

$$\dot{m} = \rho u_m A \qquad \text{[5-4]}$$

Figure 5-3 | Velocity profile for (*a*) laminar flow in a tube and (*b*) turbulent tube flow.

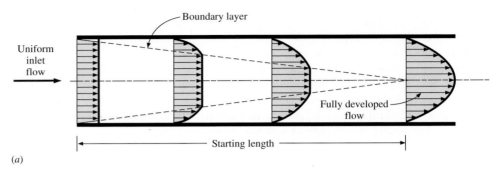

(*a*)

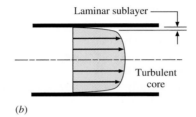

(*b*)

where

$\dot{m}$ = mass rate of flow

u_m = mean velocity

A = cross-sectional area

We define the mass velocity as

$$\text{Mass velocity} = G = \frac{\dot{m}}{A} = \rho u_m \qquad \textbf{[5-5]}$$

so that the Reynolds number may also be written

$$\text{Re}_d = \frac{Gd}{\mu} \qquad \textbf{[5-6]}$$

Equation (5-6) is sometimes more convenient to use than Equation (5-3).

5-3 | INVISCID FLOW

Although no real fluid is inviscid, in some instances the fluid may be treated as such, and it is worthwhile to present some of the equations that apply in these circumstances. For example, in the flat-plate problem discussed above, the flow at a sufficiently large distance from the plate will behave as a nonviscous flow system. The reason for this behavior is that the velocity gradients normal to the flow direction are very small, and hence the viscous-shear forces are small.

If a balance of forces is made on an element of incompressible fluid and these forces are set equal to the change in momentum of the fluid element, the Bernoulli equation for flow along a streamline results:

$$\frac{p}{\rho} + \frac{1}{2}\frac{V^2}{g_c} = \text{const} \qquad \textbf{[5-7a]}$$

or, in differential form,

$$\frac{dp}{\rho} + \frac{V\,dV}{g_c} = 0 \qquad\qquad \textbf{[5-7b]}$$

where

ρ = fluid density, kg/m^3

p = pressure at particular point in flow, Pa

V = velocity of flow at that point, m/s

The Bernoulli equation is sometimes considered an energy equation because the $V^2/2g_c$ term represents kinetic energy and the pressure represents potential energy; however, it must be remembered that these terms are derived on the basis of a dynamic analysis, so that the equation is fundamentally a dynamic equation. In fact, the concept of kinetic energy is based on a dynamic analysis.

When the fluid is compressible, an energy equation must be written that will take into account changes in internal thermal energy of the system and the corresponding changes in temperature. For a one-dimensional flow system this equation is the steady-flow energy equation for a control volume,

$$i_1 + \frac{1}{2g_c}V_1^2 + Q = i_2 + \frac{1}{2g_c}V_2^2 + Wk \qquad\qquad \textbf{[5-8]}$$

where i is the enthalpy defined by

$$i = e + pv \qquad\qquad \textbf{[5-9]}$$

and where

e = internal energy

Q = heat added to control volume

Wk = net external work done in the process

v = specific volume of fluid

(The symbol i is used to denote the enthalpy instead of the customary h to avoid confusion with the heat-transfer coefficient.) The subscripts 1 and 2 refer to entrance and exit conditions to the control volume. To calculate pressure drop in compressible flow, it is necessary to specify the equation of state of the fluid, for example, for an ideal gas,

$$p = \rho RT \quad \Delta e = c_v \Delta T \quad \Delta i = c_p \Delta T$$

The gas constant for a particular gas is given in terms of the universal gas constant $\Re$ as

$$R = \frac{\Re}{M}$$

where M is the molecular weight and $\Re = 8314.5$ J/kg · mol · K. For air, the appropriate ideal-gas properties are

$$R_{\text{air}} = 287 \text{ J/kg} \cdot \text{K} \quad c_{p,\text{air}} = 1.005 \text{ kJ/kg} \cdot °\text{C} \quad c_{v,\text{air}} = 0.718 \text{ kJ/kg} \cdot °\text{C}$$

To solve a particular problem, we must also specify the process. For example, reversible adiabatic flow through a nozzle yields the following familiar expressions relating the properties at some point in the flow to the Mach number and the stagnation properties, i.e., the properties where the velocity is zero:

$$\frac{T_0}{T} = 1 + \frac{\gamma - 1}{2} M^2$$

$$\frac{p_0}{p} = \left(1 + \frac{\gamma - 1}{2} M^2\right)^{\gamma/(\gamma-1)}$$

$$\frac{\rho_0}{\rho} = \left(1 + \frac{\gamma - 1}{2} M^2\right)^{1/(\gamma-1)}$$

where

$T_0, p_0, \rho_0 =$ stagnation properties
$\gamma =$ ratio of specific heats c_p/c_v
$M =$ Mach number

$$M = \frac{V}{a}$$

where a is the local velocity of sound, which may be calculated from

$$a = \sqrt{\gamma g_c R T} \qquad \text{[5-10]}$$

for an ideal gas.[†] For air behaving as an ideal gas this equation reduces to

$$a = 20.045\sqrt{T} \quad \text{m/s} \qquad \text{[5-11]}$$

where T is in degrees Kelvin.

| EXAMPLE 5-1 | Water Flow in a Diffuser |

Water at 20°C flows at 8 kg/s through the diffuser arrangement shown in Figure Example 5-1. The diameter at section 1 is 3.0 cm, and the diameter at section 2 is 7.0 cm. Determine the increase in static pressure between sections 1 and 2. Assume frictionless flow.

Figure Example 5-1

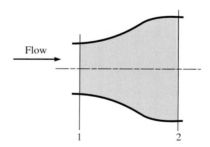

■ Solution
The flow cross-sectional areas are

$$A_1 = \frac{\pi d_1^2}{4} = \frac{\pi (0.03)^2}{4} = 7.069 \times 10^{-4} \text{ m}^2$$

$$A_2 = \frac{\pi d_2^2}{4} = \frac{\pi (0.07)^2}{4} = 3.848 \times 10^{-3} \text{ m}^2$$

[†]The isentropic flow formulas are derived in Reference 7, p. 629.

The density of water at 20°C is 1000 kg/m³, and so we may calculate the velocities from the mass-continuity relation

$$u = \frac{\dot{m}}{\rho A}$$

$$u_1 = \frac{8.0}{(1000)(7.069 \times 10^{-4})} = 11.32 \text{ m/s} \quad [37.1 \text{ ft/s}]$$

$$u_2 = \frac{8.0}{(1000)(3.848 \times 10^{-3})} = 2.079 \text{ m/s} \quad [6.82 \text{ ft/s}]$$

The pressure difference is obtained from the Bernoulli equation (5-7a):

$$\frac{p_2 - p_1}{\rho} = \frac{1}{2g_c}(u_1^2 - u_2^2)$$

$$p_2 - p_1 = \frac{1000}{2}[(11.32)^2 - (2.079)^2]$$

$$= 61.91 \text{ kPa} \quad [8.98 \text{ lb/in}^2 \text{ abs}]$$

Isentropic Expansion of Air **EXAMPLE 5-2**

Air at 300°C and 0.7 MPa pressure is expanded isentropically from a tank until the velocity is 300 m/s. Determine the static temperature, pressure, and Mach number of the air at the high-velocity condition. $\gamma = 1.4$ for air.

■ **Solution**

We may write the steady-flow energy equation as

$$i_1 = i_2 + \frac{u_2^2}{2g_c}$$

because the initial velocity is small and the process is adiabatic. In terms of temperature,

$$c_p(T_1 - T_2) = \frac{u_2^2}{2g_c}$$

$$(1005)(300 - T_2) = \frac{(300)^2}{(2)(1.0)}$$

$$T_2 = 255.2°C = 528.2 \text{ K} \quad [491.4°F]$$

We may calculate the pressure from the isentropic relation

$$\frac{p_2}{p_1} = \left(\frac{T_2}{T_1}\right)^{\gamma/(\gamma-1)}$$

$$p_2 = (0.7)\left(\frac{528.2}{573}\right)^{3.5} = 0.526 \text{ MPa} \quad [76.3 \text{ lb/in}^2 \text{ abs}]$$

The velocity of sound at condition 2 is

$$a_2 = (20.045)(528.2)^{1/2} = 460.7 \text{ m/s} \quad [1511 \text{ ft/s}]$$

so that the Mach number is

$$M_2 = \frac{u_2}{a_2} = \frac{300}{460.7} = 0.651$$

5-4 | LAMINAR BOUNDARY LAYER ON A FLAT PLATE

Consider the elemental control volume shown in Figure 5-4. We derive the equation of motion for the boundary layer by making a force-and-momentum balance on this element. To simplify the analysis we assume:

1. The fluid is incompressible and the flow is steady.
2. There are no pressure variations in the direction perpendicular to the plate.
3. The viscosity is constant.
4. Viscous-shear forces in the y direction are negligible.

We apply Newton's second law of motion,

$$\sum F_x = \frac{d(mV)_x}{d\tau}$$

The above form of Newton's second law of motion applies to a system of constant mass. In fluid dynamics it is not usually convenient to work with elements of mass; rather, we deal with elemental control volumes such as that shown in Figure 5-4, where mass may flow in or out of the different sides of the volume, which is fixed in space. For this system the force balance is then written

$$\sum F_x = \text{increase in momentum flux in } x \text{ direction}$$

The momentum flux in the x direction is the product of the mass flow through a particular side of the control volume and the x component of velocity at that point.

The mass entering the left face of the element per unit time is

$$\rho u \, dy$$

Figure 5-4 | Elemental control volume for force balance on laminar boundary layer.

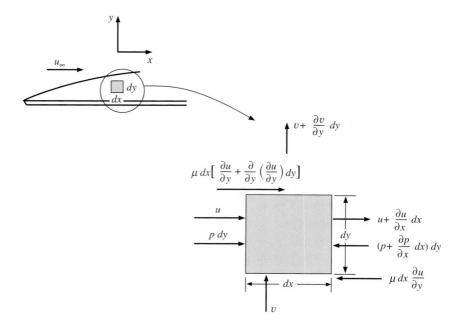

if we assume unit depth in the z direction. Thus the momentum flux entering the left face per unit time is

$$\rho u \, dy \, u = \rho u^2 \, dy$$

The mass flow leaving the right face is

$$\rho \left(u + \frac{\partial u}{\partial x} \, dx \right) dy$$

and the momentum flux leaving the right face is

$$\rho \left(u + \frac{\partial u}{\partial x} \, dx \right)^2 dy$$

The mass flow entering the bottom face is

$$\rho v \, dx$$

and the mass flow leaving the top face is

$$\rho \left(v + \frac{\partial v}{\partial y} \, dy \right) dx$$

A mass balance on the element yields

$$\rho u \, dy + \rho v \, dx = \rho \left(u + \frac{\partial u}{\partial x} \, dx \right) dy + \rho \left(v + \frac{\partial v}{\partial y} \, dy \right) dx$$

or

$$\frac{\partial u}{\partial x} + \frac{\partial v}{\partial y} = 0 \qquad\qquad \textbf{[5-12]}$$

This is the mass continuity equation for the boundary layer.

Returning to the momentum-and-force analysis, the momentum flux *in the x direction* that enters the bottom face is

$$\rho v u \, dx$$

and the momentum *in the x direction* that leaves the top face is

$$\rho \left(v + \frac{\partial v}{\partial y} \, dy \right) \left(u + \frac{\partial u}{\partial y} \, dy \right) dx$$

We are interested only in the momentum in the x direction because the forces considered in the analysis are those in the x direction. These forces are those due to viscous shear and the pressure forces on the element. The pressure force on the left face is $p \, dy$, and that on the right is $-[p + (\partial p / \partial x) \, dx] \, dy$, so that the net pressure force in the direction of motion is

$$-\frac{\partial p}{\partial x} \, dx \, dy$$

The viscous-shear force on the bottom face is

$$-\mu \frac{\partial u}{\partial y} \, dx$$

and the shear force on the top is

$$\mu \, dx \left[\frac{\partial u}{\partial y} + \frac{\partial}{\partial y} \left(\frac{\partial u}{\partial y} \right) dy \right]$$

The net viscous-shear force in the direction of motion is the sum of the two terms:

$$\text{Net viscous-shear force} = \mu \frac{\partial^2 u}{\partial y^2} \, dx \, dy$$

Equating the sum of the viscous-shear and pressure forces to the net momentum transfer in the x direction, we have

$$\mu \frac{\partial^2 u}{\partial y^2} \, dx \, dy - \frac{\partial p}{\partial x} \, dx \, dy = \rho \left(u + \frac{\partial u}{\partial x} dx \right)^2 dy - \rho u^2 \, dy$$
$$+ \rho \left(v + \frac{\partial v}{\partial y} dy \right) \left(u + \frac{\partial u}{\partial y} dy \right) dx - \rho vu \, dx$$

Clearing terms, making use of the continuity relation (5-12), and neglecting second-order differentials, gives

$$\rho \left(u \frac{\partial u}{\partial x} + v \frac{\partial u}{\partial y} \right) = \mu \frac{\partial^2 u}{\partial y^2} - \frac{\partial p}{\partial x} \qquad \text{[5-13]}$$

This is the momentum equation of the laminar boundary layer with constant properties. The equation may be solved exactly for many boundary conditions, and the reader is referred to the treatise by Schlichting [1] for details of the various methods employed in the solutions. In Appendix B we have included the classical method for obtaining an exact solution to Equation (5-13) for laminar flow over a flat plate. For the development in this chapter we shall be satisfied with an approximate analysis that furnishes an easier solution without a loss in physical understanding of the processes involved. The approximate method is due to von Kármán [2].

Consider the boundary-layer flow system shown in Figure 5-5. The free-stream velocity outside the boundary layer is u_∞, and the boundary-layer thickness is δ. We wish to make a momentum-and-force balance on the control volume bounded by the planes 1, 2, A-A, and the solid wall. The velocity components normal to the wall are neglected, and only those in the x direction are considered. We assume that the control volume is sufficiently high that it always encloses the boundary layer; that is, $H > \delta$.

Figure 5-5 | Elemental control volume for integral momentum analysis of laminar boundary layer.

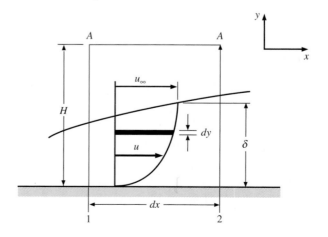

The mass flow through plane 1 is

$$\int_0^H \rho u \, dy \tag{a}$$

and the momentum flow through plane 1 is

$$\int_0^H \rho u^2 \, dy \tag{b}$$

The momentum flow through plane 2 is

$$\int_0^H \rho u^2 \, dy + \frac{d}{dx} \left(\int_0^H \rho u^2 \, dy \right) dx \tag{c}$$

and the mass flow through plane 2 is

$$\int_0^H \rho u \, dy + \frac{d}{dx} \left(\int_0^H \rho u \, dy \right) dx \tag{d}$$

Considering the conservation of mass and the fact that no mass can enter the control volume through the solid wall, the additional mass flow in expression (d) over that in (a) must enter through plane A-A. This mass flow carries with it a momentum in the x direction equal to

$$u_\infty \frac{d}{dx} \left(\int_0^H \rho u \, dy \right) dx$$

The net momentum flow out of the control volume is therefore

$$\frac{d}{dx} \left(\int_0^H \rho u^2 \, dy \right) dx - u_\infty \frac{d}{dx} \left(\int_0^H \rho u \, dy \right) dx$$

This expression may be put in a somewhat more useful form by recalling the product formula from the differential calculus:

$$d(\eta \phi) = \eta \, d\phi + \phi \, d\eta$$

or

$$\eta \, d\phi = d(\eta \phi) - \phi \, d\eta$$

In the momentum expression given above, the integral

$$\int_0^H \rho u \, dy$$

is the ϕ function and u_∞ is the η function. Thus

$$u_\infty \frac{d}{dx} \left(\int_0^H \rho u \, dy \right) dx = \frac{d}{dx} \left(u_\infty \int_0^H \rho u \, dy \right) dx - \frac{du_\infty}{dx} \left(\int_0^H \rho u \, dy \right) dx$$

$$\tag{5-14}$$

$$= \frac{d}{dx} \left(\int_0^H \rho u u_\infty \, dy \right) dx - \frac{du_\infty}{dx} \left(\int_0^H \rho u \, dy \right) dx$$

The u_∞ may be placed inside the integral since it is not a function of y and thus may be treated as a constant insofar as an integral with respect to y is concerned.

Returning to the analysis, the force on plane 1 is the pressure force pH and that on plane 2 is $[p + (dp/dx)\,dx]H$. The shear force at the wall is

$$-\tau_w\,dx = -\mu\,dx\,\left.\frac{\partial u}{\partial y}\right]_{y=0}$$

There is no shear force at plane A-A since the velocity gradient is zero outside the boundary layer. Setting the forces on the element equal to the net increase in momentum and collecting terms gives

$$-\tau_w - \frac{dp}{dx}H = -\rho\frac{d}{dx}\int_0^H (u_\infty - u)u\,dy + \frac{du_\infty}{dx}\int_0^H \rho u\,dy \qquad \textbf{[5-15]}$$

This is the integral momentum equation of the boundary layer. If the pressure is constant throughout the flow,

$$\frac{dp}{dx} = 0 = -\rho u_\infty\frac{du_\infty}{dx} \qquad \textbf{[5-16]}$$

since the pressure and free-stream velocity are related by the Bernoulli equation. For the constant-pressure condition, the integral boundary-layer equation becomes

$$\rho\frac{d}{dx}\int_0^\delta (u_\infty - u)u\,dy = \tau_w = \mu\,\left.\frac{\partial u}{\partial y}\right]_{y=0} \qquad \textbf{[5-17]}$$

The upper limit on the integral has been changed to δ because the integrand is zero for $y > \delta$ since $u = u_\infty$ for $y > \delta$.

If the velocity profile were known, the appropriate function could be inserted in Equation (5-17) to obtain an expression for the boundary-layer thickness. For our approximate analysis we first write down some conditions that the velocity function must satisfy:

$$u = 0 \qquad \text{at } y = 0 \qquad \textbf{[a]}$$
$$u = u_\infty \qquad \text{at } y = \delta \qquad \textbf{[b]}$$
$$\frac{\partial u}{\partial y} = 0 \qquad \text{at } y = \delta \qquad \textbf{[c]}$$

For a constant-pressure condition Equation (5-13) yields

$$\frac{\partial^2 u}{\partial y^2} = 0 \qquad \text{at } y = 0 \qquad \textbf{[d]}$$

since the velocities u and v are zero at $y = 0$. We assume that the velocity profiles at various x positions are similar; that is, they have the same functional dependence on the y coordinate. There are four conditions to satisfy. The simplest function that we can choose to satisfy these conditions is a polynomial with four arbitrary constants. Thus

$$u = C_1 + C_2 y + C_3 y^2 + C_4 y^3 \qquad \textbf{[5-18]}$$

Applying the four conditions (a) to (d),

$$\frac{u}{u_\infty} = \frac{3}{2}\frac{y}{\delta} - \frac{1}{2}\left(\frac{y}{\delta}\right)^3 \qquad \textbf{[5-19]}$$

Inserting the expression for the velocity into Equation (5-17) gives

$$\frac{d}{dx}\left\{\rho u_\infty^2\int_0^\delta \left[\frac{3}{2}\frac{y}{\delta} - \frac{1}{2}\left(\frac{y}{\delta}\right)^3\right]\left[1 - \frac{3}{2}\frac{y}{\delta} + \frac{1}{2}\left(\frac{y}{\delta}\right)^3\right]dy\right\} = \mu\,\left.\frac{\partial u}{\partial y}\right]_{y=0}$$

$$= \frac{3}{2}\frac{\mu u_\infty}{\delta}$$

Carrying out the integration leads to

$$\frac{d}{dx}\left(\frac{39}{280}\rho u_\infty^2 \delta\right) = \frac{3}{2}\frac{\mu u_\infty}{\delta}$$

Since ρ and u_∞ are constants, the variables may be separated to give

$$\delta\, d\delta = \frac{140}{13}\frac{\mu}{\rho u_\infty}\, dx = \frac{140}{13}\frac{\nu}{u_\infty}\, dx$$

and

$$\frac{\delta^2}{2} = \frac{140}{13}\frac{\nu x}{u_\infty} + \text{const}$$

At $x = 0$, $\delta = 0$, so that

$$\delta = 4.64\sqrt{\frac{\nu x}{u_\infty}} \qquad\qquad \textbf{[5-20]}$$

This may be written in terms of the Reynolds number as

$$\frac{\delta}{x} = \frac{4.64}{\text{Re}_x^{1/2}}$$

where

$$\text{Re}_x = \frac{u_\infty x}{\nu} \qquad\qquad \textbf{[5-21]}$$

The exact solution of the boundary-layer equations as given in Appendix B yields

$$\frac{\delta}{x} = \frac{5.0}{\text{Re}_x^{1/2}} \qquad\qquad \textbf{[5-21a]}$$

Mass Flow and Boundary-Layer Thickness   EXAMPLE 5-3

Air at 27°C and 1 atm flows over a flat plate at a speed of 2 m/s. Calculate the boundary-layer thickness at distances of 20 cm and 40 cm from the leading edge of the plate. Calculate the mass flow that enters the boundary layer between $x = 20$ cm and $x = 40$ cm. The viscosity of air at 27°C is 1.85×10^{-5} kg/m · s. Assume unit depth in the z direction.

■ **Solution**
The density of air is calculated from

$$\rho = \frac{p}{RT} = \frac{1.0132 \times 10^5}{(287)(300)} = 1.177 \text{ kg/m}^3 \quad [0.073 \text{ lb}_m/\text{ft}^3]$$

The Reynolds number is calculated as

$$\text{At } x = 20 \text{ cm:} \qquad \text{Re} = \frac{(1.177)(2.0)(0.2)}{1.85 \times 10^{-5}} = 25{,}448$$

$$\text{At } x = 40 \text{ cm:} \qquad \text{Re} = \frac{(1.177)(2.0)(0.4)}{1.85 \times 10^{-5}} = 50{,}897$$

The boundary-layer thickness is calculated from Equation (5-21):

$$\text{At } x = 20 \text{ cm:} \qquad \delta = \frac{(4.64)(0.2)}{(25{,}448)^{1/2}} = 0.00582 \text{ m} \quad [0.24 \text{ in}]$$

$$\text{At } x = 40 \text{ cm:} \qquad \delta = \frac{(4.64)(0.4)}{(50{,}897)^{1/2}} = 0.00823 \text{ m} \quad [0.4 \text{ in}]$$

To calculate the mass flow that enters the boundary layer from the free stream between $x = 20$ cm and $x = 40$ cm, we simply take the difference between the mass flow in the boundary layer at these two x positions. At any x position the mass flow in the boundary layer is given by the integral

$$\int_0^\delta \rho u \, dy$$

where the velocity is given by Equation (5-19),

$$u = u_\infty \left[\frac{3}{2} \frac{y}{\delta} - \frac{1}{2} \left(\frac{y}{\delta} \right)^3 \right]$$

Evaluating the integral with this velocity distribution, we have

$$\int_0^\delta \rho u_\infty \left[\frac{3}{2} \frac{y}{\delta} - \frac{1}{2} \left(\frac{y}{\delta} \right)^3 \right] dy = \frac{5}{8} \rho u_\infty \delta$$

Thus the mass flow entering the boundary layer is

$$\Delta m = \frac{5}{8} \rho u_\infty (\delta_{40} - \delta_{20})$$
$$= (\tfrac{5}{8})(1.177)(2.0)(0.0082 - 0.0058)$$
$$= 3.531 \times 10^{-3} \text{ kg/s} \quad [7.78 \times 10^{-3} \text{ lb}_m/\text{s}]$$

5-5 | ENERGY EQUATION OF THE BOUNDARY LAYER

The foregoing analysis considered the fluid dynamics of a laminar-boundary-layer flow system. We shall now develop the energy equation for this system and then proceed to an integral method of solution.

Consider the elemental control volume shown in Figure 5-6. To simplify the analysis we assume

1. Incompressible steady flow
2. Constant viscosity, thermal conductivity, and specific heat
3. Negligible heat conduction in the direction of flow (x direction), i.e.,

$$\frac{\partial T}{\partial x} \ll \frac{\partial T}{\partial y}$$

Then, for the element shown, the energy balance may be written

Energy convected in left face + energy convected in bottom face

+ heat conducted in bottom face

+ net viscous work done on element

= energy convected out right face + energy convected out top face
+ heat conducted out top face

The convective and conduction energy quantities are indicated in Figure 5-6, and the energy term for the viscous work may be derived as follows. The viscous work may be computed as a product of the net viscous-shear force and the distance this force moves in unit time. The viscous-shear force is the product of the shear-stress and the area dx,

$$\mu \frac{\partial u}{\partial y} dx$$

Figure 5-6 | Elemental volume for energy analysis of laminar boundary layer.

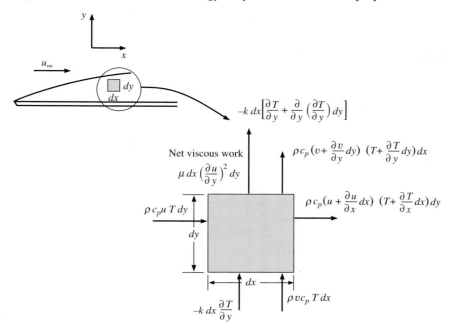

and the distance through which it moves per unit time in respect to the elemental control volume $dx\, dy$ is

$$\frac{\partial u}{\partial y}\, dy$$

so that the net viscous energy delivered to the element is

$$\mu \left(\frac{\partial u}{\partial y}\right)^2 dx\, dy$$

Writing the energy balance corresponding to the quantities shown in Figure 5-6, assuming unit depth in the z direction, and neglecting second-order differentials yields

$$\rho c_p \left[u\frac{\partial T}{\partial x} + v\frac{\partial T}{\partial y} + T\left(\frac{\partial u}{\partial x} + \frac{\partial v}{\partial y}\right)\right] dx\, dy = k\frac{\partial^2 T}{\partial y^2} dx\, dy + \mu \left(\frac{\partial u}{\partial y}\right)^2 dx\, dy$$

Using the continuity relation

$$\frac{\partial u}{\partial x} + \frac{\partial v}{\partial y} = 0 \qquad\qquad \text{[5-12]}$$

and dividing by ρc_p gives

$$u\frac{\partial T}{\partial x} + v\frac{\partial T}{\partial y} = \alpha\frac{\partial^2 T}{\partial y^2} + \frac{\mu}{\rho c_p}\left(\frac{\partial u}{\partial y}\right)^2 \qquad\qquad \text{[5-22]}$$

This is the energy equation of the laminar boundary layer. The left side represents the net transport of energy into the control volume, and the right side represents the sum of the net heat conducted out of the control volume and the net viscous work done on the element. The viscous-work term is of importance only at high velocities since its magnitude will be small compared with the other terms when low-velocity flow is studied. This may be shown with an order-of-magnitude analysis of the two terms on the right side of Equation (5-22). For this order-of-magnitude analysis we might consider the velocity as having the order of the free-stream velocity u_∞ and the y dimension of the order of δ. Thus

$$u \sim u_\infty \qquad \text{and} \qquad y \sim \delta$$

$$\alpha \frac{\partial^2 T}{\partial y^2} \sim \alpha \frac{T}{\delta^2}$$

so that

$$\frac{\mu}{\rho c_p} \left(\frac{\partial u}{\partial y} \right)^2 \sim \frac{\mu}{\rho c_p} \frac{u_\infty^2}{\delta^2}$$

If the ratio of these quantities is small, that is,

$$\frac{\mu}{\rho c_p \alpha} \frac{u_\infty^2}{T} \ll 1 \qquad\qquad \textbf{[5-23]}$$

then the viscous dissipation is small in comparison with the conduction term. Let us rearrange Equation (5-23) by introducing

$$\Pr = \frac{\nu}{\alpha} = \frac{c_p \mu}{k}$$

where Pr is called the Prandtl number, which we shall discuss later. Equation (5-23) becomes

$$\Pr \frac{u_\infty^2}{c_p T} \ll 1 \qquad\qquad \textbf{[5-24]}$$

As an example, consider the flow of air at

$$u_\infty = 70 \text{ m/s} \qquad T = 20°C = 293 \text{ K} \qquad p = 1 \text{ atm}$$

For these conditions $c_p = 1005 = \text{J/kg} \cdot °\text{C}$ and $\Pr = 0.7$ so that

$$\Pr \frac{u_\infty^2}{c_p T} = \frac{(0.7)(70)^2}{(1005)(293)} = 0.012 \ll 1.0$$

indicating that the viscous dissipation is small for even this rather large flow velocity of 70 m/s. Thus, for low-velocity incompressible flow, we have

$$u \frac{\partial T}{\partial x} + v \frac{\partial T}{\partial y} = \alpha \frac{\partial^2 T}{\partial y^2} \qquad\qquad \textbf{[5-25]}$$

In reality, our derivation of the energy equation has been a simplified one, and several terms have been left out of the analysis because they are small in comparison with others. In this way we immediately arrive at the boundary-layer approximation, without resorting to a cumbersome elimination process to obtain the final simplified relation. The general derivation of the boundary-layer energy equation is very involved and quite beyond the scope of our discussion. The interested reader should consult the books by Schlichting [1] and White [5] for more information.

There is a striking similarity between Equation (5-25) and the momentum equation for constant pressure,

$$u \frac{\partial u}{\partial x} + v \frac{\partial u}{\partial y} = \nu \frac{\partial^2 u}{\partial y^2} \qquad\qquad \textbf{[5-26]}$$

The solution to the two equations will have exactly the same form when $\alpha = \nu$. Thus we should expect that the relative magnitudes of the thermal diffusivity and kinematic viscosity would have an important influence on convection heat transfer since these magnitudes relate the velocity distribution to the temperature distribution. This is exactly the case, and we shall see the role that these parameters play in the subsequent discussion.

5-6 | THE THERMAL BOUNDARY LAYER

Just as the hydrodynamic boundary layer was defined as that region of the flow where viscous forces are felt, a thermal boundary layer may be defined as that region where temperature gradients are present in the flow. These temperature gradients would result from a heat-exchange process between the fluid and the wall.

Consider the system shown in Figure 5-7. The temperature of the wall is T_w, the temperature of the fluid outside the thermal boundary layer is T_∞, and the thickness of the thermal boundary layer is designated as δ_t. At the wall, the velocity is zero, and the heat transfer into the fluid takes place by conduction. Thus the local heat flux per unit area, q'', is

$$\frac{q}{A} = q'' = -k \left. \frac{\partial T}{\partial y} \right]_{\text{wall}} \qquad \textbf{[5-27]}$$

From Newton's law of cooling [Equation (1-8)],

$$q'' = h(T_w - T_\infty) \qquad \textbf{[5-28]}$$

where h is the convection heat-transfer coefficient. Combining these equations, we have

$$h = \frac{-k(\partial T/\partial y)_{\text{wall}}}{T_w - T_\infty} \qquad \textbf{[5-29]}$$

so that we need only find the temperature gradient at the wall in order to evaluate the heat-transfer coefficient. This means that we must obtain an expression for the temperature distribution. To do this, an approach similar to that used in the momentum analysis of the boundary layer is followed.

The conditions that the temperature distribution must satisfy are

$$T = T_w \qquad \text{at } y = 0 \qquad \textbf{[a]}$$

$$\frac{\partial T}{\partial y} = 0 \qquad \text{at } y = \delta_t \qquad \textbf{[b]}$$

$$T = T_\infty \qquad \text{at } y = \delta_t \qquad \textbf{[c]}$$

Figure 5-7 | Temperature profile in the thermal boundary layer.

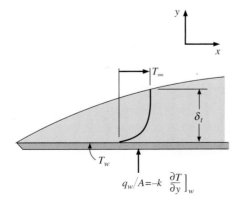

Figure 5-8 | Control volume for integral energy analysis of laminar boundary flow.

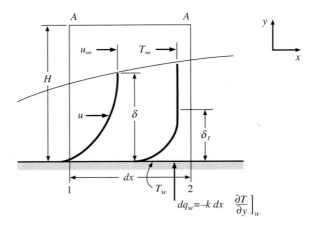

and by writing Equation (5-25) at $y=0$ with no viscous heating we find

$$\frac{\partial^2 T}{\partial y^2} = 0 \qquad \text{at } y=0 \tag{d}$$

since the velocities must be zero at the wall.

Conditions (a) to (d) may be fitted to a cubic polynomial as in the case of the velocity profile, so that

$$\frac{\theta}{\theta_\infty} = \frac{T - T_w}{T_\infty - T_w} = \frac{3}{2}\frac{y}{\delta_t} - \frac{1}{2}\left(\frac{y}{\delta_t}\right)^3 \tag{5-30}$$

where $\theta = T - T_w$. There now remains the problem of finding an expression for δ_t, the thermal-boundary-layer thickness. This may be obtained by an integral analysis of the energy equation for the boundary layer.

Consider the control volume bounded by the planes 1, 2, A-A, and the wall as shown in Figure 5-8. It is assumed that the thermal boundary layer is thinner than the hydrodynamic boundary layer, as shown. The wall temperature is T_w, the free-stream temperature is T_∞, and the heat given up to the fluid over the length dx is dq_w. We wish to make the energy balance

Energy convected in + viscous work within element

$$+ \text{heat transfer at wall} = \text{energy convected out} \tag{5-31}$$

The energy convected in through plane 1 is

$$\rho c_p \int_0^H uT\, dy$$

and the energy convected out through plane 2 is

$$\rho c_p \left(\int_0^H uT\, dy \right) + \frac{d}{dx}\left(\rho c_p \int_0^H uT\, dy \right) dx$$

The mass flow through plane A-A is

$$\frac{d}{dx}\left(\int_0^H \rho u\, dy \right) dx$$

Figure 5-9 | Hydrodynamic and thermal boundary layers on a
flat plate. Heating starts at $x = x_0$.

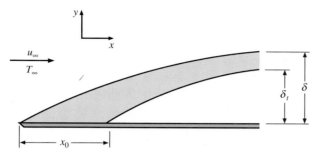

and this carries with it an energy equal to

$$c_p T_\infty \frac{d}{dx}\left(\int_0^H \rho u\, dy\right)dx$$

The net viscous work done within the element is

$$\mu\left[\int_0^H \left(\frac{du}{dy}\right)^2 dy\right]dx$$

and the heat transfer at the wall is

$$dq_w = -k\, dx\, \left.\frac{\partial T}{\partial y}\right]_w$$

Combining these energy quantities according to Equation (5-31) and collecting terms gives

$$\frac{d}{dx}\left[\int_0^H (T_\infty - T)u\, dy\right] + \frac{\mu}{\rho c_p}\left[\int_0^H \left(\frac{du}{dy}\right)^2 dy\right] = \alpha\, \left.\frac{\partial T}{\partial y}\right]_w \qquad \textbf{[5-32]}$$

This is the integral energy equation of the boundary layer for constant properties and constant free-stream temperature T_∞.

To calculate the heat transfer at the wall, we need to derive an expression for the thermal-boundary-layer thickness that may be used in conjunction with Equations (5-29) and (5-30) to determine the heat-transfer coefficient. For now, we neglect the viscous-dissipation term; this term is very small unless the velocity of the flow field becomes very large. And the calculation of high-velocity heat transfer will be considered later.

The plate under consideration need not be heated over its entire length. The situation that we shall analyze is shown in Figure 5-9, where the hydrodynamic boundary layer develops from the leading edge of the plate, while heating does not begin until $x = x_0$.

Inserting the temperature distribution Equation (5-30) and the velocity distribution Equation (5-19) into Equation (5-32) and neglecting the viscous-dissipation term, gives

$$\frac{d}{dx}\left[\int_0^H (T_\infty - T)u\, dy\right] = \frac{d}{dx}\left[\int_0^H (\theta_\infty - \theta)u\, dy\right]$$

$$= \theta_\infty u_\infty \frac{d}{dx}\left\{\int_0^H \left[1 - \frac{3}{2}\frac{y}{\delta_t} + \frac{1}{2}\left(\frac{y}{\delta_t}\right)^3\right]\left[\frac{3}{2}\frac{y}{\delta} - \frac{1}{2}\left(\frac{y}{\delta}\right)^3\right] dy\right\}$$

$$= \alpha\, \left.\frac{\partial T}{\partial y}\right]_{y=0} = \frac{3\alpha\theta_\infty}{2\delta_t}$$

Let us assume that the thermal boundary layer is thinner than the hydrodynamic boundary layer. Then we only need to carry out the integration to $y = \delta_t$ since the integrand is zero for $y > \delta_t$. Performing the necessary algebraic manipulation, carrying out the integration, and making the substitution $\zeta = \delta_t/\delta$ yields

$$\theta_\infty u_\infty \frac{d}{dx}\left[\delta\left(\frac{3}{20}\zeta^2 - \frac{3}{280}\zeta^4\right)\right] = \frac{3}{2}\frac{\alpha\theta_\infty}{\delta\zeta} \qquad \text{[5-33]}$$

Because $\delta_t < \delta$, $\zeta < 1$, and the term involving ζ^4 is small compared with the ζ^2 term, we neglect the ζ^4 term and write

$$\frac{3}{20}\theta_\infty u_\infty \frac{d}{dx}(\delta\zeta^2) = \frac{3}{2}\frac{\alpha\theta_\infty}{\zeta\delta} \qquad \text{[5-34]}$$

Performing the differentiation gives

$$\frac{1}{10}u_\infty\left(2\delta\zeta\frac{d\zeta}{dx} + \zeta^2\frac{d\delta}{dx}\right) = \frac{\alpha}{\delta\zeta}$$

or

$$\frac{1}{10}u_\infty\left(2\delta^2\zeta^2\frac{d\zeta}{dx} + \zeta^3\delta\frac{d\delta}{dx}\right) = \alpha$$

But

$$\delta\,d\delta = \frac{140}{13}\frac{\nu}{u_\infty}dx$$

and

$$\delta^2 = \frac{280}{13}\frac{\nu x}{u_\infty}$$

so that we have

$$\zeta^3 + 4x\zeta^2\frac{d\zeta}{dx} = \frac{13}{14}\frac{\alpha}{\nu} \qquad \text{[5-35]}$$

Noting that

$$\zeta^2\frac{d\zeta}{dx} = \frac{1}{3}\frac{d}{dx}\zeta^3$$

we see that Equation (5-35) is a linear differential equation of the first order in ζ^3, and the solution is

$$\zeta^3 = Cx^{-3/4} + \frac{13}{14}\frac{\alpha}{\nu}$$

When the boundary condition

$$\delta_t = 0 \qquad \text{at } x = x_0$$
$$\zeta = 0 \qquad \text{at } x = x_0$$

is applied, the final solution becomes

$$\zeta = \frac{\delta_t}{\delta} = \frac{1}{1.026}\mathrm{Pr}^{-1/3}\left[1 - \left(\frac{x_0}{x}\right)^{3/4}\right]^{1/3} \qquad \text{[5-36]}$$

where

$$\mathrm{Pr} = \frac{\nu}{\alpha} \qquad \text{[5-37]}$$

has been introduced. The ratio ν/α is called the Prandtl number after Ludwig Prandtl, the German scientist who introduced the concepts of boundary-layer theory.

When the plate is heated over the entire length, $x_0 = 0$, and

$$\frac{\delta_t}{\delta} = \zeta = \frac{1}{1.026}\mathrm{Pr}^{-1/3} \qquad [5\text{-}38]$$

In the foregoing analysis the assumption was made that $\zeta < 1$. This assumption is satisfactory for fluids having Prandtl numbers greater than about 0.7. Fortunately, most gases and liquids fall within this category. Liquid metals are a notable exception, however, since they have Prandtl numbers of the order of 0.01.

The Prandtl number ν/α has been found to be the parameter that relates the relative thicknesses of the hydrodynamic and thermal boundary layers. The kinematic viscosity of a fluid conveys information about the rate at which momentum may diffuse through the fluid because of molecular motion. The thermal diffusivity tells us the same thing in regard to the diffusion of heat in the fluid. Thus the ratio of these two quantities should express the relative magnitudes of diffusion of momentum and heat in the fluid. But these diffusion rates are precisely the quantities that determine how thick the boundary layers will be for a given external flow field; large diffusivities mean that the viscous or temperature influence is felt farther out in the flow field. The Prandtl number is thus the connecting link between the velocity field and the temperature field.

The Prandtl number is dimensionless when a consistent set of units is used:

$$\mathrm{Pr} = \frac{\nu}{\alpha} = \frac{\mu/\rho}{k/\rho c_p} = \frac{c_p \mu}{k} \qquad [5\text{-}39]$$

In the SI system a typical set of units for the parameters would be μ in kilograms per second per meter, c_p in kilojoules per kilogram per Celsius degree, and k in kilowatts per meter per Celsius degree. In the English system one would typically employ μ in pound mass per hour per foot, c_p in Btu per pound mass per Fahrenheit degree, and k in Btu per hour per foot per Fahrenheit degree.

Returning now to the analysis, we have

$$h = \frac{-k(\partial T/\partial y)_w}{T_w - T_\infty} = \frac{3}{2}\frac{k}{\delta_t} = \frac{3}{2}\frac{k}{\zeta\delta} \qquad [5\text{-}40]$$

Substituting for the hydrodynamic-boundary-layer thickness from Equation (5-21) and using Equation (5-36) gives

$$h_x = 0.332k\,\mathrm{Pr}^{1/3}\left(\frac{u_\infty}{\nu x}\right)^{1/2}\left[1 - \left(\frac{x_0}{x}\right)^{3/4}\right]^{-1/3} \qquad [5\text{-}41]$$

The equation may be nondimensionalized by multiplying both sides by x/k, producing the dimensionless group on the left side,

$$\mathrm{Nu}_x = \frac{h_x x}{k} \qquad [5\text{-}42]$$

called the Nusselt number after Wilhelm Nusselt, who made significant contributions to the theory of convection heat transfer. Finally,

$$\mathrm{Nu}_x = 0.332\mathrm{Pr}^{1/3}\,\mathrm{Re}_x^{1/2}\left[1 - \left(\frac{x_0}{x}\right)^{3/4}\right]^{-1/3} \qquad [5\text{-}43]$$

or, for the plate heated over its entire length, $x_0 = 0$ and

$$\mathrm{Nu}_x = 0.332\mathrm{Pr}^{1/3}\,\mathrm{Re}_x^{1/2} \qquad [5\text{-}44]$$

Equations (5-41), (5-43), and (5-44) express the local values of the heat-transfer coefficient in terms of the distance from the leading edge of the plate and the fluid properties. For the case where $x_0 = 0$ the average heat-transfer coefficient and Nusselt number may be obtained by integrating over the length of the plate:

$$\bar{h} = \frac{\int_0^L h_x \, dx}{\int_0^L dx} = 2h_{x=L} \qquad [\text{5-45}a]$$

For a plate where heating starts at $x = x_0$, it can be shown that the average heat transfer coefficient can be expressed as

$$\frac{\bar{h}_{x_0-L}}{h_{x=L}} = 2L \frac{1 - (x_0/L)^{3/4}}{L - x_0} \qquad [\text{5-45}b]$$

In this case, the total heat transfer for the plate would be

$$q_{\text{total}} = \bar{h}_{x_0-L}(L - x_0)(T_w - T_\infty)$$

assuming the heated section is at the constant temperature T_w. For the plate heated over the entire length,

$$\overline{\text{Nu}}_L = \frac{\bar{h}L}{k} = 2 \, \text{Nu}_{x=L} \qquad [\text{5-46}a]$$

or

$$\overline{\text{Nu}}_L = \frac{\bar{h}L}{k} = 0.664 \, \text{Re}_L^{1/2} \text{Pr}^{1/3} \qquad [\text{5-46}b]$$

where

$$\text{Re}_L = \frac{\rho u_\infty L}{\mu}$$

The reader should carry out the integrations to verify these results.

The foregoing analysis was based on the assumption that the fluid properties were constant throughout the flow. When there is an appreciable variation between wall and free-stream conditions, it is recommended that the properties be evaluated at the so-called *film temperature* T_f, defined as the arithmetic mean between the wall and free-stream temperature,

$$T_f = \frac{T_w + T_\infty}{2} \qquad [\text{5-47}]$$

An exact solution to the energy equation is given in Appendix B. The results of the exact analysis are the same as those of the approximate analysis given above.

Constant Heat Flux

The above analysis has considered the laminar heat transfer from an isothermal surface. In many practical problems the surface *heat flux* is essentially constant, and the objective is to find the distribution of the plate-surface temperature for given fluid-flow conditions. For the constant-heat-flux case it can be shown that the local Nusselt number is given by

$$\text{Nu}_x = \frac{hx}{k} = 0.453 \, \text{Re}_x^{1/2} \, \text{Pr}^{1/3} \qquad [\text{5-48}]$$

which may be expressed in terms of the wall heat flux and temperature difference as

$$\text{Nu}_x = \frac{q_w x}{k(T_w - T_\infty)} \qquad [\text{5-49}]$$

The average temperature difference along the plate, for the constant-heat-flux condition, may be obtained by performing the integration

$$\overline{T_w - T_\infty} = \frac{1}{L} \int_0^L (T_w - T_\infty)\, dx = \frac{1}{L} \int_0^L \frac{q_w x}{k\, \mathrm{Nu}_x}\, dx$$

$$= \frac{q_w L / k}{0.6795\, \mathrm{Re}_L^{1/2}\, \mathrm{Pr}^{1/3}} \qquad\qquad \textbf{[5-50]}$$

or

$$q_w = \tfrac{3}{2} h_{x=L} (\overline{T_w - T_\infty})$$

In these equations q_w is the heat flux per unit area and will have the units of watts per square meter (W/m^2) in SI units or British thermal units per hour per square foot ($Btu/h \cdot ft^2$) in the English system. Note that the heat flux $q_w = q/A$ is assumed constant over the entire plate surface.

Other Relations

Equation (5-44) is applicable to fluids having Prandtl numbers between about 0.6 and 50. It would not apply to fluids with very low Prandtl numbers like liquid metals or to high-Prandtl-number fluids like heavy oils or silicones. For a very wide range of Prandtl numbers, Churchill and Ozoe [9] have correlated a large amount of data to give the following relation for laminar flow on an isothermal flat plate:

$$\mathrm{Nu}_x = \frac{0.3387\, \mathrm{Re}_x^{1/2}\, \mathrm{Pr}^{1/3}}{\left[1 + \left(\dfrac{0.0468}{\mathrm{Pr}}\right)^{2/3}\right]^{1/4}} \qquad \text{for } \mathrm{Re}_x\, \mathrm{Pr} > 100 \qquad \textbf{[5-51]}$$

For the constant-heat-flux case, 0.3387 is changed to 0.4637 and 0.0468 is changed to 0.0207. Properties are still evaluated at the film temperature.

Isothermal Flat Plate Heated Over Entire Length   **EXAMPLE 5-4**

For the flow system in Example 5-3 assume that the plate is heated over its entire length to a temperature of 60°C. Calculate the heat transferred in (*a*) the first 20 cm of the plate and (*b*) the first 40 cm of the plate.

■ **Solution**
The total heat transfer over a certain length of the plate is desired; so we wish to calculate average heat-transfer coefficients. For this purpose we use Equations (5-44) and (5-45), evaluating the properties at the film temperature:

$$T_f = \frac{27 + 60}{2} = 43.5°C = 316.5\ K \quad [110.3°F]$$

From Appendix A the properties are

$$\nu = 17.36 \times 10^{-6}\ m^2/s \quad [1.87 \times 10^{-4}\ ft^2/s]$$
$$k = 0.02749\ W/m \cdot °C \quad [0.0159\ Btu/h \cdot ft \cdot °F]$$
$$\mathrm{Pr} = 0.7$$
$$c_p = 1.006\ kJ/kg \cdot °C \quad [0.24\ Btu/lb_m \cdot °F]$$

At $x = 20$ cm

$$\mathrm{Re}_x = \frac{u_\infty x}{\nu} = \frac{(2)(0.2)}{17.36 \times 10^{-6}} = 23{,}041$$

$$\mathrm{Nu}_x = \frac{h_x x}{k} = 0.332 \mathrm{Re}_x^{1/2} \mathrm{Pr}^{1/3}$$

$$= (0.332)(23{,}041)^{1/2}(0.7)^{1/3} = 44.74$$

$$h_x = \mathrm{Nu}_x \left(\frac{k}{x}\right) = \frac{(44.74)(0.02749)}{0.2}$$

$$= 6.15 \ \mathrm{W/m^2 \cdot {}^\circ C} \quad [1.083 \ \mathrm{Btu/h \cdot ft^2 \cdot {}^\circ F}]$$

The average value of the heat-transfer coefficient is twice this value, or

$$\overline{h} = (2)(6.15) = 12.3 \ \mathrm{W/m^2 \cdot {}^\circ C} \quad [2.17 \ \mathrm{Btu/h \cdot ft^2 \cdot {}^\circ F}]$$

The heat flow is

$$q = \overline{h} A (T_w - T_\infty)$$

If we assume unit depth in the z direction,

$$q = (12.3)(0.2)(60 - 27) = 81.18 \ \mathrm{W} \quad [277 \ \mathrm{Btu/h}]$$

At $x = 40$ cm

$$\mathrm{Re}_x = \frac{u_\infty x}{\nu} = \frac{(2)(0.4)}{17.36 \times 10^{-6}} = 46{,}082$$

$$\mathrm{Nu}_x = (0.332)(46{,}082)^{1/2}(0.7)^{1/3} = 63.28$$

$$h_x = \frac{(63.28)(0.02749)}{0.4} = 4.349 \ \mathrm{W/m^2 \cdot {}^\circ C}$$

$$\overline{h} = (2)(4.349) = 8.698 \ \mathrm{W/m^2 \cdot {}^\circ C} \quad [1.53 \ \mathrm{Btu/h \cdot ft^2 \cdot {}^\circ F}]$$

$$q = (8.698)(0.4)(60 - 27) = 114.8 \ \mathrm{W} \quad [392 \ \mathrm{Btu/h}]$$

EXAMPLE 5-5   Flat Plate with Constant Heat Flux

A 1.0-kW heater is constructed of a glass plate with an electrically conducting film that produces a constant heat flux. The plate is 60 cm by 60 cm and placed in an airstream at 27°C, 1 atm with $u_\infty = 5$ m/s. Calculate the average temperature difference along the plate and the temperature difference at the trailing edge.

■ **Solution**
Properties should be evaluated at the film temperature, but we do not know the plate temperature. So for an initial calculation, we take the properties at the free-stream conditions of

$$T_\infty = 27^\circ C = 300 \ \mathrm{K}$$

$$\nu = 15.69 \times 10^{-6} \ \mathrm{m^2/s} \qquad \mathrm{Pr} = 0.708 \qquad k = 0.02624 \ \mathrm{W/m \cdot {}^\circ C}$$

$$\mathrm{Re}_L = \frac{(0.6)(5)}{15.69 \times 10^{-6}} = 1.91 \times 10^5$$

From Equation (5-50) the average temperature difference is

$$\overline{T_w - T_\infty} = \frac{[1000/(0.6)^2](0.6)/0.02624}{0.6795(1.91 \times 10^5)^{1/2}(0.708)^{1/3}} = 240^\circ C$$

Now, we go back and evaluate properties at

$$T_f = \frac{240 + 27 + 27}{2} = 147°\text{C} = 420 \text{ K}$$

and obtain

$$\nu = 28.22 \times 10^{-6} \text{ m}^2/\text{s} \qquad \text{Pr} = 0.687 \qquad k = 0.035 \text{ W/m} \cdot °\text{C}$$

$$\text{Re}_L = \frac{(0.6)(5)}{28.22 \times 10^{-6}} = 1.06 \times 10^5$$

$$\overline{T_w - T_\infty} = \frac{[1000/(0.6)^2](0.6)/0.035}{0.6795(1.06 \times 10^5)^{1/2}(0.687)^{1/3}} = 243°\text{C}$$

At the end of the plate ($x = L = 0.6$ m) the temperature difference is obtained from Equations (5-48) and (5-50) with the constant 0.453 to give

$$(T_w - T_\infty)_{x=L} = \frac{(243.6)(0.6795)}{0.453} = 365.4°\text{C}$$

An alternate solution would be to base the Nusselt number on Equation (5-51).

Plate with Unheated Starting Length | EXAMPLE 5-6

Air at 1 atm and 300 K flows across a 20-cm-square plate at a free-stream velocity of 20 m/s. The last half of the plate is heated to a constant temperature of 350 K. Calculate the heat lost by the plate.

■ **Solution**

First we evaluate the air properties at the film temperature

$$T_f = (T_w + T_\infty)/2 = 325 \text{ K}$$

and obtain

$$\nu = 18.23 \times 10^{-6} \text{m}^2/\text{s} \quad k = 0.02814 \text{ W/m} \cdot °\text{C} \quad \text{Pr} = 0.7$$

At the trailing edge of the plate the Reynolds number is

$$\text{Re}_L = u_\infty L/\nu = (20)(0.2)/18.23 \times 10^{-6} = 2.194 \times 10^5$$

or, laminar flow over the length of the plate.

Heating does not start until the last half of the plate, or at a position $x_0 = 0.1$ m. The local heat-transfer coefficient for this condition is given by Equation (5-41):

$$h_x = 0.332k \text{ Pr}^{1/3}(u_\infty/\nu x)^{1/2}[1 - (x_0/x)^{0.75}]^{-1/3} \qquad [a]$$

Inserting the property values along with $x_0 = 0.1$ gives

$$h_x = 8.6883x^{-1/2}(1 - 0.17783x^{-0.75})^{-1/3} \qquad [b]$$

The plate is 0.2 m wide so the heat transfer is obtained by integrating over the heated length $x_0 < x < L$

$$q = (0.2)(T_w - T_\infty)\int_{x_0 = 0.1}^{L = 0.2} h_x dx \qquad [c]$$

Inserting Equation (b) in Equation (c) and performing the numerical integration gives

$$q = (0.2)(8.6883)(0.4845)(350 - 300) = 421 \text{ W} \qquad [d]$$

The average value of the heat-transfer coefficient *over the heated length* is given by

$$h = q/(T_w - T_\infty)(L - x_0)W = 421/(350 - 300)(0.2 - 0.1)(0.2) = 421 \text{ W/m}^2 \cdot {}^\circ\text{C}$$

where W is the width of the plate.

An easier calculation can be made by applying Equation (5-45b) to determine the average heat transfer coefficient over the heated portion of the plate. The result is

$$h = 425.66 \text{ W/m}^2 \cdot {}^\circ\text{C} \quad \text{and} \quad q = 425.66 \text{ W}$$

which indicates, of course, only a small error in the numerical integeration.

| **EXAMPLE 5-7** | Oil Flow Over Heated Flat Plate |

Engine oil at 20°C is forced over a 20-cm-square plate at a velocity of 1.2 m/s. The plate is heated to a uniform temperature of 60°C. Calculate the heat lost by the plate.

■ **Solution**

We first evaluate the film temperature:

$$T_f = \frac{20 + 60}{2} = 40^\circ\text{C}$$

The properties of engine oil are

$$\rho = 876 \text{ kg/m}^3 \qquad \nu = 0.00024 \text{ m}^2/\text{s}$$
$$k = 0.144 \text{ W/m} \cdot {}^\circ\text{C} \qquad \text{Pr} = 2870$$

The Reynolds number is

$$\text{Re} = \frac{u_\infty L}{\nu} = \frac{(1.2)(0.2)}{0.00024} = 1000$$

Because the Prandtl number is so large we will employ Equation (5-51) for the solution. We see that h_x varies with x in the same fashion as in Equation (5-44), that is, $h_x \propto x^{-1/2}$, so that we get the same solution as in Equation (5-45) for the average heat-transfer coefficient. Evaluating Equation (5-51) at $x = 0.2$ gives

$$\text{Nu}_x = \frac{(0.3387)(1000)^{1/2}(2870)^{1/3}}{\left[1 + \left(\dfrac{0.0468}{2870}\right)^{2/3}\right]^{1/4}} = 152.2$$

and

$$h_x = \frac{(152.2)(0.144)}{0.2} = 109.6 \text{ W/m}^2 \cdot {}^\circ\text{C}$$

The average value of the convection coefficient is

$$h = (2)(109.6) = 219.2 \text{ W/m}^2 \cdot {}^\circ\text{C}$$

so that the total heat transfer is

$$q = hA(T_w - T_\infty) = (219.2)(0.2)^2(60 - 20) = 350.6 \text{ W}$$

5-7 | THE RELATION BETWEEN FLUID FRICTION AND HEAT TRANSFER

We have already seen that the temperature and flow fields are related. Now we seek an expression whereby the frictional resistance may be directly related to heat transfer.

The shear stress at the wall may be expressed in terms of a friction coefficient C_f:

$$\tau_w = C_f \frac{\rho u_\infty^2}{2} \qquad [5\text{-}52]$$

Equation (5-52) is the defining equation for the friction coefficient. The shear stress may also be calculated from the relation

$$\tau_w = \mu \left.\frac{\partial u}{\partial y}\right]_w$$

Using the velocity distribution given by Equation (5-19), we have

$$\tau_w = \frac{3}{2}\frac{\mu u_\infty}{\delta}$$

and making use of the relation for the boundary-layer thickness gives

$$\tau_w = \frac{3}{2}\frac{\mu u_\infty}{4.64}\left(\frac{u_\infty}{\nu x}\right)^{1/2} \qquad [5\text{-}53]$$

Combining Equations (5-52) and (5-53) leads to

$$\frac{C_{fx}}{2} = \frac{3}{2}\frac{\mu u_\infty}{4.64}\left(\frac{u_\infty}{\nu x}\right)^{1/2}\frac{1}{\rho u_\infty^2} = 0.323\,\mathrm{Re}_x^{-1/2} \qquad [5\text{-}54]$$

The exact solution of the boundary-layer equations yields

$$\frac{C_{fx}}{2} = 0.332\,\mathrm{Re}_x^{-1/2} \qquad [5\text{-}54a]$$

Equation (5-44) may be rewritten in the following form:

$$\frac{\mathrm{Nu}_x}{\mathrm{Re}_x\,\mathrm{Pr}} = \frac{h_x}{\rho c_p u_\infty} = 0.332\,\mathrm{Pr}^{-2/3}\,\mathrm{Re}_x^{-1/2}$$

The group on the left is called the Stanton number,

$$\mathrm{St}_x = \frac{h_x}{\rho c_p u_\infty}$$

so that

$$\mathrm{St}_x\,\mathrm{Pr}^{2/3} = 0.332\,\mathrm{Re}_x^{1/2} \qquad [5\text{-}55]$$

Upon comparing Equations (5-54) and (5-55), we note that the right sides are alike except for a difference of about 3 percent in the constant, which is the result of the approximate nature of the integral boundary-layer analysis. We recognize this approximation and write

$$\mathrm{St}_x \mathrm{Pr}^{2/3} = \frac{C_{fx}}{2}$$ [5-56]

Equation (5-56), called the *Reynolds-Colburn analogy,* expresses the relation between fluid friction and heat transfer for laminar flow on a flat plate. The heat-transfer coefficient thus could be determined by making measurements of the frictional drag on a plate under conditions in which no heat transfer is involved.

It turns out that Equation (5-56) can also be applied to turbulent flow over a flat plate and in a modified way to turbulent flow in a tube. It does not apply to laminar tube flow. In general, a more rigorous treatment of the governing equations is necessary when embarking on new applications of the heat-transfer–fluid-friction analogy, and the results do not always take the simple form of Equation (5-56). The interested reader may consult the references at the end of the chapter for more information on this important subject. At this point, the simple analogy developed above has served to amplify our understanding of the physical processes in convection and to reinforce the notion that heat-transfer and viscous-transport processes are related at both the microscopic and macroscopic levels.

EXAMPLE 5-8	Drag Force on a Flat Plate

For the flow system in Example 5-4 compute the drag force exerted on the first 40 cm of the plate using the analogy between fluid friction and heat transfer.

■ **Solution**
We use Equation (5-56) to compute the friction coefficient and then calculate the drag force. An average friction coefficient is desired, so

$$\overline{\mathrm{St}}\, \mathrm{Pr}^{2/3} = \frac{\overline{C}_f}{2}$$ [a]

The density at 316.5 K is

$$\rho = \frac{p}{RT} = \frac{1.0132 \times 10^5}{(287)(316.5)} = 1.115 \;\mathrm{kg/m^3}$$

For the 40-cm length

$$\overline{\mathrm{St}} = \frac{\overline{h}}{\rho c_p u_\infty} = \frac{8.698}{(1.115)(1006)(2)} = 3.88 \times 10^{-3}$$

Then from Equation (a)

$$\frac{\overline{C}_f}{2} = (3.88 \times 10^{-3})(0.7)^{2/3} = 3.06 \times 10^{-3}$$

The average shear stress at the wall is computed from Equation (5-52):

$$\overline{\tau}_w = \overline{C}_f \rho \frac{u_\infty^2}{2}$$
$$= (3.06 \times 10^{-3})(1.115)(2)^2$$
$$= 0.0136 \;\mathrm{N/m^2}$$

The drag force is the product of this shear stress and the area,

$$D = (0.0136)(0.4) = 5.44 \;\mathrm{mN} \quad [1.23 \times 10^{-3} \;\mathrm{lb}_f]$$

5-8 | TURBULENT-BOUNDARY-LAYER HEAT TRANSFER

Consider a portion of a turbulent boundary layer as shown in Figure 5-10. A very thin region near the plate surface has a laminar character, and the viscous action and heat transfer take place under circumstances like those in laminar flow. Farther out, at larger y distances from the plate, some turbulent action is experienced, but the molecular viscous action and heat conduction are still important. This region is called the *buffer layer*. Still farther out, the flow is fully turbulent, and the main momentum- and heat-exchange mechanism is one involving macroscopic lumps of fluid moving about in the flow. In this fully turbulent region we speak of *eddy viscosity* and *eddy thermal conductivity*. These eddy properties may be 10 to 20 times as large as the molecular values.

The physical mechanism of heat transfer in turbulent flow is quite similar to that in laminar flow; the primary difference is that one must deal with the eddy properties instead of the ordinary thermal conductivity and viscosity. The main difficulty in an analytical treatment is that these eddy properties vary across the boundary layer, and the specific variation can be determined only from experimental data. This is an important point. All analyses of turbulent flow must eventually rely on experimental data because there is no completely adequate theory to predict turbulent-flow behavior.

If one observes the instantaneous macroscopic velocity in a turbulent-flow system, as measured with a laser anemometer or other sensitive device, significant fluctuations about the mean flow velocity are observed as indicated in Figure 5-11, where $\bar{u}$ is designated as the mean velocity and u' is the *fluctuation* from the mean. The instantaneous velocity is therefore

$$u = \bar{u} + u' \qquad [5\text{-}57]$$

The mean value of the fluctuation u' must be zero over an extended period for steady flow conditions. There are also fluctuations in the y component of velocity, so we would write

$$v = \bar{v} + v' \qquad [5\text{-}58]$$

The fluctuations give rise to a turbulent-shear stress that may be analyzed by referring to Figure 5-12.

For a unit area of the plane $P\text{-}P$, the instantaneous turbulent mass-transport rate across the plane is $\rho v'$. Associated with this mass transport is a change in the x component of

Figure 5-10 | Velocity profile in turbulent boundary layer on a flat plate.

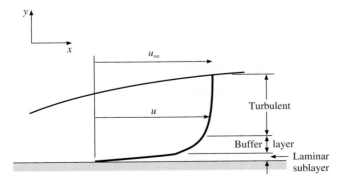

Figure 5-11 | Turbulent fluctuations with time.

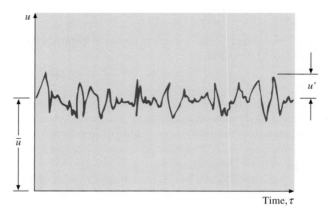

Figure 5-12 | Turbulent shear stress and mixing length.

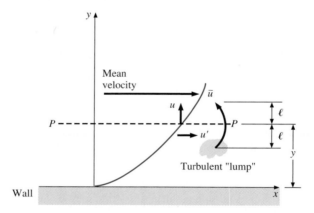

velocity u'. The net momentum flux per unit area, in the x direction, represents the turbulent-shear stress at the plane P-P, or $\rho v' u'$. When a turbulent lump moves upward ($v' > 0$), it enters a region of higher $\bar{u}$ and is therefore likely to effect a slowing-down fluctuation in u', that is, $u' < 0$. A similar argument can be made for $v' < 0$, so that the average turbulent-shear stress will be given as

$$\tau_t = -\overline{\rho v' u'} \qquad \textbf{[5-59]}$$

We must note that even though $\overline{v'} = \overline{u'} = 0$, the average of the *fluctuation product* $\overline{u'v'}$ is *not* zero.

Eddy Viscosity and the Mixing Length

Let us define an eddy viscosity or eddy diffusivity for momentum ϵ_M such that

$$\tau_t = -\overline{\rho v' u'} = \rho \epsilon_M \frac{du}{dy} \qquad \textbf{[5-60]}$$

We have already likened the macroscopic transport of heat and momentum in turbulent flow to their molecular counterparts in laminar flow, so the definition in Equation (5-60) is a natural consequence of this analogy. To analyze molecular-transport problems one normally introduces the concept of *mean free path,* or the average distance a particle travels between collisions. Prandtl introduced a similar concept for describing turbulent-flow phenomena. The *Prandtl mixing length* is the distance traveled, on the average, by the turbulent lumps of fluid in a direction normal to the mean flow.

Let us imagine a turbulent lump that is located a distance ℓ above or below the plane *P-P,* as shown in Figure 5-12. These lumps of fluid move back and forth across the plane and give rise to the eddy or turbulent-shear-stress effect. At $y + \ell$ the velocity would be approximately

$$u(y + \ell) \approx u(y) + \ell \frac{\partial u}{\partial y}$$

while at $y - \ell$,

$$u(y - \ell) \approx u(y) - \ell \frac{\partial u}{\partial y}$$

Prandtl postulated that the turbulent fluctuation u' is proportional to the mean of the above two quantities, or

$$u' \approx \ell \frac{\partial u}{\partial y} \qquad\qquad \textbf{[5-61]}$$

The distance ℓ is called the Prandtl mixing length. Prandtl also postulated that v' would be of the same order of magnitude as u' so that the turbulent-shear stress of Equation (5-60) could be written

$$\tau_t = -\overline{\rho u' v'} = \rho \ell^2 \left(\frac{\partial u}{\partial y} \right)^2 = \rho \epsilon_M \frac{\partial u}{\partial y} \qquad\qquad \textbf{[5-62]}$$

The eddy viscosity ϵ_M thus becomes

$$\epsilon_M = \ell^2 \frac{\partial u}{\partial y} \qquad\qquad \textbf{[5-63]}$$

We have already noted that the eddy properties, and hence the mixing length, vary markedly through the boundary layer. Many analysis techniques have been applied over the years to take this variation into account. Prandtl's hypothesis was that the mixing length is proportional to distance from the wall, or

$$\ell = Ky \qquad\qquad \textbf{[5-64]}$$

where K is the proportionality constant. The additional assumption was made that in the near-wall region the shear stress is approximately constant so that $\tau_t \approx \tau_w$. When this assumption is used along with Equation (5-64), Equation (5-62) yields

$$\tau_w = \rho K^2 y^2 \left(\frac{\partial u}{\partial y} \right)^2$$

Taking the square root and integrating with respect to y gives

$$u = \frac{1}{K} \sqrt{\frac{\tau_w}{\rho}} \ln y + C \qquad\qquad \textbf{[5-65]}$$

where C is the constant of integration. Equation (5-65) matches very well with experimental data except in the region very close to the wall, where the laminar sublayer is present. In the sublayer the velocity distribution is essentially linear.

Let us now quantify our earlier qualitative description of a turbulent boundary layer by expressing the shear stress as the sum of a molecular and turbulent part:

$$\frac{\tau}{\rho} = (\nu + \epsilon_M)\frac{\partial u}{\partial y} \qquad\qquad \textbf{[5-66]}$$

The so-called universal velocity profile is obtained by introducing two nondimensional coordinates

$$u^+ = \frac{u}{\sqrt{\tau_w/\rho}} \qquad\qquad \textbf{[5-67]}$$

$$y^+ = \frac{\sqrt{\tau_w/\rho}\,y}{\nu} \qquad\qquad \textbf{[5-68]}$$

Using these parameters and assuming $\tau \approx$ constant, we can rewrite Equation (5-66) as

$$du^+ = \frac{dy^+}{1 + \epsilon_M/\nu} \qquad\qquad \textbf{[5-69]}$$

In terms of our previous qualitative discussion, the laminar sublayer is the region where $\epsilon_M \sim 0$, the buffer layer has $\epsilon_M \sim \nu$, and the turbulent layer has $\epsilon_M \gg \nu$. Therefore, taking $\epsilon_M = 0$ in Equation (5-69) and integrating yields

$$u^+ = y^+ + c$$

At the wall, $u^+ = 0$ for $y^+ = 0$ so that $c = 0$ and

$$u^+ = y^+ \qquad\qquad \textbf{[5-70]}$$

is the velocity relation (a linear one) for the laminar sublayer. In the fully turbulent region $\epsilon_M/\nu \gg 1$. From Equation (5-65)

$$\frac{\partial u}{\partial y} = \frac{1}{K}\sqrt{\frac{\tau_w}{\rho}}\frac{1}{y}$$

Substituting this relation along with Equation (5-64) into Equation (5-63) gives

$$\epsilon_M = K\sqrt{\frac{\tau_w}{\rho}}\,y$$

or

$$\frac{\epsilon_m}{\nu} = Ky^+ \qquad\qquad \textbf{[5-71]}$$

Substituting this relation in Equation (5-69) for $\epsilon_M/\nu \gg 1$ and integrating gives

$$u^+ = \frac{1}{K}\ln y^+ + c \qquad\qquad \textbf{[5-72]}$$

This same *form* of equation will also be obtained for the buffer region. The limits of each region are obtained by comparing the above equations with experimental velocity measurements, with the following generally accepted constants:

$$\begin{array}{lll}
\text{Laminar sublayer: } 0 < y^+ < 5 & u^+ = y^+ & \\
\text{Buffer layer: } 5 < y^+ < 30 & u^+ = 5.0\ln y^+ - 3.05 & \textbf{[5-73]} \\
\text{Turbulent layer: } 30 < y^+ < 400 & u^+ = 2.5\ln y^+ + 5.5 &
\end{array}$$

The equation set (5-73) is called the *universal velocity profile* and matches very well with experimental data; however, we should note once again that the constants in the equations must be determined from experimental velocity measurements. The satisfying point is that the simple Prandtl mixing-length model yields an equation form that fits the data so well.

Turbulent heat transfer is analogous to turbulent momentum transfer. The turbulent momentum flux postulated by Equation (5-59) carries with it a turbulent energy fluctuation proportional to the temperature gradient. We thus have, in analogy to Equation (5-62),

$$\left(\frac{q}{A}\right)_{\text{turb}} = -\rho c_p \epsilon_H \frac{\partial T}{\partial y} \qquad [5\text{-}74]$$

or, for regions where both molecular and turbulent energy transport are important,

$$\frac{q}{A} = -\rho c_p (\alpha + \epsilon_H) \frac{\partial T}{\partial y} \qquad [5\text{-}75]$$

Turbulent Heat Transfer Based on Fluid-Friction Analogy

Various analyses, similar to the one for the universal velocity profile above, have been performed to predict turbulent-boundary-layer heat transfer. The analyses have met with good success, but for our purposes the Colburn analogy between fluid friction and heat transfer is easier to apply and yields results that are in agreement with experiment and of simpler form.

In the turbulent-flow region, where $\epsilon_M \gg \nu$ and $\epsilon_H \gg \alpha$, we define the turbulent Prandtl number as

$$\text{Pr}_t = \frac{\epsilon_M}{\epsilon_H} \qquad [5\text{-}76]$$

If we can expect that the eddy momentum and energy transport will both be increased in the same proportion compared with their molecular values, we might anticipate that heat-transfer coefficients can be calculated by Equation (5-56) with the ordinary molecular Prandtl number used in the computation. In the turbulent core of the boundary layer the eddy viscosity may be as high as 100 times the molecular value experienced in the laminar sublayer, and a similar behavior is experienced for the eddy diffusivity for heat compared to the molecular diffusivity. To account for the Prandtl number effect over the entire boundary layer a weighted average is needed, and it turns out that use of $\text{Pr}^{2/3}$ works very well and matches with the laminar heat-transfer–fluid-friction analogy. We thus will base our calculations on this analogy, and we need experimental values for C_f for turbulent boundary layer flows to carry out these computations.

Schlichting [1] has surveyed experimental measurements of friction coefficients for turbulent flow on flat plates. We present the results of that survey so that they may be employed in the calculation of turbulent heat transfer with the fluid-friction–heat-transfer analogy. The *local* skin-friction coefficient is given by

$$C_{fx} = 0.0592 \, \text{Re}_x^{-1/5} \qquad [5\text{-}77]$$

for Reynolds numbers between 5×10^5 and 10^7. At higher Reynolds numbers from 10^7 to 10^9 the formula of Schultz-Grunow [8] is recommended:

$$C_{fx} = 0.370 (\log \text{Re}_x)^{-2.584} \qquad [5\text{-}78]$$

The *average-friction coefficient* for a flat plate with a laminar boundary layer up to $\mathrm{Re_{crit}}$ and turbulent thereafter can be calculated from

$$\overline{C}_f = \frac{0.455}{(\log \mathrm{Re}_L)^{2.584}} - \frac{A}{\mathrm{Re}_L} \qquad \mathrm{Re}_L < 10^9 \qquad \textbf{[5-79]}$$

where the constant A depends on $\mathrm{Re_{crit}}$ in accordance with Table 5-1. A somewhat simpler formula can be obtained for lower Reynolds numbers as

$$\overline{C}_f = \frac{0.074}{\mathrm{Re}_L^{1/5}} - \frac{A}{\mathrm{Re}_L} \qquad \mathrm{Re}_L < 10^7 \qquad \textbf{[5-80]}$$

Table 5-1

$\mathrm{Re_{crit}}$	3×10^5	5×10^5	10^6	3×10^6
A	1055	1742	3340	8940

Equations (5-79) and (5-80) are in agreement within their common range of applicability, and the one to be used in practice will depend on computational convenience.

Applying the fluid-friction analogy $\mathrm{St}\, \mathrm{Pr}^{2/3} = C_f/2$, we obtain the local turbulent heat transfer as:

$$\mathrm{St}_x\, \mathrm{Pr}^{2/3} = 0.0296\, \mathrm{Re}_x^{-1/5} \qquad 5 \times 10^5 < \mathrm{Re}_x < 10^7 \qquad \textbf{[5-81]}$$

or

$$\mathrm{St}_x\, \mathrm{Pr}^{2/3} = 0.185(\log \mathrm{Re}_x)^{-2.584} \qquad 10^7 < \mathrm{Re}_x < 10^9 \qquad \textbf{[5-82]}$$

The average heat transfer over the entire laminar-turbulent boundary layer is

$$\overline{\mathrm{St}}\, \mathrm{Pr}^{2/3} = \frac{\overline{C}_f}{2} \qquad \textbf{[5-83]}$$

For $\mathrm{Re_{crit}} = 5 \times 10^5$ and $\mathrm{Re}_L < 10^7$, Equation (5-80) can be used to obtain

$$\overline{\mathrm{St}}\, \mathrm{Pr}^{2/3} = 0.037\, \mathrm{Re}_L^{-1/5} - 871\, \mathrm{Re}_L^{-1} \qquad \textbf{[5-84]}$$

Recalling that $\overline{\mathrm{St}} = \overline{\mathrm{Nu}}/(\mathrm{Re}_L\, \mathrm{Pr})$, we can rewrite Equation (5-84) as

$$\overline{\mathrm{Nu}}_L = \frac{\overline{h}L}{k} = \mathrm{Pr}^{1/3}(0.037\, \mathrm{Re}_L^{0.8} - 871) \qquad \textbf{[5-85]}$$

The average heat-transfer coefficient can also be obtained by integrating the local values over the entire length of the plate. Thus,

$$h = \frac{1}{L}\left(\int_0^{x_{\mathrm{crit}}} h_{\mathrm{lam}}\, dx + \int_{x_{\mathrm{crit}}}^{L} h_{\mathrm{turb}}\, dx \right)$$

Using Equation (5-55) for the laminar portion, $\mathrm{Re_{crit}} = 5 \times 10^5$, and Equation (5-81) for the turbulent portion gives the same result as Equation (5-85). For higher Reynolds numbers the friction coefficient from Equation (5-79) may be used, so that

$$\mathrm{Nu}_L = \frac{\overline{h}L}{k} = [0.228\mathrm{Re}_L\,(\log\mathrm{Re}_L)^{-2.584} - 871]\mathrm{Pr}^{1/3} \qquad \textbf{[5-85a]}$$

for $10^7 < \mathrm{Re}_L < 10^9$ and $\mathrm{Re}_{\mathrm{crit}} = 5 \times 10^5$.

The reader should note that if a transition Reynolds number different from 500,000 is chosen, then Equations (5-84) and (5-85) must be changed accordingly. An alternative equation is suggested by Whitaker [10] that may give better results with some liquids because of the viscosity-ratio term:

$$\overline{\mathrm{Nu}}_L = 0.036\,\mathrm{Pr}^{0.43}\,(\mathrm{Re}_L^{0.8} - 9200)\left(\frac{\mu_\infty}{\mu_w}\right)^{1/4} \qquad \textbf{[5-86]}$$

for

$$0.7 < \mathrm{Pr} < 380$$
$$2 \times 10^5 < \mathrm{Re}_L < 5.5 \times 10^6$$
$$0.26 < \frac{\mu_\infty}{\mu_w} < 3.5$$

All properties except μ_w are evaluated at the free-stream temperature. For gases the viscosity ratio is dropped and the properties are evaluated at the film temperature.

Constant Heat Flux

For constant-wall-heat flux in turbulent flow it is shown in Reference 11 that the local Nusselt number is only about 4 percent higher than for the isothermal surface; that is,

$$\mathrm{Nu}_x = 1.04\,\mathrm{Nu}_x\Big]_{T_w=\mathrm{const}} \qquad \textbf{[5-87]}$$

Some more comprehensive methods of correlating turbulent-boundary-layer heat transfer are given by Churchill [11].

Turbulent Heat Transfer from
Isothermal Flat Plate   **EXAMPLE 5-9**

Air at 20°C and 1 atm flows over a flat plate at 35 m/s. The plate is 75 cm long and is maintained at 60°C. Assuming unit depth in the z direction, calculate the heat transfer from the plate.

■ **Solution**
We evaluate properties at the film temperature:

$$T_f = \frac{20 + 60}{2} = 40°\mathrm{C} = 313\ \mathrm{K}$$

$$\rho = \frac{p}{RT} = \frac{1.0132 \times 10^5}{(287)(313)} = 1.128\ \mathrm{kg/m}^3$$

$$\mu = 1.906 \times 10^{-5}\ \mathrm{kg/m \cdot s}$$

$$\mathrm{Pr} = 0.7 \qquad k = 0.02723\ \mathrm{W/m \cdot °C} \qquad c_p = 1.007\ \mathrm{kJ/kg \cdot °C}$$

The Reynolds number is

$$\mathrm{Re}_L = \frac{\rho u_\infty L}{\mu} = \frac{(1.128)(35)(0.75)}{1.906 \times 10^{-5}} = 1.553 \times 10^6$$

and the boundary layer is turbulent because the Reynolds number is greater than 5×10^5. Therefore, we use Equation (5-85) to calculate the average heat transfer over the plate:

$$\overline{\mathrm{Nu}}_L = \frac{\overline{h}L}{k} = \mathrm{Pr}^{1/3}(0.037\,\mathrm{Re}_L^{0.8} - 871)$$

$$= (0.7)^{1/3}[(0.037)(1.553 \times 10^6)^{0.8} - 871] = 2180$$

$$\overline{h} = \overline{\mathrm{Nu}}_L \frac{k}{L} = \frac{(2180)(0.02723)}{0.75} = 79.1 \ \mathrm{W/m^2 \cdot {}^\circ C} \quad [13.9 \ \mathrm{Btu/h \cdot ft^2 \cdot {}^\circ F}]$$

$$q = \overline{h}A(T_w - T_\infty) = (79.1)(0.75)(60 - 20) = 2373 \ \mathrm{W} \quad [8150 \ \mathrm{Btu/h}]$$

5-9 | TURBULENT-BOUNDARY-LAYER THICKNESS

A number of experimental investigations have shown that the velocity profile in a turbulent boundary layer, outside the laminar sublayer, can be described by a one-seventh-power relation

$$\frac{u}{u_\infty} = \left(\frac{y}{\delta}\right)^{1/7} \qquad \textbf{[5-88]}$$

where δ is the boundary-layer thickness as before. For purposes of an integral analysis the momentum integral can be evaluated with Equation (5-88) because the laminar sublayer is so thin. However, the wall shear stress cannot be calculated from Equation (5-88) because it yields an infinite value at $y = 0$.

To determine the turbulent-boundary-layer thickness we employ Equation (5-17) for the integral momentum relation and evaluate the wall shear stress from the empirical relations for skin friction presented previously. According to Equation (5-52),

$$\tau_w = \frac{C_f \rho u_\infty^2}{2}$$

and so for $\mathrm{Re}_x < 10^7$ we obtain from Equation (5-77)

$$\tau_w = 0.0296 \left(\frac{\nu}{u_\infty x}\right)^{1/5} \rho u_\infty^2 \qquad \textbf{[5-89]}$$

Now, using the integral momentum equation for zero pressure gradient [Equation (5-17)] along with the velocity profile and wall shear stress, we obtain

$$\frac{d}{dx} \int_0^\delta \left[1 - \left(\frac{y}{\delta}\right)^{1/7}\right] \left(\frac{y}{\delta}\right)^{1/7} dy = 0.0296 \left(\frac{\nu}{u_\infty x}\right)^{1/5}$$

Integrating and clearing terms gives

$$\frac{d\delta}{dx} = \frac{72}{7}(0.0296) \left(\frac{\nu}{u_\infty}\right)^{1/5} x^{-1/5} \qquad \textbf{[5-90]}$$

We shall integrate this equation for two physical situations:

1. The boundary layer is fully turbulent from the leading edge of the plate.

2. The boundary layer follows a laminar growth pattern up to $Re_{crit} = 5 \times 10^5$ and a turbulent growth thereafter.

For the first case, we integrate Equation (5-89) with the condition that $\delta = 0$ at $x = 0$ to obtain

$$\frac{\delta}{x} = 0.381 \, Re_x^{-1/5} \tag{5-91}$$

For case 2 we have the condition

$$\delta = \delta_{lam} \qquad \text{at } x_{crit} = 5 \times 10^5 \frac{\nu}{u_\infty} \tag{5-92}$$

Now, δ_{lam} is calculated from the exact relation of Equation (5-21a):

$$\delta_{lam} = 5.0 x_{crit} (5 \times 10^5)^{-1/2} \tag{5-93}$$

Integrating Equation (5-89) gives

$$\delta - \delta_{lam} = \frac{72}{7}(0.0296)\left(\frac{\nu}{u_\infty}\right)^{1/5}\frac{5}{4}\left(x^{4/5} - x_{crit}^{4/5}\right) \tag{5-94}$$

Combining the various relations above gives

$$\frac{\delta}{x} = 0.381 \, Re_x^{-1/5} - 10{,}256 \, Re_x^{-1} \tag{5-95}$$

This relation applies only for the region $5 \times 10^5 < Re_x < 10^7$.

Turbulent-Boundary-Layer Thickness | EXAMPLE 5-10

Calculate the turbulent-boundary-layer thickness at the end of the plate for Example 5-9, assuming that it develops (a) from the leading edge of the plate and (b) from the transition point at $Re_{crit} = 5 \times 10^5$.

■ **Solution**
Since we have already calculated the Reynolds number as $Re_L = 1.553 \times 10^6$, it is a simple matter to insert this value in Equations (5-91) and (5-95) along with $x = L = 0.75$ m to give

$$(a) \ \delta = (0.75)(0.381)(1.553 \times 10^6)^{-0.2} = 0.0165 \text{ m} = 16.5 \text{ mm } [0.65 \text{ in}]$$

$$(b) \ \delta = (0.75)[(0.381)(1.553 \times 10^6)^{-0.2} - 10{,}256(1.553 \times 10^6)^{-1}]$$
$$= 0.0099 \text{ m} = 9.9 \text{ mm } [0.39 \text{ in}]$$

The two values differ by 40 percent.

An overall perspective of the behavior of the local and average heat-transfer coefficients is indicated in Figure 5-13. The fluid is atmospheric air flowing across an isothermal flat plate at $u_\infty = 30$ m/s, and the calculations were made with Equations (5-55), (5-81), and (5-85), which assume a value of $Re_{crit} = 5 \times 10^5$. The corresponding value of x_{crit} is 0.2615 m and the plate length is 5.23 m at $Re = 10^7$. The corresponding boundary-layer thickness is plotted in Figure 5-14. As we have noted before, the heat-transfer coefficient varies inversely with the boundary-layer thickness, and an increase in heat transfer is experienced when turbulence begins.

Figure 5-13 | Local and average heat-transfer coefficient for atmospheric airflow over isothermal flat plate at $u_\infty = 30$ m/s (*a*) semilog scale (*b*) log scale.

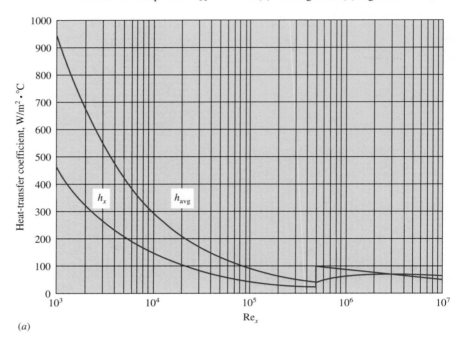

(*a*)

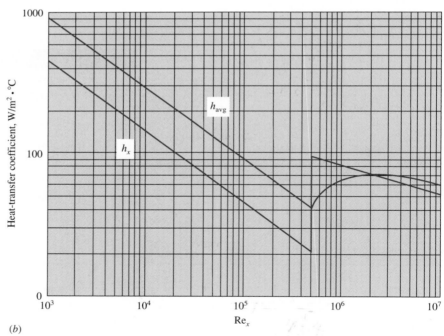

(*b*)

Figure 5-14 | Boundary-layer thickness for atmospheric air at $u_\infty = 30$ m/s.

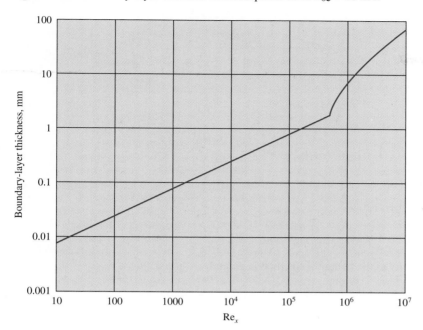

5-10 | HEAT TRANSFER IN LAMINAR TUBE FLOW

Consider the tube-flow system in Figure 5-15. We wish to calculate the heat transfer under developed flow conditions when the flow remains laminar. The wall temperature is T_w, the radius of the tube is r_o, and the velocity at the center of the tube is u_0. It is assumed that the pressure is uniform at any cross section. The velocity distribution may be derived by considering the fluid element shown in Figure 5-16. The pressure forces are balanced by the viscous-shear forces so that

$$\pi r^2\, dp = \tau 2\pi r\, dx = 2\pi r \mu\, dx\, \frac{du}{dr}$$

or

$$du = \frac{1}{2\mu} r \frac{dp}{dx}\, dr$$

Figure 5-15 | Control volume for energy analysis in tube flow.

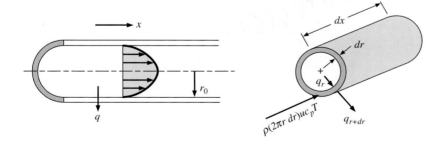

Figure 5-16 | Force balance on fluid element in tube flow.

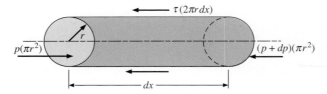

and

$$u = \frac{1}{4\mu}\frac{dp}{dx}r^2 + \text{const} \qquad\qquad \textbf{[5-96]}$$

With the boundary condition

$$u = 0 \qquad \text{at } r = r_o$$

$$u = \frac{1}{4\mu}\frac{dp}{dx}(r^2 - r_o^2)$$

the velocity at the center of the tube is given by

$$u_0 = -\frac{r_o^2}{4\mu}\frac{dp}{dx} \qquad\qquad \textbf{[5-97]}$$

so that the velocity distribution may be written

$$\frac{u}{u_0} = 1 - \frac{r^2}{r_o^2} \qquad\qquad \textbf{[5-98]}$$

which is the familiar parabolic distribution for laminar tube flow. Now consider the heat-transfer process for such a flow system. To simplify the analysis, we assume that there is a constant heat flux at the tube wall; that is,

$$\frac{dq_w}{dx} = 0$$

The heat flow conducted into the annular element is

$$dq_r = -k2\pi r\,dx\,\frac{\partial T}{\partial r}$$

and the heat conducted out is

$$dq_{r+dr} = -k2\pi(r+dr)\,dx\left(\frac{\partial T}{\partial r} + \frac{\partial^2 T}{\partial r^2}\,dr\right)$$

The net heat convected out of the element is

$$2\pi r\,dr\,\rho c_p u\,\frac{\partial T}{\partial x}\,dx$$

The energy balance is

Net energy convected out = net heat conducted in

or, neglecting second-order differentials,

$$r\rho c_p u\,\frac{\partial T}{\partial x}\,dx\,dr = k\left(\frac{\partial T}{\partial r} + r\frac{\partial^2 T}{\partial r^2}\right)dx\,dr$$

which may be rewritten

$$\frac{1}{ur}\frac{\partial}{\partial r}\left(r\frac{\partial T}{\partial r}\right) = \frac{1}{\alpha}\frac{\partial T}{\partial x} \qquad \text{[5-99]}$$

We assume that the heat flux at the wall is constant, so that the average fluid temperature must increase linearly with x, or

$$\frac{\partial T}{\partial x} = \text{const}$$

This means that the temperature profiles will be similar at various x distances along the tube. The boundary conditions on Equation (5-98) are

$$\frac{\partial T}{\partial r} = 0 \qquad \text{at } r=0$$

$$k\frac{\partial T}{\partial r}\bigg]_{r=r_o} = q_w = \text{const}$$

To obtain the solution to Equation (5-99), the velocity distribution given by Equation (5-98) must be inserted. It is assumed that the temperature and velocity fields are independent; that is, a temperature gradient does not affect the calculation of the velocity profile. This is equivalent to specifying that the properties remain constant in the flow. With the substitution of the velocity profile, Equation (5-99) becomes

$$\frac{\partial}{\partial r}\left(r\frac{\partial T}{\partial r}\right) = \frac{1}{\alpha}\frac{\partial T}{\partial x}u_0\left(1-\frac{r^2}{r_o^2}\right)r$$

Integration yields

$$r\frac{\partial T}{\partial r} = \frac{1}{\alpha}\frac{\partial T}{\partial x}u_0\left(\frac{r^2}{2}-\frac{r^4}{4r_o^2}\right)+C_1$$

and a second integration gives

$$T = \frac{1}{\alpha}\frac{\partial T}{\partial x}u_0\left(\frac{r^2}{4}-\frac{r^4}{16r_o^2}\right)+C_1\ln r+C_2$$

Applying the first boundary condition, we find that

$$C_1 = 0$$

The second boundary condition has been satisfied by noting that the axial temperature gradient $\partial T/\partial x$ is constant. The temperature distribution may finally be written in terms of the temperature at the center of the tube:

$$T = T_c \qquad \text{at } r=0 \qquad \text{so that} \qquad C_2 = T_c$$

$$T - T_c = \frac{1}{\alpha}\frac{\partial T}{\partial x}\frac{u_0 r_o^2}{4}\left[\left(\frac{r}{r_o}\right)^2 - \frac{1}{4}\left(\frac{r}{r_o}\right)^4\right] \qquad \text{[5-100]}$$

The Bulk Temperature

In tube flow the convection heat-transfer coefficient is usually defined by

$$\text{Local heat flux} = q'' = h(T_w - T_b) \qquad \text{[5-101]}$$

where T_w is the wall temperature and T_b is the so-called *bulk temperature,* or energy-average fluid temperature across the tube, which may be calculated from

$$T_b = \overline{T} = \frac{\int_0^{r_o} \rho 2\pi r\, dr\, u c_p T}{\int_0^{r_o} \rho 2\pi r\, dr\, u c_p} \qquad \text{[5-102]}$$

The reason for using the bulk temperature in the definition of heat-transfer coefficients for tube flow may be explained as follows. In a tube flow there is no easily discernible free-stream condition as is present in the flow over a flat plate. Even the centerline temperature T_c is not easily expressed in terms of the inlet flow variables and the heat transfer. For most tube- or channel-flow heat-transfer problems, the topic of central interest is the total energy transferred to the fluid in either an elemental length of the tube or over the entire length of the channel. At any x position, the temperature that is indicative of the total energy of the flow is an integrated mass-energy average temperature over the entire flow area. The numerator of Equation (5-102) represents the total energy flow through the tube, and the denominator represents the product of mass flow and specific heat integrated over the flow area. The bulk temperature is thus representative of the total energy of the flow at the particular location. For this reason, the bulk temperature is sometimes referred to as the "mixing cup" temperature, since it is the temperature the fluid would assume if placed in a mixing chamber and allowed to come to equilibrium. For the temperature distribution given in Equation (5-100), the bulk temperature is a linear function of x because the heat flux at the tube wall is constant. Calculating the bulk temperature from Equation (5-102), we have

$$T_b = T_c + \frac{7}{96} \frac{u_0 r_o^2}{\alpha} \frac{\partial T}{\partial x} \qquad \text{[5-103]}$$

and for the wall temperature

$$T_w = T_c + \frac{3}{16} \frac{u_0 r_o^2}{\alpha} \frac{\partial T}{\partial x} \qquad \text{[5-104]}$$

The heat-transfer coefficient is calculated from

$$q = hA(T_w - T_b) = kA \left(\frac{\partial T}{\partial r} \right)_{r=r_o} \qquad \text{[5-105]}$$

$$h = \frac{k(\partial T/\partial r)_{r=r_o}}{T_w - T_b}$$

The temperature gradient is given by

$$\left. \frac{\partial T}{\partial r} \right]_{r=r_o} = \frac{u_0}{\alpha} \frac{\partial T}{\partial x} \left(\frac{r}{2} - \frac{r^3}{4r_o^2} \right)_{r=r_o} = \frac{u_0 r_o}{4\alpha} \frac{\partial T}{\partial x} \qquad \text{[5-106]}$$

Substituting Equations (5-103), (5-104), and (5-106) in Equation (5-105) gives

$$h = \frac{24}{11} \frac{k}{r_o} = \frac{48}{11} \frac{k}{d_o}$$

Expressed in terms of the Nusselt number, the result is

$$\mathrm{Nu}_d = \frac{h d_o}{k} = 4.364 \qquad \text{[5-107]}$$

which is in agreement with an exact calculation by Sellars, Tribus, and Klein [3], that considers the temperature profile as it develops. Some empirical relations for calculating heat transfer in laminar tube flow will be presented in Chapter 6.

We may remark at this time that when the statement is made that a fluid enters a tube at a certain temperature, it is the bulk temperature to which we refer. The bulk temperature is used for overall energy balances on systems.

5-11 | TURBULENT FLOW IN A TUBE

The developed velocity profile for turbulent flow in a tube will appear as shown in Figure 5-17. A laminar sublayer, or "film," occupies the space near the surface, while the central core of the flow is turbulent. To determine the heat transfer analytically for this situation, we require, as usual, a knowledge of the temperature distribution in the flow. To obtain this temperature distribution, the analysis must take into consideration the effect of the turbulent eddies in the transfer of heat and momentum. We shall use an approximate analysis that relates the conduction and transport of heat to the transport of momentum in the flow (i.e., viscous effects).

The heat flow across a fluid element in laminar flow may be expressed by

$$\frac{q}{A} = -k\frac{dT}{dy}$$

Dividing both sides of the equation by ρc_p,

$$\frac{q}{\rho c_p A} = -\alpha\frac{dT}{dy}$$

It will be recalled that α is the molecular diffusivity of heat. In turbulent flow one might assume that the heat transport could be represented by

$$\frac{q}{\rho c_p A} = -(\alpha + \epsilon_H)\frac{dT}{dy} \qquad \text{[5-108]}$$

where ϵ_H is an eddy diffusivity of heat.

Equation (5-108) expresses the total heat conduction as a sum of the molecular conduction and the macroscopic eddy conduction. In a similar fashion, the shear stress in turbulent flow could be written

$$\frac{\tau}{\rho} = \left(\frac{\mu}{\rho} + \epsilon_M\right)\frac{du}{dy} = (\nu + \epsilon_M)\frac{du}{dy} \qquad \text{[5-109]}$$

where ϵ_M is the eddy diffusivity for momentum. We now assume that the heat and momentum are transported at the same rate; that is, $\epsilon_M = \epsilon_H$ and $\nu = \alpha$, or $\text{Pr} = 1$.

Dividing Equation (5-108) by Equation (5-109) gives

$$\frac{q}{c_p A\tau}\,du = -dT$$

An additional assumption is that the ratio of the heat transfer per unit area to the shear stress is constant across the flow field. This is consistent with the assumption that heat and

Figure 5-17 | Velocity profile in turbulent tube flow.

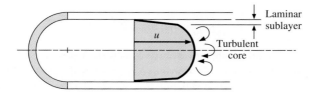

momentum are transported at the same rate. Thus

$$\frac{q}{A\tau} = \text{const} = \frac{q_w}{A_w \tau_w} \tag{5-110}$$

Then, integrating Equation (5-109) between wall conditions and mean bulk conditions gives

$$\frac{q_w}{A_w \tau_w c_p} \int_{u=0}^{u=u_m} du = \int_{T_w}^{T_b} -dT$$

$$\frac{q_w u_m}{A_w \tau_w c_p} = T_w - T_b \tag{5-111}$$

But the heat transfer at the wall may be expressed by

$$q_w = h A_w (T_w - T_b)$$

and the shear stress may be calculated from

$$\tau_w = \frac{\Delta p(\pi d_o^2)}{4\pi d_o L} = \frac{\Delta p}{4} \frac{d_o}{L}$$

The pressure drop may be expressed in terms of a friction factor f by

$$\Delta p = f \frac{L}{d_o} \rho \frac{u_m^2}{2} \tag{5-112}$$

so that

$$\tau_w = \frac{f}{8} \rho u_m^2 \tag{5-113}$$

Substituting the expressions for τ_w and q_w in Equation (5-111) gives

$$\text{St} = \frac{h}{\rho c_p u_m} = \frac{\text{Nu}_d}{\text{Re}_d \text{ Pr}} = \frac{f}{8} \tag{5-114}$$

Equation (5-114) is called the Reynolds analogy for tube flow. It relates the heat-transfer rate to the frictional loss in tube flow and is in fair agreement with experiments when used with gases whose Prandtl numbers are close to unity. (Recall that $\text{Pr} = 1$ was one of the assumptions in the analysis.)

An empirical formula for the turbulent-friction factor up to Reynolds numbers of about 2×10^5 for the flow in smooth tubes is

$$f = \frac{0.316}{\text{Re}_d^{1/4}} \tag{5-115}$$

Inserting this expression in Equation (5-113) gives

$$\frac{\text{Nu}_d}{\text{Re}_d \text{ Pr}} = 0.0395 \text{ Re}_d^{-1/4}$$

or

$$\text{Nu}_d = 0.0395 \text{ Re}_d^{3/4} \tag{5-116}$$

since we assumed the Prandtl number to be unity. This derivation of the relation for turbulent heat transfer in smooth tubes is highly restrictive because of the $\text{Pr} \approx 1.0$ assumption. The heat-transfer–fluid-friction analogy of Section 5-7 indicated a Prandtl-number dependence of $Pr^{2/3}$ for the flat-plate problem and, as it turns out, this dependence works fairly well for turbulent tube flow. Equations (5-114) and (5-116) may be modified by this factor to yield

$$St\ Pr^{2/3} = \frac{f}{8}$$

[5-114a]

$$Nu_d = 0.0395\ Re_d^{3/4}\ Pr^{1/3}$$

[5-116a]

As we shall see in Chapter 6, Equation (5-116a) predicts heat-transfer coefficients that are somewhat higher than those observed in experiments. The purpose of the discussion at this point has been to show that one may arrive at a relation for turbulent heat transfer in a fairly simple analytical fashion. As we have indicated earlier, a rigorous development of the Reynolds analogy between heat transfer and fluid friction involves considerations beyond the scope of our discussion, and the simple path of reasoning chosen here is offered for the purpose of indicating the general nature of the physical processes.

For calculation purposes, a more correct relation to use for turbulent flow in a smooth tube is Equation (6-4a), which we list here for comparison:

$$Nu_d = 0.023\ Re_d^{0.8}\ Pr^{0.4}$$

[6-4a]

All properties in Equation (6-4a) are evaluated at the bulk temperature.

5-12 | HEAT TRANSFER IN HIGH-SPEED FLOW

Our previous analysis of boundary-layer heat transfer (Section 5-6) neglected the effects of viscous dissipation within the boundary layer. When the free-stream velocity is very high, as in high-speed aircraft, these dissipation effects must be considered. We begin our analysis by considering the adiabatic case, i.e., a perfectly insulated wall. In this case the wall temperature may be considerably higher than the free-stream temperature even though no heat transfer takes place. This high temperature results from two situations: (1) the increase in temperature of the fluid as it is brought to rest at the plate surface while the kinetic energy of the flow is converted to internal thermal energy, and (2) the heating effect due to viscous dissipation. Consider the first situation. The kinetic energy of the gas is converted to thermal energy as the gas is brought to rest, and this process is described by the steady-flow energy equation for an adiabatic process:

$$i_0 = i_\infty + \frac{1}{2g_c}u_\infty^2$$

[5-117]

where i_0 is the stagnation enthalpy of the gas. This equation may be written in terms of temperature as

$$c_p(T_0 - T_\infty) = \frac{1}{2g_c}u_\infty^2$$

where T_0 is the stagnation temperature and T_∞ is the static free-stream temperature. Expressed in terms of the free-stream Mach number, it is

$$\frac{T_0}{T_\infty} = 1 + \frac{\gamma - 1}{2}M_\infty^2$$

[5-118]

where M_∞ is the Mach number, defined as $M_\infty = u_\infty/a$, and a is the acoustic velocity, which for an ideal gas may be calculated with

$$a = \sqrt{\gamma g_c R T}$$

[5-119]

where R is the gas constant for the particular gas.

In the actual case of a boundary-layer flow problem, the fluid is not brought to rest reversibly because the viscous action is basically an irreversible process in a thermodynamic sense. In addition, not all the free-stream kinetic energy is converted to thermal energy—part is lost as heat, and part is dissipated in the form of viscous work. To take into account the irreversibilities in the boundary-layer flow system, a *recovery factor* is defined by

$$r = \frac{T_{aw} - T_{\infty}}{T_0 - T_{\infty}} \qquad \text{[5-120]}$$

where T_{aw} is the actual adiabatic wall temperature and T_{∞} is the static temperature of the free stream. The recovery factor may be determined experimentally, or, for some flow systems, analytical calculations may be made.

The boundary-layer energy equation

$$u\frac{\partial T}{\partial x} + v\frac{\partial T}{\partial y} = \alpha\frac{\partial^2 T}{\partial y^2} + \frac{\mu}{\rho c_p}\left(\frac{\partial u}{\partial y}\right)^2$$

has been solved for the high-speed-flow situation, taking into account the viscous-heating term. Although the complete solution is somewhat tedious, the final results are remarkably simple. For our purposes we present only the results and indicate how they may be applied. The reader is referred to Appendix B for an exact solution to Equation (5-22). An excellent synopsis of the high-speed heat-transfer problem is given in a report by Eckert [4]. Some typical boundary-layer temperature profiles for an adiabatic wall in high-speed flow are given in Figure B-3.

The essential result of the high-speed heat-transfer analysis is that heat-transfer rates may generally be calculated with the same relations used for low-speed incompressible flow when the average heat-transfer coefficient is redefined with the relation

$$q = \overline{h}A(T_w - T_{aw}) \qquad \text{[5-121]}$$

Notice that the difference between the adiabatic wall temperature and the actual wall temperature is used in the definition so that the expression will yield a value of zero heat flow when the wall is at the adiabatic wall temperature. For gases with Prandtl numbers near unity, the following relations for the recovery factor have been derived:

Laminar flow: $\qquad\qquad\qquad\qquad\qquad r = \text{Pr}^{1/2} \qquad\qquad\qquad\qquad$ [5-122]

Turbulent flow: $\qquad\qquad\qquad\qquad\qquad r = \text{Pr}^{1/3} \qquad\qquad\qquad\qquad$ [5-123]

These recovery factors may be used in conjunction with Equation (5-120) to obtain the adiabatic wall temperature.

In high-velocity boundary layers substantial temperature gradients may occur, and there will be correspondingly large property variations across the boundary layer. The constant-property heat-transfer equations may still be used if the properties are introduced at a reference temperature T^* as recommended by Eckert:

$$T^* = T_{\infty} + 0.50(T_w - T_{\infty}) + 0.22(T_{aw} - T_{\infty}) \qquad \text{[5-124]}$$

The analogy between heat transfer and fluid friction [Equation (5-56)] may also be used when the friction coefficient is known. Summarizing the relations used for high-speed heat-transfer calculations:

Laminar boundary layer ($\text{Re}_x < 5 \times 10^5$):

$$\text{St}_x^* \, \text{Pr}^{*2/3} = 0.332 \, \text{Re}_x^{*-1/2} \qquad \text{[5-125]}$$

Turbulent boundary layer ($5 \times 10^5 < \mathrm{Re}_x < 10^7$):

$$\mathrm{St}_x^* \, \mathrm{Pr}^{*2/3} = 0.0296 \, \mathrm{Re}_x^{*-1/5} \qquad \textbf{[5-126]}$$

Turbulent boundary layer ($10^7 < \mathrm{Re}_x < 10^9$):

$$\mathrm{St}_x^* \, \mathrm{Pr}^{*2/3} = 0.185(\log \mathrm{Re}_x^*)^{-2.584} \qquad \textbf{[5-127]}$$

The superscript * in the above equations indicates that the properties are evaluated at the reference temperature given by Equation (5-124).

To obtain an average heat-transfer coefficient, the above expressions must be integrated over the length of the plate. If the Reynolds number falls in a range such that Equation (5-127) must be used, the integration cannot be expressed in closed form, and a numerical integration must be performed. Care must be taken in performing the integration for the high-speed heat-transfer problem since the reference temperature is different for the laminar and turbulent portions of the boundary layer. This results from the different value of the recovery factor used for laminar and turbulent flow as given by Equations (5-122) and (5-123).

When very high flow velocities are encountered, the adiabatic wall temperature may become so high that dissociation of the gas will take place and there will be a very wide variation of the properties in the boundary layer. Eckert [4] recommends that these problems be treated on the basis of a heat-transfer coefficient defined in terms of *enthalpy* difference:

$$q = h_i A(i_w - i_{aw}) \qquad \textbf{[5-128]}$$

The enthalpy recovery factor is then defined as

$$r_i = \frac{i_{aw} - i_\infty}{i_0 - i_\infty} \qquad \textbf{[5-129]}$$

where i_{aw} is the enthalpy at the adiabatic wall conditions. The same relations as before are used to calculate the recovery factor and heat-transfer except that all properties are evaluated at a reference enthalpy i^* given by

$$i^* = i_\infty + 0.5(i_w - i_\infty) + 0.22(i_{aw} - i_\infty) \qquad \textbf{[5-130]}$$

The Stanton number is redefined as

$$\mathrm{St}_i = \frac{h_i}{\rho u_\infty} \qquad \textbf{[5-131]}$$

This Stanton number is then used in Equation (5-125), (5-126), or (5-127) to calculate the heat-transfer coefficient. When calculating the enthalpies for use in the above relations, the *total* enthalpy must be used; that is chemical energy of dissociation as well as internal thermal energy must be included. The reference-enthalpy method has proved successful for calculating high-speed heat-transfer with an accuracy of better than 10 percent.

High-Speed Heat Transfer for a Flat Plate   EXAMPLE 5-11

A flat plate 70 cm long and 1.0 m wide is placed in a wind tunnel where the flow conditions are $M = 3$, $p = \frac{1}{20}$ atm, and $T = -40°C$. How much cooling must be used to maintain the plate temperature at $35°C$?

■ **Solution**

We must consider the laminar and turbulent portions of the boundary layer separately because the recovery factors, and hence the adiabatic wall temperatures, used to establish the heat flow will be different for each flow regime. It turns out that the difference is rather small in this problem, but we shall follow a procedure that would be used if the difference were appreciable, so that the general method of solution may be indicated. The free-stream acoustic velocity is calculated from

$$a = \sqrt{\gamma g_c R T_\infty} = [(1.4)(1.0)(287)(233)]^{1/2} = 306 \text{ m/s} \quad [1003 \text{ ft/s}]$$

so that the free-stream velocity is

$$u_\infty = (3)(306) = 918 \text{ m/s} \quad [3012 \text{ ft/s}]$$

The maximum Reynolds number is estimated by making a computation based on properties evaluated at free-stream conditions:

$$\rho_\infty = \frac{(1.0132 \times 10^5)(\frac{1}{20})}{(287)(233)} = 0.0758 \text{ kg/m}^3 \quad [4.73 \times 10^{-3} \text{ lb}_m/\text{ft}^3]$$

$$\mu_\infty = 1.434 \times 10^{-5} \text{ kg/m} \cdot \text{s} \quad [0.0347 \text{ lb}_m/\text{h} \cdot \text{ft}]$$

$$\text{Re}_{L,\infty} = \frac{(0.0758)(918)(0.70)}{1.434 \times 10^{-5}} = 3.395 \times 10^6$$

Thus we conclude that both laminar and turbulent-boundary-layer heat transfer must be considered. We first determine the reference temperatures for the two regimes and then evaluate properties at these temperatures.

Laminar portion

$$T_0 = T_\infty \left(1 + \frac{\gamma - 1}{2} M_\infty^2 \right) = (233)[1 + (0.2)(3)^2] = 652 \text{ K}$$

Assuming a Prandtl number of about 0.7, we have

$$r = \text{Pr}^{1/2} = (0.7)^{1/2} = 0.837$$

$$r = \frac{T_{aw} - T_\infty}{T_0 - T_\infty} = \frac{T_{aw} - 233}{652 - 233}$$

and $T_{aw} = 584 \text{ K} = 311°\text{C} \ [592°\text{F}]$. Then the reference temperature from Equation (5-123) is

$$T^* = 233 + (0.5)(308 - 233) + (0.22)(584 - 233) = 347.8 \text{ K}$$

Checking the Prandtl number at this temperature, we have

$$\text{Pr}^* = 0.697$$

so that the calculation is valid. If there were an appreciable difference between the value of Pr* and the value used to determine the recovery factor, the calculation would have to be repeated until agreement was reached.

The other properties to be used in the laminar heat-transfer analysis are

$$\rho^* = \frac{(1.0132 \times 10^5)(1/20)}{(287)(347.8)} = 0.0508 \text{ kg/m}^3$$

$$\mu^* = 2.07 \times 10^{-5} \text{ kg/m} \cdot \text{s}$$

$$k^* = 0.03 \text{ W/m} \cdot °\text{C} \quad [0.0173 \text{ Btu/h} \cdot \text{ft} \cdot °\text{F}]$$

$$c_p^* = 1.009 \text{ kJ/kg} \cdot °\text{C}$$

Turbulent portion

Assuming $Pr = 0.7$ gives

$$r = Pr^{1/3} = 0.888 = \frac{T_{aw} - T_\infty}{T_0 - T_\infty} = \frac{T_{aw} - 233}{652 - 233}$$

$$T_{aw} = 605 \text{ K} = 332°C$$

$$T^* = 233 + (0.5)(308 - 233) + (0.22)(605 - 233) = 352.3 \text{ K}$$

$$Pr^* = 0.695$$

The agreement between Pr^* and the assumed value is sufficiently close. The other properties to be used in the turbulent heat-transfer analysis are

$$\rho^* = \frac{(1.0132 \times 10^5)(1/20)}{(287)(352.3)} = 0.0501 \text{ kg/m}^3$$

$$\mu^* = 2.09 \times 10^{-5} \text{ kg/m} \cdot \text{s}$$

$$k^* = 0.0302 \text{ W/m} \cdot °C \qquad c_p^* = 1.009 \text{ kJ/kg} \cdot °C$$

Laminar heat transfer

We assume

$$Re_{crit}^* = 5 \times 10^5 = \frac{\rho^* u_\infty x_c}{\mu^*}$$

$$x_c = \frac{(5 \times 10^5)(2.07 \times 10^{-5})}{(0.0508)(918)} = 0.222 \text{ m}$$

$$\overline{Nu^*} = \frac{\bar{h} x_c}{k^*} = 0.664 \left(Re_{crit}^*\right)^{1/2} Pr^{*1/3}$$

$$= (0.664)(5 \times 10^5)^{1/2}(0.697)^{1/3} = 416.3$$

$$\bar{h} = \frac{(416.3)(0.03)}{0.222} = 56.25 \text{ W/m}^2 \cdot °C \quad [9.91 \text{ Btu/h} \cdot \text{ft}^2 \cdot °F]$$

This is the average heat-transfer coefficient for the laminar portion of the boundary layer, and the heat transfer is calculated from

$$q = \bar{h} A (T_w - T_{aw})$$

$$= (56.26)(0.222)(308 - 584)$$

$$= -3445 \text{ W} \quad [-11,750 \text{ Btu/h}]$$

so that 3445 W of cooling is required in the laminar region of the plate per meter of depth in the z direction.

Turbulent heat transfer

To determine the turbulent heat transfer we must obtain an expression for the local heat-transfer coefficient from

$$St_x^* Pr^{*2/3} = 0.0296 Re_x^{*-1/5}$$

and then integrate from $x = 0.222$ m to $x = 0.7$ m to determine the total heat transfer:

$$h_x = Pr^{*-2/3} \rho^* u_\infty c_p (0.0296) \left(\frac{\rho^* u_\infty x}{\mu^*}\right)^{-1/5}$$

Inserting the numerical values for the properties gives

$$h_x = 94.34 x^{-1/5}$$

The average heat-transfer coefficient in the turbulent region is determined from

$$\bar{h} = \frac{\int_{0.222}^{0.7} h_x \, dx}{\int_{0.222}^{0.7} dx} = 111.46 \text{ W/m}^2 \cdot °\text{C} \quad [19.6 \text{ Btu/h} \cdot \text{ft}^2 \cdot °\text{F}]$$

Using this value we may calculate the heat transfer in the turbulent region of the flat plate:

$$q = \bar{h} A (T_w - T_{aw})$$
$$= (111.46)(0.7 - 0.222)(308 - 605)$$
$$= -15{,}823 \text{ W} \quad [-54{,}006 \text{ Btu/h}]$$

The total amount of cooling required is the sum of the heat transfers for the laminar and turbulent portions:

$$\text{Total cooling} = 3445 + 15{,}823 = 19{,}268 \text{ W} \quad [65{,}761 \text{ Btu/h}]$$

These calculations assume unit depth of 1 m in the z direction.

5-13 | SUMMARY

Most of this chapter has been concerned with flow over flat plates and the associated heat transfer. For convenience of the reader we have summarized the heat-transfer, boundary-layer thickness, and friction-coefficient equations in Table 5-2 along with the restrictions that apply. Our presentation of convection heat transfer is incomplete at this time and will be developed further in Chapters 6 and 7. Even so, we begin to see the structure of a procedure for solution of convection problems:

1. Establish the geometry of the situation; for now we are mainly restricted to flow over flat plates.
2. Determine the fluid involved and evaluate the fluid properties. This will usually be at the film temperature.
3. Establish the boundary conditions (i.e., constant temperature or constant heat flux).
4. Establish the flow regime as determined by the Reynolds number.
5. Select the appropriate equation, taking into account the flow regime and any fluid property restrictions which may apply.
6. Calculate the value(s) of the convection heat-transfer coefficient and/or heat transfer.

At the conclusion of Chapter 7 we shall give a general procedure for all convection problems and the information contained in Table 5-2 will comprise one ingredient in the overall recipe. The interested reader may wish to consult Section 7-14 and Figure 7-15 for a preview of this information and some perspective of the way the material in the present chapter fits in.

REVIEW QUESTIONS

1. What is meant by a hydrodynamic boundary level?
2. Define the Reynolds number. Why is it important?
3. What is the physical mechanism of viscous action?
4. Distinguish between laminar and turbulent flow in a physical sense.

Table 5-2 | Summary of equations for flow over flat plates. Properties evaluated at $T_f = (T_w + T_\infty)/2$ unless otherwise noted.

Flow regime	Restrictions	Equation	Equation number
		Heat transfer	
Laminar, local	$T_w = \text{const}, \text{Re}_x < 5 \times 10^5,$ $0.6 < \text{Pr} < 50$	$\text{Nu}_x = 0.332\,\text{Pr}^{1/3}\text{Re}_x^{1/2}$	(5-44)
Laminar, local	$T_w = \text{const}, \text{Re}_x < 5 \times 10^5,$ $\text{Re}_x\,\text{Pr} > 100$	$\text{Nu}_x = \dfrac{0.3387\,\text{Re}_x^{1/2}\,\text{Pr}^{1/3}}{\left[1 + \left(\dfrac{0.0468}{\text{Pr}}\right)^{2/3}\right]^{1/4}}$	(5-51)
Laminar, local	$q_w = \text{const}, \text{Re}_x < 5 \times 10^5,$ $0.6 < \text{Pr} < 50$	$\text{Nu}_x = 0.453\,\text{Re}_x^{1/2}\,\text{Pr}^{1/3}$	(5-48)
Laminar, local	$q_w = \text{const}, \text{Re}_x < 5 \times 10^5$	$\text{Nu}_x = \dfrac{0.4637\,\text{Re}_x^{1/2}\,\text{Pr}^{1/3}}{\left[1 + \left(\dfrac{0.0207}{\text{Pr}}\right)^{2/3}\right]^{1/4}}$	(5-51)
Laminar, average	$\text{Re}_L < 5 \times 10^5,\ T_w = \text{const}$	$\overline{\text{Nu}}_L = 2\,\text{Nu}_{x=L} = 0.664\,\text{Re}_L^{1/2}\,\text{Pr}^{1/3}$	(5-46)
Laminar, local	$T_w = \text{const}, \text{Re}_x < 5 \times 10^5,$ $\text{Pr} \ll 1$ (liquid metals)	$\text{Nu}_x = 0.564(\text{Re}_x\,\text{Pr})^{1/2}$	
Laminar, local	$T_w = \text{const},$ starting at $x = x_0, \text{Re}_x < 5 \times 10^5,$ $0.6 < \text{Pr} < 50$	$\text{Nu}_x = 0.332\,\text{Pr}^{1/3}\,\text{Re}_x^{1/2}\left[1 - \left(\dfrac{x_0}{x}\right)^{3/4}\right]^{-1/3}$	(5-43)
Turbulent, local	$T_w = \text{const}, 5 \times 10^5 < \text{Re}_x < 10^7$	$\text{St}_x\,\text{Pr}^{2/3} = 0.0296\,\text{Re}_x^{-0.2}$	(5-81)
Turbulent, local	$T_w = \text{const}, 10^7 < \text{Re}_x < 10^9$	$\text{St}_x\,\text{Pr}^{2/3} = 0.185(\log\text{Re}_x)^{-2.584}$	(5-82)
Turbulent, local	$q_w = \text{const}, 5 \times 10^5 < \text{Re}_x < 10^7$	$\text{Nu}_x = 1.04\,\text{Nu}_{x\,Tw=\text{const}}$	(5-87)
Laminar-turbulent, average	$T_w = \text{const}, \text{Re}_x < 10^7,$ $\text{Re}_{\text{crit}} = 5 \times 10^5$	$\overline{\text{St}}\,\text{Pr}^{2/3} = 0.037\,\text{Re}_L^{-0.2} - 871\,\text{Re}_L^{-1}$ $\overline{\text{Nu}}_L = \text{Pr}^{1/3}(0.037\,\text{Re}_L^{0.8} - 871)$	(5-84) (5-85)
Laminar-turbulent, average	$T_w = \text{const}, \text{Re}_x < 10^7,$ liquids, μ at $T_\infty,$ μ_w at T_w	$\overline{\text{Nu}}_L = 0.036\,\text{Pr}^{0.43}(\text{Re}_L^{0.8} - 9200)\left(\dfrac{\mu_\infty}{\mu_w}\right)^{1/4}$	(5-86)
High-speed flow	$T_w = \text{const},$ $q = hA(T_w - T_{aw})$ $r = (T_{aw} - T_\infty)/(T_o - T_\infty)$ = recovery factor = $\text{Pr}^{1/2}$ (laminar) = $\text{Pr}^{1/3}$ (turbulent)	Same as for low-speed flow with properties evaluated at $T^* = T_\infty + 0.5(T_w - T_\infty) + 0.22(T_{aw} - T_\infty)$	(5-124)
		Boundary-layer thickness	
Laminar	$\text{Re}_x < 5 \times 10^5$	$\dfrac{\delta}{x} = 5.0\,\text{Re}_x^{-1/2}$	(5-21a)
Turbulent	$\text{Re}_x < 10^7,$ $\delta = 0$ at $x = 0$	$\dfrac{\delta}{x} = 0.381\,\text{Re}_x^{-1/5}$	(5-91)
Turbulent	$5 \times 10^5 < \text{Re}_x < 10^7,$ $\text{Re}_{\text{crit}} = 5 \times 10^5,$ $\delta = \delta_{\text{lam}}$ at Re_{crit}	$\dfrac{\delta}{x} = 0.381\,\text{Re}_x^{-1/5} - 10,256\,\text{Re}_x^{-1}$	(5-95)
		Friction coefficients	
Laminar, local	$\text{Re}_x < 5 \times 10^5$	$C_{fx} = 0.332\,\text{Re}_x^{-1/2}$	(5-54)
Turbulent, local	$5 \times 10^5 < \text{Re}_x < 10^7$	$C_{fx} = 0.0592\,\text{Re}_x^{-1/5}$	(5-77)
Turbulent, local	$10^7 < \text{Re}_x < 10^9$	$C_{fx} = 0.37(\log\text{Re}_x)^{-2.584}$	(5-78)
Turbulent, average	$\text{Re}_{\text{crit}} < \text{Re}_x < 10^9$	$\overline{C}_f = \dfrac{0.455}{(\log\text{Re}_L)^{2.584}} - \dfrac{A}{\text{Re}_L}$ A from Table 5-1	(5-79)

5. What is the momentum equation for the laminar boundary layer on a flat plate? What assumptions are involved in the derivation of this equation?

6. How is the boundary-layer thickness defined?

7. What is the energy equation for the laminar boundary layer on a flat plate? What assumptions are involved in the derivation of this equation?

8. What is meant by a thermal boundary layer?

9. Define the Prandtl number. Why is it important?

10. Describe the physical mechanism of convection. How is the convection heat-transfer coefficient related to this mechanism?

11. Describe the relation between fluid friction and heat transfer.

12. Define the bulk temperature. How is it used?

13. How is the heat-transfer coefficient defined for high-speed heat-transfer calculations?

LIST OF WORKED EXAMPLES

5-1 Water flow in a diffuser

5-2 Isentropic expansion of air

5-3 Mass flow and boundary-layer thickness

5-4 Isothermal flat plate heated over entire length

5-5 Flat plate with constant heat flux

5-6 Plate with unheated starting length

5-7 Oil flow over heated flat plate

5-8 Drag force on a flat plate

5-9 Turbulent heat transfer from isothermal flat plate

5-10 Turbulent-boundary-layer thickness

5-11 High-speed heat transfer for a flat plate

PROBLEMS

5-1 A certain nozzle is designed to expand air from stagnation conditions of 1.38 MPa and 200°C to 0.138 MPa. The mass rate of flow is designed to be 4.5 kg/s. Suppose this nozzle is used in conjunction with a blowdown wind-tunnel facility so that the nozzle is suddenly allowed to discharge into a perfectly evacuated tank. What will the temperature of the air in the tank be when the pressure in the tank equals 0.138 MPa? Assume that the tank is perfectly insulated and that air behaves as a perfect gas. Assume that the expansion in the nozzle is isentropic.

5-2 Using a linear velocity profile

$$\frac{u}{u_\infty} = \frac{y}{\delta}$$

for a flow over a flat plate, obtain an expression for the boundary-layer thickness as a function of x.

5-3 Using the continuity relation

$$\frac{\partial u}{\partial x} + \frac{\partial v}{\partial y} = 0$$

along with the velocity distribution

$$\frac{u}{u_\infty} = \frac{3}{2}\frac{y}{\delta} - \frac{1}{2}\left(\frac{y}{\delta}\right)^3$$

and the expression for the boundary-layer thickness

$$\frac{\delta}{x} = \frac{4.64}{\sqrt{Re_x}}$$

derive an expression for the y component of velocity v as a function of x and y. Calculate the value of v at the outer edge of the boundary layer at distances of 6 and 12 in from the leading edge for the conditions of Example 5-3.

5-4 Repeat Problem 5-3 for the linear velocity profile of Problem 5-2.

5-5 Using the linear-velocity profile in Problem 5-2 and a cubic-parabola temperature distribution [Equation (5-30)], obtain an expression for heat-transfer coefficient as a function of the Reynolds number for a laminar boundary layer on a flat plate.

5-6 Air at 20 kPa and 5°C enters a 2.5-cm-diameter tube at a velocity of 1.5 m/s. Using a flat-plate analysis, estimate the distance from the entrance at which the flow becomes fully developed.

5-7 Oxygen at a pressure of 2 atm and 27°C blows across a 50-cm-square plate at a velocity of 30 m/s. The plate temperature is maintained constant at 127°C. Calculate the total heat lost by the plate.

5-8 A fluid flows between two large parallel plates. Develop an expression for the velocity distribution as a function of distance from the centerline between the two plates under developed flow conditions.

5-9 Using the energy equation given by Equation (5-32), determine an expression for heat-transfer coefficient under the conditions

$$u = u_\infty = \text{const}\qquad \frac{T - T_w}{T_\infty - T_w} = \frac{y}{\delta_t}$$

where δ_t is the thermal-boundary-layer thickness.

5-10 Derive an expression for the heat transfer in a laminar boundary layer on a flat plate under the condition $u = u_\infty = $ constant. Assume that the temperature distribution is given by the cubic-parabola relation in Equation (5-30). This solution approximates the condition observed in the flow of a liquid metal over a flat plate.

5-11 Show that $\partial^3 u/\partial y^3 = 0$ at $y = 0$ for an incompressible laminar boundary layer on a flat plate with zero-pressure gradient.

5-12 Review the analytical developments of this chapter and list the restrictions that apply to the following equations: (5-25), (5-26), (5-44), (5-46), (5-85), and (5-107).

5-13 Calculate the ratio of thermal-boundary-layer thickness to hydrodynamic-boundary-layer thickness for the following fluids: air at 1 atm and 20°C, water at 20°C, helium at 1 atm and 20°C, liquid ammonia at 20°C, glycerine at 20°C.

5-14 For water flowing over a flat plate at 15°C and 3 m/s, calculate the mass flow through the boundary layer at a distance of 5 cm from the leading edge of the plate.

5-15 Air at 90°C and 1 atm flows over a flat plate at a velocity of 30 m/s. How thick is the boundary layer at a distance of 2.5 cm from the leading edge of the plate?

5-16 Air flows over a flat plate at a constant velocity of 20 m/s and ambient conditions of 20 kPa and 20°C. The plate is heated to a constant temperature of 75°C, starting at a distance of 7.5 cm from the leading edge. What is the total heat transfer from the leading edge to a point 35 cm from the leading edge?

5-17 Water at 15°C flows between two large parallel plates at a velocity of 1.5 m/s. The plates are separated by a distance of 15 mm. Estimate the distance from the leading edge where the flow becomes fully developed.

5-18 Air at standard conditions of 1 atm and 27°C flows over a flat plate at 20 m/s. The plate is 60 cm square and is maintained at 97°C. Calculate the heat transfer from the plate.

5-19 Air at 7 kPa and 35°C flows across a 30-cm-square flat plate at 7.5 m/s. The plate is maintained at 65°C. Estimate the heat lost from the plate.

5-20 Air at 90°C and atmospheric pressure flows over a horizontal flat plate at 60 m/s. The plate is 60 cm square and is maintained at a uniform temperature of 10°C. What is the total heat transfer?

5-21 Nitrogen at 2 atm and 500 K flows across a 40-cm-square plate at a velocity of 25 m/s. Calculate the cooling required to maintain the plate surface at a constant temperature of 300 K.

5-22 Plot the heat-transfer coefficient versus length for flow over a 1-m-long flat plate under the following conditions: (a) helium at 1 lb/in^2 abs, 80°F, $u_\infty = 10$ ft/s [3.048 m/s]; (b) hydrogen at 1 lb/in^2 abs, 80°F, $u_\infty = 10$ ft/s; (c) air at 1 lb/in^2 abs, 80°F, $u_\infty = 10$ ft/s; (d) water at 80°F, $u_\infty = 10$ ft/s; (e) helium at 20 lb/in^2 abs, 80°F, $u_\infty = 10$ ft/s.

5-23 Calculate the heat transfer from a 20-cm-square plate over which air flows at 35°C and 14 kPa. The plate temperature is 250°C, and the free-stream velocity is 6 m/s.

5-24 Air at 20 kPa and 20°C flows across a flat plate 60 cm long. The free-stream velocity is 30 m/s, and the plate is heated over its total length to a temperature of 55°C. For $x = 30$ cm, calculate the value of y for which u will equal 22.5 m/s.

5-25 For the flow system in Problem 5-24, calculate the value of the friction coefficient at a distance of 15 cm from the leading edge.

5-26 Air at a pressure of 200 kPa and free-stream temperature of 27°C flows over a square flat plate at a velocity of 30 m/s. The Reynolds number is 10^6 at the edge of the plate. Calculate the heat transfer for an isothermal plate maintained at 57°C.

5-27 Calculate the boundary layer thickness at the edge of the plate for the flow system in Problem 5-26. State the assumptions.

5-28 Air at 5°C and 70 kPa flows over a flat plate at 6 m/s. A heater strip 2.5 cm long is placed on the plate at a distance of 15 cm from the leading edge. Calculate the heat lost from the strip per unit depth of plate for a heater surface temperature of 65°C.

5-29 Air at 1 atm and 27°C blows across a large concrete surface 15 m wide maintained at 55°C. The flow velocity is 4.5 m/s. Calculate the convection heat loss from the surface.

5-30 Air at 300 K and 75 kPa flows over a 1-m-square plate at a velocity of 45 m/s. The plate is maintained at a constant temperature of 400 K. Calculate the heat lost by the plate.

5-31 A horizontal flat plate is maintained at 50°C and has dimensions of 50 cm by 50 cm. Air at 50 kPa and 10°C is blown across the plate at 20 m/s. Calculate the heat lost from the plate.

5-32 Air flows across a 20-cm-square plate with a velocity of 5 m/s. Free-stream conditions are 10°C and 0.2 atm. A heater in the plate surface furnishes a constant heat-flux condition at the wall so that the average wall temperature is 100°C. Calculate the surface heat flux and the value of h at an x position of 10 cm.

5-33 Calculate the flow velocity necessary to produce a Reynolds number of 10^7 for flow across a 1-m-square plate with the following fluids: (*a*) water at 20°C, (*b*) air at 1 atm and 20°C, (*c*) Freon 12 at 20°C, (*d*) ammonia at 20°C, and (*e*) helium at 20°C.

5-34 Calculate the average heat-transfer coefficient for each of the cases in Problem 5-31 assuming all properties are evaluated at 20°C.

5-35 Calculate the boundary-layer thickness at the end of the plate for each case in Problem 5-33.

5-36 A blackened plate is exposed to the sun so that a constant heat flux of 800 W/m^2 is absorbed. The back side of the plate is insulated so that all the energy absorbed is dissipated to an airstream that blows across the plate at conditions of 25°C, 1 atm, and 3 m/s. The plate is 25 cm square. Estimate the average temperature of the plate. What is the plate temperature at the trailing edge?

5-37 Air at 0.5 atm pressure and 27°C flows across a 34-cm-square plate at a velocity of 20 m/s. The plate temperature is maintained at 127°C. Calculate the heat lost by the plate.

5-38 Helium at 3 atm and 73°C flows across a 35-cm-square plate that is maintained at a surface temperature of 113°C. The free-stream velocity is 50 m/s. Calculate the convection heat lost by the plate.

5-39 Air at 1 atm and 300 K blows across a 50-cm-square flat plate at a velocity such that the Reynolds number at the downstream edge of the plate is 1.1×10^5. Heating does not begin until halfway along the plate and then the surface temperature is 400 K. Calculate the heat transfer from the plate.

5-40 Air at 20°C and 14 kPa flows at a velocity of 150 m/s past a flat plate 1 m long that is maintained at a constant temperature of 150°C. What is the average heat-transfer rate per unit area of plate?

5-41 Derive equations equivalent to Equation (5-85) for critical Reynolds numbers of 3×10^5, 10^6, and 3×10^6.

5-42 Assuming that the local heat-transfer coefficient for flow on a flat plate can be represented by Equation (5-81) and that the boundary layer starts at the leading edge of the plate, determine an expression for the average heat-transfer coefficient.

5-43 A 10-cm-square plate has an electric heater installed that produces a constant heat flux. Water at 10°C flows across the plate at a velocity of 3 m/s. What is the total heat which can be dissipated if the plate temperature is not to exceed 80°C?

5-44 Repeat Problem 5-41 for air at 1 atm and 300 K.

5-45 Helium at 1 atm and 300 K is used to cool a 1-m-square plate maintained at 500 K. The flow velocity is 50 m/s. Calculate the total heat loss from the plate. What is the boundary-layer thickness as the flow leaves the plate?

5-46 A light breeze at 10 mi/h blows across a metal building in the summer. The height of the building wall is 3.7 m, and the width is 6.1 m. A net energy flux of 347 W/m^2 from the sun is absorbed in the wall and subsequently dissipated to the surrounding air by convection. Assuming that the air is 27°C and 1 atm and blows across the wall as on a flat plate, estimate the average temperature the wall will attain for equilibrium conditions.

5-47 The bottom of a corn-chip fryer is 10 ft long by 3 ft wide and is maintained at a temperature of 420°F. Cooking oil flows across this surface at a velocity of 1 ft/s and has a free-stream temperature of 400°F. Calculate the heat transfer to the oil and estimate the maximum boundary-layer thickness. Properties of the oil may be taken as $\nu = 2 \times 10^{-6}$ m^2/s, $k = 0.12$ W/m $\cdot$ °C, and Pr $= 40$.

5-48 Air at 27°C and 1 atm blows over a 4.0-m-square flat plate at a velocity of 40 m/s. The plate temperature is 77°C. Calculate the total heat transfer.

5-49 The roof of a building is 30 m by 60 m, and because of heat loading by the sun it attains a temperature of 300 K when the ambient air temperature is 0°C. Calculate the heat loss from the roof for a mild breeze blowing at 5 mi/h across the roof ($L = 30$ m).

5-50 Air at 1 atm and 30°C flows over a 15-cm-square plate at a velocity of 10 m/s. Calculate the maximum boundary layer thickness.

5-51 Nitrogen at 1 atm and 300 K blows across a horizontal flat plate at a velocity of 33 m/s. The plate has a constant surface temperature of 400 K. Calculate the heat lost by the plate if the plate dimensions are 60 cm by 30 cm with the longer dimension in the direction of flow. Express in watts.

5-52 Helium at atmospheric pressure and 30°C flows across a square plate at a free-stream velocity of 15 m/s. Calculate the boundary layer thickness at a position where Re$_x = 250,000$.

5-53 Suppose the Reynolds number in problem 5-52 is attained at the edge of the plate, and the plate is maintained at a constant temperature of 60°C. Calculate the heat lost by the plate.

5-54 Air at 0.2 MPa and 25°C flows over a square flat plate at a velocity of 60 m/s. The plate is 0.5 m on a side and is maintained at a constant temperature of 150°C. Calculate the heat lost from the plate.

5-55 Helium at a pressure of 150 kPa and a temperature of 20°C flows across a 1-m-square plate at a velocity of 50 m/s. The plate is maintained at a constant temperature of 100°C. Calculate the heat lost by the plate.

5-56 Air at 50 kPa and 250 K flows across a 2-m-square plate at a velocity of 20 m/s. The plate is maintained at a constant temperature of 350 K. Calculate the heat lost by the plate.

5-57 Nitrogen at 50 kPa and 300 K flows over a flat plate at a velocity of 100 m/s. The length of the plate is 1.2 m and the plate is maintained at a constant temperature of 400 K. Calculate the heat lost by the plate.

5-58 Hydrogen at 2 atm and 15°C flows across a 1-m-square flat plate at a velocity of 6 m/s. The plate is maintained at a constant temperature of 139°C. Calculate the heat lost by the plate.

5-59 Liquid ammonia at 10°C is forced across a square plate 40 cm on a side at a velocity of 5 m/s. The plate is maintained at 50°C. Calculate the heat lost by the plate.

5-60 Helium flows across a 1.0-m-square plate at a velocity of 50 m/s. The helium is at a pressure of 45 kPa and a temperature of 50°C. The plate is maintained at a constant temperature of 136°C. Calculate the heat lost by the plate.

5-61 Air at 0.1 atm flows over a flat plate at a velocity of 300 m/s. The plate temperature is maintained constant at 100°C and the free-stream air temperature is 10°C. Calculate the heat transfer for a plate that is 80 cm square.

5-62 Water at 21°C flows across a 30-cm-square flat plate at a velocity of 6 m/s. The plate is maintained at a constant temperature of 54°F. Calculate the heat lost by the plate.

5-63 Plot h_x versus x for air at 1 atm and 300 K flowing at a velocity of 30 m/s across a flat plate. Take Re$_{crit} = 5 \times 10^5$ and use semilog plotting paper. Extend the plot to an x value equivalent to Re $= 10^9$. Also plot the average heat-transfer coefficient over this same range.

5-64 Air flows over a flat plate at 1 atm and 350 K with a velocity of 30 m/s. Calculate the mass flow through the boundary layer at x locations where $Re_x = 10^6$ and 10^7.

5-65 Air flows with a velocity of 6 m/s across a 20-cm-square plate at 50 kPa and 300 K. An electrical heater is installed in the plate such that it produces a constant heat flux. What is the total heat that can be dissipated if the plate temperature cannot exceed 600 K?

5-66 "Slug" flow in a tube may be described as that flow in which the velocity is constant across the entire flow area of the tube. Obtain an expression for the heat-transfer coefficient in this type of flow with a constant-heat-flux condition maintained at the wall. Compare the results with those of Section 5-10. Explain the reason for the difference in answers on a physical basis.

5-67 Compare the average heat transfer coefficients for the following three situations: (a) airflow at 1 atm and 300 K across a flat plate such that $Re_L = 10^5$; (b) helium flow at 1 atm and 300 K across a flat plate with the same values of ρu_∞ and L as in (a); (c) Flow of liquid water at 300 K across a flat plate with the same values of ρu_∞ and L as in (a).

  Evaluate all properties at $T = 300$ K. What do you conclude from this comparison?

5-68 Air at 1.2 atm and 27°C flows across a 60-cm-square plate at a free-stream velocity of 40 m/s. The plate is maintained at a constant temperature of 177°C. Calculate the heat lost by one side of the plate.

5-69 Air at 50.66 kPa and −23°C blows across a square flat plate that is maintained at a constant temperature of 77°C. The free-stream velocity is 30 m/s and the plate is 50 cm on each side. Calculate the heat lost from the plate, expressed in watts.

5-70 Helium at a pressure of 200 kPa and −18°C flows over a flate plate at a velocity of 20 m/s. The plate is maintained at a constant temperature of 93°C. If the plate is 30 cm square, calculate the heat loss expressed in watts.

5-71 Assume that the velocity distribution in the turbulent core for tube flow may be represented by

$$\frac{u}{u_c} = \left(1 - \frac{r}{r_o}\right)^{1/7}$$

where u_c is the velocity at the center of the tube and r_o is the tube radius. The velocity in the laminar sublayer may be assumed to vary linearly with the radius. Using the friction factor given by Equation (5-115), derive an equation for the thickness of the laminar sublayer. For this problem the average flow velocity may be calculated using only the turbulent velocity distribution.

5-72 Using the velocity profile in Problem 5-71, obtain an expression for the eddy diffusivity of momentum as a function of radius.

5-73 In heat-exchanger applications, it is frequently important to match heat-transfer requirements with pressure-drop limitations. Assuming a fixed total heat-transfer requirement and a fixed temperature difference between wall and bulk conditions as well as a fixed pressure drop through the tube, derive expressions for the length and diameter of the tube, assuming turbulent flow of a gas with the Prandtl number near unity.

5-74 Water flows in a 2.5-cm-diameter pipe so that the Reynolds number based on diameter is 1500 (laminar flow is assumed). The average bulk temperature is 35°C. Calculate the maximum water velocity in the tube. (Recall that $u_m = 0.5u_0$.) What would the heat-transfer coefficient be for such a system if the tube wall was subjected to a constant heat flux and the velocity and temperature profiles were completely developed? Evaluate properties at bulk temperature.

5-75 A slug flow is encountered in an annular-flow system that is subjected to a constant heat flux at both the inner and outer surfaces. The temperature is the same at both inner and outer surfaces at identical x locations. Derive an expression for the temperature distribution in such a flow system, assuming constant properties and laminar flow.

5-76 Air at Mach 4 and 3 lb/in^2 abs, 0°F, flows past a flat plate. The plate is to be maintained at a constant temperature of 200°F. If the plate is 18 in long, how much cooling will be required to maintain this temperature?

5-77 Air flows over an isothermal flat plate maintained at a constant temperature of 65°C. The velocity of the air is 600 m/s at static properties of 15°C and 7 kPa. Calculate the average heat-transfer coefficient for a plate 1 m long.

5-78 Air at 7 kPa and −40°C flows over a flat plate at Mach 4. The plate temperature is 35°C, and the plate length is 60 cm. Calculate the adiabatic wall temperature for the laminar portion of the boundary layer.

5-79 A wind tunnel is to be constructed to produce flow conditions of Mach 2.8 at $T_\infty = -40°C$ and $p = 0.05$ atm. What is the stagnation temperature for these conditions? What would be the adiabatic wall temperature for the laminar and turbulent portions of a boundary layer on a flat plate? If a flat plate were installed in the tunnel such that $\text{Re}_L = 10^7$, what would the heat transfer be for a constant wall temperature of 0°C?

5-80 Compute the drag force exerted on the plate by each of the systems in Problem 5-22.

5-81 Glycerin at 30°C flows past a 30-cm-square flat plate at a velocity of 1.5 m/s. The drag force is measured as 8.9 N (both sides of the plate). Calculate the heat-transfer coefficient for such a flow system.

5-82 Calculate the drag (viscous-friction) force on the plate in Problem 5-23 under the conditions of no heat transfer. Do not use the analogy between fluid friction and heat transfer for this calculation; that is, calculate the drag directly by evaluating the viscous-shear stress at the wall.

5-83 Nitrogen at 1 atm and 20°C is blown across a 130-cm-square flat plate at a velocity of 3.0 m/s. The plate is maintained at a constant temperature of 100°C. Calculate the average-friction coefficient and the heat transfer from the plate.

5-84 Using the velocity distribution for developed laminar flow in a tube, derive an expression for the friction factor as defined by Equation 5-112.

5-85 Engine oil at 10°C flows across a 15-cm-square plate upon which is imposed a constant heat flux of 10 kW/m^2. Determine (*a*) the average temperature difference, (*b*) the temperature difference at the trailing edge, and (*c*) the average heat-transfer coefficient. Use the Churchill relation [Equation (5-51)]. $u_\infty = 0.5$ m/s.

5-86 Work Problem 5-85 for a constant plate surface temperature equal to that at the trailing edge, and determine the total heat transfer.

5-87 For air at 25°C and 1 atm, with a free-stream velocity of 45 m/s, calculate the length of a flat plate to produce Reynolds numbers of 5×10^5 and 10^8. What are the boundary-layer thicknesses at these Reynolds numbers?

5-88 Determine the boundary-layer thickness at $\text{Re} = 5 \times 10^5$ for the following fluids flowing over a flat plate at 20 m/s: (*a*) air at 1 atm and 10°C, (*b*) saturated liquid water at 10°C, (*c*) hydrogen at 1 atm and 10°C, (*d*) saturated liquid ammonia at 10°C, and (*e*) saturated liquid Freon 12 at 10°C.

5-89 Many of the heat-transfer relations for flow over a flat plate are of the form

$$\mathrm{Nu}_x = \frac{h_x x}{k} = C\,\mathrm{Re}_x^n\, f(\mathrm{Pr})$$

Obtain an expression for $\overline{h}_L / h_{x=L}$ in terms of the constants C and n.

5-90 Compare Equations (5-51) and (5-44) for engine oil at 20°C and a Reynolds number of 10,000.

5-91 Air at 1 atm and 300 K blows across a square plate 75 cm on a side that is maintained at 350 K. The free-stream velocity is 45 m/s. Calculate the heat transfer and drag force on one side of the plate. Also calculate the heat transfer for just the laminar portion of the boundary layer.

5-92 Taking the critical Reynolds number as 5×10^5 for Problem 5-87, calculate the boundary-layer thickness at this point and at the trailing edge of the plate assuming (a) laminar flow to $\mathrm{Re}_{\mathrm{crit}}$ and turbulent thereafter and (b) turbulent flow from the leading edge.

5-93 If the plate temperature in Problem 5-91 is raised to 500 K while keeping the free-stream conditions the same, calculate the total heat transfer evaluating properties at (a) free-stream conditions, (b) film temperature, and (c) wall temperature. Comment on the results.

5-94 Air at 250 K and 1 atm blows across a 30-cm-square plate at a velocity of 10 m/s. The plate maintains a constant heat flux of 700 W/m^2. Determine the plate temperatures at x locations of 1, 5, 10, 20, and 30 cm.

5-95 Engine oil at 20°C is forced across a 20-cm-square plate at 10 m/s. The plate surface is maintained at 40°C. Calculate the heat lost by the plate and the drag force for one side of an unheated plate.

5-96 A large flat plate 4.0 m long and 1.0 m wide is exposed to an atmospheric air at 27°C with a velocity of 30 mi/h in a direction parallel to the 4.0-m dimension. If the plate is maintained at 77°C, calculate the total heat loss. Also calculate the heat flux in watts per square meter at x locations of 3 cm, 50 cm, 1.0 m, and 4.0 m.

5-97 Air at 1 atm and 300 K blows across a 10-cm-square plate at 30 m/s. Heating does not begin until $x = 5.0$ cm, after which the plate surface is maintained at 400 K. Calculate the total heat lost by the plate.

5-98 For the plate and flow conditions of Problem 5-97, only a 0.5-cm strip centered at $x = 5.0$ cm is heated to 400 K. Calculate the heat lost by this strip.

5-99 Two 20-cm-square plates are separated by a distance of 3.0 cm. Air at 1 atm, 300 K, and 15 m/s enters the space separating the plates. Will there be interference between the two boundary layers?

5-100 Water at 15.6°C flows across a 20-cm-square plate with a velocity of 2 m/s. A thin strip, 5 mm wide, is placed on the plate at a distance of 10 cm from the leading edge. If the strip is heated to a temperature of 37.8°C, calculate the heat lost from the strip.

5-101 Air at 300 K and 4 atm blows across a 10-cm-square plate at a velocity of 35 m/s. The plate is maintained at a constant temperature of 400 K. Calculate the heat lost from the plate.

5-102 An electric heater is installed on the plate of Problem 5-97 that will produce a constant heat flux of 1000 W/m^2 for the same airflow conditions across the plate. What is the maximum temperature that will be experienced by the plate surface?

5-103 In a certain application the critical Reynolds number for flow over a flat plate is 10^6. Air at 1 atm, 300 K, and 10 m/s flows across an isothermal plate with this critical

Reynolds number, and with a plate temperature of 400 K. The Reynolds number at the end of the plate is 5×10^6. What will the average heat transfer coefficient be for this system? How long is the plate? What is the heat lost from the plate?

5-104 Calculate the average heat transfer coefficient for the flow conditions of Problem 5-103, but with a critical Reynolds number of 5×10^5. What is the heat lost by the plate in this circumstance?

5-105 Glycerin at 10°C flows across a 30-cm-square plate with a velocity of 2 m/s. The plate surface is isothermal at 30°C. Calculate the heat lost by the plate.

5-106 Ethylene glycol at 20°C flows across an isothermal plate maintained at 0°C. The plate is 20 cm square and the Reynolds number at the end of the plate is 100,000. Calculate the heat gained by the plate.

Design-Oriented Problems

5-107 A low-speed wind tunnel is to be designed to study boundary layers up to $\mathrm{Re}_x = 10^7$ with air at 1 atm and 25°C. The maximum flow velocity that can be expected from an existing fan system is 30 m/s. How long must the flat-plate test-section be to produce the required Reynolds numbers? What will the maximum boundary-layer thickness be under these conditions? What would the maximum boundary-layer thicknesses be for flow velocities at 7 m/s and 12 m/s?

5-108 Using Equations (5-55), (5-81), and (5-82) for the local heat transfer in their respective ranges, obtain an expression for the average heat transfer coefficient, or Nusselt number, over the range $5 \times 10^5 < \mathrm{Re}_L < 10^9$ with $\mathrm{Re}_{\mathrm{crit}} = 5 \times 10^5$. Use a numerical technique to perform the necessary integration and a curve fit to simplify the results.

5-109 An experiment is to be deligned to demonstrate measurement of heat loss for water flowing over a flat plate. The plate is 30 cm square and it will be maintained nearly constant in temperature at 50°C while the water temperature will be about 10°C. (*a*) Calculate the flow velocities necessary to study a range of Reynolds numbers from 10^4 to 10^7. (*b*) Estimate the heat-transfer coefficients and heat-transfer rates for several points in the specified range.

5-110 Consider the flow of air over a flat plate under laminar flow conditions at 1 atm. Investigate the influence of temperature on the heat transfer coefficient by examining five cases with constant free-stream temperature of 20°C, constant free-stream velocity, and surface temperatures of 50, 100, 150, 250, and 350°C. What do you conclude from this analysis? From the results, determine an approximate variation of the heat-transfer coefficient with absolute temperature for air at 1 atm.

REFERENCES

1. Schlichting, H. *Boundary Layer Theory*, 7th ed., McGraw-Hill, 1979.
2. von Kármán, T. "Über laminaire und turbulente Reibung," *Angew. Math. Mech.,* vol. 1, pp. 233–52, 1921; also *NACA Tech. Mem.* 1092, 1946.
3. Sellars, J. R., M. Tribus, and J. S. Klein: "Heat Transfer to Laminar Flows in a Round Tube or Flat Conduit: The Graetz Problem Extended," *Trans. ASME,* vol. 78, p. 441, 1956.
4. Eckert, E. R. G. "Survey of Boundary Layer Heat Transfer at High Velocities and High Temperatures," *WADC Tech. Rep.,* pp. 59–624, April 1960.
5. White, F. M. *Viscous Fluid Flow*, New York McGraw-Hill, 1974.
6. Knudsen, J. D., and D. L. Katz: *Fluid Dynamics and Heat Transfer,* New York: McGraw-Hill, 1958.

7. Holman, J. P. *Thermodynamics,* 4th ed., New York, McGraw-Hill, 1988.

8. Schultz-Grunow, F. "Neues Widerstandsgesetz für glatte Platten," *Lüfifahrtforschüng,* vol. 17, p. 239, 1940; also *NACA Tech. Mem.* 986, 1941.

9. Churchill, S. W., and H. Ozoe. "Correlations for Laminar Forced Convection in Flow over an Isothermal Flat Plate and in Developing and Fully Developed Flow in an Isothermal Tube," *J. Heat Transfer,* vol. 95, p. 46, 1973.

10. Whitaker, S. "Forced Convection Heat Transfer Correlation for Flow in Pipes, Past Flat Plates, Single Cylinders, Single Spheres, and for Flow in Packed Beds and Tube Bundles," *AIChEJ.,* vol. 18, p. 361, 1972.

11. Churchill, S. W. "A Comprehensive Correlating Equation for Forced Convection from Flat Plates," *AIChE J.,* vol. 22, p. 264, 1976.

CHAPTER

6

Empirical and Practical Relations for Forced-Convection Heat Transfer

6-1 | INTRODUCTION

The discussion and analyses of Chapter 5 have shown how forced-convection heat transfer may be calculated for several cases of practical interest; the problems considered, however, were those that could be solved in an analytical fashion. In this way, the principles of the convection process and their relation to fluid dynamics were demonstrated, with primary emphasis being devoted to a clear understanding of physical mechanism. Regrettably, it is not always possible to obtain analytical solutions to convection problems, and the individual is forced to resort to experimental methods to obtain design information, as well as to secure the more elusive data that increase the physical understanding of the heat-transfer processes.

Results of experimental data are usually expressed in the form of either empirical formulas or graphical charts so that they may be utilized with a maximum of generality. It is in the process of trying to generalize the results of one's experiments, in the form of some empirical correlation, that difficulty is encountered. If an analytical solution is available for a similar problem, the correlation of data is much easier, since one may guess at the functional form of the results, and hence use the experimental data to obtain values of constants or exponents for certain significant parameters such as the Reynolds or Prandtl numbers. If an analytical solution for a similar problem is not available, the individual must resort to intuition based on physical understanding of the problem, or shrewd inferences that one may be able to draw from the differential equations of the flow processes based upon dimensional or order-of-magnitude estimates. In any event, there is no substitute for physical insight and understanding.

To show how one might proceed to analyze a new problem to obtain an important functional relationship from the differential equations, consider the problem of determining the hydrodynamic-boundary-layer thickness for flow over a flat plate. This problem was solved in Chapter 5, but we now wish to make an order-of-magnitude analysis of the differential equations to obtain the functional form of the solution. The momentum equation

$$u\frac{\partial u}{\partial x} + v\frac{\partial u}{\partial y} = v\frac{\partial^2 u}{\partial y^2}$$

must be solved in conjunction with the continuity equation

$$\frac{\partial u}{\partial x} + \frac{\partial v}{\partial y} = 0$$

Within the boundary layer we may say that the velocity u is of the order of the free-stream velocity u_∞. Similarly, the y dimension is of the order of the boundary-layer thickness δ. Thus

$$u \sim u_\infty$$
$$y \sim \delta$$

and we might write the continuity equation in an approximate form as

$$\frac{\partial u}{\partial x} + \frac{\partial v}{\partial y} = 0$$
$$\frac{u_\infty}{x} + \frac{v}{\delta} \approx 0$$

or

$$v \sim \frac{u_\infty \delta}{x}$$

Then, by using this order of magnitude for v, the analysis of the momentum equation would yield

$$u \frac{\partial u}{\partial x} + v \frac{\partial u}{\partial y} = v \frac{\partial^2 u}{\partial y^2}$$
$$u_\infty \frac{u_\infty}{x} + \frac{u_\infty \delta}{x} \frac{u_\infty}{\delta} \approx v \frac{u_\infty}{\delta^2}$$

or

$$\delta^2 \sim \frac{vx}{u_\infty}$$
$$\delta \sim \sqrt{\frac{vx}{u_\infty}}$$

Dividing by x to express the result in dimensionless form gives

$$\frac{\delta}{x} \sim \sqrt{\frac{v}{u_\infty x}} = \frac{1}{\sqrt{\mathrm{Re}_x}}$$

This functional variation of the boundary-layer thickness with the Reynolds number and x position is precisely that which was obtained in Section 5-4. Although this analysis is rather straightforward and does indeed yield correct results, the order-of-magnitude analysis may not always be so fortunate when applied to more complex problems, particularly those involving turbulent- or separated-flow regions. Nevertheless, one may often obtain valuable information and physical insight by examining the order of magnitude of various terms in a governing differential equation for the particular problem at hand.

A conventional technique used in correlation of experimental data is that of dimensional analysis, in which appropriate dimensionless groups such as the Reynolds and Prandtl numbers are derived from purely dimensional and functional considerations. There is, of course, the assumption of flow-field and temperature-profile similarity for geometrically similar heating surfaces. Generally speaking, the application of dimensional analysis to any new problem is extremely difficult when a previous analytical solution of some sort is not available. It is usually best to attempt an order-of-magnitude analysis such as the one above if the governing differential equations are known. In this way it may be possible to determine the significant dimensionless variables for correlating experimental data. In some complex flow and heat-transfer problems a clear physical model of the processes may not be available, and the engineer must first try to establish this model before the experimental data can be correlated.

Schlichting [6], Giedt [7], and Kline [28] discuss similarity considerations and their use in boundary-layer and heat-transfer problems.

The purpose of the foregoing discussion has not been to emphasize or even to imply any new method for solving problems, but rather to indicate the necessity of applying intuitive physical reasoning to a difficult problem and to point out the obvious advantage of using any and all information that may be available. When the problem of correlation of experimental data for a previously unsolved situation is encountered, one must frequently adopt devious methods to accomplish the task.

6-2 | EMPIRICAL RELATIONS FOR PIPE AND TUBE FLOW

The analysis of Section 5-10 has shown how one might analytically attack the problem of heat transfer in fully developed laminar tube flow. The cases of undeveloped laminar flow, flow systems where the fluid properties vary widely with temperature, and turbulent-flow systems are considerably more complicated but are of very important practical interest in the design of heat exchangers and associated heat-transfer equipment. These more complicated problems may sometimes be solved analytically, but the solutions, when possible, are very tedious. For design and engineering purposes, empirical correlations are usually of greatest practical utility. In this section we present some of the more important and useful empirical relations and point out their limitations.

The Bulk Temperature

First let us give some further consideration to the bulk-temperature concept that is important in all heat-transfer problems involving flow inside closed channels. In Chapter 5 we noted that the bulk temperature represents energy average or "mixing cup" conditions. Thus, for the tube flow depicted in Figure 6-1 the total energy added can be expressed in terms of a bulk-temperature difference by

$$q = \dot{m}c_p(T_{b_2} - T_{b_1}) \qquad \textbf{[6-1]}$$

provided c_p is reasonably constant over the length. In some differential length dx the heat added dq can be expressed either in terms of a bulk-temperature difference or in terms of

Figure 6-1 | Total heat transfer in terms of bulk-temperature difference.

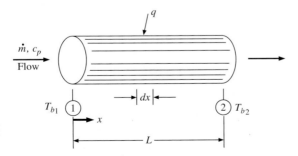

the heat-transfer coefficient

$$dq = \dot{m}c_p dT_b = h(2\pi r)dx(T_w - T_b) \qquad \textbf{[6-2]}$$

where T_w and T_b are the wall and bulk temperatures at the particular x location. The total heat transfer can also be expressed as

$$q = hA(T_w - T_b)_{\text{av}} \qquad \textbf{[6-3]}$$

where A is the total surface area for heat transfer. Because both T_w and T_b can vary along the length of the tube, a suitable averaging process must be adopted for use with Equation (6-3). In this chapter most of our attention will be focused on methods for determining h, the convection heat-transfer coefficient. Chapter 10 will discuss different methods for taking proper account of temperature variations in heat exchangers.

A traditional expression for calculation of heat transfer in fully developed turbulent flow in smooth tubes is that recommended by Dittus and Boelter [1]:*

$$\mathrm{Nu}_d = 0.023\,\mathrm{Re}_d^{0.8}\mathrm{Pr}^n \qquad \textbf{[6-4a]}$$

The properties in this equation are evaluated at the average fluid bulk temperature, and the exponent n has the following values:

$$n = \begin{cases} 0.4 & \text{for heating of the fluid} \\ 0.3 & \text{for cooling of the fluid} \end{cases}$$

Equation (6-4) is valid for fully developed turbulent flow in smooth tubes for fluids with Prandtl numbers ranging from about 0.6 to 100 and with moderate temperature differences between wall and fluid conditions. More recent information by Gnielinski [45] suggests that better results for turbulent flow in smooth tubes may be obtained from the following:

$$\mathrm{Nu} = 0.0214\left(\mathrm{Re}^{0.8} - 100\right)\mathrm{Pr}^{0.4} \qquad \textbf{[6-4b]}$$

for $0.5 < \mathrm{Pr} < 1.5$ and $10^4 < \mathrm{Re} < 5 \times 10^6$ or

$$\mathrm{Nu} = 0.012\left(\mathrm{Re}^{0.87} - 280\right)\mathrm{Pr}^{0.4} \qquad \textbf{[6-4c]}$$

for $1.5 < \mathrm{Pr} < 500$ and $3000 < \mathrm{Re} < 10^6$.

One may ask the reason for the functional form of Equation (6-4). Physical reasoning, based on the experience gained with the analyses of Chapter 5, would certainly indicate a dependence of the heat-transfer process on the flow field, and hence on the Reynolds number. The relative rates of diffusion of heat and momentum are related by the Prandtl number, so that the Prandtl number is expected to be a significant parameter in the final solution. We can be rather confident of the dependence of the heat transfer on the Reynolds and Prandtl numbers. But the question arises as to the correct functional form of the relation; that is, would one necessarily expect a product of two power functions of the Reynolds and Prandtl numbers? The answer is that one might expect this functional form since it appears in the flat-plate analytical solutions of Chapter 5, as well as the Reynolds analogy for turbulent flow in tubes. In addition, this type of functional relation is convenient to use in correlating experimental data, as described below.

---

*Equation (6-4a) is a rather famous equation in the annals of convection heat transfer, but it appears [47] that the constant 0.023 and exponents 0.4 and 0.3 were actually recommended by McAdams [10, 2nd ed., 1942] as a meld between the values given in Reference 1 and those suggested by Colburn [15].

Suppose a number of experiments are conducted with measurements taken of heat-transfer rates of various fluids in turbulent flow inside smooth tubes under different temperature conditions. Different-diameter tubes may be used to vary the range of the Reynolds number in addition to variations in the mass-flow rate. We wish to generalize the results of these experiments by arriving at one empirical equation that represents all the data. As described above, we may anticipate that the heat-transfer data will be dependent on the Reynolds and Prandtl numbers. A power function for each of these parameters is a simple type of relation to use, so we assume

$$\mathrm{Nu}_d = C\mathrm{Re}_d{}^m\mathrm{Pr}^n$$

where C, m, and n are constants to be determined from the experimental data.

A log-log plot of Nu_d versus Re_d is first made for one fluid to estimate the dependence of the heat transfer on the Reynolds number (i.e., to find an approximate value of the exponent m). This plot is made for one fluid at a constant temperature, so that the influence of the Prandtl number will be small, since the Prandtl number will be approximately constant for the one fluid. By using this first estimate for the exponent m, the data for all fluids are plotted as $\log(\mathrm{Nu}_d/\mathrm{Re}_d{}^m)$ versus $\log \mathrm{Pr}$, and a value for the exponent n is determined. Then, by using this value of n, all the data are plotted again as $\log(\mathrm{Nu}_d/\mathrm{Pr}^n)$ versus $\log \mathrm{Re}_d$, and a final value of the exponent m is determined as well as a value for the constant C. An example of this final type of data plot is shown in Figure 6-2. The final correlation equation may represent the data within ± 25 percent.

Readers should recognize that obtaining empirical correlations for convection heat transfer phenomena is not as simple as the preceding discussion might imply. The "data

Figure 6-2 | Typical data correlation for forced convection in smooth tubes, turbulent flow.

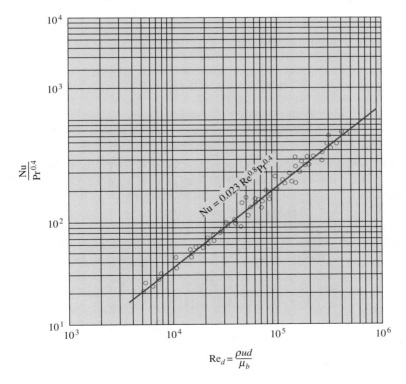

Figure 6-3 | Influence of heating on velocity profile in laminar tube flow.

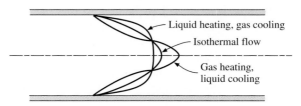

points" shown in Figure 6-2 are entirely fictitious and more consistent than normally might be encountered. Careful attention must be given to experiment design to minimize experimental uncertainties that can creep into the final data correlation. A very complete discussion of the design of an experiment for measurement of convection heat transfer in smooth tubes is given in Reference 51, along with an extensive discussion of techniques for estimating the propagation of experimental uncertainties and the methods for obtaining the best correlation equation for the available data.

If wide temperature differences are present in the flow, there may be an appreciable change in the fluid properties between the wall of the tube and the central flow. These property variations may be evidenced by a change in the velocity profile as indicated in Figure 6-3. The deviations from the velocity profile for isothermal flow as shown in this figure are a result of the fact that the viscosity of gases increases with an increase in temperature, while the viscosities of liquids decrease with an increase in temperature.

To take into account the property variations, Sieder and Tate [2] recommend the following relation:

$$\mathrm{Nu}_d = 0.027\,\mathrm{Re}_d^{0.8}\mathrm{Pr}^{1/3}\left(\frac{\mu}{\mu_w}\right)^{0.14} \qquad \textbf{[6-5]}$$

All properties are evaluated at bulk-temperature conditions, except μ_w, which is evaluated at the wall temperature.

Equations (6-4) and (6-5) apply to fully developed turbulent flow in tubes. In the entrance region the flow is not developed, and Nusselt [3] recommended the following equation:

$$\mathrm{Nu}_d = 0.036\,\mathrm{Re}_d^{0.8}\mathrm{Pr}^{1/3}\left(\frac{d}{L}\right)^{0.055} \qquad \text{for } 10 < \frac{L}{d} < 400 \qquad \textbf{[6-6]}$$

where L is the length of the tube and d is the tube diameter. The properties in Equation (6-6) are evaluated at the mean bulk temperature. Hartnett [24] has given experimental data on the thermal entrance region for water and oils. Definitive studies of turbulent transfer with water in smooth tubes and at uniform heat flux have been presented by Allen and Eckert [25].

The above equations offer simplicity in computation, but uncertainties on the order of ±25 percent are not uncommon. Petukhov [42] has developed a more accurate, although more complicated, expression for fully developed turbulent flow in smooth tubes:

$$\mathrm{Nu}_d = \frac{(f/8)\,\mathrm{Re}_d\mathrm{Pr}}{1.07 + 12.7(f/8)^{1/2}\left(\mathrm{Pr}^{2/3} - 1\right)}\left(\frac{\mu_b}{\mu_w}\right)^n \qquad \textbf{[6-7]}$$

where $n = 0.11$ for $T_w > T_b$, $n = 0.25$ for $T_w < T_b$, and $n = 0$ for constant heat flux or for gases. All properties are evaluated at $T_f = (T_w + T_b)/2$ except for μ_b and μ_w. The friction factor may be obtained either from Figure 6-4 or from the following for smooth tubes:

$$f = (1.82\,\log_{10}\mathrm{Re}_d - 1.64)^{-2} \qquad \textbf{[6-8]}$$

Figure 6-4 | Friction factors for pipes, from Reference 5.

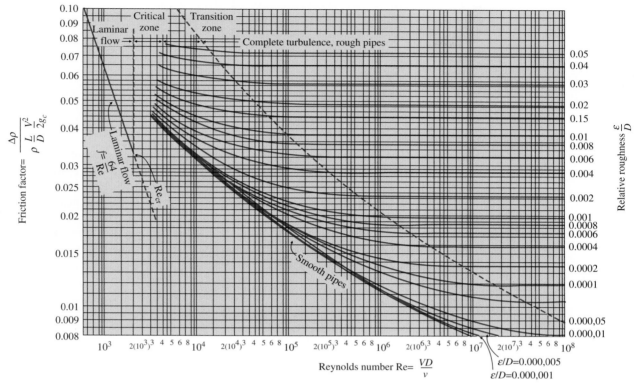

Equation (6-7) is applicable for the following ranges:

$$0.5 < Pr < 200 \qquad \text{for 6 percent accuracy}$$
$$0.5 < Pr < 2000 \qquad \text{for 10 percent accuracy}$$
$$10^4 < Re_d < 5 \times 10^6$$
$$0.8 < \mu_b/\mu_w < 40$$

Hausen [4] presents the following empirical relation for fully developed laminar flow in tubes at constant wall temperature:

$$\overline{Nu}_d = 3.66 + \frac{0.0668(d/L)\,Re_d Pr}{1 + 0.04[(d/L)\,Re_d Pr]^{2/3}} \qquad \textbf{[6-9]}$$

The heat-transfer coefficient calculated from this relation is the average value over the entire length of tube. Note that the Nusselt number approaches a constant value of 3.66 when the tube is sufficiently long. This situation is similar to that encountered in the constant-heat-flux problem analyzed in Chapter 5 [Equation (5-107)], except that in this case we have a constant wall temperature instead of a linear variation with length. The temperature profile is fully developed when the Nusselt number approaches a constant value.

A somewhat simpler empirical relation was proposed by Sieder and Tate [2] for laminar heat transfer in tubes:

$$\overline{Nu}_d = 1.86(Re_d Pr)^{1/3} \left(\frac{d}{L}\right)^{1/3} \left(\frac{\mu}{\mu_w}\right)^{0.14} \qquad \textbf{[6-10]}$$

In this formula the average heat-transfer coefficient is based on the arithmetic average of the inlet and outlet temperature differences, and all fluid properties are evaluated at the mean bulk temperature of the fluid, except μ_w, which is evaluated at the wall temperature. Equation (6-10) obviously cannot be used for extremely long tubes since it would yield a zero heat-transfer coefficient. A comparison by Knudsen and Katz [9, p. 377] of Equation (6-10) with other relationships indicates that it is valid for

$$\mathrm{Re}_d \mathrm{Pr}\frac{d}{L} > 10$$

The product of the Reynolds and Prandtl numbers that occurs in the laminar-flow correlations is called the Peclet number.

$$\mathrm{Pe} = \frac{du\rho c_p}{k} = \mathrm{Re}_d\mathrm{Pr} \qquad\qquad \textbf{[6-11]}$$

The calculation of laminar heat-transfer coefficients is frequently complicated by the presence of natural-convection effects that are superimposed on the forced-convection effects. The treatment of combined forced- and free-convection problems is discussed in Chapter 7.

The empirical correlations presented above, with the exception of Equation (6-7), apply to smooth tubes. Correlations are, in general, rather sparse where rough tubes are concerned, and it is sometimes appropriate that the Reynolds analogy between fluid friction and heat transfer be used to effect a solution under these circumstances. Expressed in terms of the Stanton number,

$$\mathrm{St}_b\,\mathrm{Pr}_f^{2/3} = \frac{f}{8} \qquad\qquad St_b = \frac{h}{c_p\rho u_m} \qquad\qquad \textbf{[6-12]}$$

The friction coefficient f is defined by

$$\Delta p = f\frac{L}{d}\rho\frac{u_m^2}{2g_c} \qquad\qquad \textbf{[6-13]}$$

where u_m is the mean flow velocity. Values of the friction coefficient for different roughness conditions are shown in Figure 6-4. An empirical relation for the friction factor for rough tubes is given in References [49, 50] as

$$f = 1.325 / \left[\ln(\varepsilon/3.7d) + 5.74/\mathrm{Re}_d^{0.9}\right]^2 \qquad\qquad \textbf{[6-13a]}$$

for $10^{-6} < \varepsilon/d < 10^{-3}$ and $5000 < \mathrm{Re}_d < 10^8$.

Note that the relation in Equation (6-12) is the same as Equation (5-114), except that the Stanton number has been multiplied by $\mathrm{Pr}^{2/3}$ to take into account the variation of the thermal properties of different fluids. This correction follows the recommendation of Colburn [15], and is based on the reasoning that fluid friction and heat transfer in tube flow are related to the Prandtl number in the same way as they are related in flat-plate flow [Equation (5-56)]. In Equation (6-12) the Stanton number is based on bulk temperature, while the Prandtl number and friction factor are based on properties evaluated at the film temperature. Further information on the effects of tube roughness on heat transfer is given in References 27, 29, 30, and 31.

If the channel through which the fluid flows is not of circular cross section, it is recommended that the heat-transfer correlations be based on the hydraulic diameter D_H, defined by

$$D_H = \frac{4A}{P} \qquad\qquad \textbf{[6-14]}$$

where A is the cross-sectional area of the flow and P is the wetted perimeter. This particular grouping of terms is used because it yields the value of the physical diameter when applied to a circular cross section. The hydraulic diameter should be used in calculating the Nusselt and Reynolds numbers, and in establishing the friction coefficient for use with the Reynolds analogy.

Although the hydraulic-diameter concept frequently yields satisfactory relations for fluid friction and heat transfer in many practical problems, there are some notable exceptions where the method does not work. Some of the problems involved in heat transfer in noncircular channels have been summarized by Irvine [20] and Knudsen and Katz [9]. The interested reader should consult these discussions for additional information.

Shah and London [40] have compiled the heat-transfer and fluid-friction information for fully developed laminar flow in ducts with a variety of flow cross sections, and some of the resulting relations are shown in Table 6-1. In this table the following nomenclature applies, with the Nusselt and Reynolds numbers based on the hydraulic diameter of the flow cross-section area:

$\overline{\text{Nu}}_H$ = average Nusselt number for uniform heat flux in flow direction and
       uniform wall temperature at particular flow cross section

$\overline{\text{Nu}}_T$ = average Nusselt number for uniform wall temperature

$f\text{Re}_{D_H}/4$ = product of friction factor and Reynolds number based on hydraulic diameter

Table 6-1 | Heat transfer and fluid friction for fully developed laminar flow in ducts of various cross sections. Average Nusselt numbers based on hydraulic diameters of cross sections.

Geometry $(L/D_h > 100)$	Nu_H Constant axial wall heat flux	Nu_T Constant axial wall temperature	$f\,\text{Re}_{D_H}/4$
$\dfrac{b}{a} = \dfrac{\sqrt{3}}{2}$ (triangle, 60°)	3.111	2.47	13.333
$\dfrac{b}{a} = 1$ (square)	3.608	2.976	14.227
(hexagon)	4.002	3.34	15.054
$\dfrac{b}{a} = \dfrac{1}{2}$	4.123	3.391	15.548
(circle)	4.364	3.657	16.000
$\dfrac{b}{a} = \dfrac{1}{4}$	5.331	4.44	18.23
$\dfrac{b}{a} = \dfrac{1}{3}$	4.79	3.96	17.25
$\dfrac{b}{a} = \dfrac{1}{8}$	6.490	5.597	20.585
$\dfrac{b}{a} = 0$	8.235	7.541	24.000
Heated / Insulated, $\dfrac{b}{a} = 0$	5.385	4.861	24.000

Figure 6-5 | Local and average Nusselt numbers for circular tube thermal entrance regions in fully developed laminar flow.

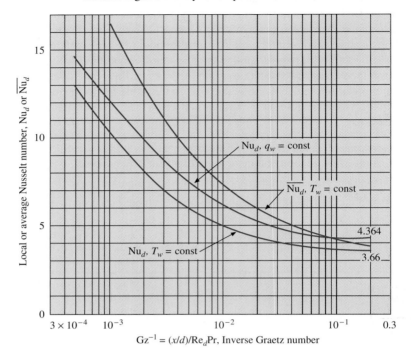

Figure 6-6 | Turbulent thermal entry Nusselt numbers for circular tubes with q_w = constant.

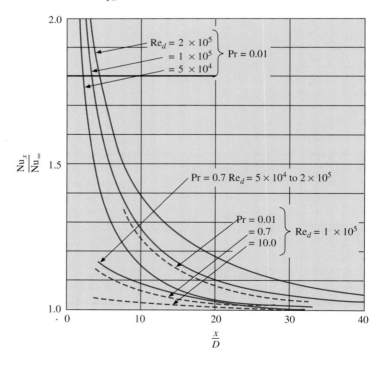

Kays [36] and Sellars, Tribus, and Klein (Reference 3, Chapter 5) have calculated the local and average Nusselt numbers for laminar entrance regions of circular tubes for the case of a fully developed velocity profile. Results of these analyses are shown in Figure 6-5 in terms of the inverse Graetz number, where

$$\text{Graetz number} = \text{Gz} = \text{Re Pr}\frac{d}{x} \qquad \textbf{[6-15]}$$

Entrance Effects in Turbulent Flow

Entrance effects for turbulent flow in tubes are more complicated than for laminar flow and cannot be expressed in terms of a simple function of the Graetz number. Kays [36] has computed the influence for several values of Re and Pr with the results summarized in Figure 6-6. The ordinate is the ratio of the local Nusselt number to that a long distance from the inlet, or for fully developed thermal conditions. In general, the higher the Prandtl number, the shorter the entry length. We can see that the thermal entry lengths are much shorter for turbulent flow than for the laminar counterpart.

A very complete survey of the many heat-transfer correlations that are available for tube and channel flow is given by Kakac, Shah, and Aung [46].

Turbulent Heat Transfer in a Tube	EXAMPLE 6-1

Air at 2 atm and 200°C is heated as it flows through a tube with a diameter of 1 in (2.54 cm) at a velocity of 10 m/s. Calculate the heat transfer per unit length of tube if a constant-heat-flux condition is maintained at the wall and the wall temperature is 20°C above the air temperature, all along the length of the tube. How much would the bulk temperature increase over a 3-m length of the tube?

■ **Solution**
We first calculate the Reynolds number to determine if the flow is laminar or turbulent, and then select the appropriate empirical correlation to calculate the heat transfer. The properties of air at a bulk temperature of 200°C are

$$\rho = \frac{p}{RT} = \frac{(2)(1.0132 \times 10^5)}{(287)(473)} = 1.493 \text{ kg/m}^3 \qquad [0.0932 \text{ lb}_m/\text{ft}^3]$$

$$\text{Pr} = 0.681$$

$$\mu = 2.57 \times 10^{-5} \text{ kg/m} \cdot \text{s} \qquad [0.0622 \text{ lb}_m/\text{h} \cdot \text{ft}]$$

$$k = 0.0386 \text{ W/m} \cdot °C \qquad [0.0223 \text{ Btu/h} \cdot \text{ft} \cdot °F]$$

$$c_p = 1.025 \text{ kJ/kg} \cdot °C$$

$$\text{Re}_d = \frac{\rho u_m d}{\mu} = \frac{(1.493)(10)(0.0254)}{2.57 \times 10^{-5}} = 14,756$$

so that the flow is turbulent. We therefore use Equation (6-4a) to calculate the heat-transfer coefficient.

$$\text{Nu}_d = \frac{hd}{k} = 0.023 \text{Re}_d^{0.8} \text{Pr}^{0.4} = (0.023)(14,756)^{0.8}(0.681)^{0.4} = 42.67$$

$$h = \frac{k}{d}\text{Nu}_d = \frac{(0.0386)(42.67)}{0.0254} = 64.85 \text{ W/m}^2 \cdot °C \qquad [11.42 \text{ Btu/h} \cdot \text{ft}^2 \cdot °F]$$

The heat flow per unit length is then

$$\frac{q}{L} = h\pi d(T_w - T_b) = (64.85)\pi(0.0254)(20) = 103.5 \text{ W/m} \qquad [107.7 \text{ Btu/ft}]$$

We can now make an energy balance to calculate the increase in bulk temperature in a 3.0-m length of tube:

$$q = \dot{m}c_p\Delta T_b = L\left(\frac{q}{L}\right)$$

We also have

$$\dot{m} = \rho u_m \frac{\pi d^2}{4} = (1.493)(10)\pi \frac{(0.0254)^2}{4}$$

$$= 7.565 \times 10^{-3} \text{ kg/s} \quad [0.0167 \text{ lb}_m/\text{s}]$$

so that we insert the numerical values in the energy balance to obtain

$$(7.565 \times 10^{-3})(1025)\Delta T_b = (3.0)(103.5)$$

and

$$\Delta T_b = 40.04°\text{C} \quad [104.07°\text{F}]$$

EXAMPLE 6-2   Heating of Water in Laminar Tube Flow

Water at 60°C enters a tube of 1-in (2.54-cm) diameter at a mean flow velocity of 2 cm/s. Calculate the exit water temperature if the tube is 3.0 m long and the wall temperature is constant at 80°C.

■ **Solution**
We first evaluate the Reynolds number at the inlet bulk temperature to determine the flow regime. The properties of water at 60°C are

$$\rho = 985 \text{ kg/m}^3 \qquad c_p = 4.18 \text{ kJ/kg} \cdot °\text{C}$$
$$\mu = 4.71 \times 10^{-4} \text{ kg/m} \cdot \text{s} \quad [1.139 \text{ lb}_m/\text{h} \cdot \text{ft}]$$
$$k = 0.651 \text{ W/m} \cdot °\text{C} \qquad \text{Pr} = 3.02$$
$$\text{Re}_d = \frac{\rho u_m d}{\mu} = \frac{(985)(0.02)(0.0254)}{4.71 \times 10^{-4}} = 1062$$

so the flow is laminar. Calculating the additional parameter, we have

$$\text{Re}_d\text{Pr}\frac{d}{L} = \frac{(1062)(3.02)(0.0254)}{3} = 27.15 > 10$$

so Equation (6-10) is applicable. We do not yet know the mean bulk temperature to evaluate properties so we first make the calculation on the basis of 60°C, determine an exit bulk temperature, and then make a second iteration to obtain a more precise value. When inlet and outlet conditions are designated with the subscripts 1 and 2, respectively, the energy balance becomes

$$q = h\pi dL\left(T_w - \frac{T_{b_1} + T_{b_2}}{2}\right) = \dot{m}c_p(T_{b_2} - T_{b_1}) \qquad [a]$$

At the wall temperature of 80°C we have

$$\mu_w = 3.55 \times 10^{-4} \text{ kg/m} \cdot \text{s}$$

From Equation (6-10)

$$\text{Nu}_d = (1.86)\left[\frac{(1062)(3.02)(0.0254)}{3}\right]^{1/3}\left(\frac{4.71}{3.55}\right)^{0.14} = 5.816$$

$$h = \frac{k\text{Nu}_d}{d} = \frac{(0.651)(5.816)}{0.0254} = 149.1 \text{ W/m}^2 \cdot °\text{C} \quad [26.26 \text{ Btu/h} \cdot \text{ft}^2 \cdot °\text{F}]$$

The mass flow rate is

$$\dot{m} = \rho \frac{\pi d^2}{4} u_m = \frac{(985)\pi(0.0254)^2(0.02)}{4} = 9.982 \times 10^{-3} \text{ kg/s}$$

Inserting the value for h into Equation (a) along with $\dot{m}$ and $T_{b_1} = 60°C$ and $T_w = 80°C$ gives

$$(149.1)\pi(0.0254)(3.0)\left(80 - \frac{T_{b_2} + 60}{2}\right) = (9.982 \times 10^{-3})(4180)(T_{b_2} - 60) \qquad [b]$$

This equation can be solved to give

$$T_{b_2} = 71.98°C$$

Thus, we should go back and evaluate properties at

$$T_{b,\text{mean}} = \frac{71.98 + 60}{2} = 66°C$$

We obtain

$$\rho = 982 \text{ kg/m}^3 \qquad c_p = 4185 \text{ J/kg} \cdot °C \qquad \mu = 4.36 \times 10^{-4} \text{ kg/m} \cdot \text{s}$$

$$k = 0.656 \text{ W/m} \cdot °C \qquad \text{Pr} = 2.78$$

$$\text{Re}_d = \frac{(1062)(4.71)}{4.36} = 1147$$

$$\text{Re Pr} \frac{d}{L} = \frac{(1147)(2.78)(0.0254)}{3} = 27.00$$

$$\text{Nu}_d = (1.86)(27.00)^{1/3}\left(\frac{4.36}{3.55}\right)^{0.14} = 5.743$$

$$h = \frac{(0.656)(5.743)}{0.0254} = 148.3 \text{ W/m}^2 \cdot °C$$

We insert this value of h back into Equation (a) to obtain

$$T_{b_2} = 71.88°C \qquad [161.4°F]$$

The iteration makes very little difference in this problem. If a large bulk-temperature difference had been encountered, the change in properties could have had a larger effect.

Heating of Air in Laminar Tube Flow for Constant Heat Flux   EXAMPLE 6-3

Air at 1 atm and 27°C enters a 5.0-mm-diameter smooth tube with a velocity of 3.0 m/s. The length of the tube is 10 cm. A constant heat flux is imposed on the tube wall. Calculate the heat transfer if the exit bulk temperature is 77°C. Also calculate the exit wall temperature and the value of h at exit.

■ **Solution**

We first must evaluate the flow regime and do so by taking properties at the average bulk temperature

$$\overline{T}_b = \frac{27 + 77}{2} = 52°C = 325 \text{ K}$$

$$\nu = 18.22 \times 10^{-6} \text{ m}^2/\text{s} \qquad \text{Pr} = 0.703 \qquad k = 0.02814 \text{ W/m} \cdot °C$$

$$\text{Re}_d = \frac{ud}{\nu} = \frac{(3)(0.005)}{18.22 \times 10^{-6}} = 823 \qquad [a]$$

so that the flow is laminar. The tube length is rather short, so we expect a thermal entrance effect and shall consult Figure 6-5. The inverse Graetz number is computed as

$$Gz^{-1} = \frac{1}{Re_d Pr} \frac{x}{d} = \frac{0.1}{(823)(0.703)(0.005)} = 0.0346$$

Therefore, for $q_w = $ constant, we obtain the Nusselt number at exit from Figure 6-5 as

$$Nu = \frac{hd}{k} = 4.7 = \frac{q_w d}{(T_w - T_b)k} \qquad [b]$$

The total heat transfer is obtained in terms of the overall energy balance:

$$q = \dot{m} c_p \left(T_{b_2} - T_{b_1} \right)$$

At entrance $\rho = 1.1774$ kg/m^3, so the mass flow is

$$\dot{m} = (1.1774)\pi(0.0025)^2(3.0) = 6.94 \times 10^{-5} \text{ kg/s}$$

and

$$q = (6.94 \times 10^{-5})(1006)(77 - 27) = 3.49 \text{ W}$$

Thus we may find the heat transfer without actually determining wall temperatures or values of h. However, to determine T_w we must compute q_w for insertion in Equation (b). We have

$$q = q_w \pi d L = 3.49 \text{ W}$$

and

$$q_w = 2222 \text{ W/m}^2$$

Now, from Equation (b)

$$(T_w - T_b)_{x=L} = \frac{(2222)(0.005)}{(4.7)(0.02814)} = 84°C$$

The wall temperature at exit is thus

$$T_w]_{x=L} = 84 + 77 = 161°C$$

and the heat-transfer coefficient is

$$h_{x=L} = \frac{q_w}{(T_w - T_b)_{x=L}} = \frac{2222}{84} = 26.45 \text{ W/m}^2 \cdot °C$$

| **EXAMPLE 6-4** | Heating of Air with Isothermal Tube Wall |

Repeat Example 6-3 for the case of constant wall temperature.

■ **Solution**

We evaluate properties as before and now enter Figure 6-5 to determine $\overline{Nu}_d$ for $T_w = $ constant. For $Gz^{-1} = 0.0346$ we read

$$Nu_d = 5.15$$

We thus calculate the average heat-transfer coefficient as

$$h = (5.15)\left(\frac{k}{d}\right) = \frac{(5.15)(0.02814)}{0.005} = 29.98 \text{ W/m}^2 \cdot °C$$

We base the heat transfer on a mean bulk temperature of 52°C, so that

$$q = \bar{h}\pi dL(T_w - \overline{T_b}) = 3.49 \text{ W}$$

and

$$T_w = 76.67 + 52 = 128.67°C$$

Heat Transfer in a Rough Tube   EXAMPLE 6-5

A 2.0-cm-diameter tube having a relative roughness of 0.001 is maintained at a constant wall temperature of 90°C. Water enters the tube at 40°C and leaves at 60°C. If the entering velocity is 3 m/s, calculate the length of tube necessary to accomplish the heating.

■ **Solution**
We first calculate the heat transfer from

$$q = \dot{m}c_p \Delta T_b = (989)(3.0)\pi(0.01)^2(4174)(60 - 40) = 77{,}812 \text{ W}$$

For the rough-tube condition, we may employ the Petukhov relation, Equation (6-7). The mean film temperature is

$$T_f = \frac{90 + 50}{2} = 70°C$$

and the fluid properties are

$$\rho = 978 \text{ kg/m}^3 \qquad \mu = 4.0 \times 10^{-4} \text{ kg/m} \cdot \text{s}$$
$$k = 0.664 \text{ W/m} \cdot °C \qquad Pr = 2.54$$

Also,

$$\mu_b = 5.55 \times 10^{-4} \text{ kg/m} \cdot \text{s}$$
$$\mu_w = 2.81 \times 10^{-4} \text{ kg/m} \cdot \text{s}$$

The Reynolds number is thus

$$Re_d = \frac{(978)(3)(0.02)}{4 \times 10^{-4}} = 146{,}700$$

Consulting Figure 6-4, we find the friction factor as

$$f = 0.0218 \qquad f/8 = 0.002725$$

Because $T_w > T_b$, we take $n = 0.11$ and obtain

$$Nu_d = \frac{(0.002725)(146{,}700)(2.54)}{1.07 + (12.7)(0.002725)^{1/2}(2.54^{2/3} - 1)}\left(\frac{5.55}{2.81}\right)^{0.11}$$
$$= 666.8$$

$$h = \frac{(666.8)(0.664)}{0.02} = 22138 \text{ W/m}^2 \cdot °C$$

The tube length is then obtained from the energy balance

$$q = \bar{h}\pi dL(T_w - \overline{T_b}) = 77{,}812 \text{ W}$$
$$L = 1.40 \text{ m}$$

| EXAMPLE 6-6 | Turbulent Heat Transfer in a Short Tube |

Air at 300 K and 1 atm enters a smooth tube having a diameter of 2 cm and length of 10 cm. The air velocity is 40 m/s. What constant heat flux must be applied at the tube surface to result in an air temperature rise of 5°C? What average wall temperature would be necessary for this case?

■ **Solution**

Because of the relatively small value of $L/d = 10/2 = 5$ we may anticipate that thermal entrance effects will be present in the flow. First, we determine the air properties at 300 K as

$$v = 15.69 \times 10^{-6} \text{m}^2/\text{s} \qquad k = 0.02624 \text{ W/m} \cdot \text{°C} \qquad \text{Pr} = 0.7$$
$$c_p = 1006 \text{ J/kg} \cdot \text{°C} \qquad \rho = 1.18 \text{ kg/m}^3$$

We calculate the Reynolds number as

$$\text{Re}_d = ud/v = (40)(0.02)/15.69 \times 10^{-6} = 50{,}988$$

so the flow is turbulent. Consulting Figure 6-6 for this value of Re_d, Pr = 0.7, and $L/d = 5$ we find

$$\text{Nu}_x/\text{Nu}_\infty \cong 1.15$$

or the heat-transfer coefficient is about 15 percent higher that it would be for thermally developed flow. We calculate the heat-transfer coefficient for developed flow using

$$\text{Nu}_d = 0.023 \text{ Re}_d^{0.8} \text{Pr}^{0.4}$$
$$= 0.023(50988)^{0.8}(0.7)^{0.4} = 116.3$$

and

$$h = k\text{Nu}_d/d = (0.02624)(116.3)/0.02 = 152.6 \text{ W/m}^2 \cdot \text{°C}$$

Increasing this value by 15 percent,

$$h = (1.15)(152.6) = 175.5 \text{ W/m}^2 \cdot \text{°C}$$

The mass flow is

$$\dot{m} = \rho u A_c = (1.18)(40)\pi(0.02)^2/4 = 0.0148 \text{ kg/s}$$

so the total heat transfer is

$$q = \dot{m}c_p \Delta T_b = (0.0148)(1006)(5) = 74.4 \text{ W}$$

This heat flow is convected from a tube surface area of

$$A = \pi dL = \pi(0.02)(0.1) = 0.0628 \text{ m}^2$$

so the heat flux is

$$q/A = 74.4/0.0628 = 11841 \text{ W/m}^2 = h(T_w - T_b)$$

We have

$$\overline{T}_b = (300 + 305)/2 = 302.5 \text{ K}$$

so that

$$\overline{T}_w = \overline{T}_b + 11841/175.5 = 302.5 + 67.5 = 370 \text{ K}$$

6-3 | FLOW ACROSS CYLINDERS AND SPHERES

While the engineer may frequently be interested in the heat-transfer characteristics of flow systems inside tubes or over flat plates, equal importance must be placed on the heat transfer that may be achieved by a cylinder in cross flow, as shown in Figure 6-7. As would be expected, the boundary-layer development on the cylinder determines the heat-transfer characteristics. As long as the boundary layer remains laminar and well behaved, it is possible to compute the heat transfer by a method similar to the boundary-layer analysis of Chapter 5. It is necessary, however, to include the pressure gradient in the analysis because this influences the boundary-layer velocity profile to an appreciable extent. In fact, it is this pressure gradient that causes a separated flow region to develop on the back side of the cylinder when the free-stream velocity is sufficiently large.

The phenomenon of boundary-layer separation is indicated in Figure 6-8. The physical reasoning that explains the phenomenon in a qualitative way is as follows: Consistent with boundary-layer theory, the pressure through the boundary layer is essentially constant at any x position on the body. In the case of the cylinder, one might measure x distance from the front stagnation point of the cylinder. Thus the pressure in the boundary layer should follow that of the free stream for potential flow around a cylinder, provided this behavior would not contradict some basic principle that must apply in the boundary layer. As the flow progresses along the front side of the cylinder, the pressure would decrease and then increase along the back side of the cylinder, resulting in an increase in free-stream velocity on the front side of the cylinder and a decrease on the back side. The transverse velocity (that velocity parallel to the surface) would decrease from a value of u_∞ at the outer edge of the boundary layer to zero at the surface. As the flow proceeds to the back side of the cylinder, the pressure increase causes a reduction in velocity in the free stream and throughout the

Figure 6-7 | Cylinder in cross flow.

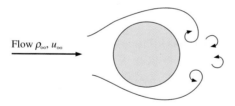

Flow ρ_∞, u_∞

Figure 6-8 | Velocity distributions indicating flow separation on a cylinder in cross flow.

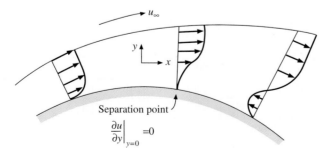

u_∞

y

x

Separation point

$$\left.\frac{\partial u}{\partial y}\right|_{y=0} = 0$$

boundary layer. The pressure increase and reduction in velocity are related through the Bernoulli equation written along a streamline:

$$\frac{dp}{\rho} = -d\left(\frac{u^2}{2g_c}\right)$$

Since the pressure is assumed constant throughout the boundary layer, we note that reverse flow may begin in the boundary layer near the surface; that is, the momentum of the fluid layers near the surface is not sufficiently high to overcome the increase in pressure. When the velocity gradient at the surface becomes zero, the flow is said to have reached a separation point:

$$\text{Separation point at } \quad \frac{\partial u}{\partial y}\Bigg]_{y=0} = 0$$

This separation point is indicated in Figure 6-8. As the flow proceeds past the separation point, reverse-flow phenomena may occur, as also shown in Figure 6-8. Eventually, the separated-flow region on the back side of the cylinder becomes turbulent and random in motion.

The drag coefficient for bluff bodies is defined by

$$\text{Drag force} = F_D = C_D A \frac{\rho u_\infty^2}{2g_c} \qquad \qquad \textbf{[6-16]}$$

where C_D is the drag coefficient and A is the *frontal area* of the body exposed to the flow, which, for a cylinder, is the product of diameter and length. The values of the drag coefficient for cylinders and spheres are given as a function the Reynolds number in Figures 6-9 and 6-10.

The drag force on the cylinder is a result of a combination of frictional resistance and so-called form, or pressure drag, resulting from a low-pressure region on the rear of the cylinder created by the flow-separation process. At low Reynolds numbers of the order

Figure 6-9 | Drag coefficient for circular cylinders as a function of the Reynolds number, from Reference 6.

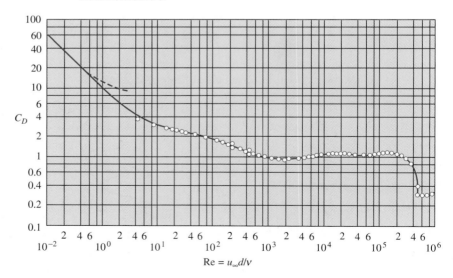

Figure 6-10 | Drag coefficient for spheres as a function of the Reynolds number, from Reference 6.

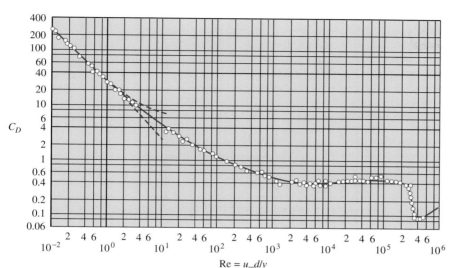

of unity, there is no flow separation, and all the drag results from viscous friction. At Reynolds numbers of the order of 10, the friction and form drag are of the same order, while the form drag resulting from the turbulent separated-flow region predominates at Reynolds numbers greater than 1000. At Reynolds numbers of approximately 10^5, based on diameter, the boundary-layer flow may become turbulent, resulting in a steeper velocity profile and extremely late flow separation. Consequently, the form drag is reduced, and this is represented by the break in the drag-coefficient curve at about $Re = 3 \times 10^5$. The same reasoning applies to the sphere as to the circular cylinder. Similar behavior is observed with other bluff bodies, such as elliptic cylinders and airfoils.

The flow processes discussed above obviously influence the heat transfer from a heated cylinder to a fluid stream. The detailed behavior of the heat transfer from a heated cylinder to air has been investigated by Giedt [7], and the results are summarized in Figure 6-11. At the lower Reynolds numbers (70,800 and 101,300) a minimum point in the heat-transfer coefficient occurs at approximately the point of separation. There is a subsequent increase in the heat-transfer coefficient on the rear side of the cylinder, resulting from the turbulent eddy motion in the separated flow. At the higher Reynolds numbers two minimum points are observed. The first occurs at the point of transition from laminar to turbulent boundary layer, and the second minimum occurs when the turbulent boundary layer separates. There is a rapid increase in heat transfer when the boundary layer becomes turbulent and another when the increased eddy motion at separation is encountered.

Because of the complicated nature of the flow-separation processes, it is not possible to calculate analytically the average heat-transfer coefficients in cross flow; however, McAdams [10] was able to correlate the data of a number of investigators for heating and cooling of air as shown in the plot of Figure 6-12. The data points have been omitted, but scatter of ± 20 percent is not uncommon. The Prandtl number was not included in the original correlation plot because it is essentially constant at about 0.72 for all the data. Following the reasoning of Prandtl number dependence indicated by Equation (5-85), Knudsen and Katz [9] suggested that the correlation be extended to liquids by inclusion of $Pr^{1/3}$.

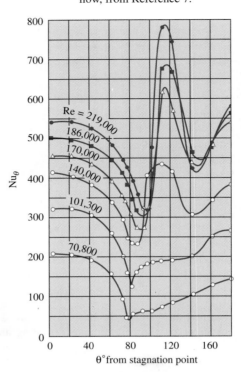

Figure 6-11 | Local Nusselt number for heat transfer from a cylinder in cross flow, from Reference 7.

The resulting correlation for average heat-transfer coefficients in cross flow over circular cylinders is

$$\mathrm{Nu}_{df} = \frac{hd}{k_f} = C\left(\frac{u_\infty d}{\nu_f}\right)^n \mathrm{Pr}_f^{1/3} \qquad \textbf{[6-17]}$$

where the constants C and n are tabulated in Table 6-2. Properties for use with Equation (6-17) are evaluated at the film temperature as indicated by the subscript f.

In obtaining the correlation constants for Table 6-2, the original calculations were based on air data alone, fitting straight-line segments to a log–log plot like that of Figure 6-12. For such data the Prandtl number is very nearly constant at about 0.72. It was reasoned in Reference 9 that the same correlation might be employed for liquids by introducing the factor $\mathrm{Pr}^{1/3}$ and dividing out $(0.72)^{1/3}$, or multiplying by 1.11. This reasoning has been borne out in practice.

Figure 6-13 shows the temperature field around heated cylinders placed in a transverse airstream. The dark lines are lines of constant temperature, made visible through the use of an interferometer. Note the separated-flow region that develops on the back side of the cylinder at the higher Reynolds numbers and the turbulent field that is present in that region.

Note also the behavior at the lowest Reynolds number of 23. The wake rises because of thermal buoyancy effects. At this point, we are observing a behavior resulting from the superposition of free-convection currents of the same order as the forced-convection-flow

Figure 6-12 | Correlation for heating and cooling in cross flow over circular cylinders.

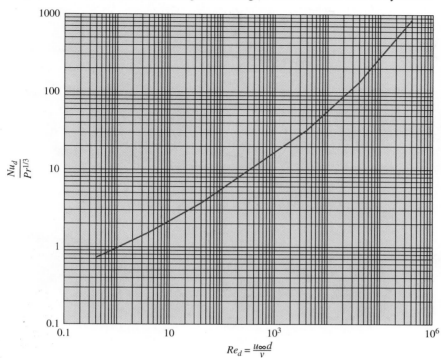

$$Re_d = \frac{u_\infty d}{\nu}$$

Table 6-2 | Constants for use with Equation (6-17), based on References 8 and 9.

Re_{df}	C	n
0.4–4	0.989	0.330
4–40	0.911	0.385
40–4000	0.683	0.466
4000–40,000	0.193	0.618
40,000–400,000	0.0266	0.805

velocities. In this regime the heat transfer is also dependent on a parameter called the Grashof number, which we shall describe in detail in Chapter 7. For higher Reynolds numbers the heat transfer is predominately by forced convection.

Fand [21] has shown that the heat-transfer coefficients from liquids to cylinders in cross flow may be better represented by the relation

$$Nu_f = (0.35 + 0.56\,Re_f^{0.52})Pr_f^{0.3} \qquad \text{[6-18]}$$

This relation is valid for $10^{-1} < Re_f < 10^5$ provided excessive free-stream turbulence is not encountered.

In some instances, particularly those involving calculations on a computer, it may be more convenient to utilize a more complicated expression than Equation (6-17) if it can be applied over a wider range of Reynolds numbers. Eckert and Drake [34] recommend the

Figure 6-13 | Interferometer photograph showing isotherms around heated horizontal cylinders placed in a transverse airstream. $Re = \rho u_\infty d/\mu$. (*Photograph courtesy E. Soehngen.*)

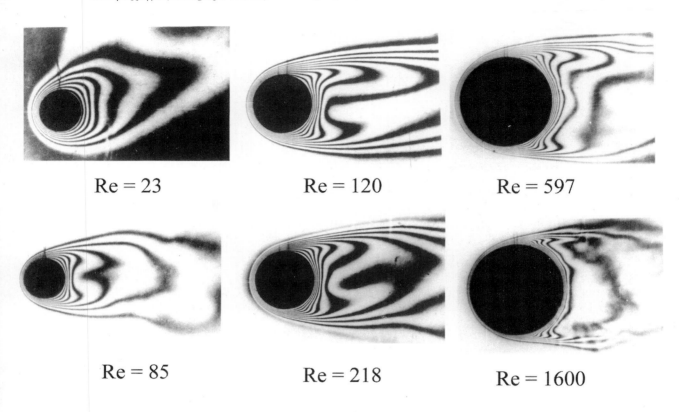

following relations for heat transfer from tubes in cross flow, based on the extensive study of References 33 and 39:

$$\mathrm{Nu} = (0.43 + 0.50\,\mathrm{Re}^{0.5})\mathrm{Pr}^{0.38}\left(\frac{\mathrm{Pr}_f}{\mathrm{Pr}_w}\right)^{0.25} \quad \text{for } 1 < \mathrm{Re} < 10^3 \qquad \textbf{[6-19]}$$

$$\mathrm{Nu} = 0.25\,\mathrm{Re}^{0.6}\mathrm{Pr}^{0.38}\left(\frac{\mathrm{Pr}_f}{\mathrm{Pr}_w}\right)^{0.25} \quad \text{for } 10^3 < \mathrm{Re} < 2 \times 10^5 \qquad \textbf{[6-20]}$$

For gases the Prandtl number ratio may be dropped, and fluid properties are evaluated at the film temperature. For liquids the ratio is retained, and fluid properties are evaluated at the free-stream temperature. Equations (6-19) and (6-20) are in agreement with results obtained using Equation (6-17) within 5 to 10 percent.

Still a more comprehensive relation is given by Churchill and Bernstein [37] that is applicable over the complete range of available data:

$$\mathrm{Nu}_d = 0.3 + \frac{0.62\,\mathrm{Re}^{1/2}\mathrm{Pr}^{1/3}}{[1 + (0.4/\mathrm{Pr})^{2/3}]^{1/4}}\left[1 + \left(\frac{\mathrm{Re}}{282{,}000}\right)^{5/8}\right]^{4/5}$$

$$\text{for } 10^2 < \mathrm{Re}_d < 10^7;\ \mathrm{Pe}_d > 0.2 \qquad \textbf{[6-21]}$$

This relation underpredicts the data somewhat in the midrange of Reynolds numbers between 20,000 and 400,000, and it is suggested that the following be employed for this range:

$$\text{Nu}_d = 0.3 + \frac{0.62 \, \text{Re}_d^{1/2} \text{Pr}^{1/3}}{\left[1 + (0.4/\text{Pr})^{2/3}\right]^{1/4}} \left[1 + \left(\frac{\text{Re}_d}{282,000}\right)^{1/2}\right]$$

$$\text{for } 20,000 < \text{Re}_d < 400,000; \, \text{Pe}_d > 0.2 \qquad \textbf{[6-22]}$$

The heat-transfer data that were used to arrive at Equations (6-21) and (6-22) include fluids of air, water, and liquid sodium. Still another correlation equation is given by Whitaker [35] as

$$\text{Nu} = \frac{\overline{h}d}{k} = (0.4 \, \text{Re}^{0.5} + 0.06 \, \text{Re}^{2/3})\text{Pr}^{0.4} \left(\frac{\mu_\infty}{\mu_w}\right)^{0.25} \qquad \textbf{[6-23]}$$

for $40 < \text{Re} < 10^5$, $0.65 < \text{Pr} < 300$, and $0.25 < \mu_\infty/\mu_w < 5.2$. All properties are evaluated at the free-stream temperature except that μ_w is at the wall temperature.

Below $\text{Pe}_d = 0.2$, Nakai and Okazaki [38] present the following relation:

$$\text{Nu}_d = \left[0.8237 - \ln\left(\text{Pe}_d^{1/2}\right)\right]^{-1} \quad \text{for } \text{Pe}_d < 0.2 \qquad \textbf{[6-24]}$$

Properties in Equations (6-21), (6-22), and (6-24) are evaluated at the film temperature.

Choice of Equation for Cross Flow Over Cylinders

The choice of equation to use for cross flow over cylinders is subject to some conjecture. Clearly, Equation (6-17) is easiest to use from a computational standpoint, and Equation (6-21) is the most comprehensive. The more comprehensive relations are preferable for computer setups because of the wide range of fluids and Reynolds numbers covered. For example, Equation (6-21) has been successful in correlating data for fluids ranging from air to liquid sodium. Equation (6-17) could not be used for liquid metals. If one were making calculations for air, either relation would be satisfactory.

Noncircular Cylinders

Jakob [22] has summarized the results of experiments with heat transfer from noncircular cylinders. Equation (6-17) is employed in order to obtain an empirical correlation for gases, and the constants for use with this equation are summarized in Table 6-3. The data upon

Table 6-3 | Constants for heat transfer from noncircular cylinders according to Reference 22, for use with Equation (6-17).

Geometry	Re_{df}	C	n
$u_\infty \rightarrow \diamondsuit \updownarrow d$	$5 \times 10^3 - 10^5$	0.246	0.588
$u_\infty \rightarrow \square \updownarrow d$	$5 \times 10^3 - 10^5$	0.102	0.675
$u_\infty \rightarrow \hexagon \updownarrow d$	$5 \times 10^3 - 1.95 \times 10^4$	0.160	0.638
	$1.95 \times 10^4 - 10^5$	0.0385	0.782
$u_\infty \rightarrow \hexagon \updownarrow d$	$5 \times 10^3 - 10^5$	0.153	0.638
$u_\infty \rightarrow \vert \updownarrow d$	$4 \times 10^3 - 1.5 \times 10^4$	0.228	0.731

which Table 6-3 is based were for gases with $Pr \sim 0.7$ and were modified by the same $1.11\,Pr^{1/3}$ factor employed for the information presented in Table 6-2.

Spheres

McAdams [10] recommends the following relation for heat transfer from spheres to a flowing gas:

$$\frac{hd}{k_f} = 0.37 \left(\frac{u_\infty d}{v_f}\right)^{0.6} \quad \text{for } 17 < Re_d < 70{,}000 \tag{6-25}$$

Achenbach [43] has obtained relations applicable over a still wider range of Reynolds numbers for air with $Pr = 0.71$:

$$Nu = 2 + (0.25\,Re + 3 \times 10^{-4}Re^{1.6})^{1/2} \quad \text{for } 100 < Re < 3 \times 10^5 \tag{6-26}$$

$$Nu = 430 + a\,Re + b\,Re^2 + c\,Re^3 \quad \text{for } 3 \times 10^5 < Re < 5 \times 10^6 \tag{6-27}$$

with

$$a = 0.5 \times 10^{-3} \qquad b = 0.25 \times 10^{-9} \qquad c = -3.1 \times 10^{-17}$$

For flow of liquids past spheres, the data of Kramers [11] may be used to obtain the correlation

$$\frac{hd}{k_f} Pr_f^{-0.3} = 0.97 + 0.68 \left(\frac{u_\infty d}{v_f}\right)^{0.5} \quad \text{for } 1 < Re_d < 2000 \tag{6-28}$$

Vliet and Leppert [19] recommend the following expression for heat transfer from spheres to oil and water over a more extended range of Reynolds numbers from 1 to 200,000:

$$Nu\,Pr^{-0.3}\left(\frac{\mu_w}{\mu}\right)^{0.25} = 1.2 + 0.53\,Re_d^{0.54} \tag{6-29}$$

where all properties are evaluated at free-stream conditions, except μ_w, which is evaluated at the surface temperature of the sphere. Equation (6-26) represents the data of Reference 11, as well as the more recent data of Reference 19.

All the above data have been brought together by Whitaker [35] to develop a single equation for gases and liquids flowing past spheres:

$$Nu = 2 + \left(0.4\,Re_d^{1/2} + 0.06\,Re_d^{2/3}\right)Pr^{0.4}\,(\mu_\infty/\mu_w)^{1/4} \tag{6-30}$$

which is valid for the range $3.5 < Re_d < 8 \times 10^4$ and $0.7 < Pr < 380$. Properties in Equation (6-30) are evaluated at the free-stream temperature.

| **EXAMPLE 6-7** | Airflow Across Isothermal Cylinder |

Air at 1 atm and 35°C flows across a 5.0-cm-diameter cylinder at a velocity of 50 m/s. The cylinder surface is maintained at a temperature of 150°C. Calculate the heat loss per unit length of the cylinder.

■ **Solution**

We first determine the Reynolds number and then find the applicable constants from Table 6-2 for use with Equation (6-17). The properties of air are evaluated at the film temperature:

$$T_f = \frac{T_w + T_\infty}{2} = \frac{150 + 35}{2} = 92.5°C = 365.5 \text{ K}$$

$$\rho_f = \frac{p}{RT} = \frac{1.0132 \times 10^5}{(287)(365.5)} = 0.966 \text{ kg/m}^3 \quad [0.0603 \text{ lb}_m/\text{ft}^3]$$

$$\mu_f = 2.14 \times 10^{-5} \text{ kg/m} \cdot \text{s} \quad [0.0486 \text{ lb}_m/\text{h} \cdot \text{ft}]$$

$$k_f = 0.0312 \text{ W/m} \cdot °C \quad [0.018 \text{ Btu/h} \cdot \text{ft} \cdot °F]$$

$$\text{Pr}_f = 0.695$$

$$\text{Re}_f = \frac{\rho u_\infty d}{\mu} = \frac{(0.966)(50)(0.05)}{2.14 \times 10^{-5}} = 1.129 \times 10^5$$

From Table 6-2

$$C = 0.0266 \qquad n = 0.805$$

so from Equation (6-17)

$$\frac{hd}{k_f} = (0.0266)(1.129 \times 10^5)^{0.805}(0.695)^{1/3} = 275.1$$

$$h = \frac{(275.1)(0.0312)}{0.05} = 171.7 \text{ W/m}^2 \cdot °C \quad [30.2 \text{ Btu/h} \cdot \text{ft}^2 \cdot °F]$$

The heat transfer per unit length is therefore

$$\frac{q}{L} = h\pi d(T_w - T_\infty)$$

$$= (171.7)\pi(0.05)(150 - 35)$$

$$= 3100 \text{ W/m} \quad [3226 \text{ Btu/ft}]$$

Heat Transfer from Electrically Heated Wire   EXAMPLE 6-8

A fine wire having a diameter of 3.94×10^{-5} m is placed in a 1-atm airstream at 25°C having a flow velocity of 50 m/s perpendicular to the wire. An electric current is passed through the wire, raising its surface temperature to 50°C. Calculate the heat loss per unit length.

■ **Solution**

We first obtain the properties at the film temperature:

$$T_f = (25 + 50)/2 = 37.5°C = 310 \text{ K}$$

$$\nu_f = 16.7 \times 10^{-6} \text{ m}^2/\text{s} \qquad k = 0.02704 \text{ W/m} \cdot °C$$

$$\text{Pr}_f = 0.706$$

The Reynolds number is

$$\text{Re}_d = \frac{u_\infty d}{\nu_f} = \frac{(50)(3.94 \times 10^{-5})}{16.7 \times 10^{-6}} = 118$$

The Peclet number is $\text{Pe} = \text{Re Pr} = 83.3$, and we find that Equations (6-17), (6-21), or (6-19) apply. Let us make the calculation with both the simplest expression, (6-17), and the most complex, (6-21), and compare results.

Using Equation (6-17) with $C = 0.683$ and $n = 0.466$, we have

$$\mathrm{Nu}_d = (0.683)(118)^{0.466}(0.705)^{1/3} = 5.615$$

and the value of the heat-transfer coefficient is

$$h = \mathrm{Nu}_d \left(\frac{k}{d}\right) = 5.615 \frac{0.02704}{3.94 \times 10^{-5}} = 3854 \ \mathrm{W/m^2 \cdot {}^\circ C}$$

The heat transfer per unit length is then

$$q/L = \pi \, dh(T_w - T_\infty) = \pi(3.94 \times 10^{-5})(3854)(50 - 25)$$
$$= 11.93 \ \mathrm{W/m}$$

Using Equation (6-21), we calculate the Nusselt number as

$$\mathrm{Nu}_d = 0.3 + \frac{(0.62)(118)^{1/2}(0.705)^{1/3}}{\left[1 + (0.4/0.705)^{2/3}\right]^{1/4}} \left[1 + (118/282{,}000)^{5/8}\right]^{4/5}$$
$$= 5.593$$

and

$$h = \frac{(5.593)(0.02704)}{3.94 \times 10^{-5}} = 3838 \ \mathrm{W/m^2 \cdot {}^\circ C}$$

and

$$q/L = (3838)\pi(3.94 \times 10^{-5})(50 - 25) = 11.88 \ \mathrm{W/m}$$

Here, we find the two correlations differing by 0.4 percent if the value from Equation (6-21) is taken as correct, or 0.2 percent from the mean value. Data scatter of ± 15 percent is not unusual for the original experiments.

EXAMPLE 6-9   Heat Transfer from Sphere

Air at 1 atm and 27°C blows across a 12-mm-diameter sphere at a free-stream velocity of 4 m/s. A small heater inside the sphere maintains the surface temperature at 77°C. Calculate the heat lost by the sphere.

■ **Solution**

Consulting Equation (6-30) we find that the Reynolds number is evaluated at the free-stream temperature. We therefore need the following properties: at $T_\infty = 27°C = 300 \ \mathrm{K}$,

$$\nu = 15.69 \times 10^{-6} \ \mathrm{m^2/s} \qquad k = 0.02624 \ \mathrm{W/m \cdot {}^\circ C},$$

$$\mathrm{Pr} = 0.708 \qquad \mu_\infty = 1.8462 \times 10^{-5} \ \mathrm{kg/m \cdot s}$$

At $T_w = 77°C = 350 \ \mathrm{K}$,

$$\mu_w = 2.075 \times 10^{-5}$$

The Reynolds number is thus

$$\mathrm{Re}_d = \frac{(4)(0.012)}{15.69 \times 10^{-6}} = 3059$$

From Equation (6-30),

$$\overline{\mathrm{Nu}} = 2 + [(0.4)(3059)^{1/2} + (0.06)(3059)^{2/3}](0.708)^{0.4} \left(\frac{1.8462}{2.075}\right)^{1/4}$$

$$= 31.40$$

and

$$\bar{h} = \overline{\mathrm{Nu}}\left(\frac{k}{d}\right) = \frac{(31.4)(0.02624)}{0.012} = 68.66 \text{ W/m}^2 \cdot {}^\circ\text{C}$$

The heat transfer is then

$$q = \bar{h}A(T_w - T_\infty) = (68.66)(4\pi)(0.006)^2(77 - 27) = 1.553 \text{ W}$$

For comparison purposes let us also calculate the heat-transfer coefficient using Equation (6-25). The film temperature is $T_f = (350 + 300)/2 = 325$ K so that

$$\nu_f = 18.23 \times 10^{-6} \text{ m}^2/\text{s} \qquad k_f = 0.02814 \text{ W/m} \cdot {}^\circ\text{C}$$

and the Reynolds number is

$$\mathrm{Re}_d = \frac{(4)(0.012)}{18.23 \times 10^{-6}} = 2633$$

From Equation (6-25)

$$\mathrm{Nu}_f = (0.37)(2633)^{0.6} = 41.73$$

and $\bar{h}$ is calculated as

$$\bar{h} = \mathrm{Nu}\left(\frac{k_f}{d}\right) = \frac{(41.73)(0.02814)}{0.012} = 97.9 \text{ W/m}^2 \cdot {}^\circ\text{C}$$

or about 42 percent higher than the value calculated before.

6-4 | FLOW ACROSS TUBE BANKS

Because many heat-exchanger arrangements involve multiple rows of tubes, the heat-transfer characteristics for tube banks are of important practical interest. The heat-transfer characteristics of staggered and in-line tube banks were studied by Grimson [12], and on the basis of a correlation of the results of various investigators, he was able to represent data in the form of Equation (6-17). The original data were for gases with $\mathrm{Pr} \sim 0.7$. To extend the use to liquids, the present writer has modified the constants by the same $1.11\mathrm{Pr}^{1/3}$ factor employed in Tables 6-2 and 6-3. The values of the constant C and the exponent n are given in Table 6-4 in terms of the geometric parameters used to describe the tube-bundle arrangement. The Reynolds number is based on the maximum velocity occurring in the tube bank; that is, the velocity through the minimum-flow area. This area will depend on the geometric tube arrangement. The nomenclature for use with Table 6-4 is shown in Figure 6-14. The data of Table 6-4 pertain to tube banks having 10 or more rows of tubes in the direction of flow. For fewer rows the ratio of h for N rows deep to that for 10 rows is given in Table 6-5.

Determination of Maximum Flow Velocity

For flows normal to in-line tube banks the maximum flow velocity will occur through the minimum frontal area $(S_n - d)$ presented to the incoming free stream velocity u_∞. Thus,

$$u_{\max} = u_\infty[S_n/(S_n - d)] \qquad \text{(in-line arrangement)}$$

for this configuration. For the staggered arrangement the same maximum flow velocity will be experienced *if the normal area at the entrance to the tube bank is the minimum flow area*. This may not be the case for close spacing in the parallel direction, as when S_p is small. For the staggered case, the flow enters the tube bank through the area $S_n - d$ and then splits

Table 6-4 | Modified correlation of Grimson for heat transfer in tube banks of 10 rows or more, from Reference 12, for use with Equation (6-17).

$\dfrac{S_p}{d}$	$\dfrac{S_n}{d}$ 1.25		1.5		2.0		3.0	
	C	n	C	n	C	n	C	n
				In line				
1.25	0.386	0.592	0.305	0.608	0.111	0.704	0.0703	0.752
1,5	0.407	0.586	0.278	0.620	0.112	0.702	0.0753	0.744
2.0	0.464	0.570	0.332	0.602	0.254	0.632	0.220	0.648
3.0	0.322	0.601	0.396	0.584	0.415	0.581	0.317	0.608
				Staggered				
0.6	—	—	—	—	—	—	0.236	0.636
0.9	—	—	—	—	0.495	0.571	0.445	0.581
1.0	—	—	0.552	0.558	—	—	—	—
1.125	—	—	—	—	0.531	0.565	0.575	0.560
1.25	0.575	0.556	0.561	0.554	0.576	0.556	0.579	0.562
1.5	0.501	0.568	0.511	0.562	0.502	0.568	0.542	0.568
2.0	0.448	0.572	0.462	0.568	0.535	0.556	0.498	0.570
3.0	0.344	0.592	0.395	0.580	0.488	0.562	0.467	0.574

Figure 6-14 | Nomenclature for use with Table 6-4: (*a*) in-line tube rows; (*b*) staggered tube rows.

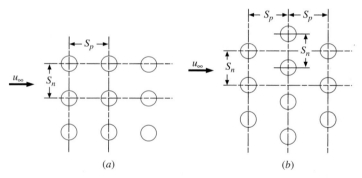

(a)   (b)

Table 6-5 | Ratio of h for N rows deep to that for 10 rows deep, for use with Equation (6-17).

N	1	2	3	4	5	6	7	8	9	10
Ratio for staggered tubes	0.68	0.75	0.83	0.89	0.92	0.95	0.97	0.98	0.99	1.0
Ratio for in-line tubes	0.64	0.80	0.87	0.90	0.92	0.94	0.96	0.98	0.99	1.0

From Reference 17.

into the two areas $[(S_n/2)^2 + S_p^2]^{1/2} - d$. If the sum of these two areas is less than $S_n - d$, then they will represent the minimum flow area and the maximum velocity in the tube bank will be

$$u_{max} = \frac{u_\infty (S_n/2)}{[(S_n/2)^2 + S_p^2]^{1/2} - d}$$

where, again, u_∞ is the free-stream velocity entering the tube bank.

Pressure drop for flow of gases over a bank of tubes may be calculated with Equation (6-31), expressed in pascals:

$$\Delta p = \frac{2 f' G_{max}^2 N}{\rho} \left(\frac{\mu_w}{\mu_b} \right)^{0.14} \qquad \text{[6-31]}$$

where

G_{max} = mass velocity at minimum flow area, kg/m$^2 \cdot$ s

ρ = density evaluated at free-stream conditions, kg/m^3

N = number of transverse rows

μ_b = average free-stream viscosity, N $\cdot$ s/m^2

The empirical friction factor f' is given by Jakob [18] as

$$f' = \left\{ 0.25 + \frac{0.118}{[(S_n - d)/d]^{1.08}} \right\} \text{Re}_{max}^{-0.16} \qquad \text{[6-32]}$$

for staggered tube arrangements, and

$$f' = \left\{ 0.044 + \frac{0.08 S_p/d}{[(S_n - d)/d]^{0.43 + 1.13 d/S_p}} \right\} \text{Re}_{max}^{-0.15} \qquad \text{[6-33]}$$

for in-line arrangements.

Zukauskas [39] has presented additional information for tube bundles that takes into account wide ranges of Reynolds numbers and property variations. The correlating equation takes the form

$$\text{Nu} = \frac{\overline{h} d}{k} = C \text{Re}_{d,max}^n \text{Pr}^{0.36} \left(\frac{\text{Pr}}{\text{Pr}_w} \right)^{1/4} \qquad \text{[6-34]}$$

where all properties except Pr_w are evaluated at T_∞ and the values of the constants are given in Table 6-6 for greater than 20 rows of tubes. This equation is applicable for $0.7 < \text{Pr} < 500$ and $10 < \text{Re}_{d,max} < 10^6$. For gases the Prandtl number ratio has little influence and is dropped. Once again, note that the Reynolds number is based on the maximum velocity in the tube bundle. For less than 20 rows in the direction of flow the correction factor in Table 6-7 should be applied. It is essentially the same as for the Grimson correlation.

Table 6-6 | Constants for Zukauskas correlation [Equation (6-34)] for heat transfer in tube banks of 20 rows or more.

Geometry	Re$_{d,max}$	C	n
In-line	10–100	0.8	0.4
	100–10^3	Treat as individual tubes	
	10^3 – 2 × 10^5	0.27	0.63
	> 2 × 10^5	0.21	0.84
Staggered	10–100	0.9	0.4
	100–10^3	Treat as individual tubes	
	10^3 – 2 × 10^5	$0.35 \left(\frac{S_n}{S_L} \right)^{0.2}$ for $\frac{S_n}{S_L} < 2$	0.60
	10^3 – 2 × 10^5	0.40 for $\frac{S_n}{S_L} > 2$	0.60
	> 2 × 10^5	0.022	0.84

From Reference 39.

Table 6-7 | Ratio of h for N rows deep to that for 20 rows deep according to Reference 39 and for use with Equation (6-34).

N	2	3	4	5	6	8	10	16	20
Staggered	0.77	0.84	0.89	0.92	0.94	0.97	0.98	0.99	1.0
In-line	0.70	0.80	0.90	0.92	0.94	0.97	0.98	0.99	1.0

Additional information is given by Morgan [44]. Further information on pressure drop is given in Reference 39.

The reader should keep in mind that these relations correlate experimental data with an uncertainty of about ± 25 percent.

EXAMPLE 6-10	Heating of Air with In-Line Tube Bank

Air at 1 atm and 10°C flows across a bank of tubes 15 rows high and 5 rows deep at a velocity of 7 m/s measured at a point in the flow before the air enters the tube bank. The surfaces of the tubes are maintained at 65°C. The diameter of the tubes is 1 in [2.54 cm]; they are arranged in an in-line manner so that the spacing in both the normal and parallel directions to the flow is 1.5 in [3.81 cm]. Calculate the total heat transfer per unit length for the tube bank and the exit air temperature.

■ **Solution**

The constants for use with Equation (6-17) may be obtained from Table 6-4, using

$$\frac{S_p}{d} = \frac{3.81}{2.54} = 1.5 \qquad \frac{S_n}{d} = \frac{3.81}{2.54} = 1.5$$

so that

$$C = 0.278 \qquad n = 0.620$$

The properties of air are evaluated at the film temperature, which at entrance to the tube bank is

$$T_{f_1} = \frac{T_w + T_\infty}{2} = \frac{65 + 10}{2} = 37.5°C = 310.5 \text{ K} \quad [558.9°R]$$

Then

$$\rho_f = \frac{p}{RT} = \frac{1.0132 \times 10^5}{(287)(310.5)} \simeq 1.137 \text{ kg/m}^3$$

$\mu_f = 1.894 \times 10^{-5} \text{ kg/m} \cdot \text{s}$

$k_f = 0.027 \text{ W/m} \cdot °C \quad [0.0156 \text{ Btu/h} \cdot \text{ft} \cdot °F]$

$c_p = 1007 \text{ J/kg} \cdot °C \quad [0.24 \text{ Btu/lb}_m \cdot °F]$

$\text{Pr} = 0.706$

To calculate the maximum velocity, we must determine the minimum flow area. From Figure 6-14 we find that the ratio of the minimum flow area to the total frontal area is $(S_n - d)/S_n$. The maximum velocity is thus

$$u_{max} = u_\infty \frac{S_n}{S_n - d} = \frac{(7)(3.81)}{3.81 - 2.54} = 21 \text{ m/s} \quad [68.9 \text{ ft/s}] \qquad [a]$$

where u_∞ is the incoming velocity before entrance to the tube bank. The Reynolds number is computed by using the maximum velocity.

$$\text{Re} = \frac{\rho u_{max} d}{\mu} = \frac{(1.137)(21)(0.0254)}{1.894 \times 10^{-5}} = 32,020 \qquad [b]$$

The heat-transfer coefficient is then calculated with Equation (6-17):

$$\frac{hd}{k_f} = (0.278)(32,020)^{0.62}(0.706)^{1/3} = 153.8 \qquad [c]$$

$$h = \frac{(153.8)(0.027)}{0.0254} = 164 \text{ W/m}^2 \cdot {}^\circ\text{C} \quad [28.8 \text{ Btu/h} \cdot \text{ft}^2 \cdot {}^\circ\text{F}] \qquad [d]$$

This is the heat-transfer coefficient that would be obtained if there were 10 rows of tubes in the direction of the flow. Because there are only 5 rows, this value must be multiplied by the factor 0.92, as determined from Table 6-5.

The total surface area for heat transfer, considering unit length of tubes, is

$$A = N\pi d(1) = (15)(5)\pi(0.0254) = 5.985 \text{ m}^2/\text{m}$$

where N is the total number of tubes.

Before calculating the heat transfer, we must recognize that the air temperature increases as the air flows through the tube bank. Therefore, this must be taken into account when using

$$q = hA(T_w - T_\infty) \qquad [e]$$

As a good approximation, we can use an arithmetic average value of T_∞ and write for the energy balance*

$$q = hA\left(T_w - \frac{T_{\infty,1} + T_{\infty,2}}{2}\right) = \dot{m}c_p(T_{\infty,2} - T_{\infty,1}) \qquad [f]$$

where now the subscripts 1 and 2 designate entrance and exit to the tube bank. The mass flow at entrance to the 15 tubes is

$$\dot{m} = \rho_\infty u_\infty(15)S_n$$

$$\rho_\infty = \frac{p}{RT_\infty} = \frac{1.0132 \times 10^5}{(287)(283)} = 1.246 \text{ kg/m}^3 \qquad [g]$$

$$\dot{m} = (1.246)(7)(15)(0.0381) = 4.99 \text{ kg/s} \quad [11.0 \text{ lb}_m/\text{s}]$$

so that Equation (f) becomes

$$(0.92)(164)(5.985)\left(65 - \frac{10 + T_{\infty,2}}{2}\right) = (4.99)(1006)(T_{\infty,2} - 10)$$

that may be solved to give

$$T_{\infty,2} = 19.08{}^\circ\text{C}$$

The heat transfer is then obtained from the right side of Equation (f):

$$q = (4.99)(1006)(19.08 - 10) = 45.6 \text{ kW/m}$$

This answer could be improved somewhat by recalculating the air properties based on a mean value of T_∞, but the improvement would be small and well within the accuracy of the empirical heat-transfer correlation of Equation (6-17).

---

*A better approach may be to base the heat transfer on a so-called log mean temperature difference (LMTD). This method is discussed in detail in Section 10-5 in connection with heat exchangers. In the present problem, the arithmetic temperature difference is satisfactory.

EXAMPLE 6-11	Alternate Calculation Method

Compare the heat-transfer coefficient calculated with Equation (6-34) with the value obtained in Example 6-10.

■ **Solution**

Properties for use in Equation (6-34) are evaluated at free-stream conditions of 10°C, so we have

$$\nu = 14.2 \times 10^{-6} \qquad \text{Pr} = 0.712 \qquad k = 0.0249 \qquad \text{Pr}_w = 0.70$$

The Reynolds number is

$$\text{Re}_{d,\max} = \frac{(21)(0.0254)}{14.2 \times 10^{-6}} = 37{,}563$$

so that the constants for Equation (6-34) are $C = 0.27$ and $n = 0.63$.

Inserting values, we obtain

$$\frac{\bar{h}d}{k} = (0.27)(37{,}563)^{0.63}(0.712/0.7)^{1/4} = 206.5$$

and

$$h = \frac{(206.5)(0.0249)}{0.0254} = 202.4 \text{ W/m}^2 \cdot {}^\circ\text{C}$$

Multiplying by a factor of 0.92 from Table 6-7 to correct for only 5 tube rows gives

$$h = (0.92)(202.4) = 186.3 \text{ W/m}^2 \cdot {}^\circ\text{C}$$

or a value about 13 percent higher than in Example 6-10. Both values are within the accuracies of the correlations.

6-5 | LIQUID-METAL HEAT TRANSFER

Considerable interest has been placed on liquid-metal heat transfer because of the high heat-transfer rates that may be achieved with these media. These high heat-transfer rates result from the high thermal conductivities of liquid metals as compared with other fluids; as a consequence, they are particularly applicable to situations where large energy quantities must be removed from a relatively small space, as in a nuclear reactor. In addition, the liquid metals remain in the liquid state at higher temperatures than conventional fluids like water and various organic coolants. This also makes more compact heat-exchanger design possible. Liquid metals are difficult to handle because of their corrosive nature and the violent action that may result when they come into contact with water or air; even so, their advantages in certain heat-transfer applications have overshadowed their shortcomings, and suitable techniques for handling them have been developed.

Let us first consider the simple flat plate with a liquid metal flowing across it. The Prandtl number for liquid metals is very low, of the order of 0.01, so that the thermal-boundary-layer thickness should be substantially larger than the hydrodynamic-boundary-layer thickness. The situation results from the high values of thermal conductivity for liquid metals and is depicted in Figure 6-15. Since the ratio of δ/δ_t is small, the velocity profile has a very blunt shape over most of the thermal boundary layer. As a first approximation, then, we might assume a slug-flow model for calculation of the heat transfer; that is we take

$$u = u_\infty \qquad\qquad \textbf{[6-35]}$$

Figure 6-15 | Boundary-layer regimes for analysis of liquid-metal heat transfer.

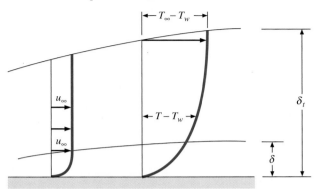

throughout the thermal boundary layer for purposes of computing the energy-transport term in the integral energy equation (Section 5-6):

$$\frac{d}{dx}\left[\int_0^{\delta_t}(T_\infty - T)u\,dy\right] = \alpha\frac{\partial T}{\partial y}\bigg]_w \qquad [6\text{-}36]$$

The conditions on the temperature profile are the same as those in Section 5-6, so that we use the cubic parabola as before:

$$\frac{\theta}{\theta_\infty} = \frac{T - T_w}{T_\infty - T_w} = \frac{3}{2}\frac{y}{\delta_t} - \frac{1}{2}\left(\frac{y}{\delta_t}\right)^3 \qquad [6\text{-}37]$$

Inserting Equations (6-35) and (6-37) in (6-36) gives

$$\theta_\infty u_\infty \frac{d}{dx}\left\{\int_0^{\delta_t}\left[1 - \frac{3}{2}\frac{y}{\delta_t} + \frac{1}{2}\left(\frac{u}{\delta_t}\right)^3\right]dy\right\} = \frac{3\alpha\theta_\infty}{2\delta_t} \qquad [6\text{-}38]$$

that may be integrated to give

$$2\delta_t d\delta_t = \frac{8\alpha}{u_\infty}dx \qquad [6\text{-}39]$$

The solution to this differential equation is

$$\delta_t = \sqrt{\frac{8\alpha x}{u_\infty}} \qquad [6\text{-}40]$$

for a plate heated over its entire length.

The heat-transfer coefficient may be expressed by

$$h_x = \frac{-k(\partial T/\partial y)_w}{T_w - T_\infty} = \frac{3k}{2\delta_t} = \frac{3\sqrt{2}}{8}k\sqrt{\frac{u_\infty}{\alpha x}} \qquad [6\text{-}41]$$

This relationship may be put in dimensionless form as

$$\mathrm{Nu}_x = \frac{h_x x}{k} = 0.530(\mathrm{Re}_x\mathrm{Pr})^{1/2} = 0.530\mathrm{Pe}^{1/2} \qquad [6\text{-}42]$$

Using Equation (5-21) for the hydrodynamic-boundary-layer thickness,

$$\frac{\delta}{x} = \frac{4.64}{\mathrm{Re}_x^{1/2}}$$ [6-43]

we may compute the ratio δ/δ_t:

$$\frac{\delta}{\delta_t} = \frac{4.64}{\sqrt{8}}\sqrt{\mathrm{Pr}} = 1.64\sqrt{\mathrm{Pr}}$$ [6-44]

Using $\mathrm{Pr} \sim 0.01$, we obtain

$$\frac{\delta}{\delta_t} \sim 0.16$$

which is in reasonable agreement with our slug-flow model.

The flow model for liquid metals discussed above illustrates the general nature of liquid-metal heat transfer, and it is important to note that the heat transfer is dependent on the Peclet number. Empirical correlations are usually expressed in terms of this parameter, four of which we present below.

Extensive data on liquid metals are given in Reference 13, and the heat-transfer characteristics are summarized in Reference 23. Lubarsky and Kaufman [14] recommended the following relation for calculation of heat-transfer coefficients in fully developed turbulent flow of liquid metals in smooth tubes with uniform heat flux at the wall:

$$\mathrm{Nu}_d = \frac{hd}{k} = 0.625(\mathrm{Re}_d\mathrm{Pr})^{0.4}$$ [6-45]

All properties for use in Equation (6-45) are evaluated at the bulk temperature. Equation (6-45) is valid for $10^2 < \mathrm{Pe} < 10^4$ and for $L/d > 60$. Seban and Shimazaki [16] propose the following relation for calculation of heat transfer to liquid metals in tubes with constant wall temperature:

$$\mathrm{Nu}_d = 5.0 + 0.025(\mathrm{Re}_d\mathrm{Pr})^{0.8}$$ [6-46]

where all properties are evaluated at the bulk temperature. Equation (6-42) is valid for $\mathrm{Pe} > 10^2$ and $L/d > 60$.

More recent data by Skupinshi, Tortel, and Vautrey [26] with sodium-potassium mixtures indicate that the following relation may be preferable to that of Equation (6-45) for constant-heat-flux conditions:

$$\mathrm{Nu} = 4.82 + 0.0185\,\mathrm{Pe}^{0.827}$$ [6-47]

This relation is valid for $3.6 \times 10^3 < \mathrm{Re} < 9.05 \times 10^5$ and $10^2 < \mathrm{Pe} < 10^4$. Additional correlations are given by Sleicher and Rouse [48].

Witte [32] has measured the heat transfer from a sphere to liquid sodium during forced convection, with the data being correlated by

$$\mathrm{Nu} = 2 + 0.386(\mathrm{Re}\,\mathrm{Pr})^{0.5}$$ [6-48]

for the Reynolds number range $3.56 \times 10^4 < \mathrm{Re} < 1.525 \times 10^5$.

Kalish and Dwyer [41] have presented information on liquid-metal heat transfer in tube bundles.

In general, there are many open questions concerning liquid-metal heat transfer, and the reader is referred to References 13 and 23 for more information.

Heating of Liquid Bismuth in Tube	**EXAMPLE 6-12**

Liquid bismuth flows at a rate of 4.5 kg/s through a 5.0-cm-diameter stainless-steel tube. The bismuth enters at 415°C and is heated to 440°C as it passes through the tube. If a constant heat flux is maintained along the tube and the tube wall is at a temperature 20°C higher than the bismuth bulk temperature, calculate the length of tube required to effect the heat transfer.

■ **Solution**

Because a constant heat flux is maintained, we may use Equation (6-47) to calculate the heat-transfer coefficient. The properties of bismuth are evaluated at the average bulk temperature of $(415 + 440)/2 = 427.5°C$

$$\mu = 1.34 \times 10^{-3} \text{ kg/m} \cdot \text{s} \quad [3.242 \text{ lb}_m/\text{h} \cdot \text{ft}]$$
$$c_p = 0.149 \text{ kJ/kg} \cdot °\text{C} \quad [0.0356 \text{ Btu/lb}_m \cdot °\text{F}]$$
$$k = 15.6 \text{ W/m} \cdot °\text{C} \quad [9.014 \text{ Btu/h} \cdot \text{ft} \cdot °\text{F}]$$
$$\text{Pr} = 0.013$$

The total transfer is calculated from

$$q = \dot{m}c_p \Delta T_b = (4.5)(149)(440 - 415) = 16.76 \text{ kW} \quad [57,186 \text{ Btu/h}] \qquad [a]$$

We calculate the Reynolds and Peclet numbers as

$$\text{Re}_d = \frac{dG}{\mu} = \frac{(0.05)(4.5)}{[\pi(0.05)^2/4](1.34 \times 10^{-3})} = 85,520 \qquad [b]$$
$$\text{Pe} = \text{Re} \, \text{Pr} = (85,520)(0.013) = 1111$$

The heat-transfer coefficient is then calculated from Equation 6-47

$$\text{Nu}_d = 4.82 + (0.0185)(1111)^{0.827} = 10.93 \qquad [c]$$
$$h = \frac{(10.93)(15.6)}{0.05} = 3410 \text{ W/m}^2 \cdot °\text{C} \quad [600 \text{ Btu/h} \cdot \text{ft}^2 \cdot °\text{F}]$$

The total required surface area of the tube may now be computed from

$$q = hA(T_w - T_b) \qquad [d]$$

where we use the temperature difference of 20°C;

$$A = \frac{16,760}{(3410)(20)} = 0.246 \text{ m}^2 \quad [2.65 \text{ ft}^2]$$

This area in turn can be expressed in terms of the tube length

$$A = \pi dL \quad \text{and} \quad L = \frac{0.246}{\pi(0.05)} = 1.57 \text{ m} \quad [5.15 \text{ ft}]$$

6-6 | SUMMARY

In contrast to Chapter 5, which was mainly analytical in character, this chapter has dealt almost entirely with empirical correlations that may be used to calculate convection heat transfer. The general calculation procedure is as follows:

1. Establish the geometry of the situation.
2. Make a preliminary determination of appropriate fluid properties.
3. Establish the flow regime by calculating the Reynolds or Peclet number.

Table 6-8 | Summary of forced-convection relations. (See text for property evaluation.)

Subscripts: b = bulk temperature, f = film temperature, ∞ = free stream temperature, w = wall temperature

Geometry	Equation	Restrictions	Equation number
Tube flow	$Nu_d = 0.023\, Re_d^{0.8} Pr^n$	Fully developed turbulent flow, $n = 0.4$ for heating, $n = 0.3$ for cooling, $0.6 < Pr < 100$, $2500 < Re_d < 1.25 \times 10^5$	(6-4a)
Tube flow	$Nu_d = 0.0214(Re_d^{0.8} - 100)Pr^{0.4}$	$0.5 < Pr < 1.5$, $10^4 < Re_d < 5 \times 10^6$	(6-4b)
	$Nu_d = 0.012(Re_d^{0.87} - 280)Pr^{0.4}$	$1.5 < Pr < 500$, $3000 < Re_d < 10^6$	(6-4c)
Tube flow	$Nu_d = 0.027\, Re_d^{0.8} Pr^{1/3} \left(\dfrac{\mu}{\mu_w}\right)^{0.14}$	Fully developed turbulent flow	(6-5)
Tube flow, entrance region	$Nu_d = 0.036\, Re_d^{0.8} Pr^{1/3} \left(\dfrac{d}{L}\right)^{0.055}$ See also Figures 6-5 and 6-6	Turbulent flow $10 < \dfrac{L}{d} < 400$	(6-6)
Tube flow	Petukov relation	Fully developed turbulent flow, $0.5 < Pr < 2000$, $10^4 < Re_d < 5 \times 10^6$, $0 < \dfrac{\mu_b}{\mu_w} < 40$	(6-7)
Tube flow	$Nu_d = 3.66 + \dfrac{0.0668(d/L)\, Re_d Pr}{1 + 0.04[(d/L)\, Re_d Pr]^{2/3}}$	Laminar, T_w = const.	(6-9)
Tube flow	$Nu_d = 1.86(Re_d Pr)^{1/3} \left(\dfrac{d}{L}\right)^{1/3} \left(\dfrac{\mu}{\mu_w}\right)^{0.14}$	Fully developed laminar flow, T_w = const. $Re_d Pr \dfrac{d}{L} > 10$	(6-10)
Rough tubes	$St_b Pr_f^{2/3} = \dfrac{f}{8}$ or Equation (6-7)	Fully developed turbulent flow	(6-12)
Noncircular ducts	Reynolds number evaluated on basis of hydraulic diameter $D_H = \dfrac{4A}{P}$ A = flow cross-section area, P = wetted perimeter	Same as particular equation for tube flow	(6-14)
Flow across cylinders	$Nu_f = C\, Re_{df}^n\, Pr^{1/3}$ C and n from Table 6-2	$0.4 < Re_{df} < 400{,}000$	(6-17)
Flow across cylinders	$Nu_{df} =$ $0.3 + \dfrac{0.62\, Re_f^{1/2}\, Pr^{1/3}}{\left[1 + \left(\dfrac{0.4}{Pr}\right)^{2/3}\right]^{1/4}} \left[1 + \left(\dfrac{Re_f}{282{,}000}\right)^{5/8}\right]^{4/5}$	$10^2 < Re_f < 10^7$, $Pe > 0.2$	(6-21)
Flow across cylinders		See text	(6-18) to (6-20) (6-22) to (6-24)
Flow across noncircular cylinders	$Nu = C\, Re_{df}^n\, Pr^{1/3}$ See Table 6-3 for values of C and n.		(6-17)

Table 6-8 | *(Continued).*

Subscripts: b = bulk temperature, f = film temperature, ∞ = free stream temperature, w = wall temperature

Geometry	Equation	Restrictions	Equation number
Flow across spheres	$\mathrm{Nu}_{df} = 0.37\,\mathrm{Re}_{df}^{0.6}$	$\mathrm{Pr} \sim 0.7$ (gases), $17 < \mathrm{Re} < 70{,}000$	(6-25)
	$\mathrm{Nu}_d\,\mathrm{Pr}^{-0.3}(\mu_w/\mu)^{0.25} = 1.2 + 0.53\,\mathrm{Re}_d^{0.54}$	Water and oils $1 < \mathrm{Re} < 200{,}000$ Properties at T_∞	(6-29)
	$\mathrm{Nu}_d = 2 + \left(0.4\,\mathrm{Re}_d^{1/2} + 0.06\,\mathrm{Re}_d^{2/3}\right)\mathrm{Pr}^{0.4}(\mu_\infty/\mu_w)^{1/4}$	$0.7 < \mathrm{Pr} < 380,\ 3.5 < \mathrm{Re}_d < 80{,}000$, Properties at T_∞	(6-30)
Flow across tube banks	$\mathrm{Nu}_f = C\,\mathrm{Re}_{f,\max}^{n}\,\mathrm{Pr}_f^{1/3}$ C and n from Table 6-4	See text	(6-17)
Flow across tube banks	$\mathrm{Nu}_d = C\,\mathrm{Re}_{d,\max}^{n}\,\mathrm{Pr}^{0.36}\left(\dfrac{\mathrm{Pr}}{\mathrm{Pr}_w}\right)^{1/4}$	$0.7 < \mathrm{Pr} < 500,\ 10 < \mathrm{Re}_{d,\max} < 10^6$	(6-34)
Liquid metals		See text	(6-37) to (6-48)
Friction factor	$\Delta p = f(L/d)\rho u_m^2/2g_c$, $u_m = \dot{m}/\rho A_c$		(6-13)

4. Select an equation that fits the geometry and flow regime and reevaluate properties, if necessary, in accordance with stipulations and the equation.

5. Proceed to calculate the value of h and/or the heat-transfer rate.

We should note that the data upon which the empirical equations are based are most often taken under laboratory conditions where it is possible to exert careful control over temperature and flow variables. In a practical application such careful control may not be present and there may be deviations from heat-transfer rates calculated from the equations given here. Our purpose is not to discourage the reader by this remark, but rather to indicate that sometimes it will be quite satisfactory to use a simple correlation over a more elaborate expression even if the simple relation has a larger scatter in its data representation. Our purpose has been to present a variety of expressions (where available) so that some choices can be made.

Finally, the most important relations of this chapter are listed in Table 6-8 for quick reference purposes.

Our presentation of convection is not yet complete. Chapter 7 will discuss the relations that are used for calculation of free convection heat transfer as well as combined free and forced convection. At the conclusion of that chapter we will present a general procedure to follow in *all* convection problems that will extend the outline given in the five steps above. This procedure will make use of the correlation summary Tables 5-2 and 6-8 along with a counterpart presented in Table 7-5 for free convection systems.

REVIEW QUESTIONS

1. What is the Dittus-Boelter equation? When does it apply?

2. How may heat-transfer coefficients be calculated for flow in rough pipes?

3. What is the hydraulic diameter? When is it used?

4. What is the form of equation used to calculate heat transfer for flow over cylinders and bluff bodies?

5. Why does a slug-flow model yield reasonable results when applied to liquid-metal heat transfer?
6. What is the Peclet number?
7. What is the Graetz number?

LIST OF WORKED EXAMPLES

PROBLEMS

6-1 Engine oil enters a 5.0-mm-diameter tube at 120°C. The tube wall is maintained at 50°C, and the inlet Reynolds number is 1000. Calculate the heat transfer, average heat-transfer coefficient, and exit oil temperature for tube lengths of 10, 20, and 50 cm.

6-2 Water at an average bulk temperature of 80°F flows inside a horizontal smooth tube with wall temperature maintained at 180°F. The tube length is 2 m, and diameter is 3 mm. The flow velocity is 0.04 m/s. Calculate the heat-transfer rate.

6-3 Calculate the flow rate necessary to produce a Reynolds number of 15,000 for the flow of air at 1 atm and 300 K in a 2.5-cm-diameter tube. Repeat for liquid water at 300 K.

6-4 Liquid ammonia flows in a duct having a cross section of an equilateral triangle 1.0 cm on a side. The average bulk temperature is 20°C, and the duct wall temperature is 50°C. Fully developed laminar flow is experienced with a Reynolds number of 1000. Calculate the heat transfer per unit length of duct.

6-5 Water flows in a duct having a cross section 5 × 10 mm with a mean bulk temperature of 20°C. If the duct wall temperature is constant at 60°C and fully developed laminar flow is experienced, calculate the heat transfer per unit length.

6-6 Water at the rate of 3 kg/s is heated from 5 to 15°C by passing it through a 5-cm-ID copper tube. The tube wall temperature is maintained at 90°C. What is the length of the tube?

6-7 Water at the rate of 0.8 kg/s is heated from 35 to 40°C in a 2.5-cm-diameter tube whose surface is at 90°C. How long must the tube be to accomplish this heating?

6-8 Water flows through a 2.5-cm-ID pipe 1.5 m long at a rate of 1.0 kg/s. The pressure drop is 7 kPa through the 1.5-m length. The pipe wall temperature is maintained at a constant temperature of 50°C by a condensing vapor, and the inlet water temperature is 20°C. Estimate the exit water temperature.

6-9 Water at the rate of 1.3 kg/s is to be heated from 60°F to 100°F in a 2.5-cm-diameter tube. The tube wall is maintained at a constant temperature of 40°C. Calculate the length of tube required for the heating process.

6-10 Water at the rate of 1 kg/s is forced through a tube with a 2.5-cm ID. The inlet water temperature is 15°C, and the outlet water temperature is 50°C. The tube wall temperature is 14°C higher than the water temperature all along the length of the tube. What is the length of the tube?

6-11 Engine oil enters a 1.25-cm-diameter tube 3 m long at a temperature of 38°C. The tube wall temperature is maintained at 65°C, and the flow velocity is 30 cm/s. Estimate the total heat transfer to the oil and the exit temperature of the oil.

6-12 Air at 1 atm and 15°C flows through a long rectangular duct 7.5 cm by 15 cm. A 1.8-m section of the duct is maintained at 120°C, and the average air temperature at exit from this section is 65°C. Calculate the airflow rate and the total heat transfer.

6-13 Water at the rate of 0.5 kg/s is forced through a smooth 2.5-cm-ID tube 15 m long. The inlet water temperature is 10°C, and the tube wall temperature is 15°C higher than the water temperature all along the length of the tube. What is the exit water temperature?

6-14 Water at an average temperature of 300 K flows at 0.7 kg/s in a 2.5-cm-diameter tube 6 m long. The pressure drop is measured as 2 kPa. A constant heat flux is imposed, and the average wall temperature is 55°C. Estimate the exit temperature of the water.

6-15 An oil with $Pr = 1960$, $\rho = 860 \text{ kg/m}^3$, $\nu = 1.6 \times 10^{-4} \text{ m}^2/\text{s}$, and $k = 0.14 \text{ W/m} \cdot \text{°C}$ enters a 2.5-mm-diameter tube 60 cm long. The oil entrance temperature is 20°C, the mean flow velocity is 30 cm/s, and the tube wall temperature is 120°C. Calculate the heat-transfer rate.

6-16 Liquid ammonia flows through a 2.5-cm-diameter smooth tube 2.5 m long at a rate of 0.4 kg/s. The ammonia enters at 10°C and leaves at 38°C, and a constant heat flux is imposed on the tube wall. Calculate the average wall temperature necessary to effect the indicated heat transfer.

6-17 Liquid Freon 12 (CCl_2F_2) flows inside a 1.25-cm-diameter tube at a velocity of 3 m/s. Calculate the heat-transfer coefficient for a bulk temperature of 10°C. How does this compare with water at the same conditions?

6-18 Water at an average temperature of 10°C flows in a 2.5-cm-diameter tube 6 m long at a rate of 0.4 kg/s. The pressure drop is measured as 3 kPa. A constant heat flux is imposed, and the average wall temperature is 50°C. Estimate the exit temperature of the water.

6-19 Water at the rate of 0.4 kg/s is to be cooled from 71 to 32°C. Which would result in less pressure drop—to run the water through a 12.5-mm-diameter pipe at a constant temperature of 4°C or through a constant-temperature 25-mm-diameter pipe at 20°C?

6-20 Air at 1400 kPa enters a duct 7.5 cm in diameter and 6 m long at a rate of 0.5 kg/s. The duct wall is maintained at an average temperature of 500 K. The average air temperature in the duct is 550 K. Estimate the decrease in temperature of the air as it passes through the duct.

6-21 Air flows at 100°C and 300 kPa in a 1.2-cm-(inside)-diameter tube at a velocity such that a Reynolds number of 15,000 is obtained. The outside of the tube is subjected to a crossflow of air at 100 kPa, 30°C, and a free-stream velocity of 20 m/s. The tube wall thickness is 1.0 mm. Calculate the overall heat transfer coefficient for this system. What would be the temperature drop of the air inside the tube per centimeter of length.

6-22 Liquid water is to be heated from 60°F to 120°F in a smooth tube. The tube has an electric heat supplied that provides a constant heat flux such that the tube wall temperature is always 30°F above the water bulk temperature. The Reynolds number used for calculating the heat-transfer coefficient is 100,000. Calculate the length of tube required for heating, expressed in meters, if the tube has a diameter of 0.5 cm.

6-23 An annulus consists of the region between two concentric tubes having diameters of 4 cm and 5 cm. Ethylene glycol flows in this space at a velocity of 6.9 m/s. The entrance temperature is 20°C, and the exit temperature is 40°C. Only the inner tube is a heating surface, and it is maintained constant at 80°C. Calculate the length of annulus necessary to effect the heat transfer.

6-24 An air-conditioning duct has a cross section of 45 cm by 90 cm. Air flows in the duct at a velocity of 7.5 m/s at conditions of 1 atm and 300 K. Calculate the heat-transfer coefficient for this system and the pressure drop per unit length.

6-25 Water flows in a 3.0-cm-diameter tube having a relative roughness of 0.002 with a constant wall temperature of 80°C. If the water enters at 20°C, estimate the convection coefficient for a Reynolds number of 10^5.

6-26 Liquid Freon 12 (CCl_2F_2) enters a 3.5-mm-diameter tube at 0°C and with a flow rate such that the Reynolds number is 700 at entrance conditions. Calculate the length of tube necessary to raise the fluid temperature to 20°C if the tube wall temperature is constant at 40°C.

6-27 Air enters a small duct having a cross section of an equilateral triangle, 3.0 mm on a side. The entering temperature is 27°C and the exit temperature is 77°C. If the flow rate is 5×10^{-5} kg/s and the tube length is 30 cm, calculate the tube wall temperature necessary to effect the heat transfer. Also calculate the pressure drop. The pressure is 1 atm.

6-28 Air at 90 kPa and 27°C enters a 4.0-mm-diameter tube with a mass flow rate of 7×10^{-5} kg/s. A constant heat flux is imposed at the tube surface so that the tube wall temperature is 70°C above the fluid bulk temperature. Calculate the exit air temperature for a tube length of 12 cm.

6-29 Air at 110 kPa and 40°C enters a 6.0-mm-diameter tube with a mass flow rate of 8×10^{-5} kg/s. The tube wall temperature is maintained constant at 140°C. Calculate the exit air temperature for a tube length of 14 cm.

6-30 Engine oil at 40°C enters a 1-cm-diameter tube at a flow rate such that the Reynolds number at entrance is 50. Calculate the exit oil temperature for a tube length of 8 cm and a constant tube wall temperature of 80°C.

6-31 Water flows in a 2-cm-diameter tube at an average flow velocity of 8 m/s. If the water enters at 20°C and leaves at 30°C and the tube length is 10 m, estimate the average wall temperature necessary to effect the required heat transfer.

6-32 Engine oil at 20°C enters a 2.0-mm-diameter tube at a velocity of 1.2 m/s. The tube wall temperature is constant at 60°C and the tube is 1.0 m long. Calculate the exit oil temperature.

6-33 Water flows inside a smooth tube at a mean flow velocity of 3.0 m/s. The tube diameter is 25 mm and a constant heat flux condition is maintained at the tube wall such that the tube temperature is always 20°C above the water temperature. The water enters the tube at 30°C and leaves at 50°C. Calculate the tube length necessary to accomplish the indicated heating.

6-34 Liquid ammonia at 10°C and 1 atm flows across a horizontal cylinder at a velocity of 5 m/s. The cylinder has a diameter of 2.5 cm and length of 125 cm and is maintained at a temperature of 30°C. Calculate the heat lost by the cylinder.

6-35 Water enters a 3-mm-diameter tube at 21°C and leaves at 32°C. The flow rate is such that the Reynolds number is 600. The tube length is 10 cm and is maintained at a constant temperature of 60°C. Calculate the water flow rate.

6-36 Water enters a 3.0-cm-diameter tube at 15°C and leaves at 38°C. The flow rate is 1.0 kg/s and the tube wall temperature is 60°C. Calculate the length of the tube.

6-37 Glycerin flows in a 5-mm-diameter tube at such a rate that the Reynolds number is 10. The glycerine enters at 10°C and leaves at 30°C. The tube wall is maintained constant at 40°C. Calculate the length of the tube.

6-38 A 5-cm-diameter cylinder maintained at 80°C is placed in a nitrogen flow stream at 2 atm pressure and 10°C. The nitrogen flows across the cylinder with a velocity of 5 m/s. Calculate the heat lost by the cylinder per meter of length.

6-39 Air at 1 atm and 10°C blows across a 4-cm-diameter cylinder maintained at a surface temperature of 54°C. The air velocity is 25 m/s. Calculate the heat loss from the cylinder per unit length.

6-40 Air at 200 kPa blows across a 20-cm-diameter cylinder at a velocity of 25 m/s and temperature of 10°C. The cylinder is maintained at a constant temperature of 80°C. Calculate the heat transfer and drag force per unit length.

6-41 Water at 43°C enters a 5-cm-ID pipe having a relative roughness of 0.002 at a rate of 6 kg/s. If the pipe is 9 m long and is maintained at 71°C, calculate the exit water temperature and the total heat transfer.

6-42 A short tube is 6.4 mm in diameter and 15 cm long. Water enters the tube at 1.5 m/s and 38°C, and a constant-heat-flux condition is maintained such that the tube wall temperature remains 28°C above the water bulk temperature. Calculate the heat-transfer rate and exit water temperature.

6-43 Ethylene glycol is to be cooled from 65 to 40°C in a 3.0-cm-diameter tube. The tube wall temperature is maintained constant at 20°C. The glycol enters the tube with a velocity of 10 m/s. Calculate the length of tube necessary to accomplish this cooling.

6-44 Air at 70 kPa and 20°C flows across a 5-cm-diameter cylinder at a velocity of 15 m/s. Compute the drag force exerted on the cylinder.

6-45 A heated cylinder at 450 K and 2.5 cm in diameter is placed in an atmospheric airstream at 1 atm and 325 K. The air velocity is 30 m/s. Calculate the heat loss per meter of length for the cylinder.

6-46 Assuming that a human can be approximated by a cylinder 30 cm in diameter and 1.1 m high with a surface temperature of 24°C, calculate the heat the person would lose while standing in a 30-mi/h wind whose temperature is 0°C.

6-47 Assume that one-half the heat transfer from a cylinder in cross flow occurs on the front half of the cylinder. On this assumption, compare the heat transfer from a cylinder in cross flow with the heat transfer from a flat plate having a length equal to the distance from the stagnation point on the cylinder. Discuss this comparison.

6-48 Water at 15.56°C is to be heated in a 2-mm-ID tube until the exit temperature reaches 26.67°C. The tube wall temperature is maintained at 48.99°C and the inlet flow velocity is 0.3 m/s. Calculate the length of tube required in meters to accomplish this heating. Also calculate the total heating required, expressed in watts.

6-49 An isothermal cylinder having a diameter of 2.0 cm and maintained at 50°C is placed in a helium flow system having free-stream conditions of 200 kPa, 20°C, and $u_\infty = 25$ m/s. Calculate the heat lost for a cylinder length of 50 cm.

6-50 A 0.13-mm-diameter wire is exposed to an airstream at -30°C and 54 kPa. The flow velocity is 230 m/s. The wire is electrically heated and is 12.5 mm long. Calculate the electric power necessary to maintain the wire surface temperature at 175°C.

6-51 Air at 90°C and 1 atm flows past a heated 1.5-mm-diameter wire at a velocity of 6 m/s. The wire is heated to a temperature of 150°C. Calculate the heat transfer per unit length of wire.

6-52 A fine wire 0.025 mm in diameter and 15 cm long is to be used to sense flow velocity by measuring the electrical heat that can be dissipated from the wire when placed in an airflow stream. The resistivity of the wire is 70 $\mu\Omega \cdot$ cm. The temperature of the wire is determined by measuring its electric resistance relative to some reference temperature T_0 so that

$$R = R_0[1 + a(T - T_0)]$$

For this particular wire the value of the temperature coefficient a is 0.006°C^{-1}. The resistance can be determined from measurements of the current and voltage impressed on the wire, and

$$R = \frac{E}{I}$$

Suppose a measurement is made for air at 20°C with a flow velocity of 10 m/s and the wire temperature is 40°C. What values of voltage and current would be measured for these conditions if R_0 is evaluated at $T_0 = 20$°C? What values of voltage and current would be measured for the same wire temperature but flow velocities of 15 m/s and 20 m/s?

6-53 Helium at 1 atm and 325 K flows across a 3-mm-diameter cylinder that is heated to 425 K. The flow velocity is 9 m/s. Calculate the heat transfer per unit length of wire. How does this compare with the heat transfer for air under the same conditions?

6-54 Calculate the heat-transfer rate per unit length for flow over a 0.025-mm-diameter cylinder maintained at 65°C. Perform the calculation for (a) air at 20°C and 1 atm and (b) water at 20°C; $u_\infty = 6$ m/s.

6-55 Compare the heat-transfer results of Equations (6-17) and (6-18) for water at Reynolds numbers of 10^3, 10^4, and 10^5 and a film temperature of 90°C.

6-56 A pipeline in the Arctic carries hot oil at 50°C. A strong arctic wind blows across the 50-cm-diameter pipe at a velocity of 13 m/s and a temperature of -35°C. Estimate the heat loss per meter of pipe length.

6-57 Two tubes are available, a 4.0-cm-diameter tube and a 4.0-cm-square tube. Air at 1 atm and 27°C is blown across the tubes with a velocity of 20 m/s. Calculate the heat transfer in each case if the tube wall temperature is maintained at 50°C.

6-58 A 3.0-cm-diameter cylinder is subjected to a cross flow of carbon dioxide at 200°C and a pressure of 1 atm. The cylinder is maintained at a constant temperature of 50°C and the carbon dioxide velocity is 40 m/s. Calculate the heat transfer to the cylinder per meter of length.

6-59 Water having an average bulk temperature of 100°F flows in a smooth tube with a diameter of 1.25 cm. The flow rate is such that a Reynolds number of 100,000 is experienced, and the tube wall is maintained at an average temperature of 160°F.

If the tube length is 1.5 m calculate the exit bulk temperature of the water. Express in °C.

6-60 Using a suitable computer software package, integrate the local heat transfer coefficient results of Figure 6-11 to obtain average values of h for each Reynolds number shown. Subsequently, compare the results with values calculated from the information in Table 6-2. If needed, consult Reference 7 for additional information.

6-61 Helium at 150 kPa and 20°C is forced at 50 m/s across a horizontal cylinder having a diameter of 30 cm and a length of 6 m. Calculate the heat lost by the cylinder if its surface temperature is maintained constant at 100°C.

6-62 A 0.25-inch-diameter cylinder is maintained at a constant temperature of 300°C and placed in a cross flow of CO_2 at $p = 100$ kPa and $T = 30$°C. Calculate the heat loss for a 4.5-m length of the cylinder if the CO_2 velocity is 35 m/s.

6-63 A 20-cm-diameter cylinder is placed in a cross-flow CO_2 stream at 1 atm and 300 K. The cylinder is maintained at a constant temperature of 400 K and the CO_2 velocity is 50 m/s. Calculate the heat lost by the cylinder per meter of length.

6-64 Air flows across a 4-cm-square cylinder at a velocity of 12 m/s. The surface temperature is maintained at 85°C. Free-stream air conditions are 20°C and 0.6 atm. Calculate the heat loss from the cylinder per meter of length.

6-65 Water flows over a 3-mm-diameter sphere at 5 m/s. The free-stream temperature is 38°C, and the sphere is maintained at 93°C. Calculate the heat-transfer rate.

6-66 A spherical water droplet having a diameter of 1.3 mm is allowed to fall from rest in atmospheric air at 1 atm and 20°C. Estimate the velocities the droplet will attain after a drop of 30, 60, and 300 m.

6-67 A spherical tank having a diameter of 4.0 m is maintained at a surface temperature of 40°C. Air at 1 atm and 20°C blows across the tank at 6 m/s. Calculate the heat loss.

6-68 A heated sphere having a diameter of 3 cm is maintained at a constant temperature of 90°C and placed in a water flow stream at 20°C. The water flow velocity is 3.5 m/s. Calculate the heat lost by the sphere.

6-69 A small sphere having a diameter of 6 mm has an electric heating coil inside, which maintains the outer surface temperature at 220°C. The sphere is exposed to an airstream at 1 atm and 20°C with a velocity of 20 m/s. Calculate the heating rate which must be supplied to the sphere.

6-70 Air at a pressure of 3 atm blows over a flat plate at a velocity of 75 m/s. The plate is maintained at 200°C and the free-stream temperature is 30°C. Calculate the heat loss for a plate which is 1 m square.

6-71 Air at 3.5 MPa and 38°C flows across a tube bank consisting of 400 tubes of 1.25-cm OD arranged in a staggered manner 20 rows high; $S_p = 3.75$ cm and $S_n = 2.5$ cm. The incoming-flow velocity is 9 m/s, and the tube-wall temperatures are maintained constant at 20°C by a condensing vapor on the inside of the tubes. The length of the tubes is 1.5 m. Estimate the exit air temperature as it leaves the tube bank.

6-72 A tube bank uses an in-line arrangement with $S_n = S_p = 1.9$ cm and 6.33-mm-diameter tubes. Six rows of tubes are employed with a stack 50 tubes high. The surface temperature of the tubes is constant at 90°C, and atmospheric air at 20°C is forced across them at an inlet velocity of 4.5 m/s before the flow enters the tube bank. Calculate the total heat transfer per unit length for the tube bank. Estimate the pressure drop for this arrangement.

6-73 Air at 1 atm and 300 K flows across an in-line tube bank having 10 vertical and 10 horizontal rows. The tube diameter is 2 cm and the center-to-center spacing is 4 cm in both the normal and parallel directions. Calculate the convection heat-transfer coefficient for this situation if the entering free-stream velocity is 10 m/s and properties may be evaluated at free-stream conditions.

6-74 Repeat Problem 6-73 for a staggered-tube arrangement with the same values of S_p and S_n.

6-75 Condensing steam at 150°C is used on the inside of a bank of tubes to heat a cross-flow stream of CO_2 that enters at 3 atm, 35°C, and 5 m/s. The tube bank consists of 100 tubes of 1.25-cm OD in a square in-line array with $S_n = S_p = 1.875$ cm. The tubes are 60 cm long. Assuming the outside-tube-wall temperature is constant at 150°C, calculate the overall heat transfer to the CO_2 and its exit temperature.

6-76 An in-line tube bank is constructed of 2.5-cm-diameter tubes with 15 rows high and 7 rows deep. The tubes are maintained at 90°C, and atmospheric air is blown across them at 20°C and $u_\infty = 12$ m/s. The arrangement has $S_p = 3.75$ and $S_n = 5.0$ cm. Calculate the heat transfer from the tube bank per meter of length. Also calculate the pressure drop.

6-77 Air at 300 K and 1 atm enters an in-line tube bank consisting of five rows of 10 tubes each. The tube diameter is 2.5 cm and $S_n = S_p = 5.0$ cm. The incoming velocity is 10 m/s and the tube wall temperatures are constant at 350 K. Calculate the exit air temperature.

6-78 Atmospheric air at 20°C flows across a 5-cm-square rod at a velocity of 15 m/s. The velocity is normal to one of the faces of the rod. Calculate the heat transfer per unit length for a surface temperature of 90°C.

6-79 A certain home electric heater uses thin metal strips to dissipate heat. The strips are 6 mm wide and are oriented normal to the airstream, which is produced by a small fan. The air velocity is 2 m/s, and seven 35-cm strips are employed. If the strips are heated to 870°C, estimate the total convection heat transfer to the room air at 20°C. (Note that in such a heater, much of the *total* transfer will be by thermal radiation.)

6-80 A square duct, 30 cm by 30 cm, is maintained at a constant temperature of 30°C and an airstream of 50°C and 1 atm is forced across it with a velocity of 6 m/s. Calculate the heat gained by the duct. How much would the heat flow be reduced if the flow velocity were reduced in half?

6-81 Using the slug-flow model, show that the boundary-layer energy equation reduces to the same form as the transient-conduction equation for the semi-infinite solid of Section 4-3. Solve this equation and compare the solution with the integral analysis of Section 6-5.

6-82 Liquid bismuth enters a 2.5-cm-diameter stainless-steel pipe at 400°C at a rate of 1 kg/s. The tube wall temperature is maintained constant at 450°C. Calculate the bismuth exit temperature if the tube is 60 cm long.

6-83 Liquid sodium is to be heated from 120 to 149°C at a rate of 2.3 kg/s. A 2.5-cm-diameter electrically heated tube is available (constant heat flux). If the tube wall temperature is not to exceed 200°C, calculate the minimum length required.

6-84 Determine an expression for the average Nusselt number for liquid metals flowing over a flat plate. Use Equation (6-42) as a starting point.

6-85 Water at the rate of 0.8 kg/s at 93°C is forced through a 5-cm-ID copper tube at a suitable velocity. The wall thickness is 0.8 mm. Air at 15°C and atmospheric pressure is forced over the outside of the tube at a velocity of 15 m/s in a direction normal to the axis of the tube. What is the heat loss per meter of length of the tube?

6-86 Air at 1 atm and 350 K enters a 1.25-cm-diameter tube with a flow rate of 35 g/s. The surface temperature of the tube is 300 K, and its length is 12 m. Calculate the heat lost by the air and the exit air temperature.

6-87 Air flows across a 5.0-cm-diameter smooth tube with free-stream conditions of 20°C, 1 atm, and $u_\infty = 25$ m/s. If the tube surface temperature is 120°C, calculate the heat loss per unit length.

6-88 Engine oil enters an 8-m-long tube at 20°C. The tube diameter is 20 mm, and the flow rate is 0.4 kg/s. Calculate the outlet temperature of the oil if the tube surface temperature is maintained at 80°C.

6-89 Air at 1 atm and 300 K with a flow rate of 0.2 kg/s enters a rectangular 10-by-20-cm duct that is 250 cm long. If the duct surface temperature is maintained constant at 400 K, calculate the heat transfer to the air and the exit air temperature.

6-90 Air at 1 atm and 300 K flows inside a 1.5-mm-diameter smooth tube such that the Reynolds number is 1200. Calculate the heat-transfer coefficients for tube lengths of 1, 10, 20, and 100 cm.

6-91 Water at an average bulk temperature of 10°C flows inside a channel shaped like an equilateral triangle 2.5 cm on a side. The flow rate is such that a Reynolds number of 50,000 is obtained. If the tube-wall temperature is maintained 15°C higher than the water bulk temperature, calculate the length of tube needed to effect a 10°C increase in bulk temperature. What is the total heat transfer under this condition?

6-92 Air at 1 atm and 300 K flows normal to a square noncircular cylinder such that the Reynolds number is 10^4. Compare the heat transfer for this system with that for a circular cylinder having diameter equal to a side of the square. Repeat the calculation for the first, third, and fourth entries of Table 6-3.

6-93 Air at 1 atm and 300 K flows across a sphere such that the Reynolds number is 50,000. Compare Equations (6-25) and (6-26) for these conditions. Also compare with Equation (6-30).

6-94 Water at 10°C flows across a 2.5-cm-diameter sphere at a free-stream velocity of 4 m/s. If the surface temperature of the sphere is 60°C, calculate the heat loss.

6-95 A tube bank consists of a square array of 144 tubes arranged in an in-line position. The tubes have a diameter of 1.5 cm and length of 1.0 m; the center-to-center tube spacing is 2.0 cm. If the surface temperature of the tubes is maintained at 350 K and air enters the tube bank at 1 atm, 300 K, and $u_\infty = 6$ m/s, calculate the total heat lost by the tubes.

6-96 Though it may be classified as a rather simple mistake, a frequent cause for substantial error in convection calculations is failure to select the correct geometry for the problem. Consider the following three geometries for flow of air at 1 atm, 300 K, and a Reynolds number of 50,000: (a) flow across a cylinder with diameter of 10 cm, (b) flow inside a tube with diameter of 10 cm, and (c) flow along a flat plate of length 10 cm. Calculate the average heat-transfer coefficient for each of these geometries and comment on the results.

6-97 Water flows at an average flow velocity of 10 ft/s in a smooth tube at an average temperature of 60°F. The tube diameter is 2.5 cm. Calculate the length of tube required to cause the bulk temperature of the water to rise 10°C if the tube wall temperature is maintained at 150°F.

6-98 It has been noted that convection heat transfer is dependent on fluid properties, which in turn are dependent on temperature. Consider flow of atmospheric air at 0.012 kg/s in a smooth 2.5-cm-diameter tube. Assuming that the Dittus-Boelter relation [Equation (6-4a)] applies, calculate the average heat-transfer coefficient for properties evaluated at 300, 400, 500, and 800 K. Comment on the results.

6-99 Repeat Problem 6-98 for the same mass flow of atmospheric helium with properties evaluated at 255, 477, and 700 K and comment on the results.

6-100 Air at 300 K flows in a 5-mm-diameter tube at a flow rate such that the Reynolds number is 50,000. The tube length is 50 mm. Estimate the average heat-transfer coefficient for a constant heat flux at the wall.

6-101 Water at 15.6°C flows in a 5-mm-diameter tube having a length of 50 mm. The flow rate is such that the Peclet number is 1000. If the tube wall temperature is constant at 49°C, what temperature increase will be experienced by the water?

6-102 Air at 1 atm flows in a rectangular duct having dimensions of 30 cm by 60 cm. The mean flow velocity is 7.5 m/s at a mean bulk temperature of 300 K. If the duct wall temperature is constant at 325 K, estimate the air temperature increase over a duct length of 30 m.

6-103 Glycerin at 10°C flows in a rectangular duct 1 cm by 8 cm and 1 m long. The flow rate is such that the Reynolds number is 250. Estimate the average heat transfer coefficient for an isothermal wall condition.

6-104 Air at 300 K blows normal to a 6-mm heated strip maintained at 600 K. The air velocity is such that the Reynolds number is 15,000. Calculate the heat loss for a 50-cm-long strip.

6-105 Repeat Problem 6-104 for flow normal to a square rod 6 mm on a side.

6-106 Repeat Problem 6-104 for flow parallel to a 6-mm strip. (Calculate heat transfer for both sides of the strip.)

6-107 Air at 1 atm flows normal to a square in-line bank of 400 tubes having diameters of 6 mm and lengths of 50 cm. $S_n = S_d = 9$ mm. The air enters the tube bank at 300 K and at a velocity such that the Reynolds number based on inlet properties and the maximum velocity at inlet is 50,000. If the outside wall temperature of the tubes is 400 K, calculate the air temperature rise as it flows through the tube bank.

6-108 Repeat Problem 6-107 for a tube bank with a staggered arrangement, the same dimensions, and the same free-stream inlet velocity to the tube bank.

6-109 Compare the Nusselt number results for heating air in a smooth tube at 300 K and Reynolds numbers of 50,000 and 100,000, as calculated from Equation (6-4a), (6-4b), and (6-4c). What do you conclude from these results?

6-110 Repeat Problem 6-109 for heating water at 21°C.

6-111 Compare the results obtained from Equations (6-17), (6-21), (6-22), and (6-23) for air at 1 atm and 300 K flowing across a cylinder maintained at 400 K, with Reynolds numbers of 50,000 and 100,000. What do you conclude from these results?

6-112 Repeat Problem 6-111 for flow or water at 21°C across a cylinder maintained at 32.2°C. What do you conclude from the results?

Design-Oriented Problems

6-113 Using the values of the local Nusselt number given in Figure 6-11, obtain values for the average Nusselt number as a function of the Reynolds number. Plot the results as log Nu versus log Re, and obtain an equation that represents all the data. Compare this correlation with that given by Equation (6-17) and Table 6-2.

6-114 A heat exchanger is constructed so that hot flue gases at 700 K flow inside a 2.5-cm-ID copper tube with 1.6-mm wall thickness. A 5.0-cm tube is placed around the 2.5-cm-diameter tube, and high-pressure water at 150°C flows in the annular space between the tubes. If the flow rate of water is 1.5 kg/s and the total heat transfer is 17.5 kW, estimate the length of the heat exchanger for a gas mass flow of 0.8 kg/s. Assume that the properties of the flue gas are the same as those of air at atmospheric pressure and 700 K.

6-115 Compare Equations (6-19), (6-20), and (6-21) with Equation (6-17) for a gas with $Pr = 0.7$ at the following Reynolds numbers: (a) 500, (b) 1000, (c) 2000, (d) 10,000, (e) 100,000.

6-116 A more compact version of the tube bank in Problem 6-72 can be achieved by reducing the S_p and S_n dimensions while still retaining the same number of tubes. Investigate the effect of reducing S_p and S_n in half, that is, $S_p = S_n = 0.95$ cm. Calculate the heat transfer and pressure drop for this new arrangement.

6-117 The drag coefficient for a sphere at Reynolds numbers less than 100 may be approximated by $C_D = b\mathrm{Re}^{-1}$, where b is a constant. Assuming that the Colburn analogy between heat transfer and fluid friction applies, derive an expression for the heat loss from a sphere of diameter d and temperature T_s, released from rest and allowed to fall in a fluid of temperature T_∞. (Obtain an expression for the heat lost between the time the sphere is released and the time it reaches some velocity v. Assume that the Reynolds number is less than 100 during this time and that the sphere remains at a constant temperature.)

6-118 Consider the application of the Dittus-Boelter relation [Equation (6-4a)] to turbulent flow of air in a smooth tube under developed turbulent flow conditions. For a fixed mass flow rate and tube diameter (selected at your discretion) investigate the effect of bulk temperature on the heat-transfer coefficient by calculating values of h for average bulk temperatures of 20, 50, 100, 200, and 300°C. What do you conclude from this calculation? From the results, estimate the dependence of the heat-transfer coefficient for air on absolute temperature.

6-119 A *convection* electric oven is one that employs a fan to force air across the food in addition to radiant heat from electric heating elements. Consider two oven temperature settings at 175°C and 230°C. Make assumptions regarding airflow velocities in order to estimate oven heating performance with and without convection under these two temperature conditions. Make your own assumptions as to the type of food to be cooked. Enthusiasts claim that the convection oven will cook in half the time of the all-radiant model. How do you evaluate this claim? What would you recommend as a prudent claim for the manufacturer of the oven to make? As a concrete example consider cooking a 25-pound turkey at Thanksgiving. Consult whatever sources (cookbooks) you think appropriate to check your calculations. Make recommendations that you feel would be acceptable to a typical home chef who is fussy about such matters.

6-120 A smooth glass plate is coated with a special electrically conductive film that may be used to produce a constant heat flux on the plate. Estimate the airflow velocity that must be used to remove 850 W from a 0.5-m-square plate, maintained at an average temperature of 65°C and dissipating heat to air at 1 atm and 20°C. Suppose the plate also radiates like a black surface to the surroundings at 20°C. What flow velocity would be necessary to dissipate the 850 W under these conditions?

REFERENCES

1. Dittus, F. W., and L. M. K. Boelter. *Univ. Calif. (Berkeley) Pub. Eng.,* vol. 2, p. 443, 1930.

2. Sieder, E. N., and C. E. Tate. "Heat Transfer and Pressure Drop of Liquids in Tubes," *Ind. Eng. Chem.,* vol. 28, p. 1429, 1936.

3. Nusselt, W. "Der Wärmeaustausch zwischen Wand und Wasser im Rohr," *Forsch. Geb. Ingenieurwes.,* vol. 2, p. 309, 1931.

4. Hausen, H. "Darstellung des Wärmeuberganges in Rohren durch verallgemeinerte Potenzbeziehungen," *VDIZ.,* no. 4, p. 91, 1943.

5. Moody, F. F. "Friction Factors for Pipe Flow," *Trans. ASME,* vol. 66, p. 671, 1944.

6. Schlichting, H. *Boundary Layer Theory,* 7th ed. New York: McGraw-Hill, 1979.

7. Giedt, W. H. "Investigation of Variation of Point Unit-Heat-Transfer Coefficient around a Cylinder Normal to an Air Stream," *Trans. ASME,* vol. 71, pp. 375–81, 1949.

8. Hilpert, R. "Wärmeabgabe von geheizen Drahten und Rohren," *Forsch. Geb. Ingenieurwes.,* vol. 4, p. 220, 1933.

9. Knudsen, J. D., and D. L. Katz. *Fluid Dynamics and Heat Transfer.* New York: McGraw-Hill, 1958.

10. McAdams, W. H. *Heat Transmission,* 3d ed. New York: McGraw-Hill, 1954.

11. Kramers, H. "Heat Transfer from Spheres to Flowing Media," *Physica,* vol. 12, p. 61, 1946.

12. Grimson, E. D. "Correlation and Utilization of New Data on Flow Resistance and Heat Transfer for Cross Flow of Gases over Tube Banks," *Trans. ASME,* vol. 59, pp. 583–94, 1937.

13. Lyon, R. D. (Ed.). *Liquid Metals Handbook,* 3d ed. Washington, D.C.: Atomic Energy Commission and U.S. Navy Department, 1952.

14. Lubarsky, B., and S. J. Kaufman. "Review of Experimental Investigations of Liquid-Metal Heat Transfer," *NACA Tech. Note* 3336, 1955.

15. Colburn, A. P. "A Method of Correlating Forced Convection Heat Transfer Data and a Comparison with Fluid Friction," *Trans. AIChE,* vol. 29, p. 174, 1933.

16. Seban, R. A., and T. T. Shimazaki. "Heat Transfer to a Fluid Flowing Turbulently in a Smooth Pipe with Walls at Constant Temperature," *Trans. ASME,* vol. 73, p. 803, 1951.

17. Kays, W. M., and R. K. Lo. "Basic Heat Transfer and Flow Friction Data for Gas Flow Normal to Banks of Staggered Tubes: Use of a Transient Technique," *Stanford Univ. Tech. Rep.* 15, Navy Contract N6-ONR251 T.O. 6, 1952.

18. Jakob, M. "Heat Transfer and Flow Resistance in Cross Flow of Gases over Tube Banks," *Trans. ASME,* vol. 60, p. 384, 1938.

19. Vliet, G. C., and G. Leppert. "Forced Convection Heat Transfer from an Isothermal Sphere to Water," *J. Heat Transfer,* serv. C, vol. 83, p. 163, 1961.

20. Irvine, T. R. "Noncircular Duct Convective Heat Transfer," in W. Ibele (ed.), *Modern Developments in Heat Transfer.* New York: Academic Press, 1963.

21. Fand, R. M. "Heat Transfer by Forced Convection from a Cylinder to Water in Crossflow," *Int. J. Heat Mass Transfer,* vol. 8, p. 995, 1965.

22. Jakob, M. *Heat Transfer,* vol. 1. New York: John Wiley, 1949.

23. Stein, R. "Liquid Metal Heat Transfer," *Adv. Heat Transfer,* vol. 3, 1966.

24. Hartnett, J. P. "Experimental Determination of the Thermal Entrance Length for the Flow of Water and of Oil in Circular Pipes," *Trans. ASME,* vol. 77, p. 1211, 1955.

25. Allen, R. W., and E. R. G. Eckert. "Friction and Heat Transfer Measurements to Turbulent Pipe Flow of Water (Pr = 7 and 8) at Uniform Wall Heat Flux," *J. Heat Transfer,* ser. C, vol. 86, p. 301, 1964.

26. Skupinshi, E., J. Tortel, and L. Vautrey. "Détermination des coéfficients de convection a'un alliage sodium-potassium dans un tube circulaire," *Int. J. Heat Mass Transfer,* vol. 8, p. 937, 1965.

27. Dipprey, D. F., and R. H. Sabersky. "Heat and Momentum Transfer in Smooth and Rough Tubes at Various Prandtl Numbers," *Int. J. Heat Mass Transfer,* vol. 6, p. 329, 1963.

28. Kline, S. J. *Similitude and Approximation Theory.* New York: McGraw-Hill, 1965.

29. Townes, H. W., and R. H. Sabersky. "Experiments on the Flow over a Rough Surface," *Int. J. Heat Mass Transfer,* vol. 9, p. 729, 1966.

30. Gowen, R. A., and J. W. Smith. "Turbulent Heat Transfer from Smooth and Rough Surfaces," *Int. J. Heat Mass Transfer,* vol. 11, p. 1657, 1968.

31. Sheriff, N., and P. Gumley. "Heat Transfer and Friction Properties of Surfaces with Discrete Roughness," *Int. J. Heat Mass Transfer,* vol. 9, p. 1297, 1966.

32. Witte, L. C. "An Experimental Study of Forced-Convection Heat Transfer from a Sphere to Liquid Sodium," *J. Heat Transfer,* vol. 90, p. 9, 1968.

33. Zukauskas, A. A., V. Makarevicius, and A. Schlanciauskas: *Heat Transfer in Banks of Tubes in Crossflow of Fluid.* Vilnius, Lithuania: Mintis, 1968.

34. Eckert, E. R. G., and R. M. Drake. *Analysis of Heat and Mass Transfer.* New York: McGraw-Hill, 1972.

35. Whitaker, S. "Forced Convection Heat-Transfer Correlations for Flow in Pipes, Past Flat Plates, Single Cylinders, Single Spheres, and Flow in Packed Bids and Tube Bundles," *AIChE J.,* vol. 18, p. 361, 1972.

36. Kays, W. M. *Convective Heat and Mass Transfer,* pp. 187–90. New York: McGraw-Hill, 1966.

37. Churchill, S. W., and M. Bernstein. "A Correlating Equation for Forced Convection from Gases and Liquids to a Circular Cylinder in Crossflow," *J. Heat Transfer,* vol. 99, pp. 300–306, 1977.

38. Nakai, S., and T. Okazaki. "Heat Transfer from a Horizontal Circular Wire at Small Reynolds and Grashof Numbers—1 Pure Convection," *Int. J. Heat Mass Transfer,* vol. 18, p. 387, 1975.

39. Zukauskas, A. "Heat Transfer from Tubes in Cross Flow," *Adv. Heat Transfer,* vol. 8, pp. 93–160, 1972.

40. Shah, R. K., and A. L. London. *Laminar Flow: Forced Convection in Ducts.* New York: Academic Press, 1978.

41. Kalish, S., and O. E. Dwyer. "Heat Transfer to NaK Flowing through Unbaffled Rod Bundles," *Int. J. Heat Mass Transfer,* vol. 10, p. 1533, 1967.

42. Petukhov, B. S. "Heat Transfer and Friction in Turbulent Pipe Flow with Variable Physical Properties," in J. P. Hartnett and T. F. Irvine, (eds.). *Advances in Heat Transfer.* New York: Academic Press, pp. 504–64, 1970.

43. Achenbach, E. "Heat Transfer from Spheres up to $Re = 6 \times 10^6$," *Proc. Sixth Int. Heat Trans. Conf.,* vol. 5, Washington, D.C.: Hemisphere Pub. Co., pp. 341–46, 1978.

44. Morgan, V. T. "The Overall Convective Heat Transfer from Smooth Circular Cylinders," in T. F. Irvine and J. P. Hartnett, (eds.). *Advances in Heat Transfer,* vol. 11, New York : Academic Press, 1975.

45. Gnielinski, V. "New Equations for Heat and Mass Transfer in Turbulent Pipe and Channel Flow," *Int. Chem. Engng.,* vol. 16, pp. 359–68, 1976.

46. Kakac, S., R. K. Shah, and W. Aung. *Handbook of Single-Phase Convection Heat Transfer.* New York: John Wiley, 1987.

47. Winterton, R. H. S., "Where Did the Dittus and Boelter Equation Come From?," *Int. J. Heat and Mass Transfer,* vol. 41, p. 809, 1998.

48. Sleicher, C. A., and M. W. Rouse, "A Convenient Correlation for Heat Transfer in Constant and Variable Property Fluids in Turbulent Flow," *Int. J. Heat and Mass Transfer,* vol. 18, p. 677, 1975.

49. Churchill, S. W., "Empirical Expressions for the Shear Stress in Turbulent Flow in Commercial Pipe," *AIChEJ.,* vol. 19, p. 375, 1973.

50. Swanee, P. K., and A. K. Jain, "Explicit Equations for Pipe Flow Problems," *J. H. Proc. ASCE,* p. 657, May 1976.

51. Holman, J. P. *Experimental Methods for Engineers*, 7th ed. chapters 3 and 9. McGraw-Hill, New York, 2001.

CHAPTER 7

Natural Convection Systems

7-1 | INTRODUCTION

Our previous discussions of convection heat transfer have considered only the calculation of forced-convection systems where the fluid is forced by or through the heat-transfer surface. Natural, or free, convection is observed as a result of the motion of the fluid due to density changes arising from the heating process. A hot radiator used for heating a room is one example of a practical device that transfers heat by free convection. The movement of the fluid in free convection, whether it is a gas or a liquid, results from the buoyancy forces imposed on the fluid when its density in the proximity of the heat-transfer surface is decreased as a result of the heating process. The buoyancy forces would not be present if the fluid were not acted upon by some external force field such as gravity, although gravity is not the only type of force field that can produce the free-convection currents; a fluid enclosed in a rotating machine is acted upon by a centrifugal force field, and thus could experience free-convection currents if one or more of the surfaces in contact with the fluid were heated. The buoyancy forces that give rise to the free-convection currents are called *body forces*.

7-2 | FREE-CONVECTION HEAT TRANSFER ON A VERTICAL FLAT PLATE

Consider the vertical flat plate shown in Figure 7-1. When the plate is heated, a free-convection boundary layer is formed, as shown. The velocity profile in this boundary layer is quite unlike the velocity profile in a forced-convection boundary layer. At the wall the velocity is zero because of the no-slip condition; it increases to some maximum value and then decreases to zero at the edge of the boundary layer since the "free-stream" conditions are at rest in the free-convection system. The initial boundary-layer development is laminar; but at some distance from the leading edge, depending on the fluid properties and the temperature difference between wall and environment, turbulent eddies are formed, and transition to a turbulent boundary layer begins. Farther up the plate the boundary layer may become fully turbulent.

To analyze the heat-transfer problem, we must first obtain the differential equation of motion for the boundary layer. For this purpose we choose the x coordinate along the plate and the y coordinate perpendicular to the plate as in the analyses of Chapter 5. The only new force that must be considered in the derivation is the weight of the element of fluid.

Figure 7-1 | Boundary layer on a vertical flat plate.

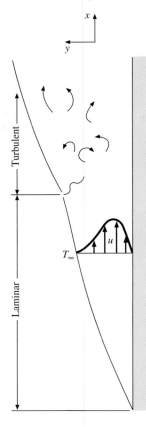

As before, we equate the sum of the external forces in the x direction to the change in momentum flux through the control volume $dx\,dy$. There results

$$\rho\left(u\frac{\partial u}{\partial x} + v\frac{\partial u}{\partial y}\right) = -\frac{\partial p}{\partial x} - \rho g + \mu\frac{\partial^2 u}{\partial y^2} \qquad [7\text{-}1]$$

where the term $-\rho g$ represents the weight force exerted on the element. The pressure gradient in the x direction results from the change in elevation up the plate. Thus

$$\frac{\partial p}{\partial x} = -\rho_\infty g \qquad [7\text{-}2]$$

In other words, the change in pressure over a height dx is equal to the weight per unit area of the fluid element. Substituting Equation (7-2) into Equation (7-1) gives

$$\rho\left(u\frac{\partial u}{\partial x} + v\frac{\partial u}{\partial y}\right) = g(\rho_\infty - \rho) + \mu\frac{\partial^2 u}{\partial y^2} \qquad [7\text{-}3]$$

The density difference $\rho_\infty - \rho$ may be expressed in terms of the volume coefficient of expansion β, defined by

$$\beta = \frac{1}{V}\left(\frac{\partial V}{\partial T}\right)_p = \frac{1}{V_\infty}\frac{V - V_\infty}{T - T_\infty} = \frac{\rho_\infty - \rho}{\rho(T - T_\infty)}$$

so that

$$\rho\left(u\frac{\partial u}{\partial x} + v\frac{\partial u}{\partial y}\right) = g\rho\beta(T - T_\infty) + \mu\frac{\partial^2 u}{\partial y^2} \qquad [7\text{-}4]$$

This is the equation of motion for the free-convection boundary layer. Notice that the solution for the velocity profile demands a knowledge of the temperature distribution. The energy equation for the free-convection system is the same as that for a forced-convection system at low velocity:

$$\rho c_p\left(u\frac{\partial T}{\partial x} + v\frac{\partial T}{\partial y}\right) = k\frac{\partial^2 T}{\partial y^2} \qquad [7\text{-}5]$$

The volume coefficient of expansion β may be determined from tables of properties for the specific fluid. For ideal gases it may be calculated from (see Problem 7-3)

$$\beta = \frac{1}{T}$$

where T is the absolute temperature of the gas.

Even though the fluid motion is the result of density variations, these variations are quite small, and a satisfactory solution to the problem may be obtained by assuming incompressible flow, that is, $\rho = $ constant. To effect a solution of the equation of motion, we use the integral method of analysis similar to that used in the forced-convection problem of Chapter 5. Detailed boundary-layer analyses have been presented in References 13, 27, and 32.

For the free-convection system, the integral momentum equation becomes

$$\frac{d}{dx}\left(\int_0^\delta \rho u^2\,dy\right) = -\tau_w + \int_0^\delta \rho g\beta(T - T_\infty)\,dy$$

$$= -\mu\left.\frac{\partial u}{\partial y}\right]_{y=0} + \int_0^\delta \rho g\beta(T - T_\infty)\,dy \qquad [7\text{-}6]$$

and we observe that the functional form of both the velocity and the temperature distributions must be known in order to arrive at the solution. To obtain these functions, we proceed in much the same way as in Chapter 5. The following conditions apply for the temperature distribution:

$$T = T_w \quad \text{at } y = 0$$
$$T = T_\infty \quad \text{at } y = \delta$$
$$\frac{\partial T}{\partial y} = 0 \quad \text{at } y = \delta$$

so that we obtain for the temperature distribution

$$\frac{T - T_\infty}{T_w - T_\infty} = \left(1 - \frac{y}{\delta}\right)^2 \tag{7-7}$$

Three conditions for the velocity profile are

$$u = 0 \quad \text{at } y = 0$$
$$u = 0 \quad \text{at } y = \delta$$
$$\frac{\partial u}{\partial y} = 0 \quad \text{at } y = \delta$$

An additional condition may be obtained from Equation (7-4) by noting that

$$\frac{\partial^2 u}{\partial y^2} = -g\beta \frac{T_w - T_\infty}{\nu} \quad \text{at } y = 0$$

As in the integral analysis for forced-convection problems, we assume that the velocity profiles have geometrically similar shapes at various x distances along the plate. For the free-convection problem, we now assume that the velocity may be represented as a polynomial function of y multiplied by some arbitrary function of x. Thus,

$$\frac{u}{u_x} = a + by + cy^2 + dy^3$$

where u_x is a fictitious velocity that is a function of x. The cubic-polynomial form is chosen because there are four conditions to satisfy, and this is the simplest type of function that may be used. Applying the four conditions to the velocity profile listed above, we have

$$\frac{u}{u_x} = \frac{\beta \delta^2 g(T_w - T_\infty)}{4u_x \nu} \frac{y}{\delta} \left(1 - \frac{y}{\delta}\right)^2$$

The term involving the temperature difference, δ^2, and u_x may be incorporated into the function u_x so that the final relation to be assumed for the velocity profile is

$$\frac{u}{u_x} = \frac{y}{\delta} \left(1 - \frac{y}{\delta}\right)^2 \tag{7-8}$$

A plot of Equation (7-8) is given in Figure 7-2. Substituting Equations (7-7) and (7-8) into Equation (7-6) and carrying out the integrations and differentiations yields

$$\frac{1}{105} \frac{d}{dx} (u_x^2 \delta) = \frac{1}{3} g\beta(T_w - T_\infty)\delta - \nu \frac{u_x}{\delta} \tag{7-9}$$

The integral form of the energy equation for the free-convection system is

$$\frac{d}{dx} \left[\int_0^\delta u(T - T_\infty) dy \right] = -\alpha \frac{dT}{dy} \bigg]_{y=0} \tag{7-10}$$

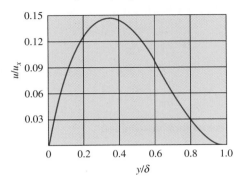

Figure 7-2 | Free-convection velocity profile given by Equation (7-8).

and when the assumed velocity and temperature distributions are inserted into this equation and the operations are performed, there results

$$\frac{1}{30}(T_w - T_\infty)\frac{d}{dx}(u_x\delta) = 2\alpha\frac{T_w - T_\infty}{\delta}$$ [7-11]

It is clear from the reasoning that led to Equation (7-8) that

$$u_x \sim \delta^2$$ [7-12]

Inserting this type of relation in Equation (7-9) yields the result that

$$\delta \sim x^{1/4}$$ [7-13]

We therefore assume the following exponential functional variations for u_x and δ:

$$u_x = C_1 x^{1/2}$$ [7-14]

$$\delta = C_2 x^{1/4}$$ [7-15]

Introducing these relations into Equations (7-9) and (7-11) gives

$$\frac{5}{420}C_1^2 C_2 x^{1/4} = g\beta(T_w - T_\infty)\frac{C_2}{3}x^{1/4} - \frac{C_1}{C_2}\nu x^{1/4}$$ [7-16]

and

$$\frac{1}{40}C_1 C_2 x^{-1/4} = \frac{2\alpha}{C_2}x^{-1/4}$$ [7-17]

These two equations may be solved for the constants C_1 and C_2 to give

$$C_1 = 5.17\nu\left(\frac{20}{21} + \frac{\nu}{\alpha}\right)^{-1/2}\left[\frac{g\beta(T_w - T_\infty)}{\nu^2}\right]^{1/2}$$ [7-18]

$$C_2 = 3.93\left(\frac{20}{21} + \frac{\nu}{\alpha}\right)^{1/4}\left[\frac{g\beta(T_w - T_\infty)}{\nu^2}\right]^{-1/4}\left(\frac{\nu}{\alpha}\right)^{-1/2}$$ [7-19]

The resultant expressions for the boundary layer thickness and fictitious velocity u_x are

$$\frac{\delta}{x} = 3.93\,\mathrm{Pr}^{-1/2}(0.952 + \mathrm{Pr})^{1/4}\mathrm{Gr}_x^{-1/4}$$ [7-20a]

$$u_x \frac{x}{\nu} = 5.17(0.952 + \text{Pr})^{-1/2} \text{Gr}_x^{1/2} \qquad [7\text{-}20b]$$

The velocity profile shown in Figure 7-2 has its maximum value at $y/\delta = 1/3$, giving $u_{max} = (4/27)u_x = 0.148u_x$. The mass flow through the boundary layer at any x position may be determined by evaluating the integral

$$\dot{m} = \int \rho u \, dy = \int_0^\delta \rho u_x \frac{y}{\delta} \left(1 - \frac{y}{\delta}\right)^2 dy = \frac{1}{12} \rho u_x \delta = 0.083 \rho u_x \delta = \frac{9}{16} \rho u_{max} \delta \qquad [7\text{-}20c]$$

The respective values of δ and u_x determined from Equations (7-20a) and (7-20b) may be inserted to obtain the mass flow values.

The Prandtl number $\text{Pr} = \nu/\alpha$ has been introduced in the above expressions along with a new dimensionless group called the *Grashof number* Gr_x:

$$\text{Gr}_x = \frac{g\beta(T_w - T_\infty)x^3}{\nu^2} \qquad [7\text{-}21]$$

The heat-transfer coefficient may be evaluated from

$$q_w = -kA \left. \frac{dT}{dy} \right]_w = hA(T_w - T_\infty)$$

Using the temperature distribution of Equation (7-7), one obtains

$$h = \frac{2k}{\delta} \qquad \text{or} \qquad \frac{hx}{k} = \text{Nu}_x = 2\frac{x}{\delta}$$

so that the dimensionless equation for the heat-transfer coefficient becomes

$$\text{Nu}_x = 0.508 \, \text{Pr}^{1/2}(0.952 + \text{Pr})^{-1/4} \text{Gr}_x^{1/4} \qquad [7\text{-}22]$$

Equation (7-22) gives the variation of the local heat-transfer coefficient along the vertical plate. The average heat-transfer coefficient may then be obtained by performing the integration

$$\bar{h} = \frac{1}{L} \int_0^L h_x dx \qquad [7\text{-}23]$$

For the variation given by Equation (7-22), the average coefficient is

$$\bar{h} = \tfrac{4}{3} h_{x=L} \qquad [7\text{-}24]$$

The Grashof number may be interpreted physically as a dimensionless group representing the ratio of the buoyancy forces to the viscous forces in the free-convection flow system. It has a role similar to that played by the Reynolds number in forced-convection systems and is the primary variable used as a criterion for transition from laminar to turbulent boundary-layer flow. For air in free convection on a vertical flat plate, the critical Grashof number has been observed by Eckert and Soehngen [1] to be approximately 4×10^8. Values ranging between 10^8 and 10^9 may be observed for different fluids and environment "turbulence levels."

A very complete survey of the stability and transition of free-convection boundary layers has been given by Gebhart et al. [13–15].

The foregoing analysis of free-convection heat transfer on a vertical flat plate is the simplest case that may be treated mathematically, and it has served to introduce the new

Figure 7-3 | Pulsed free-convection boundary layer on vertical flat plate. Distance between letters = 5 cm.

dimensionless variable, the Grashof number,[†] which is important in all free-convection problems. But as in some forced-convection problems, experimental measurements must be relied upon to obtain relations for heat transfer in other circumstances. These circumstances are usually those in which it is difficult to predict temperature and velocity profiles analytically. Turbulent free convection is an important example, just as is turbulent forced convection, of a problem area in which experimental data are necessary; however, the problem is more acute with free-convection flow systems than with forced-convection systems because the velocities are usually so small that they are very difficult to measure. For example, the maximum free-convection velocity experienced by a vertical plate heated to 45°C and exposed to atmospheric room air at 25°C is only about 350 mm/s. Despite the experimental difficulties, velocity measurements have been performed using hydrogen-bubble techniques [26], hot-wire anemometry [28], and quartz-fiber anemometers. Temperature field measurements have been obtained through the use of the Zehnder-Mach interferometer. The laser anemometer [29] is particularly useful for free-convection measurements because it does not disturb the flow field.

An interferometer indicates lines of constant density in a fluid flow field. For a gas in free convection at low pressure these lines of constant density are equivalent to lines of constant temperature. Once the temperature field is obtained, the heat transfer from a surface in free convection may be calculated by using the temperature gradient at the surface and the thermal conductivity of the gas. Several interferometric studies of free convection have been made [1–3], and Figure 7-3 indicates the isotherms in a free-convection boundary layer on a vertical flat plate with $T_W = 48$°C and $T_\infty = 20$°C in room air. The spacing between the horizontal markers is about 2.5 cm, indicating a boundary-layer thickness of about that same value. The letter A corresponds to the leading edge of the plate. Note that the isotherms are more closely spaced near the plate surface, indicating a higher temperature gradient in that region. The oscillatory or "wave" shape of the boundary layer isotherms is caused by a heat pulse from a fine wire located at $x = 2.5$ cm and having a frequency of about 2.5 Hz. The pulse moves up the plate at about the boundary layer velocity, so an indication of the velocity profile may be obtained by connecting the maximum points in the isotherms. Such a profile is indicated in Figure 7-4. Eventually, at about Gr = 10^8–10^9 small oscillations in the boundary layer become amplified and transition to turbulence begins. The region shown in Figure 7-3 is all laminar.

A number of references treat the various theoretical and empirical aspects of free-convection problems. One of the most extensive discussions is given by Gebhart et. al. [13], and the interested reader may wish to consult this reference for additional information.

7-3 | EMPIRICAL RELATIONS FOR FREE CONVECTION

Over the years it has been found that average free-convection heat-transfer coefficients can be represented in the following functional form for a variety of circumstances:

$$\overline{\mathrm{Nu}}_f = C(\mathrm{Gr}_f\,\mathrm{Pr}_f)^m \qquad\qquad \textbf{[7-25]}$$

[†]History is not clear on the point, but it appears that the Grashof number was named for Franz Grashof, a professor of applied mechanics at Karlsruhe around 1863 and one of the founding directors of *Verein deutscher Ingenieure* in 1855. He developed some early steam-flow formulas but made no significant contributions to free convection [36].

Figure 7-4 | Free-convection velocity profile indicated by connecting maximum points in boundary-layer isotherms of Figure 7-3.

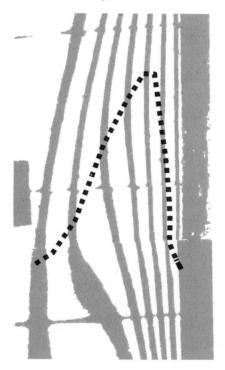

where the subscript f indicates that the properties in the dimensionless groups are evaluated at the film temperature

$$T_f = \frac{T_\infty + T_w}{2}$$

The product of the Grashof and Prandtl numbers is called the Rayleigh number:

$$\text{Ra} = \text{Gr Pr} \qquad\qquad\qquad \textbf{[7-26]}$$

Characteristic Dimensions

The characteristic dimension to be used in the Nusselt and Grashof numbers depends on the geometry of the problem. For a vertical plate it is the height of the plate L; for a horizontal cylinder it is the diameter d; and so forth. Experimental data for free-convection problems appear in a number of references, with some conflicting results. The purpose of the sections that follow is to give these results in a summary form that may be easily used for calculation purposes. The functional form of Equation (7-25) is used for many of these presentations, with the values of the constants C and m specified for each case. Table 7-1 provides a summary of the values of these correlation constants for different geometries, and the sections that follow discuss the correlations in more detail.

For convenience of the reader, the present author has presented a graphical meld of the correlations for the isothermal vertical plate and horizontal cylinder configurations in the form of Figures 7-5 and 7-6. These figures may be used in lieu of the formulas when a quick estimate of performance is desired.

Table 7-1 | Constants for use with Equation (7-25) for isothermal surfaces.

Geometry	$Gr_f\,Pr_f$	C	m	Reference(s)
Vertical planes and cylinders	$10^{-1}-10^4$	Use Fig. 7-5	Use Fig. 7-5	4
	10^4-10^9	0.59	$\frac{1}{4}$	4
	10^9-10^{13}	0.021	$\frac{2}{5}$	30
	10^9-10^{13}	0.10	$\frac{1}{3}$	22, 16[†]
Horizontal cylinders	$0-10^{-5}$	0.4	0	4
	$10^{-5}-10^4$	Use Fig. 7-6	Use Fig. 7-6	4
	10^4-10^9	0.53	$\frac{1}{4}$	4
	10^9-10^{12}	0.13	$\frac{1}{3}$	4
	$10^{-10}-10^{-2}$	0.675	0.058	76[†]
	$10^{-2}-10^2$	1.02	0.148	76[†]
	10^2-10^4	0.850	0.188	76
	10^4-10^7	0.480	$\frac{1}{4}$	76
	10^7-10^{12}	0.125	$\frac{1}{3}$	76
Upper surface of heated plates or lower surface of cooled plates	$2\times10^4-8\times10^6$	0.54	$\frac{1}{4}$	44, 52
Upper surface of heated plates or lower surface of cooled plates	$8\times10^6-10^{11}$	0.15	$\frac{1}{3}$	44, 52
Lower surface of heated plates or upper surface of cooled plates	10^5-10^{11}	0.27	$\frac{1}{4}$	44, 37, 75
Vertical cylinder, height = diameter characteristic length = diameter	10^4-10^6	0.775	0.21	77
Irregular solids, characteristic length = distance fluid particle travels in boundary layer	10^4-10^9	0.52	$\frac{1}{4}$	78

[†] Preferred.

7-4 | FREE CONVECTION FROM VERTICAL PLANES AND CYLINDERS

Isothermal Surfaces

For vertical surfaces, the Nusselt and Grashof numbers are formed with L, the height of the surface as the characteristic dimension. If the boundary-layer thickness is not large compared with the diameter of the cylinder, the heat transfer may be calculated with the same relations used for vertical plates. The general criterion is that a vertical cylinder may be treated as a vertical flat plate [13] when

$$\frac{D}{L} \geq \frac{35}{Gr_L^{1/4}} \qquad [7\text{-}27]$$

where D is the diameter of the cylinder. For vertical cylinders too small to meet this criteria, the analysis of Reference [84] for gases with $Pr = 0.7$ indicates that the flat plate results for the average heat-transfer coefficient should be multiplied by a factor F to account for the curvature, where

$$F = 1.3[(L/D)/Gr_D]^{1/4} + 1.0 \qquad [7\text{-}27a]$$

For *isothermal* surfaces, the values of the constants C and m are given in Table 7-1 with the appropriate references noted for further consultation. The reader's attention is directed

Figure 7-5 | Free-convection heat transfer from vertical isothermal plates.

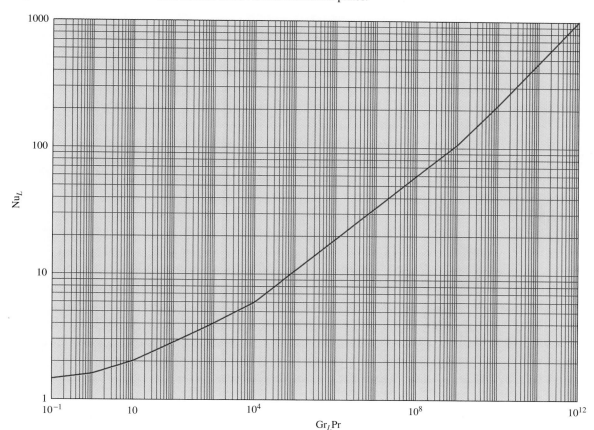

to the two sets of constants given for the turbulent case ($\mathrm{Gr}_f \, \mathrm{Pr}_f > 10^9$). Although there may appear to be a decided difference in these constants, a comparison by Warner and Arpaci [22] of the two relations with experimental data indicates that both sets of constants fit available data. There are some indications from the analytical work of Bayley [16], as well as heat flux measurements of Reference 22, that the relation

$$\mathrm{Nu}_f = 0.10(\mathrm{Gr}_f \, \mathrm{Pr}_f)^{1/3}$$

may be preferable.

More complicated relations have been provided by Churchill and Chu [71] that are applicable over wider ranges of the Rayleigh number:

$$\overline{\mathrm{Nu}} = 0.68 + \frac{0.670 \, \mathrm{Ra}^{1/4}}{[1 + (0.492/\mathrm{Pr})^{9/16}]^{4/9}} \qquad \text{for } \mathrm{Ra}_L < 10^9 \qquad \textbf{[7-28]}$$

$$\overline{\mathrm{Nu}}^{1/2} = 0.825 + \frac{0.387 \, \mathrm{Ra}^{1/6}}{[1 + (0.492/\mathrm{Pr})^{9/16}]^{8/27}} \qquad \text{for } 10^{-1} < \mathrm{Ra}_L < 10^{12} \qquad \textbf{[7-29]}$$

Equation (7-28) is also a satisfactory representation for constant heat flux. Properties for these equations are evaluated at the film temperature.

Figure 7-6 | Free-convection heat transfer from horizontal isothermal cylinders.

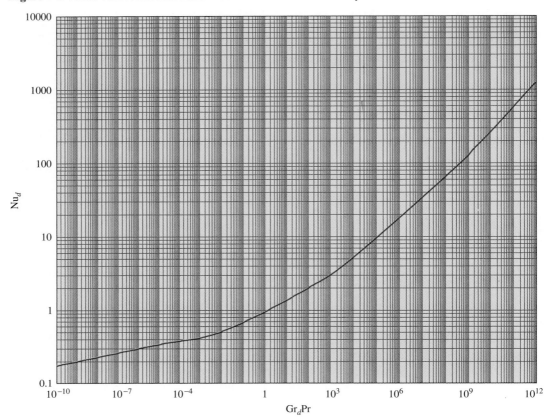

Constant-Heat-Flux Surfaces

Extensive experiments have been reported in References 25, 26, and 39 for free convection from vertical and inclined surfaces to water under constant-heat-flux conditions. In such experiments, the results are presented in terms of a modified Grashof number, Gr^*:

$$Gr_x^* = Gr_x \, Nu_x = \frac{g\beta q_w x^4}{k\nu^2} \qquad [7\text{-}30]$$

where $q_w = q/A$ is the heat flux per unit area and is assumed constant over the entire plate surface area.

The *local* heat-transfer coefficients were correlated by the following relation for the laminar range:

$$Nu_{xf} = \frac{hx}{k_f} = 0.60(Gr_x^* \, Pr_f)^{1/5} \qquad 10^5 < Gr_x^* \, Pr < 10^{11}; \, q_w = const \qquad [7\text{-}31]$$

It is to be noted that the criterion for laminar flow expressed in terms of Gr_x^* is not the same as that expressed in terms of Gr_x. Boundary-layer transition was observed to begin between $Gr_x^* \, Pr = 3 \times 10^{12}$ and 4×10^{13} and to end between 2×10^{13} and 10^{14}. Fully developed turbulent flow was present by $Gr_x^* \, Pr = 10^{14}$, and the experiments were extended up to $Gr_x^* \, Pr = 10^{16}$. For the turbulent region, the local heat-transfer coefficients are correlated with

$$Nu_x = 0.17(Gr_x^* \, Pr)^{1/4} \qquad 2 \times 10^{13} < Gr_x^* \, Pr < 10^{16}; \, q_w = const \qquad [7\text{-}32]$$

All properties in Equations (7-31) and (7-32) are evaluated at the local film temperature. Although these experiments were conducted for water, the resulting correlations are shown to work for air as well. The average heat-transfer coefficient for the constant-heat-flux case may not be evaluated from Equation (7-24) but must be obtained through a separate application of Equation (7-23). Thus, for the laminar region, using Equation (7-31) to evaluate h_x,

$$\bar{h} = \frac{1}{L} \int_0^L h_x dx$$

$$\bar{h} = \tfrac{5}{4} h_{x=L} \qquad q_w = \text{const}$$

At this point we may note the relationship between the correlations in the form of Equation (7-25) and those just presented in terms of $\text{Gr}_x^* = \text{Gr}_x \, \text{Nu}_x$. Writing Equation (7-25) as a *local* heat-transfer form gives

$$\text{Nu}_x = C(\text{Gr}_x \, \text{Pr})^m \qquad\qquad \textbf{[7-33]}$$

Inserting $\text{Gr}_x = \text{Gr}_x^*/\text{Nu}_x$ gives

$$\text{Nu}_x^{1+m} = C(\text{Gr}_x^* \, \text{Pr})^m$$

or

$$\text{Nu}_x = C^{1/(1+m)}(\text{Gr}_x^* \, \text{Pr})^{m/(1+m)} \qquad\qquad \textbf{[7-34]}$$

Thus, when the "characteristic" values of m for laminar and turbulent flow are compared to the exponents on Gr_x^*, we obtain

Laminar, $m = \dfrac{1}{4}$: $\qquad\qquad\qquad\qquad \dfrac{m}{1+m} = \dfrac{1}{5}$

Turbulent, $m = \dfrac{1}{3}$: $\qquad\qquad\qquad\quad \dfrac{m}{1+m} = \dfrac{1}{4}$

While the Gr* formulation is easier to employ for the constant-heat-flux case, we see that the characteristic exponents fit nicely into the scheme that is presented for the isothermal surface correlations.

It is also interesting to note the variation of h_x with x in the two characteristic regimes. For the laminar range $m = \tfrac{1}{4}$, and from Equation (7-25)

$$h_x \sim \frac{1}{x}(x^3)^{1/4} = x^{-1/4}$$

In the turbulent regime $m = \tfrac{1}{3}$, and we obtain

$$h_x \sim \frac{1}{x}(x^3)^{1/3} = \text{const with } x$$

So when turbulent free convection is encountered, the local heat-transfer coefficient is essentially constant with x.

Churchill and Chu [71] show that Equation (7-28) may be modified to apply to the constant-heat-flux case if the average Nusselt number is based on the wall heat flux and the temperature difference at the center of the plate ($x = L/2$). The result is

$$\overline{\text{Nu}}_L^{1/4}(\overline{\text{Nu}}_L - 0.68) = \frac{0.67(\text{Gr}_L^* \, \text{Pr})^{1/4}}{[1 + (0.492/\text{Pr})^{9/16}]^{4/9}} \qquad\qquad \textbf{[7-35]}$$

where $\overline{\text{Nu}}_L = q_w L/(k\overline{\Delta T})$ and $\overline{\Delta T} = T_w - T_\infty$ at $L/2 - T_\infty$.

| **EXAMPLE 7-1** | Constant Heat Flux from Vertical Plate |

In a plant location near a furnace, a net radiant energy flux of 800 W/m^2 is incident on a vertical metal surface 3.5 m high and 2 m wide. The metal is insulated on the back side and painted black so that all the incoming radiation is lost by free convection to the surrounding air at 30°C. What average temperature will be attained by the plate?

■ **Solution**

We treat this problem as one with constant heat flux on the surface. Since we do not know the surface temperature, we must make an estimate for determining T_f and the air properties. An *approximate* value of h for free-convection problems is 10 W/m$^2 \cdot$ °C, and so, *approximately*,

$$\Delta T = \frac{q_w}{h} \approx \frac{800}{10} = 80°C$$

Then

$$T_f \approx \frac{80}{2} + 30 = 70°C = 343 \text{ K}$$

At 70°C the properties of air are

$$\nu = 2.043 \times 10^{-5} \text{ m}^2/\text{s} \qquad \beta = \frac{1}{T_f} = 2.92 \times 10^{-3} \text{ K}^{-1}$$

$$k = 0.0295 \text{ W/m} \cdot °C \qquad \text{Pr} = 0.7$$

From Equation (7-30), with $x = 3.5$ m,

$$\text{Gr}_x^* = \frac{g\beta q_w x^4}{k\nu^2} = \frac{(9.8)(2.92 \times 10^{-3})(800)(3.5)^4}{(0.0295)(2.043 \times 10^{-5})^2} = 2.79 \times 10^{14}$$

We may therefore use Equation (7-32) to evaluate h_x:

$$h_x = \frac{k}{x}(0.17)(\text{Gr}_x^* \text{Pr})^{1/4}$$

$$= \frac{0.0295}{3.5}(0.17)(2.79 \times 10^{14} \times 0.7)^{1/4}$$

$$= 5.36 \text{ W/m}^2 \cdot °C \text{ [0.944 Btu/h} \cdot \text{ft}^2 \cdot °\text{F]}$$

In the turbulent heat transfer governed by Equation (7-32), we note that

$$\text{Nu}_x = \frac{hx}{k} \sim (\text{Gr}_x^*)^{1/4} \sim (x^4)^{1/4}$$

or h_x does not vary with x, and we may take this as the average value. The value of $h = 5.41$ W/m$^2 \cdot$ °C is less than the approximate value we used to estimate T_f. Recalculating ΔT, we obtain

$$\Delta T = \frac{q_w}{h} = \frac{800}{5.36} = 149°C$$

Our new film temperature would be

$$T_f = 30 + \frac{149}{2} = 104.5°C$$

At 104.5°C the properties of air are

$$\nu = 2.354 \times 10^{-5} \text{ m}^2/\text{s} \qquad \beta = \frac{1}{T_f} = 2.65 \times 10^{-3}/\text{K}$$

$$k = 0.0320 \text{ W/m} \cdot °C \qquad \text{Pr} = 0.695$$

Then

$$\mathrm{Gr}_x^* = \frac{(9.8)(2.65 \times 10^{-3})(800)(3.5)^4}{(0.0320)(2.354 \times 10^{-5})^2} = 1.75 \times 10^{14}$$

and h_x is calculated from

$$h_x = \frac{k}{x}(0.17)(\mathrm{Gr}_x^* \, \mathrm{Pr})^{1/4}$$

$$= \frac{(0.0320)(0.17)}{3.5}[(1.758 \times 10^{14})(0.695)]^{1/4}$$

$$= 5.17 \; \mathrm{W/m^2 \cdot {}^\circ C} \; [-0.91 \; \mathrm{Btu/h \cdot ft^2 \cdot {}^\circ F}]$$

Our new temperature difference is calculated as

$$\Delta T = (T_w - T_\infty)_{\mathrm{av}} = \frac{q_w}{h} = \frac{800}{5.17} = 155^\circ C$$

The average wall temperature is therefore

$$T_{w,\mathrm{av}} = 155 + 30 = 185^\circ C$$

Another iteration on the value of T_f is not warranted by the improved accuracy that would result.

Heat Transfer from Isothermal Vertical Plate   **EXAMPLE 7-2**

A large vertical plate 4.0 m high is maintained at 60°C and exposed to atmospheric air at 10°C. Calculate the heat transfer if the plate is 10 m wide.

■ **Solution**

We first determine the film temperature as

$$T_f = \frac{60 + 10}{2} = 35^\circ C = 308 \; K$$

The properties of interest are thus

$$\beta = \frac{1}{308} = 3.25 \times 10^{-3} \qquad k = 0.02685$$

$$\nu = 16.5 \times 10^{-6} \qquad \mathrm{Pr} = 0.7$$

and

$$\mathrm{Gr} \, \mathrm{Pr} = \frac{(9.8)(3.25 \times 10^{-3})(60 - 10)(4)^3}{(16.5 \times 10^{-6})^2}0.7$$

$$= 2.62 \times 10^{11}$$

We then may use Equation (7-29) to obtain

$$\overline{\mathrm{Nu}}^{1/2} = 0.825 + \frac{(0.387)(2.62 \times 10^{11})^{1/6}}{[1 + (0.492/0.7)^{9/16}]^{8/27}}$$

$$= 26.75$$

$$\overline{\mathrm{Nu}} = 716$$

The heat-transfer coefficient is then

$$\overline{h} = \frac{(716)(0.02685)}{4.0} = 4.80 \; \mathrm{W/m^2 \cdot {}^\circ C}$$

The heat transfer is

$$q = \overline{h} A (T_w - T_\infty)$$
$$= (4.80)(4)(10)(60 - 10) = 9606 \text{ W}$$

As an alternative, we could employ the simpler relation

$$\text{Nu} = 0.10 (\text{Gr Pr})^{1/3}$$
$$= (0.10)(2.62 \times 10^{11})^{1/3} = 639.9$$

which gives a value about 10 percent lower than Equation (7-29).

7-5 | FREE CONVECTION FROM HORIZONTAL CYLINDERS

The values of the constants C and m are given in Table 7-1 according to References 4 and 76. The predictions of Morgan (Reference 76 in Table 7-1) are the most reliable for Gr Pr of approximately 10^{-5}. A more complicated expression for use over a wider range of Gr Pr is given by Churchill and Chu [70]:

$$\overline{\text{Nu}}^{1/2} = 0.60 + 0.387 \left\{ \frac{\text{Gr Pr}}{[1 + (0.559/\text{Pr})^{9/16}]^{16/9}} \right\}^{1/6} \quad \text{for } 10^{-5} < \text{Gr Pr}$$
$$< 10^{12} \quad \textbf{[7-36]}$$

A simpler equation is available from Reference 70 but is restricted to the laminar range of $10^{-6} < \text{Gr Pr} < 10^9$:

$$\text{Nu}_d = 0.36 + \frac{0.518 (\text{Gr}_d \, \text{Pr})^{1/4}}{[1 + (0.559/\text{Pr})^{9/16}]^{4/9}} \quad \textbf{[7-37]}$$

Properties in Equations (7-36) and (7-37) are evaluated at the film temperature.

Heat transfer from horizontal cylinders to liquid metals may be calculated from Reference 46:

$$\text{Nu}_d = 0.53 (\text{Gr}_d \, \text{Pr}^2)^{1/4} \quad \textbf{[7-38]}$$

EXAMPLE 7-3   Heat Transfer from Horizontal Tube in Water

A 2.0-cm-diameter horizontal heater is maintained at a surface temperature of 38°C and submerged in water at 27°C. Calculate the free-convection heat loss per unit length of the heater.

■ **Solution**
The film temperature is

$$T_f = \frac{38 + 27}{2} = 32.5°C$$

From Appendix A the properties of water are

$$k = 0.630 \text{ W/m} \cdot °C$$

and the following term is particularly useful in obtaining the Gr Pr product when it is multiplied by $d^3 \Delta T$:

$$\frac{g \beta \rho^2 c_p}{\mu k} = 2.48 \times 10^{10} \quad [1/\text{m}^3 \cdot °C]$$

$$\text{Gr Pr} = (2.48 \times 10^{10})(38 - 27)(0.02)^3 = 2.18 \times 10^6$$

Using Table 7-1, we get $C = 0.53$ and $m = \frac{1}{4}$, so that

$$\text{Nu} = (0.53)(2.18 \times 10^6)^{1/4} = 20.36$$

$$h = \frac{(20.36)(0.63)}{0.02} = 642 \text{ W/m}^2 \cdot {}^\circ\text{C}$$

The heat transfer is thus

$$\frac{q}{L} = h\pi d(T_w - T_\infty)$$

$$= (642)\pi(0.02)(38 - 27) = 443 \text{ W/m}$$

Heat Transfer from Fine Wire in Air EXAMPLE 7-4

A fine wire having a diameter of 0.02 mm is maintained at a constant temperature of 54°C by an electric current. The wire is exposed to air at 1 atm and 0°C. Calculate the electric power necessary to maintain the wire temperature if the length is 50 cm.

■ **Solution**

The film temperature is $T_f = (54 + 0)/2 = 27°\text{C} = 300$ K, so the properties are

$$\beta = 1/300 = 0.00333 \qquad \nu = 15.69 \times 10^{-6} \text{ m}^2/\text{s}$$
$$k = 0.02624 \text{ W/m} \cdot {}^\circ\text{C} \qquad \text{Pr} = 0.708$$

The Gr Pr product is then calculated as

$$\text{Gr Pr} = \frac{(9.8)(0.00333)(54 - 0)(0.02 \times 10^{-3})^3}{(15.69 \times 10^{-6})^2}(0.708) = 4.05 \times 10^{-5}$$

From Table 7-1 we find $C = 0.675$ and $m = 0.058$ so that

$$\overline{\text{Nu}} = (0.675)(4.05 \times 10^{-5})^{0.058} = 0.375$$

and

$$\overline{h} = \overline{\text{Nu}}\left(\frac{k}{d}\right) = \frac{(0.375)(0.02624)}{0.02 \times 10^{-3}} = 492.6 \text{ W/m}^2 \cdot {}^\circ\text{C}$$

The heat transfer or power required is then

$$q = \overline{h}A\,(T_w - T_\infty) = (492.6)\pi(0.02 \times 10^{-3})(0.5)(54 - 0) = 0.836 \text{ W}$$

Heated Horizontal Pipe in Air EXAMPLE 7-5

A horizontal pipe 1 ft (0.3048 m) in diameter is maintained at a temperature of 250°C in a room where the ambient air is at 15°C. Calculate the free-convection heat loss per meter of length.

■ **Solution**

We first determine the Grashof-Prandtl number product and then select the appropriate constants from Table 7-1 for use with Equation (7-25). The properties of air are evaluated at the film temperature:

$$T_f = \frac{T_w + T_\infty}{2} = \frac{250 + 15}{2} = 132.5°\text{C} = 405.5 \text{ K}$$

$$k = 0.03406 \text{ W/m} \cdot {}^\circ\text{C} \qquad \beta = \frac{1}{T_f} = \frac{1}{405.5} = 2.47 \times 10^{-3} \text{ K}^{-1}$$

$$\nu = 26.54 \times 10^{-6} \text{ m}^2/\text{s} \qquad \text{Pr} = 0.687$$

$$\begin{aligned}
\text{Gr}_d \, \text{Pr} &= \frac{g\beta(T_w - T_\infty)d^3}{\nu^2}\text{Pr} \\
&= \frac{(9.8)(2.47 \times 10^{-3})(250 - 15)(0.3048)^3(0.687)}{(26.54 \times 10^{-6})^2} \\
&= 1.571 \times 10^8
\end{aligned}$$

From Table 7-1, $C = 0.53$ and $m = \frac{1}{4}$, so that

$$\text{Nu}_d = 0.53(\text{Gr}_d \, \text{Pr})^{1/4} = (0.53)(1.571 \times 10^8)^{1/4} = 59.4$$

$$h = \frac{k \text{Nu}_d}{d} = \frac{(0.03406)(59.4)}{0.3048} = 6.63 \text{ W/m}^2 \cdot {}^\circ\text{C} \quad [1.175 \text{ Btu/h} \cdot \text{ft}^2 \cdot {}^\circ\text{F}]$$

The heat transfer per unit length is then calculated from

$$\frac{q}{L} = h\pi \, d(T_w - T_\infty) = 6.63\pi(0.3048)(250 - 15) = 1.49 \text{ kW/m} \quad [1560 \text{ Btu/h} \cdot \text{ft}]$$

As an alternative, we could employ the more complicated expression, Equation (7-36), for solution of the problem. The Nusselt number thus would be calculated as

$$\overline{\text{Nu}}^{1/2} = 0.60 + 0.387 \left\{ \frac{1.571 \times 10^8}{[1 + (0.559/0.687)^{9/16}]^{16/9}} \right\}^{1/6}$$

$$\text{Nu} = 64.7$$

or a value about 8 percent higher.

7-6 | FREE CONVECTION FROM HORIZONTAL PLATES

Isothermal Surfaces

The average heat-transfer coefficient from horizontal flat plates is calculated with Equation (7-25) and the constants given in Table 7-1. The characteristic dimension for use with these relations has traditionally [4] been taken as the length of a side for a square, the mean of the two dimensions for a rectangular surface, and $0.9d$ for a circular disk. References 52 and 53 indicate that better agreement with experimental data can be achieved by calculating the characteristic dimension with

$$L = \frac{A}{P} \tag{7-39}$$

where A is the area and P is the perimeter of the surface. This characteristic dimension is also applicable to unsymmetrical planforms.

Constant Heat Flux

The experiments of Reference 44 have produced the following correlations for constant heat flux on a horizontal plate. For the heated surface facing upward,

$$\overline{\text{Nu}}_L = 0.13(\text{Gr}_L \, \text{Pr})^{1/3} \qquad \text{for } \text{Gr}_L \, \text{Pr} < 2 \times 10^8 \tag{7-40}$$

and

$$\overline{\mathrm{Nu}}_L = 0.16(\mathrm{Gr}_L\,\mathrm{Pr})^{1/3} \qquad \text{for } 2 \times 10^8 < \mathrm{Gr}_L\,\mathrm{Pr} < 10^{11} \qquad \textbf{[7-41]}$$

For the heated surface facing downward,

$$\overline{\mathrm{Nu}}_L = 0.58(\mathrm{Gr}_L\,\mathrm{Pr})^{1/5} \qquad \text{for } 10^6 < \mathrm{Gr}_L\,\mathrm{Pr} < 10^{11} \qquad \textbf{[7-42]}$$

In these equations all properties except β are evaluated at a temperature T_e defined by

$$T_e = T_w - 0.25(T_w - T_\infty)$$

and T_w is the *average* wall temperature related, as before, to the heat flux by

$$\overline{h} = \frac{q_w}{T_w - T_\infty}$$

The Nusselt number is formed as before:

$$\overline{\mathrm{Nu}}_L = \frac{\overline{h}L}{k} = \frac{q_w L}{(T_w - T_\infty)k}$$

Section 7-7 discusses an extension of these equations to inclined surfaces.

Irregular Solids

There is no general correlation which can be applied to irregular solids. The results of Reference 77 indicate that Equation (7-25) may be used with $C = 0.775$ and $m = 0.208$ for a vertical cylinder with height equal to diameter. Nusselt and Grashof numbers are evaluated by using the diameter as characteristic length. Lienhard [78] offers a prescription that takes the characteristic length as the distance a fluid particle travels in the boundary layer and uses values of $C = 0.52$ and $m = \frac{1}{4}$ in Equation (7-25) in the laminar range. This may serve as an estimate for calculating the heat-transfer coefficient in the absence of specific information on a particular geometric shape. Bodies of unity aspect ratio are studied extensively in Reference 81.

Cube Cooling in Air	**EXAMPLE 7-6**

A cube, 20 cm on a side, is maintained at 60°C and exposed to atmospheric air at 10°C. Calculate the heat transfer.

■ **Solution**
This is an irregular solid so we use the information in the last entry of Table 7-1 in the absence of a specific correlation for this geometry. The properties were evaluated in Example 7-2 as

$$\beta = 3.25 \times 10^{-3} \quad k = 0.02685$$
$$\nu = 17.47 \times 10^{-6} \quad \mathrm{Pr} = 0.7$$

The characteristic length is the distance a particle travels in the boundary layer, which is $L/2$ along the bottom plus L along the side plus $L/2$ on the top, or $2L = 40$ cm. The Gr Pr product is thus:

$$\mathrm{Gr\,Pr} = \frac{(9.8)(3.25 \times 10^{-3})(60 - 10)(0.4)^3}{(17.47 \times 10^{-6})^2}(0.7) = 2.34 \times 10^8$$

From the last entry in Table 7-1 we find $C = 0.52$ and $n = 1/4$ and calculate the Nusselt number as

$$\mathrm{Nu} = (0.52)(2.34 \times 10^8)^{1/4} = 64.3$$

and

$$\bar{h} = \mathrm{Nu}\frac{k}{L} = \frac{(64.3)(0.02685)}{(0.4)} = 4.32 \text{ W/m}^2 \cdot {}^\circ\text{C}$$

The cube has six sides so the area is $6(0.2)^2 = 0.24 \text{ m}^2$ and the heat transfer is

$$q = \bar{h}A(T_w - T_\infty) = (4.32)(0.24)(60 - 10) = 51.8 \text{ W}$$

7-7 | FREE CONVECTION FROM INCLINED SURFACES

Figure 7-7 | Coordinate system for inclined plates.

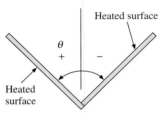

Extensive experiments have been conducted by Fujii and Imura [44] for heated plates in water at various angles of inclination. The angle that the plate makes with the vertical is designated θ, with positive angles indicating that the heater surface faces downward, as shown in Figure 7-7. For the inclined plate facing downward with approximately constant heat flux, the following correlation was obtained for the average Nusselt number:

$$\overline{\mathrm{Nu}}_e = 0.56(\mathrm{Gr}_e\,\mathrm{Pr}_e\cos\theta)^{1/4} \qquad \theta < 88^\circ;\ 10^5 < \mathrm{Gr}_e\,\mathrm{Pr}_e\cos\theta < 10^{11} \qquad \textbf{[7-43]}$$

In Equation (7-43) all properties except β are evaluated at a reference temperature T_e defined by

$$T_e = T_w - 0.25(T_w - T_\infty) \qquad \textbf{[7-44]}$$

where T_w is the *mean* wall temperature and T_∞ is the free-stream temperature; β is evaluated at a temperature of $T_\infty + 0.50(T_w - T_\infty)$. For almost-horizontal plates facing downward, that is, $88^\circ < \theta < 90^\circ$, an additional relation was obtained as

$$\overline{\mathrm{Nu}}_e = 0.58(\mathrm{Gr}_e\,\mathrm{Pr}_e)^{1/5} \qquad 10^6 < \mathrm{Gr}_e\,\mathrm{Pr}_e < 10^{11} \qquad \textbf{[7-45]}$$

For an inclined plate with heated surface facing upward the empirical correlations become more complicated. For angles between -15 and -75° a suitable correlation is

$$\overline{\mathrm{Nu}}_e = 0.14[(\mathrm{Gr}_e\,\mathrm{Pr}_e)^{1/3} - (\mathrm{Gr}_c\,\mathrm{Pr}_e)^{1/3}] + 0.56(\mathrm{Gr}_e\,\mathrm{Pr}_e\cos\theta)^{1/4} \qquad \textbf{[7-46]}$$

for the range $10^5 < \mathrm{Gr}_e\,\mathrm{Pr}_e\cos\theta < 10^{11}$. The quantity Gr_c is a critical Grashof relation indicating when the Nusselt number starts to separate from the laminar relation of Equation (7-43) and is given in the following tabulation:

θ, degrees	Gr_c
-15	5×10^9
-30	2×10^9
-60	10^8
-75	10^6

For $\mathrm{Gr}_e < \mathrm{Gr}_c$ the first term of Equation (7-46) is dropped out. Additional information is given by Vliet [39] and Pera and Gebhart [45]. There is some evidence to indicate that the above relations may also be applied to constant-temperature surfaces.

Experimental measurements with air on constant-heat-flux surfaces [51] have shown that Equation (7-31) may be employed for the laminar region if we replace Gr_x^* by $\mathrm{Gr}_x^* \cos\theta$

for both upward- and downward-facing heated surfaces. In the turbulent region with air, the following empirical correlation was obtained:

$$\text{Nu}_x = 0.17(\text{Gr}_x^* \, \text{Pr})^{1/4} \qquad 10^{10} < \text{Gr}_x^* \, \text{Pr} < 10^{15} \qquad \textbf{[7-47]}$$

where the Gr_x^* is the same as for the vertical plate when the heated surface faces upward. When the heated surface faces downward, Gr_x^* is replaced by $\text{Gr}^* \cos^2 \theta$. Equation (7-47) reduces approximately to the relation recommended in Table 7-1 for an isothermal vertical plate.

For inclined cylinders the data of Reference 73 indicate that laminar heat transfer under constant-heat-flux conditions may be calculated with the following relation:

$$\text{Nu}_L = [0.60 - 0.488(\sin\theta)^{1.03}](\text{Gr}_L \, \text{Pr})^{\frac{1}{4} + \frac{1}{12}(\sin\theta)^{1.75}} \qquad \text{for } \text{Gr}_L \, \text{Pr} < 2 \times 10^8 \quad \textbf{[7-48]}$$

where θ is the angle the cylinder makes with the vertical; that is, $0°$ corresponds to a vertical cylinder. Properties are evaluated at the film temperature except β, which is evaluated at ambient conditions.

Uncertainties still remain in the prediction of free convection from inclined surfaces, and an experimental-data scatter of ± 20 percent is not unusual for the empirical relations presented above.

7-8 | NONNEWTONIAN FLUIDS

When the shear-stress viscosity relation of the fluid does not obey the simple newtonian expression of Equation (5-1), the above equations for free-convection heat transfer do not apply. Extremely viscous polymers and lubricants are examples of fluids with nonnewtonian behavior. Successful analytical and experimental studies have been carried out with such fluids, but the results are very complicated. The interested reader should consult References 48 to 50 for detailed information on this subject.

7-9 | SIMPLIFIED EQUATIONS FOR AIR

Simplified equations for the heat-transfer coefficient from various surfaces to air at atmospheric pressure and moderate temperatures are given in Table 7-2. These relations may be extended to higher or lower pressures by multiplying by the following factors:

$$\left(\frac{p}{101.32}\right)^{1/2} \qquad \text{for laminar cases}$$

$$\left(\frac{p}{101.32}\right)^{2/3} \qquad \text{for turbulent cases}$$

where p is the pressure in kilopascals. Due caution should be exercised in the use of these simplified relations because they are only approximations of the more precise equations stated earlier.

The reader will note that the use of Table 7-2 requires a knowledge of the value of the Grashof-Prandtl number product. This might seem to be self-defeating, in that another calculation is required. However, with a bit of experience one learns the range of Gr Pr to be expected in various geometrical-physical situations, and thus the simplified expressions can be an expedient for quick problem solving. As we have noted, they are not a substitute for the more comprehensive expressions.

Table 7-2 | Simplified equations for free convection from various surfaces to air at atmospheric pressure, adapted from Table 7-1.

Surface	Laminar, $10^4 < \mathrm{Gr}_f \mathrm{Pr}_f < 10^9$	Turbulent, $\mathrm{Gr}_f \mathrm{Pr}_f > 10^9$
Vertical plane or cylinder	$h = 1.42 \left(\dfrac{\Delta T}{L} \right)^{1/4}$	$h = 1.31 (\Delta T)^{1/3}$
Horizontal cylinder	$h = 1.32 \left(\dfrac{\Delta T}{d} \right)^{1/4}$	$h = 1.24 (\Delta T)^{1/3}$
Horizontal plate:		
Heated plate facing upward or cooled plate facing downward	$h = 1.32 \left(\dfrac{\Delta T}{L} \right)^{1/4}$	$h = 1.52 (\Delta T)^{1/3}$
Heated plate facing downward or cooled plate facing upward	$h = 0.59 \left(\dfrac{\Delta T}{L} \right)^{1/4}$	
Heated cube; L = length of side, Area $= 6L^2$	$h = 1.052 \left(\dfrac{\Delta T}{L} \right)^{1/4}$	

where h = heat-transfer coefficient, W/m$^2 \cdot$ °C
$\Delta T = T_w - T_\infty$, °C
L = vertical or horizontal dimension, m
d = diameter, m

| **EXAMPLE 7-7** | Calculation with Simplified Relations |

Compute the heat transfer for the conditions of Example 7-5 using the simplified relations of Table 7-2.

■ **Solution**
In Example 7-5 we found that a rather large pipe with a substantial temperature difference between the surface and air still had a Gr Pr product of $1.57 \times 10^8 < 10^9$, so a laminar equation is selected from Table 7-2. The heat-transfer coefficient is given by

$$h = 1.32 \left(\frac{\Delta T}{d} \right)^{1/4} = 1.32 \left(\frac{250 - 15}{0.3048} \right)^{1/4}$$

$$= 6.96 \ \mathrm{W/m^2 \cdot {}^\circ C}$$

The heat transfer is then

$$\frac{q}{L} = (6.96)\pi(0.3048)(250 - 15) = 1.57 \ \mathrm{kW/m}$$

Note that the simplified relation gives a value approximately 4 percent higher than Equation (7-25).

7-10 | FREE CONVECTION FROM SPHERES

Yuge [5] recommends the following empirical relation for free-convection heat transfer from spheres to air:

$$\mathrm{Nu}_f = \frac{\overline{h}d}{k_f} = 2 + 0.392 \ \mathrm{Gr}_f^{1/4} \qquad \text{for } 1 < \mathrm{Gr}_f < 10^5 \tag{7-49}$$

This equation may be modified by the introduction of the Prandtl number to give

$$\mathrm{Nu}_f = 2 + 0.43 (\mathrm{Gr}_f \ \mathrm{Pr}_f)^{1/4} \tag{7-50}$$

Properties are evaluated at the film temperature, and it is expected that this relation would be primarily applicable to calculations for free convection in gases. However, in the absence of more specific information it may also be used for liquids. We may note that for very low values of the Grashof-Prandtl number product the Nusselt number approaches a value of 2.0. This is the value that would be obtained for pure conduction through an infinite stagnant fluid surrounding the sphere, as obtained from Table 3-1.

For higher ranges of the Rayleigh number the experiments of Amato and Tien [79] with water suggest the following correlation:

$$\text{Nu}_f = 2 + 0.50(\text{Gr}_f \, \text{Pr}_f)^{1/4} \qquad \text{[7-51]}$$

for $3 \times 10^5 < \text{Gr} \, \text{Pr} < 8 \times 10^8$.

Churchill [83] suggests a more general formula for spheres, applicable over a wider range of Rayleigh numbers:

$$\text{Nu} = 2 + \frac{0.589 \text{Ra}_d{}^{1/4}}{[1 + (0.469/\text{Pr})^{9/16}]^{4/9}} \qquad \text{[7-52]}$$

for $\text{Ra}_d < 10^{11}$ and $\text{Pr} > 0.5$.

7-11 | FREE CONVECTION IN ENCLOSED SPACES

The free-convection flow phenomena inside an enclosed space are interesting examples of very complex fluid systems that may yield to analytical, empirical, and numerical solutions. Consider the system shown in Figure 7-8, where a fluid is contained between two vertical plates separated by the distance δ. As a temperature difference $\Delta T_w = T_1 - T_2$ is impressed on the fluid, a heat transfer will be experienced with the approximate flow regions shown in Figure 7-9, according to MacGregor and Emery [18]. In this figure, the Grashof number

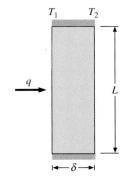

Figure 7-8 | Nomenclature for free convection in enclosed vertical spaces.

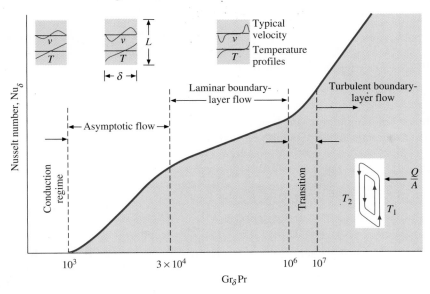

Figure 7-9 | Schematic diagram and flow regimes for the vertical convection layer, according to Reference 18.

is calculated as

$$\text{Gr}_\delta = \frac{g\beta(T_1 - T_2)\delta^3}{\nu^2} \qquad \text{[7-53]}$$

At very low Grashof numbers, there are very minute free-convection currents and the heat transfer occurs mainly by conduction across the fluid layer. As the Grashof number is increased, different flow regimes are encountered, as shown, with a progressively increasing heat transfer as expressed through the Nusselt number

$$\text{Nu}_\delta = \frac{h\delta}{k}$$

Although some open questions still remain, the experiments of Reference 18 may be used to predict the heat transfer to a number of liquids under constant-heat-flux conditions. The empirical correlations obtained were:

$$\text{Nu}_\delta = 0.42(\text{Gr}_\delta \, \text{Pr})^{1/4}\text{Pr}^{0.012}\left(\frac{L}{\delta}\right)^{-0.30} \qquad q_w = \text{const} \qquad \text{[7-54]}$$

$$10^4 < \text{Gr}_\delta \, \text{Pr} < 10^7$$
$$1 < \text{Pr} < 20,000$$
$$10 < L/\delta < 40$$

$$\text{Nu}_\delta = 0.46 \, (\text{Gr}_\delta \, \text{Pr})^{1/3} \qquad q_w = \text{const} \qquad \text{[7-55]}$$
$$10^6 < \text{Gr}_\delta \, \text{Pr} < 10^9$$
$$1 < \text{Pr} < 20$$
$$1 < L/\delta < 40$$

The heat flux is calculated as

$$\frac{q}{A} = q_w = h(T_1 - T_2) = \text{Nu}_\delta \frac{k}{\delta}(T_1 - T_2) \qquad \text{[7-56]}$$

The results are sometimes expressed in the alternate form of an *effective* or *apparent thermal conductivity* k_e, defined by

$$\frac{q}{A} = k_e \frac{T_1 - T_2}{\delta} \qquad \text{[7-57]}$$

By comparing Equations (7-56) and (7-57), we see that

$$\text{Nu}_\delta \equiv \frac{k_e}{k} \qquad \text{[7-58]}$$

In the building industry the heat transfer across an air gap is sometimes expressed in terms of the R values (see Section 2-3), so that

$$\frac{q}{A} = \frac{\Delta T}{R}$$

In terms of the above discussion, the R value would be

$$R = \frac{\delta}{k_e} \qquad \text{[7-59]}$$

Heat transfer in horizontal enclosed spaces involves two distinct situations. If the upper plate is maintained at a higher temperature than the lower plate, the lower-density fluid is above the higher-density fluid and no convection currents will be experienced. In this case

Figure 7-10 | Benard-cell pattern in enclosed fluid layer heated from below, from Reference 33.

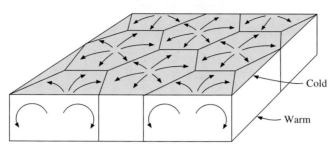

the heat transfer across the space will be by conduction alone and $Nu_\delta = 1.0$, where δ is still the separation distance between the plates. The second, and more interesting, case is experienced when the lower plate has a higher temperature than the upper plate. For values of Gr_δ below about 1700, pure conduction is still observed and $Nu_\delta = 1.0$. As convection begins, a pattern of hexagonal cells is formed as shown in Figure 7-10. These patterns are called Benard cells [33]. Turbulence begins at about $Gr_\delta = 50,000$ and destroys the cellular pattern.

Free convection in inclined enclosures is discussed by Dropkin and Somerscales [12]. Evans and Stefany [9] have shown that transient natural-convection heating or cooling in closed vertical or horizontal cylindrical enclosures may be calculated with

$$Nu_f = 0.55(Gr_f \, Pr_f)^{1/4} \qquad\qquad \textbf{[7-60]}$$

for the range $0.75 < L/d < 2.0$. The Grashof number is formed with the length of the cylinder L.

The analysis and experiments of Reference 43 indicate that it is possible to represent the effective thermal conductivity for fluids between concentric spheres with the relation

$$\frac{k_e}{k} = 0.228(Gr_\delta \, Pr)^{0.226} \qquad\qquad \textbf{[7-61]}$$

where now the gap spacing is $\delta = r_o - r_i$. The effective thermal conductivity given by Equation (7-61) is to be used with the conventional relation for steady-state conduction in a spherical shell:

$$q = \frac{4\pi k_e r_i r_o \Delta T}{r_o - r_i} \qquad\qquad \textbf{[7-62]}$$

Equation (7-61) is valid for $0.25 \leq \delta/r_i \leq 1.5$ and

$$1.2 \times 10^2 < Gr \, Pr < 1.1 \times 10^9 \qquad 0.7 < Pr < 4150$$

Properties are evaluated at a volume mean temperature T_m defined by

$$T_m = \frac{(r_m^3 - r_i^3)T_i + (r_o^3 - r_m^3)T_o}{r_o^3 - r_i^3} \qquad\qquad \textbf{[7-63]}$$

where $r_m = (r_i + r_o)/2$. Equation (7-61) may also be used for eccentric spheres with a coordinate transformation as described in Reference 43.

Experimental results for free convection in enclosures are not always in agreement, but we can express them in a general form as

$$\frac{k_e}{k} = C(Gr_\delta \, Pr)^n \left(\frac{L}{\delta}\right)^m \qquad\qquad \textbf{[7-64]}$$

Table 7-3 | Summary of empirical relations for free convection in enclosures in the form of Equation (7-61), correlation constants adjusted by Holman [74].

Fluid	Geometry	Gr_δ Pr	Pr	$\dfrac{L}{\delta}$	C	n	m	Reference(s)
Gas	Vertical plate, isothermal	< 2000	$k_e/k = 1.0$					6, 7, 55, 59
		6000–200,000	0.5–2	11–42	0.197	$\frac{1}{4}$	$-\frac{1}{9}$	
		$200{,}000\text{–}1.1 \times 10^7$	0.5–2	11–42	0.073	$\frac{1}{3}$	$-\frac{1}{9}$	
	Horizontal plate, isothermal heated from	< 1700	$k_e/k = 1.0$					
		1700–7000	0.5–2	—	0.059	0.4	0	6, 7, 55, 59, 62, 63
	below	$7000\text{–}3.2 \times 10^5$	0.5–2	—	0.212	$\frac{1}{4}$	0	66
		$> 3.2 \times 10^5$	0.5–2	—	0.061	$\frac{1}{3}$	0	
Liquid	Vertical plate, constant heat flux or isothermal	< 2000	$k_e/k = 1.0$					
		$10^4\text{–}10^7$	1–20,000	10–40	Eq. 7-52	—	—	18, 61
		$10^6\text{–}10^9$	1–20	1–40	0.046	$\frac{1}{3}$	0	
	Horizontal plate, isothermal, heated from	< 1700	$k_e/k = 1.0$	—				7, 8, 58, 63, 66
		1700–6000	1–5000	—	0.012	0.6	0	
	below	6000–37,000	1–5000	—	0.375	0.2	0	
		$37{,}000\text{–}10^8$	1–20	—	0.13	0.3	0	
		$> 10^8$	1–20	—	0.057	$\frac{1}{3}$	0	
Gas or liquid	Vertical annulus	Same as vertical plates						
	Horizontal annulus, isothermal	$6000\text{–}10^6$	1–5000	—	0.11	0.29	0	56, 57, 60
		$10^6\text{–}10^8$	1–5000	—	0.40	0.20	0	
	Spherical annulus	$120\text{–}1.1 \times 10^9$	0.7–4000	—	0.228	0.226	0	43

Table 7-3 lists values of the constants C, n, and m for a number of physical circumstances. These values may be used for design purposes in the absence of specific data for the geometry or fluid being studied. We should remark that some of the data correlations represented by Table 7-3 have been artificially adjusted by Holman [74] to give the characteristic exponents of $\frac{1}{4}$ and $\frac{1}{3}$ for the laminar and turbulent regimes of free convection. However, it appears that the error introduced by this adjustment is not significantly greater than the disagreement between different experimental investigations. The interested reader may wish to consult the specific references for more details.

For the annulus space the heat transfer is based on

$$q = \frac{2\pi k_e L \Delta T}{\ln(r_o/r_i)} \qquad \text{[7-65]}$$

where L is the length of the annulus and the gap spacing is $\delta = r_o - r_i$.

Extensive correlations for free convection between cylindrical, cubical, and spherical bodies and various enclosure geometries are given by Warrington and Powe [80]. The correlations cover a wide range of fluids.

Free convection through vertical plane layers of nonnewtonian fluids is discussed in Reference 38, but the results are too complicated to present here.

In the absence of more specific design information, the heat transfer for inclined enclosures may be calculated by substituting g' for g in the Grashof number, where

$$g' = g \cos \theta \qquad \text{[7-66]}$$

and θ is the angle that the heater surface makes with the horizontal. This transformation may be expected to hold up to inclination angles of $60°$ and applies *only* to those cases where the hotter surface is facing upward. Further information is available from Hollands et al. [66, 67, 69, 82].

Radiation R-Value for a Gap

As we have seen in conduction heat transfer, radiation boundary conditions may play an important role in the overall heat-transfer problem. This is particularly true in free-convection situations because free-convection heat-transfer rates are typically small. We will show in Section 8-7, Equation (8-42), that the radiant transfer across a gap separating two large parallel planes may be calculated with

$$q/A = \frac{\sigma \left(T_1^4 - T_2^4 \right)}{1/\epsilon_1 + 1/\epsilon_2 - 1} \qquad \text{[7-67]}$$

where the temperatures are in degrees Kelvin and the ϵ's are the respective emissivities of the surfaces. Using the concept of the R-value discussed in Section 2-3, we could write

$$(q/A)_{\text{rad}} = \Delta T / R_{\text{rad}}$$

and thus could determine an R-value for the radiation heat transfer in conjunction with Equation (7-67). That value would be strongly temperature-dependent and would operate in parallel with the R-value for the convection across the space, which could be obtained from

$$(q/A)_{\text{conv}} = k_e \Delta T / \delta = \Delta T / R_{\text{conv}}$$

so that

$$R_{\text{conv}} = \delta / k_e$$

The total R-value for the combined radiation and convection across the space would be written as

$$R_{\text{tot}} = \frac{1}{1/R_{\text{rad}} + 1/R_{\text{conv}}}$$

The concept of combined radiation and convection in confined spaces is important in building applications.

Heat Transfer Across Vertical Air Gap   EXAMPLE 7-8

Air at atmospheric pressure is contained between two 0.5-m-square vertical plates separated by a distance of 15 mm. The temperatures of the plates are 100 and 40°C, respectively. Calculate the free-convection heat transfer across the air space. Also calculate the radiation heat transfer across the air space if both surfaces have $\epsilon = 0.2$.

■ **Solution**
We evaluate the air properties at the mean temperature between the two plates:

$$T_f = \frac{100 + 40}{2} = 70°\text{C} = 343 \text{ K}$$

$$\rho = \frac{p}{RT} = \frac{1.0132 \times 10^5}{(287)(343)} = 1.029 \text{ kg/m}^3$$

$$\beta = \frac{1}{T_f} = \frac{1}{343} = 2.915 \times 10^{-3} \text{ K}^{-1}$$

$$\mu = 2.043 \times 10^{-5} \text{ kg/m} \cdot \text{s} \qquad k = 0.0295 \text{ W/m} \cdot °\text{C} \qquad \text{Pr} = 0.7$$

The Grashof-Prandtl number product is now calculated as

$$\mathrm{Gr}_\delta\,\mathrm{Pr} = \frac{(9.8)(1.029)^2(2.915\times10^{-3})(100-40)(15\times10^{-3})^3}{(2.043\times10^{-5})^2}0.7$$

$$= 1.027\times10^4$$

We may now use Equation (7-64) to calculate the effective thermal conductivity, with $L=0.5$ m, $\delta=0.015$ m, and the constants taken from Table 7-3:

$$\frac{k_e}{k} = (0.197)(1.027\times10^4)^{1/4}\left(\frac{0.5}{0.015}\right)^{-1/9} = 1.343$$

The heat transfer may now be calculated with Equation (7-54). The area is $(0.5)^2=0.25$ m^2, so that

$$q = \frac{(1.343)(0.0295)(0.25)(100-40)}{0.015} = 39.62\text{ W}\qquad[135.2\text{ Btu/h}]$$

The radiation heat flux is calculated with Equation (7-67), taking $T_1=373$ K, $T_2=313$ K, and $\epsilon_1=\epsilon_2=0.2$. Thus, with $\sigma=5.669\times10^{-8}$ W/m$^2\cdot$K^4,

$$(q/A)_{\mathrm{rad}} = \frac{(5.669\times10^{-8})(373^4-313^4)}{[1/0.2+1/0.2-1]} = 61.47\text{ W/m}^2$$

and

$$q_{\mathrm{rad}} = (0.5)^2(61.47) = 15.37\text{ W}$$

or about half the value of the convection transfer across the space. Further calculation would show that for a smaller value of $\epsilon=0.05$, the radiation transfer is reduced to 3.55 W or, for a larger value of $\epsilon=0.8$, the transfer is 92.2 W. In any event, radiation heat transfer can be an important factor in such problems.

EXAMPLE 7-9   Heat Transfer Across Horizontal Air Gap

Two horizontal plates 20 cm on a side are separated by a distance of 1 cm with air at 1 atm in the space. The temperatures of the plates are 100°C for the lower and 40°C for the upper plate. Calculate the heat transfer across the air space.

■ **Solution**
The properties are the same as given in Example 7-8:

$$\rho = 1.029\text{ kg/m}^3\qquad \beta=2.915\times10^{-3}\text{ K}^{-1}$$
$$\mu = 2.043\times10^{-5}\text{ kg/m}\cdot\text{s}\qquad k=0.0295\text{ W/m}\cdot°\text{C}$$
$$\mathrm{Pr} = 0.7$$

The Gr Pr product is evaluated on the basis of the separating distance, so we have

$$\mathrm{Gr\,Pr} = \frac{(9.8)(1.029)^2(2.915\times10^{-3})(100-40)(0.01)^3}{(2.043\times10^{-5})^2}(0.7) = 3043$$

Consulting Table 7-3, we find $C=0.059$, $n=0.4$, and $m=0$ so that

$$\frac{k_e}{k} = (0.059)(3043)^{0.4}\left(\frac{0.2}{0.01}\right)^0 = 1.46$$

and

$$q = \frac{k_e A(T_1-T_2)}{\delta} = \frac{(1.460)(0.0295)(0.2)^2(100-40)}{0.01} = 10.34\text{ W}$$

Heat Transfer Across Water Layer   **EXAMPLE 7-10**

Two 50-cm horizontal square plates are separated by a distance of 1 cm. The lower plate is maintained at a constant temperature of 100°F and the upper plate is constant at 80°F. Water at atmospheric pressure occupies the space between the plates. Calculate the heat lost by the lower plate.

■ **Solution**

We evaluate properties at the mean temperature of 90°F and obtain, for water,

$$k = 0.623 \text{ W/m} \cdot °\text{C} \qquad \frac{g\beta\rho^2 c_p}{\mu k} = 2.48 \times 10^{10}$$

The Grashof-Prandtl number product is now evaluated using the plate spacing of 1 cm as the characteristic dimension.

$$\text{Gr Pr} = (2.48 \times 10^{10})(0.01)^3(100 - 80)(5/9) = 2.76 \times 10^5$$

Now, using Equation (7-64) and consulting Table 7-3 we obtain

$$C = 0.13 \qquad n = 0.3 \qquad m = 0$$

Therefore, Equation (7-64) becomes

$$\frac{k_e}{k} = (0.13)(2.76 \times 10^5)^{0.3} = 5.57$$

The effective thermal conductivity is thus

$$k_e = (0.623)(5.57) = 3.47 \text{ W/m} \cdot °\text{C}$$

and the heat transfer is

$$q = k_e A \Delta T / \delta = \frac{(3.47)(0.5)^2(100 - 80)(5/9)}{0.01} = 964 \text{ W}$$

We see, of course, that the heat transfer across a water gap is considerably larger than for an air gap [Example 7-9] because of the larger thermal conductivity.

Reduction of Convection in Air Gap   **EXAMPLE 7-11**

A vertical air gap between two glass plates is to be evacuated so that the convective currents are essentially eliminated, that is, the air behaves as a pure conductor. For air at a mean temperature of 300 K and a temperature difference of 20°C, calculate the vacuum necessary for glass spacings of 1 and 2 cm.

■ **Solution**

Consulting Table 7-3, we find that for gases, a value of $\text{Gr}_\delta \text{ Pr} < 2000$ is necessary to reduce the system to one of pure conduction. At 300 K the propertiues of air are

$$k = 0.02624 \text{ W/m} \cdot °\text{C} \quad \text{Pr} = 0.7 \quad \mu = 1.846 \times 10^{-5} \text{ kg/m} \cdot \text{s} \quad \beta = 1/300$$

and

$$\rho = p/RT = p/(287)(300)$$

We have

$$\text{Gr}_\delta \text{ Pr} = g\beta\rho^2 \Delta T \delta^3 \text{Pr}/\mu^2 = 2000$$

$$= (9.8)(1/300)[p/(287)(300)]^2(20)\delta^3(0.7)/(1.846 \times 10^{-5})^2$$

and $p^2 \delta^3 = 7773$. Therefore, for a plate spacing of $\delta = 1$ cm we have

$$p = [7773/(0.01)^3]^{1/2} = 88200 \text{ Pa}$$

or, vacuum $= p_{\text{atm}} - p = 101320 - 88200 = 13120$ Pa. For a spacing of 2 cm,

$$p = 31190 \text{ Pa and vacuum} = 70130 \text{ Pa}$$

Both vacuum figures are modest and easily achieved in practice.

Evacuated (Low-Density) Spaces

In the equations presented for free convection in enclosures we have seen that when the product Gr_δ Pr is sufficiently small, usually less than about 2000, the fluid layer behaves as if pure conduction were involved and $k_e/k \to 1.0$. This means that the free-convection flow velocities are small. A small value of Gr_δ can result from either lowering the fluid pressure (density) or by reducing the spacing δ. If the pressure of a gas is reduced sufficiently, we refer to the situation as a low-density problem, which is influenced by the mean free path of the molecules and by individual molecular impacts.

A number of practical situations involve heat transfer between a solid surface and a low-density gas. In employing the term *low density*, we shall mean those circumstances where the mean free path of the gas molecules is no longer small in comparison with a characteristic dimension of the heat-transfer surface. The *mean free path* λ is the distance a molecule travels, on the average, between collisions. The larger this distance becomes, the greater the distance required to communicate the temperature of a hot surface to a gas in contact with it. This means that we shall not necessarily be able to assume that a gas in the immediate neighborhood of the surface will have the same temperature as the heated surface, as was done in the boundary-layer analyses. Evidently, the parameter that is of principal interest is a ratio of the mean free path to a characteristic body dimension. This grouping is called the Knudsen number,

$$\text{Kn} = \frac{\lambda}{L} \tag{7-68}$$

According to the kinetic theory of gases, the mean free path may be calculated from

$$\lambda = \frac{0.707}{4\pi r^2 n} \tag{7-69}$$

where r is the effective molecular radius for collisions and n is the molecular density. An approximate relation for the mean free path of air molecules is given by

$$\lambda = 2.27 \times 10^{-5} \frac{T}{p} \text{ meters} \tag{7-70}$$

where T is in degrees Kelvin and p is in pascals.

As a first example of low-density heat transfer let us consider the two parallel infinite plates shown in Figure 7-11. The plates are maintained at different temperatures and separated by a gaseous medium. Let us first consider a case where the density or plate spacing is low enough that free convection effects are negligible, but with a gas density sufficiently high so that $\lambda \to 0$ and a linear temperature profile through the gas will be

Figure 7-11 | Effect of mean free path on conduction heat transfer between parallel plates: (a) physical model; (b) anticipated temperature profiles.

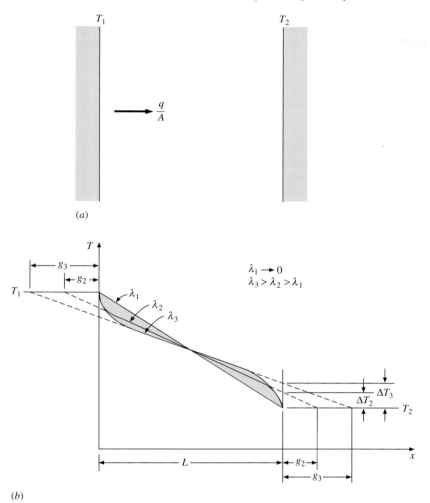

experienced, as shown for the case of λ_1. As the gas density is lowered, the larger mean free paths require a greater distance from the heat-transfer surfaces in order for the gas to accommodate to the surface temperatures. The anticipated temperature profiles are shown in Figure 7-11b. Extrapolating the straight portion of the low-density curves to the wall produces a temperature "jump" ΔT, which may be calculated by making the following energy balance:

$$\frac{q}{A} = k\frac{T_1 - T_2}{g + L + g} = k\frac{\Delta T}{g} \qquad \textbf{[7-71]}$$

In this equation we are assuming that the extrapolation distance g is the same for both plate surfaces. In general, the temperature jump will depend on the type of surface, and these extrapolation distances will not be equal unless the materials are identical. For different types of materials we should have

$$\frac{q}{A} = k\frac{T_1 - T_2}{g_1 + L + g_2} = k\frac{\Delta T_1}{g_1} = k\frac{\Delta T_2}{g_2} \qquad \text{[7-72]}$$

where now ΔT_1 and ΔT_2 are the temperature jumps at the two heat-transfer surfaces and g_1 and g_2 are the corresponding extrapolation distances. For identical surfaces the temperature jump would then be expressed as

$$\Delta T = \frac{g}{2g + L}(T_1 - T_2) \qquad \text{[7-73]}$$

Similar expressions may be developed for low-density conduction between concentric cylinders. In order to predict the heat-transfer rate it is necessary to establish relations for the temperature jump for various gas-to-solid interfaces.

We have already mentioned that the temperature-jump effect arises as a result of the failure of the molecules to "accommodate" to the surface temperature when the mean free path becomes of the order of a characteristic body dimension. The parameter that describes this behavior is called the *accommodation coefficient* α, defined by

$$\alpha = \frac{E_i - E_r}{E_i - E_w} \qquad \text{[7-74]}$$

where

$E_i = $ energy of incident molecules on a surface

$E_r = $ energy of molecules reflected from the surface

$E_w = $ energy molecules would have if they acquired energy of wall at temperature T_w

Values of the accommodation coefficient must be determined from experiment, and some typical values are given in Table 7-4.

It is possible to employ the kinetic theory of gases along with values of α to determine the temperature jump at a surface. The result of such an analysis is

$$T_{y=0} - T_w = \frac{2 - \alpha}{\alpha}\frac{2\gamma}{\gamma + 1}\frac{\lambda}{\text{Pr}}\frac{\partial T}{\partial y}\Bigg]_{y=0} \qquad \text{[7-75]}$$

Table 7-4 | Thermal accommodation coefficients for air at low pressure in contact with various surfaces.

Surface	Accommodation coefficient, α
Flat black lacquer on bronze	0.88–0.89
Bronze, polished	0.91–0.94
Machined	0.89–0.93
Etched	0.93–0.95
Cast iron, polished	0.87–0.93
Machined	0.87–0.88
Etched	0.89–0.96
Aluminum, polished	0.87–0.95
Machined	0.95–0.97
Etched	0.89–0.97

Figure 7-12 | Nomenclature for use with Equation (7-75).

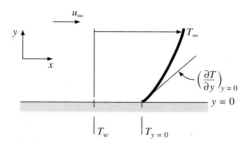

The nomenclature for Equation (7-75) is noted in Figure 7-12. This temperature jump is denoted by ΔT in Figure 7-11, and the temperature gradient for use with Figure 7-11 would be

$$\frac{T_1 - T_2 - 2\Delta T}{L}$$

For very low densities (high vacuum) the mean free path may become very large compared to the plate separation distance and the conduction-convection heat transfer will approach zero. The reader should recognize, however, that the total heat transfer across the gap-space will be the sum of conduction-convection and radiation heat transfer. We will discuss radiation heat transfer in detail in Chapter 8, but we have already provided the relation in Equation (7-67) for calculation of radiant heat transfer between two parallel plates. We note that ϵ approaches 1.0 for highly absorptive surfaces and has a small value for highly reflective surfaces. Example 7-12 illustrates the application of the low-density relations to calculation of heat transfer across a gap.

Heat Transfer Across Evacuated Space EXAMPLE 7-12

Two polished-aluminum plates ($\epsilon = 0.06$) are separated by a distance of 2.5 cm in air at a pressure of 10^{-6} atm. The plates are maintained at 100 and 30°C, respectively. Calculate the conduction heat transfer through the air gap. Compare this with the radiation heat transfer and the conduction for air at normal atmospheric pressure.

■ Solution
We first calculate the mean free path to determine if low-density effects are important. From Equation (7-70), at an average temperature of 65°C = 338 K,

$$\lambda = \frac{(2.27 \times 10^{-5})(338)}{(1.0132 \times 10^{+5})(10^{-6})} = 0.0757 \text{ m} = 7.57 \text{ cm} \quad [0.248 \text{ ft}]$$

Since the plate spacing is only 2.5 cm, we should expect low-density effects to be important. Evaluating properties at the mean air temperature of 65°C, we have

$$k = 0.0291 \text{ W/m} \cdot °C \quad [0.0168 \text{ Btu/h} \cdot \text{ft} \cdot °F]$$

$$\gamma = 1.40 \qquad \text{Pr} = 0.7 \qquad \alpha \approx 0.9 \qquad \text{from Table 7-4}$$

Combining Equation (7-75) with the central-temperature-gradient relation gives

$$\Delta T = \frac{2 - \alpha}{\alpha} \frac{2\gamma}{\gamma + 1} \frac{\lambda}{\text{Pr}} \frac{T_1 - T_2 - 2\Delta T}{L}$$

Inserting the appropriate properties gives

$$\Delta T = \frac{2 - 0.9}{0.9} \frac{2.8}{2.4} \frac{0.0757}{0.7} \frac{100 - 30 - 2\Delta T}{0.025}$$
$$= 32.38°C \quad [58.3°F]$$

The conduction heat transfer is thus

$$\frac{q}{A} = k\frac{T_1 - T_2 - 2\Delta T}{L} = \frac{(0.0291)(70 - 64.76)}{0.025}$$
$$= 6.099 \text{ W/m}^2 \quad [1.93 \text{ Btu/h} \cdot \text{ft}^2]$$

At normal atmospheric pressure the conduction would be

$$\frac{q}{A} = k\frac{T_1 - T_2}{L} = 81.48 \text{ W/m}^2 \quad [25.8 \text{ Btu/h} \cdot \text{ft}^2]$$

The radiation heat transfer is calculated with Equation (8-42), taking $\epsilon_1 = \epsilon_2 = 0.06$ for polished aluminum:

$$\left(\frac{q}{A}\right)_{rad} = \frac{\sigma(T_1^4 - T_2^4)}{2/\epsilon - 1} = \frac{(5.669 \times 10^{-8})(393^4 - 303^4)}{2/0.06 - 1}$$
$$= 27.05 \text{ W/m}^2 \quad [8.57 \text{ Btu/h} \cdot \text{ft}^2]$$

Thus, at the low-density condition the radiation heat transfer is almost 5 times as large as the conduction, even with highly polished surfaces.

7-12 | COMBINED FREE AND FORCED CONVECTION

A number of practical situations involve convection heat transfer that is neither "forced" nor "free" in nature. The circumstances arise when a fluid is forced over a heated surface at a rather low velocity. Coupled with the forced-flow velocity is a convective velocity that is generated by the buoyancy forces resulting from a reduction in fluid density near the heated surface.

A summary of combined free- and forced-convection effects in tubes has been given by Metais and Eckert [10], and Figure 7-13 presents the regimes for combined convection in vertical tubes. Two different combinations are indicated in this figure. *Aiding flow* means that the forced- and free-convection currents are in the same direction, while *opposing flow* means that they are in the opposite direction. The abbreviation UWT means uniform wall temperature, and the abbreviation UHF indicates data for uniform heat flux. It is fairly easy to anticipate the qualitative results of the figure. A large Reynolds number implies a large forced-flow velocity, and hence less influence of free-convection currents. The larger the value of the Grashof-Prandtl product, the more one would expect free-convection effects to prevail.

Figure 7-14 presents the regimes for combined convection in horizontal tubes. In this figure the Graetz number is defined as

$$\text{Gz} = \text{Re} \, \text{Pr} \frac{d}{L} \qquad\qquad \textbf{[7-76]}$$

The applicable range of Figures 7-13 and 7-14 is for

$$10^{-2} < \text{Pr}\left(\frac{d}{L}\right) < 1$$

Figure 7-13 | Regimes of free, forced, and mixed convection for flow through vertical tubes, according to Reference 10.

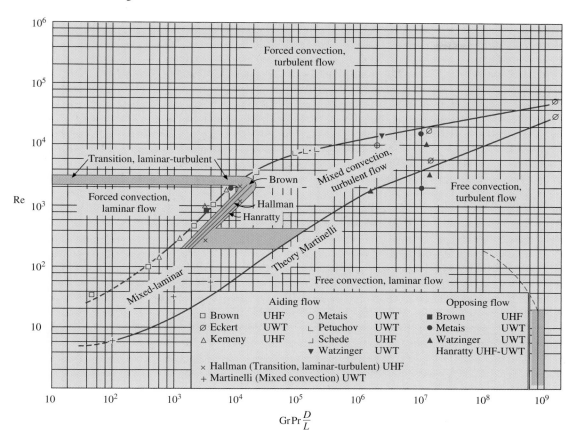

The correlations presented in the figures are for constant wall temperature. All properties are evaluated at the film temperature.

Brown and Gauvin [17] have developed a better correlation for the mixed-convection, laminar flow region of Figure 7-14:

$$\text{Nu} = 1.75 \left(\frac{\mu_b}{\mu_w} \right)^{0.14} [\text{Gz} + 0.012(\text{Gz Gr}^{1/3})^{4/3}]^{1/3} \qquad \textbf{[7-77]}$$

where μ_b is evaluated at the bulk temperature. This relation is preferred over that shown in Figure 7-14. Further information is available in Reference 68. The problem of combined free and forced convection from horizontal cylinders is treated in detail by Fand and Keswani [47].

Criterion for Free or Forced Convection

The general notion that is applied in combined-convection analysis is that the predominance of a heat-transfer mode is governed by the fluid velocity associated with that mode. A forced-convection situation involving a fluid velocity of 30 m/s, for example, would be expected to overshadow most free-convection effects encountered in ordinary gravitational fields because the velocities of the free-convection currents are small in comparison with

Figure 7-14 | Regimes of free, forced, and mixed convection for flow through horizontal tubes, according to Metais and Eckert [10].

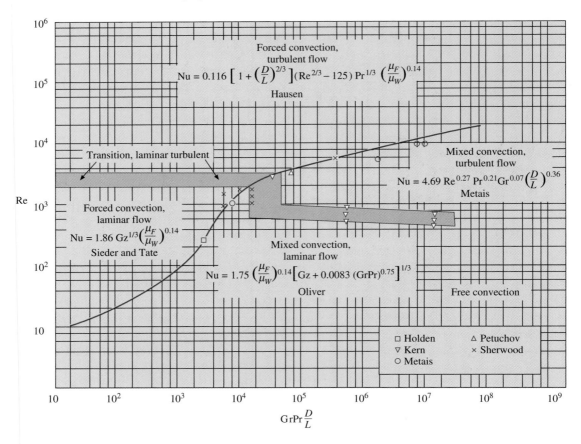

30 m/s. On the other hand, a forced-flow situation at very low velocities (~ 0.3 m/s) might be influenced appreciably by free-convection currents. An order-of-magnitude analysis of the free-convection boundary-layer equations will indicate a general criterion for determining whether free-convection effects dominate. The criterion is that when

$$Gr/Re^2 > 10 \qquad\qquad \textbf{[7-78]}$$

free convection is of primary importance. This result is in agreement with Figures 7-13 and 7-14.

| EXAMPLE 7-13 | Combined Free and Forced Convection with Air |

Air at 1 atm and 27°C is forced through a horizontal 25-mm-diameter tube at an average velocity of 30 cm/s. The tube wall is maintained at a constant temperature of 140°C. Calculate the heat-transfer coefficient for this situation if the tube is 0.4 m long.

■ **Solution**
For this calculation we evaluate properties at the film temperature:

$$T_f = \frac{140 + 27}{2} = 83.5°C = 356.5 \text{ K}$$

$$\rho_f = \frac{p}{RT} = \frac{1.0132 \times 10^5}{(287)(356.5)} = 0.99 \text{ kg/m}^3$$

$$\beta = \frac{1}{T_f} = 2.805 \times 10^{-3} \text{ K}^{-1} \qquad \mu_w = 2.337 \times 10^{-5} \text{ kg/m} \cdot \text{s}$$

$$\mu_f = 2.102 \times 10^{-5} \text{ kg/m} \cdot \text{s} \qquad k_f = 0.0305 \text{ W/m} \cdot {}^\circ\text{C} \qquad \text{Pr} = 0.695$$

Let us take the bulk temperature as 27°C for evaluating μ_b; then

$$\mu_b = 1.8462 \times 10^{-5} \text{ kg/m} \cdot \text{s}$$

The significant parameters are calculated as

$$\text{Re}_f = \frac{\rho u d}{\mu} = \frac{(0.99)(0.3)(0.025)}{2.102 \times 10^{-5}} = 3.53$$

$$\text{Gr} = \frac{\rho^2 g \beta (T_w - T_b) d^3}{\mu^2} = \frac{(0.99)^2 (9.8)(2.805 \times 10^{-3})(140 - 27)(0.025)^3}{(2.102 \times 10^{-5})^2}$$

$$= 1.007 \times 10^5$$

$$\text{Gr Pr} \frac{d}{L} = (1.077 \times 10^5)(0.695)\frac{0.025}{0.4} = 4677$$

According to Figure 7-14, the mixed-convection-flow regime is encountered. Thus we must use Equation (7-77). The Graetz number is calculated as

$$\text{Gz} = \text{Re Pr} \frac{d}{L} = \frac{(353)(0.695)(0.025)}{0.4} = 15.33$$

and the numerical calculation for Equation (7-77) becomes

$$\text{Nu} = 1.75 \left(\frac{1.8462}{2.337}\right)^{0.14} \{15.33 + (0.012)[(15.33)(1.077 \times 10^5)^{1/3}]^{4/3}\}^{1/3}$$

$$= 7.70$$

The average heat-transfer coefficient is then calculated as

$$\bar{h} = \frac{k}{d} \text{Nu} = \frac{(0.0305)(7.70)}{0.025} = 9.40 \text{ W/m}^2 \cdot {}^\circ\text{C} \quad [1.67 \text{ Btu/h} \cdot \text{ft}^2 \cdot {}^\circ\text{F}]$$

It is interesting to compare this value with that which would be obtained for strictly laminar forced convection. The Sieder-Tate relation [Equation (6-10)] applies, so that

$$\text{Nu} = 1.86(\text{Re Pr})^{1/3} \left(\frac{\mu_f}{\mu_w}\right)^{0.14} \left(\frac{d}{L}\right)^{1/3}$$

$$= 1.86 \, \text{Gz}^{1/3} \left(\frac{\mu_f}{\mu_w}\right)^{0.14}$$

$$= (1.86)(15.33)^{1/3} \left(\frac{2.102}{2.337}\right)^{0.14}$$

$$= 4.55$$

and

$$\bar{h} = \frac{(4.55)(0.0305)}{0.025} = 5.55 \text{ W/m}^2 \cdot {}^\circ\text{C} \quad [0.977 \text{ Btu/h} \cdot \text{ft}^2 \cdot {}^\circ\text{F}]$$

Thus there would be an error of −41 percent if the calculation were made strictly on the basis of laminar forced convection.

Table 7-5 | Summary of free-convection heat-transfer relations T. For most cases, properties are evaluated at $T_f = (T_w + T_\infty)/2$.

Geometry	Equation	Restrictions	Equation number
A variety of isothermal surfaces	$\mathrm{Nu}_f = C(\mathrm{Gr}_f\,\mathrm{Pr}_f)^m$ C and m from Table 7-1	See Table 7-1	(7-25)
Vertical isothermal surface	$\overline{\mathrm{Nu}}^{1/2} = 0.825 + \dfrac{0.387\,\mathrm{Ra}^{1/6}}{[1+(0.492/\mathrm{Pr})^{9/16}]^{8/27}}$	$10^{-1} < \mathrm{Ra}_L < 10^{12}$ Also see Fig. 7-5	(7-29)
Vertical surface, constant heat flux, local h	$\mathrm{Nu}_{xf} = C(\mathrm{Gr}_x^*\mathrm{Pr}_f)^m$	$C = 0.60,\ m = \frac{1}{5}$ for $10^5 < \mathrm{Gr}_x^*\mathrm{Pr} < 10^{11}$ $C = 0.17,\ m = \frac{1}{4}$ for $2 \times 10^{13} < \mathrm{Gr}^*\mathrm{Pr} < 10^{16}$	(7-31) (7-32)
Isothermal horizontal cylinders	$\overline{\mathrm{Nu}}^{1/2} = 0.60 + 0.387\left\{\dfrac{\mathrm{Gr}\,\mathrm{Pr}}{[1+(0.559/\mathrm{Pr})^{9/16}]^{16/9}}\right\}^{1/6}$	$10^{-5} < \mathrm{Gr}\,\mathrm{Pr} < 10^{13}$ Also see Fig. 7-6	(7-36)
Horizontal surface, constant heat flux		See text	(7-39) to (7-42)
Inclined surfaces	Section 7-7	See text	
Spheres	$\mathrm{Nu} = 2 + 0.43(\mathrm{Gr}\,\mathrm{Pr})^{1/4}$ $\mathrm{Nu} = 2 + 0.5(\mathrm{Gr}\,\mathrm{Pr})^{1/4}$ $\mathrm{Nu} = 2 + \dfrac{0.589(\mathrm{Gr}\,\mathrm{Pr})^{1/4}}{[1+(0.469/\mathrm{Pr})^{9/16}]^{4/9}}$	$1 < \mathrm{Gr}\,\mathrm{Pr} < 10^5$ water, $3 \times 10^5 < \mathrm{Gr}\,\mathrm{Pr} < 8 \times 10^8$ $0.5 < \mathrm{Pr}$ $\mathrm{Gr}\,\mathrm{Pr} < 10^{11}$	(7-50) (7-51) (7-52)
Enclosed spaces	$q = k_e A(\Delta T/\delta)$ $\frac{k_e}{k} = C(\mathrm{Gr}_\delta\,\mathrm{Pr})^n\,(L/\delta)^m$	Constants C, m, and n from Table 7-3 Pure conduction for $\mathrm{Gr}_\delta\,\mathrm{Pr} < 2000$	(7-57) (7-64)
Across evacuated spaces	Most transfer is by radiation		

7-13 | SUMMARY

By now the reader will have sensed that there is an abundance of empirical relations for natural convection systems. Our purposes in this section are to (1) issue a few words of caution and (2) provide a convenient table to summarize the relations.

Most free-convection data are collected under laboratory conditions in still air, still water, etc. A practical free-convection problem might not be so fortunate and the boundary layer could have a slightly added forced-convection effect. In addition, real surfaces in practice are *seldom* isothermal or constant heat flux so the correlations developed from laboratory data for these conditions may not strictly apply. The net result, of course, is that the engineer must realize that calculated values of the heat-transfer coefficient can vary ± 25 percent from what will actually be experienced.

For solution of free-convection problems one should follow a procedure similar to that given in Chapter 6 for forced-convection problems. To aid the reader, a summary of free-convection correlations is given in Table 7-5.

7-14 | SUMMARY PROCEDURE FOR ALL CONVECTION PROBLEMS

At the close of Chapter 6 we gave a brief procedure for calculation of convection heat transfer. We now are in a position to expand that discussion to include the possibility of free-convection exchange. The procedure is as follows:

1. Specify the fluid involved and be prepared to determine properties of that fluid. This may seem like a trivial step, but a surprisingly large number of errors are made in practice by choosing the *wrong* fluid, that is to say, air instead of water.

2. Specify the geometry of the problem. Again, a seemingly simple matter, but important. Is there flow *inside* a tube, or flow *across the outside* of a tube, or flow *along the length* of a tube? Is the flow internal or external?

3. Decide whether the problem involves free convection or forced convection. If there is no specification of forcing the fluid through a channel or across some heated surface, free convection may be presumed. If there is a clear specification of a flow velocity, or mass flow rate, then forced convection may be assumed. When very small forced velocities are involved, combination free convection–forced convection may be encountered and the relative magnitudes of Re and Gr may need to be examined.

4. Once steps 1–3 are accomplished, decide on a temperature for evaluating fluid properties. This will usually be some average bulk temperature for forced flow in channels, and a film temperature $T_f = (T_\infty + T_{surface})/2$ for either free or forced-convection flow over exterior surfaces. Some modification of this calculation may be needed once the final convection relation for h is determined.

5. Determine the flow regime by evaluating the Grashof-Prandtl number product for free-convection problems or the Reynolds number for forced-convection situations. *Be particularly careful to employ the correct characteristic body dimension in this calculation.* A large number of mistakes are made in practice by failing to make this calculation properly, in accordance with the findings of step 2 above. At this point, determine if an average or local heat-transfer coefficient is required in the problem. Revise the calculation of Gr Pr or Re as needed.

6. Select an appropriate correlation equation for h in terms of the findings above. Be sure the equation fits the flow situation and geometry of the problem. If the equation selected requires modification of temperature-property determinations, revise the calculations in steps 4 and 5.

7. Calculate the value of h needed for the problem. Again, check to be sure that the calculation matches the geometry, fluid, type of flow, and flow regime for the problem.

8. Determine convection heat transfer for the problem, which is usually calculated with an equation of the form

$$q = h A_{surface}(T_{surface} - T_{free\ stream})$$

for either free or forced convection over exterior surfaces, and

$$q = h A_{surface}(T_{surface} - T_{bulk})$$

for forced convection inside channels. Be careful to employ the correct value for $A_{surface}$, which is the **surface area in contact with the fluid for which h is calculated.** For forced convection inside a tube, $A_{surface}$ is $\pi d_i L$ (*not* the flow cross-sectional area $\pi d_i^2/4$), while for crossflow or free convection on the outside of a tube, $A_{surface}$ is $\pi d_o L$. The surface area for complicated fin arrangements like those illustrated in Figure 2-14 would be the total fin(s) surface area in contact with the surrounding fluid (presumably air). This procedure is summarized in Figure 7-15, which also appears in the inside back cover.

REVIEW QUESTIONS

1. Why is an analytical solution of a free-convection problem more involved than its forced-convection counterpart?

2. Define the Grashof number. What is its physical significance?

Figure 7-15 | Summary of convection calculation procedure.

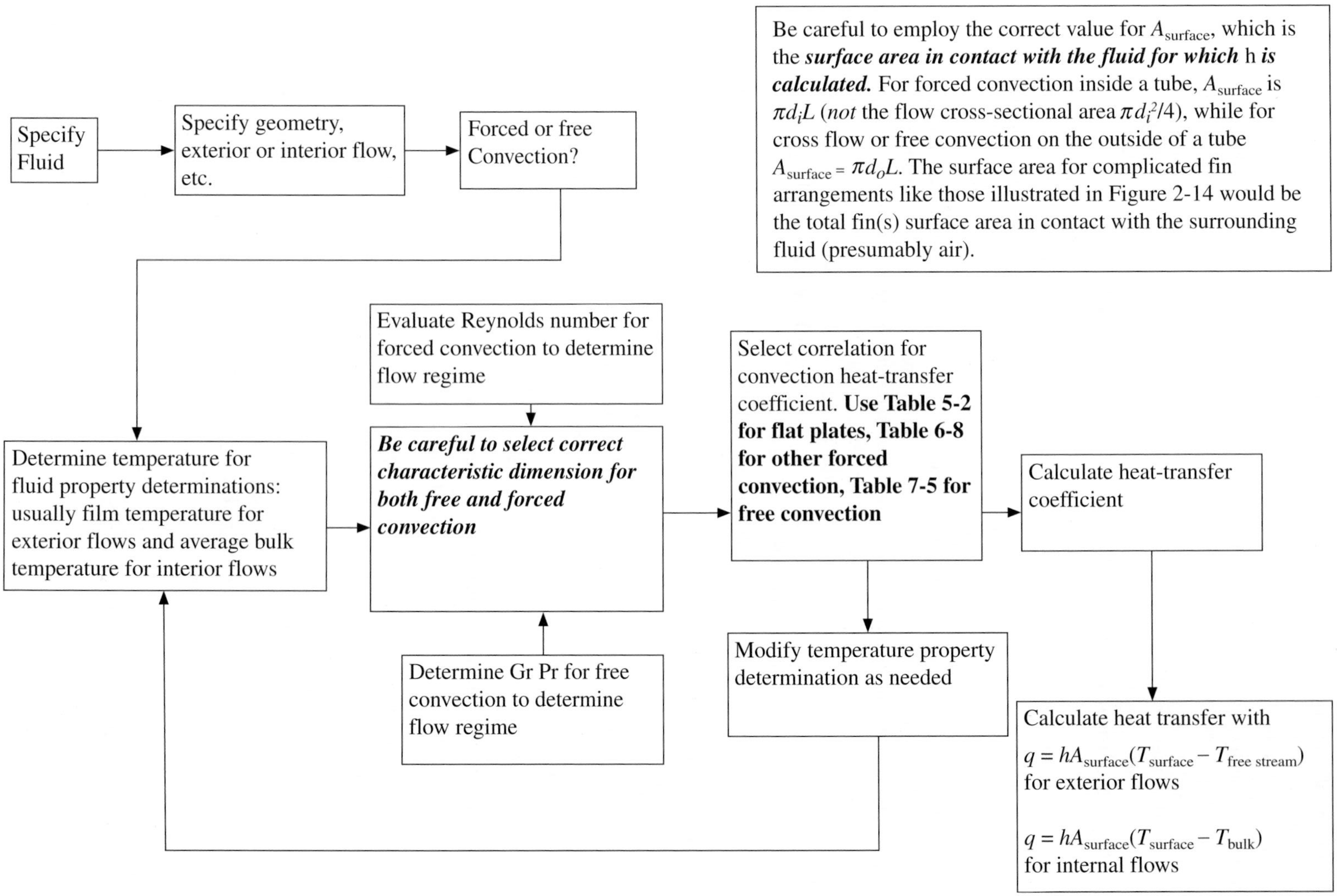

Specify Fluid

Specify geometry, exterior or interior flow, etc.

Forced or free Convection?

Be careful to employ the correct value for $A_{surface}$, which is the **surface area in contact with the fluid for which h is calculated.** For forced convection inside a tube, $A_{surface}$ is $\pi d_i L$ (*not* the flow cross-sectional area $\pi d_i^2/4$), while for cross flow or free convection on the outside of a tube $A_{surface} = \pi d_o L$. The surface area for complicated fin arrangements like those illustrated in Figure 2-14 would be the total fin(s) surface area in contact with the surrounding fluid (presumably air).

Evaluate Reynolds number for forced convection to determine flow regime

Be careful to select correct characteristic dimension for both free and forced convection

Determine temperature for fluid property determinations: usually film temperature for exterior flows and average bulk temperature for interior flows

Determine Gr Pr for free convection to determine flow regime

Select correlation for convection heat-transfer coefficient. **Use Table 5-2 for flat plates, Table 6-8 for other forced convection, Table 7-5 for free convection**

Modify temperature property determination as needed

Calculate heat-transfer coefficient

Calculate heat transfer with
$q = hA_{surface}(T_{surface} - T_{free\ stream})$ for exterior flows

$q = hA_{surface}(T_{surface} - T_{bulk})$ for internal flows

3. What is the approximate criterion for transition to turbulence in a free-convection boundary layer?

4. What functional form of equation is normally used for correlation of free-convection heat-transfer data?

5. Discuss the problem of combined free and forced convection.

6. What is the approximate criterion dividing pure conduction and free convection in an enclosed space between vertical walls?

7. How is a modified Grashof number defined for a constant-heat-flux condition on a vertical plate?

LIST OF WORKED EXAMPLES

7-1 Constant heat flux from vertical plate

7-2 Heat transfer from isothermal vertical plate

7-3 Heat transfer from horizontal tube in water

7-4 Heat transfer from fine wire in air

7-5 Heated horizontal pipe in air

7-6 Cube cooling in air

7-7 Calculation with simplified relations

7-8 Heat transfer across vertical air gap

7-9 Heat transfer across horizontal air gap

7-10 Heat transfer across water layer

7-11 Reduction of convection in air gap

7-12 Heat transfer across evacuated space

7-13 Combined free and forced convection with air

PROBLEMS

7-1 Suppose the heat-transfer coefficients for forced or free convection over vertical flat plates are to be compared. Develop an approximate relation between the Reynolds and Grashof numbers such that the heat-transfer coefficients for pure forced convection and pure free convection are equal. Assume laminar flow.

7-2 For a vertical isothermal flat plate at 93°C exposed to air at 20°C and 1 atm, plot the free-convection velocity profiles as a function of distance from the plate surface at x positions of 15, 30, and 45 cm.

7-3 Show that $\beta = 1/T$ for an ideal gas having the equation of state $p = \rho RT$.

7-4 A 1-ft-square vertical plate is maintained at 65°C and is exposed to atmospheric air at 15°C. Compare the free-convection heat transfer from this plate with that which would result from forcing air over the plate at a velocity equal to the maximum velocity that occurs in the free-convection boundary layer. Discuss this comparison.

7-5 A vertical flat plat maintained at 350 K is exposed to room air at 300 K and 1 atm. Estimate the plate height necessary to produce a free-convection boundary layer thickness of 2.0 cm.

7-6 Plot the free-convection boundary-layer thickness as a function of x for a vertical plate maintained at 80°C and exposed to air at atmospheric pressure and 15°C. Consider the laminar portion only.

7-7 Two vertical flat plates at 65°C are placed in a tank of water at 25°C. If the plates are 30 cm high, what is the minimum spacing that will prevent interference of the free-convection boundary layers?

7-8 A 2.0-cm-diameter cylinder is placed horizontally in a pool of water at 70°F. The surface of the cylinder is maintained at 130°F. Calculate the heat lost by the cylinder per meter of length.

7-9 A vertical cylinder having a length of 30 cm is maintained at 100°C and exposed to room air at 15°C. Calculate the minimum diameter the cylinder can have in order to behave as a vertical flat plate.

7-10 A 1-m-square vertical plate is heated to 300°C and placed in room air at 25°C. Calculate the heat loss from one side of the plate.

7-11 A vertical flat plate is maintained at a constant temperature of 120°F and exposed to atmospheric air at 70°F. At a distance of 14 in. from the leading edge of the plate the boundary layer thickness is 1.0 in. Estimate the thickness of the boundary layer at a distance of 24 in. from the leading edge.

7-12 Condensing steam at 1 atm is used to maintain a vertical plate 20 cm high and 3.0 m wide at a constant temperature of 100°C. The plate is exposed to room air at 20°C. What flow rate of air will result from this heating process? What is the total heating supplied to the room air?

7-13 A vertical flat plate 10 cm high and 1.0 m wide is maintained at a constant temperature of 310 K and submerged in a large pool of liquid water at 290 K. Calculate the free-convection heat lost by the plate and the free-convection flow rate induced by the heated plate.

7-14 A vertical cylinder 1.8 m high and 7.5 cm in diameter is maintained at a temperature of 93°C in an atmospheric environment of 30°C. Calculate the heat lost by free convection from this cylinder. For this calculation the cylinder may be treated as a vertical flat plate.

7-15 The outside wall of a building 6 m high receives an average radiant heat flux from the sun of 1100 W/m². Assuming that 95 W/m² is conducted through the wall, estimate the outside wall temperature. Assume the atmospheric air on the outside of the building is at 20°C.

7-16 Assuming that a human may be approximated by a vertical cylinder 30 cm in diameter and 2.0 m tall, estimate the free-convection heat loss for a surface temperature of 24°C in ambient air at 20°C.

7-17 A 30-cm-square vertical plate is heated electrically such that a constant-heat-flux condition is maintained with a total heat dissipation of 30 W. The ambient air is at 1 atm and 20°C. Calculate the value of the heat-transfer coefficient at heights of 15 and 30 cm. Also calculate the average heat-transfer coefficient for the plate.

7-18 A 0.3-m-square vertical plate is maintained at 55°C and exposed to room air at 1 atm and 20°C. Calculate the heat lost from *both* sides of the plate.

7-19 Calculate the free-convection heat loss from a 0.61-m-square vertical plate maintained at 100°C and exposed to helium at 20°C and a pressure of 2 atm.

7-20 A large vertical plate 6.1 m high and 1.22 m wide is maintained at a constant temperature of 57°C and exposed to atmospheric air at 4°C. Calculate the heat lost by the plate.

7-21 A 1-m-square vertical plate is maintained at 49°C and exposed to room air at 21°C. Calculate the heat lost by the plate.

7-22 What vertical distance is necessary to produce a Rayleigh number of 10^{12} in air at standard conditions and $\Delta T = 10°C$?

7-23 A 25-by-25-cm vertical plate is fitted with an electric heater that produces a constant heat flux of 1000 W/m^2. The plate is submerged in water at 15°C. Calculate the heat-transfer coefficient and the average temperature of the plate. How much heat would be lost by an isothermal surface at this average temperature?

7-24 Assume that one-half of the heat transfer by free convection from a horizontal cylinder occurs on each side of the cylinder because of symmetry considerations. Going by this assumption, compare the heat transfer on each side of the cylinder with that from a vertical flat plate having a height equal to the circumferential distance from the bottom stagnation point to the top stagnation point on the cylinder. Discuss this comparison.

7-25 A horizontal cylindrical heater with $d = 2$ cm is placed in a pool of sodium-potassium mixture with 22 percent sodium. The mixture is at 120°C, and the heater surface is constant at 200°C. Calculate the heat transfer for a heater 40 cm long.

7-26 A vertical flat plate 15 cm high and 50 cm wide is maintained at a constant temperature of 325 K and placed in a large tank of helium at a pressure of 2.2 atm and a temperature of 0°C. Calculate the heat lost by the plate and the free-convection flow rate induced.

7-27 A horizontal heating rod having a diameter of 3.0 cm and a length of 1 m is placed in a pool of saturated liquid ammonia at 20°C. The heater is maintained at a constant surface temperature of 70°C. Calculate the heat-transfer rate.

7-28 Condensing steam at 120°C is to be used inside a 7.5-cm-diameter horizontal pipe to provide heating for a certain work area where the ambient air temperature is 20°C. The total heating required is 29.3 kW. What length pipe would be required to accomplish this heating?

7-29 A 10-cm length of platinum wire 0.4 mm in diameter is placed horizontally in a container of water at 38°C and is electrically heated so that the surface temperature is maintained at 93°C. Calculate the heat lost by the wire.

7-30 Water at the rate of 0.8 kg/s at 90°C flows through a steel pipe with 2.5-cm ID and 3-cm OD. The outside surface temperature of the pipe is 85°C, and the temperature of the surrounding air is 20°C. The room pressure is 1 atm, and the pipe is 15 m long. How much heat is lost by free convection to the room?

7-31 A horizontal pipe 8.0 cm in diameter is located in a room where atmospheric air is at 25°C. The surface temperature of the pipe is 140°C. Calculate the free-convection heat loss per meter of pipe.

7-32 A horizontal 1.25-cm-OD tube is heated to a surface temperature of 250°C and exposed to air at room temperature of 20°C and 1 atm. What is the free-convection heat transfer per unit length of tube?

7-33 A horizontal electric heater 2.5 cm in diameter is submerged in a light-oil bath at 93°C. The heater surface temperature is maintained at 150°C. Calculate the heat lost per meter of length of the heater.

7-34 A 0.3-m-square air-conditioning duct carries air at a temperature such that the outside temperature of the duct is maintained at 15.6°C and is exposed to room air at 27°C. Estimate the heat gained by the duct per meter of length.

7-35 A fine wire having a diameter of 0.001 in (0.0254 mm) is heated by an electric current and placed horizontally in a chamber containing helium at 3 atm and 10°C. If the surface temperature of the wire is not to exceed 240°C, calculate the electric power to be supplied per unit length.

7-36 A heated horizontal cylinder having a surface temperature of 93°C, diameter of 10 cm, and length of 2.0 m is exposed to Helium at 1 atm and −18°C. Calculate the heat lost by the cylinder.

7-37 A large circular duct, 3.0 m in diameter, carries hot gases at 250°C. The outside of the duct is exposed to room air at 1 atm and 20°C. Estimate the heat loss per unit length of the duct.

7-38 A 2.0-cm-diameter cylinder is placed in a tank of glycerine at 20°C. The surface temperature of the heater is 60°C, and its length is 60 cm. Calculate the heat transfer.

7-39 A 3.5-cm-diameter cylinder contains an electric heater that maintains a constant heat flux at the surface of 1500 W/m². If the cylinder is inclined at an angle of 35° with the horizontal and exposed to room air at 20°C, estimate the average surface temperature.

7-40 A 30-cm-diameter horizontal pipe is maintained at a constant temperature of 25°C and placed in room air at 20°C. Calculate the free-convection heat loss from the pipe per unit length.

7-41 A 12.5 cm-diameter duct is maintained at a constant temperature of 260°C by hot combustion gases inside. The duct is located horizontally in a small warehouse area having an ambient temperature of 20°C. Calculate the length of the duct necessary to provide 37 kW of convection heating.

7-42 A horizontal cylinder with diameter of 5 cm and length of 3 m is maintained at 180°F and submerged in water that is at 60°F. Calculate the heat lost by the cylinder.

7-43 A 2.0-m-diameter horizontal cylinder is maintained at a constant temperature of 77°C and exposed to a large warehouse space at 27°C. The cylinder is 20 m long. Calculate the heat lost by the cylinder.

7-44 Calculate the rate of free-convection heat loss from a 30-cm-diameter sphere maintained at 90°C and exposed to atmospheric air at 20°C.

7-45 A 2.5-cm-diameter sphere at 35°C is submerged in water at 10°C. Calculate the rate of free-convection heat loss.

7-46 A spherical balloon gondola 2.4 m in diameter rises to an altitude where the ambient pressure is 1.4 kPa and the ambient temperature is −50°C. The outside surface of the sphere is at approximately 0°C. Estimate the free-convection heat loss from the outside of the sphere. How does this compare with the forced-convection loss from such a sphere with a low free-stream velocity of approximately 30 cm/s?

7-47 A 4.0-cm diameter sphere is maintained at 38°C and submerged in water at 15°C. Calculate the heat-transfer rate under these conditions.

7-48 Apply the reasoning pertaining to the last entry of Table 7-1 to free convection from a sphere and compare with Equation (7-50).

7-49 Using the information in Table 7-1 and the simplified relations of Table 7-2, devise a simplified relation that may be used as a substitute for Equation (7-50) to calculate free convection from a sphere to air at 1 atm.

7-50 A horizontal tube having a diameter of 30 cm is maintained at a constant temperature of 204°C and exposed to helium at 3 atm and 93°C. Calculate the heat lost from the tube for a tube length of 10.4 m. Express in units of watts.

7-51 A large bare duct having a diameter of 30 cm runs horizontally across a factory area having environmental conditions of 20°C and 1 atm. The length of the duct is 100 m. Inside the duct a low pressure steam flow maintains the duct wall temperature constant at 120°C. Calculate the total heat lost by convection from the duct to the room.

7-52 A circular hot plate, 15 cm in diameter, is maintained at 150°C in atmospheric air at 20°C. Calculate the free-convection heat loss when the plate is in a horizontal position.

7-53 An engine-oil heater consists of a large vessel with a square-plate electric-heater surface in the bottom of the vessel. The heater plate is 30 by 30 cm and is maintained at a constant temperature of 60°C. Calculate the heat-transfer rate for an oil temperature of 20°C.

7-54 Small electric strip heaters with a width of 6 mm are oriented in a horizontal position. The strips are maintained at 500°C and exposed to room air at 20°C. Assuming that the strips dissipate heat from both the top and the bottom surfaces, estimate the strip length required to dissipate 2 kW of heat by free convection.

7-55 The top surface of a 10-by-10-m horizontal plate is maintained at 25°C and exposed to room temperature at 28°C. Estimate the heat transfer.

7-56 A 4-by-4-m horizontal heater is placed in room air at 15°C. Both the top and the bottom surfaces are heated to 50°C. Estimate the total heat loss by free convection.

7-57 A horizontal plate, uniform in temperature at 400 K, has the shape of an equilateral triangle 45 cm on each side and is exposed to atmospheric air at 300 K. Estimate the heat lost by the plate.

7-58 A heated plate, 20 by 20 cm, is inclined at an angle of 60° with the horizontal and placed in water. Approximately constant-heat-flux conditions prevail with a mean plate temperature of 40°C and the heated surface facing downward. The water temperature is 20°C. Calculate the heat lost by the plate.

7-59 Repeat Problem 7-58 for the heated plate facing upward.

7-60 A double plate-glass window is constructed with a 1.25-cm air space. The plate dimensions are 1.2 by 1.8 m. Calculate the free-convection heat-transfer rate through the air space for a temperature difference of 30°C and $T_1 = 20$°C.

7-61 A flat-plate solar collector is 1 m square and is inclined at an angle of 20° with the horizontal. The hot surface at 160°C is placed in an enclosure that is evacuated to a pressure of 0.1 atm. Above the hot surface, and parallel to it, is the transparent window that admits the radiant energy from the sun. The hot surface and window are separated by a distance of 8 cm. Because of convection to the surroundings, the window temperature is maintained at 40°C. Calculate the free-convection heat transfer between the hot surface and the transparent window.

7-62 A flat plate 1 by 1 m is inclined at 30° with the horizontal and exposed to atmospheric air at 30°C and 1 atm. The plate receives a net radiant-energy flux from the sun of 700 W/m², which then is dissipated to the surroundings by free convection. What average temperature will be attained by the plate?

7-63 A horizontal cylinder having a diameter of 5 cm and an emissivity of 0.5 is placed in a large room, the walls of which are maintained at 35°C. The cylinder loses heat by natural convection with an h of 6.5 W/m² · °C. A sensitive thermocouple placed on the surface of the cylinder measures the temperature as 30°C. What is the temperature of the air in the room?

7-64 A 10-by-10-cm plate is maintained at 80°C and inclined at 45° with the horizontal. Calculate the heat loss from both sides of the plate to room air at 20°C.

7-65 A 5-by-5-cm plate is maintained at 50°C and inclined at 60° with the horizontal. Calculate the heat loss from both sides of the plate to water at 20°C.

7-66 Air at 1 atm and 38°C is forced through a horizontal 6.5-mm-diameter tube at an average velocity of 30 m/s. The tube wall is maintained at 540°C, and the tube is 30 cm long. Calculate the average heat-transfer coefficient. Repeat for a velocity of 30 m/s and a tube wall temperature of 800°C.

7-67 A small copper block having a square bottom 2.5 by 2.5 cm and a vertical height of 5 cm cools in room air at 1 atm and 15°C. The block is isothermal at 100°C. Calculate the heat-transfer rate.

7-68 A horizontal plate in the shape of an equilateral triangle 40 cm on a side is maintained at a constant temperature of 55°C and exposed to atmospheric air at 25°C. Calculate the heat lost by the top surface of the plate.

7-69 A small horizontal heater is in the shape of a circular disk with a diameter of 3 cm. The disk is maintained at 70°C and exposed to atmospheric air at 30°C. Calculate the heat loss.

7-70 A hot ceramic block at 400°C has dimensions of 15 by 15 by 8 cm high. It is exposed to room air at 27°C. Calculate the free-convection heat loss.

7-71 A magnetic amplifier is encased in a cubical box 15 cm on a side and must dissipate 50 W to surrounding air at 20°C. Estimate the surface temperature of the box.

7-72 A glass thermometer is placed in a large room, the walls of which are maintained at 10°C. The convection coefficient between the thermometer and the room air is 5 W/m^2 · °C, and the thermometer indicates a temperature of 30°C. Determine the temperature of the air in the room. Take $\epsilon = 1.0$.

7-73 A horizontal air-conditioning duct having a horizontal dimension of 30 cm and a vertical dimension of 15 cm is maintained at 45°C and exposed to atmospheric air at 20°C. Calculate the heat lost per unit length of duct.

7-74 Two 30-cm-square vertical plates are separated by a distance of 1.25 cm, and the space between them is filled with water. A constant-heat-flux condition is imposed on the plates such that the average temperature is 38°C for one and 60°C for the other. Calculate the heat-transfer rate under these conditions. Evaluate properties at the mean temperature.

7-75 An enclosure contains helium at a pressure of 1.3 atm and has two vertical heating surfaces, which are maintained at 80 and 20°C, respectively. The vertical surfaces are 40 by 40 cm and are separated by a gap of 2.0 cm. Calculate the free-convection heat transfer between the vertical surfaces.

7-76 A horizontal annulus with inside and outside diameters of 8 and 10 cm, respectively, contains liquid water. The inside and outside surfaces are maintained at 40 and 20°C, respectively. Calculate the heat transfer across the annulus space per meter of length.

7-77 Two concentric spheres are arranged to provide storage of brine inside the inner sphere at a temperature of −10°C. The inner-sphere diameter is 2 m, and the gap spacing is 5 cm. The outer sphere is maintained at 30°C, and the gap space is evacuated to a pressure of 0.05 atm. Estimate the free-convection heat transfer across the gap space.

7-78 A large vat used in food processing contains a hot oil at 400°F. Surrounding the vat on the vertical sides is a shell that is cooled to 140°F. The air space separating the vat and the shell is 35 cm high and 3 cm thick. Estimate the free-convection loss per square meter of surface area.

7-79 Two 30-cm-square vertical plates are separated by a distance of 2.5 cm and air at 1 atm. The two plates are maintained at temperatures of 200 and 90°C, respectively. Calculate the heat-transfer rate across the air space.

7-80 A horizontal air space is separated by a distance of 1.6 mm. Estimate the heat-transfer rate per unit area for a temperature difference of 165°C, with one plate temperature at 90°C.

7-81 Repeat Problem 7-80 for a horizontal space filled with water.

7-82 An atmospheric vertical air space 4.0 ft high has a temperature differential of 20°F at 300 K. Calculate and plot k_e/k and the R value for spacings of 0 to 10 in. At approximately what spacing is the R value a maximum?

7-83 Two vertical plates 50 by 50 cm are separated by a space of 4 cm that is filled with water. The plate temperatures are 50 and 20°C. Calculate the heat transfer across the space.

7-84 Repeat Problem 7-83 for the plates oriented in a horizontal position with the 50°C surface as the lower plate.

7-85 Two vertical plates 1.1 by 1.1 m are separated by a 4.0-cm air space. The two surface temperatures are at 300 and 350 K. The heat transfer in the space can be reduced by decreasing the pressure of the air. Calculate and plot k_e/k and the R value as a function of pressure. To what value must the pressure be reduced to make $k_e/k = 1.0$?

7-86 Repeat Problem 7-85 for two horizontal plates with the 350 K surface on the bottom.

7-87 An air space in a certain building wall is 10 cm thick and 2 m high. Estimate the free-convection heat transfer through this space for a temperature difference of 17°C.

7-88 A vertical enclosed space contains air at 2 atm. The space is 3 m high by 2 m deep and the spacing between the vertical plates is 6 cm. One plate is maintained at 300 K while the other is at 400 K. Calculate the convection heat transfer between the two vertical plates.

7-89 Develop an expression for the optimum spacing for vertical plates in air in order to achieve minimum heat transfer, assuming that the heat transfer results from pure conduction at $Gr_\delta < 2000$. Plot this optimum spacing as a function of temperature difference for air at 1 atm.

7-90 Air at atmospheric pressure is contained between two vertical plates maintained at 100°C and 20°C, respectively. The plates are 1.0 m on a side and spaced 8 cm apart. Calculate the convection heat transfer across the air space.

7-91 A special section of insulating glass is constructed of two glass plates 30 cm square separated by an air space of 1 cm. Calculate the percent reduction in heat transfer of this arrangement compared to free convection from a vertical plate with a temperature difference of 30°C.

7-92 One way to reduce the free-convection heat loss in a horizontal solar collector is to reduce the pressure in the space separating the glass admitting the solar energy and the black absorber below. Assume the bottom surface is at 120°C and the top surface is at 20°C. Calculate the pressures that are necessary to eliminate convection for spacings of 1, 2, 5, and 10 cm.

7-93 Air at 20°C and 1 atm is forced upward through a vertical 2.5-cm-diameter tube 30 cm long. Calculate the total heat-transfer rate where the tube wall is maintained at 200°C and the flow velocity is 45 cm/s.

7-94 A horizontal tube is maintained at a surface temperature of 55°C and exposed to atmospheric air at 27°C. Heat is supplied to the tube by a suitable electric heater that produces an input of 175 W for each meter of length. Find the expected power input if the surface temperature is raised to 83°C.

7-95 A large vertical plate is maintained at a surface temperature of 140°F and exposed to air at 1 atm and 70°F. Estimate the vertical position on the plate where the boundary layer becomes turbulent. What is the average q/A for the portion of the plate preceding this location? What is the maximum velocity in the boundary layer at this location?

7-96 The horizontal air space over a solar collector has a spacing of 2.5 cm. The lower plate is maintained at 70°C while the upper plate is at 30°C. Calculate the free convection across the space for air at 1 atm. 1f the spacing is reduced to 1.0 cm, by how much is the heat transfer changed?

7-97 One concept of a solar collector reduces the pressure of the air gap to a value low enough to eliminate free-convection effects. For the air space in Problem 7-96 determine the pressures to eliminate convection; that is, Gr Pr < 1700.

7-98 A 2.5-cm sphere is maintained at a surface temperature of 120°F and exposed to a fluid at 80°F. Compare the heat loss for (a) air and (b) water.

7-99 Air at 1 atm is contained between two concentric spheres having diameters of 10 and 8 cm and maintained at temperatures of 300 and 400 K. Calculate the free-convection heat transfer across the air gap.

7-100 Some canned goods are to be cooled from room temperature of 300 K by placing them in a refrigerator maintained at 275 K. The cans have diameter and height of 8.0 cm. Calculate the cooling rate. Approximately how long will it take the temperature of the can to drop to 290 K if the contents have the properties of water? Use lumped-capacity analysis.

7-101 A 5.0-cm-diameter horizontal disk is maintained at 120°F and submerged in water at 80°F. Calculate the heat lost from the top and bottom of the disk.

7-102 A 10-cm-square plate is maintained at 400 K on the bottom side, and exposed to air at 1 atm and 300 K. The plate is inclined at 45° with the vertical. Calculate the heat lost by the bottom surface of the plate.

7-103 Calculate the heat transfer for the plate of Problem 7-102 if the heated surface faces upward.

7-104 A vertical cylinder 50 cm high is maintained at 400 K and exposed to air at 1 atm and 300 K. What is the minimum diameter for which the vertical-flat-plate relations may be used to calculate the heat transfer? What would the heat transfer be for this diameter?

7-105 Derive an expression for the ratio of the heat conducted through an air layer at low density to that conducted for $\lambda = 0$. Plot this ratio versus λ/L for $\alpha = 0.9$ and air properties evaluated at 35°C.

7-106 A superinsulating material is to be constructed of polished aluminum sheets separated by a distance of 0.8 mm. The space between the sheets is sealed and evacuated to a pressure of 10^{-5} atm. Four sheets are used. The two outer sheets are maintained at 35 and 90°C and have a thickness of 0.75 mm, whereas the inner sheets have a thickness of 0.18 mm. Calculate the conduction and radiation transfer across the layered section per unit area. For this calculation, allow the inner sheets to "float" in the determination of the radiation heat transfer. Evaluate properties at 65°C.

7-107 Two large polished plates are separated by a distance of 1.3 mm, and the space between them is evacuated to a pressure of 10^{-5} atm. The surface properties of the plates are $\alpha_1 = 0.87$, $\epsilon_1 = 0.08$, $\alpha_2 = 0.95$, $\epsilon_2 = 0.23$, where α is the accommodation coefficient. The plate temperatures are $T_1 = 70°C$ and $T_2 = 4°C$. Calculate the total heat transfer between the plates by low-density conduction and radiation.

7-108 A smooth glass plate is coated with a special coating that is electrically conductive and can produce constant-heat-flux conditions. One of these surfaces, 0.5 m square, is suspended vertically in room air at 20°C. What heat flux would be experienced and what would be the electric power input to maintain an average surface temperature of 65°C on both sides of the plate? Suppose the plate surface radiates approximately as a blackbody. What amount of heat would be dissipated for the same average surface temperature?

7-109 A 20-cm-square vertical plate is heated to a temperature of 30°C and submerged in glycerin at 10°C. Calculate the heat lost from both sides of the plate.

7-110 A vertical cylinder 30 cm high and 30 cm in diameter is maintained at a surface temperature of 43.3°C while submerged in water at 10°C. Calculate the heat lost from the total surface area of the cylinder.

7-111 A 1.0-cm-diameter horizontal cylinder is maintained at a constant surface temperature of 400 K and exposed to oxygen at 300 K and 1.5 atm. The length of the cylinder is 125 cm. Calculate the heat lost by the cylinder.

7-112 Air is contained between two vertical plates spaced 2 cm apart, with the air space evacuated so that the mean free path is equal to the plate spacing. One plate is at 400 K with $\epsilon = 0.1$ while the other plate is at 300 K with $\epsilon = 0.15$. The accommodation coefficients for the surfaces are 0.9. Calculate the heat transfer between the two plates.

7-113 A 40-cm-diameter sphere is maintained at 400 K and exposed to room air at 20°C. Calculate the free convection heat loss from the sphere. If the surface has $\epsilon = 0.9$, also calculate the radiation heat lost from the sphere.

7-114 Two 20-cm-square plates are maintained at 350 and 400 K and separated by a distance of 2 cm. The space between the plates is filled with helium at 2 atm. Calculate the heat transfer through the gap space.

7-115 A 30-cm-square horizontal plate is exposed to air at 1 atm and 25°C. The plate surface is maintained at 125°C on both sides. Calculate the free convection loss from the plate.

7-116 A horizontal 1-mm-diameter stainless-steel wire having $k = 16$ W/m · °C and a resistivity of 70 $\mu\Omega \cdot$ cm is exposed to air at 1 atm and 20°C. The wire length is 1 m. What is the maximum temperature that will occur in the wire and the voltage that must be impressed on it to produce a surface temperature of 134°C?

7-117 A vertical cylindrical surface has a diameter of 10.5 cm, a height of 30 cm, and is exposed to air at 1 atm and 15°C. The cylindrical surface is maintained at 100°C. Calculate the free convection heat loss from the cylindrical surface. State your assumptions.

7-118 A wire having a diameter of 0.025 mm is placed in a horizontal position in room air at 1 atm and 300 K. A voltage is impressed on the wire, producing a surface temperature of 865 K. The surface emissivity of the wire is 0.9. Calculate the heat loss from the wire per unit length by both free convection and radiation.

7-119 A flat surface having the shape of an equilateral triangle, 20 cm on a side, is maintained at 400 K and exposed to air at 1 atm and 300 K. Calculate the heat lost from the top surface of the triangle.

7-120 A horizontal disk having a diameter of 10 cm is maintained at 49°C and submerged in water at 1 atm and 10°C. Calculate the free convection heat loss from the top surface of the disk.

Design-Oriented Problems

7-121 A free-convection heater is to be designed that will dissipate 10,000 kJ/h to room air at 300 K. The heater surface temperature must not exceed 350 K. Consider four alternatives: (*a*) a group of vertical surfaces, (*b*) a single vertical surface, (*c*) a single horizontal surface, and (*d*) a group of horizontal cylindrical surfaces. Examine these alternatives and suggest a design.

7-122 A special double-pane insulating window glass is to be constructed of two glass plates separated by an air gap. The plates are square, 60 by 60 cm, and are designed to be used with temperatures of -10 and $+20°C$ on the respective plates. Assuming the air in the gap is at 1 atm, calculate and plot the free convection across the gap as a function of gap spacing for a vertical window. What conclusions can you draw from this plot from a design standpoint?

7-123 A standing rib roast is cooked for a holiday dinner and is removed from the oven when the temperature reaches $120°C$. The roast cools by combined free convection and radiation in a room at 300 K. Using whatever reference material is necessary, estimate the time required for the temperature of the roast to reach $50°C$. Be sure to state all assumptions.

7-124 Repeat Problem 7-122 for a horizontal window with the hot surface on the lower side.

7-125 Energy-conservation advocates claim that storm windows can substantially reduce energy losses (or gains). Consider a vertical 1.0-m-square window covered by a storm window with an air gap of 2.5 cm. The inside window is at $15°C$ and the outside storm window is at $-10°C$. Calculate the R value for the gap. What would the R value be for the same thickness of fiberglass blanket?

7-126 An evacuated thermal insulation is to be designed that will incorporate multiple layers of reflective sheets ($\epsilon = 0.04$) separated by air gap spaces that are partially evacuated and have spacing δ sufficiently small that $k_e/k = 1.0$. The insulation is to be designed to operate over a temperature differential from $0°C$ to $200°C$. Investigate the possibilities of using 1, 2, 3, or 4 gap spaces and comment on the influence of different factors on the design, such as gap-space size and the evacuation pressure necessary to produce $k_e/k = 1.0$.

7-127 Aunt Maude frequently complains of a "draft" while sitting next to a window in her New York apartment in the winter, and she also says her feet get cold. She remarks that the window seems to leak cold air in the winter but not hot air in the summer. (She has air-conditioning, so her windows are closed in the summer.) Using appropriate assumptions, analyze and explain the "draft" problem and make some quantitative estimates of what the draft may be. How does the analysis account for her feet being cold?

7-128 A circular air-conditioning duct carries cool air at $5°C$ and is constructed of 1 percent carbon steel with a thickness of 0.2 mm and an outside diameter of 18 cm. The duct is in a horizontal position and gains heat from room air at $20°C$. If the average air velocity in the duct is 7.5 m/s, estimate the air temperature rise in a duct run of 30 m. Be sure to state your assumptions in arriving at an answer.

7-129 An experiment is to be designed to measure free convection heat-transfer coefficients from spheres by preheating aluminum spheres of various diameters to an initial temperature and then measuring the temperature response as each sphere cools in room air. Because of the low value of the Biot number (see Chapter 4) the sphere may be assumed to behave as a lumped capacity. The sphere is also blackened so that the radiation loss from the outer surface will be given by $q_{rad} = \sigma A_{surf}(T^4 - T_{surr}^4)$, where the temperatures are in degrees Kelvin. From information in this chapter, anticipate the cooling curve behavior for 5-mm-, 25-mm-, and 50-mm-diameter aluminum spheres cooling from $230°C$ in room air at $20°C$. How often would you advise reading the temperatures of the spheres and room? What range of Rayleigh numbers would you expect to observe in these experiments? Can you suggest a way to correlate the data in terms of the significant dimensionless groups?

7-130 In a television weather report a "wind chill factor" is frequently stated. The actual factor is based on empirical data. You are asked to come up with an expression for wind chill based on the information presented in Chapters 6 and 7. In obtaining this relation you may assume that (1) a man can be approximated as a vertical cylinder 30 cm in diameter and 1.8 m tall, (2) wind chill expresses the equivalent air temperature the cylinder would experience in free convection when losing heat by forced convection to air at the ambient temperature and velocity u_∞, (3) forced convection heat loss from the cylinder can be obtained from Equation (6-17) with the appropriate values of C and n, and (4) free convection from the vertical cylinder can be obtained from the simplified expressions of Table 7-2. Based on these assumptions, devise relationship(s) to predict the wind chill for ambient temperatures between -12 and $+10°C$ and wind velocities between 5 and 40 mi/h (1 mi/h $= 0.447$ m/s). Other assumptions must be made in addition to the ones stated. Be sure to clearly note your assumptions in arriving at the relation(s) for wind chill. If convenient, check other sources of information to verify your results. If you are currently experiencing winter weather, compare your results with a television weather report.

REFERENCES

1. Eckert, E. R. G., and E. Soehngen. "Interferometric Studies on the Stability and Transition to Turbulence of a Free Convection Boundary Layer," *Proc. Gen. Discuss. Heat Transfer ASMEIME, London,* 1951.

2. Eckert, E. R. G., and E. Soehngen. "Studies on Heat Transfer in Laminar Free Convection with the Zehnder-Mach Interferometer," *USAF Tech. Rep.* 5747, December 1948.

3. Holman, J. P., H. E. Gartrell, and E. E. Soehngen. "An Interferometric Method of Studying Boundary Layer Oscillations," *J. Heat Transfer,* ser. C, vol. 80, August 1960.

4. McAdams, W. H. *Heat Transmission,* 3d ed., New York: McGraw-Hill, 1954.

5. Yuge, T. "Experiments on Heat Transfer from Spheres Including Combined Natural and Forced Convection," *J. Heat Transfer,* ser. C, vol. 82, p. 214, 1960.

6. Jakob, M. "Free Convection through Enclosed Gas Layers," *Trans. ASME,* vol. 68, p. 189, 1946.

7. Jakob, M. *Heat Transfer,* vol. 1, New York: John Wiley, 1949.

8. Globe, S., and D. Dropkin. *J. Heat Transfer,* February 1959, pp. 24–28.

9. Evans, L. B., and N. E. Stefany. "An Experimental Study of Transient Heat Transfer to Liquids in Cylindrical Enclosures," *AIChE Pap. 4, Heat Transfer Conf Los Angeles,* August 1965.

10. Metais, B., and E. R. G. Eckert. "Forced, Mixed, and Free Convection Regimes," *J. Heat Transfer,* ser. C, vol. 86, p. 295, 1964.

11. Bishop, E. N., L. R. Mack, and J. A. Scanlan. "Heat Transfer by Natural Convection between Concentric Spheres," *Int. J. Heat Mass Transfer,* vol. 9, p. 649, 1966.

12. Dropkin, D., and E. Somerscales. "Heat Transfer by Natural Convection in Liquids Confined by Two Parallel Plates Which Are Inclined at Various Angles with Respect to the Horizontal," *J. Heat Transfer,* vol. 87, p. 71, 1965.

13. Gebhart, B., Y. Jaluria, R. L. Mahajan, and B. Sammakia. *Buoyancy Induced Flows and Transport.* New York: Hemisphere Publishing Corp., 1988.

14. Gebhart, B. "Natural Convection Flow, Instability, and Transition," *ASME Pap.* 69-HT-29, August 1969.

15. Mollendorf, J. C., and B. Gebhart. "An Experimental Study of Vigorous Transient Natural Convection," *ASME Pap.* 70-HT-2, May 1970.

16. Bayley, F. J. "An Analysis of Turbulent Free Convection Heat Transfer," *Proc. Inst. Mech. Eng.,* vol. 169, no. 20, p. 361, 1955.

17. Brown, C. K., and W. H. Gauvin. "Combined Free and Forced Convection, I, II," *Can. J. Chem. Eng.,* vol. 43, no. 6, pp. 306, 313, 1965.

18. MacGregor, R. K., and A. P. Emery. "Free Convection through Vertical Plane Layers: Moderate and High Prandtl Number Fluids," *J. Heat Transfer,* vol. 91, p. 391, 1969.

19. Newell, M. E., and F .W. Schmidt. "Heat Transfer by Laminar Natural Convection within Rectangular Enclosures," *J. Heat Transfer,* vol. 92, pp. 159–168, 1970.

20. Husar, R. B., and E. M. Sparrow. "Patterns of Free Convection Flow Adjacent to Horizontal Heated Surfaces," *Int. J. Heat Mass Trans.,* vol. 11, p. 1206, 1968.

21. Habne, E. W. P. "Heat Transfer and Natural Convection Patterns on a Horizontal Circular Plate," *Int. J. Heat Mass Transfer,* vol. 12, p. 651, 1969.

22. Warner, C. Y., and V. S. Arpaci. "An Experimental Investigation of Turbulent Natural Convection in Air at Low Pressure along a Vertical Heated Flat Plate," *Int. J. Heat Mass Transfer,* vol. 11, p. 397, 1968.

23. Gunness, R. C., Jr., and B. Gebhart. "Stability of Transient Convection," *Phys. Fluids,* vol. 12, p. 1968, 1969.

24. Rotern, Z., and L. Claassen. "Natural Convection above Unconfined Horizontal Surfaces," *J. Fluid Mech.,* vol. 39, pt. 1, p. 173, 1969.

25. Vliet, G. C. "Natural Convection Local Heat Transfer on Constant Heat Flux Inclined Surfaces," *J. Heat Transfer,* vol. 91, p. 511, 1969.

26. Vliet, G. C., and C. K. Lin. "An Experimental Study of Turbulent Natural Convection Boundary Layers," *J. Heat Transfer,* vol. 91, p. 517, 1969.

27. Ostrach, S. "An Analysis of Laminar-Free-Convection Flow and Heat Transfer about a Flat Plate Parallel to the Direction of the Generating Body Force," *NACA Tech. Rep.* 1111, 1953.

28. Cheesewright, R. "Turbulent Natural Convection from a Vertical Plane Surface," *J. Heat Transfer,* vol. 90, p. 1, February 1968.

29. Flack, R. D., and C. L. Witt. "Velocity Measurements in Two Natural Convection Air Flows Using a Laser Velocimeter," *J. Heat Transfer,* vol. 101, p. 256, 1979.

30. Eckert, E. R. G., and T. W. Jackson. "Analysis of Turbulent Free Convection Boundary Layer on a Flat Plate," *NACA Rep.* 1015, 1951.

31. King, W. J. "The Basic Laws and Data of Heat Transmission," *Mech. Eng.,* vol. 54, p. 347, 1932.

32. Sparrow, E. M., and J. L. Gregg. "Laminar Free Convection from a Vertical Flat Plate," *Trans. ASME,* vol. 78, p. 435, 1956.

33. Benard, H. "Les Tourbillons cellulaires dans une nappe liquide transportant de la chaleur par convection en régime permanent," *Ann. Chim. Phys.,* vol. 23, pp. 62–144, 1901.

34. *Progress in Heat and Mass Transfer,* vol. 2, Eckert Presentation Volume. New York: Pergamon Press, 1969.

35. Gebhart, B., T. Audunson, and L. Pera. *Fourth Int. Heat Transfer Conf., Paris,* August 1970.

36. Sanders, C. J., and J. P. Holman. "Franz Grashof and the Grashof Number," *Int. J. Heat Mass Transfer,* vol. 15, p. 562, 1972.

37. Clifton, J. V., and A. J. Chapman. "Natural Convection on a Finite-Size Horizontal Plate," *Int. J. Heat Mass Transfer,* vol. 12, p. 1573, 1969.

38. Emery, A. F., H. W. Chi, and J. D. Dale. "Free Convection through Vertical Plane Layers of Non-Newtonian Power Law Fluids," *ASME Pap.* 70-WA/HT-1.

39. Vliet, G. C. "Natural Convection Local Heat Transfer on Constant Heat Flux Inclined Surfaces," *Trans. ASME,* vol. 91C, p. 511, 1969.

40. Bergles, A. E., and R. R. Simonds. "Combined Forced and Free Convection for Laminar Flow in Horizontal Tubes with Uniform Heat Flux," *Int. J. Heat Mass Transfer,* vol. 14, p. 1989, 1971.

41. Aihara, T., Y. Yamada, and S. Endo. "Free Convection along the Downward-facing Surface of a Heated Horizontal Plate," *Int. J. Heat Mass Transfer,* vol. 15, p. 2535, 1972.

42. Saunders, O. A., M. Fishenden, and H. D. Mansion. "Some Measurement of Convection by an Optical Method," *Engineering,* p. 483, May 1935.

43. Weber, N., R. E. Rowe, E. H. Bishop, and J. A. Scanlan. "Heat Transfer by Natural Convection between Vertically Eccentric Spheres," *ASME Pap.* 72-WA/HT-2.

44. Fujii, T., and H. Imura. "Natural Convection Heat Transfer from a Plate with Arbitrary Inclination," *Int. J. Heat Mass Transfer,* vol. 15, p. 755, 1972.

45. Pera, L., and B. Gebhart. "Natural Convection Boundary Layer Flow over Horizontal and Slightly Inclined Surfaces," *Int. J. Heat Mass Transfer,* vol. 16, p. 1131, 1973.

46. Hyman, S. C., C. F. Bonilla, and S. W. Ehrlich. "Heat Transfer to Liquid Metals from Horizontal Cylinders," *AiChE Symp. Heat Transfer, Atlantic City,* 1953, p. 21.

47. Fand, R. M., and K. K. Keswani. "Combined Natural and Forced Convection Heat Transfer from Horizontal Cylinders to Water," *Int. J. Heat Mass Transfer,* vol. 16, p. 175, 1973.

48. Dale, J. D., and A. F. Emery. "The Free Convection of Heat from a Vertical Plate to Several Non-Newtonian Pseudoplastic Fluids," *ASME Pap.* 71-HT-S.

49. Fujii, T., O. Miyatake, M. Fujii, H. Tanaka, and K. Murakami. "Natural Convective Heat Transfer from a Vertical Isothermal Surface to a Non-Newtonian Sutterby Fluid," *Int. J. Heat Mass Transfer,* vol. 16, p. 2177, 1973.

50. Soehngen, E. E. "Experimental Studies on Heat Transfer at Very High Prandtl Numbers," *Prog. Heat Mass Transfer,* vol. 2, p. 125, 1969.

51. Vliet, G. C., and D. C. Ross. "Turbulent Natural Convection on Upward and Downward Facing Inclined Constant Heat Flux Surfaces," *ASME Pap.* 74-WA/HT-32.

52. Llyod, J. R., and W. R. Moran. "Natural Convection Adjacent to Horizontal Surface of Various Planforms," *ASME Pap.* 74-WA/HT-66.

53. Goldstein, R. J., E. M. Sparrow, and D.C. Jones. "Natural Convection Mass Transfer Adjacent to Horizontal Plates," *Int. J. Heat Mass Transfer,* vol. 16, p. 1025, 1973.

54. Holman, J. P., and J. H. Boggs. "Heat Transfer to Freon 12 near the Critical State in a Natural Circulation Loop," *J. Heat Transfer,* vol. 80, p. 221, 1960.

55. Mull, W., and H. Reiher. "Der Wärmeschutz von Luftschichten," *Beih. Gesund. Ing.,* ser. 1, no. 28, 1930.

56. Krasshold, H. "Wärmeabgabe von zylindrischen Flussigkeitsschichten bei natürlichen Konvektion," *Forsch. Geb. Ingenieurwes,* vol. 2, p. 165, 1931.

57. Beckmann, W. "Die Wärmeübertragung in zylindrischen Gasschichten bei natürlicher Konvektion," *Forsch. Geb. Ingenieurwes,* vol. 2, p. 186, 1931.

58. Schmidt, E. "Free Convection in Horizontal Fluid Spaces Heated from Below." *Proc. Int. Heat Transfer Conf., Boulder, Col., ASME,* 1961.

59. Graff, J. G. A., and E. F. M. Van der Held. "The Relation between the Heat Transfer and Convection Phenomena in Enclosed Plain Air Players," *Appl. Sci. Res.,* ser. A, vol. 3, p. 393, 1952.

60. Liu, C. Y., W. K. Mueller, and F. Landis. "Natural Convection Heat Transfer in Long Horizontal Cylindrical Annuli," *Int. Dev. Heat Transfer,* pt. 5, pap. 117, p. 976, 1961.

61. Emery, A., and N. C. Chu. "Heat Transfer across Vertical Layers," *J. Heat Transfer,* vol. 87, p. 110, 1965.

62. O'Toole, J., and P. L. Silveston. "Correlation of Convective Heat Transfer in Confined Horizontal Layers," *Chem. Eng. Prog. Symp.,* vol. 57, no. 32, p. 81, 1961.

63. Goldstein, R. J., and T. Y. Chu. "Thermal Convection in a Horizontal Layer of Air," *Prog. Heat Mass Transfer,* vol. 2, p. 55, 1969.

64. Singh, S. N., R. C. Birkebak, and R. M. Drake. "Laminar Free Convection Heat Transfer from Downward-facing Horizontal Surfaces of Finite Dimensions," *Prog. Heat Mass Transfer,* vol. 2, p. 87, 1969.

65. McDonald, J. S., and T. J. Connally. "Investigation of Natural Convection Heat Transfer in Liquid Sodium," *Nucl Sci. Eng.,* vol. 8, p. 369, 1960.

66. Hollands, K. G. T., G. D. Raithby, and L. Konicek. "Correlation Equations for Free Convection Heat Transfer in Horizontal Layers of Air and Water," *Int. J. Heat Mass Transfer,* vol. 18, p. 879, 1975.

67. Hollands, K. G. T., T. E. Unny, and G. D. Raithby. "Free Convective Heat Transfer across Inclined Air Layers," *ASME Pap.* 75-HT-55, August 1975.

68. Depew, C. A., J. L. Franklin, and C. H. Ito. "Combined Free and Forced Convection in Horizontal, Uniformly Heated Tubes," *ASME Pap.* 75-HT-19, August 1975.

69. Raithby, G. D., and K. G. T. Hollands. "A General Method of Obtaining Approximate Solutions to Laminar and Turbulent Free Convection Problems," *Advances in Heat Transfer,* New York: Academic Press, 1974.

70. Churchill, S. W., and H. H. S. Chu. "Correlating Equations for Laminar and Turbulent Free Convection from a Horizontal Cylinder," *Int. J. Heat Mass Transfer,* vol. 18, p. 1049, 1975.

71. Churchill, S. W., and H. H. S. Chu. "Correlating Equations for Laminar and Turbulent Free Convection from a Vertical Plate," *Int. J. Heat Mass Transfer,* vol. 18, p. 1323, 1975.

72. Churchill, S. W. "A Comprehensive Correlating Equation for Laminar, Assisting, Forced and Free Convection," *AiChE J.,* vol. 23, no. 1, p. 10, 1977.

73. Al-Arabi, M., and Y. K. Salman. "Laminar Natural Convection Heat Transfer from an Inclined Cylinder," *Int. J. Heat Mass Transfer,* vol. 23, pp. 45–51, 1980.

74. Holman, J. P. *Heat Transfer,* 4th ed. New York: McGraw-Hill, 1976.

75. Hatfield, D. W., and D. K. Edwards. "Edge and Aspect Ratio Effects on Natural Convection from the Horizontal Heated Plate Facing Downwards," *Int. J. Heat Mass Transfer,* vol. 24, p. 1019, 1981.

76. Morgan, V. T. *The Overall Convective Heat Transfer from Smooth Circular Cylinders, Advances in Heat Transfer* (T. F. Irvine and J. P. Hartnett, eds.), vol.11, New York: Academic Press, 1975.

77. Sparrow, E. M., and M. A. Ansari. "A Refutation of King's Rule for Multi-Dimensional External Natural Convection," *Int. J. Heat Mass Transfer,* vol. 26, p. 1357, 1983.

78. Lienhard, J. H. "On the Commonality of Equations for Natural Convection from Immersed Bodies," *Int. J. Heat Mass Transfer,* vol. 16, p. 2121, 1973.

79. Amato, W. S., and C. L. Tien. "Free Convection Heat Transfer from Isothermal Spheres in Water," *Int. J. Heat Mass Transfer,* vol. 15, p. 327, 1972.

80. Warrington, R. O., and R. E. Powe. "The Transfer of Heat by Natural Convection Between Bodies and Their Enclosures," *Int. J. Heat Mass Transfer,* vol. 28, p. 319, 1985.

81. Sparrow, E. M., and A. J. Stretton. "Natural Convection from Bodies of Unity Aspect Ratio," *Int. J. Heat Mass Transfer,* vol. 28, p. 741, 1985.

82. El Sherbing, S. M., G. D. Raithby, and K. G. T. Hollands. "Heat Transfer across Vertical and Inclined Air Layers," *J. Heat Transfer,* vol. 104C, p. 96, 1982.

83. Churchill, S. W. "Free Convection Around Immersed Bodies," p. 2.5.7–24, in G. F. Hewitt (ed.), *Heat Exchanger Design Handbook,* Washington, D.C.: Hemisphere Publishing Corp., 1983.

84. Minkowycz, W. J., and E. M. Sparrow. "Local Nonsimilar Solutions for Natural Convection on a Vertical Cylinder," *J. Heat Transfer,* vol. 96, p. 178, 1974.

CHAPTER 8

Radiation Heat Transfer

8-1 | INTRODUCTION

Preceding chapters have shown how conduction and convection heat transfer may be calculated with the aid of both mathematical analysis and empirical data. We now wish to consider the third mode of heat transfer—thermal radiation. *Thermal radiation* is that electromagnetic radiation emitted by a body as a result of its temperature. In this chapter, we shall first describe the nature of thermal radiation, its characteristics, and the properties that are used to describe materials insofar as the radiation is concerned. Next, the transfer of radiation through space will be considered. Finally, the overall problem of heat transfer by thermal radiation will be analyzed, including the influence of the material properties and the geometric arrangement of the bodies on the total energy that may be exchanged.

8-2 | PHYSICAL MECHANISM

There are many types of electromagnetic radiation; thermal radiation is only one. Regardless of the type of radiation, we say that it is propagated at the speed of light, 3×10^8 m/s. This speed is equal to the product of the wavelength and frequency of the radiation,

$$c = \lambda \nu$$

where

$c =$ speed of light
$\lambda =$ wavelength
$\nu =$ frequency

The unit for λ may be centimeters, angstroms (1 Å $= 10^{-8}$ cm), or micrometers (1 μm $= 10^{-6}$ m). A portion of the electromagnetic spectrum is shown in Figure 8-1. Thermal radiation lies in the range from about 0.1 to 100 μm, while the visible-light portion of the spectrum is very narrow, extending from about 0.35 to 0.75 μm.

The propagation of thermal radiation takes place in the form of discrete quanta, each quantum having an energy of

$$E = h\nu \qquad \textbf{[8-1]}$$

where h is Planck's constant and has the value

$$h = 6.625 \times 10^{-34} \text{ J} \cdot \text{s}$$

Figure 8-1 | Electromagnetic spectrum.

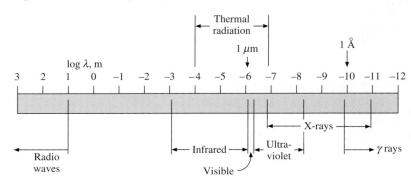

A very rough physical picture of the radiation propagation may be obtained by considering each quantum as a particle having energy, mass, and momentum, just as we considered the molecules of a gas. So, in a sense, the radiation might be thought of as a "photon gas" that may flow from one place to another. Using the relativistic relation between mass and energy, expressions for the mass and momentum of the "particles" could thus be derived; namely,

$$E = mc^2 = h\nu$$
$$m = \frac{h\nu}{c^2}$$
$$\text{Momentum} = c\frac{h\nu}{c^2} = \frac{h\nu}{c}$$

By considering the radiation as such a gas, the principles of quantum-statistical thermodynamics can be applied to derive an expression for the energy density of radiation per unit volume and per unit wavelength as[†]

$$u_\lambda = \frac{8\pi hc\lambda^{-5}}{e^{hc/\lambda kT} - 1} \tag{8-2}$$

where k is Boltzmann's constant, 1.38066×10^{-23} J/molecule · K. When the energy density is integrated over all wavelengths, the total energy emitted is proportional to absolute temperature to the fourth power:

$$E_b = \sigma T^4 \tag{8-3}$$

Equation (8-3) is called the Stefan-Boltzmann law, E_b is the energy radiated per unit time and per unit area by the ideal radiator, and σ is the Stefan-Boltzmann constant, which has the value

$$\sigma = 5.669 \times 10^{-8} \text{ W/m}^2 \cdot \text{K}^4 \quad [0.1714 \times 10^{-8} \text{ Btu/h} \cdot \text{ft}^2 \cdot {}^\circ\text{R}^4]$$

where E_b is in watts per square meter and T is in degrees Kelvin. In the thermodynamic analysis the energy density is related to the energy radiated from a surface per unit time and per unit area. Thus the heated interior surface of an enclosure produces a certain energy density of thermal radiation in the enclosure. We are interested in radiant exchange with surfaces—hence the reason for the expression of radiation from a surface in terms of its temperature. The subscript b in Equation (8-3) denotes that this is the radiation from a blackbody. We call this *blackbody radiation* because materials that obey this law appear black to the eye; they appear black because they do not reflect any radiation. Thus a blackbody is

---

[†] See, for example, J. P. Holman, *Thermodynamics*, 4th ed. New York: McGraw-Hill, 1988, p. 350.

also considered as one that absorbs all radiation incident upon it. E_b is called the *emissive power* of a blackbody.

It is important to note at this point that the "blackness" of a surface to thermal radiation can be quite deceiving insofar as visual observations are concerned. A surface coated with lampblack appears black to the eye and turns out to be black for the thermal-radiation spectrum. On the other hand, snow and ice appear quite bright to the eye but are essentially "black" for long-wavelength thermal radiation. Many white paints are also essentially black for long-wavelength radiation. This point will be discussed further in later sections.

8-3 | RADIATION PROPERTIES

When radiant energy strikes a material surface, part of the radiation is reflected, part is absorbed, and part is transmitted, as shown in Figure 8-2. We define the reflectivity ρ as the fraction reflected, the absorptivity α as the fraction absorbed, and the transmissivity τ as the fraction transmitted. Thus

$$\rho + \alpha + \tau = 1 \qquad \textbf{[8-4]}$$

Most solid bodies do not transmit thermal radiation, so that for many applied problems the transmissivity may be taken as zero. Then

$$\rho + \alpha = 1$$

Two types of reflection phenomena may be observed when radiation strikes a surface. If the angle of incidence is equal to the angle of reflection, the reflection is called *specular*. On the other hand, when an incident beam is distributed uniformly in all directions after reflection, the reflection is called *diffuse*. These two types of reflection are depicted

Figure 8-2 | Sketch showing effects of incident radiation.

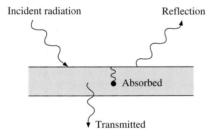

Figure 8-3 | (*a*) Specular ($\phi_1 = \phi_2$) and (*b*) diffuse reflection.

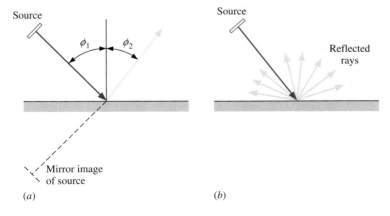

Figure 8-4 | Sketch showing model used for deriving Kirchhoff's law.

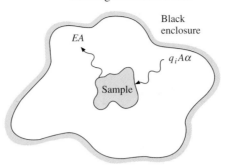

in Figure 8-3. Note that a specular reflection presents a mirror image of the source to the observer. No real surface is either specular or diffuse. An ordinary mirror is quite specular for visible light, but would not necessarily be specular over the entire wavelength range of thermal radiation. Ordinarily, a rough surface exhibits diffuse behavior better than a highly polished surface. Similarly, a polished surface is more specular than a rough surface. The influence of surface roughness on thermal-radiation properties of materials is a matter of serious concern and remains a subject for continuing research.

The emissive power of a body E is defined as the energy emitted by the body per unit area and per unit time. One may perform a thought experiment to establish a relation between the emissive power of a body and the material properties defined above. Assume that a perfectly black enclosure is available, i.e., one that absorbs all the incident radiation falling upon it, as shown schematically in Figure 8-4. This enclosure will also emit radiation according to the T^4 law. Let the radiant flux arriving at some area in the enclosure be q_i W/m². Now suppose that a body is placed inside the enclosure and allowed to come into temperature equilibrium with it. At equilibrium the energy absorbed by the body must be equal to the energy emitted; otherwise there would be an energy flow into or out of the body that would raise or lower its temperature. At equilibrium we may write

$$E A = q_i A \alpha \qquad [8\text{-}5]$$

If we now replace the body in the enclosure with a blackbody of the same size and shape and allow it to come to equilibrium with the enclosure *at the same temperature,*

$$E_b A = q_i A(1) \qquad [8\text{-}6]$$

since the absorptivity of a blackbody is unity. If Equation (8-5) is divided by Equation (8-6),

$$\frac{E}{E_b} = \alpha$$

and we find that the ratio of the emissive power of a body to the emissive power of a blackbody *at the same temperature* is equal to the absorptivity of the body. This ratio is defined as the *emissivity* ϵ of the body,

$$\epsilon = \frac{E}{E_b} \qquad [8\text{-}7]$$

so that

$$\epsilon = \alpha \qquad [8\text{-}8]$$

Equation (8-8) is called Kirchhoff's identity. At this point we note that the emissivities and absorptivities that have been discussed are the *total* properties of the particular material;

that is, they represent the integrated behavior of the material over all wavelengths. Real substances emit less radiation than ideal black surfaces as measured by the emissivity of the material. In reality, the emissivity of a material varies with temperature and the wavelength of the radiation.

The Gray Body

A *gray body* is defined such that the monochromatic emissivity ϵ_λ of the body is independent of wavelength. The *monochromatic emissivity* is defined as the ratio of the monochromatic emissive power of the body to the monochromatic emissive power of a blackbody at the same wavelength and temperature. Thus

$$\epsilon_\lambda = \frac{E_\lambda}{E_{b\lambda}}$$

The total emissivity of the body may be related to the monochromatic emissivity by noting that

$$E = \int_0^\infty \epsilon_\lambda E_{b\lambda} \, d\lambda \qquad \text{and} \qquad E_b = \int_0^\infty E_{b\lambda} \, d\lambda = \sigma T^4$$

so that

$$\epsilon = \frac{E}{E_b} = \frac{\int_0^\infty \epsilon_\lambda E_{b\lambda} \, d\lambda}{\sigma T^4} \qquad \text{[8-9]}$$

where $E_{b\lambda}$ is the emissive power of a blackbody per unit wavelength. If the gray-body condition is imposed, that is, ϵ_λ = constant, Equation (8-9) reduces to

$$\epsilon = \epsilon_\lambda \qquad \text{[8-10]}$$

The emissivities of various substances vary widely with wavelength, temperature, and surface condition. Some typical values of the total emissivity of various surfaces are given in Appendix A. We may note that the tabulated values are subject to considerable experimental uncertainty. A rather complete survey of radiation properties is given in Reference 14.

The functional relation for $E_{b\lambda}$ was derived by Planck by introducing the quantum concept for electromagnetic energy. The derivation is now usually performed by methods of statistical thermodynamics, and $E_{b\lambda}$ is shown to be related to the energy density of Equation (8-2) by

$$E_{b\lambda} = \frac{u_\lambda c}{4} \qquad \text{[8-11]}$$

or

$$E_{b\lambda} = \frac{C_1 \lambda^{-5}}{e^{C_2/\lambda T} - 1} \qquad \text{[8-12]}$$

where

λ = wavelength, μm

T = temperature, K

$C_1 = 3.743 \times 10^8 \text{ W} \cdot \mu\text{m}^4/\text{m}^2 \quad [1.187 \times 10^8 \text{ Btu} \cdot \mu\text{m}^4/\text{h} \cdot \text{ft}^2]$

$C_2 = 1.4387 \times 10^4 \, \mu\text{m} \cdot \text{K} \quad [2.5896 \times 10^4 \, \mu\text{m} \cdot °\text{R}]$

A plot of $E_{b\lambda}$ as a function of temperature and wavelength is given in Figure 8-5a. Notice that the peak of the curve is shifted to the shorter wavelengths for the higher temperatures. These maximum points in the radiation curves are related by Wien's displacement law,

$$\lambda_{\max} T = 2897.6 \, \mu\text{m} \cdot \text{K} \quad [5215.6 \, \mu\text{m} \cdot °\text{R}] \qquad \text{[8-13]}$$

Figure 8-5 | (*a*) Blackbody emissive power as a function of wavelength and temperature; (*b*) comparison of emissive power of ideal blackbodies and gray bodies with that of a real surface.

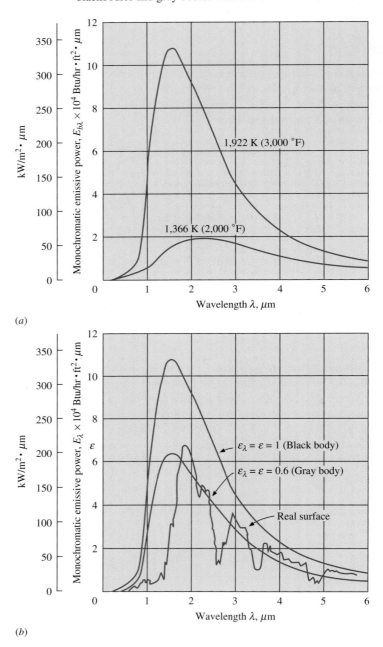

(*a*)

(*b*)

Figure 8-5*b* indicates the relative radiation spectra from a blackbody at 3000°F and a corresponding ideal gray body with emissivity equal to 0.6. Also shown is a curve indicating an approximate behavior for a real surface, which may differ considerably from that of either an ideal blackbody or an ideal gray body. For analysis purposes surfaces are usually considered as gray bodies, with emissivities taken as the integrated average value.

The shift in the maximum point of the radiation curve explains the change in color of a body as it is heated. Since the band of wavelengths visible to the eye lies between about 0.3 and 0.7 μm, only a very small portion of the radiant-energy spectrum at low temperatures is detected by the eye. As the body is heated, the maximum intensity is shifted to the shorter wavelengths, and the first visible sign of the increase in temperature of the body is a dark-red color. With further increase in temperature, the color appears as a bright red, then bright yellow, and finally white. The material also appears much brighter at higher temperatures because a larger portion of the total radiation falls within the visible range.

We are frequently interested in the amount of energy radiated from a blackbody in a certain specified wavelength range. The fraction of the total energy radiated between 0 and λ is given by

$$\frac{E_{b0-\lambda}}{E_{b0-\infty}} = \frac{\int_0^\lambda E_{b\lambda}\, d\lambda}{\int_0^\infty E_{b\lambda}\, d\lambda} \qquad\qquad \textbf{[8-14]}$$

Equation (8-12) may be rewritten by dividing both sides by T^5, so that

$$\frac{E_{b\lambda}}{T^5} = \frac{C_1}{(\lambda T)^5 (e^{C_2/\lambda T} - 1)} \qquad\qquad \textbf{[8-15]}$$

Now, for any specified temperature, the integrals in Equation (8-14) may be expressed in terms of the single variable λT. The ratio in Equation (8-14) is plotted in Figure 8-6 and tabulated in Table 8-1, along with the ratio in Equation (8-15). If the radiant energy emitted

Figure 8-6 | Fraction of blackbody radiation in wavelength interval.

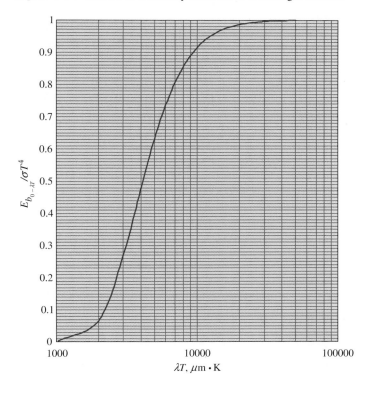

Table 8-1 | Radiation functions.

λT	$E_{b\lambda}/T^5$	$\dfrac{E_{b_0-\lambda T}}{\sigma T^4}$	λT	$E_{b\lambda}/T^5$	$\dfrac{E_{b_0-\lambda T}}{\sigma T^4}$
$\mu m \cdot K$	$\dfrac{W}{m^2 \cdot K^5 \cdot \mu m} \times 10^{11}$		$\mu m \cdot K$	$\dfrac{W}{m^2 \cdot K^5 \cdot \mu m} \times 10^{11}$	
1000	0.02110	0.00032	6300	0.42760	0.76180
1100	0.04846	0.00091	6400	0.41128	0.76920
1200	0.09329	0.00213	6500	0.39564	0.77631
1300	0.15724	0.00432	6600	0.38066	0.78316
1400	0.23932	0.00779	6700	0.36631	0.78975
1500	0.33631	0.01285	6800	0.35256	0.79609
1600	0.44359	0.01972	6900	0.33940	0.80219
1700	0.55603	0.02853	7000	0.32679	0.80807
1800	0.66872	0.03934	7100	0.31471	0.81373
1900	0.77736	0.05210	7200	0.30315	0.81918
2000	0.87858	0.06672	7300	0.29207	0.82443
2100	0.96994	0.08305	7400	0.28146	0.82949
2200	1.04990	0.10088	7500	0.27129	0.83436
2300	1.11768	0.12002	7600	0.26155	0.83906
2400	1.17314	0.14025	7700	0.25221	0.84359
2500	1.21659	0.16135	7800	0.24326	0.84796
2600	1.24868	0.18311	7900	0.23468	0.85218
2700	1.27029	0.20535	8000	0.22646	0.85625
2800	1.28242	0.22788	8100	0.21857	0.86017
2900	1.28612	0.25055	8200	0.21101	0.86396
3000	1.28245	0.27322	8300	0.20375	0.86762
3100	1.27242	0.29576	8400	0.19679	0.87115
3200	1.25702	0.31809	8500	0.19011	0.87456
3300	1.23711	0.34009	8600	0.18370	0.87786
3400	1.21352	0.36172	8700	0.17755	0.88105
3500	1.18695	0.38290	8800	0.17164	0.88413
3600	1.15806	0.40359	8900	0.16596	0.88711
3700	1.12739	0.42375	9000	0.16051	0.88999
3800	1.09544	0.44336	9100	0.15527	0.89277
3900	1.06261	0.46240	9200	0.15024	0.89547
4000	1.02927	0.48085	9300	0.14540	0.89807
4100	0.99571	0.49872	9400	0.14075	0.90060
4200	0.96220	0.51599	9500	0.13627	0.90304
4300	0.92892	0.53267	9600	0.13197	0.90541
4400	0.89607	0.54877	9700	0.12783	0.90770
4500	0.86376	0.56429	9800	0.12384	0.90992
4600	0.83212	0.57925	9900	0.12001	0.91207
4700	0.80124	0.59366	10,000	0.11632	0.91415
4800	0.77117	0.60753	10,200	0.10934	0.91813
4900	0.74197	0.62088	10,400	0.10287	0.92188
5000	0.71366	0.63372	10,600	0.09685	0.92540
5100	0.68628	0.64606	10,800	0.09126	0.92872
5200	0.65983	0.65794	11,000	0.08606	0.93184
5300	0.63432	0.66935	11,200	0.08121	0.93479
5400	0.60974	0.68033	11,400	0.07670	0.93758
5500	0.58608	0.69087	11,600	0.07249	0.94021
5600	0.56332	0.70101	11,800	0.06856	0.94270
5700	0.54146	0.71076	12,000	0.06488	0.94505
5800	0.52046	0.72012	12,200	0.06145	0.94728
5900	0.50030	0.72913	12,400	0.05823	0.94939
6000	0.48096	0.73778	12,600	0.05522	0.95139
6100	0.46242	0.74610	12,800	0.05240	0.95329
6200	0.44464	0.75410	13,000	0.04976	0.95509

Table 8-1 | (*Continued*).

λT	$E_{b\lambda}/T^5$	$\dfrac{E_{b_0-\lambda T}}{\sigma T^4}$	λT	$E_{b\lambda}/T^5$	$\dfrac{E_{b_0-\lambda T}}{\sigma T^4}$
$\mu m \cdot K$	$\dfrac{W}{m^2 \cdot K^5 \cdot \mu m} \times 10^{11}$		$\mu m \cdot K$	$\dfrac{W}{m^2 \cdot K^5 \cdot \mu m} \times 10^{11}$	
13,200	0.04728	0.95680	19,800	0.01151	0.98515
13,400	0.04494	0.95843	20,000	0.01110	0.98555
13,600	0.04275	0.95998	21,000	0.00931	0.98735
13,800	0.04069	0.96145	22,000	0.00786	0.98886
14,000	0.03875	0.96285	23,000	0.00669	0.99014
14,200	0.03693	0.96418	24,000	0.00572	0.99123
14,400	0.03520	0.96546	25,000	0.00492	0.99217
14,600	0.03358	0.96667	26,000	0.00426	0.99297
14,800	0.03205	0.96783	27,000	0.00370	0.99367
15,000	0.03060	0.96893	28,000	0.00324	0.99429
15,200	0.02923	0.96999	29,000	0.00284	0.99482
15,400	0.02794	0.97100	30,000	0.00250	0.99529
15,600	0.02672	0.97196	31,000	0.00221	0.99571
15,800	0.02556	0.97288	32,000	0.00196	0.99607
16,000	0.02447	0.97377	33,000	0.00175	0.99640
16,200	0.02343	0.97461	34,000	0.00156	0.99669
16,400	0.02245	0.97542	35,000	0.00140	0.99695
16,600	0.02152	0.97620	36,000	0.00126	0.99719
16,800	0.02063	0.97694	37,000	0.00113	0.99740
17,000	0.01979	0.97765	38,000	0.00103	0.99759
17,200	0.01899	0.97834	39,000	0.00093	0.99776
17,400	0.01823	0.97899	40,000	0.00084	0.99792
17,600	0.01751	0.97962	41,000	0.00077	0.99806
17,800	0.01682	0.98023	42,000	0.00070	0.99819
18,000	0.01617	0.98081	43,000	0.00064	0.99831
18,200	0.01555	0.98137	44,000	0.00059	0.99842
18,400	0.01496	0.98191	45,000	0.00054	0.99851
18,600	0.01439	0.98243	46,000	0.00049	0.99861
18,800	0.01385	0.98293	47,000	0.00046	0.99869
19,000	0.01334	0.98340	48,000	0.00042	0.99877
19,200	0.01285	0.98387	49,000	0.00039	0.99884
19,400	0.01238	0.98431	50,000	0.00036	0.99890
19,600	0.01193	0.98474			

between wavelengths λ_1 and λ_2 is desired, then

$$E_{b_{\lambda 1 - \lambda 2}} = E_{b_{0-\infty}} \left(\frac{E_{b_{0-\lambda 2}}}{E_{b_{0-\infty}}} - \frac{E_{b_{0-\lambda 1}}}{E_{b_{0-\infty}}} \right) \qquad \textbf{[8-16]}$$

where $E_{b_{0-\infty}}$ is the total radiation emitted over all wavelengths,

$$E_{b_{0-\infty}} = \sigma T^4 \qquad \textbf{[8-17]}$$

and is obtained by integrating the Planck distribution formula of Equation (8-12) over all wavelengths.

Solar radiation has a spectrum approximating that of a blackbody at 5800 K. Ordinary window glass transmits radiation up to about 2.5 μm. Consulting Table 8-1 for $\lambda T = (2.5)(5800) = 14,500 \ \mu m \cdot K$, we find the fraction of the solar spectrum below 2.5μm to be about 0.97. Thus the glass transmits most of the solar radiation incident upon it. In contrast, room radiation at about 300 K below 2.5 μm has $\lambda T = (2.5)(300) = 750 \ \mu m \cdot K$, and only a minute fraction (less than 0.001 percent) of this radiation would be transmitted

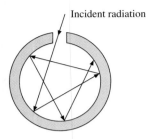

Incident radiation

through the glass. The glass, which is essentially transparent for visible light, is almost totally opaque for thermal radiation emitted at ordinary room temperatures.

Construction of a Blackbody

The concept of a blackbody is an idealization; that is, a perfect blackbody does not exist—all surfaces reflect radiation to some extent, however slight. A blackbody may be approximated very accurately, however, in the following way. A cavity is constructed, as shown in Figure 8-7, so that it is very large compared with the size of the opening in the side. An incident ray of energy is reflected many times on the inside before finally escaping from the side opening. With each reflection there is a fraction of the energy absorbed corresponding to the absorptivity of the inside of the cavity. After the many absorptions, practically all the incident radiation at the side opening is absorbed. It should be noted that the cavity of Figure 8-7 behaves approximately as a blackbody emitter as well as an absorber.

EXAMPLE 8-1   Transmission and Absorption in a Glass Plate

A glass plate 30 cm square is used to view radiation from a furnace. The transmissivity of the glass is 0.5 from 0.2 to 3.5 μm. The emissivity may be assumed to be 0.3 up to 3.5 μm and 0.9 above that. The transmissivity of the glass is zero, except in the range from 0.2 to 3.5 μm. Assuming that the furnace is a blackbody at 2000°C, calculate the energy absorbed in the glass and the energy transmitted.

■ **Solution**

$$T = 2000°C = 2273 \text{ K}$$
$$\lambda_1 T = (0.2)(2273) = 454.6 \ \mu\text{m} \cdot \text{K}$$
$$\lambda_2 T = (3.5)(2273) = 7955.5 \ \mu\text{m} \cdot \text{K}$$
$$A = (0.3)^2 = 0.09 \text{ m}^2$$

From Table 8-1

$$\frac{E_{b0-\lambda 1}}{\sigma T^4} = 0 \qquad \frac{E_{b0-\lambda 2}}{\sigma T^4} = 0.85443$$

$$\sigma T^4 = (5.669 \times 10^{-8})(2273)^4 = 1513.3 \text{ kW/m}^2$$

Total incident radiation is

$$0.2 \ \mu\text{m} < \lambda < 3.5 \ \mu\text{m} = (1.5133 \times 10^6)(0.85443 - 0)(0.3)^2$$
$$= 116.4 \text{ kW} \quad [3.97 \times 10^5 \text{ Btu/h}]$$

Total radiation transmitted $= (0.5)(116.4) = 58.2$ kW

$$\frac{\text{Radiation}}{\text{absorbed}} = \begin{cases} (0.3)(116.4) = 34.92 \text{ kW} & \text{for } 0 < \lambda < 3.5 \ \mu\text{m} \\ (0.9)(1 - 0.85443)(1513.3)(0.09) = 17.84 \text{ kW} & \text{for } 3.5 \ \mu\text{m} < \lambda < \infty \end{cases}$$

Total radiation absorbed $= 34.92 + 17.84 = 52.76$ kW   [180,000 Btu/h]

8-4 | RADIATION SHAPE FACTOR

Consider two black surfaces A_1 and A_2, as shown in Figure 8-8. We wish to obtain a general expression for the energy exchange between these surfaces when they are maintained at different temperatures. The problem becomes essentially one of determining the amount of

Figure 8-8 | Sketch showing area elements used in deriving
radiation shape factor.

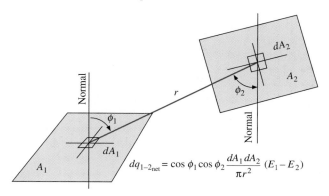

$$dq_{1-2_{net}} = \cos \phi_1 \cos \phi_2 \frac{dA_1 \, dA_2}{\pi r^2} (E_1 - E_2)$$

energy that leaves one surface and reaches the other. To solve this problem the *radiation
shape factors* are defined as

F_{1-2} = fraction of energy leaving surface 1 that reaches surface 2

F_{2-1} = fraction of energy leaving surface 2 that reaches surface 1

F_{i-j} = fraction of energy leaving surface *i* that reaches surface *j*

Other names for the radiation shape factor are *view factor, angle factor,* and *configuration
factor.* The energy leaving surface 1 and arriving at surface 2 is

$$E_{b1} A_1 F_{12}$$

and the energy leaving surface 2 and arriving at surface 1 is

$$E_{b2} A_2 F_{21}$$

Since the surfaces are black, all the incident radiation will be absorbed, and the net energy
exchange is

$$E_{b1} A_1 F_{12} - E_{b2} A_2 F_{21} = Q_{1-2}$$

If both surfaces are at the same temperature, there can be no heat exchange, that is, $Q_{1-2} = 0$.
Also, for $T_1 = T_2$

$$E_{b1} = E_{b2}$$

so that

$$A_1 F_{12} = A_2 F_{21} \qquad\qquad \textbf{[8-18]}$$

The net heat exchange is therefore

$$Q_{1-2} = A_1 F_{12}(E_{b1} - E_{b2}) = A_2 F_{21}(E_{b1} - E_{b2}) \qquad\qquad \textbf{[8-19]}$$

Equation (8-18) is known as a reciprocity relation, and it applies in a general way for
any two surfaces *i* and *j*:

$$A_i F_{ij} = A_j F_{ji} \qquad\qquad \textbf{[8-18a]}$$

Although the relation is derived for black surfaces, it holds for other surfaces also as long
as diffuse radiation is involved.

Figure 8-9 | Elevation view of area shown
in Figure 8-8.

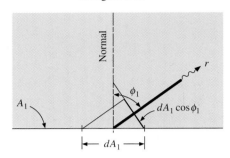

We now wish to determine a general relation for F_{12} (or F_{21}). To do this, we consider the elements of area dA_1 and dA_2 in Figure 8-8. The angles ϕ_1 and ϕ_2 are measured between a normal to the surface and the line drawn between the area elements r. The projection of dA_1 on the line between centers is

$$dA_1 \cos \phi_1$$

This may be seen more clearly in the elevation drawing shown in Figure 8-9. We assume that the surfaces are diffuse, that is, that the intensity of the radiation is the same in all directions. The intensity is the radiation emitted per unit area and per unit of solid angle in a certain specified direction. So, in order to obtain the energy emitted by the element of area dA_1 in a certain direction, we must multiply the intensity by the projection of dA_1 in the specified direction. Thus the energy leaving dA_1 in the direction given by the angle ϕ_1 is

$$I_b \, dA_1 \cos \phi_1 \qquad [a]$$

where I_b is the blackbody intensity. The radiation arriving at some area element dA_n at a distance r from A_1 would be

$$I_b \, dA_1 \cos \phi_1 \, \frac{dA_n}{r^2} \qquad [b]$$

where dA_n is constructed normal to the radius vector. The quantity dA_n/r^2 represents the solid angle subtended by the area dA_n. The intensity may be obtained in terms of the emissive power by integrating expression (b) over a hemisphere enclosing the element of area dA_1. In a spherical coordinate system like that in Figure 8-10,

$$dA_n = r^2 \sin \phi \, d\psi \, d\phi$$

Figure 8-10 | Spherical coordinate system used in derivation of radiation shape factor.

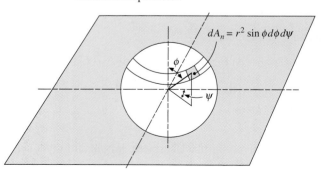

Then

$$E_b \, dA_1 = I_b \, dA_1 \int_0^{2\pi} \int_0^{\pi/2} \sin\phi \cos\phi \, d\phi \, d\psi$$
$$= \pi I_b \, dA_1$$

so that

$$E_b = \pi I_b \qquad\qquad \textbf{[8-20]}$$

We may now return to the energy-exchange problem indicated in Figure 8-8. The area element dA_n is given by

$$dA_n = \cos\phi_2 \, dA_2$$

so that the energy leaving dA_1 that arrives at dA_2 is

$$dq_{1-2} = E_{b1} \cos\phi_1 \cos\phi_2 \frac{dA_1 \, dA_2}{\pi r^2}$$

That energy leaving dA_2 and arriving at dA_1 is

$$dq_{2-1} = E_{b2} \cos\phi_2 \cos\phi_1 \frac{dA_2 \, dA_1}{\pi r^2}$$

and the net energy exchange is

$$q_{\mathrm{net}_{1-2}} = (E_{b1} - E_{b2}) \int_{A_2} \int_{A_1} \cos\phi_1 \cos\phi_2 \frac{dA_1 dA_2}{\pi r^2} \qquad\qquad \textbf{[8-21]}$$

The integral is either $A_1 F_{12}$ or $A_2 F_{21}$ according to Equation (8-19). To evaluate the integral, the specific geometry of the surfaces A_1 and A_2 must be known. We shall work out an elementary problem and then present the results for more complicated geometries in graphical and equation form.

Consider the radiation from the small area dA_1 to the flat disk A_2, as shown in Figure 8-11. The element of area dA_2 is chosen as the circular ring of radius x. Thus

$$dA_2 = 2\pi x \, dx$$

Figure 8-11 | Radiation from a small-area element to a disk.

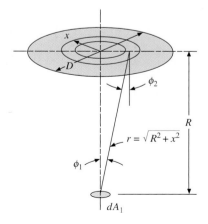

We note that $\phi_1 = \phi_2$ and apply Equation (8-21), integrating over the area A_2:

$$dA_1 \; F_{dA_1-A_2} = dA_1 \int_{A_2} \cos^2 \phi_1 \; \frac{2\pi x \; dx}{\pi r^2}$$

Making the substitutions

$$r = (R^2 + x^2)^{1/2} \quad \text{and} \quad \cos \phi_1 = \frac{R}{(R^2 + x^2)^{1/2}}$$

we have

$$dA_1 \; F_{dA_1-A_2} = dA_1 \int_0^{D/2} \frac{2R^2 x \; dx}{(R^2 + x^2)^2}$$

Performing the integration gives

$$dA_1 \; F_{dA_1-A_2} = -dA_1 \left(\frac{R^2}{R^2 + x^2} \right)\Bigg]_0^{D/2} = dA_1 \frac{D^2}{4R^2 + D^2}$$

so that

$$F_{dA_1-A_2} = \frac{D^2}{4R^2 + D^2} \tag{8-22}$$

The calculation of shape factors may be extended to more complex geometries, as described in References 3, 5, 24, and 32; 32 gives a very complete catalog of analytical relations and graphs for shape factors. For our purposes we give only the results of a few geometries as shown in Figures 8-12 through 8-16. The analytical relations for these geometries are given in Table 8-2.

Figure 8-12 | Radiation shape factor for radiation between parallel rectangles.

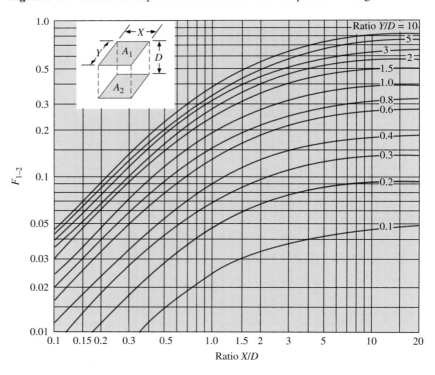

Figure 8-13 | Radiation shape factor for radiation between parallel equal coaxial disks.

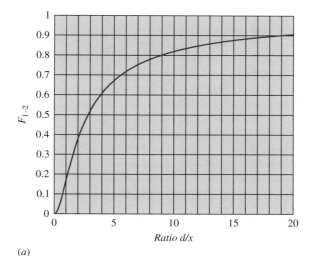

(a)

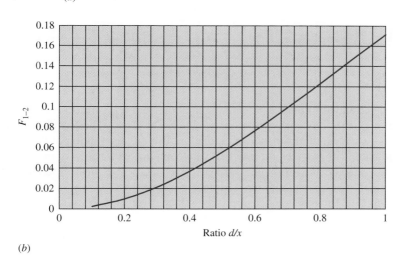

(b)

Real-Surface Behavior

Real surfaces exhibit interesting deviations from the ideal surfaces described in the preceding paragraphs. Real surfaces, for example, are not perfectly diffuse, and hence the intensity of emitted radiation is not constant over all directions. The directional-emittance characteristics of several types of surfaces are shown in Figure 8-17. These curves illustrate the characteristically different behavior of electric conductors and nonconductors. Conductors emit more energy in a direction having a large azimuth angle. This behavior may be satisfactorily explained with basic electromagnetic wave theory, and is discussed in Reference 24. As a result of this basic behavior of conductors and nonconductors, we may anticipate the appearance of a sphere which is heated to incandescent temperatures, as shown in Figure 8-18. An electric conducting sphere will appear bright around the rim since more energy is emitted at large angles ϕ. A sphere constructed of a nonconducting material will have the opposite behavior and will appear bright in the center and dark around the edge.

Figure 8-14 | Radiation shape factor for radiation between perpendicular rectangles with a common edge.

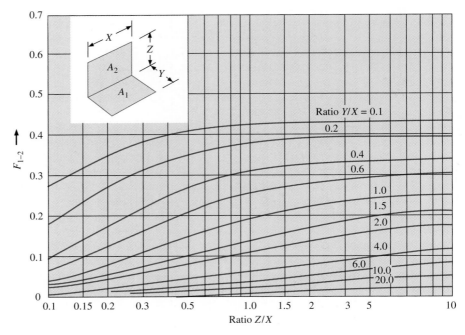

Reflectance and absorptance of thermal radiation from real surfaces are a function not only of the surface itself but also of the surroundings. These properties are dependent on the direction and wavelength of the incident radiation. But the distribution of the intensity of incident radiation with wavelength may be a very complicated function of the temperatures and surface characteristics of all the surfaces that incorporate the surroundings. Let us denote the total incident radiation on a surface per unit time, per unit area, and per unit wavelength as G_λ. Then the total absorptivity will be given as the ratio of the total energy absorbed to the total energy incident on the surface,

or

$$\alpha = \frac{\int_0^\infty \alpha_\lambda G_\lambda \, d\lambda}{\int_0^\infty G_\lambda \, d\lambda} \qquad \text{[8-23]}$$

If we are fortunate enough to have a gray body such that $\epsilon_\lambda = \epsilon = \text{constant}$, this relation simplifies considerably. It may be shown that Kirchoff's law [Equation(8-8)] may be written for monochromatic radiation as

$$\epsilon_\lambda = \alpha_\lambda \qquad \text{[8-24]}$$

Therefore, for a gray body, $\alpha_\lambda = \text{constant}$, and Equation (8-23) expresses the result that the total absorptivity is also constant and independent of the wavelength distribution of incident radiation. Furthermore, since the emissivity and absorptivity are constant over all wavelengths for a gray body, they must be independent of temperature as well. Unhappily, real surfaces are not always "gray" in nature, and significant errors may ensue by assuming gray-body behavior. On the other hand, analysis of radiation exchange using real-surface behavior is so complicated that the ease and simplification of the gray-body assumption is justified by the practical utility it affords. References 10, 11, and 24 present comparisons of heat-transfer calculations based on both gray and nongray analyses.

Figure 8-15 | Radiation shape factors for two concentric cylinders of finite length. (*a*) Outer cylinder to itself; (*b*) outer cylinder to inner cylinder.

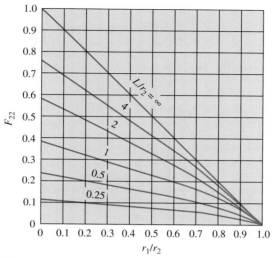

(*a*)

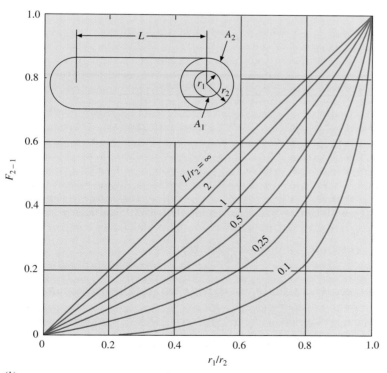

(*b*)

Figure 8-16 | Radiation shape factor for radiation between two parallel coaxial disks.

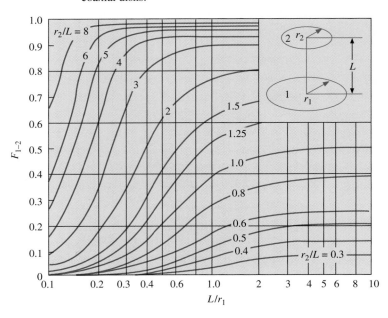

Table 8-2 | Radiation shape factor relations.

Geometry	Shape factor
1. Parallel, equal rectangles (Fig. 8-12) $x = X/D, y = Y/D$	$F_{1-2} = (2/\pi xy)\{ \ln[(1+x^2)(1+y^2)/(1+x^2+y^2)]^{1/2} + x(1+y^2)^{1/2}\tan^{-1}[x/(1+y^2)^{1/2}]$ $\quad + y(1+x^2)^{1/2}\tan^{-1}[y/(1+x^2)^{1/2}] - x\tan^{-1}x - y\tan^{-1}y\}$
2. Parallel, equal, coaxial disks (Fig. 8-13) $R = d/2x, X = (2R^2+1)/R^2$	$F_{1-2} = [X - (X^2-4)^{1/2}]/2$
3. Perpendicular rectangles with a common edge (Fig. 8-14) $H = Z/X, W = Y/X$	$F_{1-2} = (1/\pi W)\big(W\tan^{-1}(1/W) + H\tan^{-1}(1/H) - (H^2+W^2)^{1/2}\tan^{-1}[1/(H^2+W^2)^{1/2}]$ $\quad + (1/4)\ln\{[(1+W^2)(1+H^2)/(1+W^2+H^2)] \times [W^2(1+W^2+H^2)/(1+W^2)(W^2+H^2)]^{W^2}$ $\quad \times [H^2(1+H^2+W^2)/(1+H^2)(H^2+W^2)]^{H^2}\}\big)$
4. Finite, coaxial cylinders (Fig. 8-15) $X = r_2/r_1, Y = L/r_1$ $A = X^2 + Y^2 - 1$ $B = Y^2 - X^2 + 1$	$F_{2-1} = (1/X) - (1/\pi X)\{\cos^{-1}(B/A) - (1/2Y)[(A^2+4A-4X^2+4)^{1/2}\cos^{-1}(B/XA)$ $\quad + B\sin^{-1}(1/X) - \pi A/2]\}$ $F_{2-2} = 1 - (1/X) + (2/\pi X)\tan^{-1}[2(X^2-1)^{1/2}/Y]$ $\quad - (Y/2\pi X)([\sqrt{(4X^2+Y^2)}/Y]\sin^{-1}\{[4(X^2-1) + (Y/X)^2(X^2-2)]/[Y^2+4(X^2-1)]\}$ $\quad - \sin^{-1}[(X^2-2)/X^2] + (\pi/2)[(4X^2+Y^2)^{1/2}/Y - 1])$
5. Parallel, coaxial disks (Fig. 8-16) $R_1 = r_1/L$ $R_2 = r_2/L$ $X = 1 + (1+R_2^2)/R_1^2$	$F_{1-2} = \{X - [X^2 - 4(R_2/R_1)^2]^{1/2}\}/2$

Figure 8-17 | Typical directional behavior of emissivity for conductors and
nonconductors. ϵ_ϕ is emissivity at angle ϕ measured from
normal to surface. Nonconductor curves are for (a) wet ice,
(b) wood, (c) glass, (d) paper, (e) clay, (f) copper oxide, and
(g) aluminum oxide.

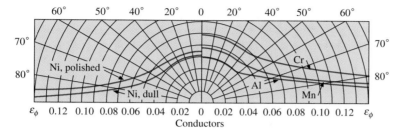

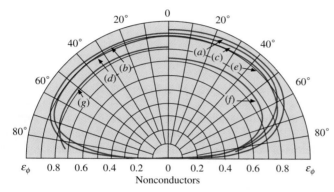

Figure 8-18 | Effect of directional emittance on appearance of an
incandescent sphere: (a) electrical conductor; (b) electrical
nonconductor.

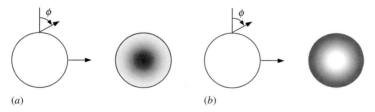

(a)   (b)

Heat Transfer Between Black Surfaces   EXAMPLE 8-2

Two parallel black plates 0.5 by 1.0 m are spaced 0.5 m apart. One plate is maintained at 1000°C
and the other at 500°C. What is the net radiant heat exchange between the two plates?

■ **Solution**
The ratios for use with Figure 8-12 are

$$\frac{Y}{D} = \frac{0.5}{0.5} = 1.0 \qquad \frac{X}{D} = \frac{1.0}{0.5} = 2.0$$

so that $F_{12} = 0.285$. The heat transfer is calculated from

$$q = A_1 F_{12}(E_{b1} - E_{b2}) = \sigma A_1 F_{12}(T_1^4 - T_2^4)$$
$$= (5.669 \times 10^{-8})(0.5)(0.285)(1273^4 - 773^4)$$
$$= 18.33 \text{ kW} \quad [62,540 \text{ Btu/h}]$$

8-5 | RELATIONS BETWEEN SHAPE FACTORS

Some useful relations between shape factors may be obtained by considering the system shown in Figure 8-19. Suppose that the shape factor for radiation from A_3 to the combined area $A_{1,2}$ is desired. This shape factor must be given very simply as

$$F_{3-1,2} = F_{3-1} + F_{3-2} \qquad \text{[8-25]}$$

that is, the total shape factor is the sum of its parts. We could also write Equation (8-25) as

$$A_3 F_{3-1,2} = A_3 F_{3-1} + A_3 F_{3-2} \qquad \text{[8-26]}$$

and making use of the reciprocity relations

$$A_3 F_{3-1,2} = A_{1,2} F_{1,2-3}$$
$$A_3 F_{3-1} = A_1 F_{1-3}$$
$$A_3 F_{3-2} = A_2 F_{2-3}$$

the expression could be rewritten

$$A_{1,2} F_{1,2-3} = A_1 F_{1-3} + A_2 F_{2-3} \qquad \text{[8-27]}$$

which simply states that the total radiation arriving at surface 3 is the sum of the radiations from surfaces 1 and 2. Suppose we wish to determine the shape factor F_{1-3} for the surfaces in Figure 8-20 in terms of known shape factors for perpendicular rectangles with a common edge. We may write

$$F_{1-2,3} = F_{1-2} + F_{1-3}$$

in accordance with Equation (8-25). Both $F_{1-2,3}$ and F_{1-2} may be determined from Figure 8-14, so that F_{1-3} is easily calculated when the dimensions are known. Now consider the somewhat more complicated situation shown in Figure 8-21. An expression for the shape factor F_{1-4} is desired in terms of known shape factors for perpendicular rectangles

Figure 8-19 | Sketch showing some relations between shape factors.

$$F_{3-1,2} = F_{3-1} + F_{3-2}$$
$$A_3 F_{3-1,2} = A_3 F_{3-1} + A_3 F_{3-2}$$
$$A_{1,2} F_{1,2-3} = A_1 F_{1-3} + A_2 F_{2-3}$$

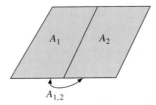

Figure 8-20

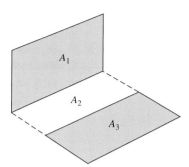

Figure 8-21

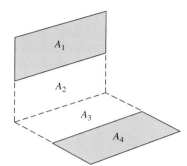

with a common edge. We write

$$A_{1,2}F_{1,2-3,4} = A_1F_{1-3,4} + A_2F_{2-3,4} \qquad [a]$$

in accordance with Equation (8-25). Both $F_{1,2-3,4}$ and $F_{2-3,4}$ can be obtained from Figure 8-14, and $F_{1-3,4}$ may be expressed

$$A_1F_{1-3,4} = A_1F_{1-3} + A_1F_{1-4} \qquad [b]$$

Also

$$A_{1,2}F_{1,2-3} = A_1F_{1-3} + A_2F_{2-3} \qquad [c]$$

Solving for A_1F_{1-3} from (c), inserting this in (b), and then inserting the resultant expression for $A_1F_{1-3,4}$ in (a) gives

$$A_{1,2}F_{1,2-3,4} = A_{1,2}F_{1,2-3} - A_2F_{2-3} + A_1F_{1-4} + A_2F_{2-3,4} \qquad [d]$$

Notice that all shape factors except F_{1-4} may be determined from Figure 8-14. Thus

$$F_{1-4} = \frac{1}{A_1}(A_{1,2}F_{1,2-3,4} + A_2F_{2-3} - A_{1,2}F_{1,2-3} - A_2F_{2-3,4}) \qquad [8\text{-}28]$$

In the foregoing discussion the tacit assumption has been made that the various bodies do not see themselves, that is,

$$F_{11} = F_{22} = F_{33} = 0 \cdots$$

To be perfectly general, we must include the possibility of concave curved surfaces, which may then see themselves. The general relation is therefore

$$\sum_{j=1}^{n} F_{ij} = 1.0 \qquad [8\text{-}29]$$

where F_{ij} is the fraction of the total energy leaving surface i that arrives at surface j. Thus for a three-surface enclosure we would write

$$F_{11} + F_{12} + F_{13} = 1.0$$

and F_{11} represents the fraction of energy leaving surface 1 that strikes surface 1. A certain amount of care is required in analyzing radiation exchange between curved surfaces.

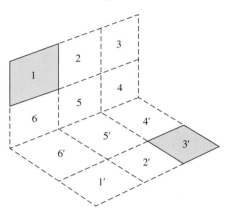

Figure 8-22 | Generalized perpendicular-rectangle arrangement.

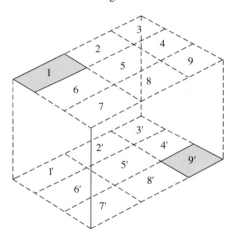

Figure 8-23 | Generalized parallel-rectangle arrangement.

Hamilton and Morgan [5] have presented generalized relations for parallel and perpendicular rectangles in terms of shape factors which may be obtained from Figures 8-12 and 8-14. The two situations of interest are shown in Figures 8-22 and 8-23. For the perpendicular rectangles of Figure 8-22 it can be shown that the following reciprocity relations apply [5]:

$$A_1 F_{13'} = A_3 F_{31'} = A_{3'} F_{3'1} = A_{1'} F_{1'3} \tag{8-30}$$

By making use of these reciprocity relations, the radiation shape factor $F_{13'}$ may be expressed by

$$A_1 F_{13'} = \tfrac{1}{2}\big[K_{(1,2,3,4,5,6)^2} - K_{(2,3,4,5)^2} - K_{(1,2,5,6)^2} + K_{(4,5,6)^2} - K_{(4,5,6)-(1',2',3',4',5',6')} $$
$$- K_{(1,2,3,4,5,6)-(4',5',6')} + K_{(1,2,5,6)-(5',6')} + K_{(2,3,4,5)-(4',5')} + K_{(5,6)-(1',2',5',6')} $$
$$+ K_{(4,5)-(2',3',4',5')} + K_{(2,5)^2} - K_{(2,5)-5'} - K_{(5,6)^2} - K_{(4,5)^2} - K_{5-(2',5')} + K_{5^2}\big] \tag{8-31}$$

where the K terms are defined by

$$K_{m-n} = A_m F_{m-n} \tag{8-32}$$

$$K_{(m)^2} = A_m F_{m-m'} \tag{8-33}$$

The generalized parallel-rectangle arrangement is depicted in Figure 8-23. The reciprocity relations that apply to this situation are given in Reference 5 as

$$A_1 F_{19'} = A_3 F_{37'} = A_9 F_{91'} = A_7 F_{73'} \tag{8-34}$$

Making use of these relations, it is possible to derive the shape factor $F_{19'}$ as

$$A_1 F_{19'} = \tfrac{1}{4} \big[K_{(1,2,3,4,5,6,7,8,9)^2} - K_{(1,2,5,6,7,8)^2} - K_{(2,3,4,5,8,9)^2} - K_{(1,2,3,4,5,6)^2}$$
$$+ K_{(1,2,5,6)^2} + K_{(2,3,4,5)^2} + K_{(4,5,8,9)^2} - K_{(4,5)^2} - K_{(5,8)^2} - K_{(5,6)^2}$$
$$- K_{(4,5,6,7,8,9)^2} + K_{(5,6,7,8)^2} + K_{(4,5,6)^2} + K_{(2,5,8)^2} - K_{(2,5)^2} + K_{(5)^2} \big] \tag{8-35}$$

The nomenclature for the K terms is the same as given in Equations (8-32) and (8-33).

Shape-Factor Algebra for Open Ends of Cylinders

EXAMPLE 8-3

Two concentric cylinders having diameters of 10 and 20 cm have a length of 20 cm. Calculate the shape factor between the open ends of the cylinders.

■ **Solution**

We use the nomenclature of Figure 8-15 for this problem and designate the open ends as surfaces 3 and 4. We have $L/r_2 = 20/10 = 2.0$ and $r_1/r_2 = 0.5$; so from Figure 8-15 or Table 8-2 we obtain

$$F_{21} = 0.4126 \qquad F_{22} = 0.3286$$

Using the reciprocity relation [Equation (8-18)] we have

$$A_1 F_{12} = A_2 F_{21} \qquad \text{and} \qquad F_{12} = (d_2/d_1) F_{21} = (20/10)(0.4126) = 0.8253$$

For surface 2 we have

$$F_{21} + F_{22} + F_{23} + F_{24} = 1.0$$

From symmetry $F_{23} = F_{24}$ so that

$$F_{23} = F_{24} = \left(\tfrac{1}{2}\right)(1 - 0.4126 - 0.3286) = 0.1294$$

Using reciprocity again,

$$A_2 F_{23} = A_3 F_{32}$$

and

$$F_{32} = \frac{\pi(20)(20)}{\pi(20^2 - 10^2)/4} 0.1294 = 0.6901$$

We observe that $F_{11} = F_{33} = F_{44} = 0$ and for surface 3

$$F_{31} + F_{32} + F_{34} = 1.0 \tag{a}$$

So, if F_{31} can be determined, we can calculate the desired quantity F_{34}. For surface 1

$$F_{12} + F_{13} + F_{14} = 1.0$$

and from symmetry $F_{13} = F_{14}$ so that

$$F_{13} = \left(\frac{1}{2}\right)(1 - 0.8253) = 0.0874$$

Using reciprocity gives

$$A_1 F_{13} = A_3 F_{31}$$

$$F_{31} = \frac{\pi(10)(20)}{\pi(20^2 - 10^2)/4}0.0874 = 0.233$$

Then, from Equation (a)

$$F_{34} = 1 - 0.233 - 0.6901 = 0.0769$$

| EXAMPLE 8-4 | Shape-Factor Algebra for Truncated Cone |

A truncated cone has top and bottom diameters of 10 and 20 cm and a height of 10 cm. Calculate the shape factor between the top surface and the side and also the shape factor between the side and itself.

■ **Solution**
We employ Figure 8-16 for solution of this problem and take the nomenclature as shown, designating the top as surface 2, the bottom as surface 1, and the side as surface 3. Thus, the desired quantities are F_{23} and F_{33}. We have $L/r_1 = 10/10 = 1.0$ and $r_2/L = 5/10 = 0.5$. Thus, from Figure 8-16

$$F_{12} = 0.12$$

From reciprocity [Equation (8-18)]

$$A_1 F_{12} = A_2 F_{21}$$
$$F_{21} = (20/10)^2(0.12) = 0.48$$

and

$$F_{22} = 0$$

so that

$$F_{21} + F_{23} = 1.0$$

and

$$F_{23} = 1 - 0.48 = 0.52$$

For surface 3,

$$F_{31} + F_{32} + F_{33} = 1.0 \qquad\qquad [a]$$

so we must find F_{31} and F_{32} in order to evaluate F_{33}. Since $F_{11} = 0$, we have

$$F_{12} + F_{13} = 1.0 \quad \text{and} \quad F_{13} = 1 - 0.12 = 0.88$$

and from reciprocity

$$A_1 F_{13} = A_3 F_{31} \qquad\qquad [b]$$

The surface area of the side is

$$A_3 = \pi(r_1 + r_2)\left[(r_1 - r_2)^2 + L^2\right]^{1/2}$$
$$= \pi(5 + 10)(5^2 + 10^2)^{1/2} = 526.9 \text{ cm}^2$$

So, from Equation (b)

$$F_{31} = \frac{\pi(10^2)}{526.9} 0.88 = 0.525$$

A similar procedure applies with surface 2 so that

$$F_{32} = \frac{\pi(5)^2}{526.9} 0.52 = 0.0775$$

Finally, from Equation (a)

$$F_{33} = 1 - 0.525 - 0.0775 = 0.397$$

Shape-Factor Algebra for Cylindrical Reflector   **EXAMPLE 8-5**

The long circular half-cylinder shown in Figure Example 8-5 has a diameter of 60 cm and a square rod 20 by 20 cm placed along the geometric centerline. Both are surrounded by a large enclosure. Find F_{12}, F_{13}, and F_{11} in accordance with the nomenclature in the figure.

Figure Example 8-5

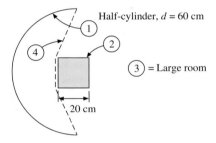

Half-cylinder, $d = 60$ cm
1
4
2
3 = Large room
20 cm

■ **Solution**

From symmetry we have

$$F_{21} = F_{23} = 0.5 \qquad [a]$$

In general, $F_{11} + F_{12} + F_{13} = 1.0$. To aid in the analysis we create the fictitious surface 4 shown as the dashed line. For this surface, $F_{41} = 1.0$. Now, all radiation leaving surface 1 will arrive either at 2 or at 3. Likewise, this radiation will arrive at the imaginary surface 4, so that

$$F_{14} = F_{12} + F_{13} \qquad [b]$$

From reciprocity,

$$A_1 F_{14} = A_4 F_{41}$$

The areas are, for unit length,

$$A_1 = \pi d/2 = \pi(0.6)/2 = 0.942$$
$$A_4 = 0.2 + (2)[(0.1)^2 + (0.2)^2]^{1/2} = 0.647$$
$$A_2 = (4)(0.2) = 0.8$$

so that

$$F_{14} = \frac{A_4}{A_1} F_{41} = \frac{(0.647)(1.0)}{0.942} = 0.686 \qquad [c]$$

We also have, from reciprocity,

$$A_2 F_{21} = A_1 F_{12}$$

so

$$F_{12} = \frac{A_2}{A_1} F_{21} = \frac{(0.8)(0.5)}{0.942} = 0.425 \qquad [d]$$

Combining (b), (c), and (d) gives

$$F_{13} = 0.686 - 0.425 = 0.261$$

Finally,

$$F_{11} = 1 - F_{12} - F_{13} = 1 - 0.425 - 0.261 = 0.314$$

This example illustrates how one may make use of clever geometric considerations to calculate the radiation shape factors.

8-6 | HEAT EXCHANGE BETWEEN NONBLACKBODIES

The calculation of the radiation heat transfer between black surfaces is relatively easy because all the radiant energy that strikes a surface is absorbed. The main problem is one of determining the geometric shape factor, but once this is accomplished, the calculation of the heat exchange is very simple. When nonblackbodies are involved, the situation is much more complex, for all the energy striking a surface will not be absorbed; part will be reflected back to another heat-transfer surface, and part may be reflected out of the system entirely. The problem can become complicated because the radiant energy can be reflected back and forth between the heat-transfer surfaces several times. The analysis of the problem must take into consideration these multiple reflections if correct conclusions are to be drawn.

We shall assume that all surfaces considered in our analysis are diffuse, gray, and uniform in temperature and that the reflective and emissive properties are constant over all the surface. Two new terms may be defined:

$G =$ irradiation

    $=$ total radiation incident upon a surface per unit time and per unit area

$J =$ radiosity

    $=$ total radiation that leaves a surface per unit time and per unit area

In addition to the assumptions stated above, we shall also assume that the radiosity and irradiation are uniform over each surface. This assumption is not strictly correct, even for ideal gray diffuse surfaces, but the problems become exceedingly complex when this analytical restriction is not imposed. Sparrow and Cess [10] give a discussion of such problems. As shown in Figure 8-24, the radiosity is the sum of the energy emitted and the energy reflected when no energy is transmitted, or

$$J = \epsilon E_b + \rho G \qquad [8-36]$$

where ϵ is the emissivity and E_b is the blackbody emissive power. Since the transmissivity is assumed to be zero, the reflectivity may be expressed as

$$\rho = 1 - \alpha = 1 - \epsilon$$

so that

$$J = \epsilon E_b + (1 - \epsilon)G \qquad [8-37]$$

Figure 8-24 | (*a*) Surface energy balance for opaque material; (*b*) element representing "surface resistance" in the radiation-network method.

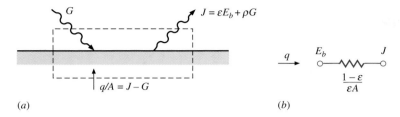

(*a*)   (*b*)

The net energy leaving the surface is the difference between the radiosity and the irradiation:

$$\frac{q}{A} = J - G = \epsilon E_b + (1 - \epsilon)G - G$$

Solving for G in terms of J from Equation (8-37),

$$q = \frac{\epsilon A}{1 - \epsilon}(E_b - J)$$

or

$$q = \frac{E_b - J}{(1 - \epsilon)/\epsilon A} \qquad \textbf{[8-38]}$$

At this point we introduce a very useful interpretation for Equation (8-38). If the denominator of the right side is considered as the surface resistance to radiation heat transfer, the numerator as a potential difference, and the heat flow as the "current," then a network element could be drawn as in Figure 8-24(*b*) to represent the physical situation. This is the first step in the network method of analysis originated by Oppenheim [20].

Now consider the exchange of radiant energy by two surfaces, A_1 and A_2, shown in Figure 8-25. Of that total radiation leaving surface 1, the amount that reaches surface 2 is

$$J_1 A_1 F_{12}$$

and of that total energy leaving surface 2, the amount that reaches surface 1 is

$$J_2 A_2 F_{21}$$

Figure 8-25 | (*a*) Spatial energy exchange between two surfaces; (*b*) element representing "space resistance" in the radiation-network method.

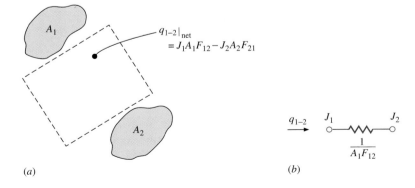

(*a*)   (*b*)

The net interchange between the two surfaces is

$$q_{1-2} = J_1 A_1 F_{12} - J_2 A_2 F_{21}$$

But

$$A_1 F_{12} = A_2 F_{21}$$

so that

$$q_{1-2} = (J_1 - J_2) A_1 F_{12} = (J_1 - J_2) A_2 F_{21}$$

or

$$q_{1-2} = \frac{J_1 - J_2}{1/A_1 F_{12}} \qquad\qquad \textbf{[8-39]}$$

We may thus construct a network element that represents Equation (8-39), as shown in Figure 8-25b. The two network elements shown in Figures 8-24 and 8-25 represent the essentials of the radiation-network method. To construct a network for a particular radiation heat-transfer problem we need only connect a "surface resistance" $(1 - \epsilon)/\epsilon A$ to each surface and a "space resistance" $1/A_i F_{ij}$ between the radiosity potentials. For example, two surfaces that exchange heat with each other *and nothing else* would be represented by the network shown in Figure 8-26. In this case the net heat transfer would be the overall potential difference divided by the sum of the resistances:

$$
\begin{aligned}
q_{net} &= \frac{E_{b1} - E_{b2}}{(1 - \epsilon_1)/\epsilon_1 A_1 + 1/A_1 F_{12} + (1 - \epsilon_2)/\epsilon_2 A_2} \\
&= \frac{\sigma(T_1^4 - T_2^4)}{(1 - \epsilon_1)/\epsilon_1 A_1 + 1/A_1 F_{12} + (1 - \epsilon_2)/\epsilon_2 A_2}
\end{aligned}
\qquad \textbf{[8-40]}
$$

A network for a three-body problem is shown in Figure 8-27. In this case each of the bodies exchanges heat with the other two. The heat exchange between body 1 and body 2 would be

$$q_{1-2} = \frac{J_1 - J_2}{1/A_1 F_{12}}$$

and that between body 1 and body 3,

$$q_{1-3} = \frac{J_1 - J_3}{1/A_1 F_{13}}$$

To determine the heat flows in a problem of this type, the values of the radiosities must be calculated. This may be accomplished by performing standard methods of analysis used in dc circuit theory. The most convenient method is an application of Kirchhoff's current law to the circuit, which states that the sum of the currents entering a node is zero. Example 8-6 illustrates the use of the method for the three-body problem.

Figure 8-26 | Radiation network for two surfaces that see each other and nothing else.

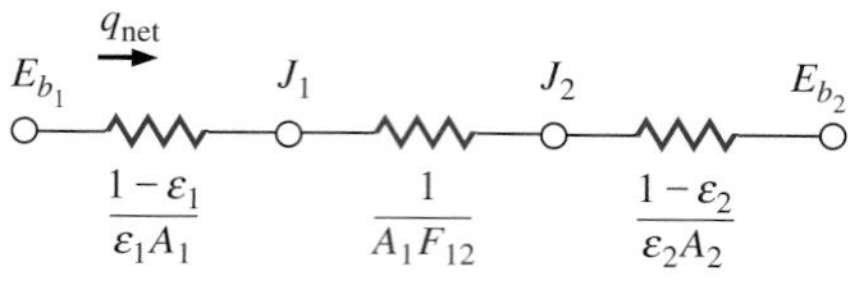

Figure 8-27 | Radiation network for three surfaces that see each other and nothing else.

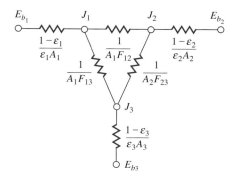

Insulated Surfaces and Surfaces with Large Areas

As we have seen, $(E_b - J)$ represents the potential difference for heat flow through the surface resistance $(1 - \epsilon)/\epsilon A$. If a surface is perfectly insulated, or re-radiates all the energy incident upon it, it has zero heat flow and the potential difference across the surface resistance is zero, resulting in $J = E_b$. But, *the insulated surface does not have zero surface resistance.* In effect, the J node in the network is *floating,* that is, it does not draw any current. On the other hand, a surface with a very large area $(A \to \infty)$ has a surface resistance approaching zero, which makes it behave like a blackbody with $\epsilon = 1.0$. It, too, will have $J = E_b$ because of the zero surface resistance. Thus, these two cases—insulated surface and surface with a large area—both have $J = E_b$, but for entirely different reasons. We will make use of these special cases in several examples.

A problem that may be easily solved with the network method is that of two flat surfaces exchanging heat with one another but connected by a third surface that does not exchange heat, i.e., one that is perfectly insulated. This third surface nevertheless influences the heat-transfer process because it absorbs and re-radiates energy to the other two surfaces that exchange heat. The network for this system is shown in Figure 8-28. Notice that node J_3 is not connected to a radiation surface resistance because surface 3 does not exchange energy. A surface resistance $(1 - \epsilon)/\epsilon A$ exists, but because there is no heat current flow there is no

Figure 8-28 | Radiation network for two plane or convex surfaces enclosed by a third surface that is nonconducting but re-radiating (insulated).

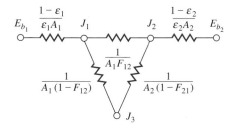

potential difference, and $J_3 = E_{b_3}$. Notice also that the values for the space resistances have been written

$$F_{13} = 1 - F_{12}$$
$$F_{23} = 1 - F_{21}$$

since surface 3 completely surrounds the other two surfaces. For the *special case where surfaces 1 and 2 are convex*, that is, they do not see themselves and $F_{11} = F_{22} = 0$, Figure 8-28 is a simple series-parallel network that may be solved for the heat flow as

$$q_{net} = \frac{\sigma A_1 (T_1^4 - T_2^4)}{\dfrac{A_1 + A_2 - 2A_1 F_{12}}{A_2 - A_1(F_{12})^2} + \left(\dfrac{1}{\epsilon_1} - 1\right) + \dfrac{A_1}{A_2}\left(\dfrac{1}{\epsilon_2} - 1\right)} \qquad \textbf{[8-41]}$$

where the reciprocity relation

$$A_1 F_{12} = A_2 F_{21}$$

has been used to simplify the expression. *It is to be noted again that Equation (8-41) applies only to surfaces that do not see themselves; that is, $F_{11} = F_{22} = 0$.* If these conditions do not apply, one must determine the respective shape factors and solve the network accordingly. Example 8-7 gives an appropriate illustration of a problem involving an insulated surface.

This network, and others that follow, assume that the only heat exchange is by radiation. Conduction and convection are neglected for now.

| **EXAMPLE 8-6** | Hot Plates Enclosed by a Room |

Two parallel plates 0.5 by 1.0 m are spaced 0.5 m apart, as shown in Figure Example 8-6. One plate is maintained at 1000°C and the other at 500°C. The emissivities of the plates are 0.2 and 0.5, respectively. The plates are located in a very large room, the walls of which are maintained at 27°C. The plates exchange heat with each other and with the room, but only the plate surfaces facing each other are to be considered in the analysis. Find the net transfer to each plate and to the room.

Figure Example 8-6 | (*a*) Schematic. (*b*) Network.

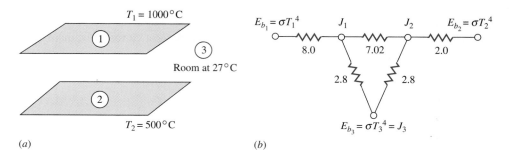

(*a*)   (*b*)

■ **Solution**
This is a three-body problem, the two plates and the room, so the radiation network is shown in Figure 8-27. From the data of the problem

$$T_1 = 1000°C = 1273 \text{ K} \qquad A_1 = A_2 = 0.5 \text{ m}^2$$
$$T_2 = 500°C = 773 \text{ K} \qquad \epsilon_1 = 0.2$$
$$T_3 = 27°C = 300 \text{ K} \qquad \epsilon_2 = 0.5$$

Because the area of the room A_3 is very large, the resistance $(1-\epsilon_3)/\epsilon_3 A_3$ may be taken as zero and we obtain $E_{b_3}=J_3$. The shape factor F_{12} was given in Example 8-2:

$$F_{12}=0.285=F_{21}$$
$$F_{13}=1-F_{12}=0.715$$
$$F_{23}=1-F_{21}=0.715$$

The resistances in the network are calculated as

$$\frac{1-\epsilon_1}{\epsilon_1 A_1}=\frac{1-0.2}{(0.2)(0.5)}=8.0 \qquad \frac{1-\epsilon_2}{\epsilon_2 A_2}=\frac{1-0.5}{(0.5)(0.5)}=2.0$$

$$\frac{1}{A_1 F_{12}}=\frac{1}{(0.5)(0.285)}=7.018 \qquad \frac{1}{A_1 F_{13}}=\frac{1}{(0.5)(0.715)}=2.797$$

$$\frac{1}{A_2 F_{23}}=\frac{1}{(0.5)(0.715)}=2.797$$

Taking the resistance $(1-\epsilon_3)/\epsilon_3 A_3$ as zero, we have the network as shown. To calculate the heat flows at each surface we must determine the radiosities J_1 and J_2. The network is solved by setting the sum of the heat currents entering nodes J_1 and J_2 to zero:

node J_1:

$$\frac{E_{b_1}-J_1}{8.0}+\frac{J_2-J_1}{7.018}+\frac{E_{b_3}-J_1}{2.797}=0 \qquad\qquad \textbf{[a]}$$

node J_2:

$$\frac{J_1-J_2}{7.018}+\frac{E_{b_3}-J_2}{2.797}+\frac{E_{b_2}-J_2}{2.0}=0 \qquad\qquad \textbf{[b]}$$

Now

$$E_{b_1}=\sigma T_1^4=148.87 \text{ kW/m}^2 \quad [47{,}190 \text{ Btu/h}\cdot\text{ft}^2]$$

$$E_{b_2}=\sigma T_2^4=20.241 \text{ kW/m}^2 \quad [6416 \text{ Btu/h}\cdot\text{ft}^2]$$

$$E_{b_3}=\sigma T_3^4=0.4592 \text{ kW/m}^2 \quad [145.6 \text{ Btu/h}\cdot\text{ft}^2]$$

Inserting the values of E_{b_1}, E_{b_2} and E_{b_3} into Equations (*a*) and (*b*), we have two equations and two unknowns J_1 and J_2 that may be solved simultaneously to give

$$J_1=33.469 \text{ kW/m}^2 \quad J_2=15.054 \text{ kW/m}^2$$

The total heat lost by plate 1 is

$$q_1=\frac{E_{b_1}-J_1}{(1-\epsilon_1)/\epsilon_1 A_1}=\frac{148.87-33.469}{8.0}=14.425 \text{ kW}$$

and the total heat lost by plate 2 is

$$q_2=\frac{E_{b_2}-J_2}{(1-\epsilon_2)/\epsilon_2 A_2}=\frac{20.241-15.054}{2.0}=2.594 \text{ kW}$$

The total heat received by the room is

$$q_3=\frac{J_1-J_3}{1/A_1 F_{13}}+\frac{J_2-J_3}{1/A_2 F_{23}}$$

$$=\frac{33.469-0.4592}{2.797}+\frac{15.054-0.4592}{2.797}=17.020 \text{ kW} \quad [58{,}070 \text{ Btu/h}]$$

From an overall-balance standpoint we must have

$$q_3=q_1+q_2$$

because the net energy lost by both plates must be absorbed by the room.

| EXAMPLE 8-7 | Surface in Radiant Balance |

Two rectangles 50 by 50 cm are placed perpendicularly with a common edge. One surface has $T_1 = 1000$ K, $\epsilon_1 = 0.6$, while the other surface is insulated and in radiant balance with a large surrounding room at 300 K. Determine the temperature of the insulated surface and the heat lost by the surface at 1000 K.

Figure Example 8-7 | (a) Schematic. (b) Network.

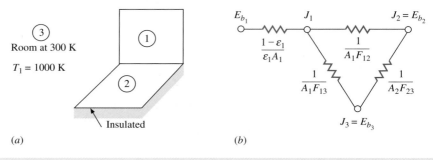

(a)   (b)

■ **Solution**

Although this problem involves two surfaces that exchange heat and one that is insulated or re-radiating, Equation (8-41) may not be used for the calculation because one of the heat-exchanging surfaces (the room) is not convex. The radiation network is shown in Figure Example 8-7 where surface 3 is the room and surface 2 is the insulated surface. Note that $J_3 = E_{b_3}$ because the room is large and $(1 - \epsilon_3)/\epsilon_3 A_3$ approaches zero. Because surface 2 is insulated it has zero heat transfer and $J_2 = E_{b_2}$. J_2 "floats" in the network and is determined from the overall radiant balance. From Figure 8-14 the shape factors are

$$F_{12} = 0.2 = F_{21}$$

Because $F_{11} = 0$ and $F_{22} = 0$ we have

$$F_{12} + F_{13} = 1.0 \qquad \text{and} \qquad F_{13} = 1 - 0.2 = 0.8 = F_{23}$$

$$A_1 = A_2 = (0.5)^2 = 0.25 \text{ m}^2$$

The resistances are

$$\frac{1 - \epsilon_1}{\epsilon_1 A_1} = \frac{0.4}{(0.6)(0.25)} = 2.667$$

$$\frac{1}{A_1 F_{13}} = \frac{1}{A_2 F_{23}} = \frac{1}{(0.25)(0.8)} = 5.0$$

$$\frac{1}{A_1 F_{12}} = \frac{1}{(0.25)(0.2)} = 20.0$$

We also have

$$E_{b_1} = (5.669 \times 10^{-8})(1000)^4 = 5.669 \times 10^4 \text{ W/m}^2$$

$$J_3 = E_{b_3} = (5.669 \times 10^{-8})(300)^4 = 459.2 \text{ W/m}^2$$

The overall circuit is a series-parallel arrangement and the heat transfer is

$$q = \frac{E_{b_1} - E_{b_3}}{R_{\text{equiv}}}$$

We have

$$R_{\text{equiv}} = 2.667 + \frac{1}{\frac{1}{5} + 1/(20 + 5)} = 6.833$$

and

$$q = \frac{56,690 - 459.2}{6.833} = 8.229 \text{ kW} \quad [28,086 \text{ Btu/h}]$$

This heat transfer can also be written

$$q = \frac{E_{b_1} - J_1}{(1 - \epsilon_1)/\epsilon_1 A_1}$$

Inserting the values we obtain

$$J_1 = 34,745 \text{ W/m}^2$$

The value of J_2 is determined from proportioning the resistances between J_1 and J_3, so that

$$\frac{J_1 - J_2}{20} = \frac{J_1 - J_3}{20 + 5}$$

and

$$J_2 = 7316 = E_{b_2} = \sigma T_2^4$$

Finally, we obtain the temperature of the insulated surface as

$$T_2 = \left(\frac{7316}{5.669 \times 10^{-8}} \right)^{1/4} = 599.4 \text{ K} \quad [619°\text{F}]$$

■ **Comment**

Note, once again, that we have made use of the $J = E_b$ relation in two instances in this example, but for two different reasons. $J_2 = E_{b_2}$ because surface 2 is insulated and there is zero current flow through the surface resistance, while $J_3 = E_{b_3}$ because the surface resistance for surface 3 approaches zero as $A_3 \rightarrow \infty$.

8-7 | INFINITE PARALLEL SURFACES

When two infinite parallel planes are considered, A_1 and A_2 are equal; and the radiation shape factor is unity since all the radiation leaving one plane reaches the other. The network is the same as in Figure 8-26, and the heat flow per unit area may be obtained from Equation (8-40) by letting $A_1 = A_2$ and $F_{12} = 1.0$. Thus

$$\frac{q}{A} = \frac{\sigma(T_1^4 - T_2^4)}{1/\epsilon_1 + 1/\epsilon_2 - 1} \qquad \textbf{[8-42]}$$

When two long concentric cylinders as shown in Figure 8-29 exchange heat we may again apply Equation (8-40). Rewriting the equation and noting that $F_{12} = 1.0$,

$$q = \frac{\sigma A_1 (T_1^4 - T_2^4)}{1/\epsilon_1 + (A_1/A_2)(1/\epsilon_2 - 1)} \qquad \textbf{[8-43]}$$

The area ratio A_1/A_2 may be replaced by the diameter ratio d_1/d_2 when cylindrical bodies are concerned.

Convex Object in Large Enclosure

Equation (8-43) is particularly important when applied to the limiting case of a convex object completely enclosed by a very large concave surface. In this instance $A_1/A_2 \rightarrow 0$

Figure 8-29 | Radiation exchange between two cylindrical surfaces.

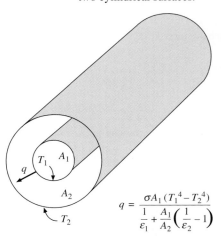

$$q = \frac{\sigma A_1 \left(T_1{}^4 - T_2{}^4 \right)}{\dfrac{1}{\varepsilon_1} + \dfrac{A_1}{A_2}\left(\dfrac{1}{\varepsilon_2} - 1\right)}$$

Figure 8-30 | Radiation heat transfer between simple two-body diffuse, gray surfaces. In all cases $F_{12} = 1.0$.

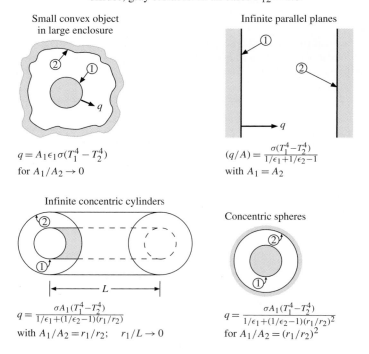

Small convex object in large enclosure

$$q = A_1 \epsilon_1 \sigma (T_1^4 - T_2^4)$$
for $A_1/A_2 \to 0$

Infinite parallel planes

$$(q/A) = \frac{\sigma(T_1^4 - T_2^4)}{1/\epsilon_1 + 1/\epsilon_2 - 1}$$
with $A_1 = A_2$

Infinite concentric cylinders

$$q = \frac{\sigma A_1 (T_1^4 - T_2^4)}{1/\epsilon_1 + (1/\epsilon_2 - 1)(r_1/r_2)}$$
with $A_1/A_2 = r_1/r_2$; $r_1/L \to 0$

Concentric spheres

$$q = \frac{\sigma A_1 (T_1^4 - T_2^4)}{1/\epsilon_1 + (1/\epsilon_2 - 1)(r_1/r_2)^2}$$
for $A_1/A_2 = (r_1/r_2)^2$

and the following simple relation results:

$$q = \sigma A_1 \epsilon_1 (T_1^4 - T_2^4) \qquad \textbf{[8-43a]}$$

This equation is readily applied to calculate the radiation-energy loss from a hot object in a large room.

Some of the radiation heat-transfer cases for simple two-body problems are summarized in Figure 8-30. In this figure, both surfaces are assumed to be gray and diffuse.

Open Hemisphere in Large Room   **EXAMPLE 8-8**

The 30-cm-diameter hemisphere in Figure Example 8-8 is maintained at a constant temperature of 500°C and insulated on its back side. The surface emissivity is 0.4. The opening exchanges radiant energy with a large enclosure at 30°C. Calculate the net radiant exchange.

Figure Example 8-8

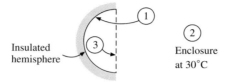

Insulated
hemisphere

Enclosure
at 30°C

■ **Solution**

This is an object completely surrounded by a large enclosure but the inside surface of the sphere is *not convex;* that is, it sees itself, and therefore we are *not* permitted to use Equation (8-43a). In the figure we take the inside of the sphere as surface 1 and the enclosure as surface 2. We also create an imaginary surface 3 covering the opening. We actually have a two-surface problem (surfaces 1 and 2) and therefore may use Equation (8-40) to calculate the heat transfer. Thus,

$$E_{b_1} = \sigma T_1^4 = \sigma (773)^4 = 20{,}241 \ \text{W/m}^2$$

$$E_{b_2} = \sigma T_2^4 = \sigma (303)^4 = 478 \ \text{W/m}^2$$

$$A_1 = 2\pi r^2 = (2)\pi (0.15)^2 = 0.1414 \ \text{m}^2$$

$$\frac{1 - \epsilon_1}{\epsilon_1 A_1} = \frac{0.6}{(0.4)(0.1414)} = 10.61$$

$$A_2 \to \infty$$

so that

$$\frac{1 - \epsilon_2}{\epsilon_2 A_2} \to 0$$

Now, at this point we recognize that all of the radiation leaving surface 1 that will eventually arrive at enclosure 2 will also hit the imaginary surface 3 (i.e., $F_{12} = F_{13}$). We also recognize that

$$A_1 F_{13} = A_3 F_{31}$$

But, $F_{31} = 1.0$ so that

$$F_{13} = F_{12} = \frac{A_3}{A_1} = \frac{\pi r^2}{2\pi r^2} = 0.5$$

Then $1/A_1 F_{12} = 1/(0.1414)(0.5) = 14.14$ and we can calculate the heat transfer by inserting the quantities in Equation (8-40):

$$q = \frac{20{,}241 - 478}{10.61 + 14.14 + 0} = 799 \ \text{W}$$

Apparent Emissivity of a Cavity

Consider the cavity shown in Figure 8-31 having an internal concave surface area A_i and emissivity ϵ_i radiating out through the opening with area A_o. The cavity exchanges radiant energy with a surrounding at T_s having an area that is large compared to the area of the opening. We want to determine a relationship for an apparent emissivity of the opening in terms of the above variables. If one considers the imaginary surface A_o covering the

Figure 8-31 | Apparent emissivity of cavity.

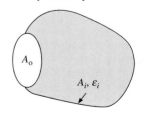

A_o

$A_i, \ \varepsilon_i$

Surrounding, T_s
$A_s \gg A_o$

opening and exchanging heat with A_i we have

$$F_{oi} = 1.0$$

and, from reciprocity,

$$A_o F_{oi} = A_i F_{io}$$

But, $F_{io} = F_{is}$ so that

$$A_i F_{is} = A_o \qquad \textbf{[8-44]}$$

The net radiant exchange of surface A_i with the large enclosure A_s is given by

$$q_{i-s} = (E_{bi} - E_{bs})/[(1 - \epsilon_i)/\epsilon_i A_i + 1/A_i F_{is}] \qquad \textbf{[8-45]}$$

and the net radiant energy exchange of an imaginary surface A_o having an apparent emissivity ϵ_a with the large surroundings is given by Equation (8-43a) as

$$q_{o-s} = \epsilon_a A_o (E_{bi} - E_{bs}) \qquad \textbf{[8-46]}$$

for A_o at the same temperature at the cavity surface A_i. Substituting (8-44) in (8-45) and equating (8-45) and (8-46) gives, after algebraic manipulation,

$$\epsilon_a = \epsilon_i A_i / [A_o + \epsilon_i (A_i - A_o)] \qquad \textbf{[8-47]}$$

We can observe the following behavior for ϵ_a in limiting cases:

$$\epsilon_a = \epsilon_i \quad \text{for } A_o = A_i \text{ or no cavity at all}$$

Figure 8-32 | Apparent emissivity of cavity.

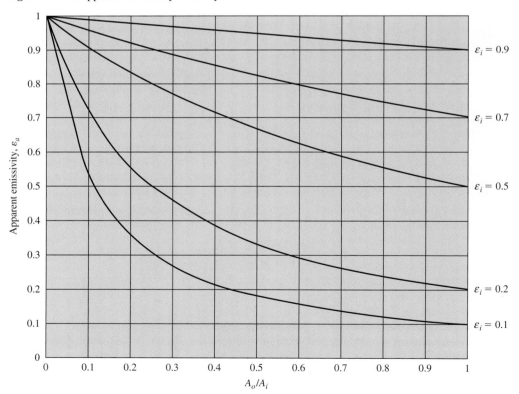

and

$$\epsilon_a \to 1.0 \quad for \ A_i \gg A_o$$

or a very large cavity. A plot of Equation (8-47) is given in Figure 8-32.

The apparent emissivity concept may also be used to analyze transient problems that admit to the lumped capacity approximation. Such an example is discussed in Appendix D, section D-6. In addition, an example is given of multiple lumped capacity formulation applied to the heating of a box of electronic components exchanging energy by convection and radiation with an enclosure.

Effective Emissivity of Finned Surface	**EXAMPLE 8-9**

A repeating finned surface having the relative dimensions shown in Figure Example 8-9 is utilized to produce a higher effective emissivity than that for a flat surface alone. Calculate the effective emissivity of the combination of fin tip and open cavity for surface emissivities of 0.2, 0.5, and 0.8.

Figure Example 8-9

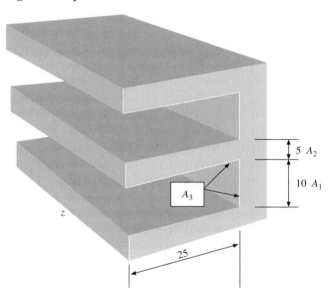

■ **Solution**

For unit depth in the z-dimension we have

$$A_1 = 10, \ A_2 = 5, \ A_3 = (2)(25) + 10 = 60$$

The apparent emissivity of the open cavity area A_1 is given by Equation (8-47) as

$$\epsilon_{a1} = \epsilon A_3 / [A_1 + \epsilon(A_3 - A_1)] = 60\epsilon/(10 + 50\epsilon) \qquad [a]$$

For constant *surface emissivity* the emitted energy from the total erea $A_1 + A_2$ is

$$(\epsilon_{a1} A_1 + \epsilon A_2) E_b \qquad [b]$$

and the energy emitted per unit area for that total area is

$$[(\epsilon_{a1} A_1 + \epsilon A_2)/(A_1 + A_2)] E_b \qquad [c]$$

The coefficient of E_b is the effective emissivity, ϵ_{eff} of the combination of the flat surface and open cavity. Inserting Equation (a) in (c) gives the following numerical values:

For $\epsilon = 0.2$ $\epsilon_{\text{eff}} = 0.4667$
For $\epsilon = 0.5$ $\epsilon_{\text{eff}} = 0.738$
For $\epsilon = 0.8$ $\epsilon_{\text{eff}} = 0.907$

One could employ these effective values to calculate the radiation performance of such a finned surface in conjunction with applicable radiation properties of surrounding surfaces.

8-8 | RADIATION SHIELDS

One way of reducing radiant heat transfer between two particular surfaces is to use materials that are highly reflective. An alternative method is to use radiation shields between the heat-exchange surfaces. These shields do not deliver or remove any heat from the overall system; they only place another resistance in the heat-flow path so that the overall heat transfer is retarded. Consider the two parallel infinite planes shown in Figure 8-33a. We have shown that the heat exchange between these surfaces may be calculated with Equation (8-42). Now consider the same two planes, but with a radiation shield placed between them, as in Figure 8-33b. The heat transfer will be calculated for this latter case and compared with the heat transfer without the shield.

Since the shield does not deliver or remove heat from the system, the heat transfer between plate 1 and the shield must be precisely the same as that between the shield and plate 2, and this is the overall heat transfer. Thus

$$\left(\frac{q}{A}\right)_{1-3} = \left(\frac{q}{A}\right)_{3-2} = \frac{q}{A}$$

$$\frac{q}{A} = \frac{\sigma(T_1^4 - T_3^4)}{1/\epsilon_1 + 1/\epsilon_3 - 1} = \frac{\sigma(T_3^4 - T_2^4)}{1/\epsilon_3 + 1/\epsilon_2 - 1} \qquad \textbf{[8-48]}$$

The only unknown in Equation (8-48) is the temperature of the shield T_3. Once this temperature is obtained, the heat transfer is easily calculated. If the emissivities of all three surfaces are equal, that is, $\epsilon_1 = \epsilon_2 = \epsilon_3$, we obtain the simple relation

$$T_3^4 = \frac{1}{2}(T_1^4 + T_2^4) \qquad \textbf{[8-49]}$$

Figure 8-33 | Radiation between parallel infinite planes with and without a radiation shield.

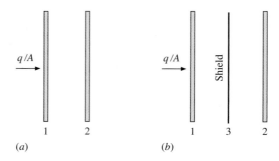

Figure 8-34 | Radiation network for two parallel planes separated by one radiation shield.

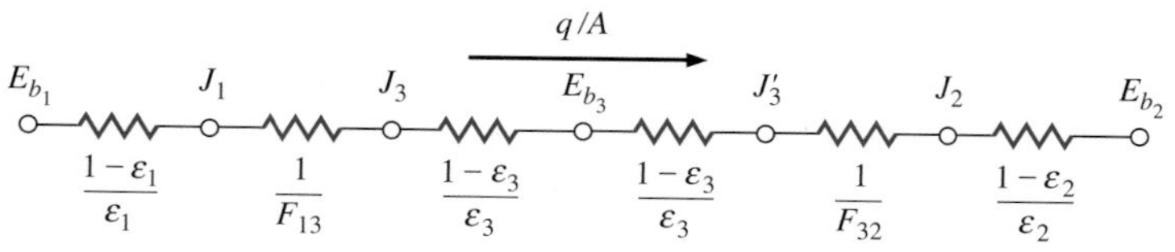

and the heat transfer is

$$\frac{q}{A} = \frac{\frac{1}{2}\sigma(T_1^4 - T_2^4)}{1/\epsilon_1 + 1/\epsilon_3 - 1}$$

But since $\epsilon_3 = \epsilon_2$, we observe that this heat flow is just one-half of that which would be experienced if there were no shield present. The radiation network corresponding to the situation in Figure 8-33b is given in Figure 8-34.

By inspecting the network in Figure 8-34, we see that the radiation heat transfer is impeded by the insertion of three resistances more than would be present with just two surfaces facing each other: an extra space resistance and two extra surface resistances for the shield. The higher the reflectivity of the shield (i.e., the smaller its emissivity), the greater will be the surface resistances inserted. Even for a black shield, with $\epsilon = 1$ and zero surface resistance, there will still be an extra space resistance inserted in the network. As a result, insertion of any surface that intercepts the radiation path will always cause some reduction in the heat-transfer rate, regardless of its surface emissive properties.

Multiple-radiation-shield problems may be treated in the same manner as that outlined above. When the emissivities of all surfaces are different, the overall heat transfer may be calculated most easily by using a series radiation network with the appropriate number of elements, similar to the one in Figure 8-34. If the emissivities of all surfaces are equal, a rather simple relation may be derived for the heat transfer when the surfaces may be considered as infinite parallel planes. Let the number of shields be n. Considering the radiation network for the system, all the "surface resistances" would be the same since the emissivities are equal. There would be two of these resistances for each shield and one for each heat-transfer surface. There would be $n + 1$ "space resistances," and these would all be unity since the radiation shape factors are unity for the infinite parallel planes. The total resistance in the network would thus be

$$R(n \text{ shields}) = (2n + 2)\frac{1 - \epsilon}{\epsilon} + (n + 1)(1) = (n + 1)\left(\frac{2}{\epsilon} - 1\right)$$

The resistance when no shield is present is

$$R(\text{no shield}) = \frac{1}{\epsilon} + \frac{1}{\epsilon} - 1 = \frac{2}{\epsilon} - 1$$

We note that the resistance with the shields in place is $n + 1$ times as large as when the shields are absent. Thus

$$\left(\frac{q}{A}\right)_{\substack{\text{with} \\ \text{shields}}} = \frac{1}{n + 1}\left(\frac{q}{A}\right)_{\substack{\text{without} \\ \text{shields}}} \qquad \qquad \textbf{[8-50]}$$

if the temperatures of the heat-transfer surfaces are maintained the same in both cases. The radiation-network method may also be applied to shield problems involving cylindrical systems. In these cases the proper area relations must be used in formulating the resistance elements.

Notice that the analyses above, dealing with infinite parallel planes, have been carried out on a per-unit-area basis because all areas are the same.

EXAMPLE 8-10	Heat-Transfer Reduction with Parallel-Plate Shield

Two very large parallel planes with emissivities 0.3 and 0.8 exchange heat. Find the percentage reduction in heat transfer when a polished-aluminum radiation shield ($\epsilon = 0.04$) is placed between them.

■ **Solution**
The heat transfer without the shield is given by

$$\frac{q}{A} = \frac{\sigma(T_1^4 - T_2^4)}{1/\epsilon_1 + 1/\epsilon_2 - 1} = 0.279\sigma(T_1^4 - T_2^4)$$

The radiation network for the problem with the shield in place is shown in Figure 8-34. The resistances are

$$\frac{1 - \epsilon_1}{\epsilon_1} = \frac{1 - 0.3}{0.3} = 2.333$$

$$\frac{1 - \epsilon_3}{\epsilon_3} = \frac{1 - 0.04}{0.04} = 24.0$$

$$\frac{1 - \epsilon_2}{\epsilon_2} = \frac{1 - 0.8}{0.8} = 0.25$$

The total resistance with the shield is

$$2.333 + (2)(24.0) + (2)(1) + 0.25 = 52.583$$

and the heat transfer is

$$\frac{q}{A} = \frac{\sigma(T_1^4 - T_2^4)}{52.583} = 0.01902\sigma(T_1^4 - T_2^4)$$

so that the heat transfer is *reduced* by 93.2 percent.

EXAMPLE 8-11	Open Cylindrical Shield in Large Room

The two concentric cylinders of Example 8-3 have $T_1 = 1000$ K, $\epsilon_1 = 0.8$, $\epsilon_2 = 0.2$ and are located in a large room at 300 K. The outer cylinder is in radiant balance. Calculate the temperature of the outer cylinder and the total heat lost by the inner cylinder.

Figure Example 8-11

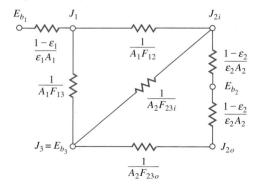

■ **Solution**

The network for this problem is shown in Figure Example 8-11. The room is designated as surface 3 and $J_3 = E_{b_3}$, because the room is very large (i.e., its surface resistance is very small). In this problem we must consider the inside and outside of surface 2 and thus have subscripts i and o to designate the respective quantities. The shape factors can be obtained from Example 8-3 as

$$F_{12} = 0.8253 \quad F_{13} = 0.1747$$
$$F_{23i} = (2)(0.1294) = 0.2588 \quad F_{23o} = 1.0$$

Also,

$$A_1 = \pi(0.1)(0.2) = 0.06283 \text{ m}^2$$
$$A_2 = \pi(0.2)(0.2) = 0.12566 \text{ m}^2$$
$$E_{b_1} = (5.669 \times 10^{-8})(1000)^4 = 5.669 \times 10^4 \text{ W/m}^2$$
$$E_{b_3} = (5.669 \times 10^{-8})(300)^4 = 459.2 \text{ W/m}^2$$

and the resistances may be calculated as

$$\frac{1 - \epsilon_1}{\epsilon_1 A_1} = 3.979 \quad \frac{1 - \epsilon_2}{\epsilon_2 A_2} = 31.83$$
$$\frac{1}{A_1 F_{12}} = 19.28 \quad \frac{1}{A_2 F_{23i}} = 30.75$$
$$\frac{1}{A_2 F_{23o}} = 7.958 \quad \frac{1}{A_1 F_{13}} = 91.1$$

The network could be solved as a series-parallel circuit to obtain the heat transfer, but we will need the radiosities anyway, so we set up three nodal equations to solve for J_1, J_{2i}, and J_{2o}. We sum the currents into each node and set them equal to zero:

node J_1:
$$\frac{E_{b_1} - J_1}{3.979} + \frac{E_{b_3} - J_1}{91.1} + \frac{J_{2i} - J_1}{19.28} = 0$$

node J_{2i}:
$$\frac{J_1 - J_{2i}}{19.28} + \frac{E_{b_3} - J_{2i}}{30.75} + \frac{J_{2o} - J_{2i}}{(2)(31.83)} = 0$$

node J_{2o}:
$$\frac{E_{b_3} - J_{2o}}{7.958} + \frac{J_{2i} - J_{2o}}{(2)(31.83)} = 0$$

These equations have the solution

$$J_1 = 49{,}732 \text{ W/m}^2$$
$$J_{2i} = 26{,}444 \text{ W/m}^2$$
$$J_{2o} = 3346 \text{ W/m}^2$$

The heat transfer is then calculated from

$$q = \frac{E_{b_1} - J_1}{(1 - \epsilon_1)/\epsilon_1 A_1} = \frac{56{,}690 - 49{,}732}{3.979} = 1749 \text{ W} \quad [5968 \text{ Btu/h}]$$

From the network we see that

$$E_{b_2} = \frac{J_{2i} + J_{2o}}{2} = \frac{26{,}444 + 3346}{2} = 14{,}895 \text{ W/m}^2$$

and

$$T_2 = \left(\frac{14{,}895}{5.669 \times 10^{-8}} \right)^{1/4} = 716 \text{ K} \quad [829°\text{F}]$$

If the outer cylinder had not been in place acting as a "shield" the heat loss from cylinder 1 could have been calculated from Equation (8-43a) as

$$q = \epsilon_1 A_1 (E_{b_1} - E_{b_3})$$
$$= (0.8)(0.06283)(56{,}690 - 459.2) = 2826 \text{ W} \quad [9644 \text{ Btu/h}]$$

8-9 | GAS RADIATION

Radiation exchange between a gas and a heat-transfer surface is considerably more complex than the situations described in the preceding sections. Unlike most solid bodies, gases are in many cases transparent to radiation. When they absorb and emit radiation, they usually do so only in certain narrow wavelength bands. Some gases, such as N_2, O_2, and others of nonpolar symmetrical molecular structure, are essentially transparent at low temperatures, while CO_2, H_2O, and various hydrocarbon gases radiate to an appreciable extent.

The absorption of radiation in gas layers may be described analytically in the following way, considering the system shown in Figure 8-35. A monochromatic beam of radiation having an intensity I_λ impinges on the gas layer of thickness dx. The decrease in intensity resulting from absorption in the layers is assumed to be proportional to the thickness of the layer and the intensity of radiation at that point. Thus

$$dI_\lambda = -a_\lambda I_\lambda \, dx \qquad \text{[8-51]}$$

where the proportionality constant a_λ is called the *monochromatic absorption coefficient*. Integrating this equation gives

$$\int_{I_{\lambda_0}}^{I_{\lambda_x}} \frac{dI_\lambda}{I_\lambda} = \int_0^x -a_\lambda \, dx$$

or

$$\frac{I_{\lambda_x}}{I_{\lambda_0}} = e^{-a_\lambda x} \qquad \text{[8-52]}$$

Equation (8-52) is called Beer's law and represents the familiar exponential-decay formula experienced in many types of radiation analyses dealing with absorption. In accordance with our definitions in Section 8-3, the monochromatic transmissivity will be given as

$$\tau_\lambda = e^{-\alpha_\lambda x} \qquad \text{[8-53]}$$

If the gas is nonreflecting, then

$$\tau_\lambda + \alpha_\lambda = 1$$

and

$$\alpha_\lambda = 1 - e^{-\alpha_\lambda x} \qquad \text{[8-54]}$$

As we have mentioned, gases frequently absorb only in narrow wavelength bands. For example, water vapor has an absorptivity of about 0.7 between 1.4 and 1.5 μm, about 0.8 between 1.6 and 1.8 μm, about 1.0 between 2.6 and 2.8 μm, and about 1.0 between 5.5 and 7.0 μm. As we have seen in Equation (8-54), the absorptivity will also be a function of the thickness of the gas layer, and there is a temperature dependence as well.

The calculation of gas-radiation properties is quite complicated, and References 23 to 25 should be consulted for detailed information.

Figure 8-35 |
Absorption in a gas layer.

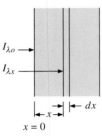

$I_{\lambda o}$

$I_{\lambda x}$

dx

x

$x = 0$

8-10 | RADIATION NETWORK FOR AN ABSORBING AND TRANSMITTING MEDIUM

The foregoing discussions have shown the methods that may be used to calculate radiation heat transfer between surfaces separated by a completely transparent medium. The radiation-network method is used to great advantage in these types of problems.

Many practical problems involve radiation heat transfer through a medium that is both absorbing and transmitting. The various glass substances are one example of this type of medium; gases are another. Some approximate transmissivities or glass substances over the wavelength range of $0.5\ \mu m < \lambda < 2.5\ \mu m$ are given in Table 8-3.

Keeping in mind the complications involved with the band absorption characteristics of gases, we shall now examine a simplified radiation network method for analyzing transmitting absorbing systems.

To begin, let us consider a simple case, that of two nontransmitting surfaces that see each other and nothing else. In addition, we let the space between these surfaces be occupied by a transmitting and absorbing medium. The practical problem might be that of two large planes separated by either an absorbing gas or a transparent sheet of glass or plastic. The situation is shown schematically in Figure 8-36. The transparent medium is designated by the subscript m. We make the assumption that the medium is nonreflecting and that Kirchhoff's identity applies, so that

$$\alpha_m + \tau_m = 1 = \epsilon_m + \tau_m \qquad \text{[8-55]}$$

The assumption that the medium is nonreflecting is a valid one when gases are considered. For glass or plastic plates this is not necessarily true, and reflectivities of the order

Table 8-3 | Approximate transmissivities for glasses at 20°C.

Glass	$\tau(0.5\ \mu m < \lambda < 2.5\ \mu m)$
Soda lime glass	
Thickness = 1.6 mm	0.9
= 6.4 mm	0.75
= 9.5 mm	0.7
= 12.7 mm	0.65
Aluminum silicate, thickness = 12.7 mm	0.85
Borosilicate = 12.7 mm	0.8
Fused silica = 12.7 mm	0.85
Pyrex = 12.7 mm	0.65

Figure 8-36 | Radiation system consisting of a transmitting medium between two planes.

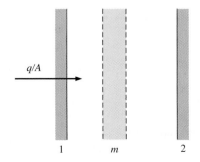

of 0.1 are common for many glass substances. In addition, the transmissive properties of glasses are usually limited to a narrow wavelength band between about 0.2 and 4 μm. Thus the analysis that follows is highly idealized and serves mainly to furnish a starting point for the solution of problems in which transmission of radiation must be considered. Other complications with gases are mentioned later in the discussion. When both reflection and transmission must be taken into account, the analysis techniques discussed in Section 8-12 must be employed.

Returning to the analysis, we note that the medium can emit and transmit radiation from one surface to the other. Our task is to determine the network elements to use in describing these two types of exchange processes. The transmitted energy may be analyzed as follows. The energy leaving surface 1 that is transmitted through the medium and arrives at surface 2 is

$$J_1 A_1 F_{12} \tau_m$$

and the energy leaving surface 2 and arrives at surface 1 is

$$J_2 A_2 F_{21} \tau_m$$

The net exchange in the transmission process is therefore

$$q_{1-2\text{transmitted}} = A_1 F_{12} \tau_m (J_1 - J_2)$$

$$q_{1-2\text{transmitted}} = \frac{J_1 - J_2}{1/A_1 F_{12}(1 - \epsilon_m)} \qquad \textbf{[8-56]}$$

Figure 8-37 | Network element for transmitted radiation through medium.

and the network element that may be used to describe this process is shown in Figure 8-37.

Now consider the exchange process between surface 1 and the transmitting medium. Since we have assumed that this medium is nonreflecting, the energy leaving the medium (other than the transmitted energy, which we have already considered) is precisely the energy emitted by the medium

$$J_m = \epsilon_m E_{bm}$$

And of the energy leaving the medium, the amount which reaches surface 1 is

$$A_m F_{m1} J_m = A_m F_{m1} \epsilon_m E_{bm}$$

Of that energy leaving surface 1, the quantity that reaches the transparent medium is

$$J_1 A_1 F_{1m} \alpha_m = J_1 A_1 F_{1m} \epsilon_m$$

At this point we note that absorption in the medium means that the incident radiation has "reached" the medium. Consistent with the above relations, the net energy exchange between the medium and surface 1 is the difference between the amount emitted by the medium toward surface 1 and that absorbed which emanated from surface 1. Thus

$$q_{m-1\text{net}} = A_m F_{m1} \epsilon_m E_{bm} - J_1 A_1 F_{1m} \epsilon_m$$

Figure 8-38 | Network element for radiation exchange between medium and surface.

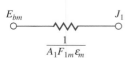

Using the reciprocity relation

$$A_1 F_{1m} = A_m F_{m1}$$

we have

$$q_{m-1\text{net}} = \frac{E_{bm} - J_1}{1/A_1 F_{1m} \epsilon_m} \qquad \textbf{[8-57]}$$

This heat-exchange process is represented by the network element shown in Figure 8-38. The total network for the physical situation of Figure 8-36 is shown in Figure 8-39.

Figure 8-39 | Total radiation network for system of Figure 8-36.

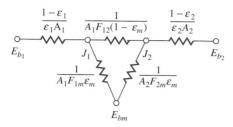

If the transport medium is maintained at some fixed temperature, then the potential E_{bm} is fixed according to

$$E_{bm} = \sigma T_m^4$$

On the other hand, if no net energy is delivered to the medium, then E_{bm} becomes a floating node, and its potential is determined by the other network elements.

In reality, the radiation shape factors F_{1-2}, F_{1-m}, and F_{2-m} are unity for this example, so that the expression for the heat flow could be simplified to some extent; however, these shape factors are included in the network resistances for the sake of generality in the analysis.

When the practical problem of heat exchange between gray surfaces through an absorbing gas is encountered, the major difficulty is that of determining the transmissivity and emissivity of the gas. These properties are functions not only of the temperature of the gas, but also of the thickness of the gas layer; that is, thin gas layers transmit more radiation than thick layers. The usual practical problem almost always involves more than two heat-transfer surfaces, as in the simple example given above. As a result, the transmissivities between the various heat-transfer surfaces can be quite different, depending on their geometric orientation. Since the temperature of the gas will vary, the transmissive and emissive properties will vary with their location in the gas. One way of handling this situation is to divide the gas body into layers and set up a radiation network accordingly, letting the potentials of the various nodes "float," and thus arriving at the gas-temperature distribution. Even with this procedure, an iterative method must eventually be employed because the radiation properties of the gas are functions of the unknown "floating potentials." Naturally, if the temperature of the gas is uniform, the solution is much easier.

We shall not present the solution of a complex gas-radiation problem since the tedious effort required for such a solution is beyond the scope of our present discussion; however, it is worthwhile to analyze a two-layer transmitting system in order to indicate the general scheme of reasoning that might be applied to more complex problems.

Consider the physical situation shown in Figure 8-40. Two radiating and absorbing surfaces are separated by two layers of transmitting and absorbing media. These two layers might represent two sheets of transparent media, such as glass, or they might represent the division of a separating gas into two parts for purposes of analysis. We designate the two transmitting and absorbing layers with the subscripts m and n. The energy exchange between surface 1 and m is given by

$$q_{1-m} = A_1 F_{1m} \epsilon_m J_1 - A_m F_{m1} \epsilon_m E_{bm} = \frac{J_1 - E_{bm}}{1/A_1 F_{1m} \epsilon_m} \qquad \textbf{[8-58]}$$

and that between surface 2 and n is

$$q_{2-n} = A_1 F_{2n} \epsilon_n J_2 - A_n F_{n2} \epsilon_n E_{bn} = \frac{J_2 - E_{bn}}{1/A_2 F_{2n} \epsilon_n} \qquad \textbf{[8-59]}$$

Figure 8-40 | Radiation system consisting of two transmitting layers between two planes.

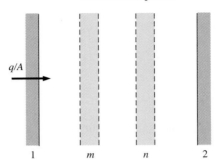

Of that energy leaving surface 1, the amount arriving at surface 2 is

$$q_{1-2} = A_1 F_{12} J_1 \tau_m \tau_n = A_1 F_{12} J_1 (1 - \epsilon_m)(1 - \epsilon_n)$$

and of that energy leaving surface 2, the amount arriving at surface 1 is

$$q_{2-1} = A_2 F_{21} J_2 \tau_n \tau_m = A_2 F_{12} J_2 (1 - \epsilon_n)(1 - \epsilon_m)$$

so that the net energy exchange by transmission between surfaces 1 and 2 is

$$q_{1-2\text{transmitted}} = A_1 F_{12}(1 - \epsilon_m)(1 - \epsilon_n)(J_1 - J_2) = \frac{J_1 - J_2}{1/A_1 F_{12}(1 - \epsilon_m)(1 - \epsilon_n)} \qquad \textbf{[8-60]}$$

Figure 8-41 | Network element for transmitted radiation between planes.

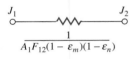

and the network element representing this transmission is shown in Figure 8-41. Of that energy leaving surface 1, the amount that is absorbed in n is

$$q_{1-n} = A_1 F_{1n} J_1 \tau_m \epsilon_n = A_1 F_{1n} J_1 (1 - \epsilon_m)\epsilon_n$$

Also,

$$q_{n-1} = A_n F_{n1} J_n \tau_m = A_n F_{n1} \epsilon_n E_{bn} (1 - \epsilon_m)$$

since

$$J_n = \epsilon_n E_{bn}$$

The net exchange between surface 1 and n is therefore

$$q_{1-n\text{net}} = A_1 F_{1n}(1 - \epsilon_m)\epsilon_n(J_1 - E_{bn}) = \frac{J_1 - E_{bn}}{1/A_1 F_{1n}(1 - \epsilon_m)\epsilon_n} \qquad \textbf{[8-61]}$$

Figure 8-42 | Network element for transmitted radiation for medium n to plane 1.

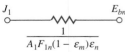

and the network element representing this situation is shown in Figure 8-42. In like manner, the net exchange between surface 2 and m is

$$q_{2-m\text{net}} = \frac{J_2 - E_{bm}}{1/A_2 F_{2m}(1 - \epsilon_n)\epsilon_m}$$

Of that radiation leaving m, the amount absorbed in n is

$$q_{m-n} = J_m A_m F_{mn} \alpha_n = A_m F_{mn} \epsilon_m \epsilon_n E_{bm}$$

and

$$q_{n-m} = A_n F_{nm} \epsilon_n \epsilon_m E_{bn}$$

Figure 8-44 | Total radiation network for system of Figure 8-40.

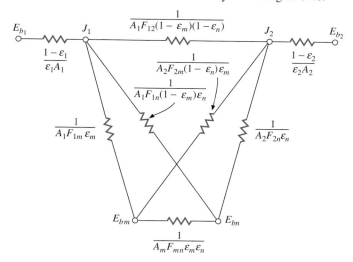

so that the net energy exchange between m and n is

$$q_{m-n_{\text{net}}} = A_m F_{mn} \epsilon_m \epsilon_n (E_{bm} - E_{bn}) = \frac{E_{bm} - E_{bn}}{1/A_m F_{mn} \epsilon_m \epsilon_n} \qquad \textbf{[8-62]}$$

and the network element representing this energy transfer is given in Figure 8-43.

The final network for the entire heat-transfer process is shown in Figure 8-44, with the surface resistances added. If the two transmitting layers m and n are maintained at given temperatures, the solution to the network is relatively easy to obtain because only two unknown potentials J_1 and J_2 need be determined to establish the various heat-flow quantities. In this case the two transmitting layers will either absorb or lose a certain quantity of energy, depending on the temperature at which they are maintained.

When no net energy is delivered to the transmitting layers, nodes E_{bm} and E_{bn} must be left "floating" in the analysis; and for this particular system four nodal equations would be required for a solution of the problem.

Figure 8-43 | Network element for radiation exchange between two transparent layers.

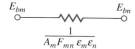

Network for Gas Radiation Between Parallel Plates   EXAMPLE 8-12

Two large parallel planes are at $T_1 = 800$ K, $\epsilon_1 = 0.3$, $T_2 = 400$ K, $\epsilon_2 = 0.7$ and are separated by a gray gas having $\epsilon_g = 0.2$, $\tau_g = 0.8$. Calculate the heat-transfer rate between the two planes and the temperature of the gas using a radiation network. Compare with the heat transfer without presence of the gas.

■ **Solution**

The network shown in Figure 8-39 applies to this problem. All the shape factors are unity for large planes and the various resistors can be computed on a unit-area basis as

$$\frac{1 - \epsilon_1}{\epsilon_1} = \frac{0.7}{0.3} = 2.333 \qquad \frac{1}{F_{12}(1 - \epsilon_g)} = \frac{1}{1 - 0.2} = 1.25$$

$$\frac{1 - \epsilon_2}{\epsilon_2} = \frac{0.3}{0.7} = 0.4286 \qquad \frac{1}{F_{1g}\epsilon_g} = \frac{1}{F_{2g}\epsilon_g} = \frac{1}{0.2} = 5.0$$

$$E_{b_1} = \sigma T_1^4 = 23{,}220 \text{ W/m}^2 \qquad E_{b_2} = \sigma T_2^4 = 1451 \text{ W/m}^2$$

The equivalent resistance of the center "triangle" is

$$R = \frac{1}{1/1.25 + 1/(5.0 + 5.0)} = 1.1111$$

The total heat transfer is then

$$\frac{q}{A} = \frac{E_{b_1} - E_{b_2}}{\sum R} = \frac{23{,}200 - 1451}{2.333 + 1.111 + 0.4286} = 5616 \text{ W/m}^2$$

If there were no gas present the heat transfer would be given by Equation (8-42):

$$\frac{q}{A} = \frac{23{,}200 - 1451}{1/0.3 + 1/0.7 - 1} = 5781 \text{ W/m}^2$$

The radiosities may be computed from

$$\frac{q}{A} = (E_{b_1} - J_1)\left(\frac{\epsilon_1}{1 - \epsilon_1}\right) = (J_2 - E_{b_2})\left(\frac{\epsilon_2}{1 - \epsilon_2}\right) = 5616 \text{ W/m}^2$$

which gives $J_1 = 10{,}096$ W/m^2 and $J_2 = 3858$ W/m^2. For the network E_{bg} is just the mean of these values

$$E_{bg} = \frac{1}{2}(10{,}096 + 3858) = 6977 = \sigma T_g^4$$

so that the temperature of the gas is

$$T_g = 592.3 \text{ K}$$

8-11 | RADIATION EXCHANGE WITH SPECULAR SURFACES

All the preceding discussions have considered radiation exchange between diffuse surfaces. In fact, the radiation shape factors defined by Equation (8-21) hold only for diffuse radiation because the radiation was assumed to have no preferred direction in the derivation of this relation. In this section we extend the analysis to take into account some simple geometries containing surfaces that may have a specular type of reflection. No real surface is completely diffuse or completely specular. We shall assume, however, that all the surfaces to be considered *emit* radiation diffusely but that they may *reflect* radiation partly in a specular manner and partly in a diffuse manner. We therefore take the reflectivity to be the sum of a specular component and a diffuse component:

$$\rho = \rho_s + \rho_D \qquad \text{[8-63]}$$

It is still assumed that Kirchhoff's identity applies so that

$$\epsilon = \alpha = 1 - \rho \qquad \text{[8-64]}$$

The net heat lost by a surface is the difference between the energy emitted and absorbed:

$$q = A(\epsilon E_b - \alpha G) \qquad \text{[8-65]}$$

We define the *diffuse radiosity* J_D as the total *diffuse* energy leaving the surface per unit area and per unit time, or

$$J_D = \epsilon E_b + \rho_D G \qquad \text{[8-66]}$$

Solving for the irradiation G from Equation (8-66) and inserting in Equation (8-65) gives

$$q = \frac{\epsilon A}{\rho_D}[E_b(\epsilon + \rho_D) - J_D]$$

or, written in a different form,

$$q = \frac{E_b - J_D/(1 - \rho_s)}{\rho_D/[\epsilon A(1 - \rho_s)]} \qquad \textbf{[8-67]}$$

where $1 - \rho_s$ has been substituted for $\epsilon + \rho_D$. It is easy to see that Equation (8-67) may be represented with the network element shown in Figure 8-45. A quick inspection will show that this network element reduces to that in Figure 8-24 for the case of a surface that reflects in only a diffuse manner (i.e., for $\rho_s = 0$).

Now let us compute the radiation exchange between two specular-diffuse surfaces. For the moment, we assume that the surfaces are oriented as shown in Figure 8-46. In this arrangement any diffuse radiation leaving surface 1 that is specularly reflected by 2 will not be reflected directly back to 1. This is an important point, for in eliminating such reflections we are considering only the *direct* diffuse exchange between the two surfaces. In subsequent paragraphs we shall show how the specular reflections must be analyzed. For the surfaces in Figure 8-46 the *diffuse* exchanges are given by

$$q_{1\to 2} = J_{1D}A_1 F_{12}(1 - \rho_{2s}) \qquad \textbf{[8-68]}$$
$$q_{2\to 1} = J_{2D}A_2 F_{21}(1 - \rho_{1s}) \qquad \textbf{[8-69]}$$

Equation (8-68) expresses the diffuse radiation leaving 1 that arrives at 2 *and* that may contribute to a diffuse radiosity of surface 2. The factor $1 - \rho_s$ represents the fraction absorbed plus the fraction reflected diffusely. The inclusion of this factor is important because we are considering only diffuse direct exchange, and thus must leave out the specular-reflection contribution for now. The net exchange is given by the difference between Equations (8-68) and (8-69), according to Reference 21.

$$q_{12} = \frac{J_{1D}/(1 - \rho_{1s}) - J_{2D}/(1 - \rho_{2s})}{1/[A_1 F_{12}(1 - \rho_{1s})(1 - \rho_{2s})]} \qquad \textbf{[8-70]}$$

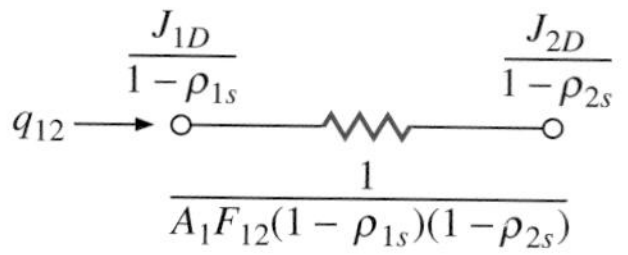

The network element representing Equation (8-70) is shown in Figure 8-47.

To analyze specular reflections we utilize a technique presented in References 12 and 13. Consider the enclosure with four long surfaces shown in Figure 8-48. Surfaces 1, 2, and 4 reflect diffusely, while surface 3 has both a specular and a diffuse component of reflection. The dashed lines represent mirror images of the surfaces 1, 2, and 4 in surface 3. (A specular reflection produces a mirror image.) The nomenclature 2 (3) designates the mirror image of surface 2 in mirror 3.

Now consider the radiation leaving 2 that arrives at 1. There is a direct diffuse radiation of

$$(q_{2\to 1})^{\text{direct}}_{\text{diffuse}} = J_2 A_2 F_{21} \qquad \textbf{[8-71]}$$

Part of the diffuse radiation from 2 is specularly reflected in 3 and strikes 1. This specularly reflected radiation acts like *diffuse* energy coming from the image surface 2 (3). Thus we may write

$$(q_{2\to 1})^{\text{specular}}_{\text{reflected}} = J_2 A_{2(3)} F_{2(3)1}\rho_{3s} \qquad \textbf{[8-72]}$$

The radiation shape factor $F_{2(3)1}$ is the one between surface 2 (3) and surface 1. The reflectivity ρ_{3s} is inserted because only this fraction of the radiation gets to 1. Of course,

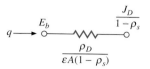

Figure 8-45 | Network element representing Equation (8-67).

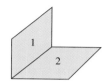

Figure 8-46

Figure 8-47 | Network element representing Equation (8-70).

Figure 8-48 | System with one specular-diffuse surface.

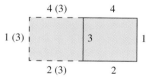

1,2,4 Diffuse reflecting
3 Specular-diffuse reflecting

$A_2 = A_{2(3)}$. We now have

$$q_{2\to1} = J_2 A_2 (F_{21} + \rho_{3s} F_{2(3)1})$$ [8-73]

Similar reasoning leads to

$$q_{1\to2} = J_1 A_1 (F_{12} + \rho_{3s} F_{1(3)2})$$ [8-74]

Combining Equations (8-73) and (8-74) and making use of the reciprocity relation $A_1 F_{12} = A_2 F_{21}$ gives

$$q_{12} = \frac{J_1 - J_2}{1/[A_1 (F_{12} + \rho_{3s} F_{1(3)2})]}$$ [8-75]

Figure 8-49 | Network element for Equation (8-75).

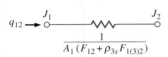

$$q_{12} \longrightarrow$$

$$\frac{1}{A_1 (F_{12} + \rho_{3s} F_{1(3)2})}$$

The network element represented by Equation (8-75) is shown in Figure 8-49.

Analogous network elements may be developed for radiation between the other surfaces in Figure 8-48, so that the final complete network becomes as shown in Figure 8-50. It is to be noted that the elements connecting to J_{3D} are simple modifications of the one shown in Figure 8-47 since $\rho_{1s} = \rho_{2s} = \rho_{4s} = 0$. An interesting observation can be made about this network for the case where $\rho_{3D} = 0$. In this instance surface 3 is completely specular and

$$J_{3D} = \epsilon_3 E_{b3}$$

so that we are left with only three unknowns, J_1, J_2, and J_4, when surface 3 is completely specular-reflecting.

Now let us complicate the problem a step further by letting the enclosure have two specular-diffuse surfaces, as shown in Figure 8-51. In this case multiple images may be formed as shown. Surface 1 (3, 2) represents the image of 1 after it is viewed first through 3 and then through 2. In other words, it is the image of surface 1 (3) in mirror 2. At the same location is surface 1 (2, 3), which is the image of surface 1 (2) in mirror 3.

This problem is complicated because multiple specular reflections must be considered. Consider the exchange between surfaces 1 and 4. Diffuse energy leaving 1 can arrive at 4 in

Figure 8-50 | Complete radiation network for system in Figure 8-48.

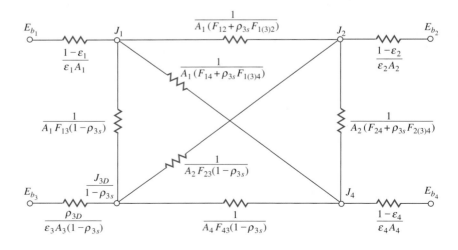

Figure 8-51 | System with two specular-diffuse
surfaces.

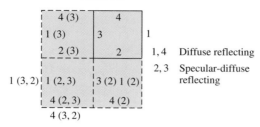

five possible ways:

direct:

$$J_1 A_1 F_{14}$$

reflection in 2 only:

$$J_1 A_1 F_{1(2)4} \rho_{2s}$$

reflection in 3 only:

$$J_1 A_1 F_{1(3)4} \rho_{3s} \qquad \textbf{[8-76]}$$

reflection first in 2 and then in 3:

$$J_1 A_1 \rho_{3s} \rho_{2s} F_{1(2,3)4}$$

reflection first in 3 and then in 2:

$$J_1 A_1 \rho_{2s} \rho_{3s} F_{1(3,2)4}$$

The last shape factor, $F_{1(3,2)4}$, is zero because surface 1 (3, 2) cannot see surface 4 when looking *through* mirror 2. On the other hand, $F_{1(2,3)4}$ is not zero because surface 1 (2, 3) can see surface 4 when looking through mirror 3. The sum of the above terms is given as

$$q_{1 \to 4} = J_1 A_1 (F_{14} + \rho_{2s} F_{1(2)4} + \rho_{3s} F_{1(3)4} + \rho_{3s} \rho_{2s} F_{1(2,3)4}) \qquad \textbf{[8-77]}$$

In a similar manner,

$$q_{4 \to 1} = J_4 A_4 (F_{41} + \rho_{2s} F_{4(2)1} + \rho_{3s} F_{4(3)1} + \rho_{3s} \rho_{2s} F_{4(3,2)1}) \qquad \textbf{[8-78]}$$

Subtracting these two equations and applying the usual reciprocity relations gives the network element shown in Figure 8-52.

Now consider the diffuse exchange between surfaces 1 and 3. Of the energy leaving 1, the amount which contributes to the diffuse radiosity of surface 3 is

$$q_{1 \to 3} = J_1 A_1 F_{13} (1 - \rho_{3s}) + J_1 A_1 \rho_{2s} F_{1(2)3} (1 - \rho_{3s}) \qquad \textbf{[8-79]}$$

The first term represents the direct exchange, and the second term represents the exchange after one specular reflection in mirror 2. As before, the factor $1 - \rho_{3s}$ is included to leave out of consideration the specular reflection from 3. This reflection, of course, is taken into account in other terms. The *diffuse* energy going from 3 to 1 is

$$q_{3 \to 1} = J_{3D} A_3 F_{31} + J_{3D} A_3 \rho_{2s} F_{3(2)1} \qquad \textbf{[8-80]}$$

The first term is the direct radiation, and the second term is that which is specularly reflected in mirror 2. Combining Equations (8-79) and (8-80) gives the network element shown in Figure 8-53.

Figure 8-52 | Network element representing
exchange between surfaces 1
and 4 of Figure 8-51.

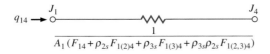

Figure 8-53 | Network element representing exchange between surfaces 1 and 3 of Figure 8-51.

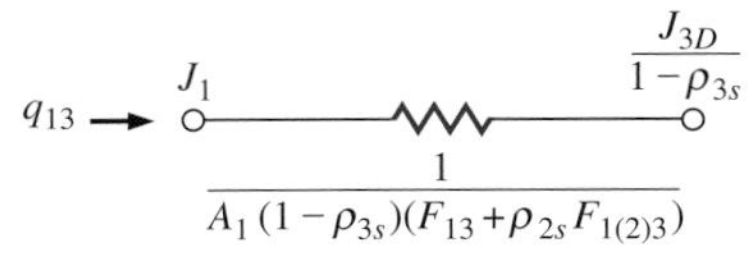

q_{13}
J_1
$\frac{J_{3D}}{1-\rho_{3s}}$
$\frac{1}{A_1(1-\rho_{3s})(F_{13}+\rho_{2s}F_{1(2)3})}$

Figure 8-54 | Complete radiation network for system in Figure 8-51.

E_{b_1}
J_1
$\frac{1}{A_1(F_{14}+\rho_{2s}F_{1(2)4}+\rho_{3s}F_{1(3)4}+\rho_{3s}\rho_{2s}F_{1(2,3)4})}$
J_4
E_{b_4}
$\frac{1-\varepsilon_1}{\varepsilon_1 A_1}$
$\frac{1}{A_4(1-\rho_{2s})(F_{42}+\rho_{3s}F_{4(3)2})}$
$\frac{1-\varepsilon_4}{\varepsilon_4 A_4}$
$\frac{1}{A_1(1-\rho_{3s})(F_{13}+\rho_{2s}F_{1(2)3})}$
$\frac{1}{A_4(1-\rho_{2s})(F_{43}+\rho_{3s}F_{4(2)3})}$
$\frac{1}{A_1(1-\rho_{2s})(F_{12}+\rho_{3s}F_{1(3)2})}$
E_{b_3}
E_{b_2}
$\frac{\rho_{3D}}{\varepsilon_3 A_3(1-\rho_{3s})}$
$\frac{J_{3D}}{1-\rho_{3s}}$
$\frac{1}{A_3F_{32}(1-\rho_{2s})(1-\rho_{3s})}$
$\frac{J_{2D}}{1-\rho_{2s}}$
$\frac{\rho_{2D}}{\varepsilon_2 A_2(1-\rho_{2s})}$

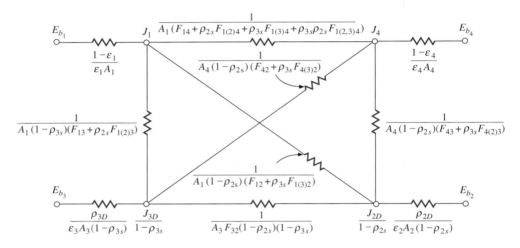

The above two elements are typical for the enclosure of Figure 8-51 and the other elements may be constructed by analogy. Thus the final complete network is given in Figure 8-54.

If both surfaces 2 and 3 are pure specular reflectors, that is,

$$\rho_{2D} = \rho_{3D} = 0$$

we have

$$J_{2D} = \epsilon_2 E_{b_2} \qquad J_{3D} = \epsilon_3 E_{b_3}$$

and the network involves only two unknowns, J_1 and J_4, under these circumstances.

We could complicate the calculation further by installing the specular surfaces opposite each other. In this case there would be an infinite number of images, and a series solution would have to be obtained; however, the series for such problems usually converge rather rapidly. The reader should consult Reference 13 for further information on this aspect of radiation exchange between specular surfaces.

8-12 | RADIATION EXCHANGE WITH TRANSMITTING, REFLECTING, AND ABSORBING MEDIA

We now consider a simple extension of the presentations in Sections 8-10 and 8-11 to analyze a medium where reflection, transmission, and absorption modes are all important.

As in Section 8-10, we shall analyze a system consisting of two parallel diffuse planes with a medium in between that may absorb, transmit, and reflect radiation. For generality we assume that the surface of the transmitting medium may have both a specular and a diffuse component of reflection. The system is shown in Figure 8-55.

For the transmitting medium m we have

$$\alpha_m + \rho_{mD} + \rho_{ms} + \tau_m = 1 \qquad \textbf{[8-81]}$$

Also

$$\epsilon_m = \alpha_m$$

The diffuse radiosity of a particular surface of the medium is defined by

$$J_{mD} = \epsilon_m E_{bm} + \rho_{mD} G \qquad \textbf{[8-82]}$$

where G is the irradiation on the particular surface. Note that J_{mD} no longer represents the total diffuse energy leaving a surface. Now it represents only emission and diffuse reflection. The transmitted energy will be analyzed with additional terms. As before, the heat exchange is written

$$q = A(\epsilon E_b - \alpha G) \qquad \textbf{[8-83]}$$

Solving for G from Equation (8-82) and making use of Equation (8-81) gives

$$q = \frac{E_{bm} - J_{mD}/(1 - \tau_m - \rho_{ms})}{\rho_{mD}/[\epsilon_m A_m (1 - \tau_m - \rho_{ms})]} \qquad \textbf{[8-84]}$$

The network element representing Equation (8-84) is shown in Figure 8-56. This element is quite similar to the one shown in Figure 8-45, except that here we must take the transmissivity into account.

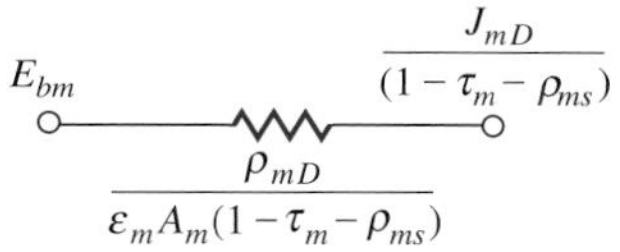

The transmitted heat exchange between surfaces 1 and 2 is the same as in Section 8-10; that is,

$$q = \frac{J_1 - J_2}{1/A_1 F_{12} \tau_m} \qquad \textbf{[8-85]}$$

The heat exchange between surface 1 and m is computed in the following way. Of that energy leaving surface 1, the amount that arrives at m and contributes to the diffuse radiosity of m is

$$q_{1 \to m} = J_1 A_1 F_{1m} (1 - \tau_m - \rho_{ms}) \qquad \textbf{[8-86]}$$

The diffuse energy leaving m that arrives at 1 is

$$q_{m \to 1} = J_{mD} A_m F_{m1} \qquad \textbf{[8-87]}$$

Subtracting (8-87) from (8-86) and using the reciprocity relation

$$A_1 F_{1m} = A_m F_{m1}$$

gives

$$q_{1m} = \frac{J_1 - J_{mD}/(1 - \tau_m - \rho_{ms})}{1/[A_1 F_{1m}(1 - \tau_m - \rho_{ms})]} \qquad \textbf{[8-88]}$$

The network element corresponding to Equation (8-89) is quite similar to the one shown in Figure 8-50. An equation similar to Equation (8-89) can be written for the radiation exchange between surface 2 and m. Finally, the complete network may be drawn as in Figure 8-57. It is to be noted that J_{mD} represents the diffuse radiosity of the left side of m, while J'_{mD} represents the diffuse radiosity of the right side of m.

Figure 8-55 | Physical system for analysis of transmitting and reflecting layers.

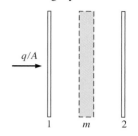

Figure 8-56 | Network element representing Equation (8-84).

Figure 8-57 | Complete radiation network for system in Figure 8-55.

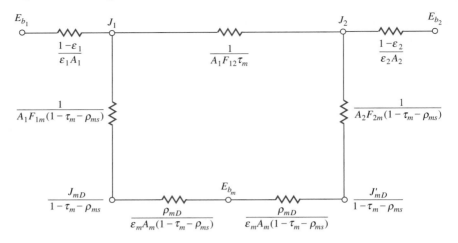

If m is maintained at a fixed temperature, then J_1 and J_2 must be obtained as a solution to nodal equations for the network. On the other hand, if no net energy is delivered to m, then E_{bm} is a floating node, and the network reduces to a simple series-parallel arrangement. In this latter case the temperature of m must be obtained by solving the network for E_{bm}.

We may extend the analysis a few steps further by distinguishing between specular and diffuse transmission. A specular transmission is one where the incident radiation goes "straight through" the material, while a diffuse transmission is encountered when the incident radiation is scattered in passing through the material, so that it emerges from the other side with a random spatial orientation. As with reflected energy, the assumption is made that the transmissivity may be represented with a specular and a diffuse component:

$$\tau = \tau_s + \tau_D \qquad \text{[8-89]}$$

The diffuse radiosity is still defined as in Equation (8-82), and the net energy exchange with a transmitting surface is given by Equation (8-84). The analysis of transmitted energy exchange with other surfaces must be handled somewhat differently, however.

Consider, for example, the arrangement in Figure 8-58. The two diffuse opaque surfaces are separated by a specular-diffuse transmitting and reflecting plane. For this example all planes are assumed to be infinite in extent. The *specular*-transmitted exchange between surfaces 1 and 3 may be calculated immediately with

$$(q_{13})_{\text{specular}-\text{transmitted}} = \frac{J_1 - J_3}{1/A_1 F_{13}\tau_{2s}} \qquad \text{[8-90]}$$

The *diffuse*-transmitted exchange between 1 and 3 is a bit more complicated. The energy leaving 1 that is transmitted diffusely through 2 is

$$J_1 A_1 F_{12}\tau_{2D}$$

Of this amount transmitted through 2, the amount that arrives at 3 is

$$(q_{13})_{\text{diffuse}-\text{transmitted}} = J_1 A_1 F_{12}\tau_{2D} F_{23} \qquad \text{[8-91]}$$

Similarly, the amount leaving 3 that is diffusely transmitted to 1 is

$$(q_{31})_{\text{diffuse}-\text{transmitted}} = J_3 A_3 F_{32}\tau_{2D} F_{21} \qquad \text{[8-92]}$$

Figure 8-58 | Radiation network for infinite parallel planes separated by a transmitting specular-diffuse plane.

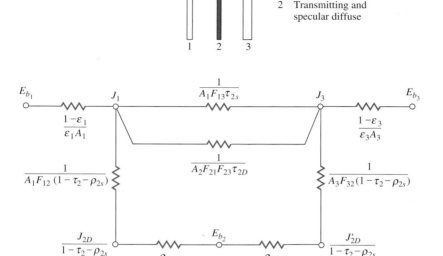

Now, by making use of the reciprocity relations, $A_1 F_{12} = A_2 F_{21}$ and $A_3 F_{32} = A_2 F_{23}$, subtraction of Equation (8-92) from Equation (8-91) gives

$$(q_{13})_{\text{net diffuse−transmitted}} = \frac{J_1 - J_3}{1/A_1 F_{21} F_{23} \tau_{2D}} \qquad \text{[8-93]}$$

Apparent Emissivity of Cavity with Transparent Cover

Using similar reasoning to that which enabled us to arrive at a relation for the apparent emissivity of a cavity in Equation (8-47), we may consider the effect a transparent covering may have on ϵ_a. The covered cavity is indicated in Figure 8-59 with the characteristics of the cover described by

$$\epsilon_2 + \rho_2 + \tau_2 = 1.0$$

The corresponding radiation network for this cavity exchanging heat with a large surrounding at T_s is shown in Figure 8-60. As in Equation (8-47), we define the apparent emissivity

Figure 8-59 | Cavity with semitransparent covering.

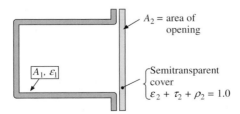

Figure 8-60 | Radiation network for cavity with partially transparent cover.

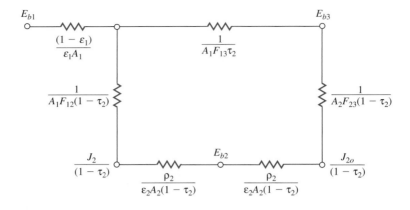

of the cavity in terms of the net radiation exchange with the surroundings as

$$q = \epsilon_a A_2 (E_{b1} - E_{bs}) \qquad \text{[8-94]}$$

The shape factors in the radiation network are determined as

$$F_{21} = 1, \quad A_1 F_{12} = A_2 F_{21} = A_2, \quad F_{23} = 1, \quad A_2 F_{23} = A_2$$

But, $F_{12} = F_{13}$ so that $A_1 F_{13} = A_2$. The heat exchange is determined from the network as

$$q = (E_{b1} - E_{bs})/\Sigma R \qquad \text{[8-95]}$$

where ΣR is the equivalent resistance for the series parallel network. Performing the necessary algebraic manipulation to evaluate ΣR, and equating the heat transfers in (8–95) and (8–94) gives the relation for the apparent emissivity as

$$\epsilon_a/(\tau_2 + \epsilon_2/2) = K/[(A_2/A_1)(1 - \epsilon_1) + K] \qquad \text{[8-96]}$$

where

$$K = \epsilon_1/(\tau_2 + \epsilon_2/2) \qquad \text{[8-96a]}$$

We may note the following behavior for three limiting conditions.

1. For $\tau_2 \to 1$, we have an open cavity and the behavior approaches that described by Equation (8-47).
2. For $\tau_2 \to 1$ and $A_2 = A_1$, we have neither cavity nor cover and $\epsilon_a \to \epsilon_1$.
3. For $A_1 \gg A_2$ we have a very large cavity with $\epsilon_a \to \tau_2 + \epsilon_2/2$.

The behavior of ϵ_a is displayed graphically in Figure 8-61.

EXAMPLE 8-13   Cavity with Transparent Cover

The rectangular cavity between the fins of Example 8-9 has $\epsilon_1 = 0.5$ along with a cover placed over the opening with the properties

$$\tau_2 = 0.5 \quad \rho_2 = 0.1 \quad \epsilon_2 = 0.4$$

Calculate the apparent emissivity of the covered opening.

Figure 8-61 | Apparent emissivity of cavity with partially transparent cover.

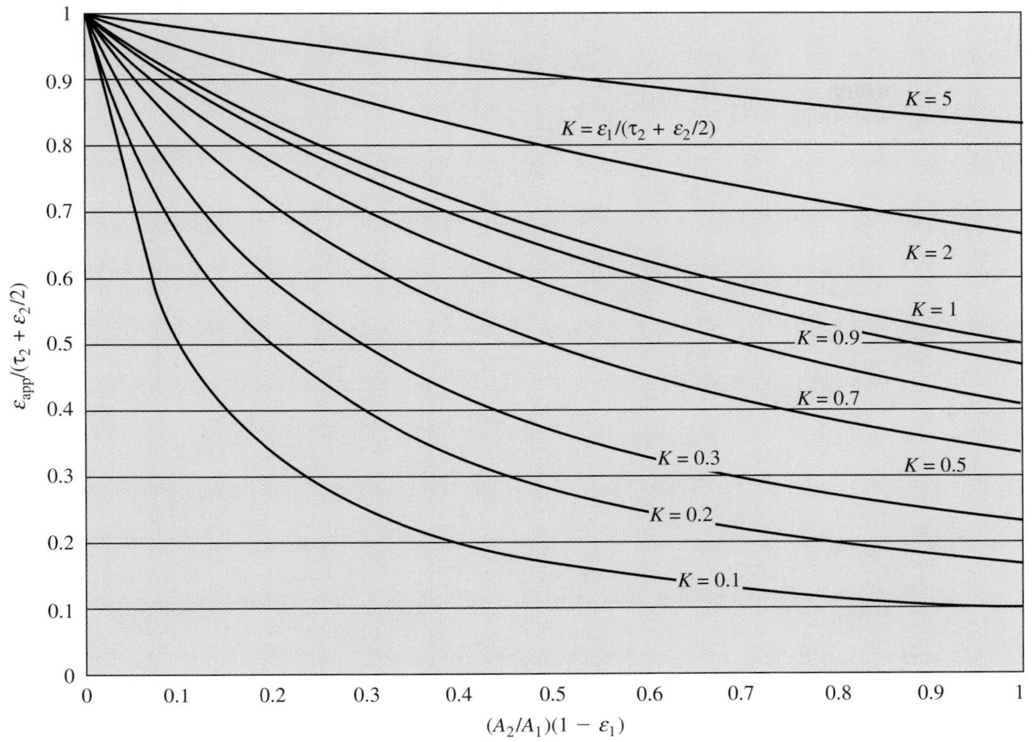

Solution

Per unit depth in the z direction we have $A_1 = 225 + 25 + 10 = 60$ and $A_2 = 10$. We may evaluate K from Equation (8-96a)

$$K = 0.5/(0.5 + 0.4/2) = 5/7$$

The value of ϵ_a is then computed from Equation (8-96) as

$$\epsilon_a = (0.5 + 0.4/2)(5/7)/[(10/60)(1 - 0.5) + 5/7] = 0.6269$$

If there were no cover present, the value of ϵ_a would be given by Equation (8-47) as

$$\epsilon_a = (0.5)(60)/[10 + (0.5)(60 - 10)] = 0.8571$$

Obviously, the presence of the cover reduces the heat transfer for values of $\tau_2 < 1.0$.

Transmitting and Reflecting System for Furnace Opening   EXAMPLE 8-14

A furnace at 1000°C has a small opening in the side that is covered with a quartz window having the following properties:

$$0 < \lambda < 4 \, \mu\text{m} \qquad \tau = 0.9 \qquad \epsilon = 0.1 \qquad \rho = 0$$
$$4 < \lambda < \infty \qquad \tau = 0 \qquad \epsilon = 0.8 \qquad \rho = 0.2$$

The interior of the furnace may be treated as a blackbody. Calculate the radiation lost through the quartz window to a room at 30°C. Diffuse surface behavior is assumed.

Figure Example 8-14

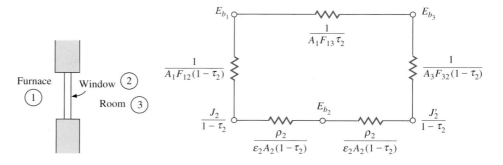

■ **Solution**
The diagram for this problem is shown in Figure Example 8-14. Because the room is large it may be treated as a blackbody also. We shall analyze the problem by calculating the heat transfer for each wavelength band and then adding them together to obtain the total. The network for each band is a modification of Figure 8-57, as shown here for black furnace and room. We shall make the calculation for unit area; then

$$A_1 = A_2 = A_3 = 1.0$$
$$F_{12} = 1.0 \qquad F_{13} = 1.0 \qquad F_{32} = 1.0$$

The total emissive powers are

$$E_{b_1} = (5.669 \times 10^{-8})(1273)^4 = 1.4887 \times 10^5 \text{ W/m}^2$$
$$E_{b_3} = (5.669 \times 10^{-8})(303)^4 = 477.8 \text{ W/m}^2$$

To determine the fraction of radiation in each wavelength band, we calculate

$$\lambda T_1 = (4)(1273) = 5092 \ \mu\text{m} \cdot \text{K}$$
$$\lambda T_3 = (4)(303) = 1212 \ \mu\text{m} \cdot \text{K}$$

Consulting Table 8-1, we find

$$E_{b_1}(0-4 \ \mu\text{m}) = 0.6450 E_{b_1} = 96{,}021 \text{ W/m}^2$$
$$E_{b_3}(0-4 \ \mu\text{m}) = 0.00235 E_{b_3} = 1.123 \text{ W/m}^2$$
$$E_{b_1}(4-\infty) = (1-0.6450) E_{b_1} = 52{,}849 \text{ W/m}^2$$
$$E_{b_3}(4-\infty) = (1-0.00235) E_{b_3} = 476.7 \text{ W/m}^2$$

We now apply these numbers to the network for the two wavelength bands, with unit areas. $0 < \lambda < 4 \ \mu\text{m}$ *band:*

$$\frac{1}{F_{13}\tau_2} = \frac{1}{0.9} \qquad \frac{1}{F_{32}(1-\tau_2)} = \frac{1}{0.1} = \frac{1}{F_{12}(1-\tau_2)}$$
$$\frac{\rho_2}{\epsilon_2(1-\tau_2)} = 0$$

The net heat transfer from the network is then

$$q = \frac{E_{b_1} - E_{b_3}}{R_{\text{equiv}}} = \frac{96,021 - 1.123}{1.0526} = 91,219 \text{ W/m}^2 \qquad 0 < \lambda < 4 \ \mu\text{m}$$

$4 \ \mu\text{m} < \lambda < +\infty$ *band:*

$$\frac{1}{F_{13}\tau_2} = \infty \qquad \frac{1}{F_{32}(1-\tau_2)} = \frac{1}{F_{12}(1-\tau_2)} = 1.0$$

$$\frac{\rho_2}{\epsilon_2(1-\tau_2)} = \frac{0.2}{0.8} = 0.25$$

The net heat transfer from the network is

$$q = \frac{E_{b_1} - E_{b_3}}{1 + 0.25 + 0.25 + 1} = \frac{52,849 - 476.7}{2.5} = 20,949 \text{ W/m}^2 \qquad 4 < \lambda < \infty$$

The total heat loss is then

$$q_{\text{total}} = 91,219 + 20,949 = 112,168 \text{ W/m}^2 \qquad [35,560 \text{ Btu/h} \cdot \text{ft}^2]$$

With no window at all, the heat transfer would have been the difference in blackbody emissive powers,

$$q - E_{b_1} - E_{b_3} = 1.4887 \times 10^5 - 477.8 = 1.4839 \times 10^5 \text{ W/m}^2 \qquad [47,040 \text{ Btu/h} \cdot \text{ft}^2]$$

8-13 | FORMULATION FOR NUMERICAL SOLUTION

The network method that we have used to analyze radiation problems is an effective artifice for visualizing radiant exchange between surfaces. For simple problems that do not involve too many surfaces, the network method affords a solution that can be obtained quite easily. When many heat-transfer surfaces are involved, it is to our advantage to formalize the procedure for writing the nodal equations. For this procedure we consider only opaque, gray, diffuse surfaces. The reader should consult Reference 10 for information on transmitting and specular surfaces. The radiant-energy balance on a particular opaque surface can be written

Net heat lost by surface = energy emitted − energy absorbed

or on a unit-area basis with the usual gray-body assumptions,

$$\frac{q}{A} = \epsilon E_b - \alpha G$$

Considering the ith surface, the total irradiation is the sum of all irradiations G_j from the other j surfaces. Thus, for $\epsilon = \alpha$,

$$\frac{q_i}{A_i} = \epsilon_i \left(E_{b_i} - \sum_j G_j \right) \qquad \text{[8-97]}$$

But, the irradiations can be expressed by

$$A_j J_j F_{ji} = G_j A_i \qquad \text{[8-98]}$$

From reciprocity, we have

$$A_j F_{ji} = A_i F_{ij}$$

so that we can combine the equations to give

$$\frac{q_i}{A_i} = \epsilon_i \left(E_{b_i} - \sum_j F_{ij} J_j \right)$$

[8-99]

The heat transfer at each surface is then evaluated in terms of the radiosities J_j. These parameters are obtained by recalling that the heat transfer can also be expressed as

$$\frac{q_i}{A_i} = J_i - G_i = J_i - \sum_j F_{ij} J_j$$

[8-100]

Combining Equations (8-99) and (8-100) gives

$$J_i - (1 - \epsilon_i) \sum_j F_{ij} J_j = \epsilon_i E_{b_i}$$

[8-101]

In the equations above it must be noted that the summations must be performed over *all* surfaces in the enclosure. For a three-surface enclosure, with $i = 1$, the summation would then become

$$\sum_j F_{ij} J_j = F_{11} J_1 + F_{12} J_2 + F_{13} J_3$$

Figure 8-62 | Radiation network balance on node J_i.

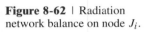

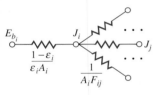

$$\frac{\epsilon_i}{1-\epsilon_i}(E_{b_i}-J_i)+\sum_j F_{ij}(J_j-J_i)=0$$

Of course, if surface 1 is convex, $F_{11} = 0$ and some simplification could be effected.

The nodal equations for the radiosities may also be derived from the nodes in the network formulation, as indicated in Figure 8-62. At each J_i node an energy balance gives

$$\frac{\epsilon_i}{1 - \epsilon_i}(E_{b_i} - J_i) + \sum_j F_{ij}(J_j - J_i) = 0$$

[8-102]

Again, an equation will be obtained for each J_i that is entirely equivalent to Equation (8-101). Once all the equations are written out they can be expressed in the matrix form

$$[A][J] = [C]$$

[8-103]

where

$$[A] = \begin{bmatrix} a_{11} & a_{12} & \cdots & a_{1i} \\ a_{21} & a_{22} & \cdots & a_{2i} \\ \cdots & \cdots & \cdots & \cdots \\ a_{i1} & a_{i2} & \cdots & a_{ii} \end{bmatrix} \qquad [J] = \begin{bmatrix} J_1 \\ J_2 \\ \vdots \\ J_i \end{bmatrix} \qquad [C] = \begin{bmatrix} C_1 \\ C_2 \\ \vdots \\ C_i \end{bmatrix}$$

The solution for the radiosities is found by obtaining the inverse to $[A]$ such that

$$[J] = [A]^{-1}[C]$$

The inverse $[A]^{-1}$ is written as

$$[A]^{-1} = \begin{bmatrix} b_{11} & b_{12} & \cdots & b_{1i} \\ b_{21} & \cdots & \cdots & \cdots \\ \cdots & \cdots & \cdots & \cdots \\ b_{i1} & b_{i2} & \cdots & b_{ii} \end{bmatrix}$$

so that the unknown radiosities are written as

$$J_1 = b_{11}C_1 + b_{12}C_2 + \cdots + b_{1i}C_i$$

$$\cdots \cdots \cdots \cdots \cdots \cdots \cdots \cdots \cdots$$

$$J_i = b_{i1}C_1 + b_{i2}C_2 + \cdots + b_{ii}C_i$$

Standard computer subroutines are available to obtain the inverse matrix and perform the final calculations of the J_i. The heat-transfer rate at each ith surface having an area A_i is then calculated from

$$\frac{q_i}{A_i} = \frac{\epsilon_i}{1 - \epsilon_i}(E_{b_i} - J_i) \qquad \textbf{[8-104]}$$

In formulating the nodal equations one must take note of the consequence of Equation (8-104) for an *insulated* surface (i.e., one for which there is no net heat transfer). Equation (8-110) thus requires that

$$E_{b_i} = J_i \qquad \text{for insulated surface} \qquad \textbf{[8-105]}$$

From a practical point of view, a Gauss-Seidel iteration scheme may be the most efficient numerical procedure to follow in solving the set of equations for the J_i's. For the Gauss-Seidel scheme the above equations must be organized in explicit form for J_i. Solving for J_i in Equation (8-101) and breaking out the F_{ii} term gives

$$J_i = (1 - \epsilon_i)\sum_{j \neq i} F_{ij}J_j + (1 - \epsilon_i)F_{ii}J_i + \epsilon_i E_{b_i}$$

$$J_i = \frac{1}{1 - F_{ii}(1 - \epsilon_i)}\left[(1 - \epsilon_i)\sum_{j \neq i} F_{ij}J_j + \epsilon_i E_{b_i}\right] \qquad \textbf{[8-106]}$$

For a surface in radiant equilibrium, $q_i/A_i = 0$ and $J_i = E_{b_i}$ may be substituted into Equation (8-106) to give

$$J_i = \frac{1}{1 - F_{ii}}\sum_{j \neq i} F_{ij}J_j \qquad \text{for } \frac{q_i}{A_i} = 0 \qquad \textbf{[8-107]}$$

If the problem formulation is to include a specified heat flux q_i/A_i at one of the ith surfaces, we can solve for E_{b_i} from Equation (8-104) to give

$$E_{b_i} = J_i + \frac{1 - \epsilon_i}{\epsilon_i}\frac{q_i}{A_i} \qquad \textbf{[8-108]}$$

Substituting this value into Equation (8-101) and then solving for J_i give

$$J_i = \frac{1}{1 - F_{ii}}\left(\sum_{j \neq i} F_{ij}J_j + \frac{q_i}{A_i}\right) \qquad \textbf{[8-109]}$$

In many cases the radiation solution must take conduction and convection into account at the ith surface. The appropriate energy balance is then, for steady state,

Heat conducted into surface + heat convected into surface

= radiant heat *lost* from surface

or

$$q_{\text{cond},i} + q_{\text{conv},i} = q_{i,\text{rad}} \qquad \textbf{[8-110]}$$

This energy balance may then be used in conjunction with Equation (8-109) to obtain the proper nodal equation for J_i.

While the above formulations may appear rather cumbersome at first glance they are easily solved by computer, with either matrix inversion or iteration. For many practical radiation problems, the number of equations is small and programmable calculators may be employed for solution. In most cases one will not know the surface properties (ϵ_i) within better than a few percent, so an iterative solution need not be carried out to unreasonable limits of precision.

In summary, we outline the computational procedure to be followed for numerical solution of radiation heat transfer between diffuse, gray surfaces.

1. Evaluate F_{ij} and ϵ_i for all surfaces.
2. Evaluate E_{b_i} for all surfaces with specified temperature.
3. Formulate nodal equations for the J_i using:
   a. Equation (8-106) for surfaces with specified T_i.
   b. Equation (8-107) for surfaces in radiant balance ($J_i = E_{b_i}$).
   c. Equation (8-109) for surfaces with specified q_i.
4. Solve the equations for the J_i's.
5. Compute the q_i's and T_i's, using:
   a. q_i from Equation (8-104) for gray surfaces and Equation (8-99) for black surfaces with specified T_i.
   b. T_i from $J_i = E_{b_i} = \sigma T_i^4$ for surfaces in radiant balance.
   c. T_i using E_{b_i} obtained from Equation (8-108) for surfaces with specified q_i.

Of course, the above equations may be put in the following form if direct matrix inversion is preferred over an iteration scheme:

$$J_i[1 - F_{ii}(1 - \epsilon_i)] - (1 - \epsilon_i) \sum_{j \neq i} F_{ij} J_j = \epsilon_i E_{b_i} \qquad \textbf{[8-106a]}$$

$$J_i(1 - F_{ii}) - \sum_{j \neq i} F_{ij} J_j = 0 \qquad \textbf{[8-107a]}$$

$$J_i(1 - F_{ii}) - \sum_{j \neq i} F_{ij} J_j = \frac{q_i}{A_i} \qquad \textbf{[8-109a]}$$

Computation of the surface temperatures and heat transfers is the same as in step 5 above.

Insulated Surfaces and Surfaces with Large Areas

We have seen in the application of the network method that an insulated surface acts as if it were perfectly reflective with $\epsilon \to 0$, and thus $J = E_b$. We may note that if one takes $\epsilon = 0$ in Equation (8-106a), Equation (8-107a) will result. When the system of equations is solved for the J's, one may thus obtain the temperature of an insulated surface from $T = (E_b/\sigma)^{1/4} = (J/\sigma)^{1/4}$. When a surface with a very large area compared to other surfaces in the system is involved, it behaves like a blackbody with $\epsilon \to 1.0$ because of its low surface resistance. If such a large surface is concave, it will behave as if $F_{ii} \to 1.0$ and all the $F_{ij} \to 0$.

Numerical Solution for Enclosure **EXAMPLE 8-15**

The geometry of Example 8-5 is used for radiant exchange with a large enclosure. Surface 2 is diffuse with $\epsilon = 0.5$ while surface 1 is perfectly insulated. $T_2 = 1000$ K, and $T_3 = 300$ K. Calculate the heat lost to the large room per unit length of surface 2, using the numerical formulation. Also calculate the temperature of the insulated surface.

■ **Solution**

For unit length we have:

$$E_{b_2} = \sigma T_2^4 = 5.669 \times 10^4 \qquad E_{b_3} = \sigma T_3^4 = 459$$

$$A_1 = (4)(0.2) = 0.8 \text{ m}^2/\text{m} \qquad A_2 = \pi(0.60)/2 = 0.94 \text{ m}^2/\text{m}$$

We will use the numerical formulation. We find from Example 8-5, using the nomenclature of the figure, $F_{11} = 0.314$, $F_{12} = 0.425$, $F_{13} = 0.261$, $F_{21} = 0.5$, $F_{22} = 0$, $F_{23} = 0.5$, $F_{31} \to 0$, $F_{32} \to 0$, $F_{33} \to 1.0$. We now write the equations. Surface 1 is insulated so we use Equation (8-107a):

$$J_1(1 - 0.314) - 0.425 J_2 - 0.261 J_3 = 0$$

Surface 2 is constant temperature so we use Equation (8-106a):

$$J_2(1 - 0) - (1 - 0.5)[0.5 J_1 + 0.5 J_3] = (0.5)(56,690)$$

Because surface 3 is so large,

$$J_3 = E_{b_3} = 459 \text{ W/m}^2$$

Rearranging the equations gives

$$0.686 J_1 - 0.425 J_2 = 119.8$$
$$-0.25 J_1 + J_2 = 28,460$$

which have the solutions

$$J_1 = 21,070 \text{ W/m}^2$$
$$J_2 = 33,727 \text{ W/m}^2$$

The heat transfer is thus

$$q = \frac{E_{b_2} - J_2}{\dfrac{1 - \epsilon_2}{\epsilon_2 A_2}} = \frac{56,690 - 33,727}{(1 - 0.5)/(0.5)(4)(0.2)} = 18,370 \text{ W/m length}$$

Because surface 1 is insulated, $J_1 = E_{b_1}$, and we could calculate the temperature as

$$T_1 = \left(\frac{21,070}{5.669 \times 10^{-8}}\right)^{1/4} = 781 \text{ K}$$

For this problem a solution using the network method might be simpler because it involves only a simple series–parallel circuit.

Numerical Solutions for Parallel Plates **EXAMPLE 8-16**

Two 1-m-square surfaces are separated by a distance of 1 m with $T_1 = 1000$ K, $T_2 = 400$ K, $\epsilon_1 = 0.8, \epsilon_2 = 0.5$. Obtain the numerical solutions for this system when (a) the plates are surrounded by a large room at 300 K and (b) the surfaces are connected by a re-radiating wall perfectly insulated on its outer surface. Part (a) of this example is identical in principle to the problem that is solved by the network method in Example 8-6.

■ **Solution**

Consulting Figure 8-12, we obtain

$$F_{12} = 0.2 \qquad F_{21} = 0.2 \qquad F_{11} = 0 = F_{22}$$
$$F_{13} = 0.8 \qquad F_{23} = 0.8$$
$$A_1 = A_2 = 1 \text{ m}^2$$

(surface 3 is the surroundings or insulated surface). For part (*a*)

$$E_{b_1} = \sigma T_1^4 = 56.69 \text{ kW/m}^2 \quad [17{,}970 \text{ Btu/h} \cdot \text{ft}^2]$$
$$E_{b_2} = \sigma T_2^4 = 1.451 \text{ kW/m}^2$$
$$E_{b_3} = \sigma T_3^4 = 0.459 \text{ kW/m}^2$$

Because $A_3 \to \infty$, F_{31} and F_{32} must approach zero since $A_1 F_{13} = A_3 F_{31}$ and $A_2 F_{23} = A_3 F_{32}$. The nodal equations are written in the form of Equation (8-107):

surface 1: $\qquad J_1 - (1 - \epsilon_1)(F_{11} J_1 + F_{12} J_2 + F_{13} J_3) = \epsilon_1 E_{b_1}$

surface 2: $\qquad J_2 - (1 - \epsilon_2)(F_{21} J_1 + F_{22} J_2 + F_{23} J_3) = \epsilon_2 E_{b_2}$ [*a*]

surface 3: $\qquad J_3 - (1 - \epsilon_3)(F_{31} J_1 + F_{32} J_2 + F_{33} J_3) = \epsilon_3 E_{b_3}$

Because F_{31} and F_{32} approach zero, F_{33} must be 1.0.

Inserting the numerical values for the various terms, we have

$$J_1 - (1 - 0.8)[(0) J_1 + (0.2) J_2 + (0.8) J_3] = (0.8)(56.69)$$
$$J_2 - (1 - 0.5)[(0.2) J_1 + (0) J_2 + (0.8) J_3] = (0.5)(1.451) \qquad [b]$$
$$J_3 - (1 - \epsilon_3)[(0) J_1 + (0) J_2 + (1.0) J_3] = \epsilon_3 (0.459)$$

The third equation yields $J_3 = 0.459 \text{ kW/m}^2 \cdot \text{K}$. Because the room is so large it acts like a hohlraum, or blackbody. But *it does not have zero heat transfer.*

Finally, the equations are written in compact form as

$$J_1 - 0.04 J_2 - 0.16 J_3 = 45.352$$
$$-0.1 J_1 + J_2 - 0.4 J_3 = 0.7255 \qquad [c]$$
$$J_3 = 0.459$$

Of course, there only remain two unknowns, J_1 and J_2, in this set.

For part (*b*), A_3 for the enclosing wall is 4.0 m^2, and we set $J_3 = E_{b_3}$ because surface 3 is insulated. From reciprocity we have

$$A_1 F_{13} = A_3 F_{31} \qquad F_{31} = \frac{(1.0)(0.8)}{4.0} = 0.2$$
$$A_2 F_{23} = A_3 F_{32} \qquad F_{32} = \frac{(1.0)(0.8)}{4.0} = 0.2$$

Then from $F_{31} + F_{32} + F_{33} = 1.0$ we have $F_{33} = 0.6$.

The set of equations in (*a*) still applies, so we insert the numerical values to obtain (with $J_3 = E_{b_3}$)

$$J_1 - (1 - 0.8)[(0) J_1 + (0.2) J_2 + (0.8) J_3] = (0.8)(56.69)$$
$$J_2 - (1 - 0.5)[(0.2) J_1 + (0) J_2 + (0.8) J_3] = (0.5)(1.451) \qquad [d]$$
$$J_3 - (1 - \epsilon_3)[(0.2) J_1 + (0.2) J_2 + (0.6) J_3] = \epsilon_3 J_3$$

Notice that the third equation of set (*d*) can be written as

$$J_3(1 - \epsilon_3) - (1 - \epsilon_3)[(0.2) J_1 + (0.2) J_2 + (0.6) J_3] = 0$$

so that the $1 - \epsilon_3$ term drops out, and we obtain our final set of equations as

$$J_1 - 0.04J_2 - 0.16J_3 = 45.352$$
$$-0.1J_1 + J_2 - 0.4J_3 = 0.7255 \qquad [e]$$
$$-0.2J_1 - 0.2J_2 + 0.4J_3 = 0$$

To obtain the heat transfers the set of equations is first solved for the radiosities. For set (c),

$$J_1 = 45.644 \text{ kW/m}^2$$
$$J_2 = 5.474 \text{ kW/m}^2$$
$$J_3 = 0.459 \text{ kW/m}^2$$

The heat transfers are obtained from Equation (8-104):

$$q_1 = \frac{A_1 \epsilon_1}{1 - \epsilon_1}(E_{b_1} - J_1) = \frac{(1.0)(0.8)}{1 - 0.8}(56.69 - 45.644) = 44.184 \text{ kW} \quad [150{,}760 \text{ Btu/h}]$$

$$q_2 = \frac{A_2 \epsilon_2}{1 - \epsilon_2}(E_{b_2} - J_2) = \frac{(1.0)(0.5)}{1 - 0.5}(1.451 - 5.474) = -4.023 \text{ kW} \quad [-13{,}730 \text{ Btu/h}]$$

The net heat *absorbed* by the room is the algebraic sum of q_1 and q_2 or

$$q_{3, \text{ absorbed}} = 44.184 - 4.023 = 40.161 \text{ kW} \quad [137{,}030 \text{ Btu/h}]$$

For part (b) the solutions to set (e) are

$$J_1 = 51.956 \text{ kW/m}^2 \qquad J_2 = 20.390 \text{ kW/m}^2 \qquad J_3 = 36.173 \text{ kW/m}^2$$

The heat transfers are

$$q_1 = \frac{A_1 \epsilon_1}{1 - \epsilon_1}(E_{b_1} - J_1) = \frac{(1.0)(0.8)}{1 - 0.8}(56.69 - 51.965) = 18.936 \text{ kW}$$

$$q_2 = \frac{A_2 \epsilon_2}{1 - \epsilon_2}(E_{b_2} - J_2) = \frac{(1.0)(0.5)}{1 - 0.5}(1.451 - 20.390) = -18.936 \text{ kW}$$

Of course, these heat transfers should be equal in magnitude with opposite sign because the insulated wall exchanges no heat. The temperature of the insulated wall is obtained from

$$J_3 = E_{b_3} = \sigma T_3^4 = 36.173 \text{ kW/m}^2$$

and

$$T_3 = 894 \text{ K} \quad [621°\text{C}, 1150°\text{F}]$$

Radiation from a Hole with Variable Radiosity   EXAMPLE 8-17

To further illustrate the radiation formulation for numerical solution we consider the circular hole 2 cm in diameter and 3 cm deep, as shown in Figure Example 8-17. The hole is machined in a large block of metal, which is maintained at 1000°C and has a surface emissivity of 0.6. The temperature of the large surrounding room is 20°C. A simple approach to this problem would assume the radiosity uniform over the entire heated internal surface. In reality, the radiosity varies over the surface, and we break it into segments 1 (bottom of the hole), 2, 3, and 4 (sides of the hole) for analysis.

The large room acts like a blackbody at 20°C, so for analysis purposes we can assume the hole is covered by an imaginary black surface 5 at 20°C. We set the problem up for a numerical solution for the radiosities and then calculate the heat-transfer rates. After that, we shall examine an insulated-surface case for this same geometry.

Figure Example 8-17

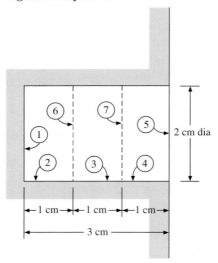

■ Solution

All the shape factors can be obtained with the aid of Figure 8-13 and the imaginary disk surfaces 6 and 7. We have

$$E_{b_1} = \sigma T_1^4 = (5.669 \times 10^{-8})(1273)^4 = 1.48874 \times 10^5 \text{ W/m}^2$$
$$= E_{b_2} = E_{b_3} = E_{b_4}$$
$$E_{b_5} = \sigma T_5^4 = (5.669 \times 10^{-8})(293)^4 = 417.8 \text{ W/m}^2$$
$$\epsilon_1 = \epsilon_2 = \epsilon_3 = \epsilon_4 = 0.6 \qquad \epsilon_5 = 1.0$$
$$A_1 = A_5 = \pi(1)^2 = \pi \text{ cm}^2 = A_6 = A_7$$
$$A_2 = A_3 = A_4 = \pi(2)(1) = 2\pi$$
$$F_{11} = F_{55} = 0 \qquad F_{16} = 0.37 \qquad F_{17} = 0.175 \qquad F_{15} = 0.1$$
$$F_{12} = 1 - F_{16} = 0.63 = F_{54}$$
$$F_{13} = F_{16} - F_{17} = 0.195 = F_{53}$$
$$F_{14} = F_{17} - F_{15} = 0.075 = F_{52}$$
$$F_{21} = F_{26} = F_{16}\frac{A_1}{A_2} = 0.315 = F_{45} = F_{36} = F_{37}$$
$$F_{22} = 1 - F_{21} - F_{26} = 0.37 = F_{33} = F_{44}$$
$$F_{31} = F_{13}\frac{A_1}{A_3} = 0.0975$$
$$F_{32} = F_{36} - F_{31} = 0.2175 = F_{34} = F_{43} = F_{23}$$
$$F_{27} = F_{26} - F_{23} = F_{21} - F_{23} = 0.0975 = F_{46}$$
$$F_{41} = F_{14}\frac{A_1}{A_4} = 0.0375 = F_{25}$$
$$F_{42} = F_{46} - F_{41} = 0.06 = F_{24}$$

The equations for the radiosities are now written in the form of Equation (8-106), noting that $F_{11} = 0$ and $J_5 = E_{b_5}$:

$$J_1 = (1 - \epsilon_1)(F_{12}J_2 + F_{13}J_3 + F_{14}J_4 + F_{15}E_{b_5}) + \epsilon_1 E_{b_1}$$
$$J_2 = \frac{1}{1 - F_{22}(1 - \epsilon_2)}[(1 - \epsilon_2)(F_{21}J_1 + F_{23}J_3 + F_{24}J_4 + F_{25}E_{b_5}) + \epsilon_2 E_{b_2}]$$

$$J_3 = \frac{1}{1 - F_{33}(1 - \epsilon_3)}[(1 - \epsilon_3)(F_{31}J_1 + F_{32}J_2 + F_{34}J_4 + F_{35}E_{b_5}) + \epsilon_3 E_{b_3}]$$

$$J_4 = \frac{1}{1 - F_{44}(1 - \epsilon_4)}[(1 - \epsilon_4)(F_{41}J_1 + F_{42}J_2 + F_{43}J_3 + F_{45}E_{b_5}) + \epsilon_4 E_{b_4}]$$

When all the numerical values are inserted, we obtain

$$J_1 = 0.252J_2 + 0.078J_3 + 0.03J_4 + 89,341$$

$$J_2 = 0.1479J_1 + 0.1021J_3 + 0.02817J_4 + 104,848$$

$$J_3 = 0.04577J_1 + 0.1021J_2 + 0.1021J_4 + 104,859$$

$$J_4 = 0.01761J_1 + 0.02817J_2 + 0.1021J_3 + 104,902$$

These equations may be solved to give

$$J_1 = 1.4003 \times 10^5 \text{ W/m}^2$$

$$J_2 = 1.4326 \times 10^5 \text{ W/m}^2$$

$$J_3 = 1.3872 \times 10^5 \text{ W/m}^2$$

$$J_4 = 1.2557 \times 10^5 \text{ W/m}^2$$

The heat transfers can be calculated from Equation (8-104):

$$q_i = \frac{\epsilon_i A_i}{1 - \epsilon_i}(E_{b_i} - J_i)$$

$$q_1 = \frac{(0.6)(\pi \times 10^{-4})}{1 - 0.6}(1.4887 - 1.4003)(10^5) = 4.1658 \text{ W}$$

$$q_2 = \frac{(0.6)(2\pi \times 10^{-4})}{1 - 0.6}(1.4887 - 1.4326)(10^5) = 5.2873 \text{ W}$$

$$q_3 = \frac{(0.6)(2\pi \times 10^{-4})}{1 - 0.6}(1.4887 - 1.3872)(10^5) = 9.5661 \text{ W}$$

$$q_4 = \frac{(0.6)(2\pi \times 10^{-4})}{1 - 0.6}(1.4887 - 1.2557)(10^5) = 21.959 \text{ W}$$

The total heat transfer is the sum of these four quantities or

$$q_{\text{total}} = 40.979 \text{ W} \quad [139.8 \text{ Btu/h}]$$

It is of interest to compare this heat transfer with the value we would obtain by assuming uniform radiosity on the hot surface. We would then have a two-body problem with

$$A_1 = \pi + 3(2\pi) = 7\pi \text{ cm}^2 \qquad A_5 = \pi \qquad F_{51} = 1.0 \qquad \epsilon_1 = 0.6 \qquad \epsilon_5 = 1.0$$

The heat transfer is then calculated from Equation (8-43), with appropriate change of nomenclature:

$$q = \frac{(E_{b_1} - E_{b_5})A_5}{1/\epsilon_5 + (A_5/A_1)(1/\epsilon_1 - 1)} = \frac{(\pi \times 10^{-4})(1.4887 \times 10^5 - 417.8)}{1 + (\frac{1}{7})(1/0.6 - 1)}$$

$$= 42.581 \text{ W} \quad [145.3 \text{ Btu/h}]$$

Thus, the simple assumption of uniform radiosity gives a heat transfer that is 3.9 percent above the value obtained by breaking the hot surface into four parts for the calculation. This indicates that the uniform-radiosity assumption we have been using is a rather good one for engineering calculations.

Let us now consider the case where surface 1 is still radiating at 1000°C with $\epsilon = 0.6$ but the side walls 2, 3, and 4 are insulated. The radiation is still to the large room at 20°C. The nodal equation for J_1 is the same as before but now the equations for J_2, J_3, and J_4 must be written in the form of Equation (8-107). When that is done and the numerical values are inserted, we obtain

$$J_1 = 0.252J_2 + 0.078J_3 + 0.03J_4 + 89,341$$
$$J_2 = 0.5J_1 + 0.3452J_3 + 0.09524J_4 + 24.869$$
$$J_3 = 0.1548J_1 + 0.3452J_2 + 0.3452J_4 + 64.66$$
$$J_4 = 0.05952J_1 + 0.0952J_2 + 0.3452J_3 + 208.9$$

When these equations are solved, we obtain

$$J_1 = 1.1532 \times 10^5 \text{ W/m}^2 \quad [36,560 \text{ Btu/h} \cdot \text{ft}^2]$$
$$J_2 = 0.81019 \times 10^5 \text{ W/m}^2$$
$$J_3 = 0.57885 \times 10^5 \text{ W/m}^2$$
$$J_4 = 0.34767 \times 10^5 \text{ W/m}^2$$

The heat transfer at surface 1 is

$$q_1 = \frac{\epsilon_1 A_1}{1 - \epsilon_1}(E_{b_1} - J_1) = \frac{(0.6)(\pi \times 10^{-4})}{1 - 0.6}(1.4887 - 1.1532)(10^5)$$
$$= 15.81 \text{ W} \quad [53.95 \text{ Btu/h}]$$

The temperatures of the insulated surface elements are obtained from

$$J_i = E_{b_i} = \sigma T_i^4$$
$$T_2 = 1093 \text{ K} = 820°C \quad [1508°F]$$
$$T_3 = 1005 \text{ K} = 732°C \quad [1350°F]$$
$$T_4 = 895 \text{ K} = 612°C \quad [1134°F]$$

It is of interest to compare the heat transfer calculated above with that obtained by assuming surfaces 2, 3, and 4 uniform in temperature and radiosity. Equation (8-41) applies for this case:

$$q = \frac{A_1(E_{b_1} - E_{b_5})}{\dfrac{A_1 + A_2 + 2A_1F_{15}}{A_5 - A_1(F_{15})^2} + \left(\dfrac{1}{\epsilon_1} - 1\right) + \dfrac{A_1}{A_5}\left(\dfrac{1}{\epsilon_5} - 1\right)}$$

and

$$q = \frac{(\pi \times 10^{-4})(1.4887 \times 10^5 - 417.8)}{\dfrac{\pi + \pi - 2\pi(0.1)}{\pi - \pi(0.1)^2} + \dfrac{1}{0.6} - 1} = 18.769 \text{ W} \quad [64.04 \text{ Btu/h}]$$

In this case the assumption of uniform radiosity at the insulated surface gives an overall heat transfer with surface 1 (bottom of hole) that is 18.7 percent too high.

Heater with Constant Heat Flux and Surrounding Shields

EXAMPLE 8-18

In Figure Example 8-18, an electric heater is installed in surface 1 such that a constant heat flux of 100 kW/m² is generated at the surface. The four surrounding surfaces are in radiant balance with surface 1 and a large surrounding room at 20°C. The surface properties are $\epsilon_1 = 0.8$ and $\epsilon_2 = \epsilon_3 = \epsilon_4 = \epsilon_5 = 0.4$. Determine the temperatures of all surfaces. The back side of surface 1 is insulated. Repeat the calculation assuming surfaces 2, 3, 4, and 5 are just one surface uniform in temperature.

Figure Example 8-18

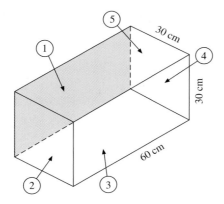

■ **Solution**

In reality, surfaces 2, 3, 4, and 5 have *two* surfaces each; an inside and an outside surface. We thus have *nine* surfaces plus the room, so a 10-body problem is involved. Of course, from symmetry we can see that $T_2 = T_4$ and $T_3 = T_5$, but we set up the problem in the general numerical formulation. We designate the large room as surface 6 and it behaves as if $\epsilon_6 = 1.0$. So, it is as if the opening were covered with a black surface at 20°C. The shape factors of the inside surfaces are obtained from Figures 8-12 and 8-14:

$$F_{16} = F_{61} = 0.285 \qquad F_{13} = F_{15} = 0.24 = F_{31} = F_{51}$$
$$F_{12} = F_{14} = 0.115 \qquad F_{24} = F_{42} = 0.068$$
$$F_{35} = F_{53} = 0.285 \qquad F_{32} = F_{52} = F_{34} = 0.115$$
$$F_{25} = F_{23} = F_{45} = F_{43} = F_{21} = F_{41} = F_{26} = F_{46} = 0.23$$
$$F_{11} = F_{22} = F_{33} = F_{44} = F_{55} = 0$$

For the *outside* surfaces,

$$F'_{26} = F'_{36} = F'_{46} = F'_{56} = 1.0$$

where the primes indicate the outside surfaces. We shall also use primes to designate the radiosities of the outside surfaces. For the room, $J_6 = E_{b_6} = (5.669 \times 10^{-8})(293)^4 = 417.8 \text{ W/m}^2$.

For surface 1 with constant heat flux, we use Equation (8-l09a) and write

$$J_1 - (F_{12}J_2 + F_{13}J_3 + F_{14}J_4 + F_{15}J_5 + F_{16}J_6) = 1.0 \times 10^5 \qquad [a]$$

Because of the radiant balance condition we have

$$(J_2 - E_{b_2})\frac{\epsilon_2 A_2}{1 - \epsilon_2} = (E_{b_2} - J'_2)\frac{\epsilon_2 A_2}{1 - \epsilon_2}$$

and

$$E_{b_2} = \frac{J_2 + J'_2}{2} \qquad [b]$$

where the prime designates the outside radiosity. A similar relation applies for surfaces 3, 4, and 5. Thus, we can use Equation (8-106a) for *inside* surface 2

$$J_2 - (1 - \epsilon_2)(F_{21}J_1 + F_{23}J_3 + F_{24}J_4 + F_{25}J_5 + F_{26}J_6) = \frac{\epsilon_2}{2}(J_2 + J_2') \qquad [c]$$

and for the *outside* surface 2

$$J_2' - (1 - \epsilon_2)(F_{26}'J_6) = \frac{\epsilon_2}{2}(J_2 + J_2') \qquad [d]$$

Equations like (c) and (d) are written for surfaces 3, 4, and 5 also, and with the shape factors and emissivities inserted the following set of equations is obtained:

$$J_1 - 0.115J_2 - 0.24J_3 - 0.115J_4 - 0.24J_5 = 1.0012 \times 10^5$$
$$-0.138J_1 + 0.8J_2 - 0.2J_2' - 0.138J_3 - 0.0408J_4 - 0.138J_5 = 57.66$$
$$0.2J_2 - 0.8J_2' = -250.68$$
$$-0.144J_1 - 0.069J_2 + 0.8J_3 - 0.2J_3' - 0.069J_4 - 0.05J_5 = 60.16$$
$$0.2J_3 - 0.8J_3' = -250.68$$
$$-0.138J_1 - 0.0408J_2 - 0.138J_3 + 0.8J_4 - 0.2J_4' - 0.138J_5 = 57.66$$
$$0.2J_4 - 0.8J_4' = -250.68$$
$$-0.144J_1 - 0.069J_2 - 0.057J_3 - 0.069J_4 + 0.8J_5 - 0.2J_5' = 60.16$$
$$0.2J_5 - 0.8J_5' = -250.68$$

We thus have nine equations and nine unknowns, which may be solved to give

$$J_1 = 1.24887 \times 10^5 \text{ W/m}^2$$
$$J_2 = J_4 = 37{,}549$$
$$J_2' = J_4' = 9701$$
$$J_3 = J_5 = 33{,}605$$
$$J_3' = J_5' = 8714$$

The temperatures are thus computed from Equation (b):

$$E_{b_2} = \frac{37{,}549 + 9701}{2} = 23{,}625 \quad T_2 = T_4 = 803.5 \text{ K}$$

$$E_{b_3} = \frac{33{,}605 + 8714}{2} = 21{,}160 \quad T_3 = T_5 = 781.6 \text{ K}$$

For surface 1 we observed that

$$\frac{q}{A} = \frac{\epsilon}{1 - \epsilon}(E_{b_1} - J_1)$$

so that

$$E_{b_1} = \frac{(1.0 \times 10^5)(1 - 0.8)}{0.8} + 1.24887 \times 10^5 = 1.49887 \times 10^5$$

and

$$T_1 = 1275 \text{ K}$$

We note again that we could have observed the symmetry of the problem and set $J_2 = J_4$, $J_2' = J_4'$, and so on. By so doing, we could have had only five equations with five unknowns.

■ **Surfaces 2, 3, 4, and 5 as one surface**

We now go back and take surfaces 2, 3, 4, and 5 as one surface, which we choose to call surface 7. The shape factors are then

$$F_{16} = F_{61} = 0.285 \qquad F_{17} = 1 - 0.285 = 0.715$$
$$A_1 = 2.0 \qquad A_7 = 6.0$$

Thus

$$F_{71} = (0.715)\left(\tfrac{2}{6}\right) = 0.2383 = F_{76}$$

$$F_{77} = 1 - (2)(0.2383) = 0.5233 \qquad F'_{76} = 1.0$$

Then for surface 1 we use Equation (8-109a) to obtain

$$J_1 - (F_{17}J_7 + F_{16}J_6) = 1.0 \times 10^5$$

Using $E_{b_7} = (J_7 + J'_7)/2$, we have for the *inside* of surface 7

$$J_7[1 - F_{77}(1 - \epsilon_7)] - (1 - \epsilon_7)(F_{71}J_1 + F_{76}J_6) = \frac{\epsilon_7}{2}(J_7 + J'_7)$$

while for the *outside* we have

$$J'_7 - (1 - \epsilon_7)F'_{76}J_6 = \frac{\epsilon_7}{2}(J_7 + J'_7)$$

When the numerical values are inserted, we obtain the set of three equations:

$$J_1 - 0.715J_7 = 1.0012 \times 10^5$$

$$-0.143J_1 + 0.486J_7 - 0.2J'_7 = 59.74$$

$$0.2J_7 - 0.8J'_7 = -250.68$$

which has the solution

$$J_1 = 1.31054 \times 10^5 \text{ W/m}^2$$

$$J_7 = 43{,}264$$

$$J'_7 = 11{,}129$$

The temperatures are then calculated as before:

$$E_{b_7} = \frac{43{,}264 + 11{,}129}{2} = 27{,}197 \qquad T_7 = 832.2 \text{ K}$$

$$E_{b_1} = \frac{(1.0 \times 10^5)(1 - 0.8)}{0.8} + 1.31054 \times 10^5 = 1.65054 \times 10^5$$

$$T_1 = 1306 \text{ K}$$

So, there is about a 30 K temperature difference between the two methods.

■ **Comment**

With such a small difference between the solutions we may conclude that the extra complexity of choosing each surface at a different radiosity is probably not worth the effort, particularly when one recognizes the uncertainties that are present in the surface emissivities. *This points out that our assumptions of uniform irradiation and radiosity, though strictly not correct, give answers that are quite satisfactory.*

Numerical Solution for Combined Convection and Radiation (Nonlinear System)   **EXAMPLE 8-19**

A 0.5 by 0.5 m plate is maintained at 1300 K and exposed to a convection and radiation surrounding at 300 K. Attached to the top are two radiation shields 0.5 by 0.5 m as shown in Figure Example 8-19(a). The convection heat-transfer coefficient for all surfaces is 50 W/m$^2 \cdot$K, and $\epsilon_1 = 0.8$, $\epsilon_2 = 0.3 = \epsilon_3$. Determine the total heat lost from the 1300 K surface and the temperature of the shields.

Figure Example 8-19

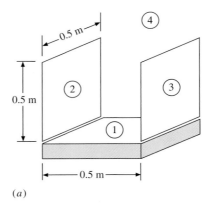

(a)

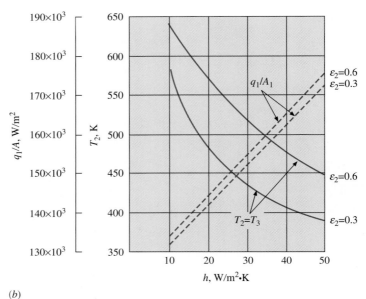

(b)

■ **Solution**

This example illustrates how it is possible to handle convection-radiation problems with the numerical formulation and an iterative computational procedure. Nomenclature is shown in the figure. Using Figures 8-12 and 8-14, we can evaluate the shape factors as

$$F_{12} = F_{13} = 0.2 \qquad F_{14} = 1 - 0.2 - 0.2 = 0.6$$
$$F_{23} = F_{32} = 0.2 \qquad F_{24L} = F_{34R} = 1.0$$
$$F_{21} = F_{12} = F_{31} = 0.2 \qquad F_{24R} = F_{34L} = 0.6$$
$$F_{11} = F_{22} = F_{33} = 0$$
$$J_{2R} = J_{3L} \qquad J_{2L} = J_{3R} \qquad \text{from symmetry}$$
$$J_4 = E_{b_4}$$

We now use Equation (8-105) to obtain a relation for J_1:

$$J_1 = (1 - \epsilon_i)[F_{12}J_{2R} + F_{13}J_{3L} + F_{14}J_4] + \epsilon_1 E_{b_1}$$

But $J_{2R} = J_{3L}$ and $F_{12} = F_{13}$ so that

$$J_1 = (1 - \epsilon_1)(2F_{13}J_{2R} + F_{14}J_4) + \epsilon_1 E_{b_1} \qquad [a]$$

We use Equation (8-108) for the overall energy balance on surface 2:

$$2h(T_\infty - T_2) = \frac{\epsilon_2}{1 - \epsilon_2}(E_{b_2} - J_{2R}) + \frac{\epsilon_2}{1 - \epsilon_2}(E_{b_2} - J_{2L})$$

$$= \frac{\epsilon_2}{1 - \epsilon_2}(2E_{b_2} - J_{2R} - J_{2L}) \qquad [b]$$

Equation (8-105) is used for surface J_{2R}.

$$J_{2R} = (1 - \epsilon_2)(F_{21}J_1 + F_{23}J_{3L} + F_{24R}J_4) + \epsilon_2 E_{b_2}$$

But $J_{2R} = J_{3L}$ so that

$$J_{2R} = \frac{1}{1 - (1 - \epsilon_2)F_{23}}[(1 - \epsilon_2)(F_{21}J_1 + F_{24R}J_4) + \epsilon_2 E_{b_2}] \qquad [c]$$

For surface J_{2L} the equation is

$$J_{2L} = (1 - \epsilon_2)(F_{24L}J_4) + \epsilon_2 E_{b_2} \qquad [d]$$

Equation (b) is nonlinear in E_b so we must use an iterative method to solve the set. Such procedures are described in Reference 34. Applying the iteration technique, we obtain the final solution set as

$$J_1 = 1.3135 \times 10^5 \quad J_{2R} = 22{,}051$$

$$J_{2L} = 710 \quad E_{b_2} = 1275 \quad T_2 = 386.6 \text{ K}$$

The total heat flux lost by surface 1 is

$$\frac{q_1}{A_1} = h(T_1 - T_\infty) + (E_{b_1} - J_1)\frac{\epsilon_1}{1 - \epsilon_1}$$

$$= 1.7226 \times 10^5 \text{ W/m}^2 \qquad [e]$$

For a 0.5 by 0.5 m surface the heat lost is thus

$$q_1 = (1.7226 \times 10^5)(0.5)^2 = 43{,}065 \text{ W}$$

Other cases may be computed, and the influence that h and ϵ_2 have on the results is shown in the accompanying figure.

■ **Comment**

This example illustrates how nonlinear equations resulting from combined convection and radiation can be solved with an iterative procedure.

8-14 | SOLAR RADIATION

Solar radiation is a form of thermal radiation having a particular wavelength distribution. Its intensity is strongly dependent on atmospheric conditions, time of year, and the angle of incidence for the sun's rays on the surface of the earth. At the outer limit of the atmosphere the total solar irradiation when the earth is at its mean distance from the sun is 1395 W/m². This number is called the *solar constant* and is subject to modification upon collection of more precise experimental data.

Not all the energy expressed by the solar constant reaches the surface of the earth, because of strong absorption by carbon dioxide and water vapor in the atmosphere. The

Figure 8-63 | Solar radiation compared to normalized blackbody spectrum with peak at 0.5 μm. Atmospheric data according to Reference 15. Peak of blackbody spectrum for 10,000°R has been normalized to match solar spectrum outside atmosphere at 0.5 μm wavelength.

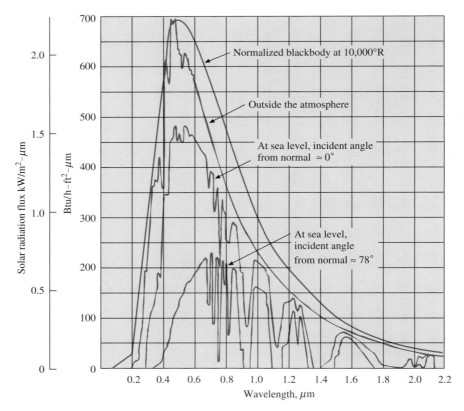

solar radiation incident on the earth's surface is also dependent on the atmospheric content of dust and other pollutants. The maximum solar energy reaches the surface of the earth when the rays are directly incident on the surface since (1) a larger view area is presented to the incoming solar flux and (2) the solar rays travel a smaller distance through the atmosphere so that there is less absorption than there would be for an incident angle tilted from the normal. Figure 8-63 indicates the atmospheric absorption effects for a sea-level location on clear days in a moderately dusty atmosphere with moderate water-vapor content.

It is quite apparent from Figure 8-63 that solar radiation that arrives at the surface of the earth does not behave like the radiation from an ideal gray body, while outside the atmosphere the distribution of energy follows more of an ideal pattern. To determine an equivalent blackbody temperature for the solar radiation, we might employ the wavelength at which the maximum in the spectrum occurs (about 0.5 μm, according to Figure 8-63) and Wien's displacement law [Equation (8-13)]. This estimate gives

$$T \approx \frac{2987.6}{0.5} = 5795 \text{ K} \qquad (10{,}431°\text{R})$$

The equivalent solar temperature for thermal radiation is therefore about 5800 K (10,000°R).

If all materials exhibited gray-body behavior, solar-radiation analysis would not present a particularly unusual problem; however, since solar radiation is concentrated at short wavelengths, as opposed to much longer wavelengths for most "earth-bound" thermal radiation,

Table 8-4 | Comparisons of absorptivities of various surfaces to solar and low-temperature thermal radiation as compiled from Reference 14.

Surface	Absorptivity	
	For solar radiation $(\lambda \sim 0.5\ \mu\text{m})$	For low-temperature radiation $\sim 25°\text{C}$ $(\lambda \sim 10\ \mu\text{m})$
Aluminum, highly polished	0.15	0.04
Copper, highly polished	0.18	0.03
Tarnished	0.65	0.75
Cast iron	0.94	0.21
Stainless steel, no. 301, polished	0.37	0.60
White marble	0.46	0.95
Asphalt	0.90	0.90
Brick, red	0.75	0.93
Gravel	0.29	0.85
Flat black lacquer	0.96	0.95
White paints, various types of pigments	0.12–0.16	0.90–0.95

a particular material may exhibit entirely different absorptance and transmittance properties for the two types of radiation. The classic example of this behavior is a greenhouse. Ordinary glass transmits radiation very readily at wavelengths below 2 μm; thus it transmits the large part of solar radiation incident upon it. This glass, however, is essentially opaque to long-wavelength radiation above 3 or 4 μm. Practically all the low-temperature radiation emitted by objects in the greenhouse is of such a long-wavelength character that it remains trapped in the greenhouse. Thus the glass allows much more radiation to come in than can escape, thereby producing the familiar heating effect. The solar radiation absorbed by objects in the greenhouse must eventually be dissipated to the surroundings by convection from the outside walls of the greenhouse.

Similar behavior is observed for the absorptance and reflectance of solar, as opposed to low-temperature, radiation from opaque metal or painted surfaces. In many instances, the total absorptivity for solar radiation can be quite different from the absorptivity for blackbody radiation at some moderate temperature like 25°C. Table 8-4 gives a brief comparison of the absorptivities for some typical surfaces for both solar and low-temperature radiation. As will be noted, rather striking differences can occur.

This brief discussion of solar radiation is not intended to be comprehensive. Rather, it has the purpose of alerting the reader to some of the property information (like that of Reference 14) when making calculations for solar radiation.

Solar–Environment Equilibrium Temperatures   EXAMPLE 8-20

Calculate the radiation equilibrium temperature for a plate exposed to a solar flux of 700 W/m^2 and a surrounding temperature of 25°C if the surface is coated with (*a*) white paint or (*b*) flat black lacquer. Neglect convection.

■ **Solution**
At radiation equilibrium the net energy absorbed from the sun must equal the long-wavelength radiation exchange with the surroundings, or

$$\left(\frac{q}{A}\right)_{\text{sun}} \alpha_{\text{sun}} = \alpha_{\text{low temp}}\sigma(T^4 - T^4_{\text{surr}}) \qquad [a]$$

For white paint we obtain from Table 8-4

$$\alpha_{sun} = 0.12 \qquad \alpha_{low\ temp} = 0.9$$

so that Equation (a) becomes

$$(700)(0.12) = (0.9)(5.669 \times 10^{-8})(T^4 - 298^4)$$

and

$$T = 312.5\ K = 39.5°C \quad [103°F]$$

For flat black lacquer we obtain

$$\alpha_{sun} = 0.96 \qquad \alpha_{low\ temp} = 0.95$$

so that Equation (a) becomes

$$(700)(0.96) = (0.95)(5.669 \times 10^{-8})(T^4 - 298^4)$$

and

$$T = 377.8\ K = 104.8°C \quad [220.6°F]$$

We conclude from this example what we may have known from the start, that some white surfaces are cooler than black surfaces in the sunlight.

Influence of Convection on Solar Equilibrium Temperatures

EXAMPLE 8-21

The surfaces in Example 8-20 are also exposed to a convection environment at 25°C having $h = 10\ W/m^2 \cdot °C$, in addition to the radiation surrounding. Calculate the radiation–convection equilibrium temperatures under these conditions.

■ Solution
In this case the solar energy absorbed must equal the sum of the radiation and convection transfers to the surroundings, or

$$(q/A)_{sun}\alpha_{sun} = \alpha_{low\ temp}\sigma(T^4 - T_{surr}^4) + h(T - T_{surr}) \qquad \textbf{[a]}$$

For the white paint, using the same surface properties as in Example 8-20 gives

$$(700)(0.12) = (0.9)(5.669 \times 10^{-8})(T^4 - 298^4) + (10)(T - 298)$$

which has the solution

$$T = 303.4\ K = 30.4°C \quad (86.7°F)$$

For the flat black surface the properties may be inserted in Equation (a) to give

$$(700)(0.96) = (0.95)(5.669 \times 10^{-8})(T^4 - 298^4) + (10)(T - 298)$$

which has a corresponding solution of

$$T = 337.8\ K = 64.8°C \quad (149°F)$$

As in Example 8-20 the black surface is hotter in the sunlight than the white surface, but the convection loss lowers the temperatures considerably.

The reader will note that the value of the convection coefficient, $h = 10\ W/m^2 \cdot °C$, is approximately what one would expect for free convection from a hot surface as calculated with the relations of Chapter 7.

A Flat-Plate Solar Collector | **EXAMPLE 8-22**

A flat-plate solar collector is constructed as shown in Figure Example 8-22a. A glass plate covers the blackened surface, which is insulated. Solar energy at the rate of 750 W/m^2 is transmitted through the glass and absorbed in the blackened surface. The surface heats up and radiates to the glass and also loses heat by convection across the air gap, which has $k_e/k = 1.4$. The outside surface of the glass loses heat by radiation and convection to the environment at 30°C with $h = 20$ W/m$^2 \cdot$°C. It is assumed that the glass does not transmit any of the thermal radiation and has $\epsilon = 0.9$. The blackened surface is assumed to have $\epsilon = 1.0$ for all radiation. Determine the temperatures of the glass and inside surface.

Figure Example 8-22a

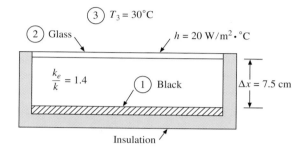

■ Solution

This is an interesting example of combined radiation and convection heat-transfer analysis. We designate the black plate as surface 1, the glass as surface 2, and the surroundings as surface 3. We assume no absorption of the solar energy in the glass.
For the black plate

$$J_1 = E_{b1}$$

The convection and solar energy delivered to surface 1 is

$$\left.\frac{q}{A}\right]_1 = \frac{k_e}{\Delta x}(T_2 - T_1) + \left.\frac{q}{A}\right]_s \qquad [a]$$

If Equation (8-109) is now applied, we have

$$E_{b1} - F_{12}J_{2i} = \frac{k_e}{\Delta x}(T_2 - T_1) + \left.\frac{q}{A}\right]_s \qquad [b]$$

where J_{2i} is the inside radiosity for the glass. The overall energy balance for surface 2 is

$$\frac{\epsilon_2}{1 - \epsilon_2}(2E_{b2} - J_{2i} - J_{2o}) = \frac{k_e}{\Delta x}(T_1 - T_2) + h(T_3 - T_2) \qquad [c]$$

where, now, J_{2o} is the outside radiosity of the glass.
    For the overall system, the solar energy absorbed must eventually be lost by convection and radiation from the outside surface of the glass. Thus,

$$\left.\frac{q}{A}\right]_s = h(T_2 - T_3) + \epsilon_2(E_{b2} - E_{b3}) \qquad [d]$$

Finally, the radiation lost from the outside of the glass can be written two ways:

$$\left.\frac{q}{A}\right]_{rad} = \epsilon_2(E_{b2} - E_{b3}) = (E_{b2} - J_{2o})\frac{\epsilon_2}{1 - \epsilon_2} \qquad [e]$$

The area of the collector is very large compared to the spacing so $F_{12} \approx 1.0$. We now have four equations and four unknowns: E_{b1}, E_{b2}, J_{2i}, and J_{2o}. Of course, T_1 and T_2 are expressed in terms of E_{b1} and E_{b2}.

The above equations may be rearranged algebraically into the set

$$E_{b1} = \sigma T_1^4$$

$$E_{b2} = \sigma T_2^4$$

$$\epsilon_2(E_{b1} + E_{b3} - 2E_{b2}) + \frac{k_e}{\Delta x}(T_1 - T_2) + h(T_3 - T_2) = 0$$

$$\epsilon_2(E_{b2} - E_{b1}) + \frac{k_e}{\Delta x}(T_2 - T_1) + \frac{q}{A}_{\text{solar}} = 0$$

This set may be solved to yield

$$T_1 = 401.4\,\text{K}$$

$$T_2 = 331.3\,\text{K}$$

A performance chart for the collector may be calculated very simply by reducing the solar input ($750\,\text{W/m}^2$) by the amount of extracted energy and recomputing the values of T_1 and T_2 under the new condition. The results are shown in Figure Example 8-22b. Note that the adiabatic condition, that is, zero load, produces a collector temperature of 401.4 K. Extensive information on solar collectors is given in Reference 31.

Figure Example 8-22b

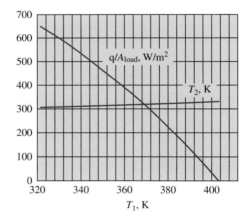

8-15 | RADIATION PROPERTIES OF THE ENVIRONMENT

We have already described the radiation spectrum of the sun and noted that the major portion of solar energy is concentrated in the short-wavelength region. It was also noted that as a consequence of this spectrum, real surfaces may exhibit substantially different absorption properties for solar radiation than for long-wavelength "earthbound" radiation.

Meteorologists and hydrologists use the term *insolation* to describe the intensity of direct solar radiation incident on a horizontal surface per unit area and per unit time, designated with the symbol I. Insolation is analogous to the term irradiation employed for incident "earthbound" radiation.

Although we shall emphasize other units, it will be helpful to mention a unit that appears in the meterological literature:

$$1 \ langley \ (Ly) = 1 \ cal/cm^2 \quad [41.86 \ kJ/m^2]$$

Insolation and radiation intensity are frequently expressed in langleys per unit time; for example, the Stefan-Boltzmann constant would be

$$\sigma = 0.826 \times 10^{-10} \ Ly/min \cdot K^4$$

Radiation heat transfer in the environment is governed by the absorption, scattering, and reflection properties of the atmosphere and natural surfaces. Two types of scattering phenomena occur in the atmosphere. *Molecular scattering* is observed because of the interaction of radiation with individual molecules. The blue color of the sky results from the scattering of the violet (short) wavelengths by the air molecules. *Particulate scattering* in the atmosphere results from the interaction of radiation with the many types of particles that may be suspended in the air. Dust, smog, and water droplets are all important types of particulate scattering centers. The scattering process is governed mainly by the size of the particle in comparison with the wavelength of radiation. Maximum scattering occurs when wavelength and particle size are equal and decreases progressively for longer wavelengths. For wavelengths smaller than the particle size, the radiation tends to be reflected.

Reflection phenomena in the atmosphere occur for wavelengths less than the particle size and are fairly independent of wavelength in this region. The term *albedo* is used to describe the reflective properties of surfaces and is defined by

$$A = albedo = \frac{reflected \ energy}{incident \ energy} \qquad \textbf{[8-111]}$$

The albedo of the surface is dependent on the angle the incoming solar rays make with the surface. The albedo could also be termed as reflectivity. The albedos of some natural surfaces are given in Table 8-5.

The atmosphere absorbs radiation quite selectively in narrow-wavelength bands. The absorption for solar radiation occurs in entirely different bands from the absorption of the radiation from the earth because of the different spectrums for the two types of radiation.

Table 8-5 | Albedos for some natural surfaces.

Surface	Albedo
Water	0.03–0.4
Black dry soil	0.14
Black moist soil	0.08
Gray dry soil	0.25–0.30
Gray moist soil	0.10–0.12
Desert loam	0.29–0.31
Bright fine sand	0.37
Snow	0.4–0.85
Sea ice	0.36–0.50
Spring wheat	0.10–0.25
Green grass	0.26
Cotton	0.20–0.22
Lettuce	0.22
Potatoes	0.19

Figure 8-64 | Thermal-radiation spectrums for the sun and earth with primary absorption bands indicated by shaded areas. Note different scales; $1 \text{ erg/cm}^2 \cdot \text{s} = 1 \text{ mW/m}^2$.

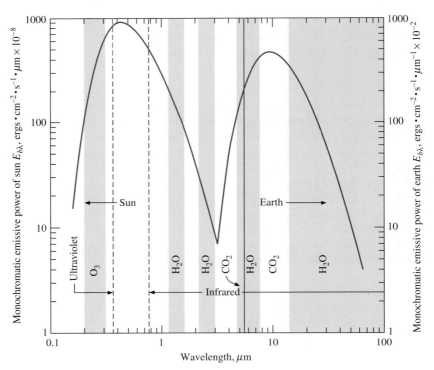

In Figure 8-64 we see the approximate spectrums for solar and earth radiation with some important absorption bands superimposed on the diagram. Note the scale differential on the two curves. A quick inspection of these curves will show that the atmosphere transmits most of the short-wavelength radiation while absorbing most of the back radiation from the earth. Therefore, the atmosphere acts very much like a greenhouse, trapping the incoming solar radiation to provide energy and warmth for humans on earth. Some concern is voiced that people may upset the energy budget of the earth through excessive contamination of the atmosphere with pollutants. Such a possibility may exist but is beyond the scope of our discussion.

The insolation at the outer edge of the atmosphere is expressed in terms of the solar constant E_{b_o}, by

$$I_o = E_{b_o} \sin \alpha \qquad \text{[8-112]}$$

where α is again the angle the rays make with the horizontal. We gave the solar constant as

$$E_{bo} = 442.4 \text{ Btu/h} \cdot \text{ft}^2$$
$$= 2.00 \text{ cal/cm}^2 \cdot \text{min}$$
$$= 1395 \text{ W/m}^2$$

Average values of incident solar radiation as a function of solar altitude angle are indicated in Table 8-6.

Table 8-6 | Average solar insolation (irradiation) on a horizontal surface under average atmospheric conditions.

Solar altitude α, deg	Average total insolation	
	Ly/h	W/m²
5	3.6	41.9
10	9.7	112.8
15	17.2	200.0
20	25.0	290.7
25	32.8	381.4
30	40.6	472.1
35	47.7	554.6
40	54.7	636.0
45	61.1	710.4
50	67.2	781.4
60	77.5	901.1
70	85.3	991.8
80	89.7	1043
90	91.4	1063

8-16 | EFFECT OF RADIATION ON TEMPERATURE MEASUREMENT

When a thermometer is placed in a gas-flow stream to measure temperature, the temperature indicated by the sensing element is determined by the overall energy balance on the element. Consider the element shown in Figure 8-65. The temperature of the gas is T_∞, the effective radiation surrounding temperature is T_s, and the temperature indicated by the thermometer is T_t. Assuming that T_∞ is greater than T_s, energy will be transferred by convection to the thermometer and then dissipated by radiation to the surroundings. Thus the energy balance becomes

$$hA(T_\infty - T_t) = \sigma A\epsilon(T_t^4 - T_s^4) \qquad \textbf{[8-113]}$$

where A is the surface area of the element and ϵ is its emissivity. Equation (8-113) assumes that the surroundings are either very large or black so that Equation (8-43a) can be applied for the radiation heat transfer.

From this energy balance we see that the temperature indicated by the thermometer is *not* the true gas temperature but some radiation-convection equilibrium temperature.

Figure 8-65 | Thermometer element in flow stream.

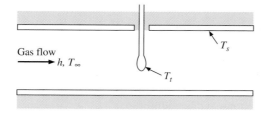

Very large errors can result in temperature measurements if this energy balance is not properly taken into account. Radiation shields are frequently employed to alleviate this difficulty.

| **EXAMPLE 8-23** | Temperature Measurement Error Caused by Radiation |

A mercury-in-glass thermometer having $\epsilon = 0.9$ hangs in a metal building and indicates a temperature of 20°C. The walls of the building are poorly insulated and have a temperature of 5°C. The value of h for the thermometer may be taken as 8.3 W/m² · °C. Calculate the true air temperature.

■ **Solution**
We employ Equation (8-119) for the solution:

$$h(T_\infty - T_t) = \sigma\epsilon(T_t^4 - T_s^4)$$

Inserting the numerical values, with $T_t = 20°C = 293$ K, $T_s = 5°C = 278$ K, gives

$$(8.3)(T_\infty - 293) = (5.669 \times 10^{-8})(0.9)(293^4 - 278^4)$$

and

$$T_\infty = 301.6 \text{ K} = 28.6°C \quad [83.5°F]$$

In this simple example the thermometer is in error by 8.6°C [15.5°F]!

8-17 | THE RADIATION HEAT-TRANSFER COEFFICIENT

In the development of convection heat transfer in the preceding chapters, we found it convenient to define a heat-transfer coefficient by

$$q_{\text{conv}} = h_{\text{conv}} A(T_w - T_\infty)$$

Since radiation heat-transfer problems are often very closely associated with convection problems, and the total heat transfer by both convection and radiation is often the objective of an analysis, it is worthwhile to put both processes on a common basis by defining a radiation heat-transfer coefficient h_r as

$$q_{\text{rad}} = h_r A_1(T_1 - T_2)$$

where T_1 and T_2 are the temperatures of the two bodies exchanging heat by radiation. The total heat transfer is then the sum of the convection and radiation,

$$q = (h_c + h_r)A_1(T_w - T_\infty) \tag{8-114}$$

if we assume that the second radiation-exchange surface is an enclosure and is at the same temperature as the fluid. For example, the heat loss by free convection and radiation from a hot steam pipe passing through a room could be calculated from Equation (8-114). Handbooks frequently tabulate "total" heat-transfer coefficients for surfaces exposed and enclosed by various environmental conditions.

In many instances the convection heat-transfer coefficient is not strongly dependent on temperature. However, this is not so with the radiation heat-transfer coefficient. The value of h_r, corresponding to Equation (8-43), could be calculated from

$$\frac{q}{A_1} = \frac{\sigma(T_1^4 - T_2^4)}{1/\epsilon_1 + (A_1/A_2)(1/\epsilon_2 - 1)} = h_r(T_1 - T_2)$$

$$h_r = \frac{\sigma(T_1^2 + T_2^2)(T_1 + T_2)}{1/\epsilon_1 + (A_1/A_2)(1/\epsilon_2 - 1)}$$

[8-115]

The corresponding R value would be

$$R_{\text{rad}} = \frac{1}{h_r}$$

and is frequently employed for calculation of radiation heat transfer across gap-spaces in building applications. Such calculations are made in combination with the information presented in Section (7-11) for free convection in enclosed spaces. Obviously, the radiation coefficient is a very strong function of temperature.

The reader may recall that we used a concept like Equation (8-115) to obtain a "radiation resistance" for numerical examples in Chapters 3 and 4.

8-18 | SUMMARY

In this chapter we have examined several means for analyzing radiation heat transfer. The gray-body assumption, although not strictly correct, is a viable method for performing heat-transfer calculations. Assumptions of uniform radiosity and irradiation over surfaces are also not strictly correct but provide an approximation that is usually well within the accuracy of knowledge of surface properties. In Table 8-7, we present a tabular summary of a few formulas that are often used.

Table 8-7 | Summary of radiation formulas.

Type	Equation	Equation number
Energy emitted by a black body	$E_b = \sigma T^4 \qquad \text{W/m}^2$	(8-3)
Basic radiation shape factor reciprocity relation	$A_i F_{ij} = A_j F_{ji}$	(8-18a)
Net energy lost by a gray surface	$q = \frac{E_b - J}{(1-\epsilon)/\epsilon A} \qquad \text{W}$	(8-38)
Net radiant exchange between convex surface 1 and large enclosure 2	$q = \sigma A_1 \epsilon_1 (T_1^4 - T_2^4) \qquad \text{W}$	(8-43a)
Radiation-balance equation for constant surface temperature of ith surface	$J_i[1 - F_{ii}(1 - \epsilon_i)] - (1 - \epsilon_i)\sum_{j \neq i} F_{ij} J_j = \epsilon_i E_{bi}$	(8-106a)
Radiation-balance equation for surface in radiant balance; i.e., $q/A = 0$	$J_i(1 - F_{ii}) - \sum_{j \neq i} F_{ij} J_j = 0$	(8-107a)
Radiation-balance equation for surface with specified heat flux	$J_i(1 - F_{ii}) - \sum_{j \neq i} F_{ij} J_j = \frac{q_i}{A_i}$	(8-109a)

REVIEW QUESTIONS

1. How does thermal radiation differ from other types of electromagnetic radiation?
2. What is the Stefan-Boltzmann law?
3. Distinguish between specular and diffuse surfaces.
4. Define radiation intensity.
5. What is Kirchhoff's identity? When does it apply?
6. What is a gray body?
7. What is meant by the radiation shape factor?
8. Define irradiation and radiosity.
9. What is Beer's law?
10. Why do surfaces absorb differently for solar or earthbound radiation?
11. Explain the greenhouse effect.
12. Why is the sky blue?
13. Define albedo.
14. What is meant by the atmospheric greenhouse effect?

LIST OF WORKED EXAMPLES

8-1 Transmission and absorption in a glass plate
8-2 Heat transfer between black surfaces
8-3 Shape-factor algebra for open ends of cylinders
8-4 Shape-factor algebra for truncated cone
8-5 Shape-factor algebra for cylindrical reflector
8-6 Hot plates enclosed by a room
8-7 Surface in radiant balance
8-8 Open hemisphere in large room
8-9 Effective emissivity of finned surface
8-10 Heat-transfer reduction with parallel-plate shield
8-11 Open cylindrical shield in large room
8-12 Network for gas radiation between parallel plates
8-13 Cavity with transparent cover
8-14 Transmitting and reflecting system for furnace opening
8-15 Numerical solution for enclosure
8-16 Numerical solutions for parallel plates
8-17 Radiation from a hole with variable radiosity
8-18 Heater with constant heat flux and surrounding shields
8-19 Numerical solution for combined convection and radiation (nonlinear system)
8-20 Solar–environment equilibrium temperatures
8-21 Influence of convection on solar equilibrium temperatures
8-22 A flat-plate solar collector
8-23 Temperature measurement error caused by radiation

PROBLEMS

8-1 Fused quartz transmits 90 percent of the incident thermal radiation between 0.2 and 4 μm. Suppose a certain heat source is viewed through a quartz window. What heat flux in watts will be transmitted through the material from blackbody radiation sources at (*a*) 800°C, (*b*) 550°C, (*c*) 250°C, and (*d*) 70°C?

8-2 Repeat Problem 8-1 for synthetic sapphire, which has a transmissivity of 0.85 between 0.2 and 5.5 μm.

8-3 Repeat Problem 8-1 for cesium iodide, which has a transmissivity of approximately 0.92 between 0.3 and 52 μm.

8-4 Calculate the energy emitted between 4 and 15 μm by a gray body at 100°F with $\epsilon = 0.6$.

8-5 A furnace with black interior walls maintained at 1100°C has an opening in the side covered with a glass window having the following properties:

$$0 < \lambda < 3 \ \mu\text{m}: \qquad \tau = 0.8 \qquad \epsilon = 0.2 \qquad \rho = 0$$
$$3 \ \mu\text{m} < \lambda < \infty: \qquad \tau = 0 \qquad \epsilon = 0.8 \qquad \rho = 0.2$$

Assume diffuse behavior and calculate the radiation lost through the window to a large room at 25°C.

8-6 A certain surface has the following absorption properties:

$$\alpha_\lambda = 0.05 \qquad 0 < \lambda < 1.2 \ \mu\text{m}$$
$$\alpha_\lambda = 0.5 \qquad 1.2 < \lambda < 3 \ \mu\text{m}$$
$$\alpha_\lambda = 0.4 \qquad 3 < \lambda < 6 \ \mu\text{m}$$
$$\alpha_\lambda = 0.2 \qquad 6 < \lambda < 20 \ \mu\text{m}$$
$$\alpha_\lambda = 0 \qquad 20 < \lambda < \infty \ \mu\text{m}$$

Calculate the total absorptivity of the surface if it is irradiated with blackbody radiation at (*a*) 300 K, (*b*) 500 K, (*c*) 1000 K, (*d*) 2000 K, (*e*) 5000 K.

8-7 Assuming solar radiation is like a blackbody at 5795 K, calculate the fraction of energy in the following wavelength bands: (*a*) 0 to 0.2 μm, (*b*) 0.2 to 0.4 μm, (*c*) 0.4 to 1.0 μm, (*d*) 1.0 to 2.0 μm, (*e*) over 2.0 μm.

8-8 Color photographic films are designed for particular wavelength sensitivities such as "daylight" and "indoor incandescent lighting." An incandescent lightbulb radiates approximately as a blackbody at 3200 K, while daylight is approximately like a blackbody at 5800 K. Make some calculations and comment on the results that might be obtained if a film is used for other than its intended light source, that is to say, daylight film with indoor lighting.

8-9 A 2.5-cm-diameter pipe carrying condensing steam at 101 kPa passes through the center of an infinite plate having a thickness of 5 cm and $k = 0.1$ W/m · °C. The plate is exposed to room air at 27°C with a convection coefficient of $h = 5.1$ W/m^2 · °C on both sides and is sprayed with a flat black paint. Calculate the heat lost by the steam pipe per meter of length. Compare with the heat that would have been lost by a black steam pipe, not placed in the insulating slab.

8-10 A black surface is at 800°C. Calculate the fraction of the total energy emitted between (*a*) 1 and 2 μm, (*b*) 2 and 3 μm, (*c*) 3 and 4 μm, (*d*) 5 and 6 μm.

8-11 A black radiation source is at 1100°C. Calculate the upper wavelength in micrometers for emissions of (*a*) 25, (*b*) 50, (*c*) 75, and (*d*) 98 percent of the total radiation.

8-12 A certain surface maintained at 1400 K has the following spectral emissive characteristics:

$$\epsilon_\lambda = 0.08 \qquad 0 < \lambda < 0.6 \ \mu\text{m}$$
$$\epsilon_\lambda = 0.4 \qquad 0.6 < \lambda < 5 \ \mu\text{m}$$
$$\epsilon_\lambda = 0.7 \qquad 5 < \lambda < \infty$$

Calculate the emissive power of the surface.

8-13 A surface has the emissive characteristics shown in Figure P8-13. Calculate the emissive power for the surface maintained at 2000 K.

Figure P8-13

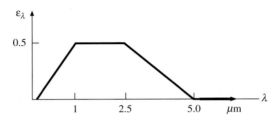

8-14 Find the radiation shape factors F_{1-2} for the situations shown in Figure P8-14.

Figure P8-14

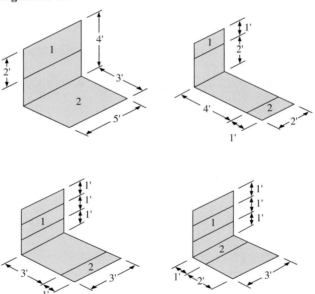

8-15 Two parallel concentric disks have $d_1 = 10$ cm, $d_2 = 5$ cm and are spaced 10 cm apart. Determine F_{12} and F_{21}.

8-16 A 10-cm-diameter disk is placed at the center and parallel to a 25-cm-diameter hemisphere. Designating the sphere as surface 1, the disk as surface 2, and a large surrounding room as surface 3, calculate all the shape factors.

8-17 Two concentric cylinders have diameters of 10 and 25 cm and a length of 7 cm. Calculate the shape factor between the open ends.

8-18 Two parallel square plates 30 cm on a side are separated by a distance of 15 cm but the edge of the lower plate lies halfway between the edges of the top plate. Calculate the shape factor for this geometry.

8-19 Two parallel disks having diameters of 50 cm are separated by a distance of 10 cm. One disk also has a 20-cm-diameter hole cut in the center. Find the shape factor from this disk to the one without the hole.

8-20 Determine the shape factors for the geometries shown in Figure P8-20a–i.

Figure P8-20a,b

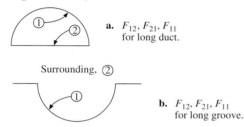

a. F_{12}, F_{21}, F_{11}
for long duct.

b. F_{12}, F_{21}, F_{11}
for long groove.

Figure P8-20c

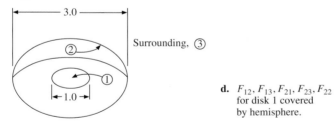

c. F_{11}, F_{13}
for short cylinder
(bottom closed).

Figure P8-20d

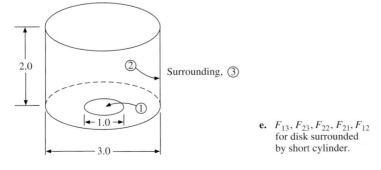

d. $F_{12}, F_{13}, F_{21}, F_{23}, F_{22}$
for disk 1 covered
by hemisphere.

Figure P8-20e

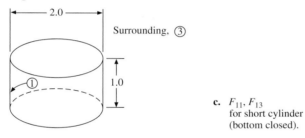

e. $F_{13}, F_{23}, F_{22}, F_{21}, F_{12}$
for disk surrounded
by short cylinder.

Figure P8-20f

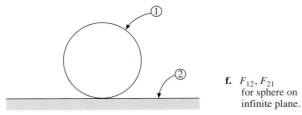

f. F_{12}, F_{21} for sphere on infinite plane.

Figure P8-20g

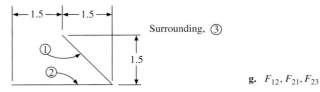

g. F_{12}, F_{21}, F_{23}

Figure P8-20h

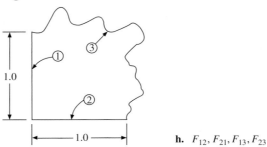

h. $F_{12}, F_{21}, F_{13}, F_{23}$

Figure P8-20i

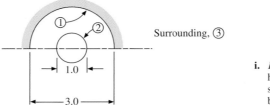

i. $F_{12}, F_{21}, F_{11}, F_{13}$; hemisphere 1, enclosing small sphere 2, surrounded by large enclosure 3.

8-21 Calculate all the shape factors for a cubical enclosure.

8-22 A regular tetrahedron has sides in the shape of equilateral triangles. Calculate all the shape factors.

8-23 A room 4×4 m square by 3 m high has one side wall maintained at 260°C; the floor is maintained at 90°C. The other four surfaces are perfectly insulated. Assume that all surfaces are black. Calculate the net heat transfer between the hot wall and the cool floor.

8-24 Two perfectly black parallel planes 1.2 by 1.2 m are separated by a distance of 1.2 m. One plane is maintained at 800 K and the other at 500 K. The plates are located in a large room whose walls are at 300 K. What is the net heat transfer between the planes?

8-25 Two parallel disks, 60 cm in diameter, are separated by a distance of 15 cm and completely surrounded by a large enclosure at 100°C. The properties of the surfaces are $T_1 = 540°C$, $\epsilon_1 = 0.7$, $T_2 = 300°C$, $\epsilon_2 = 0.5$. What is the net radiant heat transfer with each surface? (Do not include back side exchange, only that from the surfaces facing each other.)

8-26 Rework Problem 8-25 for the case where disk 2 (300°C, $\epsilon_2 = 0.5$) is perfectly specular-reflecting.

8-27 A square room 3 by 3 m has a floor heated to 300 K, a ceiling at 290 K, and walls that are assumed perfectly insulated. The height of the room is 2.5 m. The emissivity of all surfaces is 0.8. Using the network method, find the net interchange between floor and ceiling and the wall temperature.

8-28 Two parallel disks, 50 cm in diameter, are separated by a distance of 12.5 cm and enclosed by a large room at 300 K. One disk is maintained at a constant temperature of 800°C and has an emissivity of $\epsilon = 0.8$. The other disk is perfectly insulated and has an emissivity of $\epsilon = 0.1$. Calculate the net radiant energy lost by the disk maintained at 800°C.

8-29 A 50-cm-square vertical plate is maintained at a constant surface temperature of 127°C on both sides and placed in a large chamber containing air at 1.7 atm and 27°C. The walls of the large chamber are maintained at 22°C and the surface of the vertical plate is blackened on both sides so that its emissivity is 1.0. Calculate the total heat lost by both sides of the vertical plate, expressed in watts.

8-30 It is desired to transmit energy from one spaceship to another. A 1.5-m-square plate is available on each ship to accomplish this. The ships are guided so that the plates are parallel and 30 cm apart. One plate is maintained at 800°C and the other at 280°C. The emissivities are 0.5 and 0.8, respectively. Find (*a*) the net heat transferred between the spaceships in watts and (*b*) the total heat lost by the hot plate in watts. Assume that outer space is a blackbody at 0 K.

8-31 Two parallel planes 90 by 60 cm are separated by a distance of 60 cm. One plane is maintained at a temperature of 800 K and has an emissivity of 0.6. The other plane is insulated. The planes are placed in a large room that is maintained at 290 K. Calculate the temperature of the insulated plane and the energy lost by the heated plane.

8-32 A long pipe 5 cm in diameter passes through a room and is exposed to air at atmospheric pressure and temperature of 20°C. The pipe surface temperature is 93°C. Assuming that the emissivity of the pipe is 0.6, calculate the radiation heat loss per foot of length of pipe.

8-33 A 2.5-cm-diameter pipe carrying condensing steam at 101 kPa passes through the center of a square block of insulating material having $k = 0.04$ W/m · °C. The block is 5 cm on a side and 2 m long. The outside of the block is exposed to room air at 27°C and a convection coefficient of 5.1 W/m² · °C. The outside of the insulation may be assumed black. Calculate the heat lost by the steam pipe.

8-34 A vertical plate 60 cm high and 30 cm wide is maintained at a temperature of 95°C in a room where the air is 20°C and 1 atm. The walls of the rooms are also at 20°C. Assume that $\epsilon = 0.8$ for the plate. How much radiant heat is lost by the plate?

8-35 A horizontal pipe 6 m long and 12.5 cm in diameter is maintained at a temperature of 150°C in a large room where the air is 20°C and 1 atm. The walls of the room are at 38°C. Assume that $\epsilon = 0.7$ for the pipe. How much heat is lost by the pipe through both convection and radiation?

8-36 An oven with a radiant heater for drying painted metal parts on a moving conveyor belt is designed as shown in Figure P8-36. The length of the heating section is 3 m, and the heater temperature is 425°C. The sidewalls are insulating, and it is determined experimentally that the conveyor belt and parts attain a temperature of 120°C. The belt-part combination has an effective emissivity of 0.8, and the radiant-heater surface has $\epsilon = 0.7$. Calculate the energy supplied to the heater. Be sure to consider the radiation that is lost from the ends of the channel. Take the surroundings as a blackbody at 25°C.

Figure P8-36

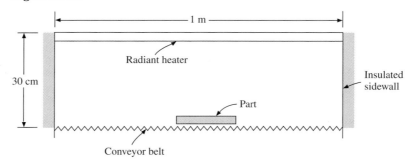

8-37 The top surface of the truncated cone in Example 8-4 has $T_1 = 1200$ K and $\epsilon_1 = 0.75$. The side surface is insulated and has $\epsilon = 0.3$. The bottom is open and exposed to a large room at 30°C. Calculate the temperature of the side surface and the heat lost to the room.

8-38 A long cylinder having a diameter of 2 cm is maintained at 600°C and has an emissivity of 0.4. Surrounding the cylinder is another long, thin-walled concentric cylinder having a diameter of 6 cm and an emissivity of 0.2 on both the inside and outside surfaces. The assembly is located in a large room having a temperature of 27°C. Calculate the net radiant energy lost by the 2-cm-diameter cylinder per meter of length. Also calculate the temperature of the 6-cm-diameter cylinder.

8-39 Two concentric cylinders have lengths of 30 cm. The inner cylinder has a diameter of 8.0 cm. What must the outer-cylinder diameter be such that $F_{12} = 0.8$ where the inner cylinder is considered as surface 1?

8-40 Two parallel 30-cm-diameter disks are separated by a distance of 10 cm and enclosed in a large room at 300 K. The disks are sprayed with a flat black paint such that the surfaces are essentially black. If the disks are maintained at 1200 K and 500 K, respectively, calculate the heat lost by each disk.

8-41 The same two disks as in Problem 8-40 are enclosed by the same large room at 300 K, but the disk previously maintained at 500 K is now allowed to exchange heat with both sides and come into radiation equilibrium with the 1200 K disk and the room. Both sides of the disk are covered with the black paint. Calculate the heat lost by the 1200 K disk under these new conditions and the temperature attained by the disk in radiant equilibrium.

8-42 The bottom surface of a cylindrical furnace has an electric heating element that produces a constant heat flux of 7 kW/m² over a diameter of 75 cm. The sidewalls are 50 cm high and may be considered as nonconducting and re-radiating. The bottom has $\epsilon = 0.8$. Over the top is placed a surface with $\epsilon = 0.6$ that is maintained at 400 K. Determine the temperature of the bottom and side surfaces.

8-43 A long rod heater with $\epsilon = 0.8$ is maintained at 980°C and is placed near a half-cylinder reflector as shown in Figure P8-43. The diameter of the rod is 7.5 cm, and the diameter of the reflector is 50 cm. The reflector is insulated, and the combined heater reflector is placed in a large room whose walls are maintained at 15°C. Calculate the radiant heat loss per unit length of the heater rod. How does this compare with the energy which would be radiated by the rod if it were used without the reflector?

Figure P8-43

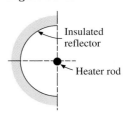

Insulated reflector

Heater rod

8-44 A conical hole is machined in a block of metal whose emissivity is 0.5. The hole is 2.5 cm in diameter at the surface and 7.5 cm deep. If the metal block is heated to 800 K, calculate the radiation emitted by the hole. Calculate the value of an *apparent* emissivity of the hole, defined as the ratio of the actual energy emitted by the hole to that energy which would be emitted by a black surface having an area equal to that of the opening and a temperature equal to that of the inside surfaces.

8-45 A hole 2.5 cm in diameter is drilled in a 7.5-cm metal plate that is maintained at 550 K. The hole is lined with a thin foil having an emissivity of 0.07. A heated surface at 700 K having an emissivity of 0.5 is placed over the hole on one side of the plate, and the hole is left open on the other side of the plate. The 700 K surface is insulated from the plate insofar as conduction is concerned. Calculate the energy emitted from the open hole.

8-46 A cylindrical hole of depth x and diameter d is drilled in a block of metal having an emissivity of ϵ. Using the definition given in Equation (8-47), plot the apparent emissivity of the hole as a function of x/d and ϵ.

8-47 A cylindrical hole of diameter d is drilled in a metal plate of thickness x. Assuming that the radiation emitted through the hole on each side of the plate is due only to the temperature of the plate, plot the apparent emissivity of the hole as a function of x/d and the emissivity of the plate material ϵ.

8-48 A 1-m-diameter cylinder, 1 m long, is maintained at 800 K and has an emissivity of 0.65. Another cylinder, 2 m in diameter and 1 m long, encloses the first cylinder and is perfectly insulated. Both cylinders are placed in a large room maintained at 300 K. Calculate the heat lost by the inner cylinder.

8-49 Two square vertical plates, 10 cm high, are coated with a flat black paint that produces a surface with essentially $\epsilon \approx 1.0$. One plate is maintained at 400 K and a large room at 300 K encloses the assembly. The plates are separated by a distance of 5 cm. Assume that the free convection heat-transfer coefficients may be calculated with the simplified relation in Table 7-2, the room air is at 1 atm and 300 K, and the second plate "floats" in radiant and convective equilibrium. Calculate the heat lost by the 400 K plate and the temperature of the other plate.

8-50 A heated plate with $T = 700$°C and emissivity of 0.8 is placed as shown in Figure P8-50. The plate is 2 by 3 m and 2-m-high walls are placed on each side. Each of these walls is insulated. The whole assembly is placed in a large room at 30°C. Draw the network for this problem assuming that the four walls act as one surface (insulated). Then calculate the heat transfer to the large room.

Figure P8-50

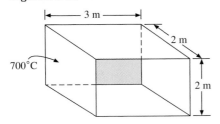

3 m

2 m

700°C

2 m

8-51 A 3.0-cm-diameter sphere is maintained at 1000°C and has an emissivity of 0.6. It is enclosed by another sphere having a diameter of 9.0 cm with an emissivity of 0.3 (inside and outside). Both spheres are enclosed in a large room at 30°C. Calculate the heat lost from the small sphere.

8-52 Two 60 by 60 cm plates are perpendicular with a common edge and are placed in a large room at 30°C. One plate has $T = 600°C$ and an emissivity of 0.65. The other plate is insulated with an emissivity of 0.45. Calculate the temperature of the insulated plate and the heat lost by the 600°C plate.

8-53 Two concentric cylinders having diameters of 10 cm and 20 cm are placed in a large room maintained at 30°C. The length of the cylinders is 10 cm and the inner cylinder is maintained at 700°C with an emissivity of 0.6. The outer cylinder is perfectly insulated and has an emissivity of 0.7. Calculate the heat lost by the inner cylinder.

8-54 A 5-m-square room has a ceiling maintained at 28°C and a floor maintained at 20°C. The connecting walls are 4 m high and perfectly insulated. Emissivity of the ceiling is 0.62 and that of the floor is 0.75. Calculate the heat transfer from ceiling to floor, and the temperature of the connecting walls.

8-55 The inside temperature of a half cylinder is maintained at 1000 K and is enclosed by a large room. The surface emissivity is 0.7. The diameter is 50 cm and the length may be assumed to be very long. The back side of the cylinder is insulated. Calculate the heat loss per unit length if the room temperature is 300 K.

8-56 Two large parallel plates are maintained at 400 and 600°C, respectively, and have surface emissivities of 0.6. A radiation shield having $\epsilon = 0.1$ is placed equidistant between the two heated plates. Calculate the radiant heat transfer through the plate-shield arrangement per m^2 of surface area.

8-57 Two parallel plates 1 m by 1 m are separated by a distance of 1 m. One plate is maintained at a temperature of 800 K and has an emissivity of 0.5. The other plate has an emissivity of 0.2 on both sides and is in radiant balance with the first plate and a large enclosure maintained at 300 K. Draw the radiation network for this problem, and determine all resistances for the network. Note the values on the network. Write any equations necessary to solve for the heat lost by the hot surface. Then solve for the heat lost, expressed in watts.

8-58 Two parallel disks having diameters of 50 cm are separated by a distance of 12.5 cm and placed in a large room at 300 K. One disk is at 1000 K and the other is maintained at 500 K. Both have emissivities of 0.8. Calculate the heat-transfer rate for each disk.

Figure P8-59

8-59 A 50-cm-disk is maintained at a temperature of 600°C with an emissivity of 0.55. Extending from the disk is a radiation shield having an emissivity of 0.1 as shown in Figure P8-59. The arrangement is placed in a large room maintained at 30°C. Calculate the heat lost by the disk and the temperature of the shield.

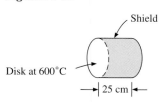

Shield

Disk at 600°C

25 cm

8-60 Two concentric cylinders have properties of $d_1 = 20$ cm, $\epsilon_1 = 0.6$, $T_1 = 500°C$ and $d_2 = 40$ cm, $\epsilon_2 = 0.7$, $T_2 = 100°C$. Both are enclosed in a large room at 20°C. Calculate the length of the cylinder that would enable one to use Equation (8-43) for calculating the heat transfer with no more than a 10 percent error.

8-61 A 30 by 30 cm plate whose emissivity is 0.5 is attached to the side of a spaceship so that it is perfectly insulated from the inside of the ship. Assuming that outer space is a blackbody at 0 K, determine the equilibrium temperature for the plate at a point in space where the radiant heat flux from the sun is 1500 W/m^2. Assume gray-body behavior.

8-62 An artificial satellite 1 m in diameter circles the earth at an altitude of 400 km. Assuming that the diameter of the earth is 12.9 Mm and the outer surface of the satellite is polished aluminum, calculate the radiation equilibrium temperature of the satellite when it is on the "dark" side of the earth. Take the earth as a blackbody at 15°C and outer space as a blackbody at 0 K. The geometric shape factor from the satellite to the earth may be taken as the ratio of the solid angle subtended by the earth to the total solid angle for radiation from the satellite. When the satellite is on the "bright" side of the earth, it is irradiated with a heat flux of approximately 1400 W/m² from the sun. Recalculate the equilibrium temperature of the satellite under these conditions, assuming gray-body behavior. Assume that the satellite receives radiation from the sun as a disk and radiates to space as a sphere.

8-63 A 2-cm-diameter hole, 3 cm deep, is drilled in a block of cast iron. Estimate the apparent emissivity of the freshly drilled hole at room temperature (20°C). Suppose the block is then heated to a temperature of 930°C. Estimate the apparent emissivity of the hole under these new conditions.

8-64 Suppose the freshly drilled hole in Problem 8-63 is sprayed with black shiny lacquer. Calculate the apparent emissivity of the hole under these conditions.

8-65 Two 30 by 30 cm vertical plates are separated by a distance of 10 cm and placed in room air at 20°C. One plate is maintained at 150°C while the other plate attains a temperature in accordance with its radiant and convection energy exchange with the 150°C plate and the surroundings. Both plates have $\epsilon = 0.8$. Using the approximate free-convection relations of Chapter 7, calculate the temperature of the other plate.

8-66 A heated rod protrudes from a spaceship. The rod loses heat to outer space by radiation. Assuming that the emissivity of the rod is ϵ and that none of the radiation leaving the rod is reflected, set up the differential equation for the temperature distribution in the rod. Also set up the boundary conditions that the differential equation must satisfy. The length of the rod is L, its cross-sectional area is A, its perimeter is P, and its base temperature is T_0. Assume that outer space is a blackbody at 0 K.

8-67 Three infinite parallel plates are arranged as shown in Figure P8-67. Plate 1 is maintained at 1200 K, and plate 3 is maintained at 300 K; $\epsilon_1 = 0.2$, $\epsilon_2 = 0.5$, and $\epsilon_3 = 0.8$. Plate 2 receives no heat from external sources. What is the temperature of plate 2?

Figure P8-67

1   2   3

8-68 Two large parallel planes having emissivities of 0.3 and 0.5 are maintained at temperatures of 900 K and 400 K, respectively. A radiation shield having an emissivity of 0.05 on both sides is placed between the two planes. Calculate (*a*) the heat-transfer rate per unit area if the shield were not present, (*b*) the heat-transfer rate per unit area with the shield present, and (*c*) the temperature of the shield.

8-69 Two parallel planes 1.2 by 1.2 m are separated by a distance of 1.2 m. The emissivities of the planes are 0.4 and 0.6, and the temperatures are 760 and 300°C, respectively. A 1.2 by 1.2 m radiation shield having an emissivity of 0.05 on both sides is located equidistant between the two planes. The combined arrangement is placed in a large room that is maintained at 40°C. Calculate (*a*) the heat-transfer rate from each of the two planes if the shield were not present, (*b*) the heat-transfer rate from each of the two planes with the shield present, (*c*) the temperature of the shield.

8-70 A long cylindrical heater 2.5 cm in diameter is maintained at 500°C and has a surface emissivity of 0.8. The heater is located in a large room whose walls are at 25°C. How much will the radiant heat transfer from the heater be reduced if it is surrounded by a 30-cm-diameter radiation shield of aluminum having an emissivity of 0.2? What is the temperature of the shield?

8-71 Two long concentric cylinders have diameters of 4 and 8 cm, respectively. The inside cylinder is at 800°C and the outer cylinder is at 100°C. The inside and outside emissivities are 0.8 and 0.4, respectively. Calculate the percent reduction in heat transfer if a cylindrical radiation shield having a diameter of 6 cm and emissivity of 0.3 is placed between the two cylinders.

8-72 Two finite-length concentric cylinders are placed in a large room maintained at 20°C. The inner cylinder has a diameter of 5.0 cm and the outer cylinder has a diameter of 10 cm. The length of the cylinders is 10 cm. The inner cylinder is newly turned cast iron and maintained at a temperature of 400°C. The outer cylinder is Monel metal oxidized at 1110°F. Calculate the heat lost by the inner cylinder.

8-73 Two parallel disks having diameters of 50 and 25 cm are separated by a distance of 25 cm. The larger disk has a temperature of 800°C and emissivity of 0.6 while the smaller disk is at 400°C and has an emissivity of 0.4. The disks are enclosed by a large evacuated chamber with walls maintained at a low temperature of −250°C (23 K). Calculate the radiant energy lost by the larger disk.

8-74 Suppose a man can be approximated by a vertical cylinder 1.8 m high and 30 cm in diameter. Conduct your own experiments to estimate the body surface temperature, and then estimate the radiant heat transfer between the body and a large black surrounding at 0, 10, 20, 30, and 40°C. What do you assume for the surface emissivity of skin?

8-75 Two parallel equal-size disks having diameters of 3 cm are separated by a distance of 60 cm. Both disks are black and enclosed by a large room at −10°C. One disk is maintained at 100°C, while the other disk is at 0°C. The net radiation exchange from (or to) the cooler disk is measured by a suitable technique. How sensitive must this measurement be, expressed in watts, to be accurate within 1.0 percent?

8-76 Repeat Problem 7-75 for a separation distance between the disks of 6 cm; of 120 cm.

8-77 An annular space is filled with a gas whose emissivity and transmissivity are 0.3 and 0.7, respectively. The inside and outside diameters of the annular space are 30 and 60 cm, and the emissivities of the surface are 0.5 and 0.3, respectively. The inside surface is maintained at 760°C, while the outside surface is maintained at 370°C. Calculate the net heat transfer per unit length from the hot surface to the cooler surface. What is the temperature of the gas? Neglect convection heat transfer.

8-78 For the conditions in Problem 8-77, plot the net heat transfer per unit length of the annulus as a function of the gas emissivity, assuming that $\epsilon_m + \tau_m = 1$.

8-79 Repeat Problem 8-77 for two infinite parallel planes with the same temperatures and emissivities. Calculate the heat-transfer rates per unit area of the parallel planes.

8-80 The gas of Problem 8-77 is forced through the annular space at a velocity of 6.0 m/s and is maintained at a temperature of 1100°C. The properties of the gas are

$$\rho = 1.6 \text{ kg/m}^3 \qquad c_p = 1.67 \text{ kJ/kg} \cdot °\text{C}$$
$$\mu = 5.4 \times 10^{-5} \text{ kg/m} \cdot \text{s} \qquad k = 0.11 \text{ W/m} \cdot °\text{C}$$

Assuming the same temperatures and emissivities of the surfaces as in Problem 8-68, estimate the heating or cooling required for the inner and outer surfaces to maintain them at these temperatures. Assume that the convection heat-transfer coefficient may be estimated with the Dittus-Boelter equation (6-4).

8-81 Two long concentric cylinders have $T_1 = 900$ K, $\epsilon_1 = 0.4$, $d_1 = 5$ cm, and $T_2 = 400$ K, $\epsilon_2 = 0.6$, and $d_2 = 10$ cm. They are separated by a gray gas having $\epsilon_g = 0.15$, $\tau_g = 0.85$. Calculate the heat-transfer rate between the two cylinders and the gas temperature using a radiation-network approach.

8-82 A square array of pin fins having diameters of 2 mm and lengths of 12 mm are fixed to a base such that there are inline center-to-center spacings of 4 mm. If the emissivities of all surfaces are 0.3, estimate the apparent emissivity of the arrangement for radiation exchange with a large surrounding environment. State assumptions.

8-83 A square honeycomb assembly consists of 400 cavities of very thin polished aluminum sheets with openings of 6 by 6 mm and depths of 10 mm. If the surface emissivity is 0.1, calculate the apparent emissivity of the assembly for radiation exchange with a large surrounding environment.

8-84 A cavity is to be designed to serve as a blackbody source. The proposed shape is that of a truncated conical hole with an opening of 25 mm and diameter at the bottom of 6 mm. If the surface emissivity of the cavity is 0.9, estimate the cavity depth necessary to produce an apparent emissivity of 0.98.

8-85 A 30-cm-diameter sphere radiates from a 6-cm-diameter hole in its side. If the surface emissivity inside the sphere is 0.5, calculate the apparent emissivity of the hole.

8-86 A cavity 25 by 25 cm, and 10 cm deep, is covered with a semitransparent material having $\tau = 0.5$, $\epsilon = 0.4$, and $\rho = 0.1$. The inside surface has $\epsilon = 0.6$. Calculate the apparent emissivity of the cavity for radiant exchange with a large surrounding environment.

8-87 Two parallel disks 10 cm in diameter are separated by a distance of 2.5 cm. One disk is maintained at 540°C, is completely diffuse-reflecting, and has an emissivity of 0.3. The other disk is maintained at 260°C but is a specular-diffuse reflector such that $\rho_D = 0.2$, $\rho_s = 0.4$. The surroundings are maintained at 20°C. Calculate the heat lost by the inside surface of each disk.

8-88 Rework Problem 8-87, assuming that the 540°C plane reflects in only a specular manner. The insulated plane is diffuse.

8-89 Draw the radiation network for a specular-diffuse surface losing heat to a large enclosure. Obtain an expression for the heat transfer under these circumstances. How does this heat transfer compare with the heat that would be lost by a completely diffuse surface with the same emissivity?

8-90 A 30 by 60 cm plate with $\epsilon = 0.6$ is placed in a large room and heated to 370°C. Only one side of the plate exchanges heat with the room. A highly reflecting plate ($\rho_s = 0.7$, $\rho_D = 0.1$) of the same size is placed perpendicular to the heated plate within the room. The room temperature is 90°C. Calculate the energy lost by the hot plate both with and without the reflector. What is the temperature of the reflector? Neglect convection.

8-91 A 5 by 5 by 2.5 cm cavity is constructed of stainless steel ($\epsilon = 0.6$) and heated to 260°C. Over the top is placed a special ground-glass window ($\rho_s = 0.1$, $\rho_D = 0.1$, $\tau_D = 0.3$, $\tau_s = 0.3$, $\epsilon = 0.2$) that is 5 by 5 cm. Calculate the heat lost to a very large room at 20°C, and compare with the energy that would be lost to the room if the glass window were not in place.

8-92 Repeat Problem 8-91 for the case of a window that is all diffuse-reflecting and all specular-transmitting, that is, $\rho_D = 0.2$, $\tau_s = 0.6$, $\epsilon = 0.2$.

8-93 The cavity of Problem 8-91 has a fused-quartz window placed over it, and the cavity is assumed to be perfectly insulated with respect to conduction and convection loss to the surroundings. The cavity is exposed to a solar irradiation flux of 900 W/m^2. Assuming that the quartz is nonreflecting and $\tau = 0.9$, calculate the equilibrium temperature of the inside surface of the cavity. Recall that the transmission range for quartz is 0.2 to 4 μm. Neglect convection loss from the window. The surroundings may be assumed to be at 20°C.

8-94 A circular cavity, 5 cm in diameter and 1.4 cm deep, is formed in a material with an emissivity of 0.8. The cavity is maintained at 200°C, and the opening is covered with a transparent material having $\tau = 0.7, \epsilon = 0.3, \rho = 0$. The outside surface of the transparent material experiences a convection heat-transfer coefficient of 17 W/m$^2 \cdot$°C. The surrounding air and room are at 20°C. Calculate the net heat lost by the cavity and the temperature of the transparent covering.

8-95 Two parallel infinite plates are maintained at 800 and 35°C with emissivities of 0.5 and 0.8, respectively. To reduce the heat-transfer rate, a radiation shield is placed between the two plates. Both sides of the shield are specular-diffuse-reflecting and have $\rho_D = 0.4$, $\rho_s = 0.4$. Calculate the heat-transfer rate with and without the shield. Compare this result with that obtained when the shield is completely diffuse-reflecting with $\rho = 0.8$.

8-96 Apply Equations (8-99) and (8-101) to the problem described by Equation (8-43). Apply the equations directly to obtain Equation (8-43a) for a convex object completely enclosed by a very large concave surface.

8-97 Rework Example 8-10 using the formulation of Equations (8-106) and (8-109).

8-98 Rework Problem 8-27 using the formulation of Equations (8-106) and (8-109). Recall that $J = E_b$ for the insulated surface.

8-99 Two 10 by 30 cm rectangular plates are spaced 10 cm apart and connected by four insulated and re-radiating walls. The plate temperatures are uniform at 1000 and 300°C, and their emissivities are 0.6 and 0.4, respectively. Using the numerical method, determine the net heat transfer under the assumptions that (a) the four re-radiating surfaces act as one surface and have uniform radiosity and (b) the four re-radiating surfaces have radiosities determined from the radiant balance with all other surfaces. Assume that the 1000 and 300°C surfaces have uniform radiosity. Also calculate the temperatures for the re-radiating surfaces for each case above.

8-100 Two parallel disks 30 cm in diameter are separated by a distance of 5 cm in a large room at 20°C. One disk contains an electric heater that produces a constant heat flux of 100 kW/m^2 and $\epsilon = 0.9$ on the surface facing the other disk. Its back surface is insulated. The other disk has $\epsilon = 0.5$ on both sides and is in radiant balance with the other disk and room. Calculate the temperatures of both disks.

Figure P8-101

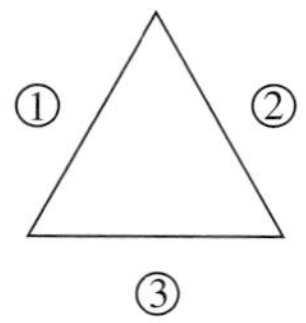

① ② ③

8-101 A long duct has an equilateral triangle shape as shown in Figure P8-101. The surface conditions are $T_1 = 900$ K, $\epsilon_1 = 0.6$, $T_2 = 1500$ K, $\epsilon_2 = 0.8$, $(q/A)_3 = 1000$ W/m^2, $\epsilon_3 = 0.7$. Calculate the heat fluxes for surfaces 1 and 2 and the temperature of surface 3.

8-102 Two 1-m-square plates are placed in a perpendicular position and joined together. One is maintained at 300°C with $\epsilon = 0.5$ while the other is perfectly insulated with $\epsilon = 0.7$. Both plates are placed in a large room maintained at 30°C. Calculate the heat lost by the 300°C plate and the temperature of the insulated plate using the numerical method.

8-103   Solve Problem 8-25 using the numerical formulation.

8-104   Two parallel disks, each 1 m in diameter, are spaced 25 cm apart. One disk is maintained at 300°C while the other disk is insulated on the back side. Both disks have an emissivity of 0.5 and are placed in a large room that is maintained at 30°C. Calculate the radiation energy lost by the 300°C disk.

8-105   Two parallel disks 50 cm in diameter are separated by a distance of 10 cm. An electric heater supplies a constant heat flux of 10 kW/m² to one disk while the other disk is maintained at a constant temperature of 350 K. The two disks are connected by an insulated surface. Both disks have surface emissivity of 0.6. Using the numerical formulation set up the equations for the problem and solve for the temperatures of the hot disk and surrounding insulated surface.

8-106   In Figure P8-106, calculate the heat loss by radiation from disk surface 1 and the irradiation on spherical surface 2. Surface 3 is insulated.

Figure P8-106

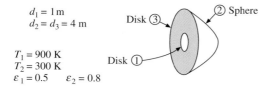

$d_1 = 1$ m
$d_2 = d_3 = 4$ m

$T_1 = 900$ K
$T_2 = 300$ K
$\varepsilon_1 = 0.5$ $\varepsilon_2 = 0.8$

8-107   A 10-cm black disk maintained at 500°C is installed at the end of a 10-cm-diameter cylinder having a length of 10 cm. The cylinder has an emissivity of 0.3 on both the inside and outside surfaces. The open end of the cylinder is exposed to a large room at 40°C. Calculate the energy lost by the 500°C disk.

8-108   A "focusing" radiation heater is constructed as shown in Figure P8-108 with the hot surface at 1000°C having an emissivity of 0.55 and a diameter of 0.5 m. The walls surrounding the heater surface are insulated and have $\epsilon = 0.4$. The arrangement is open to a large room at 20°C. Calculate the heat transfer to the room.

Figure P8-108

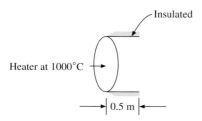

Insulated

Heater at 1000°C

0.5 m

8-109   A hollow metal shell is constructed of a material having $\epsilon = 0.45$ on the interior side that is maintained at a temperature of 800 K. The shell has a paraboloid shape with an interior area of 156 cm² and a circular opening with a diameter of 7.1 cm. The hot interior of the shell radiates to a large room maintained at 300 K. Calculate the net radiant energy lost by the shell, expressed in watts.

8-110   A tapered cylindrical hole is machined in a block of insulation such that the depth is 3 cm. The diameter at the bottom is 1 cm and the diameter at the opening is 2 cm. An electric heater element is installed at the bottom of the hole such that the temperature is maintained at 500°C with a metal surface emissivity of 0.6. Outside the hole the large surroundings are at 30°C. Assuming the insulation is "perfect," calculate the power that must be supplied to the heater. The emissivity for the insulation is 0.3.

8-111 A cubical box is 1 m on a side. The top surface has $T = 100°C$, $\epsilon = 0.7$ while the bottom surface has $T = 40°C$, $\epsilon = 0.5$. The four sides have $\epsilon = 0.6$. Two sides opposite each other have $T = 70°C$ while the other two are at 60°C. Calculate the net radiant heat transfer for each side using the numerical method.

8-112 Two parallel concentric disks have diameters of 10 and 20 cm and are separated by a distance of 10 cm. The smaller disk has $T = 700$ K and $\epsilon = 0.8$ while the larger disk has $\epsilon = 0.4$ on both sides and is in radiant balance with the small disk and a large surrounding room at 25°C. Calculate the heat lost by the small disk and the temperature of the large disk.

8-113 Two parallel disks have diameters of 50 cm and are separated by a distance of 12.5 cm. One disk is maintained at a constant temperature of 600°C and has $\epsilon = 0.63$ while the other generates a constant heat flux of 80 kW/m^2 and has $\epsilon = 0.75$. The disks are surrounded by a large room at 30°C. Calculate the heat absorbed by the room and the temperature of the constant-heat-flux disk.

8-114 Rework Problem 8-113 such that the 600°C disk is replaced with a disk having constant heat flux of 100 kW/m^2 and $\epsilon = 0.63$. What are the temperatures of the two disks under these conditions?

8-115 Two concentric cylinders have $d_1 = 10$ cm, $\epsilon_1 = 0.4$ and $d_2 = 20$ cm, $\epsilon_2 = 0.6$, respectively. The length is 10 cm. At one end of the cylinders a heater is placed that has $\epsilon = 0.8$ and covers the open space between the cylinders. The heater has a constant heat flux of 90 kW/m^2. The entire assembly is placed in a large room at 30°C. Using both the network and numerical methods, calculate the temperatures of the two cylinders.

8-116 Rework Problem 8-61 for a polished-aluminum plate having the radiation characteristics given in Table 8-4.

8-117 A slab of white marble is exposed to a solar radiation flux of 1070 W/m^2. Assuming the effective radiation temperature of the sky is $-70°C$, calculate the radiation equilibrium temperature of the slab, using the properties given in Table 8-4. For this calculation neglect all conduction and convection losses.

8-118 A large surface has $\alpha = 0.6$ for long-wavelength radiation and $\alpha = 0.95$ for solar radiation. Calculate the radiation equilibrium temperature of the surface if its back side is insulated and top side is exposed to a solar radiation flux of 950 W/m^2 and an environment at 300 K with $h = 12$ W/m$^2 \cdot °C$.

8-119 A cast-iron plate is placed in an environment at 25°C and exposed to a solar flux of 800 W/m^2. Calculate the radiation equilibrium temperature of the plate neglecting convection.

8-120 A thermometer is encased in metal shell having an exterior emissivity of 0.3. The thermometer is exposed to a transparent fluid having $h = 25$ W/m $\cdot °C$ and $T_\infty = 105°C$. The thermometer indicates a temperature of 98°C when placed in a certain large enclosure. Calculate the temperature of the enclosure expressed in °C.

8-121 A thermocouple is placed in a large heated duct to measure the temperature of the gas flowing through the duct. The duct walls are at 425°C, and the thermocouple indicates a temperature of 170°C. The heat-transfer coefficient from the gas to the thermocouple is 150 W/m$^2 \cdot °C$. The emissivity of the thermocouple material is 0.43. What is the temperature of the gas?

8-122 Suppose a flat plate is exposed to a high-speed environment. We define the radiation equilibrium temperature of the plate as that temperature it would attain if insulated so that the energy received by aerodynamic heating is just equal to the heat lost by radiation to the surroundings, that is,

$$hA(T_w - T_{aw}) = -\sigma A \epsilon (T_w^4 - T_s^4)$$

where the surroundings are supposed to be infinite and at the temperature T_s and ϵ is the emissivity of the plate surface. Assuming an emissivity of 0.8 for the surface, calculate the radiation equilibrium temperature for the flow conditions of Example 5-9. Assume that the effective radiation temperature for the surroundings is $-40°C$.

8-123 On a clear night the effective radiation temperature of the sky may be taken as $-70°C$. Assuming that there is no wind and the convection heat-transfer coefficient from the air to the dew that has collected on the grass is 28 W/m² · °C, estimate the minimum temperature that the air must have to prevent formation of frost. Neglect evaporation of the dew, and assume that the grass is insulated from the ground insofar as conduction is concerned. Take the emissivity as unity for the water.

8-124 A thermocouple enclosed in a 3.2-mm stainless-steel sheath ($\epsilon = 0.6$) is inserted horizontally into a furnace to measure the air temperature inside. The walls of the furnace are at 650°C, and the true air temperature is 560°C. What temperature will be indicated by the thermocouple? Assume free convection from the thermocouple.

8-125 The thermocouple of Problem 8-124 is placed horizontally in an air-conditioned room. The walls of the room are at 32°C, and the air temperature in the room is 20°C. What temperature is indicated by the thermocouple? What would be the effect on the reading if the thermocouple were enclosed by a polished-aluminum radiation shield?

8-126 A horizontal metal thermocouple with a diameter of 3 mm and an emissivity of 0.6 is inserted in a duct to measure the temperature of an airstream flowing at 7 m/s. The walls of the duct are at 400°C, and the thermocouple indicates a temperature of 100°C. Using the convection relations of Chapter 6, calculate the true gas temperature.

8-127 Suppose the outer cylinder of Example 8-11 loses heat by convection to the room with $h = 50$ W/m² · °C on its outside surface. Neglecting convection on other surfaces determine the temperature of the outer cylinder and the total heat lost by the inner cylinder.

8-128 A thermocouple is used to measure the air temperature inside an electrically heated metallurgical furnace. The surface of the thermocouple has an emissivity of 0.7 and a convection heat-transfer coefficient of 20 W/m² · °C. The thermocouple indicates a temperature of 700°C. The true temperature of the air is only 650°C. Estimate the temperature of the walls of the furnace.

8-129 A mercury-in-glass thermometer is inserted in a duct to measure the temperature of an air-flow stream. The thermometer indicates a temperature of 55°C and the temperature of the walls of the duct is measured with a thermocouple as 90°C. By the methods of Chapter 6 the convection heat-transfer coefficient from the thermometer to the air is calculated as 30 W/m² · °C. Calculate the temperature of the air.

8-130 Air at 20°C flows across a 50-cm-diameter cylinder at a velocity of 25 m/s. The cylinder is maintained at a temperature of 150°C and has a surface emissivity of 0.7. Calculate the total heat loss from the cylinder per unit length if the effective radiation temperature of the surroundings is 20°C.

8-131 A tungsten light filament operates at a temperature of 3400 K. What fraction of the radiation lies in the visible range, 0.4 to 0.7 μm?

8-132 Suppose the tungsten light filament in Problem 8-131 consumes 400 W of electric power and all the visible light is transmitted through the enclosing glass bulb. Calculate the area of the filament to dissipate the total of 400 W of radiant energy. What is the efficiency of the lightbulb?

8-133 A certain material has an emissivity of 0.6 between 0 and 2 μm and 0.2 between 2 and 8 μm. Calculate the emissive power of such a material maintained at 3000 K.

8-134 Show that the Stefan-Boltzmann constant can be expressed in terms of the radiation constants C_1 and C_2 by the relation

$$\sigma = \left(\frac{\pi}{C_2}\right)^4 \left(\frac{C_1}{15}\right)$$

8-135 A plastic tinted coating is frequently applied to glass windows to reduce the transmission of solar energy. The transmissivity for the uncoated glass is 0.9 from about 0.25 to 2.5 μm, while the tinted glass has the same transmissivity from 0.5 to 1.5 μm. Assuming solar radiation to be that of a blackbody at 5800 K, calculate the fraction of the incident solar energy transmitted through each glass. Also calculate the visible energy (0.4 to 0.7 μm) transmitted through each glass.

8-136 Air flows across a 0.2-m-square plate producing a convection heat-transfer coefficient of 12 W/m$^2 \cdot$°C. The air temperature is 400 K. The plate is insulated on the back side and exposed to an irradiation of 2200 W/m^2, of which 450 W/m^2 is reflected. When the surface temperature is 510 K, its emissive power is 1100 W/m^2. Calculate the emissivity and radiosity of the plate surface.

8-137 A cubical container, 3 m on a side, has the bottom surface at 550 K with $\epsilon = 0.5$ and the top surface at 300 K with $\epsilon = 0.7$. The other four sides are perfectly insulated. Calculate the heat transfer from the bottom surface.

8-138 A 10-cm-diameter disk at 700 K with $\epsilon = 0.6$ is placed at the bottom of an insulated cylinder 15 cm high. On top of the cylinder is a hemisphere 10 cm in diameter with $T = 300$ K and $\epsilon = 0.3$. Calculate the heat lost by the disk.

8-139 A 5.0-cm-diameter long cylinder with $T = 800$ K and $\epsilon = 0.7$ is enclosed by a large room at 300 K. How much will the radiant heat loss from the cylinder be decreased if it is enclosed by a cylindrical radiation shield having $d = 10$ cm and $\epsilon = 0.2$ on the inner and outer surfaces? Express as W/m of length.

8-140 A cylindrical furnace has diameter and lengths of 50 cm. The bottom disk surface is at 450 K with $\epsilon = 0.5$, the top disk is at 600 K with $\epsilon = 0.6$, and the vertical cylinder surface is at 1000 K with $\epsilon = 0.7$. Calculate the emissive power, radiosity, irradiation, and heat loss for the three surfaces.

8-141 Repeat Problem 8-140 with the top disk surface removed and the open cylindrical cavity exposed to a large room at 300 K.

8-142 An evacuated insulating material consists of outer layers of aluminum with emissivity of 0.11 and three inner polished aluminum shields having emissivity of 0.04 and spaced 2 mm apart. The total thickness, neglecting the thickness of the outer layers, is 8 mm. Calculate the radiation heat transfer across the assembly for temperatures of 300 and 85 K. Also calculate an effective thermal conductivity and R value for these temperatures.

8-143 A cylinder 40 cm in diameter and 30 cm high has the bottom disk surface maintained at 750 K with $\epsilon = 0.75$. The vertical cylindrical surface is perfectly insulated. The top disk has a 20-cm-diameter hole in the center with a surface temperature of 600 K and $\epsilon = 0.4$. The open hole is exposed to a large room at 300 K. Calculate the heat loss, emissive power, radiosity, and irradiation for all three surfaces and the temperature of the vertical cylindrical surface.

8-144 The "E factor" is sometimes defined such that the heat transfer per unit area between large parallel surfaces is given by $q/A = E\sigma(T_1^4 - T_2^4)$. Calculate the value of the E factor for two surfaces, each having $\epsilon = 0.9, 0.8, 0.5, 0.2, 0.1$, and 0.05.

8-145 Show that Equation (8-41) for radiation between plates connected by insulating surfaces reduces to Equation (8-42) for parallel infinite planes when the surfaces are very large.

8-146 Two concentric cylinders having diameters of 30 and 100 cm, respectively, have lengths of 2 m. The inside and outside emissivities are 0.2 and 0.4 and the surface temperatures are 800 K and 400 K, respectively. Calculate the net radiant heat loss from each cylinder for (*a*) enclosure by a large room at 300 K, and (*b*) the open ends of the cylinders are covered with insulating material.

8-147 A truncated cone like that in Example 8-4 is constructed of thin copper sheet having $\epsilon = 0.8$ on both sides. The top (small) opening is covered with a surface having $\epsilon = 0.2$ and $T = 700$ K, while the bottom (large) opening is covered with a surface having $\epsilon = 0.6$ and $T = 500$ K. The entire assembly is placed in a large room at 300 K. Calculate the heat lost by each of the two hot surfaces.

8-148 Suppose the outer copper surface of Problem 8-147 also loses heat by convection, with $h = 25$ W/m$^2 \cdot$°C. Calculate the heat lost by each of the hot surfaces for this condition.

8-149 A sphere having a diameter of 10 cm has one hemisphere with an emissivity of 0.5 and the other hemisphere with an emissivity of 0.8. The entire sphere is heated to a uniform temperature of 800°C and enclosed by a large room maintained at 20°C. Calculate the total heat lost by the sphere.

8-150 Calculate the radiation equilibrium temperature for a cast-iron plate exposed to a solar flux of 700 W/m^2 and a surrounding temperature of 25°C. Assume $h = 0$.

8-151 Repeat Problem 8-150 for convection loss from the surface with $h = 7.5$ W/m$^2 \cdot$°C.

8-152 A copper sphere having $\epsilon = 0.8$ and a diameter of 5 cm is allowed to cool from 300° to 50°C when enclosed by a large evacuated enclosure maintained at 0°C. Using a numerical analysis based on lumped capacity, determine the time for the sphere temperature to reach (*a*) 200°C, (*b*) 100°C, and (*c*) 50°C. (See Section D-6, Appendix D.)

8-153 A mercury-in-glass thermometer is placed vertically outdoors to measure the local air temperature. It is placed in a shaded area near a large vertical metal wall that is heated on the reverse side by solar radiation to a temperature of 40°C. A light breeze blows across the thermometer at a velocity of 3 m/s. The open-air surroundings are at the ambient air temperature. If the true air temperature is 20°C, estimate the temperature indicated by the thermometer. State assumptions.

Design-Oriented Problems

8-154 One way of constructing a blackbody cavity is to drill a hole in a metal plate. As a result of multiple reflections on the inside of the hole, the interior walls appear to have a higher emissivity than a flat surface in free space would have. A strict analysis of the cavity must take into account the fact that the irradiation is nonuniform over the interior surface. Thus the specific location at which a radiant-flux-measuring device sights on this surface must be known in order to say how "black" the surface may be. Consider a 1.25-cm-diameter hole 2.5 cm deep. Divide the interior surface

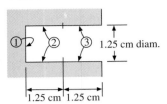

1.25 cm diam.

1.25 cm   1.25 cm

into three sections, as shown in Figure P8-154. Assume, as an approximation, that the irradiation is uniform over each of these three surfaces and that the temperature and emissivity are uniform inside the entire cavity. A radiometer will detect the total energy leaving a surface (radiosity). Calculate the ratio J/E_b for each of the three surfaces, assuming $\epsilon = 0.6$ and no appreciable radiation from the exterior surroundings.

8-155 Repeat Problem 8-154, assuming a cavity temperature of 120°C and a surrounding temperature of 93°C.

8-156 Two infinite strips, 30 cm wide, are separated by a distance of 10 cm. The strips have surface emissivities of 0.25 and 0.5, and the respective temperatures are constant over the surfaces at 200 and 1000°C. The strips are completely enclosed by a large surrounding at 0°C. Using the numerical method, divide each strip into three equal segments and calculate the net heat transfer from each strip. What would be the heat transfer of each strip if uniform radiosity were assumed for each? Assume gray-body behavior.

8-157 Repeat Problem 8-156, using one strip at 1000°C with $\epsilon = 0.6$ and the other strip perfectly insulated (radiant balance) with $\epsilon = 0.25$.

8-158 Verify the results of Example 8-18 for $\epsilon_2 = \epsilon_3 = 0.6$ and $h = 25$ W/m² · °C.

8-159 Two long concentric cylinders have the properties of $d_1 = 2$ cm, $\epsilon_1 = 0.75$, $T_1 = 600$ K and $d_2 = 5$ cm, $\epsilon_2 = 0.8$ on both inner and outer surfaces. An airstream is blown across the outer cylinder with $T_\infty = 35$°C and $h = 180$ W/m² · °C. The assembly is located in a large room at 25°C. Calculate the heat lost by the inner cylinder per unit length and the temperature of the outer cylinder.

8-160 Surface 1 in Figure P8-160 has $T_1 = 500$ K and $\epsilon_1 = 0.8$. Surface 2 is very thin, has $\epsilon = 0.4$, and is in radiant balance with surface 1 and a large enclosure at 300 K. Determine the temperature of surface 2 assuming it is uniform in temperature.

Figure P8-160

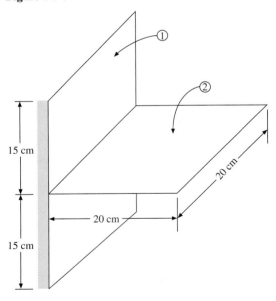

15 cm

15 cm

20 cm

20 cm

8-161 Suppose a convection coefficient of $h = 10$ W/m² · °C at $T = 300$ K was experienced on surface 2 (top and bottom) of Problem 8-160. Obtain the temperature distribution in this circumstance. Repeat for $h = 50$ W/m² · °C. Assume the surface is so thin that conduction is negligible.

8-162 Suppose surface 2 in Problem 8-160 is oxidized copper with $\epsilon = 0.78$ and has a thickness of 0.5 mm. Surface 1 has $T_1 = 1000$ K and $\epsilon_1 = 0.5$. The convection and radiation environment is at 300 K and $h = 75$ W/m² · °C. Calculate the temperatures along the fin, taking into account conduction. Consult Chapter 3 for setting up conduction heat-transfer terms.

8-163 The plate of Example 5-4 is 30 cm by 30 cm square and is sprayed with a white paint having a solar absorptivity of 0.16 and a low-temperature absorptivity of 0.09. The plate is exposed to a solar radiation flux of 1100 W/m² and allowed to reach equilibrium with the convection surroundings. Assuming that the underside of the plate is insulated, calculate the equilibrium temperature of the plate.

8-164 A thin copper disk is placed on a thick slab of insulating material coated with a highly reflective material. The outside surface of the copper disk is sprayed with a flat black paint, producing a surface that is almost a perfect absorber of thermal radiation. Embedded in the bottom of the disk is a tiny thermocouple used to measure the temperature of the disk. The entire assembly is employed as a sensing device to measure thermal radiation impressed on the face. Using whatever references are necessary discuss this application and make any calculations that may be needed for your evaluation.

8-165 In the early days of the space program, testing of astronauts' space suits and other survival gear was accomplished by sending the astronauts to high altitudes (24,000 to 30,000 m) in specially constructed balloon gondolas. Using whatever sources of information that may be required, investigate the heat transfer problems that may be involved in such an endeavor. The investigation should consider, but not be restricted to, the following factors:

1. The paint on the outside of the gondola will probably have different absorptive characteristics for solar radiation than for long-wavelength earthbound radiation, that is, it will not behave as a gray body over all wavelengths and $\alpha \neq \epsilon$.
2. The paint on the inside of the gondola will likely exhibit gray-body behavior for radiant exchange with the astronaut (inside a space suit) in the gondola.
3. At the altitudes involved, there will likely be no clouds to interfere with radiant exchange with the sun.
4. Use data for "standard atmosphere" to determine the temperature and pressure at various altitudes from 24,000 to 30,000 m.
5. Though the air temperature at high altitude is quite low, the density is also low, so that the convection heat loss from the outside of the space suit may not be enough to counteract the heat gained by radiation from the walls of the gondola. In this case, the trip may "cook" the astronaut unless provision is made to reduce the radiant transfer to the space suit, and the human inhabitant. One way to alleviate the problem is to provide radiation shields to intercept the radiation from the inside of the gondola surface to the space suit. Suggest ways this radiation shielding might be accomplished.

8-166 A solar collector consists of an insulated black surface covered by a glass plate having the following properties:

$$0 < \lambda < 2.5 \ \mu\text{m}: \quad \tau = 0.9 \quad \epsilon = 0.1 \quad \rho = 0$$
$$2.5 \ \mu\text{m} < \lambda < \infty: \quad \tau = 0 \quad \epsilon = 0.95 \quad \rho = 0.05$$

Assume that the radiant flux from the sun is 700 W/m^2 and lies entirely below 2.5 μm. Calculate the radiation equilibrium temperature of the black surface neglecting convection and with a surrounding environment at 35°C.

8-167 Refer to Problem 1-37. By now, you should know that the person is cooled by exchanging radiation with the inside of the open refrigerator. Make some suitable assumptions that will enable you to estimate the magnitude of this cooling effect. What might be the equivalent lowering of the effective temperature of the room? Use whatever convection heat transfer relations that may be appropriate from Chapters 5, 6, and 7.

8-168 Refer to Problem 1-39. The difference in "temperature comfort" results from the radiant exchange between the students and the walls. The walls are warmer in the summer than in the winter. Make some suitable assumptions regarding radiation shape factors, emissive properties, wall temperatures and geometry, and convection coefficients to analyze the situation and obtain a satisfactory explanation of the experience.

8-169 The side of a building adjoins a concrete slab and is exposed to sunlight in the summer. Experience shows that the concrete will be heated by the sunlight and will attain a higher temperature than a grassy area. As a result, it will re-radiate energy to the adjoining building wall and thereby produce a higher irradiation than would be present when the building is surrounded by a grassy area. Make appropriate assumptions regarding the radiation properties, solar irradiation, and convection to arrive at estimates of the increase in irradiation resulting from the concrete. Assume that the concrete slab extends out a distance of at least twice the height of the building.

8-170 A radiant heater for a warehouse space is to be designed to provide total heating of about 15 kW to a space at 20°C. The heater surface will be maintained at a temperature of about 900°C by an appropriate gas flame, and the surface emissivities of the heated surface(s) may range from 0.5 to 0.8. The heater surface will be rectangular in shape surrounded by four reasonably well-insulated walls similar to the arrangement shown in Figure P8-50. The "box" is open to the room or warehouse space. Consider several different configurations and/or surface conditions to determine a geometry which will accomplish the stated heating objective of 15 kW. Recall that an insulated surface is handled in the program by taking ϵ = a small value for that surface. For the geometric configuration selected, compute an estimated value for the temperatures of the insulated surfaces.

8-171 A box contains an electronic package that dissipates 200 W of heat resulting from the electrical device contained therein. The maximum temperature that may be tolerated by the electronics is 80°C and the outside of the box is exposed to room air at 20°C. Determine appropriate dimensions for a rectangular box to dissipate the required heat. Make assumptions regarding the box material, wall thickness, size of electronic device in the box, temperature drop from the device to box wall, and any other factors that may be appropriate. Assume that the total energy is dissipated from the outside of the box by a combination of free convection and radiation from the outside surface of the box. The radiation is calculated by application of Equation (8-43a) with values of the emissivity taken from Table A-10 for the box material selected.

8-172 A radiant panel heater is to be designed with a variable radiation shield as shown in Figure P8-172. The blackened panel surface is maintained at a constant temperature of 100°C by condensing steam on the reverse side. The shield sections may be rotated to vary the radiation that may reach a surrounding room at 25°C. The shield sections have $\epsilon = 0.1$ on both sides. Using whatever approximations and analysis techniques that seem appropriate estimate the heat exchange rate from the black panel to the room as a function of the rotation angle θ of the shield sections. Assume that the black panel is very large compared to the distance between it and the shield sections. Perform the design both with and without free convection from the panel and shield sections. State conclusions regarding the advisability of this design.

Figure P8-172

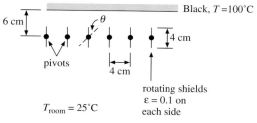

8-173 A solar water desalinization system is proposed, using an arrangement like that shown in Figure P8-173. Solar energy is transmitted through the glass plates, which transmit up to 2.5 μm and are opaque at longer wavelengths. The seawater flows through the bottom section, which is well insulated. As a result of the "trapped" radiation, the water evaporates and condenses when it comes into contact with the slanted glass surfaces exposed to the surrounding air. It then runs down the surfaces and is collected in the troughs on the sides. You are asked to evaluate this proposed design. Using appropriate values for water-condensation coefficients, forced convection outside the glass, and any other parameters you feel appropriate, estimate the yield of distilled water that might be achieved per hour with the device. Take the incoming water temperature as 25°C and the ambient air temperature as 32°C. Comment on your assumptions and results.

Figure P8-173

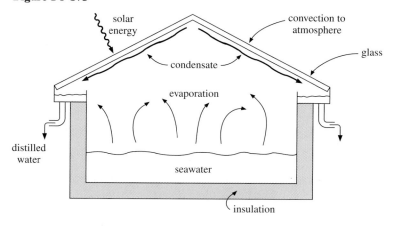

8-174 The performance of fins like those shown in Figure 2-14 for cooling electronic devices must usually be determined by experiment. If such fins are constructed of aluminum, they usually behave as a lumped capacity when cooling by free convection in ambient air. You are to design an experiment to determine the free-convection-cooling behavior of such fins by first heating the fins to some uniform temperature and then allowing them to cool in room air. Suggest how the radiation from the fins may be accounted for in the experiment. How would you attempt to correlate the experimental data to obtain some kind of *generalized* relation that describes the behavior?

8-175 A condominium in Crested Butte, Colorado, uses electric resistance elements in the ceilings for heating the interior space. A person situated in the room may be assumed to "view" each of the six surfaces equally. Make assumptions regarding acceptable levels of comfort and estimate the heating temperature that would be required in a ceiling having dimensions of 4 by 5 meters. For this calculation assume that the same comfort could be achieved with room air at 20°C and an appropriate free-convection heat-transfer coefficient from the person to the room air. Also assume that two of the room walls are outside walls with appropriate temperatures. What do you conclude from this study?

8-176 A partially transparent covering is placed over the square cavity as shown in Figure P8-176. The inside temperature of the cavity can be varied from 350 to 500 K, and the assembly exchanges heat with a large room at 300 K. Suppose it is possible to fabricate the covering material such that a variable transmissivity can be tailored to produce a constant net radiant exchange between the assembly and room for the temperature range indicated. Perform an appropriate analysis to design a suitable set of radiation properties for the covering material. Be sure to state clearly all assumptions.

Figure P8-176

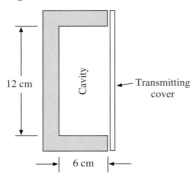

12 cm   Cavity   ← Transmitting cover

6 cm

8-177 Insulation contractors claim that the use of reflective coatings on attic insulation materials can substantially reduce the heat load on the building (house). You are asked to evaluate this claim. Choose appropriate insulating materials and thicknesses and your own estimates of attic temperatures to evaluate the claim. Consider two cases: (1) a single layer of insulation placed on top of the ceiling, with and without a reflective coating, and (2) two layers of insulation, one on top of the ceiling and one under the roof; again, with and without reflective coatings. If you do indeed find that the reflective-coated material is better, how much higher price might be justified for the material? What energy cost saving is the basis for this higher price? What are any other bases for the price estimate? Discuss assumptions and uncertainties in your analysis. If you were the marketing manager for the insulation material manufacturer, how would you present the results? Would it make sense to represent the reflective-coated material as having a larger R value?

8-178 Two parallel square plates are maintained at $T_1 = 800$ K and $T_2 = 500$ K with $\epsilon_1 = 0.6$ and $\epsilon_2 = 0.4$. They are surrounded by a large room at 300 K. Caclulate the heat transfer from each plate as a function of a non-dimensional plate spacing. Also determine a radiation transfer efficiency defined as

$$\text{Eff} = \frac{\text{net heat received by cool plate}}{\text{net heat lost by hot plate}}$$

as a function of the same nondimensional parameter.

8-179 What would the relation for the efficiency parameter of Problem 8-178 be if both plates are black?

8-180 Thin, electrically heated metal strips are arranged horizontally as shown in Figure P8-180. A fan blows air at 20°C across the strips with a free-stream velocity of 10 m/s. The assembly radiates to a room environment also at 20°C. Devise a suitable configuration to dissipate 3 kW by both convection and radiation. Specify strip length, number of heater strips, heater temperatures, and radiative properties. Be sure to state all assumptions.

Figure P8-180

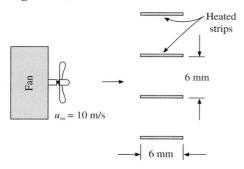

8-181 A man is enclosed by a small room maintained at an air temperature of 20°C, with a convection heat transfer coefficient of $h = 8$ W/m² · °C. The room walls are at 0°C. Compute the energy lost by the man by combined convection and radiation if his body surface temperature is 25°C. Repeat for a wall temperature of 20°C. Suppose the wall is coated with a highly reflective material. How would that influence the calculation? Alternately, suppose the man's clothing is coated with a reflective material. How would that affect the calculation? Be sure to state all assumptions clearly.

REFERENCES

1. Sears, F. W. *Introduction to Thermodynamics, Kinetic Theory, and Statistical Mechanics,* pp. 123–124, Reading, Mass.: Addison-Wesley, 1953.

2. Dunkle, R. V. "Thermal Radiation Tables and Applications," *Trans. ASME,* vol. 76, p. 549, 1954.

3. Mackey, C. O., L. T. Wright, Jr., R. E. Clark, and N. R. Gay. "Radiant Heating and Cooling, part I," *Cornell Univ. Eng. Exp. Stn. Bull.,* vol. 32, 1943.

4. Chapman, A. J. *Heat Transfer,* pp. 319–23, New York: The Macmillan Company, 1960.

5. Hamilton, D. C., and W. R. Morgan. "Radiant Interchange Configuration Factors," *NACA Tech. Note* 2836, 1952.

6. Eckert, E. R. G., and R. M. Drake. *Heat and Mass Transfer,* 2d ed., pp. 381–93. New York: McGraw-Hill, 1959.

7. McAdams, W. H. *Heat Transmission,* 3d ed., chap. 2. New York: McGraw-Hill, 1954.

8. Schmidt, E. "Messung der Gesamtstrahlung das Wasserdampfes bei Temperaturen bis 1000C," *Forsch. Geb. Ingenieurwes.,* vol. 3, p. 57, 1932.

9. Schmidt, E., and E. R. G. Eckert. "Über die Richtungsverteilung der Wärmestrahlung von Oberflachen," *Forsch. Geb. Ingenieurwes.,* vol. 6, p. 175, 1935.

10. Sparrow, E. M., and R. D. Cess. *Radiation Heat Transfer.* Englewood Cliffs, N.J.: Wadsworth Publishing Co., 1966.

11. Wiebelt, J. A. *Engineering Radiation Heat Transfer.* New York: Holt, Rinehart and Winston, 1966.

12. Eckert, E. R. G., and E. M. Sparrow. "Radiative Heat Exchange between Surfaces with Specular Reflection," *Int. J. Heat Mass Transfer,* vol. 3, pp. 43–54, 1961.

13. Sparrow, E. M., E. R. G. Eckert, and V. K. Jonsson. "An Enclosure Theory for Radiative Exchange between Specular and Diffusely Reflecting Surfaces," *J. Heat Transfer,* ser. C., vol. 84, pp. 294–299, 1962.

14. Gubareff, G. G., J. E. Janssen, and R. H. Torborg. *Thermal Radiation Properties Survey,* 2d ed. Minneapolis, Minn.: Minneapolis Honeywell Regulator Co., 1960.

15. Threlkeld, J. L., and R. C. Jordan. "Direct Solar Radiation Available on Clear Days," *ASHAE Trans.,* vol. 64, pp. 45–56, 1958.

16. Jakob, M. *Heat Transfer,* vol. 2. New York: John Wiley & Sons, 1957.

17. Birkebak, R. D., and E. R. G. Eckert. "Effects of Roughness of Metal Surfaces on Angular Distribution of Monochromatic Radiation," *J. Heat Transfer,* vol. 87, p. 85, 1965.

18. Torrance, K. E., and E. M. Sparrow. "Off Specular Peaks in the Directional Distribution of Reflected Thermal Radiation," *J. Heat Transfer,* vol. 88, p. 223, 1966.

19. Hering, R. G., and T. F. Smith. "Surface Roughness Effects on Radiant Transfer between Surfaces," *Int. J. Heat Mass Transfer,* vol. 13, p. 725, 1970.

20. Oppenheim, A. K. "Radiation Analysis by the Network Method," *Trans. ASME,* vol. 78, pp. 725–35, 1956.

21. Holman, J. P. "Radiation Networks for Specular-Diffuse Transmitting and Reflecting Surfaces," *ASME Pap.* 66 WA/HT-9, December 1966.

22. Hottel, H. C. "Radiant Heat Transmission," chap. 4, in W. H. McAdams, *Heat Transmission,* 3d ed. New York: McGraw-Hill, 1954.

23. Hottel, H. C., and A. F. Sarofim. *Radiative Transfer.* New York: McGraw-Hill, 1967.

24. Siegel, R., and J. R. Howell. *Thermal Radiation Heat Transfer,* 2d ed. New York: McGraw-Hill, 1980.

25. Eckert, E. R. G., and R. M. Drake. *Analysis of Heat and Mass Transfer.* New York: McGraw-Hill, 1972.

26. Dunkle, R. V. "Geometric Mean Beam Lengths for Radiant Heat Transfer Calculations," *J. Heat Transfer,* vol. 86, p. 75, February 1964.

27. Edwards, D. K., and K. E. Nelson. "Rapid Calculation of Radiant Energy Transfer between Nongray Walls and Isothermal H_2O and CO_2 Gas," *J. Heat Transfer,* vol. 84, p. 273, 1962.

28. Edwards, D. K. "Radiation Interchange in a Nongray Enclosure Containing an Isothermal Carbon Dioxide–Nitrogen Gas Mixture," *J. Heat Transfer,* vol. 84, p. 1, 1962.

29. Hottel, H. C., and R. B. Egbert. "Radiant Heat Transmission from Water Vapor," *Trans. AIChE,* vol. 38, p. 531, 1942.

30. Kondratyev, K. Y. *Radiative Heat Exchange in the Atmosphere.* New York: Pergamon Press, 1965.

31. Duffie, J. A., and W. A. Beckman. *Solar Energy Thermal Process,* 2nd ed. New York: JohnWiley & Sons, 1991.

32. Howell, J. R. *A Catalog of Radiation Configuration Factors.* New York: McGraw-Hill, 1982.

33. Modest, M. F. *Radiative Heat Transfer.* New York: McGraw-Hill, 1993.

34. Holman, J. P. *What Every Engineer Should Know About EXCEL,* Chap. 5. Boca Raton, FL: CRC Press, 2006.

9

Condensation and Boiling Heat Transfer

9-1 | INTRODUCTION

Our preceding discussions of convection heat transfer have considered homogeneous single-phase systems. Of equal importance are the convection processes associated with a change of phase of a fluid. The two most important examples are condensation and boiling phenomena, although heat transfer with solid-gas changes is important in a number of applications.

In many types of power or refrigeration cycles one is interested in changing a vapor to a liquid, or a liquid to a vapor, depending on the particular part of the cycle under study. These changes are accomplished by boiling or condensation, and the engineer must understand the processes involved in order to design the appropriate heat-transfer equipment. High heat-transfer rates are usually involved in boiling and condensation, and this fact has also led designers of compact heat exchangers to utilize the phenomena for heating or cooling purposes not necessarily associated with power cycles.

9-2 | CONDENSATION HEAT-TRANSFER PHENOMENA

Consider a vertical flat plate exposed to a condensable vapor. If the temperature of the plate is below the saturation temperature of the vapor, condensate will form on the surface and under the action of gravity will flow down the plate. If the liquid wets the surface, a smooth film is formed, and the process is called *film condensation*. If the liquid does not wet the surface, droplets are formed that fall down the surface in some random fashion. This process is called *dropwise condensation*. In the film-condensation process the surface is blanketed by the film, which grows in thickness as it moves down the plate. A temperature gradient exists in the film, and the film represents a thermal resistance to heat transfer. In dropwise condensation a large portion of the area of the plate is directly exposed to the vapor; there is no film barrier to heat flow, and higher heat-transfer rates are experienced. In fact, heat-transfer rates in dropwise condensation may be as much as 10 times higher than in film condensation.

Because of the higher heat-transfer rates, dropwise condensation would be preferred to film condensation, but it is extremely difficult to maintain since most surfaces become wetted after exposure to a condensing vapor over an extended period of time. Various surface coatings and vapor additives have been used in attempts to maintain dropwise condensation,

Figure 9-1 | (*a*) Steam condensation on a copper plate. The right side of the plate is clean copper where steam condenses as a continuous film. The left side has a coating of cupric oleate, which promotes dropwise condensation. The horizontal object is a thermocouple probe with a diameter of 1.7 mm. In this case the heat-transfer coefficient for the dropwise condensation is about seven times that for the filmwise condensation. (*b*) Dropwise condensation of steam at atmospheric pressure on vertical gold-plated copper surface. The surface diameter is 25 mm and the gold thickness is 0.2 μm. Note clearing of central portion following coalescence and runoff of drops into puddle at bottom center.

(*a*)

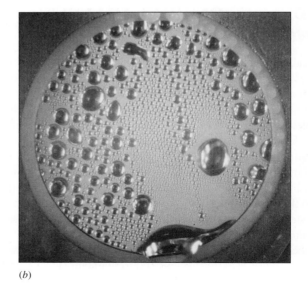

(*b*)

Source: Photographs courtesy of Professor J. W. Westwater, University of Illinois, Urbana.

Figure 9-2 | Film condensation on a vertical flat plate.

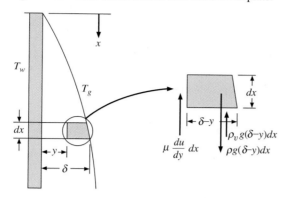

but these methods have not met with general success. Some of the pioneer work on drop condensation was conducted by Schmidt [26] and a good summary of the overall problem is presented in Reference 27. Measurements of Reference 35 indicate that the drop conduction is the main resistance to heat flow for atmospheric pressure and above. Nucleation site density on smooth surfaces can be of the order of 10^8 sites per square centimeter, and heat-transfer coefficients in the range of 170 to 290 kW/m$^2 \cdot$ °C [30,000 to 50,000 Btu/h $\cdot$ ft$^2 \cdot$ °F] have been reported by a number of investigators.

Figure 9-1 is a photograph illustrating the different appearances of dropwise and film-wise condensation. The figure caption explains the phenomena.

Film condensation on a vertical plate may be analyzed in a manner first proposed by Nusselt [1]. Consider the coordinate system shown in Figure 9-2. The plate temperature is maintained at T_w, and the vapor temperature at the edge of the film is the saturation temperature T_g. The film thickness is represented by δ, and we choose the coordinate system with the positive direction of x measured downward, as shown. It is assumed that the viscous shear of the vapor on the film is negligible at $y = \delta$. It is further assumed that a linear temperature distribution exists between wall and vapor conditions. The weight of the fluid element of thickness dx between y and δ is balanced by the viscous shear force at y and the buoyancy force due to the displaced vapor. Thus

$$\rho g(\delta - y)\, dx = \mu \frac{du}{dy}\, dx + \rho_v g(\delta - y)\, dx \qquad \text{[9-1]}$$

Integrating and using the boundary condition that $u = 0$ at $y = 0$ gives

$$u = \frac{(\rho - \rho_v)g}{\mu}\left(\delta y - \tfrac{1}{2}y^2\right) \qquad \text{[9-2]}$$

The mass flow of condensate through any x position of the film is thus given by

$$\text{Mass flow} = \dot{m} = \int_0^\delta \rho\left[\frac{(\rho - \rho_v)g}{\mu}\left(\delta y - \tfrac{1}{2}y^2\right)\right] dy$$

$$= \frac{\rho(\rho - \rho_v)g\, \delta^3}{3\mu} \qquad \text{[9-3]}$$

when unit depth is assumed. The heat transfer at the wall in the area dx is

$$q_x = -k\, dx\, \frac{\partial T}{\partial y}\bigg]_{y=0} = k\, dx\, \frac{T_g - T_w}{\delta} \qquad \text{[9-4]}$$

since a linear temperature profile was assumed. As the flow proceeds from x to $x + dx$, the film grows from δ to $\delta + d\delta$ as a result of the influx of additional condensate. The amount of condensate added between x and $x + dx$ is

$$\frac{d}{dx}\left[\frac{\rho(\rho-\rho_v)g\,\delta^3}{3\mu}\right]dx = \frac{d}{d\delta}\left[\frac{\rho(\rho-\rho_v)g\,\delta^3}{3\mu}\right]\frac{d\delta}{dx}\,dx$$
$$= \frac{\rho(\rho-\rho_v)g\,\delta^2\,d\delta}{\mu}$$

The heat removed by the wall must equal this incremental mass flow times the latent heat of condensation of the vapor. Thus

$$\frac{\rho(\rho-\rho_v)g\,\delta^2\,d\delta}{\mu}h_{fg} = k\,dx\frac{T_g - T_w}{\delta} \tag{9-5}$$

Equation (9-5) may be integrated with the boundary condition $\delta = 0$ at $x = 0$ to give

$$\delta = \left[\frac{4\mu kx(T_g - T_w)}{gh_{fg}\rho(\rho-\rho_v)}\right]^{1/4} \tag{9-6}$$

The heat-transfer coefficient is now written

$$h\,dx(T_w - T_g) = -k\,dx\frac{T_g - T_w}{\delta}$$

or

$$h = \frac{k}{\delta}$$

so that

$$h_x = \left[\frac{\rho(\rho-\rho_v)gh_{fg}k^3}{4\mu x(T_g - T_w)}\right]^{1/4} \tag{9-7}$$

Expressed in dimensionless form in terms of the Nusselt number, this is

$$\mathrm{Nu}_x = \frac{hx}{k} = \left[\frac{\rho(\rho-\rho_v)gh_{fg}x^3}{4\mu k(T_g - \overline{T}_w)}\right]^{1/4} \tag{9-8}$$

The average value of the heat-transfer coefficient is obtained by integrating over the length of the plate:

$$\overline{h} = \frac{1}{L}\int_0^L h_x\,dx = \frac{4}{3}h_{x=L} \tag{9-9}$$

or

$$\overline{h} = 0.943\left[\frac{\rho(\rho-\rho_v)gh_{fg}k_f^3}{L\mu_f(T_g - T_w)}\right]^{1/4} \tag{9-10}$$

More refined analyses of film condensation are presented in detail by Rohsenow [37]. The most significant refinements take into account a nonlinear temperature profile in the film and modifications to the energy balance to include additional energy to cool the film below the saturation temperature. Both effects can be handled by replacing h_{fg} with h'_{fg}, defined by

$$h'_{fg} = h_{fg} + 0.68c(T_g - T_w) \tag{9-11}$$

where c is the specific heat of the liquid. Otherwise, properties in Equations (9-7) and (9-10) should be evaluated at the film temperature

$$T_f = \frac{T_g + T_w}{2}$$

With these substitutions Equation (9-10) may be used for vertical plates and cylinders and fluids with Pr > 0.5 and $cT/h_{fg} \leq 1.0$.

For laminar film condensation on horizontal tubes, Nusselt obtained the relation

$$\bar{h} = 0.725 \left[\frac{\rho(\rho - \rho_v)gh_{fg}k_f^3}{\mu_f d(T_g - T_w)} \right]^{1/4} \tag{9-12}$$

where d is the diameter of the tube. When condensation occurs on a horizontal tube bank with n tubes placed directly over one another in the vertical direction, the heat-transfer coefficient may be calculated by replacing the diameter in Equation (9-12) with nd. The analysis of Reference 48 has shown that Equation (9-12) can be used for an isothermal sphere if the constant is changed to 0.815.

When a plate on which condensation occurs is sufficiently large or there is a sufficient amount of condensate flow, turbulence may appear in the condensate film. This turbulence results in higher heat-transfer rates. As in forced-convection flow problems, the criterion for determining whether the flow is laminar or turbulent is the Reynolds number, and for the condensation system it is defined as

$$\mathrm{Re}_f = \frac{D_H \rho V}{\mu_f} = \frac{4A\rho V}{P\mu_f}$$

where

$\quad D_H = $ hydraulic diameter
$\quad A = $ flow area
$\quad P = $ shear, or "wetted," perimeter
$\quad V = $ average velocity in flow

But

$$\dot{m} = \rho A V$$

so that

$$\mathrm{Re}_f = \frac{4\dot{m}}{P\mu_f} \tag{9-13}$$

where $\dot{m}$ is the mass flow through the particular section of the condensate film. For a vertical plate of unit depth, $P = 1$; for a vertical tube, $P = \pi d$. The critical Reynolds number is approximately 1800, and turbulent correlations for heat transfer must be used at Reynolds numbers greater than this value. The Reynolds number is sometimes expressed in terms of the mass flow per unit depth of plate Γ, so that

$$\mathrm{Re}_f = \frac{4\Gamma}{\mu_f} \tag{9-14}$$

In calculating the Reynolds numbers the mass flow may be related to the total heat transfer and the heat-transfer coefficient by

$$q = \bar{h}A(T_{\mathrm{sat}} - T_w) = \dot{m}h_{fg} \tag{9-15}$$

where A is the total surface area for heat transfer. Thus

$$\dot{m} = \frac{q}{h_{fg}} = \frac{\overline{h}A(T_{\text{sat}} - T_w)}{h_{fg}}$$

$$\text{Re}_f = \frac{4\overline{h}A(T_{\text{sat}} - T_w)}{h_{fg}P\mu_f} \qquad \textbf{[9-16]}$$

But

$$A = LW \qquad \text{and} \qquad P = W$$

where L and W are the height and width of the plate, respectively, so that

$$\text{Re}_f = \frac{4\overline{h}L(T_{\text{sat}} - T_w)}{h_{fg}\mu_f} \qquad \textbf{[9-17]}$$

The laminar condensation equations presented above match experimental data very well as long as the film remains smooth and well behaved. In practice, it has been found that ripples will develop in the film for Reynolds numbers as low as 30 or 40. When this occurs, the experimental values of $\overline{h}$ can be 20 percent higher than predicted by Equation (9-12). Because this is such a common occurrence, McAdams [3] was prompted to suggest that the 20 percent increase be adopted for design purposes. For our discussions here we shall use Equation (9-10) without the increase, recognizing that this is a conservative approach that provides a safety factor in design problems. If one wishes to employ the 20 percent higher coefficient, the resulting equation for vertical plates is

$$\overline{h} = 1.13\left[\frac{\rho(\rho - \rho_v)gh_{fg}k^3}{L\mu(T_g - T_w)}\right]^{1/4} \qquad \textbf{[9-18]}$$

If the vapor to be condensed is superheated, the preceding equations may be used to calculate the heat-transfer coefficient, provided the heat flow is calculated on the basis of the temperature difference between the surface and the saturation temperature corresponding to the system pressure. When a noncondensable gas is present along with the vapor, there may be an impediment of the heat transfer since the vapor must diffuse through the gas before it can condense on the surface. The reader should consult References 3 and 4 for more information on this subject.

Inclined Surfaces

If a plate or cylinder is inclined at an angle ϕ with the horizontal, the net effect on the above analysis is to replace the gravitational force with its component parallel to the heat-transfer surface, or

$$g' = g \sin \phi \qquad \textbf{[9-19]}$$

Therefore, for laminar flow, we suggest that inclined surfaces be treated with the simple substitution indicated in Equation (9-19).

9-3 | THE CONDENSATION NUMBER

Because the film Reynolds number is so important in determining condensation behavior, it is convenient to express the heat-transfer coefficient directly in terms of Re. We include the effect of inclination and write the heat-transfer equations in the form

$$\overline{h} = C\left[\frac{\rho(\rho - \rho_v)k^3 g \sin \phi \, h_{fg}}{\mu L(T_g - T_w)}\right]^{1/4} \qquad \textbf{[9-20]}$$

where the constant is evaluated for a plate or cylindrical geometry. From Equation (9-15) we can solve for $T_g - T_w$ as

$$T_g - T_w = \frac{\dot{m} h_{fg}}{\bar{h} A} \qquad \text{[9-21]}$$

where A, once again, is the surface area for heat transfer. Substituting Equation (9-21) in (9-20) and solving for $\bar{h}$ gives

$$\bar{h}^{3/4} = C \left[\frac{\rho(\rho - \rho_v)g \sin\phi k^3\; A/L}{\mu \dot{m}} \right]^{1/4} \qquad \text{[9-22]}$$

This may be restructured as

$$\bar{h}^{3/4} = C \left[\frac{\rho(\rho - \rho_v)gk^3}{\mu^2} \frac{\mu P}{4\dot{m}} \frac{4 \sin\phi\; A/P}{L} \right]^{1/4}$$

and we may solve for $\bar{h}$ as

$$\bar{h} = C^{4/3} \left[\frac{\rho(\rho - \rho_v)gk^3}{\mu^2} \frac{\mu P}{4\dot{m}} \frac{4 \sin\phi\; A/P}{L} \right]^{1/3} \qquad \text{[9-23]}$$

We now define a new dimensionless group, the *condensation number* Co, as

$$\text{Co} = \bar{h} \left[\frac{\mu^2}{k^3 \rho(\rho - \rho_v)g} \right]^{1/3} \qquad \text{[9-24]}$$

so that Equation (9-23) can be expressed in the form

$$\text{Co} = C^{4/3} \left(\frac{4 \sin\phi\; A/P}{L} \right)^{1/3} \text{Re}_f^{-1/3} \qquad \text{[9-25]}$$

For a vertical plate $A/PL = 1.0$, and we obtain, using the constant from Equation (9-10),

$$\text{Co} = 1.47\, \text{Re}_f^{-1/3} \qquad \text{for } \text{Re}_f < 1800 \qquad \text{[9-26]}$$

For a horizontal cylinder $A/PL = \pi$ and

$$\text{Co} = 1.514\, \text{Re}_f^{-1/3} \qquad \text{for } \text{Re}_f < 1800 \qquad \text{[9-27]}$$

When turbulence is encountered in the film, an empirical correlation by Kirkbride [2] may be used:

$$\text{Co} = 0.0077\, \text{Re}_f^{0.4} \qquad \text{for } \text{Re}_f > 1800 \qquad \text{[9-28]}$$

9-4 | FILM CONDENSATION INSIDE HORIZONTAL TUBES

Our discussion of film condensation so far has been limited to *exterior surfaces,* where the vapor and liquid condensate flows are not restricted by some overall flow-channel dimensions. Condensation inside tubes is of considerable practical interest because of applications to condensers in refrigeration and air-conditioning systems, but unfortunately these phenomena are quite complicated and not amenable to a simple analytical treatment. The overall flow rate of vapor strongly influences the heat-transfer rate in the forced convection-condensation system, and this in turn is influenced by the rate of liquid accumulation on the walls. Because of the complicated flow phenomena involved we shall present only two empirical relations for heat transfer and refer the reader to Rohsenow [37] for more complete information.

Chato [38] obtained the following expression for condensation of refrigerants at low vapor velocities inside horizontal tubes:

$$\bar{h} = 0.555 \left[\frac{\rho(\rho - \rho_v) g k^3 h'_{fg}}{\mu d(T_g - T_w)} \right]^{1/4} \tag{9-29}$$

where the modified enthalpy of vaporization is given by

$$h'_{fg} = h_{fg} + 0.375 c_{p,l}(T_g - T_w)$$

Liquid properties in Equation (9-29) are evaluated at the film temperature while h_{fg} and ρ_v are evaluated at the saturation temperature T_g.

Equation (9-29) is restricted to low vapor Reynolds numbers such that

$$\mathrm{Re}_v = \frac{d G_v}{\mu_v} < 35{,}000 \tag{9-30}$$

where Re_v is evaluated at *inlet* conditions to the tube. For higher flow rates an approximate empirical expression is given by Akers, Deans, and Crosser [39] as

$$\frac{\bar{h} d}{k_f} = 0.026 \, \mathrm{Pr}_f^{1/3} \, \mathrm{Re}_m^{0.8} \tag{9-31}$$

where now Re_m is a mixture Reynolds number, defined as

$$\mathrm{Re}_m = \frac{d}{\mu_f} \left[G_f + G_v \left(\frac{\rho_f}{\rho_v} \right)^{1/2} \right] \tag{9-32}$$

The mass velocities for the liquid G_f and vapor G_v are calculated as if each occupied the entire flow area. Equation (9-31) correlates experimental data within about 50 percent when

$$\mathrm{Re}_v = \frac{d G_v}{\mu_v} > 20{,}000 \qquad \mathrm{Re}_f = \frac{d G_f}{\mu_f} > 5000$$

EXAMPLE 9-1	Condensation on Vertical Plate

A vertical square plate, 30 by 30 cm, is exposed to steam at atmospheric pressure. The plate temperature is 98°C. Calculate the heat transfer and the mass of steam condensed per hour.

■ **Solution**

The Reynolds number must be checked to determine if the condensate film is laminar or turbulent. Properties are evaluated at the film temperature:

$$T_f = \frac{100 + 98}{2} = 99°C \qquad \rho_f = 960 \text{ kg/m}^3$$

$$\mu_f = 2.82 \times 10^{-4} \text{ kg/m} \cdot \text{s} \qquad k_f = 0.68 \text{ W/m} \cdot °C$$

For this problem the density of the vapor is very small in comparison with that of the liquid, and we are justified in making the substitution

$$\rho_f(\rho_f - \rho_v) \approx \rho_f^2$$

In trying to calculate the Reynolds number we find that it is dependent on the mass flow of condensate. But this is dependent on the heat-transfer coefficient, which is dependent on the Reynolds number. To solve the problem we assume either laminar or turbulent flow, calculate the heat-transfer coefficient, and then check the Reynolds number to see if our assumption was correct. Let us assume laminar film condensation. At atmospheric pressure we have

$$T_{sat} = 100°C \qquad h_{fg} = 2255 \text{ kJ/kg}$$

$$\bar{h} = 0.943 \left[\frac{\rho_f^2 g h_{fg} k_f^3}{L \mu_f (T_g - T_w)} \right]^{1/4}$$

$$= 0.943 \left[\frac{(960)^2 (9.8)(2.255 \times 10^6)(0.68)^3}{(0.3)(2.82 \times 10^{-4})(100 - 98)} \right]^{1/4}$$

$$= 13,150 \text{ W/m}^2 \cdot °C \quad [2316 \text{ Btu/h} \cdot \text{ft}^2 \cdot °F]$$

Checking the Reynolds number with Equation (9-17), we have

$$\text{Re}_f = \frac{4 \bar{h} L (T_{sat} - T_w)}{h_{fg} \mu f}$$

$$= \frac{(4)(13,150)(0.3)(100 - 98)}{(2.255 \times 10^6)(2.82 \times 10^{-4})} = 49.6$$

so that the laminar assumption was correct. The heat transfer is now calculated from

$$q = \bar{h} A (T_{sat} - T_w) = (13,150)(0.3)^2 (100 - 98) = 2367 \text{ W} \quad [8079 \text{ Btu/h}]$$

The total mass flow of condensate is

$$\dot{m} = \frac{q}{h_{fg}} = \frac{2367}{2.255 \times 10^6} = 1.05 \times 10^{-3} \text{ kg/s} = 3.78 \text{ kg/h} \quad [8.33 \text{ lb}_m/\text{h}]$$

Condensation on Tube Bank   **EXAMPLE 9-2**

One hundred tubes of 0.50-in (1.27-cm) diameter are arranged in a square array and exposed to atmospheric steam. Calculate the mass of steam condensed per unit length of tubes for a tube wall temperature of 98°C.

■ Solution

The condensate properties are obtained from Example 9-1. We employ Equation (9-12) for the solution, replacing d by nd, where $n = 10$. Thus,

$$\bar{h} = 0.725 \left[\frac{\rho_f^2 g h_{fg} k_f^3}{\mu_f nd (T_g - T_w)} \right]^{1/4}$$

$$= 0.725 \left[\frac{(960)^2 (9.8)(2.255 \times 10^6)(0.68)^3}{(2.82 \times 10^{-4})(10)(0.0127)(100 - 98)} \right]^{1/4}$$

$$= 12,540 \text{ W/m}^2 \cdot °C \quad [2209 \text{ Btu/h} \cdot \text{ft}^2 \cdot °F]$$

The total surface area is

$$\frac{A}{L} = n \pi d = (100)\pi(0.0127) = 3.99 \text{ m}^2/\text{m}$$

so the heat transfer is

$$\frac{q}{L} = h \frac{A}{L}(T_g - T_w)$$

$$= (12,540)(3.99)(100 - 98) = 100.07 \text{ kW/m}$$

The total mass flow of condensate is then

$$\frac{\dot{m}}{L} = \frac{q/L}{h_{fg}} = \frac{1.0007 \times 10^5}{2.255 \times 10^6} = 0.0444 \text{ kg/s} = 159.7 \text{ kg/h} \quad [352 \text{ lb}_m/\text{h}]$$

9-5 | BOILING HEAT TRANSFER

When a surface is exposed to a liquid and is maintained at a temperature above the saturation temperature of the liquid, boiling may occur, and the heat flux will depend on the difference in temperature between the surface and the saturation temperature. When the heated surface is submerged below a free surface of liquid, the process is referred to as *pool boiling*. If the temperature of the liquid is below the saturation temperature, the process is called *subcooled,* or *local, boiling.* If the liquid is maintained at saturation temperature, the process is known as *saturated,* or *bulk, boiling.*

The different regimes of boiling are indicated in Figure 9-3, where heat-flux data from an electrically heated platinum wire submerged in water are plotted against temperature excess $T_w - T_{sat}$. In region I, free-convection currents are responsible for motion of the fluid near the surface. In this region the liquid near the heated surface is superheated slightly, and it subsequently evaporates when it rises to the surface. The heat transfer in this region can be calculated with the free-convection relations presented in Chapter 7. In region II, bubbles begin to form on the surface of the wire and are dissipated in the liquid after breaking away from the surface. This region indicates the beginning of *nucleate boiling.* As the temperature excess is increased further, bubbles form more rapidly and rise to the surface of the liquid, where they are dissipated. This is indicated in region III. Eventually, bubbles are formed so rapidly that they blanket the heating surface and prevent the inflow of fresh liquid from taking their place. At this point the bubbles coalesce and form a vapor film that covers the

Figure 9-3 | Heat-flux data from an electrically heated platinum wire, from Farber and Scorah [9].

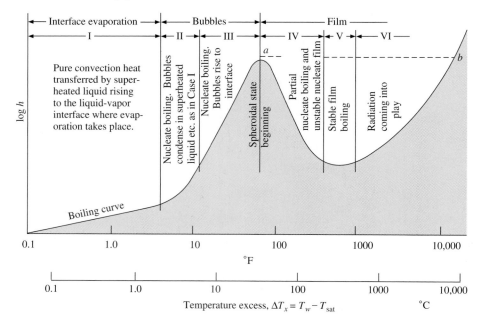

surface. The heat must be conducted through this film before it can reach the liquid and effect the boiling process. The thermal resistance of this film causes a reduction in heat flux, and this phenomenon is illustrated in region IV, the *film-boiling* region. This region represents a transition from nucleate boiling to film boiling and is unstable. Stable film boiling is eventually encountered in region V. The surface temperatures required to maintain stable film boiling are high, and once this condition is attained, a significant portion of the heat lost by the surface may be the result of thermal radiation, as indicated in region VI.

An electrically heated wire is unstable at point a since a small increase in ΔT_x at this point results in a decrease in the boiling heat flux. But the wire still must dissipate the same heat flux, or its temperature will rise, resulting in operation farther down to the boiling curve. Eventually, equilibrium may be reestablished only at point b in the film-boiling region. This temperature usually exceeds the melting temperature of the wire, so that burnout results. If the electric-energy input is quickly reduced when the system attains point a, it may be possible to observe the partial nucleate boiling and unstable film region.

In nucleate boiling, bubbles are created by the expansion of entrapped gas or vapor at small cavities in the surface. The bubbles grow to a certain size, depending on the surface tension at the liquid-vapor interface and the temperature and pressure. Depending on the temperature excess, the bubbles may collapse on the surface, may expand and detach from the surface to be dissipated in the body of the liquid, or at sufficiently high temperatures may rise to the surface of the liquid before being dissipated. When local boiling conditions are observed, the primary mechanism of heat transfer is thought to be the intense agitation at the heat-transfer surface, which creates the high heat-transfer rates observed in boiling. In saturated, or bulk, boiling the bubbles may break away from the surface because of the buoyancy action and move into the body of the liquid. In this case the heat-transfer rate is influenced by both the agitation caused by the bubbles and the vapor transport of energy into the body of the liquid.

Experiments have shown that the bubbles are not always in thermodynamic equilibrium with the surrounding liquid (i.e., the vapor inside the bubble is not necessarily at the same temperature as the liquid). Considering a spherical bubble as shown in Figure 9-4, the pressure forces of the liquid and vapor must be balanced by the surface-tension force at the vapor-liquid interface. The pressure force acts on an area of πr^2, and the surface tension acts on the interface length of $2\pi r$. The force balance is

$$\pi r^2 (p_v - p_l) = 2\pi r\sigma$$

or

$$p_v - p_l = \frac{2\sigma}{r} \qquad \text{[9-32a]}$$

Figure 9-4 | Force balance on a vapor bubble.

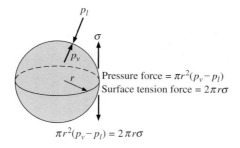

Pressure force $= \pi r^2 (p_v - p_l)$
Surface tension force $= 2\pi r\sigma$

$$\pi r^2 (p_v - p_l) = 2\pi r\sigma$$

where

$p_v =$ vapor pressure inside bubble

$p_l =$ liquid pressure

$\sigma =$ surface tension of vapor-liquid interface

Now, suppose we consider a bubble in pressure equilibrium (i.e., one that is not growing or collapsing). Let us assume that the temperature of the vapor inside the bubble is the saturation temperature corresponding to the pressure p_v. If the liquid is at the saturation temperature corresponding to the pressure p_l, it is below the temperature inside the bubble. Consequently, heat must be conducted out of the bubble, the vapor inside must condense, and the bubble must collapse. This is the phenomenon that occurs when the bubbles collapse on the heating surface or in the body of the liquid. In order for the bubbles to grow and escape to the surface, they must receive heat from the liquid. This requires that the liquid be in a superheated condition so that the temperature of the liquid is greater than the vapor temperature inside the bubble. This is a metastable thermodynamic state, but it is observed experimentally and accounts for the growth of bubbles after leaving the surface in some regions of nucleate boiling.

A number of photographic studies of boiling phenomena have been presented by Westwater et al. [17, 40, 41] that illustrate the various boiling regimes.

Figure 9-5 is a photograph illustrating several boiling regimes operating at once. The horizontal 6.1-mm-diameter copper rod is heated from the right side and immersed in

Figure 9-5 | A copper rod (6.1 mm in diameter) heated on the right side and immersed in isopropanol. Boiling regimes progress from free-convection boiling at the cooler end of the rod (left) to nucleate, transition, and finally film boiling at the right end.

Source: Photograph courtesy of Professor J. W. Westwater, University of Illinois, Urbana.

Figure 9-6 | Methanol boiling on a horizontal 9.53-mm-diameter
copper tube heated internally by condensing steam.
(a) $\Delta T_x = 37°C$, $q/A = 242.5$ kW/m^2, nucleate boiling;
(b) $\Delta T_x = 62°C$, $q/A = 217.6$ kW/m^2, transition boiling;
(c) $\Delta T_x = 82°C$, $q/A = 40.9$ kW/m^2, film boiling.

(a)

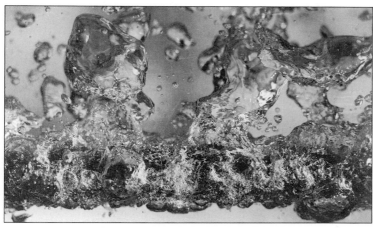

(b)

(c)

Source: Photographs courtesy of Professor J. W. Westwater, University of
Illinois, Urbana.

isopropanol. As a result of the temperature gradient along the rod, it was possible to observe the different regimes simultaneously. At the left end of the rod, the surface temperature is only slightly greater than the bulk fluid temperature, so that free-convection boiling is observed. Farther to the right, higher surface temperatures are experienced, and nucleate boiling is observed. Still farther to the right, transition boiling takes place; finally, film boiling is observed at the wall. Note the blanketing action of the vapor film on the right-hand portion of the rod.

More detailed photographs of the different boiling regimes using methanol are given in Figure 9-6. The vigorous action of nucleate boiling is illustrated in Figure 9-6a. At higher surface temperatures the bubbles start to coalesce, and transition boiling is observed, as in Figure 9-6b. Finally, at still higher temperatures the heat-transfer surface is completely covered by a vapor film, and large vapor bubbles break away from the surface. A more vigorous film-boiling phenomenon is illustrated in Figure 9-7 for methanol on a *vertical* tube. The vapor film rises up to the surface and develops into very active turbulent behavior at the top.

Figure 9-7 | A steam-heated vertical 19.05-mm-diameter copper tube displaying turbulent film boiling in methanol.
$\Delta T_x = 138°C$
$q/A = 38.8 \text{ kW/m}^2$.

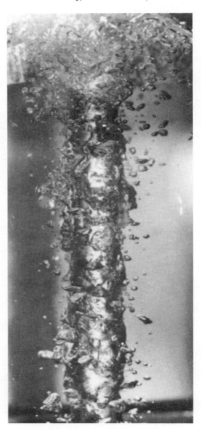

Source: Photograph courtesy of Professor J. W. Westwater, University of Illinois, Urbana.

The process of bubble growth is a complex one, but a simple qualitative explanation of the physical mechanism may be given. Bubble growth takes place when heat is conducted to the liquid-vapor interface from the liquid. Evaporation then takes place at the interface, thereby increasing the total vapor volume. Assuming that the liquid pressure remains constant, Equation (9-32a) requires that the pressure inside the bubble be reduced. Corresponding to a reduction in pressure inside the bubble will be a reduction in the vapor temperature and a larger temperature difference between the liquid and vapor if the bubble stays at its same spatial position in the liquid. However, the bubble will likely rise from the heated surface, and the farther away it moves, the lower the liquid temperature will be. Once the bubble moves into a region where the liquid temperature is below that of the vapor, heat will be conducted out, and the bubble will collapse. Hence the bubble growth process may reach a balance at some location in the liquid, or if the liquid is superheated enough, the bubbles may rise to the surface before being dissipated.

There is considerable controversy as to exactly how bubbles are initially formed on the heat-transfer surface. Surface conditions—both roughness and type of material—can play a central role in the bubble formation-and-growth drama. The mystery has not been completely solved and remains a subject of intense research. Excellent summaries of the status of knowledge of boiling heat transfer are presented in References 18, 23, 49, and 50. The interested reader is referred to these discussions for more extensive information than is presented in this chapter. Heat-transfer problems in two-phase flow are discussed by Wallis [28] and Tong [23].

Before specific relations for calculating boiling heat transfer are presented, it is suggested that the reader review the discussion of the last few pages and correlate it with some simple experimental observations of boiling. For this purpose a careful visual observation of the boiling process in a pan of water on the kitchen stove can be quite enlightening.

Rohsenow [5] correlated experimental data for nucleate pool boiling with the following relation.

$$\frac{C_l \Delta T_x}{h_{fg} \, \mathrm{Pr}_l^s} = C_{sf} \left[\frac{q/A}{\mu_l h_{fg}} \sqrt{\frac{g_c \sigma}{g(\rho_l - \rho_v)}} \right]^{0.33} \qquad \textbf{[9-33]}$$

where

C_l = specific heat of saturated liquid, $\mathrm{Btu/lb}_m \cdot {}^\circ\mathrm{F}$ or $\mathrm{J/kg} \cdot {}^\circ\mathrm{C}$
ΔT_x = temperature excess = $T_w - T_{\mathrm{sat}}$, ${}^\circ\mathrm{F}$ or ${}^\circ\mathrm{C}$
h_{fg} = enthalpy of vaporization, $\mathrm{Btu/lb}_m$ or $\mathrm{J/kg}$
Pr_l = Prandtl number of saturated liquid
q/A = heat flux per unit area, $\mathrm{Btu/h} \cdot \mathrm{ft}^2$ or $\mathrm{W/m}^2$
μ_l = liquid viscosity, $\mathrm{lb}_m/\mathrm{h} \cdot \mathrm{ft}$, or $\mathrm{kg/m} \cdot \mathrm{s}$
σ = surface tension of liquid-vapor interface, $\mathrm{lb}_f/\mathrm{ft}$ or $\mathrm{N/m}$
g = gravitational acceleration, $\mathrm{ft/s}^2$ or $\mathrm{m/s}^2$
ρ_l = density of saturated liquid, $\mathrm{lb}_m/\mathrm{ft}^3$ or $\mathrm{kg/m}^3$
ρ_v = density of saturated vapor, $\mathrm{lb}_m/\mathrm{ft}^3$ or $\mathrm{kg/m}^3$
C_{sf} = constant, determined from experimental data
s = 1.0 for water and 1.7 for other liquids

Values of the surface tension are given in Reference 10, and a brief tabulation of the vapor-liquid surface tension for water is given in Table 9-1.

The functional form of Equation (9-33) was determined by analyzing the significant parameters in bubble growth and dissipation. Experimental data for nucleate boiling of water on a platinum wire are shown in Figure 9-8, and a correlation of these data by the Rohsenow equation is shown in Figure 9-9, indicating good agreement. The value of the

Table 9-1 | Vapor-liquid surface tension for water.

Saturation temperature		Surface tension	
°F	°C	σ, mN/m	$\sigma \times 10^4$, lb_f/ft
32	0	75.6	51.8
60	15.56	73.3	50.2
100	37.78	69.8	47.8
140	60	66.0	45.2
200	93.33	60.1	41.2
212	100	58.8	40.3
320	160	46.1	31.6
440	226.67	32.0	21.9
560	293.33	16.2	11.1
680	360	1.46	1.0
705.4	374.1	0	0

Figure 9-8 | Heat-flux data for water boiling on a platinum wire $d = 0.6$ mm, from Reference 3. Numbers in parentheses are pressure in MPa.

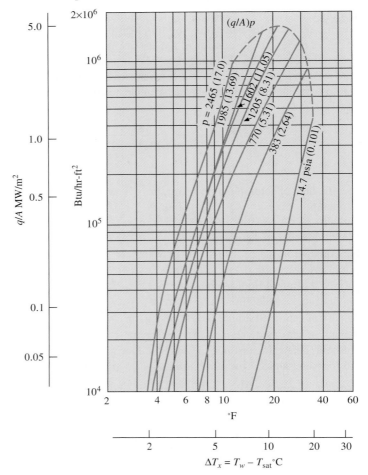

Figure 9-9 | Correlation of pool-boiling data by Equation (9-33), from Rohsenow [5].

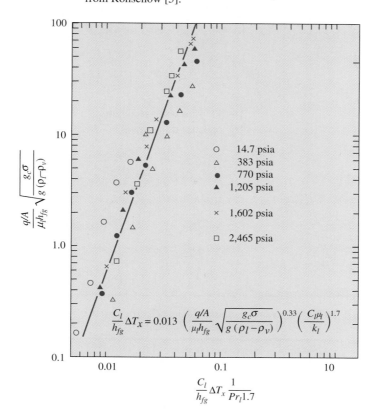

constant C_{sf} for the water-platinum combination is 0.013. Values for other fluid-surface combinations are given in Table 9-2. Equation (9-33) may be used for geometries other than horizontal wires, and in general it is found that geometry is not a strong factor in determining heat transfer for pool boiling. This would be expected because the heat transfer is primarily dependent on bubble formation and agitation, which is dependent on surface area, and not surface shape. Vachon, Nix, and Tanger [29] have determined values of the constants in the Rohsenow equation for a large number of surface-fluid combinations. There are several extenuating circumstances that influence the determination of the constants. Additional information given in Reference [57] indicates that depth of fluid as well as surface size and shape may have an effect on the values of the constants used in the Rohsenow equation.

Boiling on Brass Plate	**EXAMPLE 9-3**

A heated brass plate is submerged in a container of water at atmospheric pressure. The plate temperature is 242°F. Calculate the heat transfer per unit area of plate.

■ **Solution**

We could solve this problem by determining all the properties for use in Equation (9-33) and subsequently determining the heat flux. An alternative method is to use the data of Figure 9-8 in conjunction with Table 9-2. Upon writing Equation (9-33), we find that if the heat flux for *one* particular water-surface combination is known, the heat flux for some other surface may easily be determined in terms of the constants C_{sf} for the two surfaces since the fluid properties at any given temperature and pressure are the same regardless of the surface material. From Figure 9-8 the heat flux for the water-platinum combination is

$$\frac{q}{A} = 3 \times 10^5 \ \text{Btu/h} \cdot \text{ft}^2 \quad [946.1 \ \text{kW/m}^2]$$

since

$$T_w - T_{\text{sat}} = 242 - 212 = 30°\text{F} \quad [16.7°\text{C}]$$

From Table 9-2

$$C_{sf} = \begin{cases} 0.013 & \text{for water-platinum} \\ 0.006 & \text{for water-brass} \end{cases}$$

Accordingly,

$$\frac{(q/A)_{\text{water-brass}}}{(q/A)_{\text{water-platinum}}} = \left(\frac{C_{sf}\,\text{water-platinum}}{C_{sf}\,\text{water-brass}} \right)^3$$

and

$$\left(\frac{q}{A} \right)_{\text{water-brass}} = (3 \times 10^5) \left(\frac{0.013}{0.006} \right)^3$$

$$= 3.4 \times 10^6 \ \text{Btu/h} \cdot \text{ft}^2 \quad [1.072 \times 10^7 \ \text{W/m}^2]$$

Table 9-2 | Values of the coefficient C_{sf} for various liquid-surface combinations.

Fluid-heating-surface combination	C_{sf}
Water-copper [11][†]	0.013
Water-platinum [12]	0.013
Water-brass [13]	0.0060
Water–emery-polished copper [29]	0.0128
Water–ground and polished stainless steel [29]	0.0080
Water–chemically etched stainless steel [29]	0.0133
Water–mechanically polished stainless steel [29]	0.0132
Water–emery-polished and paraffin-treated copper [29]	0.0147
Water–scored copper [29]	0.0068
Water–Teflon pitted stainless steel [29]	0.0058
Carbon tetrachloride–copper [11]	0.013
Carbon tetrachloride–emery-polished copper [29]	0.0070
Benzene-chromium [14]	0.010
n-Butyl alcohol–copper [11]	0.00305
Ethyl alcohol–chromium [14]	0.027
Isopropyl alcohol–copper [11]	0.00225
n-Pentane–chromium [14]	0.015
n-Pentane–emery-polished copper [29]	0.0154
n-Pentane–emery-polished nickel [29]	0.0127
n-Pentane–lapped copper [29]	0.0049
n-Pentane–emery-rubbed copper [29]	0.0074
35% K_2CO_3–copper [11]	0.0054
50% K_2CO_3–copper [11]	0.0027

[†]Numbers in brackets refer to source of data.

When a liquid is forced through a channel or over a surface maintained at a temperature greater than the saturation temperature of the liquid, forced-convection boiling may result. For forced-convection boiling in smooth tubes Rohsenow and Griffith [6] recommended that the forced-convection effect be computed with the Dittus-Boelter relation of Chapter 6 [Equation (6-4)] and that this effect be added to the boiling heat flux computed from Equation (9-33). Thus

$$\left(\frac{q}{A}\right)_{total} = \left(\frac{q}{A}\right)_{boiling} + \left(\frac{q}{A}\right)_{forced\ convection}$$ **[9-34]**

For computing the forced-convection effect, it is recommended that the coefficient 0.023 be replaced by 0.019 in the Dittus-Boelter equation. The temperature difference between wall and liquid bulk temperature is used to compute the forced-convection effect.

The concept of adding the forced convection and boiling heat fluxes has been developed further in Reference 46 with good results; however, the terms in the equations are much more complicated and too elaborate to present here. An individual working in this field should consult this reference.

Forced-convection boiling is not necessarily as simple as might be indicated by Equation (9-34). This equation is generally applicable to forced-convection situations where the bulk liquid temperature is subcooled, in other words, for *local* forced-convection boiling. Once saturated or bulk boiling conditions are reached, the situation changes rapidly. A fully developed nucleate boiling phenomenon is eventually encountered that is independent of the flow velocity or forced-convection effects. Various relations have been presented for calculating the heat flux in the fully developed boiling state. McAdams et al. [21] suggested the following empirical relation for low-pressure boiling water:

$$\frac{q}{A} = 2.253(\Delta T_x)^{3.96}\ \text{W/m}^2 \qquad \text{for } 0.2 < p < 0.7 \text{ MPa}$$ **[9-35]**

For higher pressures Levy [22] recommends the relation

$$\frac{q}{A} = 283.2 p^{4/3}(\Delta T_x)^3\ \text{W/m}^2 \qquad \text{for } 0.7 < p < 14 \text{ MPa}$$ **[9-36]**

In these equations ΔT_x is in degrees Celsius and p is in megapascals.

If boiling is maintained for a sufficiently long length of tube, the majority of the flow area will be occupied by vapor. In this instance the vapor may flow rapidly in the central portion of the tube while a liquid film is vaporized along the outer surface. This situation is called *forced-convection vaporization* and is normally treated as a subject in two-phase flow and heat transfer. Several complications arise in this interesting subject, many of which are summarized by Tong [23] and Wallis [28].

The peak heat flux for nucleate pool boiling is indicated as point a in Figure 9-3 and by a dashed line in Figure 9-8. Zuber [7] has developed an analytical expression for the peak heat flux in nucleate boiling by considering the stability requirements of the interface between the vapor film and liquid. This relation is

$$\left(\frac{q}{A}\right)_{max} = \frac{\pi}{24} h_{fg}\rho_v \left[\frac{\sigma g(\rho_l - \rho_v)}{\rho_v^2}\right]^{1/4} \left(1 + \frac{\rho_v}{\rho_l}\right)^{1/2}$$ **[9-37]**

where σ is the vapor-liquid surface tension. This relation is in good agreement with experimental data. In general, the type of surface material does not affect the peak heat flux, although surface cleanliness can be an influence, dirty surfaces causing increases of approximately 15 percent in the peak value.

The peak heat flux in flow boiling is a more complicated situation because the rapid generation of vapor produces a complex two-phase flow system that strongly influences the maximum heat flux that may be attained at the heat-transfer surface. Near the heated surface a thin layer of superheated liquid is formed, followed by a layer containing both bubbles and liquid. The core of the flow is occupied, for the most part, by vapor. The heat transfer at the wall is influenced by the boundary-layer development in that region and also by the rate at which diffusion of vapor and bubbles can proceed radially. Still further complications may arise from flow oscillations that are generated under certain conditions. Gambill [24] has suggested that the critical heat flux in flow boiling may be calculated by a superposition of the critical heat flux for pool boiling [Equation (9-37)] and a forced-convection effect similar to the technique employed in Equation (9-34). Levy [25] has considered the effects of vapor diffusion on the peak heat flux in flow boiling, and Tong [23] presents a summary of available data on the subject.

An interesting peak heat-flux phenomenon is observed when liquid droplets impinge on hot surfaces. Experiments with water, acetone, alcohol, and some of the Freons indicate that the maximum heat transfer is observed for temperature excesses of about 165°C, for all the fluids. The peak flux is a function of the fluid properties and the normal component of the impact velocity. A correlation of experimental data is given in Reference 30 as

$$\frac{Q_{max}}{\rho_L d^3 \lambda} = 1.83 \times 10^{-3} \left(\frac{\rho_L{}^2 V^2 d}{\rho_{vf} \sigma g_c} \right)^{0.341} \tag{9-38}$$

where

Q_{max} = maximum heat transfer per drop
ρ_L = density of liquid droplet
V = normal component of impact velocity
ρ_{vf} = vapor density evaluated at film temperature $(T_w + T_{sat})/2$
σ = surface tension
d = drop diameter
λ = modified heat of vaporization, defined by

$$\lambda = h_{fg} + c_{pv} \left(\frac{T_w - T_{sat}}{2} \right)$$

While not immediately apparent from this equation, the heat-transfer rates in droplet impingement are quite high, and as much as 50 percent of the droplet is evaporated during the short time interval of impact and bouncing. The case of zero impact velocity is of historical note and is called the *Leidenfrost phenomenon* [31]. This latter case can be observed by watching water droplets sizzle and dance about on a hot plate. Very high heat-transfer rates are also experienced when a liquid jet impinges on a hot surface maintained at temperatures significantly greater than the saturation temperature. Both nucleate- and film-boiling phenomena can be observed, and relations for calculating the heat-transfer rates are presented in Reference 36.

Sun and Lienhard [34] have presented a relation for the peak boiling heat flux on horizontal cylinders that is in good agreement with experimental data. The relation is

$$\frac{q''_{max}}{q''_{maxF}} = 0.89 + 2.27 \exp(-3.44\sqrt{R'}) \qquad \text{for } 0.15 < R' \tag{9-39}$$

where R' is a dimensionless radius defined by

$$R' = R \left[\frac{g(\rho_l - \rho_v)}{\sigma} \right]^{1/2}$$

and q''_{maxF} is the peak heat flux on an infinite horizontal plate derived in Reference 33 as

$$q''_{maxF} = 0.131 \sqrt{\rho_v} \, h_{fg} [\sigma g (\rho_l - \rho_v)]^{1/4} \tag{9-40}$$

Here, σ is the surface tension.

Bromley [8] suggests the following relation for calculation of heat-transfer coefficients in the stable film-boiling region on a horizontal tube:

$$h_b = 0.62 \left[\frac{k_g^3 \rho_v (\rho_l - \rho_v) g (h_{fg} + 0.4 c_{pv} \Delta T_x)}{d \mu_v \Delta T_x} \right]^{1/4} \qquad \textbf{[9-41]}$$

where d is the tube diameter. This heat-transfer coefficient considers only the conduction through the film and does not include the effects of radiation. The total heat-transfer coefficient may be calculated from the empirical relation

$$h = h_b \left(\frac{h_b}{h} \right)^{1/3} + h_r \qquad \textbf{[9-42]}$$

where h_r is the radiation heat-transfer coefficient and is calculated by assuming an emissivity of unity for the liquid. Thus

$$h_r = \frac{\sigma \epsilon (T_w^4 - T_{sat}^4)}{T_w - T_{sat}} \qquad \textbf{[9-43]}$$

where σ is the Stefan-Boltzmann constant and ϵ is the emissivity of the surface. Note that Equation (9-42) will require an iterative solution for the total heat-transfer coefficient.

The properties of the vapor in Equation (9-41) are to be evaluated at the film temperature defined by

$$T_f = \frac{1}{2}(T_w + T_{sat})$$

while the enthalpy of vaporization h_{fg} is to be evaluated at the saturation temperature.

9-6 | SIMPLIFIED RELATIONS FOR BOILING HEAT TRANSFER WITH WATER

Many empirical relations have been developed to estimate the boiling heat-transfer coefficients for water. Some of the simplest relations are those presented by Jakob and Hawkins [15] for water boiling on the outside of submerged surfaces at atmospheric pressure (Table 9-3). These heat-transfer coefficients may be modified to take into account the influence of pressure by using the empirical relation

$$h_p = h_1 \left(\frac{p}{p_1} \right)^{0.4} \qquad \textbf{[9-44]}$$

Table 9-3 | Simplified relations for boiling heat-transfer coefficients to water at atmospheric pressure, $\Delta T_x = T_w - T_{sat}$, °C.

Surface	$\frac{q}{A}$, kW/m²	h, W/m² · °C	Approximate range of ΔT, °C	Approximate range of h, W/m² · °C
Horizontal	$\frac{q}{A} < 16$	$1042(\Delta T_x)^{1/3}$	0–7.76	0–2060
	$16 < \frac{q}{A} < 240$	$5.56(\Delta T_x)^3$	7.32–14.4	2180–16,600
Vertical	$\frac{q}{A} < 3$	$537(\Delta T_{ax})^{1/7}$	0–4.51	0–670
	$3 < \frac{q}{A} < 63$	$7.96(\Delta T_x)^3$	4.41–9.43	680–6680

where

h_p = heat-transfer coefficient at some pressure p
h_1 = heat-transfer coefficient at atmospheric pressure as determined from Table 9-3
p = system pressure
p_1 = standard atmospheric pressure

For forced-convection local boiling inside vertical tubes the following relation is recommended [16]:

$$h = 2.54(\Delta T_x)^3 e^{p/1.551} \text{ W/m}^2 \cdot {}^\circ\text{C} \qquad \textbf{[9-45]}$$

where ΔT_x is the temperature difference between the surface and saturated liquid in degrees Celsius and p is the pressure in MPa. The heat-transfer coefficient has the units of watts per square meter per degree Celsius. Equation (9-45) is valid over a pressure range of 5 to 170 atm.

EXAMPLE 9-4   Flow Boiling

Water at 5 atm flows inside a tube of 1-in [2.54-cm] diameter under local boiling conditions where the tube wall temperature is 10°C above the saturation temperature. Estimate the heat transfer in a 1.0-m length of tube.

■ **Solution**
For this calculation we use Equation (9-45), noting that

$$\Delta T_x = 10^\circ\text{C}$$

$$p = (5)(1.0132 \times 10^5 \text{ N/m}^2) = 0.5066 \text{ MPa}$$

The heat-transfer coefficient is then calculated as

$$h = (2.54)(10)^3 e^{0.5066/1.551}$$

$$= 3521 \text{ W/m}^2 \cdot {}^\circ\text{C} \quad [620 \text{ Btu/h} \cdot \text{ft}^2 \cdot {}^\circ\text{F}]$$

The surface area for a 1-m length of tube is

$$A = \pi dL = \pi(0.0254)(1.0) = 0.0798 \text{ m}^2$$

so the heat transfer is

$$q = hA(T_w - T_{\text{sat}})$$

$$= (3521)(0.0798)(10) = 2810 \text{ W/m}$$

EXAMPLE 9-5   Water Boiling in a Pan

Water at 1 atm boils in a stainless-steel kitchen pan with $\Delta T_x = 8^\circ\text{C}$. Estimate the heat flux that will be obtained. If the same pan operates as a pressure cooker at 0.17 MPa, what percent increase in heat flux might be expected?

■ **Solution**
We will use the simplified relation of Table 9-3 for the estimates. We do not know the value of q/A and so must choose one of the two relations for a horizontal surface from the table. We anticipate nucleate boiling, so choose

$$h = 5.56(\Delta T_x)^3 = 5.56(8)^3 = 2847 \text{ W/m}^2 \cdot {}^\circ\text{C}$$

and the heat flux is

$$q/A = h\,\Delta T = (2847)(8) = 22{,}770 \text{ W/m}^2 = 22.8 \text{ kW/m}^2$$

This value falls within the applicable range of the equation used ($16 < q/A < 240$ kW/m^2). We may note from Table 9-2 that the values of C_{sf} for platinum–water or stainless steel–water combinations are approximately equal at 0.013, so one might anticipate that the heat flux displayed for Figure 9-8 would be approximately the same for a stainless-steel surface. At $\Delta T_x = 8°C$ an extrapolation of the figure does give a value for the heat flux that is in approximate agreement with the 22.8 kW/m^2 value calculated above.

For operation as a pressure cooker we obtain the value of h from Equation (9-44)

$$h_p = h_1(p/p_1)^{0.4} = (2847)(0.17/0.101)^{0.4} = 3506 \text{ W/m}^2 \cdot °C$$

or a value 23 percent higher than that obtained at 1 atm. The corresponding heat flux is

$$q/A = h\,\Delta T = (3506)(8) = 28 \text{ kW/m}^2$$

9-7 | THE HEAT PIPE

A device that makes use of change-of-phase heat transfer in a novel way is the heat pipe, illustrated in Figure 9-10 for a horizontal position. A circular pipe has a layer of wicking material covering the inside surface, as shown, with a hollow core in the center. A condensable fluid is also contained in the pipe, and the liquid permeates the wicking material by capillary action. When heat is added to one end of the pipe (the evaporator), liquid is vaporized in the wick and the vapor moves to the central core. At the other end of the pipe, heat is removed (the condenser) and the vapor condenses back into the wick. Liquid is replenished in the evaporator section by capillary action.

Although the basic concept will work in the absence of gravity, the heat pipe may be tilted so that the condenser is at a higher elevation than the evaporator. In this case, the action of gravity serves to speed the flow of liquid down through the wicking material. This is called a *favorable tilt*. In contrast, when the condenser is placed at a lower elevation than the evaporator the action of gravity will impede the flow of liquid in the wick and the heat pipe is said to have an *adverse tilt*.

Figure 9-10 | Basic heat pipe configuration in horizontal position. Heat addition in the evaporator causes vapor release from the wick. Heat removal in the condenser causes condensation of vapor and return of liquid through the wick by capillary action.

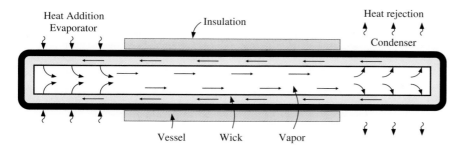

Several combinations of fluid and pipe materials have been used for heat-pipe construction, and some typical operating characteristics are summarized in Table 9-4. Very high heat fluxes are obtained; for this reason research efforts are being devoted to optimum wick designs, novel configurations for specialized applications, etc. A number of wick materials have been employed in practice, ranging from mesh screens, stainless steel fibers, and sintered fibers and powders, to etched microgrooves/channels.

The heat pipe illustrated in Figure 9-10 is called a *fixed-conductance* heat pipe because its thermal resistance is not a strong function of heat load when a single condensing fluid is employed. The basic design may be modified to operate as a *variable-conductance* heat pipe as shown in Figure 9-11.

A reservoir containing a noncondensable gas is connected to the heat-removal end of the heat pipe. This gas may then form an interface with the vapor and "choke off" part of the condensation to the wick. With increased heat addition, more vapor is generated with an increase in vapor pressure, and the noncondensable gas is forced back into the reservoir, thereby opening up additional condenser area to remove the additional heat. For a reduction in heat addition, just the reverse operation is observed. If the heat-source temperature drops below a certain minimum value, depending on the specific fluid and gas combinations in the heat pipe, a complete shutoff can occur. So the control feature can be particularly useful for fast warm-up applications in addition to its value as a temperature leveler for variable-load conditions.

Heat pipes are particularly useful in energy-conservation equipment where it is desired to recover heat from hot gases for air-preheat or supplemental heating applications. In some cases the heat pipe can take the place of more costly combinations of pumps, piping, and dual-heat-exchanger configurations. Further information on heat pipe theory and design is given in References 51–56.

Table 9-4 | Typical axial heat fluxes for heat pipes.

Temperature, °C	Fluid	Pipe material	Axial heat flux, kW/cm^2
−160	Liquid nitrogen	Stainless steel	0.067
−70 to +60	Liquid NH_3	Stainless steel	0.3
−100	Methanol	Copper	0.45
200	Water	Copper	0.67
750	Potassium	Nickel, stainless steel	5.6
850	Sodium	Nickel, stainless steel	9.3

Figure 9-11 | Variable-conductance heat pipe configuration. The noncondensable-gas reservoir may be used to provide a temperature-control device.

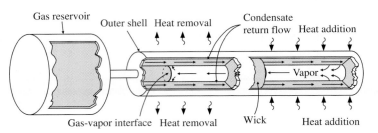

<table>
<tr><th>Heat-Flux Comparisons</th><th>EXAMPLE 9-6</th></tr>
</table>

Using the data of Table 9-4, compare the axial heat flux in a heat pipe using water as the working fluid (at about 200°C) with the heat flux in a solid copper bar 8 cm long experiencing a temperature differential of 100°C.

■ **Solution**

The heat flux per unit area is expressed as

$$q/A = -k(\Delta T/\Delta x)$$

From Table A-2 the thermal conductivity of copper is 374 W/m · °C, so that for an 8-cm length

$$q/A = -(374)(-100)/0.08 = 467.5 \text{ kW/m}^2 = 0.04675 \text{ kW/cm}^2$$

From Table 9-4 the typical axial heat flux for a water heat pipe is

$$(q/A)_{axial} = 0.67 \text{ kW/cm}^2$$

Thus the heat pipe transfers more than 10 times the heat of a pure copper rod with a substantial temperature gradient. This example illustrates why the heat pipe enjoys wide application possibilities.

9-8 | SUMMARY AND DESIGN INFORMATION

Boiling and condensation phenomena are very complicated, as we have shown in the preceding sections. The equations presented in these sections may be used to calculate heat-transfer coefficients for various geometries and fluid-surface combinations. For many preliminary design applications only approximate values of heat flux or heat-transfer coefficient are required, and Tables 9-5 to 9-7 give summaries of such information. Of course, more accurate values should be obtained for the final design of heat-transfer equipment.

Table 9-5 | Approximate values of condensation heat-transfer coefficients for vapors at 1 atm, according to References 3 and 45.

Fluid	Geometry	h W/m^2 · °C	Btu/h · ft^2 · °F	$T_g - T_w$, °C
Steam	Vertical surface	4000–11,300	700–2000	22–3
	Horizontal tubes, 15- to 76-mm diameter	9600–24,400	1700–4300	20–2
Diphenyl	Vertical surface, turbulent, 3.66 m	680–2400	120–430	72–13
	Horizontal tube, 43-mm diameter	1280–2270	225–400	15–5
Dowtherm A	Vertical surface 3.66 m turbulent	680–3060	120–540	40–20
Ethanol	Vertical surface, 152 mm	1130–2000	200–350	55–11
	Horizontal tube, 51-mm diameter	1800–2550	320–450	22–6
Propanol	Horizontal tube, 51-mm diameter	1400–1700	250–300	26–13
Butanol	Horizontal tube, 51-mm diameter	1400–1700	250–300	26–13
Benzene	Horizontal tubes, 15- to 33-mm diameter	1300–2150	230–380	45–13

Table 9-6 | Relative magnitudes of nucleate boiling heat-transfer coefficients at 1 atm relative to value for water.

Fluid	h_{fluid}/h_{water}
Water	1.0
20% sugar	0.87
10% Na_2SO_4	0.94
26% glycerin	0.83
55% glycerin	0.75
24% NaCl	0.61
Isopropanol	0.70
Methanol	0.53
Toluene	0.36
Carbon tetrachloride	0.35
n-Butanol	0.32

Table 9-7 | Approximate burnout heat flux at 1 atm, according to References 3, 43, and 44.

Fluid-surface combination	$(q/A)_{max}$		ΔT_x,
	kW/m^2	Btu/h·ft^2 $\times 10^{-3}$	°C
Water, copper	620–850	200–270	
Copper-chrome plated	940–1260	300–400	23–28
Steel	1290	410	30
Benzene, copper	130	43.5	
Aluminum	160	50.5	
Propanol, nickel-plated copper	210–340	67–110	42–50
Butanol, nickel-plated copper	250–330	79–105	33–39
Ethanol, aluminum	170	55	
Copper	250	80.5	
Methanol, copper	390	125	
Chrome-plated copper	350	111	
Steel	390	125	
Liquid H_2	30	9.53	2
Liquid N_2	100	31.7	11
Liquid O_2	150	47.5	11

REVIEW QUESTIONS

1. Why are higher heat-transfer rates experienced in dropwise condensation than in film condensation?
2. How is the Reynolds number defined for film condensation?
3. What is meant by subcooled and saturated boiling?
4. Distinguish between nucleate and film boiling.
5. How is forced-convection boiling calculated?
6. Why does radiation play a significant role in film-boiling heat transfer?

LIST OF WORKED EXAMPLES

9-1 Condensation on vertical plate
9-2 Condensation on tube bank
9-3 Boiling on brass plate
9-4 Flow boiling
9-5 Water boiling in a pan
9-6 Heat-flux comparisons

PROBLEMS

9-1 Using Equation (9-28) as a starting point, develop an expression for the average heat-transfer coefficient in turbulent condensation as a function of only the fluid properties, length of the plate, and temperature difference; that is, eliminate the Reynolds number from Equation (9-28) to obtain a relation similar to Equation (9-10) for laminar condensation.

9-2 Show that the condensation Reynolds number for laminar condensation on a vertical plate may be expressed as

$$\mathrm{Re}_f = 3.77 \left[\frac{L^3(T_g - T_w)^3 \rho_f(\rho_f - \rho_v)g k_f^3}{\mu_f^5 h_{fg}^3} \right]^{1/4}$$

9-3 Develop an expression for the total condensate flow in a turbulent film in terms of the fluid properties, the temperature difference, and the dimensions of the plate.

9-4 Plot Equations (9-26) and (9-28) as

$$\log \left\{ \bar{h} \left[\frac{\mu_f^2}{k_f^3 \rho_f(\rho_f - \rho_v)g} \right]^{1/3} \right\}$$

versus $\log \mathrm{Re}_f$. Discuss this plot.

9-5 A vertical plate 30 cm wide and 1.2 m high is maintained at 80°C and exposed to saturated steam at 1 atm. Calculate the heat transfer and the total mass of steam condensed per hour.

9-6 A 40 by 40 cm plate is inclined at an angle of 30° with the vertical and exposed to water vapor at 1 atm. The plate is maintained at 98°C. Calculate the heat-transfer and mass-flow rate of condensate.

9-7 A 50 by 50 cm square vertical plate is maintained at 95°C and exposed to saturated steam at 1 atm pressure. Calculate the amount of steam condensed per hour.

9-8 Calculate the rate of condensation on a 1.5 by 1.5 m vertical plate maintained at 40°F and exposed to saturated water vapor at 55°F; $h_{fg} = 2376$ kJ/kg at 55°F.

9-9 A vertical plate 40 by 40 cm is exposed to saturated vapor ammonia at 38°C and the plate surface is maintained constant at 30°C. Calculate the condensation rate if $h_{fg} = 1111$ kJ/kg at 38°C.

9-10 Compare the results of Equation (9-36) with that of Table 9-3.

9-11 Saturated steam at 100 lb/in^2 abs condenses on the outside of a horizontal 1-in-diameter tube. The tube wall temperature is maintained at 280°F. Calculate the heat-transfer coefficient and the condensate flow per unit length of tube. Take $T_{sat} = 328°F$ and $h_{fg} = 889$ Btu/lb.

9-12 An uninsulated, chilled water pipe carrying water at 2°C passes through a hot, humid factory area where the temperature is 35°C and the relative humidity is 80 percent because of steam-operated equipment in the factory. If the pipe is 5 cm in diameter and the exposed length is 7.5 m, estimate the condensate that will drip off the pipe. For this estimate assume that the pipe is exposed to saturated vapor at the partial pressure of the water vapor in the air.

9-13 A certain pressure cooker is designed to operate at 20 lb/in^2 gauge. It is well known that an item of food will cook faster in such a device because of the higher steam temperature at the higher pressure. Consider a certain item of food as a horizontal 4-in-diameter cylinder at a temperature of 95°F when placed in the cooker. Calculate the percentage increase in heat transfer to this cylinder for the 20-lb/in^2-gauge condition compared with condensation on the cylinder at standard atmospheric pressure.

9-14 Saturated steam at 690 kPa abs condenses on the outside of a horizontal 1-in-diameter tube. The tube wall temperature is maintained at 138°C. Calculate the heat-transfer coefficient and the condensate flow per unit length of tube.

9-15 Saturated steam at 1 atm pressure condenses on the outside of a 30-cm-diameter tube whose surface is maintained at 95°C. The tube is 15 m long. Calculate the amount of steam condensed per hour.

9-16 Condensing carbon dioxide at 20°C is in contact with a horizontal 10-cm-diameter tube maintained at 15°C. Calculate the condensation rate per meter of length if $h_{fg} = 153.2$ kJ/kg at 20°C.

9-17 A square array of four hundred 6.35-mm tubes is used to condense steam at atmospheric pressure. The tube walls are maintained at 88°C by a coolant flowing inside the tubes. Calculate the amount of steam condensed per hour per unit length of the tubes.

9-18 Saturated water vapor at 1 atm enters a horizontal 5-cm-diameter tube 1.5 m long. Estimate the condensation for a tube wall temperature of 98°C.

9-19 Steam at 1 atm is to be condensed on the outside of a bank of 10×10 horizontal tubes 2.54 cm in diameter. The tube surface temperature is maintained at 95°C. Calculate the quantity of steam condensed for a tube length of 0.61 m.

9-20 A square array of 2.54-cm-diameter tubes contains 100 tubes each having a length of 0.91 m. The distance between centers is 45.7 mm and the tube wall temperature is 97°C. The tubes are exposed to steam at 1 atm. Calculate the condensation rate in kilograms per hour.

9-21 In a large cold-storage plant, ammonia is used as the refrigerant and in one application 1.2×10^7 kJ/h must be removed by condensing the ammonia at 29.4°C with an array of tubes with walls maintained at 25.6°C. Select several tube sizes, lengths, and array dimensions that might be used to accomplish the task. $h_{fg} = 1148$ kJ/kg at 29°C.

9-22 An ammonia condenser uses a 20 by 20 array of 6.35-mm-diameter tubes 0.305 m long. The ammonia condenses at 32.2°C and the tube walls are maintained at 27.8°C by a water flow inside. Calculate the rate of condensation of ammonia. h_{fg} at 32.2°C is 1135 kJ/kg.

9-23 A condenser is to be designed to condense 10,000 kg/h of refrigerant 12 (CCl_2F_2) at 37.8°C. A square 25 by 25 array of 12-mm-diameter tubes is to be used, with water flow inside the tubes maintaining the wall temperature at 32.2°C. Calculate the length of the tubes. $h_{fg} = 130$ kJ/kg at 37.8°C.

9-24 Refrigerant 12 (CCl_2F_2) is condensed inside a horizontal 12-mm-diameter tube at a low vapor velocity. The condensing temperature is 32.2°C and the tube wall is at 26.7°C. Calculate the mass condensed per meter of tube length. $h_{fg} = 133.5$ kJ/kg at 32.2°C.

9-25 A heated vertical plate at a temperature of 107°C is immersed in a tank of water exposed to atmospheric pressure. The temperature of the water is 100°C, and boiling occurs at the surface of the plate. The area of the plate is 0.3 m². What is the heat lost from the plate in watts?

9-26 The surface tension of water at 212°F is 58.8 dyn/cm for the vapor in contact with the liquid. Assuming that the saturated vapor inside a bubble is at 213°F while the surrounding liquid is saturated at 212°F, calculate the size of the bubble.

9-27 Assuming that the bubble in Problem 9-26 moves through the liquid at a velocity of 4.5 m/s, estimate the time required to cool the bubble 0.3°C by calculating the heat-transfer coefficient for flow over a sphere and using this in a lumped-capacity analysis as described in Chapter 4.

9-28 A heated 30 by 30 cm square copper plate serves as the bottom for a pan of water at 1 atm pressure. The temperature of the plate is maintained at 119°C. Estimate the heat transferred per hour by the plate.

9-29 Compare the heat flux calculated from the simple relations of Table 9-3 with the curve for atmospheric pressure in Figure 9-8. Make the comparisons for two or three values of the temperature excess.

9-30 Water at 4 atm pressure flows inside a 2-cm-diameter tube under local boiling conditions where the tube wall temperature is 12°C above the saturation temperature. Estimate the heat transfer in a 60-cm length of tube.

9-31 Compare the heat-transfer coefficients for nucleate boiling of water, as shown in Figure 9-8, with the simplified relations given in Table 9-3.

9-32 Using Equations (9-14) and (9-7), develop Equation (9-26).

9-33 A platinum wire is submerged in saturated water at 5.3 MPa. What is the heat flux for a temperature excess of 10°C?

9-34 Water at 1 atm flows in a 1.25-cm-diameter brass tube at a velocity of 1.2 m/s. The tube wall is maintained at 110°C, and the average bulk temperature of the water is 96°C. Calculate the heat-transfer rate per unit length of tube.

9-35 A kettle with a flat bottom 30 cm in diameter is available. It is desired to boil 2.3 kg/h of water at atmospheric pressure in this kettle. At what temperature must the bottom surface of the kettle be maintained to accomplish this?

9-36 A 5-mm-diameter copper heater rod is submerged in water at 1 atm. The temperature excess is 11°C. Estimate the heat loss per unit length of the rod.

9-37 Compare the heat-transfer coefficients for boiling water and condensing steam on a horizontal tube for normal atmospheric pressure.

9-38 A certain boiler employs one hundred 2-cm-diameter tubes 1 m long. The boiler is designed to produce local forced-convection boiling of water at 3 MPa pressure with $\Delta T_x = 10$°C. Estimate the total heat-transfer rate and the amount of saturated vapor that can be produced at 3 MPa.

9-39 A horizontal tube 3 mm in diameter and 7.5 cm long is submerged in water at 1.6 atm. Calculate the surface temperature necessary to generate a heat flux of 0.2 MW/m^2.

9-40 Copper electric heating rods 2.5 cm in diameter are used to produce steam at 34-kPa-gauge pressure in a nucleate-pool-boiling arrangement where $\Delta T_x = 4°C$. Estimate the length of rod necessary to produce 908 kg/h of saturated vapor steam.

9-41 Estimate the nucleate-pool-boiling heat-transfer coefficient for a water–26% glycerin mixture at 1 atm in contact with a copper surface and $\Delta T_x = 15°C$.

9-42 Compare Equations (9-35) and (9-36) with Equation (9-45).

9-43 Estimate the nucleate-pool-boiling heat flux for water at 1 atm in contact with ground and polished stainless steel and $\Delta T_x = 17°C$.

9-44 Calculate the peak heat flux for boiling water at atmospheric pressure on a horizontal cylinder of 1.25-cm OD. Use the Lienhard relation.

9-45 How much heat would be lost from a horizontal platinum wire, 1.0 mm in diameter and 12 cm long, submerged in water at atmospheric pressure if the surface temperature of the wire is 110°C?

9-46 A horizontal pipe at 94°C is exposed to steam at atmospheric pressure and 100°C. The pipe is 4.0 cm in diameter. Calculate the condensation rate per meter of length.

9-47 Liquid water at 1 atm and 98°C flows in a horizontal 2.5-cm-diameter brass tube maintained at 110°C. Calculate the heat-transfer coefficient if the Reynolds number based on liquid bulk conditions is 40,000.

9-48 A steel bar 1.25 cm in diameter and 5 cm long is removed from a 1200°C furnace and placed in a container of water at atmospheric pressure. Estimate the heat-transfer rate from the bar when it is first placed in the water.

9-49 Estimate the peak heat flux for boiling water at normal atmospheric pressure.

9-50 Heat-transfer coefficients for boiling are usually large compared with those for ordinary convection. Estimate the flow velocity that would be necessary to produce a value of h for forced convection through a smooth 6.5-mm-diameter brass tube comparable with that which could be obtained by pool boiling with $\Delta T_x = 16.7°C$, $p = 690$ kPa, and water as the fluid. See Problem 9-11 for data on properties.

9-51 A horizontal tube having a 1.25-cm OD is submerged in water at 1 atm and 100°C. Calculate the heat flux for surface temperatures of (*a*) 540°C, (*b*) 650°C, and (*c*) 800°C. Assume $\epsilon = 0.8$, and use Equation (9-41).

9-52 Water at a rate of 1.0 liter/h in the form of 0.4-mm droplets at 25°C is sprayed on a hot surface at 280°C with an impact velocity of 3 m/s. Estimate the maximum heat transfer that can be achieved with this arrangement.

9-53 A square array of 196 tubes 1.25 cm in diameter is used to condense steam at 1 atm pressure. Water flowing inside the tubes maintains the outside surface temperature at 92°C. Calculate the condensation rate for tube lengths of 2.0 m.

9-54 A vertical plate maintained at 91°C is exposed to saturated steam at 1 atm pressure. Determine the height of plate to just produce a Reynolds number of 1800. What would the condensation rate be under these conditions?

9-55 Determine the boiling heat flux for a water–ground and polished stainless-steel combination for a temperature excess of 15°C at 1 atm.

9-56 Calculate the heat-transfer coefficient for forced-convection local boiling of water at 5 atm in a vertical tube. The temperature excess is 10°C.

9-57 Calculate the condensation rate for saturated steam at 1 atm exposed to a horizontal 30-cm-diameter cylinder maintained at 94°C.

9-58 A 20-cm-square vertical plate is maintained at 93°C and exposed to saturated water vapor at 1 atm pressure. Calculate the condensation rate and film thickness at the bottom of the plate.

9-59 Consider the copper-finned tube of Problem 2-121. Estimate the heat transfer that would result from submerging this tube in water at 1 atm with base tube temperatures at 108°C, 111°C, and 117°C.

9-60 Water is placed in a stainless-steel pan on a kitchen stove. Estimate the heat transfer rate from a 25-cm-diameter pan when the bottom surface is maintained at 104°C.

9-61 A gas flame is placed around a shallow stainless steel pan having a bottom diameter of 12.5 cm and walls 5 cm high. The gas flame may be assumed to produce uniform heating, and the pan is filled with water to a depth of 2.5 cm. Estimate the heat transfer rate to the water if the surface temperature of the pan is maintained at 104°C.

9-62 Estimate the heat transfer coefficient from a horizontal Teflon-plated stainless-steel tube to water at atmospheric pressure with a temperature excess of 17°C. What would the heat transfer be for a 1-cm-diameter tube, 1 m long?

9-63 A horizontal tube bank employs 10 columns of tubes, with each column 15 rows high. The 12-mm-diameter tubes are exposed to condensing steam at 1 atm. If the tube wall temperature is maintained at 86°C, calculate the condensation rate per meter of tube length.

9-64 A vertical plate 25 cm wide by 50 cm high at 85°C is exposed to steam at 1 atm. Estimate the condensation rate in kg/h. Also estimate the film thickness at the bottom of the plate.

Design-Oriented Problems

9-65 A condenser is to be designed to condense 1.3 kg/s of steam at atmospheric pressure. A square array of 1.25-cm-OD tubes is to be used with the outside tube walls maintained at 93°C. The spacing of the tubes is to be 1.9 cm between centers, and their length is 3 times the square dimension. How many tubes are required for the condenser, and what are the outside dimensions?

9-66 A heat exchanger is to be designed to condense 600 kg/h of steam at atmospheric pressure. A square array of four hundred 1.0-cm-diameter tubes is available for the task, and the tube wall temperature is to be maintained at 97°C. Estimate the length of tubes required.

9-67 The design situation for the desalinization system of Problem 8-173 obviously depends on the condensation heat-transfer coefficient from the transmitting material to the air-vapor mixture inside. Assuming an inclination angle of 40 degrees, estimate the condensation coefficient for several values of interior and surface temperatures. Assume that the condensation process is that of saturated vapor at the mixture temperature and neglect the presence of the air. (This is not a particularly good assumption because the condensation is influenced by diffusion through the air.) Assume different values for the length of the inclined surface.

REFERENCES

1. Nusselt, W. "Die Oberflachenkondensation des Wasserdampfes." *VDI Z.,* vol. 60, p. 541, 1916.
2. Kirkbride, C. G. "Heat Transfer by Condensing Vapors on Vertical Tubes," *Trans. AIChE,* vol. 30, p. 170, 1934.
3. McAdams, W. H. *Heat Transmission,* 3rd ed. New York: McGraw-Hill, 1954.

4. Kern, D. Q. *Process Heat Transfer.* New York: McGraw-Hill, 1950.

5. Rohsenow, W. M. "A Method of Correlating Heat Transfer Data for Surface Boiling Liquids," *Trans. ASME,* vol. 74, p. 969, 1952.

6. Rohsenow, W. M., and P. Griffith. "Correlation of Maximum Heat Flux Data for Boiling of Saturated Liquids," *AIChE-ASME Heat Transfer Symp.,* Louisville, Ky., 1955.

7. Zuber, N. "On the Stability of Boiling Heat Transfer," *Trans. ASME,* vol. 80, p. 711, 1958.

8. Bromley, L. A. "Heat Transfer in Stable Film Boiling," *Chem. Eng. Prog.,* vol. 46, p. 221, 1950.

9. Farber, E. A., and E. L. Scorah. "Heat Transfer to Water Boiling under Pressure," *Trans. ASME,* vol. 70, p. 369, 1948.

10. *Handbook of Chemistry and Physics.* Cleveland, Ohio: Chemical Rubber Publishing Company, 1960.

11. Piret, E. L., and H. S. Isbin. "Natural Circulation Evaporation Two-Phase Heat Transfer," *Chem. Eng. Prog.,* vol. 50, p. 305, 1954.

12. Addoms, J. N. "Heat Transfer at High Rates to Water Boiling outside Cylinders," Sc.D. thesis, Massachusetts Institute of Technology, Cambridge, Mass., 1948.

13. Cryder, D. S., and A. C. Finalbargo. "Heat Transmission from Metal Surfaces to Boiling Liquids: Effect of Temperature of the Liquid on Film Coefficient," *Trans. AIChE,* vol. 33, p. 346, 1937.

14. Cichelli, M. T., and C. F. Bonilla. "Heat Transfer to Liquids Boiling under Pressure," *Trans. AIChE,* vol. 41, p. 755, 1945.

15. Jakob, M., and G. Hawkins. *Elements of Heat Transfer,* 3rd ed. New York: John Wiley & Sons, 1957.

16. Jakob, M. *Heat Transfer,* vol. 2, p. 584. New York: John Wiley & Sons, 1957.

17. Haley, K. W., and J. W. Westwater. "Heat Transfer from a Fin to a Boiling Liquid," *Chem. Eng. Sci.,* vol. 20, p. 711, 1965.

18. Rohsenow, W. M. (ed.). *Developments in Heat Transfer.* Cambridge, Mass.: The M.I.T. Press, 1964.

19. Leppert, G., and C. C. Pitts. "Boiling," *Adv. Heat Transfer,* vol. 1, 1964.

20. Gebhart, C. *Heat Transfer.* New York: McGraw-Hill, 1961.

21. McAdams, W. H., et al. "Heat Transfer at High Rates to Water with Surface Boiling," *Ind. Eng. Chem.,* vol. 41, pp. 1945–55, 1949.

22. Levy, S. "Generalized Correlation of Boiling Heat Transfer," *J. Heat Transfer,* vol. 81C, pp. 37–42, 1959.

23. Tong, L. S. *Boiling Heat Transfer and Two-Phase Flow,* 2nd ed. New York: John Wiley & Sons, 1997.

24. Gambill, W. R. "Generalized Prediction of Burnout Heat Flux for Flowing, Sub-cooled, Wetting Liquids," *AIChE Rep.* 17, *5th Nat. Heat Transfer Conf., Houston,* 1962.

25. Levy, S. "Prediction of the Critical Heat Flux in Forced Convection Flow," *USAEC Rep.* 3961, 1962.

26. Schmidt, E., W. Schurig, and W. Sellschop. "Versuche über die Kondensation von Wasserdampf in Film- und Tropfenform," *Tech. Mech. Thermodyn. Bull.,* vol. 1, p. 53, 1930.

27. Citakoglu, E., and J. W. Rose. "Dropwise Condensation: Some Factors Influencing the Validity of Heat Transfer Measurements," *Int. J. Heat Mass Transfer,* vol. 11, p. 523, 1968.

28. Wallis, G. B. *One-Dimensional Two-Phase Flow.* New York: McGraw-Hill, 1969.

29. Vachon, R. I., G. H. Nix, and G. E. Tanger. "Evaluation of Constants for the Rohsenow Pool-Boiling Correlation," *J. Heat Transfer,* vol. 90, p. 239, 1968.

30. McGinnis, F. K., and J. P. Holman. "Individual Droplet Heat Transfer Rates for Splattering on Hot Surfaces," *Int. J. Heat Mass Transfer,* vol. 12, p. 95, 1969.

31. Bell, K. J. "The Leidenfrost Phenomenon: A Survey," *Chem. Eng. Prog. Symp. Ser.,* no. 79, p. 73, 1967.

32. Rohsenow, W. M. "Nucleation with Boiling Heat Transfer," *ASME Pap.* 70-HT-18.

33. Zuber, N., M. Tribus, and J. W. Westwater. "The Hydrodynamic Crises in Pool Boiling of Saturated and Subcooled Liquids," *Int. Dev. Heat Transfer,* pp. 230–35, 1963.

34. Sun, K. H., and J. H. Lienhard. "The Peak Boiling Heat Flux on Horizontal Cylinders," *Int. J. Heat Mass Trans.,* vol. 13, p. 1425, 1970.

35. Graham, C., and P. Griffith. "Drop Size Distributions and Heat Transfer in Dropwise Condensation," *Int. J. Heat Mass Transfer,* vol. 16, p. 337, 1973.

36. Ruch, M. A., and J. P. Holman. "Boiling Heat Transfer to a Freon-113 Jet Impinging Upward onto a Flat Heated Surface," *Int. J. Heat Mass Transfer,* vol. 18, p. 51, 1974.

37. Rohsenow, W. M. "Film Condensation," chap. 12 in *Handbook of Heat Transfer.* New York: McGraw-Hill, 1973.

38. Chato, J. C. *J. Am. Soc. Refrig. Air Cond. Eng.,* February 1962, p. 52.

39. Akers, W. W., H. A. Deans, and O. K. Crosser. "Condensing Heat Transfer within Horizontal Tubes," *Chem. Eng. Prog. Symp. Ser.,* vol. 55, no. 29, p. 171, 1958.

40. Westwater, J. W., and J. G. Santangelo. *Ind. Eng. Chem.,* vol. 47, p. 1605, 1955.

41. Westwater, J. W. *Am. Sci.,* vol. 47, p. 427, 1959, photo by Y. Y. Hsu.

42. Fritz, W. Verdampfen und Kondensieren, *Z. VDI, Beih, Verfahrenstech.,* no. 1, 1943.

43. Sauer, E. T., H. B. Cooper, and W. H. McAdams. "Heat Transfer to Boiling Liquids," *Mech. Eng.,* vol. 60, p. 669, 1938.

44. Weil, L. "Échanges thermiques dans les liquides bouillants," *Fourth Int. Congr. Ind. Heating,* group I, sec. 13, Rep. 210, Paris, 1952.

45. Chilton, T. H., A. P. Colburn, R. P. Genereaux, and H. C. Vernon: *Trans. ASME, Pet. Mech. Eng.,* vol. 55, p. 7, 1933.

46. Bjorge, R. W., G. R. Hall, and W. M. Rohsenow. "Correlations of Forced Convection Boiling Heat Transfer Data," *Int. J. Heat Mass Transfer,* vol. 25, p. 753, 1982.

47. Westwater, J. W. "Gold Surfaces for Condensation Heat Transfer," *Gold Bulletin,* vol. 14, pp. 95–101, 1981.

48. Dhir, V. K., and J. H. Lienhard. "Laminar Film Condensation on Plane and Axisymmetric Bodies in Non-Uniform Gravity," *J. Heat Trans.,* vol. 93, p. 97, 1971.

49. Rohsenow, W. M., and J. P. Hartnett (eds.). *Handbook of Heat Transfer,* chap. 13. New York: McGraw-Hill, 1973.

50. Lienhard, J. H. *A Heat Transfer Textbook,* chap. 10. Englewood Cliffs, N.J.: Prentice-Hall, 1981.

51. Peterson, G. P. *Introduction to Heat Pipes.* New York: John Wiley, 1994.

52. Feldman, K. T., and G. H. Whiting. "Applications of the Heat Pipe," *Mech. Eng.,* vol. 90, p. 48, November 1968.

53. Dutcher, C. H., and M. R. Burke. "Heat Pipes: A Cool Way of Cooling Circuits," *Electronics,* pp. 93–100, February 16, 1970.

54. Grover, G. M., T. P. Cotter, and G. F. Erickson. "Structures of Very High Thermal Conductance," *J. Appl. Phys.,* vol. 35, p. 1990, 1964.

55. Tien, C. L. "Fluid Mechanics of Heat Pipes," *Ann. Rev. Fluid Mechanics,* vol. 7, p. 167, 1975.

56. Chi, S. W. *Heat Pipe Theory and Practice.* New York: Hemisphere Publishing Co., 1976.

57. Pioro, I. L., "Experimental evaluation of constants for the Rohsenow pool boiling correlation," *Int. J. Heat and Mass Transfer,* vol. 42, pp. 2003–13, 1999.

Heat Exchangers

$$y = y_1 + \left(\frac{x - x_1}{x_2 - x_1}\right)(y_2 - y_1)$$
$$\dot{m} = \rho A u_m$$

10-1 | INTRODUCTION

The application of the principles of heat transfer to the design of equipment to accomplish a certain engineering objective is of extreme importance, for in applying the principles to design, the individual is working toward the important goal of product development for economic gain. Eventually, economics plays a key role in the design and selection of heat-exchange equipment, and the engineer should bear this in mind when embarking on any new heat-transfer design problem. The weight and size of heat exchangers used in space or aeronautical applications are very important parameters, and in these cases cost considerations are frequently subordinated insofar as material and heat-exchanger construction costs are concerned; however, the weight and size are important cost factors in the overall application in these fields and thus may still be considered as economic variables.

A particular application will dictate the rules that one must follow to obtain the best design commensurate with economic considerations, size, weight, etc. An analysis of all these factors is beyond the scope of our present discussion, but it is well to remember that they all must be considered in practice. Our discussion of heat exchangers will take the form of technical analysis; that is, the methods of predicting heat-exchanger performance will be outlined, along with a discussion of the methods that may be used to estimate the heat-exchanger size and type necessary to accomplish a particular task. In this respect, we limit our discussion to heat exchangers where the primary modes of heat transfer are conduction and convection. This is not to imply that radiation is not important in heat-exchanger design, for in many space applications it is the predominant means available for effecting an energy transfer. The reader is referred to the discussions by Siegal and Howell [1] and Sparrow and Cess [7] for detailed consideration of radiation heat-exchanger design.

10-2 | THE OVERALL HEAT-TRANSFER COEFFICIENT

We have already discussed the overall heat-transfer coefficient in Section 2-4 with the heat transfer through the plane wall of Figure 10-1 expressed as

$$q = \frac{T_A - T_B}{1/h_1 A + \Delta x/kA + 1/h_2 A} \qquad \text{[10-1]}$$

Figure 10-1 | Overall heat transfer through a plane wall.

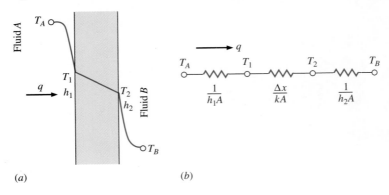

(a)   (b)

where T_A and T_B are the fluid temperatures on each side of the wall. The overall heat-transfer coefficient U is defined by the relation

$$q = UA \, \Delta T_{\text{overall}}$$ **[10-2]**

From the standpoint of heat-exchanger design, the plane wall is of infrequent application; a more important case for consideration would be that of a double-pipe heat exchanger, as shown in Figure 10-2. In this application one fluid flows on the inside of the smaller tube while the other fluid flows in the annular space between the two tubes. The convection coefficients are calculated by the methods described in previous chapters, and the overall heat transfer is obtained from the thermal network of Figure 10-2b as

$$q = \frac{T_A - T_B}{\dfrac{1}{h_i A_i} + \dfrac{\ln(r_o/r_i)}{2\pi k L} + \dfrac{1}{h_o A_o}}$$ **[10-3]**

Figure 10-2 | Double-pipe heat exchange: (a) schematic; (b) thermal-resistance network for overall heat transfer.

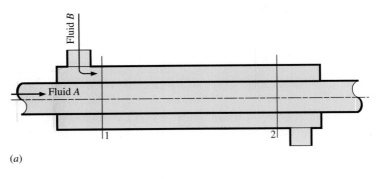

(a)

(b)

Table 10-1 | Approximate values of overall heat-transfer coefficients.

Physical situation	U	
	Btu/h · ft^2 · °F	W/m^2 · °C
Brick exterior wall, plaster interior, uninsulated	0.45	2.55
Frame exterior wall, plaster interior: uninsulated	0.25	1.42
with rock-wool insulation	0.07	0.4
Plate-glass window	1.10	6.2
Double plate-glass window	0.40	2.3
Steam condenser	200–1000	1100–5600
Feedwater heater	200–1500	1100–8500
Freon-12 condenser with water coolant	50–150	280–850
Water-to-water heat exchanger	150–300	850–1700
Finned-tube heat exchanger, water in tubes, air across tubes	5–10	25–55
Water-to-oil heat exchanger	20–60	110–350
Steam to light fuel oil	30–60	170–340
Steam to heavy fuel oil	10–30	56–170
Steam to kerosone or gasoline	50–200	280–1140
Finned-tube heat exchanger, steam in tubes, air over tubes	5–50	28–280
Ammonia condenser, water in tubes	150–250	850–1400
Alcohol condenser, water in tubes	45–120	255–680
Gas-to-gas heat exchanger	2–8	10–40

where the subscripts i and o pertain to the inside and outside of the smaller inner tube. The overall heat-transfer coefficient may be based on either the inside or outside area of the tube at the discretion of the designer. Accordingly,

$$U_i = \frac{1}{\dfrac{1}{h_i} + \dfrac{A_i \ln(r_o/r_i)}{2\pi k L} + \dfrac{A_i}{A_o}\dfrac{1}{h_o}} \qquad \textbf{[10-4a]}$$

$$U_o = \frac{1}{\dfrac{A_o}{A_i}\dfrac{1}{h_i} + \dfrac{A_o \ln(r_o/r_i)}{2\pi k L} + \dfrac{1}{h_o}} \qquad \textbf{[10-4b]}$$

Although final heat-exchanger designs will be made on the basis of careful calculations of U, it is helpful to have a tabulation of values of the overall heat-transfer coefficient for various situations that may be encountered in practice. Comprehensive information of this sort is available in References 5 and 6, and an abbreviated list of values of U is given in Table 10-1. We should remark that the value of U is governed in many cases by only one of the convection heat-transfer coefficients. In most practical problems the conduction resistance is small compared with the convection resistances. Then, if one value of h is markedly lower than the other value, it will tend to dominate the equation for U. Examples 10-1 and 10-2 illustrate this concept.

Overall Heat-Transfer Coefficient for Pipe in Air   EXAMPLE 10-1

Hot water at 98°C flows through a 2-in schedule 40 horizontal steel pipe [$k = 54$ W/m · °C] and is exposed to atmospheric air at 20°C. The water velocity is 25 cm/s. Calculate the overall heat-transfer coefficient for this situation, based on the outer area of pipe.

■ **Solution**

From Appendix A the dimensions of 2-in schedule 40 pipe are

$$ID = 2.067 \text{ in} = 0.0525 \text{ m}$$
$$OD = 2.375 \text{ in} = 0.06033 \text{ m}$$

The heat-transfer coefficient for the water flow on the inside of the pipe is determined from the flow conditions with properties evaluated at the bulk temperature. The free-convection heat-transfer coefficient on the outside of the pipe depends on the temperature difference between the surface and ambient air. This temperature difference depends on the overall energy balance. First, we evaluate h_i and then formulate an iterative procedure to determine h_o.

The properties of water at 98°C are

$$\rho = 960 \text{ kg/m}^3 \qquad \mu = 2.82 \times 10^{-4} \text{ kg/m} \cdot \text{s}$$
$$k = 0.68 \text{ W/m} \cdot {}°\text{C} \qquad \text{Pr} = 1.76$$

The Reynolds number is

$$\text{Re} = \frac{\rho u d}{\mu} = \frac{(960)(0.25)(0.0525)}{2.82 \times 10^{-4}} = 44{,}680 \qquad [a]$$

and since turbulent flow is encountered, we may use Equation (6-4):

$$\text{Nu} = 0.023 \, \text{Re}^{0.8} \, \text{Pr}^{0.4}$$
$$= (0.023)(44{,}680)^{0.8}(1.76)^{0.4} = 151.4$$
$$h_i = \text{Nu} \frac{k}{d} = \frac{(151.4)(0.68)}{0.0525} = 1961 \text{ W/m}^2 \cdot {}°\text{C} \quad [345 \text{ Btu/h} \cdot \text{ft}^2 \cdot {}°\text{F}] \qquad [b]$$

For unit length of the pipe the thermal resistance of the steel is

$$R_s = \frac{\ln(r_o/r_i)}{2\pi k} = \frac{\ln(0.06033/0.0525)}{2\pi(54)} = 4.097 \times 10^{-4} \qquad [c]$$

Again, on a unit-length basis the thermal resistance on the inside is

$$R_i = \frac{1}{h_i A_i} = \frac{1}{h_i 2\pi r_i} = \frac{1}{(1961)\pi(0.0525)} = 3.092 \times 10^{-3} \qquad [d]$$

The thermal resistance for the outer surface is as yet unknown but is written, for unit lengths,

$$R_o = \frac{1}{h_o A_o} = \frac{1}{h_o 2\pi r_o} \qquad [e]$$

From Table 7-2, for laminar flow, the simplified relation for h_o is

$$h_o = 1.32 \left(\frac{\Delta T}{d} \right)^{1/4} = 1.32 \left(\frac{T_o - T_\infty}{d} \right)^{1/4} \qquad [f]$$

where T_o is the unknown outside pipe surface temperature. We designate the inner pipe surface as T_i and the water temperature as T_w; then the energy balance requires

$$\frac{T_w - T_i}{R_i} = \frac{T_i - T_o}{R_s} = \frac{T_o - T_\infty}{R_o} \qquad [g]$$

$q = \dot{m} C_p (T_{co} - T_a)$

Combining Equations (e) and (f) gives

$$\frac{T_o - T_\infty}{R_o} = 2\pi r_o \frac{1.32}{d^{1/4}}(T_o - T_\infty)^{5/4} \qquad [h]$$

This relation may be introduced into Equation (g) to yield two equations with the two unknowns T_i and T_o:

$$\frac{98 - T_i}{3.092 \times 10^{-3}} = \frac{T_i - T_o}{4.097 \times 10^{-4}}$$

$$\frac{T_i - T_o}{4.097 \times 10^{-4}} = \frac{(\pi)(0.06033)(1.32)(T_o - 20)^{5/4}}{(0.06033)^{1/4}}$$

This is a nonlinear set that may be solved by iteration to give

$$T_o = 97.6°C \qquad T_i = 97.65°C$$

As a result, the outside heat-transfer coefficient and thermal resistance are

$$h_o = \frac{(1.32)(97.6 - 20)^{1/4}}{(0.06033)^{1/4}} = 7.91 \text{ W/m}^2 \cdot °C \quad [1.39 \text{ Btu/h} \cdot \text{ft}^2 \cdot °F]$$

$$R_o = \frac{1}{(0.06033)(7.91)\pi} = 0.667$$

The calculation clearly illustrates the fact that the free convection controls the overall heat-transfer because R_o is much larger than R_i or R_s. The overall heat-transfer coefficient based on the outer area is written in terms of these resistances as

$$U_o = \frac{1}{A_o(R_i + R_s + R_o)} \qquad [i]$$

With numerical values inserted,

$$U_o = \frac{1}{\pi(0.06033)(3.092 \times 10^{-3} + 4.097 \times 10^{-4} + 0.667)}$$
$$= 7.87 \text{ W/Area} \cdot °C$$

In this calculation we used the outside area for a 1.0-m length as

$$A_o = \pi(0.06033) = 0.1895 \text{ m}^2/\text{m}$$
$$U_o = 7.87 \text{ W/m}^2 \cdot °C$$

Thus, we find that the overall heat-transfer coefficient is almost completely controlled by the value of h_o. We might have expected this result strictly on the basis of our experience with the relative magnitude of convection coefficients; free-convection values for air are very low compared with forced convection with liquids.

Overall Heat-Transfer Coefficient for Pipe Exposed to Steam

EXAMPLE 10-2

The pipe and hot-water system of Example 10-1 is exposed to steam at 1 atm and 100°C. Calculate the overall heat-transfer coefficient for this situation based on the outer area of pipe.

■ **Solution**

We have already determined the inside convection heat-transfer coefficient in Example 10-1 as

$$h_i = 1961 \text{ W/m}^2 \cdot °C$$

The convection coefficient for condensation on the outside of the pipe is obtained by using Equation (9-12),

$$h_o = 0.725 \left[\frac{\rho(\rho - \rho_v)gh_{fg}k_f^3}{\mu_f d(T_g - T_o)} \right]^{1/4} \tag{a}$$

where T_o is the outside pipe-surface temperature. The water film properties are

$$\rho = 960 \text{ kg/m}^3 \qquad \mu_f = 2.82 \times 10^{-4} \text{ kg/m} \cdot \text{s}$$
$$k_f = 0.68 \text{ W/m} \cdot °\text{C} \qquad h_{fg} = 2255 \text{ kJ/kg}$$

so Equation (a) becomes

$$h_o = 0.725 \left[\frac{(960)^2(9.8)(2.255 \times 10^6)(0.68)^3}{(2.82 \times 10^{-4})(0.06033)(100 - T_o)} \right]^{1/4}$$
$$= 17,960(100 - T_o)^{-1/4} \tag{b}$$

and the outside thermal resistance per unit length is

$$R_o = \frac{1}{h_o A_o} = \frac{(100 - T_o)^{1/4}}{(17,960)\pi(0.06033)} = \frac{(100 - T_o)^{1/4}}{3403} \tag{c}$$

The energy balance requires

$$\frac{100 - T_o}{R_o} = \frac{T_o - T_i}{R_s} = \frac{T_i - T_w}{R_i} \tag{d}$$

From Example 10-1

$$R_i = 3.092 \times 10^{-3} \qquad R_s = 4.097 \times 10^{-4} \qquad T_w = 98°\text{C}$$

and Equations (c) and (d) may be combined to give

$$3403(100 - T_o)^{3/4} = \frac{(T_o - T_i)}{4.097 \times 10^{-4}}$$

$$\frac{T_o - T_i}{4.097 \times 10^{-4}} = \frac{T_i - 98}{3.092 \times 10^{-3}}$$

This is a nonlinear set of equations that may be solved to give

$$T_o = 99.91°\text{C} \qquad T_i = 99.69°\text{C}$$

The exterior heat-transfer coefficient and thermal resistance then become

$$h_o = 17,960(100 - 99.91)^{-1/4} = 32,790 \text{ W/m}^2 \cdot °\text{C} \tag{e}$$

$$R_o = \frac{(100 - 99.91)^{1/4}}{3403} = 1.610 \times 10^{-4} \tag{f}$$

Based on unit length of pipe, the overall heat-transfer coefficient is

$$U_o = \frac{1}{A_o(R_i + R_s + R_o)}$$
$$= \frac{1}{\pi(0.06033)(3.092 \times 10^{-3} + 4.097 \times 10^{-4} + 1.610 \times 10^{-4})} \tag{g}$$
$$= 1441 \text{ W/}°\text{C} \cdot \text{Area}$$

Since A_o and the R's were both per unit length,

$$U_o = 1441 \text{ W/m}^2 \cdot {}^\circ\text{C} \qquad [254 \text{ Btu/h} \cdot \text{h} \cdot \text{ft}^2 \cdot {}^\circ\text{F}]$$

In this problem the water-side convection coefficient is the main controlling factor because h_o is so large for a condensation process. In fact, the outside thermal resistance is smaller than the conduction resistance of the steel. The approximate *relative* magnitudes of the resistances are

$$R_o \sim 1 \qquad R_s \sim 2.5 \qquad R_i \sim 19$$

10-3 | FOULING FACTORS

After a period of operation the heat-transfer surfaces for a heat exchanger may become coated with various deposits present in the flow systems, or the surfaces may become corroded as a result of the interaction between the fluids and the material used for construction of the heat exchanger. In either event, this coating represents an additional resistance to the heat flow, and thus results in decreased performance. The overall effect is usually represented by a *fouling factor*, or fouling resistance, R_f, which must be included along with the other thermal resistances making up the overall heat-transfer coefficient.

Fouling factors must be obtained experimentally by determining the values of U for both clean and dirty conditions in the heat exchanger. The fouling factor is thus defined as

$$R_f = \frac{1}{U_{\text{dirty}}} - \frac{1}{U_{\text{clean}}}$$

An abbreviated list of recommended values of the fouling factor for various fluids is given in Table 10-2, and a very complete treatment of the subject is available in Reference [9].

Table 10-2 | Table of selected fouling factors, according to Reference 2.

Type of fluid	Fouling factor, $h \cdot \text{ft}^2 \cdot {}^\circ\text{F/Btu}$	$\text{m}^2 \cdot {}^\circ\text{C/W}$
Seawater, below 125°F	0.0005	0.00009
Above 125°F	0.001	0.002
Treated boiler feedwater above 125°F	0.001	0.0002
Fuel oil	0.005	0.0009
Quenching oil	0.004	0.0007
Alcohol vapors	0.0005	0.00009
Steam, non-oil-bearing	0.0005	0.00009
Industrial air	0.002	0.0004
Refrigerating liquid	0.001	0.0002

Influence of Fouling Factor   EXAMPLE 10-3

Suppose the water in Example 10-2 is seawater above 125°F and a fouling factor of $0.0002 \text{ m}^2 \cdot {}^\circ\text{C/W}$ is experienced. What is the percent reduction in the convection heat-transfer coefficient?

■ **Solution**

The fouling factor influences the heat-transfer coefficient on the inside of the pipe. We have

$$R_f = 0.0002 = \frac{1}{h_{\text{dirty}}} - \frac{1}{h_{\text{clean}}}$$

Using $h_{\text{clean}} = 1961$ W/m$^2 \cdot$ °C we obtain

$$h_i = 1409 \text{ W/m}^2 \cdot \text{°C}$$

This is a 28 percent reduction because of the fouling factor.

10-4 | TYPES OF HEAT EXCHANGERS

One type of heat exchanger has already been mentioned, that of a double-pipe arrangement as shown in Figure 10-2. Either counterflow or parallel flow may be used in this type of exchanger, with either the hot or cold fluid occupying the annular space and the other fluid occupying the inside of the inner pipe.

A type of heat exchanger widely used in the chemical-process industries is that of the shell-and-tube arrangement shown in Figure 10-3. One fluid flows on the inside of the tubes, while the other fluid is forced through the shell and over the outside of the tubes. To ensure that the shell-side fluid will flow across the tubes and thus induce higher heat transfer, baffles are placed in the shell as shown in the figure. Depending on the head arrangement at the ends of the exchanger, one or more tube passes may be utilized. In Figure 10-3*a* one tube pass is used, and the head arrangement for two tube passes is shown in Figure 10-3*b*.

Shell and tube exchangers my also be employed in miniature form for specialized applications in biotechnology fields. Such an exchanger with one shell pass and one tube pass is illustrated in Figure 10-3*c* and the internal tube construction in Figure 10-3*d*. Small double pipe or tube-in-tube exchangers may also be constructed in a coiled configuration as shown in Figure 10-3*e* with an enlarged view of the inlet-outlet flow connections shown in Figure 10-3*f*.

Cross-flow exchangers are commonly used in air or gas heating and cooling applications. An example of such an exchanger is shown in Figure 10-4, where a gas may be forced across a tube bundle, while another fluid is used inside the tubes for heating or cooling purposes. In this exchanger the gas flowing across the tubes is said to be a *mixed* stream, while the fluid in the tubes is said to be *unmixed*. The gas is mixed because it can move about freely in the exchanger as it exchanges heat. The other fluid is confined in separate tubular channels while in the exchanger so that it cannot mix with itself during the heat-transfer process.

A different type of cross-flow exchanger is shown in Figure 10-5. In this case the gas flows across finned-tube bundles and thus is *unmixed* since it is confined in separate channels between the fins as it passes through the exchanger. This exchanger is typical of the types used in air-conditioning applications.

If a fluid is unmixed, there can be a temperature gradient both parallel and normal to the flow direction, whereas when the fluid is mixed, there will be a tendency for the fluid temperature to equalize in the direction normal to the flow as a result of the mixing. An approximate temperature profile for the gas flowing in the exchanger of Figure 10-5 is indicated in Figure 10-6, assuming that the gas is being heated as it passes through the exchanger. The fact that a fluid is mixed or unmixed influences the overall heat transfer in the exchanger because this heat transfer is dependent on the temperature difference between the hot and cold fluids. Although beyond the scope of our discussion, there are cases where

Figure 10-3 | Photos of commercial heat exchangers. (*a*) Shell-and-tube heat exchanger with one tube pass; (*b*) head arrangement for exchanger with two tube passes; (*c*) miniature shell-and-tube exchanger with one shell pass and one tube pass; (*d*) internal construction of miniature exchanger; (*e*) miniature coiled tube-in-tube exchanger; (*f*) detail of inlet–outlet fluid connections for miniature tube-in-tube exchanger.

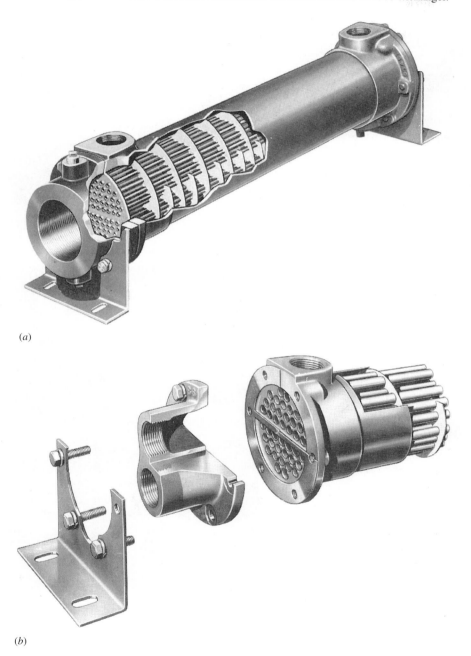

(*a*)

(*b*)

Source: (*a*), (*b*) Courtesy Young Radiator Company, Racine, Wisconsin; (*c*)–(*f*) Courtesy Exergy, Inc., Hanson, MA.

Figure 10-3 | (*Continued*).

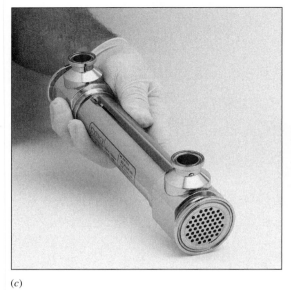

(*c*)

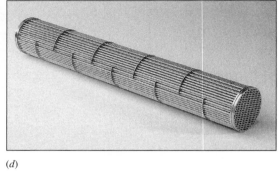

(*d*)

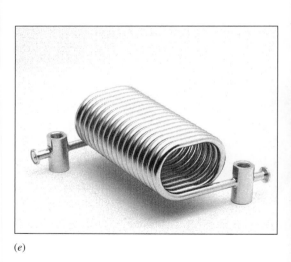

(*e*)

(*f*)

Figure 10-4 | Cross-flow heat exchanger, one fluid
mixed and one unmixed.

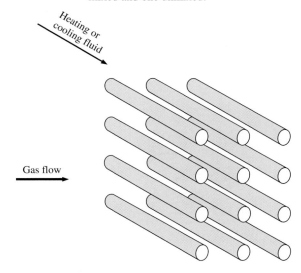

Figure 10-5 | Cross-flow heat exchanger, both fluids unmixed.

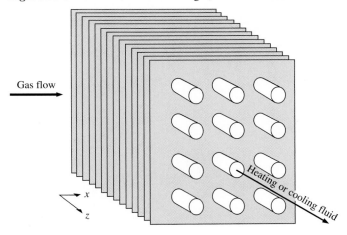

the heat exchanger flows should be considered as only "partially" mixed. Such cases are discussed in Reference 11.

There are a number of other configurations called *compact heat exchangers* that are primarily used in gas-flow systems where the overall heat-transfer coefficients are low and it is desirable to achieve a large surface area in a small volume. These exchangers generally have surface areas of greater than 650 m^2 per cubic meter of volume and will be given a fuller discussion in Section 10-7.

10-5 | THE LOG MEAN TEMPERATURE DIFFERENCE

Consider the double-pipe heat exchanger shown in Figure 10-2. The fluids may flow in either parallel flow or counterflow, and the temperature profiles for these two cases are indicated in Figure 10-7. We propose to calculate the heat transfer in this double-pipe arrangement

Figure 10-6 | Typical temperature profile for cross-flow heat exchanger of Figure 10-5.

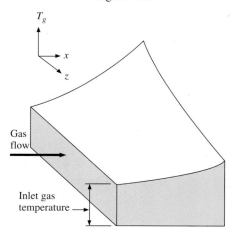

Figure 10-7 | Temperature profiles for (a) parallel flow and (b) counterflow in double-pipe heat exchanger.

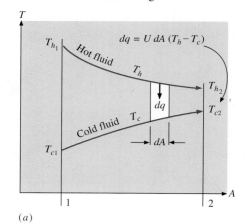

(a)

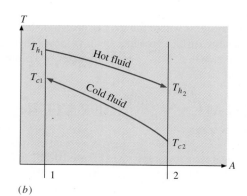

(b)

with

$$q = UA \, \Delta T_m \qquad \text{[10-5]}$$

where

U = overall heat-transfer coefficient

A = surface area for heat transfer consistent with definition of U

ΔT_m = suitable mean temperature difference across heat exchanger

An inspection of Figure 10-7 shows that the temperature difference between the hot and cold fluids varies between inlet and outlet, and we must determine the average value for use in Equation (10-5). For the parallel-flow heat exchanger shown in Figure 10-7, the heat transferred through an element of area dA may be written

$$dq = -\dot{m}_h c_h dT_h = \dot{m}_c c_c dT_c \qquad \text{[10-6]}$$

where the subscripts h and c designate the hot and cold fluids, respectively. The heat transfer could also be expressed

$$dq = U(T_h - T_c)dA \qquad \text{[10-7]}$$

From Equation (10-6)

$$dT_h = \frac{-dq}{\dot{m}_h c_h}$$

$$dT_c = \frac{dq}{\dot{m}_c c_c}$$

where $\dot{m}$ represents the mass-flow rate and c is the specific heat of the fluid. Thus

$$dT_h - dT_c = d(T_h - T_c) = -dq \left(\frac{1}{\dot{m}_h c_h} + \frac{1}{\dot{m}_c c_c} \right) \qquad \text{[10-8]}$$

Solving for dq from Equation (10-7) and substituting into Equation (10-8) gives

$$\frac{d(T_h - T_c)}{T_h - T_c} = -U \left(\frac{1}{\dot{m}_h c_h} + \frac{1}{\dot{m}_c c_c} \right) dA \qquad \text{[10-9]}$$

This differential equation may now be integrated between conditions 1 and 2 as indicated in Figure 10-7. The result is

$$\ln \frac{T_{h2} - T_{c2}}{T_{h1} - T_{c1}} = -UA \left(\frac{1}{\dot{m}_h c_h} + \frac{1}{\dot{m}_c c_c} \right) \qquad \text{[10-10]}$$

Returning to Equation (10-6), the products $\dot{m}_c c_c$ and $\dot{m}_h c_h$ may be expressed in terms of the total heat transfer q and the overall temperature differences of the hot and cold fluids. Thus

$$\dot{m}_h c_h = \frac{q}{T_{h1} - T_{h2}}$$

$$\dot{m}_c c_c = \frac{q}{T_{c2} - T_{c1}}$$

Substituting these relations into Equation (10-10) gives

$$q = UA \frac{(T_{h2} - T_{c2}) - (T_{h1} - T_{c1})}{\ln[(T_{h2} - T_{c2})/(T_{h1} - T_{c1})]} \qquad \text{[10-11]}$$

Comparing Equation (10-11) with Equation (10-5), we find that the mean temperature difference is the grouping of terms in the brackets. Thus

$$\Delta T_m = \frac{(T_{h2} - T_{c2}) - (T_{h1} - T_{c1})}{\ln[(T_{h2} - T_{c2})/(T_{h1} - T_{c1})]} \qquad \text{[10-12]}$$

Figure 10-8 | Correction-factor plot for exchanger with one shell pass and two, four, or any multiple of tube passes.

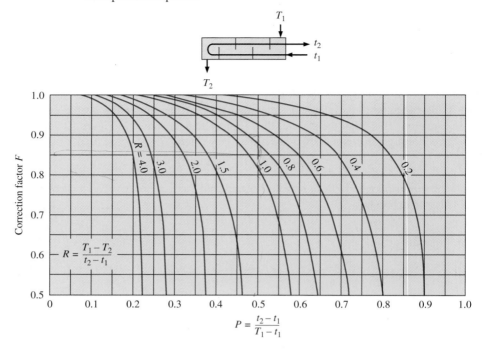

Figure 10-9 | Correction-factor plot for exchanger with two shell passes and four, eight, or any multiple of tube passes.

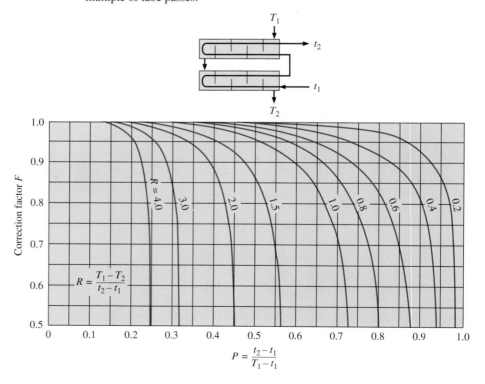

Figure 10-10 | Correction-factor plot for single-pass cross-flow exchanger, both fluids unmixed.

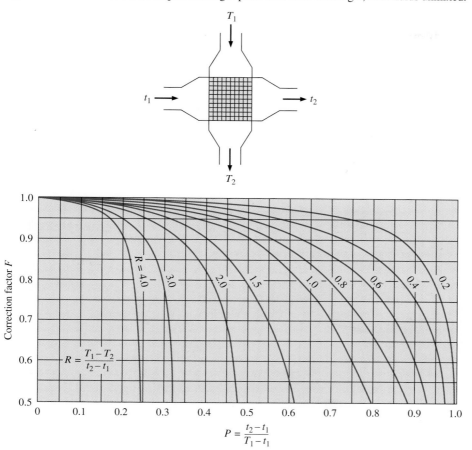

This temperature difference is called the *log mean temperature difference* (LMTD). Stated verbally, it is the temperature difference at one end of the heat exchanger less the temperature difference at the other end of the exchanger divided by the natural logarithm of the ratio of these two temperature differences. It is left as an exercise for the reader to show that this relation may also be used to calculate the LMTDs for counterflow conditions.

The above derivation for LMTD involves two important assumptions: (1) the fluid specific heats do not vary with temperature, and (2) the convection heat-transfer coefficients are constant throughout the heat exchanger. The second assumption is usually the more serious one because of entrance effects, fluid viscosity, and thermal-conductivity changes, etc. Numerical methods must normally be employed to correct for these effects. Section 10-8 describes one way of performing a variable-properties analysis.

If a heat exchanger other than the double-pipe type is used, the heat transfer is calculated by using a correction factor applied to the LMTD *for a counterflow double-pipe arrangement with the same hot and cold fluid temperatures.* The heat-transfer equation then takes the form

$$q = UAF\Delta T_m$$ **[10-13]**

Values of the correction factor F according to Reference 4 are plotted in Figures 10-8 to 10-11 for several different types of heat exchangers. When a phase change is involved, as

Figure 10-11 | Correction-factor plot for single-pass cross-flow exchanger, one fluid mixed, the other unmixed.

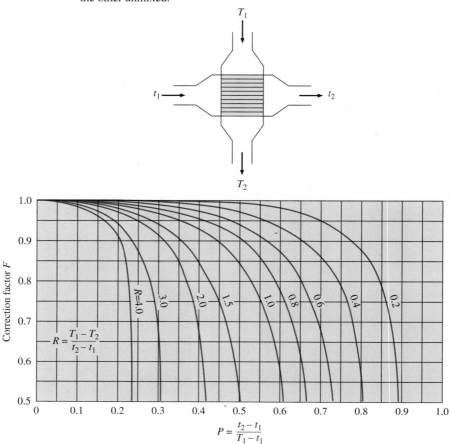

in condensation or boiling (evaporation), the fluid normally remains at essentially constant temperature and the relations are simplified. For this condition, P or R becomes zero and we obtain

$$F = 1.0 \quad \text{for boiling or condensation}$$

Examples 10-4 to 10-8 illustrate the use of the LMTD method for calculation of heat-exchanger performance.

Calculation of Heat-Exchanger Size from Known Temperatures

EXAMPLE 10-4

Water at the rate of 68 kg/min is heated from 35 to 75°C by an oil having a specific heat of 1.9 kJ/kg · °C. The fluids are used in a counterflow double-pipe heat exchanger, and the oil enters the exchanger at 110°C and leaves at 75°C. The overall heat-transfer coefficient is 320 W/m² · °C. Calculate the heat-exchanger area.

■ **Solution**

The total heat transfer is determined from the energy absorbed by the water:

$$q = \dot{m}_w c_w \Delta T_w = (68)(4180)(75 - 35) = 11.37 \text{ MJ/min} \qquad [a]$$

$$= 189.5 \text{ kW} \quad [6.47 \times 10^5 \text{ Btu/h}]$$

Since all the fluid temperatures are known, the LMTD can be calculated by using the temperature scheme in Figure 10-7b:

$$\Delta T_m = \frac{(110-75)-(75-35)}{\ln[(110-75)/(75-35)]} = 37.44°C \qquad [b]$$

Then, since $q = UA\,\Delta T_m$,

$$A = \frac{1.895 \times 10^5}{(320)(37.44)} = 15.82 \text{ m}^2 \quad [170 \text{ ft}^2]$$

Shell-and-Tube Heat Exchanger | EXAMPLE 10-5

Instead of the double-pipe heat exchanger of Example 10-4, it is desired to use a shell-and-tube exchanger with the water making one shell pass and the oil making two tube passes. Calculate the area required for this exchanger, assuming that the overall heat-transfer coefficient remains at 320 W/m$^2 \cdot$°C.

■ **Solution**

To solve this problem, we determine a correction factor from Figure 10-8 to be used with the LMTD calculated on the basis of a counterflow exchanger. The parameters according to the nomenclature of Figure 10-8 are

$$T_1 = 35°C \qquad T_2 = 75°C \qquad t_1 = 110°C \qquad t_2 = 75°C$$

$$P = \frac{t_2 - t_1}{T_1 - t_1} = \frac{75-110}{35-110} = 0.467$$

$$R = \frac{T_1 - T_2}{t_2 - t_1} = \frac{35-75}{75-110} = 1.143$$

T = oil
t = H₂O

so the correction factor is

$$F = 0.81$$

and the heat transfer is

$$q = UAF\,\Delta T_m$$

so that

$$A = \frac{1.895 \times 10^5}{(320)(0.81)(37.44)} = 19.53 \text{ m}^2 \quad [210 \text{ ft}^2]$$

Design of Shell-and-Tube Heat Exchanger | EXAMPLE 10-6

Water at the rate of 30,000 lb$_m$/h [3.783 kg/s] is heated from 100 to 130°F [37.78 to 54.44°C] in a shell-and-tube heat exchanger. On the shell side one pass is used with water as the heating fluid, 15,000 lb$_m$/h [1.892 kg/s], entering the exchanger at 200°F [93.33°C]. The overall heat-transfer coefficient is 250 Btu/h $\cdot$ ft$^2 \cdot$ °F [1419 W/m$^2 \cdot$ °C], and the average water velocity in the $\frac{3}{4}$-in [1.905-cm] diameter tubes is 1.2 ft/s [0.366 m/s]. Because of space limitations, the tube length must not be longer than 8 ft [2.438 m]. Calculate the number of tube passes, the number of tubes per pass, and the length of the tubes, consistent with this restriction.

■ **Solution**

We first assume one tube pass and check to see if it satisfies the conditions of this problem. The exit temperature of the hot water is calculated from

$$q = \dot{m}_c c_c\,\Delta T_c = \dot{m}_h c_h\,\Delta T_h$$

$$\Delta T_h = \frac{(30,000)(1)(130-100)}{(15,000)(1)} = 60°F = 33.33°C \qquad [a]$$

so

$$T_{h,\text{exit}} = 93.33 - 33.33 = 60°C$$

The total required heat transfer is obtained from Equation (a) for the cold fluid:

$$q = (3.783)(4182)(54.44 - 37.78) = 263.6 \text{ kW} \quad [8.08 \times 10^5 \text{ Btu/h}]$$

For a counterflow exchanger, with the required temperature

$$\text{LMTD} = \Delta T_m = \frac{(93.33 - 54.44) - (60 - 37.78)}{\ln[(93.33 - 54.44)/(60 - 37.78)]} = 29.78°C$$

$$q = UA \, \Delta T_m$$

$$A = \frac{2.636 \times 10^5}{(1419)(29.78)} = 6.238 \text{ m}^2 \quad [67.1 \text{ ft}^2] \qquad [b]$$

Using the average water velocity in the tubes and the flow rate, we calculate the total flow area with

$$\dot{m}_c = \rho A u$$

$$A = \frac{3.783}{(1000)(0.366)} = 0.01034 \text{ m}^2 \qquad [c]$$

This area is the product of the number of tubes and the flow area per tube:

$$0.01034 = n \frac{\pi d^2}{4}$$

$$n = \frac{(0.01034)(4)}{\pi(0.01905)^2} = 36.3$$

or $n = 36$ tubes. The surface area per tube per meter of length is

$$\pi d = \pi(0.01905) = 0.0598 \text{ m}^2/\text{tube} \cdot \text{m}$$

We recall that the total surface area required for a one-tube-pass exchanger was calculated in Equation (b) as 6.238 m². We may thus compute the length of tube for this type of exchanger from

$$n \pi d \, L = 6.238$$

$$L = \frac{6.238}{(36)(0.0598)} = 2.898 \text{ m}$$

This length is greater than the allowable 2.438 m, so we must use more than one tube pass. When we increase the number of passes, we correspondingly increase the total surface area required because of the reduction in LMTD caused by the correction factor F. We next try two tube passes. From Figure 10-8, $F = 0.88$, and thus

$$A_{\text{total}} = \frac{q}{UF\Delta T_m} = \frac{2.636 \times 10^5}{(1419)(0.88)(29.78)} = 7.089 \text{ m}^2$$

The number of tubes per pass is still 36 because of the velocity requirement. For the two-tube-pass exchanger the total surface area is now related to the length by

$$A_{\text{total}} = 2n \pi d \, L$$

so that

$$L = \frac{7.089}{(2)(36)(0.0598)} = 1.646 \text{ m} \quad [5.4 \text{ ft}]$$

This length is within the 2.438-m requirement, so the final design choice is

   Number of tubes per pass = 36

   Number of passes = 2

   Length of tube per pass = 1.646 m   [5.4 ft]

Cross-Flow Exchanger with One Fluid Mixed   EXAMPLE 10-7

A heat exchanger like that shown in Figure 10-4 is used to heat an oil in the tubes ($c = 1.9$ kJ/kg · °C) from 15°C to 85°C. Blowing across the outside of the tubes is steam that enters at 130°C and leaves at 110°C with a mass flow of 5.2 kg/sec. The overall heat-transfer coefficient is 275 W/m^2 · °C and c for steam is 1.86 kJ/kg · °C. Calculate the surface area of the heat exchanger.

Figure Example 10-7

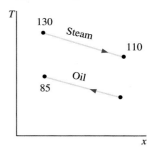

■ **Solution**

The total heat transfer may be obtained from an energy balance on the steam

$$q = \dot{m}_s c_s \, \Delta T_s = (5.2)(1.86)(130 - 110) = 193 \text{ kW}$$

We can solve for the area from Equation (10-13). The value of ΔT_m is calculated *as if the exchanger were counterflow double pipe* (i.e., as shown in Figure Example 10-7). Thus,

$$\Delta T_m = \frac{(130 - 85) - (110 - 15)}{\ln\left(\dfrac{130 - 85}{110 - 15}\right)} = 66.9°C$$

Now, from Figure 10-11, t_1 and t_2 will represent the unmixed fluid (the oil) and T_1 and T_2 will represent the mixed fluid (the steam) so that

$$T_1 = 130 \qquad T_2 = 110 \qquad t_1 = 15 \qquad t_2 = 85°C$$

and we calculate

$$R = \frac{130 - 110}{85 - 15} = 0.286 \qquad P = \frac{85 - 15}{130 - 15} = 0.609$$

Consulting Figure 10-11 we find

$$F = 0.97$$

so the area is calculated from

$$A = \frac{q}{UF \, \Delta T_m} = \frac{193{,}000}{(275)(0.97)(66.9)} = 10.82 \text{ m}^2$$

Effects of Off-Design Flow Rates for Exchanger in Example 10-7   EXAMPLE 10-8

Investigate the heat-transfer performance of the exchanger in Example 10-7 if the oil flow rate is reduced in half while the steam flow remains the same. Assume U remains constant at 275 W/m^2 · °C.

■ **Solution**

We did not calculate the oil flow in Example 10-7 but can do so now from

$$q = \dot{m}_o c_o \, \Delta T_o$$

$$\dot{m}_o = \frac{193}{(1.9)(85 - 15)} = 1.45 \text{ kg/s}$$

The new flow rate will be half this value or 0.725 kg/s. We are assuming the inlet temperatures remain the same at 130°C for the steam and 15°C for the oil. The new relation for the heat transfer is

$$q = \dot{m}_o c_o (T_{e,o} - 15) = \dot{m}_s c_{p_s} (130 - T_{e,s})$$ [a]

but the exit temperatures, $T_{e,o}$ and $T_{e,s}$ are unknown. Furthermore, ΔT_m is unknown without these temperatures, as are the values of R and P from Figure 10-11. This means we must use an iterative procedure to solve for the exit temperatures using Equation (a) and

$$q = UAF\Delta T_m$$ [b]

The general procedure is to assume values of the exit temperatures until the q's agree between Equations (a) and (b).

The objective of this example is to show that an iterative procedure is required when the inlet and outlet temperatures are not known or easily calculated. There is no need to go through this iteration because it can be avoided by using the techniques described in Section 10-6.

10-6 | EFFECTIVENESS-NTU METHOD

The LMTD approach to heat-exchanger analysis is useful when the inlet and outlet temperatures are known or are easily determined. The LMTD is then easily calculated, and the heat flow, surface area, or overall heat-transfer coefficient may be determined. When the inlet or exit temperatures are to be evaluated for a given heat exchanger, the analysis frequently involves an iterative procedure because of the logarithmic function in the LMTD. In these cases the analysis is performed more easily by utilizing a method based on the effectiveness of the heat exchanger in transferring a given amount of heat. The effectiveness method also offers many advantages for analysis of problems in which a comparison between various types of heat exchangers must be made for purposes of selecting the type best suited to accomplish a particular heat-transfer objective.

We define the heat-exchanger effectiveness as

$$\text{Effectiveness} = \epsilon = \frac{\text{actual heat transfer}}{\text{maximum possible heat transfer}}$$

The actual heat transfer may be computed by calculating either the energy lost by the hot fluid or the energy gained by the cold fluid. Consider the parallel-flow and counterflow heat exchangers shown in Figure 10-7. For the parallel-flow exchanger

$$q = \dot{m}_h c_h (T_{h_1} - T_{h_2}) = \dot{m}_c c_c (T_{c_2} - T_{c_1})$$ [10-14]

and for the counterflow exchanger

$$q = \dot{m}_h c_h (T_{h_1} - T_{h_2}) = \dot{m}_c c_c (T_{c_1} - T_{c_2})$$ [10-15]

To determine the maximum possible heat transfer for the exchanger, we first recognize that this maximum value could be attained if one of the fluids were to undergo a temperature change equal to the maximum temperature difference present in the exchanger, which is the difference in the entering temperatures for the hot and cold fluids. The fluid that might undergo this maximum temperature difference is the one having the *minimum* value of $\dot{m}c$ because the energy balance requires that the energy received by one fluid be equal to that given up by the other fluid; if we let the fluid with the larger value of $\dot{m}c$ go through the maximum temperature difference, this would require that the other fluid undergo a temperature difference greater than the maximum, and this is impossible. So, maximum possible heat transfer is expressed as

$$q_{\max} = (\dot{m}c)_{\min} (T_{h_{\text{inlet}}} - T_{c_{\text{inlet}}})$$ [10-16]

The minimum fluid may be either the hot or cold fluid, depending on the mass-flow rates and specific heats. For the parallel-flow exchanger

$$\epsilon_h = \frac{\dot{m}_h c_h (T_{h_1} - T_{h_2})}{\dot{m}_h c_h (T_{h_1} - T_{c_1})} = \frac{T_{h_1} - T_{h_2}}{T_{h_1} - T_{c_1}} \qquad \text{[10-17]}$$

$$\epsilon_c = \frac{\dot{m}_c c_c (T_{c_2} - T_{c_1})}{\dot{m}_c c_c (T_{h_1} - T_{c_1})} = \frac{T_{c_2} - T_{c_1}}{T_{h_1} - T_{c_1}} \qquad \text{[10-18]}$$

The subscripts on the effectiveness symbols designate the fluid that has the minimum value of $\dot{m}c$. For the counterflow exchanger:

$$\epsilon_h = \frac{\dot{m}_h c_h (T_{h_1} - T_{h_2})}{\dot{m}_h c_h (T_{h_1} - T_{c_2})} = \frac{T_{h_1} - T_{h_2}}{T_{h_1} - T_{c_2}} \qquad \text{[10-19]}$$

$$\epsilon_c = \frac{\dot{m}_c c_c (T_{c_1} - T_{c_2})}{\dot{m}_c c_c (T_{h_1} - T_{c_2})} = \frac{T_{c_1} - T_{c_2}}{T_{h_1} - T_{c_2}} \qquad \text{[10-20]}$$

In a general way the effectiveness is expressed as

$$\epsilon = \frac{\Delta T(\text{minimum fluid})}{\text{Maximum temperature difference in heat exchanger}} \qquad \text{[10-21]}$$

The minimum fluid is always the one experiencing the larger temperature difference in the heat exchanger, and the maximum temperature difference in the heat exchanger is always the difference in inlet temperatures of the hot and cold fluids.

We may derive an expression for the effectiveness in parallel flow double-pipe as follows. Rewriting Equation (10-10), we have

$$\ln \frac{T_{h_2} - T_{c_2}}{T_{h_1} - T_{c_1}} = -UA \left(\frac{1}{\dot{m}_h c_h} + \frac{1}{\dot{m}_c c_c} \right) = \frac{-UA}{\dot{m}_c c_c} \left(1 + \frac{\dot{m}_c c_c}{\dot{m}_h c_h} \right) \qquad \text{[10-22]}$$

or

$$\frac{T_{h_2} - T_{c_2}}{T_{h_1} - T_{c_1}} = \exp \left[\frac{-UA}{\dot{m}_c c_c} \left(1 + \frac{\dot{m}_c c_c}{\dot{m}_h c_h} \right) \right] \qquad \text{[10-23]}$$

If the cold fluid is the minimum fluid,

$$\epsilon = \frac{T_{c_2} - T_{c_1}}{T_{h_1} - T_{c_1}}$$

Rewriting the temperature ratio in Equation (10-23) gives

$$\frac{T_{h_2} - T_{c_2}}{T_{h_1} - T_{c_1}} = \frac{T_{h_1} + (\dot{m}_c c_c / \dot{m}_h c_h)(T_{c_1} - T_{c_2}) - T_{c_2}}{T_{h_1} - T_{c_1}} \qquad \text{[10-24]}$$

when the substitution

$$T_{h_2} = T_{h_1} + \frac{\dot{m}_c c_c}{\dot{m}_h c_h}(T_{c_1} - T_{c_2})$$

is made from Equation (10-6). Equation (10-24) may now be rewritten

$$\frac{(T_{h_1} - T_{c_1}) + (\dot{m}_c c_c / \dot{m}_h c_h)(T_{c_1} - T_{c_2}) + (T_{c_1} - T_{c_2})}{T_{h_1} - T_{c_1}} = 1 - \left(1 + \frac{\dot{m}_c c_c}{\dot{m}_h c_h} \right) \epsilon$$

Inserting this relation back in Equation (10-23) gives for the effectiveness

$$\epsilon = \frac{1 - \exp[(-UA/\dot{m}_c c_c)(1 + \dot{m}_c c_c / \dot{m}_h c_h)]}{1 + \dot{m}_c c_c / \dot{m}_h c_h} \qquad \text{[10-25]}$$

It may be shown that the same expression results for the effectiveness when the hot fluid is the minimum fluid, except that $\dot{m}_c c_c$ and $\dot{m}_h c_h$ are interchanged. As a consequence, the effectiveness is usually written

$$\epsilon = \frac{1 - \exp[(-UA/C_{min})(1 + C_{min}/C_{max})]}{1 + C_{min}/C_{max}} \qquad \textbf{[10-26]}$$

where $C = \dot{m}c$ is defined as the capacity rate.

A similar analysis may be applied to the counterflow case, and the following relation for effectiveness results:

$$\epsilon = \frac{1 - \exp[(-UA/C_{min})(1 - C_{min}/C_{max})]}{1 - (C_{min}/C_{max})\exp[(-UA/C_{min})(1 - C_{min}/C_{max})]} \qquad \textbf{[10-27]}$$

The grouping of terms UA/C_{min} is called the *number of transfer units* (NTU) since it is indicative of the size of the heat exchanger.

Kays and London [3] have presented effectiveness ratios for various heat-exchanger arrangements, and some of the results of their analyses are available in chart form in Figures 10-12 to 10-17. Examples 10-9 to 10-14 illustrate the use of the effectiveness-NTU method in heat-exchanger analysis.

While the effectiveness-NTU charts can be of great practical utility in design problems, there are applications where more precision is desired than can be obtained by reading the graphs. In addition, more elaborate design procedures may be computer-based, requiring analytical expressions for these curves. Table 10-3 summarizes the effectiveness relations. In some cases the objective of the analysis is a determination of NTU, and it is possible to give an explicit relation for NTU in terms of effectiveness and capacity ratio. Some of these relations are listed in Table 10-4.

The formulas for one shell pass and 2, 4, 6 tube passes are strictly correct for 2 tube passes but may produce a small error for higher multiples. The error is usually less than 1 percent for C less than 0.5 and N less than 3.0. The formulas may overpredict by about 6.5 percent at $N = 6.0$ and $C = 1.0$. Further information is given by Kraus and Kern [10].

Boilers and Condensers

We noted earlier that in a boiling or condensation process the fluid temperature stays essentially constant, or the fluid acts as if it had infinite specific heat. In these cases $C_{min}/C_{max} \rightarrow 0$ and all the heat-exchanger effectiveness relations approach a single simple equation,

$$\epsilon = 1 - e^{-NTU}$$

The equation is shown as the last entry in Table 10-3. For this case,

$$q = C_{min}(T_{h,inlet} - T_{c,inlet})[1 - \exp(UA/C_{min})]$$

where

$$C_{min} = \dot{m}_c c_c \text{ for a condenser (condensing fluid is } losing \text{ heat)}$$
$$= \dot{m}_h c_h \text{ for a boiler (boiling fluid is } gaining \text{ heat)}$$

Effectiveness and Heat-Transfer Rates

We must caution the reader about misinterpreting the physical meaning of heat-exchanger effectiveness. Just because a heat exchanger has a high *effectiveness* at a certain flow condition does not mean that it will have a higher heat-transfer *rate* than at some low effectiveness condition. High values of ε correspond to small temperature differences between the hot and cold fluid, while higher heat-transfer *rates* result from larger temperature differences

Figure 10-12 | Effectiveness for parallel-flow exchanger performance.

Figure 10-13 | Effectiveness for counterflow exchanger performance.

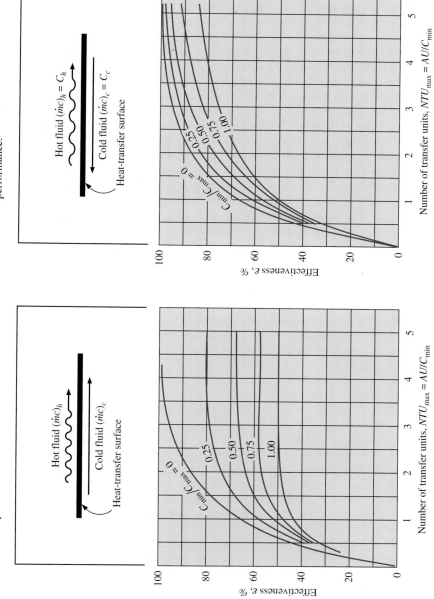

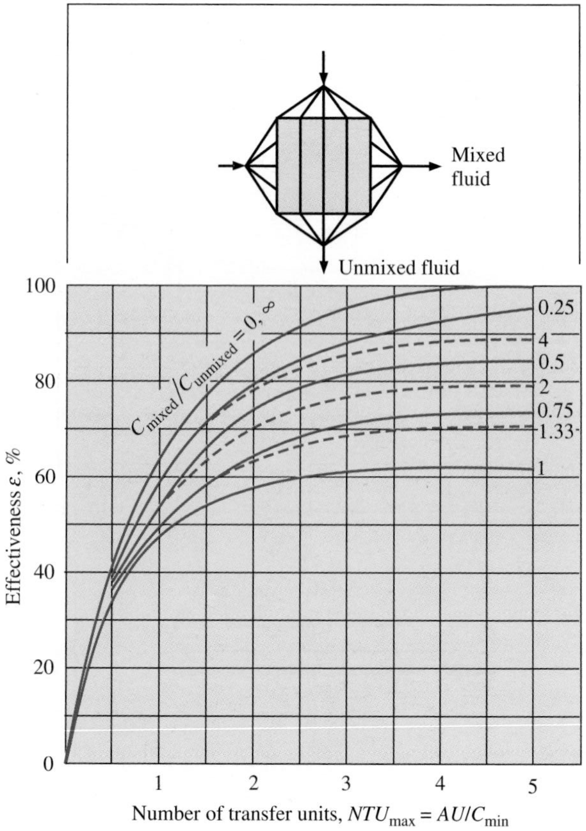

Figure 10-15 | Effectiveness for cross-flow exchanger with fluids unmixed.

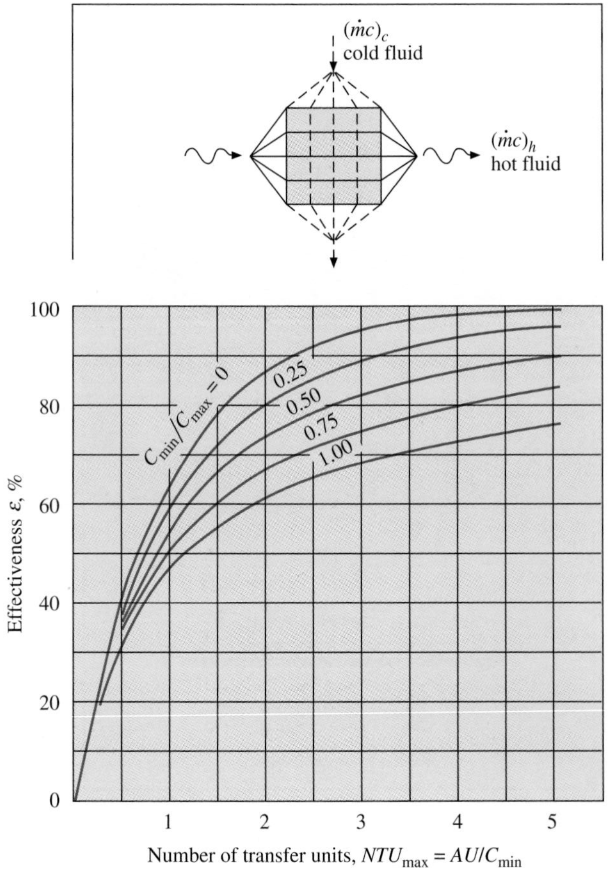

Figure 10-14 | Effectiveness for cross-flow exchanger with one fluid mixed.

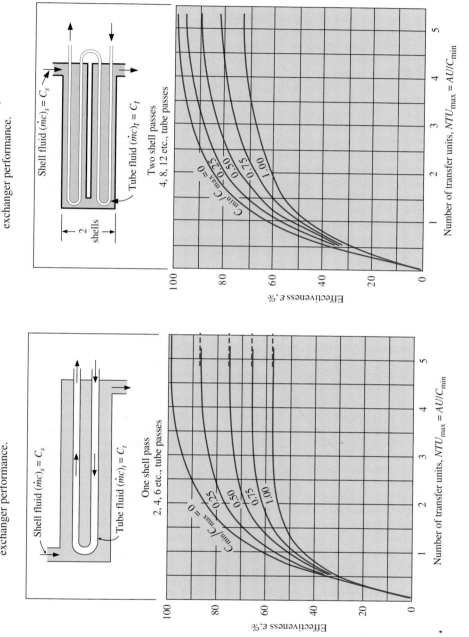

Figure 10-17 | Effectiveness for 2-4 multipass counterflow exchanger performance.

Figure 10-16 | Effectiveness for 1-2 parallel counterflow exchanger performance.

545

Table 10-3 | Heat-exchanger effectiveness relations.

$$N = \text{NTU} = \frac{UA}{C_{min}} \qquad C = \frac{C_{min}}{C_{max}}$$

Flow geometry	Relation
Double pipe:	
Parallel flow	$\epsilon = \dfrac{1 - \exp[-N(1+C)]}{1+C}$
Counterflow	$\epsilon = \dfrac{1 - \exp[-N(1-C)]}{1 - C\exp[-N(1-C)]}$
Counterflow, $C = 1$	$\epsilon = \dfrac{N}{N+1}$
Cross flow:	
Both fluids unmixed	$\epsilon = 1 - \exp\left[\dfrac{\exp(-NCn) - 1}{Cn}\right]$ where $n = N^{-0.22}$
Both fluids mixed	$\epsilon = \left[\dfrac{1}{1 - \exp(-N)} + \dfrac{C}{1 - \exp(-NC)} - \dfrac{1}{N}\right]^{-1}$
C_{max} mixed, C_{min} unmixed	$\epsilon = (1/C)\{1 - \exp[-C(1 - e^{-N})]\}$
C_{max} unmixed, C_{min} mixed	$\epsilon = 1 - \exp\{-(1/C)[1 - \exp(-NC)]\}$
Shell and tube:	
One shell pass, 2, 4, 6, tube passes	$\epsilon = 2\left\{1 + C + (1+C^2)^{1/2}\right.$ $\left. \times \dfrac{1 + \exp[-N(1+C^2)^{1/2}]}{1 - \exp[-N(1+C^2)^{1/2}]}\right\}^{-1}$

Multiple shell passes, $2n$, $4n$, $6n$ tube passes
    ($\epsilon_p = $ effectiveness of each shell pass,
    $n = $ number of shell passes)
	$\epsilon = \dfrac{[(1 - \epsilon_p C)/(1 - \epsilon_p)]^n - 1}{[(1 - \epsilon_p C)/(1 - \epsilon_p)]^n - C}$

Special case for $C = 1$	$\epsilon = \dfrac{n\epsilon_p}{1 + (n-1)\epsilon_p}$
All exchangers with $C = 0$	$\epsilon = 1 - e^{-N}$

Table 10-4 | NTU relations for heat exchangers.

$$C = C_{min}/C_{max} \qquad \epsilon = \text{effectiveness} \qquad N = \text{NTU} = UA/C_{min}$$

Flow geometry	Relation
Double pipe:	
Parallel flow	$N = \dfrac{-\ln[1 - (1+C)\epsilon]}{1+C}$
Counterflow	$N = \dfrac{1}{C-1}\ln\left(\dfrac{\epsilon - 1}{C\epsilon - 1}\right)$
Counterflow, $C = 1$	$N = \dfrac{\epsilon}{1 - \epsilon}$
Cross flow:	
C_{max} mixed, C_{min} unmixed	$N = -\ln\left[1 + \dfrac{1}{C}\ln(1 - C\epsilon)\right]$
C_{max} unmixed, C_{min} mixed	$N = \dfrac{-1}{C}\ln[1 + C\ln(1 - \epsilon)]$
Shell and tube:	
One shell pass, 2, 4, 6, tube passes	$N = -(1+C^2)^{-1/2}$ $\times \ln\left[\dfrac{2/\epsilon - 1 - C - (1+C^2)^{1/2}}{2/\epsilon - 1 - C + (1+C^2)^{1/2}}\right]$
All exchangers, $C = 0$	$N = -\ln(1 - \epsilon)$

(i.e., a greater driving potential). In a thermodynamic sense, higher effectiveness values correspond to reduced values of thermodynamic irreversibility and smaller entropy generation. To achieve *both* high heat transfer *and* high effectiveness one must increase the value of the *UA* product, either by increasing the size (and cost) of the exchanger or by forcing the fluid(s) through the heat exchanger at higher velocities to produce increased convection coefficients. Or, one may employ so-called heat-transfer augmentation techniques to increase the value of *UA*. Such techniques are discussed extensively by Bergles [12–14] and Webb [17]. In many cases pressure drop and pumping costs become important factors in the eventual design and/or selection of the heat exchanger.

Off-Design Calculation Using ϵ-NTU Method   EXAMPLE 10-9

Complete Example 10-8 using the effectiveness method.

■ **Solution**

For the steam

$$C_s = \dot{m}_s c_s = (5.2)(1.86) = 9.67 \text{ kW/°C}$$

and for the oil

$$C_o = \dot{m}_o c_o = (0.725)(1.9) = 1.38 \text{ kW/°C}$$

so the oil is the minimum fluid. We thus have

$$C_{min}/C_{max} = 1.38/9.67 = 0.143$$

and

$$\text{NTU} = UA/C_{min} = (275)(10.82)/1380 = 2.156$$

We choose to use Table 10-3 and note that C_{min} (oil) is unmixed and C_{max} (steam) is mixed so that the first relation in the table applies. We therefore calculate ϵ as

$$\epsilon = (1/0.143)\{1 - \exp[-(0.143)(1 - e^{-2.156})]\} = 0.831$$

If we were using Figure 10-14 we would have to evaluate

$$C_{mixed}/C_{unmixed} = 9.67/1.38 = 7.01$$

and would still determine $\epsilon \approx 0.8$. Now, using the effectiveness we can determine the temperature difference of the minimum fluid (oil) as

$$\Delta T_o = \epsilon(\Delta T_{max}) = (0.831)(130 - 15) = 95.5 \text{°C}$$

so that the heat transfer is

$$q = \dot{m}_o c_o \,\Delta T_o = (1.38)(95.5) = 132 \text{ kW}$$

and we find a reduction in the oil flow rate of 50 percent causes a reduction in heat transfer of only 32 percent.

Off-Design Calculation of Exchanger in Example 10-4   EXAMPLE 10-10

The heat exchanger of Example 10-4 is used for heating water as described in the example. Using the same entering-fluid temperatures, calculate the exit water temperature when only 40 kg/min of water is heated but the same quantity of oil is used. Also calculate the total heat transfer under these new conditions.

■ **Solution**

The flow rate of oil is calculated from the energy balance for the original problem:

$$\dot{m}_h c_h \, \Delta T_h = \dot{m}_c c_c \, \Delta T_c \qquad\qquad [a]$$

$$\dot{m}_h = \frac{(68)(4180)(75-35)}{(1900)(110-75)} = 170.97 \text{ kg/min}$$

The capacity rates for the new conditions are now calculated as

$$\dot{m}_h c_h = \frac{170.97}{60}(1900) = 5414 \text{ W/°C}$$

$$\dot{m}_c c_c = \frac{40}{60}(4180) = 2787 \text{ W/°C}$$

so that the water (cold fluid) is the minimum fluid, and

$$\frac{C_{\min}}{C_{\max}} = \frac{2787}{5414} = 0.515$$

$$\text{NTU}_{\max} = \frac{UA}{C_{\min}} = \frac{(320)(15.82)}{2787} = 1.816 \qquad\qquad [b]$$

where the area of 15.82 m^2 is taken from Example 10-4. From Figure 10-13 or Table 10-3 the effectiveness is

$$\epsilon = 0.744$$

and because the cold fluid is the minimum, we can write

$$\epsilon = \frac{\Delta T_{\text{cold}}}{\Delta T_{\max}} = \frac{\Delta T_{\text{cold}}}{110-35} = 0.744 \qquad\qquad [c]$$

$$\Delta T_{\text{cold}} = 55.8°\text{C}$$

and the exit water temperature is

$$T_{w,\text{exit}} = 35 + 55.8 = 90.8°\text{C}$$

The total heat transfer under the new flow conditions is calculated as

$$q = \dot{m}_c c_c \Delta T_c = \frac{40}{60}(4180)(55.8) = 155.5 \text{ kW} \quad [5.29 \times 10^5 \text{ Btu/h}] \qquad [d]$$

Notice that although the flow rate has been reduced by 41 percent (68 to 40 kg/min), the heat transfer is reduced by only 18 percent (189.5 to 155.5 kW) because the exchanger is more effective at the lower flow rate.

	Cross-Flow Exchanger with Both
EXAMPLE 10-11	Fluids Unmixed

A finned-tube heat exchanger like that shown in Figure 10-5 is used to heat 5000 ft^3/min [2.36 m^3/s] of air at 1 atm from 60 to 85°F (15.55 to 29.44°C). Hot water enters the tubes at 180°F [82.22°C], and the air flows across the tubes, producing an average overall heat-transfer coefficient of 40 Btu/h · ft^2 · °F [227 W/m^2 · °C]. The total surface area of the exchanger is 100 ft^2 [9.29 m^2]. Calculate the exit water temperature and the heat-transfer rate.

■ **Solution**

The heat transfer is calculated from the energy balance on the air. First, the inlet air density is

$$\rho = \frac{p}{RT} = \frac{1.0132 \times 10^5}{(287)(288.7)} = 1.223 \text{ kg/m}^3$$

so the mass flow of air (the cold fluid) is

$$\dot{m}_c = (2.36)(1.223) = 2.887 \text{ kg/s}$$

The heat transfer is then

$$q = \dot{m}_c c_c \, \Delta T_c = (2.887)(1006)(29.44 - 15.55)$$
$$= 40.34 \text{ kW} \quad [1.38 \times 10^5 \text{ Btu/h}] \tag{a}$$

From the statement of the problem we do not know whether the air or water is the minimum fluid. If the air is the minimum fluid, we may immediately calculate NTU and use Figure 10-15 to determine the water-flow rate and hence the exit water temperature. If the water is the minimum fluid, a trial-and-error procedure must be used with Figure 10-15 or Table 10-3. We assume that the air is the minimum fluid and then check our assumption. Then

$$\dot{m}_c c_c = (2.887)(1006) = 2904 \text{ W/}^\circ\text{C}$$

and

$$\text{NTU}_{max} = \frac{UA}{C_{min}} = \frac{(227)(9.29)}{2904} = 0.726$$

and the effectiveness based on the air as the minimum fluid is

$$\epsilon \frac{\Delta T_{air}}{\Delta T_{max}} = \frac{29.44 - 15.55}{82.22 - 15.55} = 0.208 \tag{b}$$

Entering Figure 10-15, we are unable to match these quantities with the curves. This requires that the hot fluid be the minimum. We must therefore assume values for the water-flow rate until we are able to match the performance as given by Figure 10-15 or Table 10-3. We first note that

$$C_{max} = \dot{m}_c c_c = 2904 \text{ W/}^\circ\text{C} \tag{c}$$

$$\text{NTU}_{max} = \frac{UA}{C_{min}} \tag{d}$$

$$\epsilon = \frac{\Delta T_h}{\Delta T_{max}} = \frac{\Delta T_h}{82.22 - 15.55} \tag{e}$$

$$\Delta T_h = \frac{4.034 \times 10^4}{C_{min}} = \frac{4.034 \times 10^4}{C_h} \tag{f}$$

The iterations are:

$\frac{C_{min}}{C_{max}}$	$C_{min} = \dot{m}_h c_h$	NTU_{max}	ΔT_h	ϵ From Figure 10-15 or Table 10-3	Calculated from Equation (e)
0.5	1452	1.452	27.78	0.65	0.417
0.25	726	2.905	55.56	0.89	0.833
0.22	639	3.301	63.13	0.92	0.947

We thus estimate the water-flow rate as about

$$\dot{m}_h c_h = 660 \text{ W/}^\circ\text{C}$$

and

$$\dot{m}_h = \frac{660}{4180} = 0.158 \text{ kg/s}$$

The exit water temperature is accordingly

$$T_{w,\text{exit}} = 82.22 - \frac{4.034 \times 10^4}{660} = 21.1°C$$

Alternatively, Equations (c, d, e, f) may be rearranged to give

$$N = 0.7762/C \qquad [g]$$

$$\epsilon = 0.22084/C \qquad [h]$$

where N and C are defined as in Table 10-3. The appropriate effectiveness equation from Table 10-3 (cross flow, both fluids unmixed) is

$$\epsilon = 1 - \exp\{[\exp(-NCn) - 1]/Cn\} \qquad [i]$$

where $n = N^{-0.22}$
Substituting Equations (g) and (h) in Equation (i) gives a single equation in terms of the capacity ratio C, which may be solved numerically to yield

$$C = 0.23$$

The value of $C_{\min}$ is then

$$C_{\min} = 2904 \times C = (2904)(0.23) = 668 \text{ W/°C}$$

Or, a slightly different value from the above iteration. The resulting exit water temperature is thus

$$T_{w,\text{exit}} = 82.22 - 40,340/668 = 21.8°C$$

| EXAMPLE 10-12 | Comparison of Single- or Two-Exchanger Options |

A counterflow double-pipe heat exchanger is used to heat 1.25 kg/s of water from 35 to 80°C by cooling an oil [$c_p = 2.0$ kJ/kg · °C] from 150 to 85°C. The overall heat-transfer coefficient is 150 Btu/h · ft^2 · °F. A similar arrangement is to be built at another plant location, but it is desired to compare the performance of the single counterflow heat exchanger with two smaller counterflow heat exchangers connected in series on the water side and in parallel on the oil side, as shown in Figure Example 10-12. The oil flow is split equally between the two exchangers, and it may be assumed that the overall heat-transfer coefficient for the smaller exchangers is the same as for the large exchanger. If the smaller exchangers cost 20 percent more per unit surface area, which would be the most economical arrangement—the one large exchanger or two equal-sized small exchangers?

Figure Example 10-12

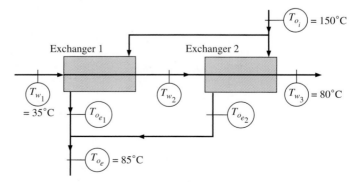

■ **Solution**

We calculate the surface area required for both alternatives and then compare costs. For the one large exchanger

$$q = \dot{m}_c c_c \, \Delta T_c = \dot{m}_h c_h \, \Delta T_h$$

$$= (1.25)(4180)(80 - 35) = \dot{m}_h c_h (150 - 85)$$

$$= 2.351 \times 10^5 \text{ W} \quad [8.02 \times 10^5 \text{ Btu/h}]$$

$$\dot{m}_c c_c = 5225 \text{ W/}°\text{C} \qquad \dot{m}_h c_h = 3617 \text{ W/}°\text{C}$$

so that the oil is the minimum fluid:

$$\epsilon_h = \frac{\Delta T_h}{150 - 35} = \frac{150 - 85}{150 - 35} = 0.565$$

$$\frac{C_{min}}{C_{max}} = \frac{3617}{5225} = 0.692$$

From Figure 10-13 or Table 10-4, $\text{NTU}_{max} = 1.09$, so that

$$A = \text{NTU}_{max} \frac{C_{min}}{U} = \frac{(1.09)(3617)}{850} = 4.649 \text{ m}^2 \quad [50.04 \text{ ft}^2]$$

We now wish to calculate the surface-area requirement for the two small exchangers shown in the sketch. We have

$$\dot{m}_h c_h = \frac{3617}{2} = 1809 \text{ W/}°\text{C}$$

$$\dot{m}_c c_c = 5225 \text{ W/}°\text{C}$$

$$\frac{C_{min}}{C_{max}} = \frac{1809}{5225} = 0.347$$

The number of transfer units is the same for each heat exchanger because UA and C_{min} are the same for each exchanger. This requires that the effectiveness be the same for each exchanger. Thus,

$$\epsilon_1 = \frac{T_{oi} - T_{oe,1}}{T_{oi} - T_{w,1}} = \epsilon_2 = \frac{T_{oi} - T_{oe,2}}{T_{oi} - T_{w,2}}$$

$$\epsilon_1 = \frac{150 - T_{oe,1}}{150 - 35} = \epsilon_2 = \frac{150 - T_{oe,2}}{150 - T_{w,2}} \qquad [a]$$

where the nomenclature for the temperatures is indicated in the sketch. Because the oil flow is the same in each exchanger and the average exit oil temperature must be 85°C, we may write

$$\frac{T_{oe,1} + T_{oe,2}}{2} = 85 \qquad [b]$$

An energy balance on the second heat exchanger gives

$$(5225)(T_{w3} - T_{w2}) = (1809)(T_{oi} - T_{oe,2})$$

$$(5225)(80 - T_{w2}) = (1809)(150 - T_{oe,2}) \qquad [c]$$

We now have the three equations (a), (b), and (c) that may be solved for the three unknowns $T_{oe,1}$, $T_{oe,2}$, and T_{w2}. The solutions are

$$T_{oe,1} = 76.98°\text{C}$$

$$T_{oe,2} = 93.02°\text{C}$$

$$T_{w2} = 60.26°\text{C}$$

The effectiveness can then be calculated as

$$\epsilon_1 = \epsilon_2 = \frac{150 - 76.98}{150 - 35} = 0.635$$

From Figure 10-13 or Table 10-4, we obtain $\text{NTU}_{max} = 1.16$, so that

$$A = \text{NTU}_{max} \frac{C_{min}}{U} = \frac{(1.16)(1809)}{850} = 2.47 \text{ m}^2$$

We thus find that 2.47 m^2 of area is required for each of the small exchangers, or a total of 4.94 m^2. This is greater than the 4.649 m^2 required in the one larger exchanger; in addition, the cost per unit area is greater so that the most economical choice would be the single larger exchanger. It may be noted, however, that the pumping costs for the oil would probably be less with the two smaller exchangers, so that this could precipitate a decision in favor of the smaller exchangers if pumping costs represented a sizable economic factor.

EXAMPLE 10-13	Shell-and-Tube Exchanger as Air Heater

Hot oil at 100°C is used to heat air in a shell-and-tube heat exchanger. The oil makes six tube passes and the air makes one shell pass; 2.0 kg/s of air are to be heated from 20 to 80°C. The specific heat of the oil is 2100 J/kg · °C, and its flow rate is 3.0 kg/s. Calculate the area required for the heat exchanger for $U = 200$ W/m^2 · °C.

■ **Solution**

The basic energy balance is

$$\dot{m}_o c_o \, \Delta T_o = \dot{m}_a c_{pa} \, \Delta T_a$$

or

$$(3.0)(2100)(100 - T_{oe}) = (2.0)(1009)(80 - 20)$$

$$T_{oe} = 80.78°C$$

We have

$$\dot{m}_h c_h = (3.0)(2100) = 6300 \text{ W/°C}$$

$$\dot{m}_c c_c = (2.0)(1009) = 2018 \text{ W/°C}$$

so the air is the minimum fluid and

$$C = \frac{C_{min}}{C_{max}} = \frac{2018}{6300} = 0.3203$$

The effectiveness is

$$\epsilon = \frac{\Delta T_c}{\Delta T_{max}} = \frac{80 - 20}{100 - 20} = 0.75$$

Now, we may use either Figure 10-16 or the analytical relation from Table 10-4 to obtain NTU. For this problem we choose to use the table.

$$N = -(1 + 0.3203^2)^{-1/2} \ln \left[\frac{2/0.75 - 1 - 0.3203 - (1 + 0.3203^2)^{1/2}}{2/0.75 - 1 - 0.3203 + (1 + 0.3203^2)^{1/2}} \right]$$

$$= 1.99$$

Now, with $U = 200$ we calculate the area as

$$A = \text{NTU} \frac{C_{min}}{U} = \frac{(1.99)(2018)}{200} = 20.09 \text{ m}^2$$

Ammonia Condenser
EXAMPLE 10-14

A shell-and-tube heat exchanger is used as an ammonia condenser with ammonia vapor entering the shell at 50°C as a saturated vapor. Water enters the single-pass tube arrangement at 20°C and the total heat transfer required is 200 kW. The overall heat-transfer coefficient is estimated from Table 10-1 as 1000 W/m² · °C. Determine the area to achieve a heat exchanger effectiveness of 60 percent with an exit water temperature of 40°C. What percent reduction in heat transfer would result if the water flow is reduced in half while keeping the heat exchanger area and U the same?

■ Solution
The mass flow can be calculated from the heat transfer with

$$q = 200 \text{ kW} = \dot{m}_w c_w \, \Delta T_w$$

so

$$\dot{m}_w = \frac{200}{(4.18)(40-20)} = 2.39 \text{ kg/s}$$

Because this is a condenser the water is the minimum fluid and

$$C_{min} = \dot{m}_w c_w = (2.39)(4.18) = 10 \text{ kW/°C}$$

The value of NTU is obtained from the last entry of Table 10-4, with $\epsilon = 0.6$:

$$N = -\ln(1-\epsilon) = -\ln(1-0.6) = 0.916$$

so that the area is calculated as

$$A = \frac{C_{min} N}{U} = \frac{(10,000)(0.916)}{1000} = 9.16 \text{ m}^2$$

When the flow rate is reduced in half the new value of NTU is

$$N = \frac{UA}{C_{min}} = \frac{(1000)(9.16)}{(10,000/2)} = 1.832$$

And the effectiveness is computed from the last entry of Table 10-3:

$$\epsilon = 1 - e^{-N} = 0.84$$

The new water temperature difference is computed as

$$\Delta T_w = \epsilon(\Delta T_{max}) = (0.84)(50-20) = 25.2°C$$

so the new heat transfer is

$$q = C_{min} \, \Delta T_w = \left(\frac{10,000}{2}\right)(25.2) = 126 \text{ kW}$$

So, by reducing the flow rate in half we have lowered the heat transfer from 200 to 126 kW, or by 37 percent.

Cross-Flow Exchanger as Energy Conservation Device
EXAMPLE 10-15

A recuperator used as an energy conservation measure employs a cross-flow heat exchanger with both fluids unmixed as shown in Figure Example 10-15. The exchanger is designed to remove 210 kW from 1200 kg/min of atmospheric air entering at 25°C. This energy is used to preheat the same quantity of air that enters from outdoor conditions at 0°C before being used for a building

Figure Example 10-15

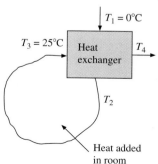

heating application. The design value of U for this flow condition is 30 W/m$^2 \cdot \,^\circ$C. The following calculations are desired:

1. The design value for the area of the heat exchanger
2. The percent reduction in heat-transfer rate if the flow rate is reduced by 50 percent while keeping the inlet temperatures and value of U constant
3. The percent reduction in heat-transfer rate if the flow rate is reduced by 50 percent and the value of U varies as mass flow to the 0.8 power, with the same inlet temperature conditions

■ **Solution**

1. The hot and cold fluids have the same flow rate with

$$\dot{m}_h = \dot{m}_c = 1200/60 = 20 \, \text{kg/s}$$

and

$$C_h = C_c = (20)(1005) = 20{,}100 \, \text{W/}^\circ\text{C}; \; C_{min}/C_{max} = 1.0 = C \text{ for use in Table 10-3}$$

The energy balance gives

$$q = 210{,}000 = C_h \, \Delta T_h = C_c \, \Delta T_c$$

and

$$\Delta T_h = \Delta T_c = 210{,}000/20{,}100 = 10.45^\circ\text{C}$$

The heat-exchanger effectiveness is

$$\varepsilon = \Delta T_{\text{min fluid}} / \Delta T_{\text{max HX}} = 10.45/(25 - 0) = 0.4179 \qquad [a]$$

Consulting Table 10-3 for a cross-flow exchanger with both fluids unmixed, and inserting the value $C = 1.0$, we have

$$\epsilon = 1 - \exp\{N^{0.22}[\exp(-N^{0.78}) - 1]\} \qquad [b]$$

Inserting $\varepsilon = 0.4179$, Equation (b) may be solved for N to yield

$$N = 0.8 = UA/C_{min} = (30)A/20{,}100$$

and

$$A = 536 \, \text{m}^2$$

This is the *design value* for the area of the heat exchanger.

2. We now examine the effect of reducing the flow rate by half, while keeping the inlet temperatures and value of U the same. Note that the flow rate of *both* fluids is reduced because they are physically the same fluid. This means that the value of C_{min}/C_{max} will remain the same at a value of 1.0, and Equation (b) above may still be used for the calculation of effectiveness. The new value of C_{min} is

$$C_{min} = (1/2)(20{,}100) = 10{,}050 \, \text{W/}^\circ\text{C}$$

so that

$$\text{NTU} = N = UA/C_{min} = (30)(536)/10050 = 1.6$$

Inserting this value in Equation (b) above gives

$$\epsilon = 0.5713$$

The temperature difference for each fluid is then

$$\Delta T = \epsilon \Delta T_{\text{max HX}} = (0.5713)(25 - 0) = 14.28^\circ\text{C}$$

The resulting heat transfer is then

$$q = \dot{m}c\Delta T = (10,050)(14.28) = 143.5\,\text{kW}$$

or a reduction of 32 percent for a reduction in flow rate of 50 percent.

3. Finally, we examine the effect of reducing the flow rate by 50 percent coupled with reduction in overall heat-transfer coefficient under the assumption that

$$U \text{ varies as } \dot{m}^{0.8} \text{ or, correspondingly, as } C_{min}^{0.8}$$

Still keeping the area constant, we would find that NTU varies as

$$\text{NTU} = N = UA/C_{min} \approx C^{0.8} \times C^{-1} = C^{-0.2}$$

Our new value of N under these conditions would be

$$N = (0.8)(10,050/20,100)^{-0.2} = 0.919$$

Inserting this value in Equation (b) above gives for the effectiveness

$$\epsilon = 0.4494$$

The corresponding temperature difference in each fluid is

$$\Delta T = \epsilon \Delta T_{\text{max HX}} = (0.4494)(25 - 0) = 11.23°\text{C}$$

The heat transfer is calculated as

$$q = \dot{m}c\Delta T = (10,050)(11.23) = 112.9\,\text{kW}$$

This is a reduction of 46 percent from the 210 kW design value at full flow. Again, we note the rather pronounced effect because *both* the hot and cold fluid flow rates are reduced, coupled with an anticipated decrease in the overall heat-transfer coefficient that may accompany the lower mass flows.

■ **Comment**

The design condition examined in step 1 resulted in a "heat recovery" of 210 kW from the warm air in the room. If the heat exchanger were not used, that energy would need to be supplied from other sources (presumably a hydrocarbon fuel) at a cost on the order of $10/10^9$ J or about $7.50 per hour of operation. The capital expense of installing the heat exchanger is usually well justified in such applications.

10-7 | COMPACT HEAT EXCHANGERS

A number of heat-exchanger surfaces do not fall into the categories discussed in the foregoing sections. Most notable are the compact exchangers that achieve a very high surface area per unit volume. These exchangers are most adaptable to applications where gas flows and low values of h are to be encountered. Kays and London [3] have studied these types of exchangers very extensively, and four typical configurations are shown in Figure 10-18. In Figure 10-18a a finned-tube exchanger is shown with flat tubes, Figure 10-18b shows a circular finned-tube array in a different configuration, and Figures 10-18c and d offer ways to achieve very high surface areas on both sides of the exchanger. These last two configurations are applicable to processes where gas-to-gas heat transfer is involved.

The heat transfer and friction factor for two typical compact exchangers are shown in Figures 10-19 and 10-20. The Stanton and Reynolds numbers are based on the mass

Figure 10-18 | Examples of compact heat-exchanger configurations.

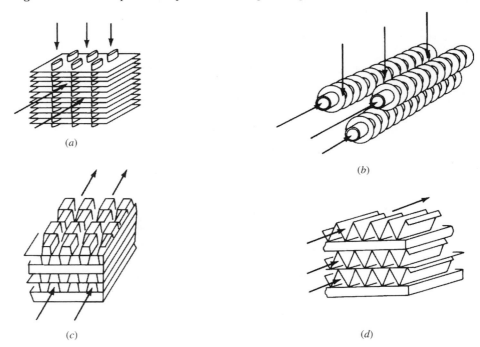

(a)

(b)

(c)

(d)

Figure 10-19 | Heat transfer and friction factor for finned flat-tube heat
exchanger according to Reference 3.

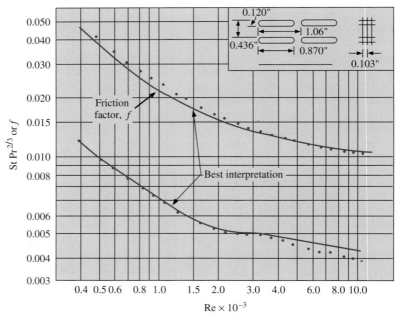

Fin pitch = 9.68 per in. (3.81/cm)
Flow passage hydraulic diameter, D_h = 0.01180 ft (3.6 mm)
Fin metal thickness = 0.004 in., (0.102 mm), copper
Free-flow area/frontal area, σ = 0.697
Total heat transfer area/total volume, α = 229 ft^2/ft^3 (751 m^2/m^3)
Fin area/total area = 0.795

Figure 10-20 │ Heat transfer and friction factor for finned circulator-tube heat exchanger according to Reference 3.

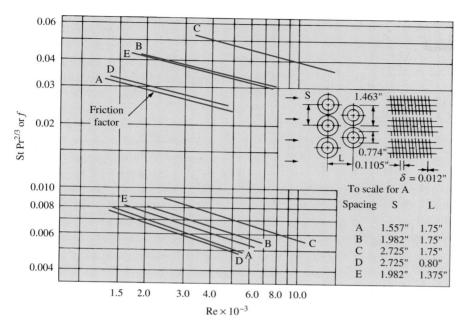

Tube outside diameter = 0.774 in. (1.93 cm)
Fin pitch = 9.05 per in. (3.56/cm)
Fin thickness = 0.012 in. (0.305/mm)
Fin area/Total area = 0.835

	A	B	C	D	E
Flow passage hydraulic diameter, D_h =	0.01681	0.02685	0.0445	0.01587	0.02108 ft
	5.12	8.18	13.56	4.84	6.43 mm
Free-flow area/frontal area, σ =	0.455	0.572	0.688	0.537	0.572
Heat trasfer area/ total volume, α =	108	85.1	61.9	135	108 ft²/ft³
	354	279	203	443	354 m²/m³

Note: Minimum free-flow area in all cases occurs in the spaces transverse to the flow, except for *D*, in which the minimum area is in the diagonals.

velocities in the minimum flow cross-sectional area and a hydraulic diameter stated in the figure.

$$G = \frac{\dot{m}}{A_c} \qquad\qquad \textbf{[10-28]}$$

The ratio of the free-flow area to frontal area

$$\sigma = \frac{A_c}{A} \qquad\qquad \textbf{[10-29]}$$

is also given in the figure. Thus,

$$\text{St} = \frac{h}{Gc_p} \qquad \text{Re} = \frac{D_h G}{\mu}$$

Fluid properties are evaluated at the average bulk temperature. Heat transfer and fluid friction *inside* the tubes are evaluated with the hydraulic diameter method discussed in Chapter 6. Pressure drop is calculated with the chart friction factor f and the following relation:

$$\Delta p = \frac{v_1 G^2}{2g_c} \left[(1 + \sigma^2) \left(\frac{v_2}{v_1} - 1 \right) + f \frac{A}{A_c} \frac{v_m}{v_1} \right] \qquad \text{[10-30]}$$

where v_1 and v_2 are the entrance and exit specific volumes, respectively, and v_m is the mean specific volume in the exchanger, normally taken as $v_m = (v_1 + v_2)/2$.

Rather meticulous design procedures are involved with compact heat exchangers, and these are given a full discussion in Reference 3.

EXAMPLE 10-16   Heat-Transfer Coefficient in Compact Exchanger

Air at 1 atm and 300 K enters an exchanger like that shown in Figure 10-19 with a velocity of 15 m/s. Calculate the heat-transfer coefficient.

■ **Solution**
We obtain the air properties from Table A-5 as

$$\rho = 1.1774 \text{ kg/m}^3 \qquad c_p = 1.0057 \text{ kJ/kg} \cdot {}^\circ\text{C}$$
$$\mu = 1.983 \times 10^{-5} \text{ kg/m} \cdot \text{s} \quad \text{Pr} = 0.708$$

From Figure 10-19 we have

$$\sigma = \frac{A_c}{A} = 0.697 \qquad D_h = 0.0118 \text{ ft} = 3.597 \text{ mm}$$

The mass velocity is thus

$$G = \frac{\dot{m}}{A_c} = \frac{\rho u_\infty A}{A_c} = \frac{(1.1774)(15)}{0.697} = 38.18 \text{ kg/m}^2 \cdot \text{s}$$

and the Reynolds number is

$$\text{Re} = \frac{D_h G}{\mu} = \frac{(3.597 \times 10^{-3})(38.18)}{1.983 \times 10^{-5}} = 6.926 \times 10^3$$

From Figure 10-19 we can read

$$\text{St Pr}^{2/3} = 0.0036 = \frac{h}{Gc_p} \text{Pr}^{2/3}$$

and the heat-transfer coefficient is

$$h = (0.0036)(38.18)(1005.7)(0.708)^{-2/3}$$
$$= 174 \text{ W/m}^2 \cdot {}^\circ\text{C} \quad [30.64 \text{ Btu/h} \cdot \text{ft}^2 \cdot {}^\circ\text{F}]$$

10-8 | ANALYSIS FOR VARIABLE PROPERTIES

The convection heat-transfer coefficient is dependent on the fluid being considered. Correspondingly, the overall heat-transfer coefficient for a heat exchanger may vary substantially through the exchanger if the fluids are such that their properties are strongly temperature-dependent. In this circumstance the analysis is best performed on a numerical or finite-difference basis. To illustrate the technique, let us consider the simple parallel-flow double-pipe heat exchanger of Section 10-5. The heat exchanger is divided into increments of surface area ΔA_j. For this incremental surface area the hot and cold temperatures are T_{hj} and T_{cj}, respectively, and we shall assume that the overall heat-transfer coefficient can be expressed as a function of these temperatures. Thus

$$U_j = U_j(T_{hj}, T_{cj})$$

The incremental heat transfer in ΔA_j is, according to Equation (10-6),

$$\Delta q_j = -(\dot{m}_h c_h)_j (T_{hj+1} - T_{hj}) = (\dot{m}_c c_c)_j (T_{cj+1} - T_{cj}) \qquad \textbf{[10-31]}$$

Also

$$\Delta q_j = U_j \Delta A_j (T_h - T_c)_j \qquad \textbf{[10-32]}$$

The finite-difference equation analogous to Equation (10-9) is

$$\frac{(T_h - T_c)_{j+1} - (T_h - T_c)_j}{(T_h - T_c)_j} = -U_j \left[\frac{1}{(\dot{m}_h c_h)_j} + \frac{1}{(\dot{m}_c c_c)_j} \right] \Delta A_j \qquad \textbf{[10-33]}$$

$$= -K_j(T_h, T_c) \Delta A_j$$

where we have introduced the indicated definition for K_j. Reducing Equation (10-33), we obtain

$$\frac{(T_h - T_c)_{j+1}}{(T_h - T_c)_j} = 1 - K_j \Delta A_j \qquad \textbf{[10-34]}$$

The numerical-analysis procedure is now clear when the inlet temperatures and flows are given:

1. Choose a convenient value of ΔA_j for the analysis.
2. Calculate the value of U for the inlet conditions and through the initial ΔA increment.
3. Calculate the value of q for this increment from Equation (10-32).
4. Calculate the values of T_h, T_c, and $T_h - T_c$ for the *next* increment, using Equations (10-31) and (10-34).
5. Repeat the foregoing steps until all the increments in ΔA are employed.

The total heat-transfer rate is then calculated from

$$q_{\text{total}} = \sum_{j=1}^{n} \Delta q_j$$

where n is the number of increments in ΔA.

A numerical analysis such as the one discussed above is easily performed with a computer. Heat-transfer rates calculated from a variable-properties analysis can frequently differ by substantial amounts from a constant-properties analysis. The most difficult part of the analysis is, of course, a determination of the values of h. The interested reader is referred to the heat-transfer literature for additional information on this complicated but important subject.

Transient Response of Thermal-Energy Storage System

EXAMPLE 10-17

A rock-bed thermal-energy storage unit is employed to remove energy from a hot airstream and store for later use. The schematic for the device is shown in Figure Example 10-17. The surface is covered with a material having an overall R value of 2 h·°F·ft^2/Btu. The inlet flow area is $5 \times 5 = 25$ ft^2, and the rock-bed length is 10 ft. Properties of the rock are

$$\rho_r = 80 \text{ lb}_m/\text{ft}^3 \quad [1281.4 \text{ kg/m}^3]$$
$$c_r = 0.21 \text{ Btu/lb}_m \cdot °\text{F} \quad [0.880 \text{ kJ/kg} \cdot °\text{C}]$$
$$k_r = 0.5 \text{ Btu/h} \cdot \text{ft} \cdot °\text{F} \quad [0.87 \text{ W/m} \cdot °\text{C}]$$

Figure Example 10-17 | (*a*) Schematic, (*b*) energy accumulation with time, (*c*) temperature-time response for $v = 1$ ft/s.

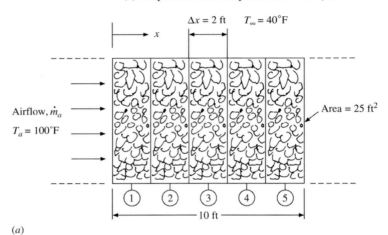

(*a*)

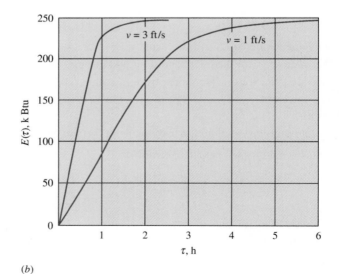

(*b*)

Figure Example 10-17 | (*Continued*).

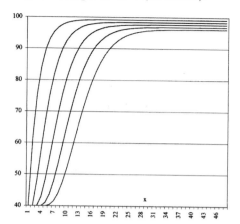

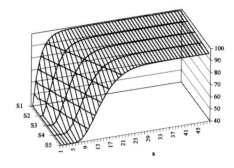

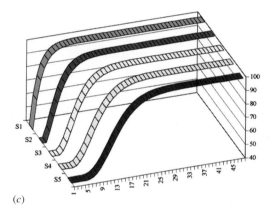

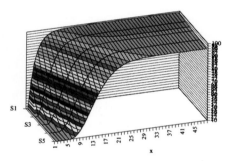

(*c*)

As the air flows through the rock, it is in such intimate contact with the rock that the air and rock temperatures may be assumed equal at any *x* position.

The rock bed is initially at 40°F and the air enters at 1 atm and 100°F. The surroundings remain at 40°F. Calculate the energy storage relative to 40°F as a function of time for inlet velocities of 1.0 and 3.0 ft/s.

■ **Solution**

It can be shown that the axial energy conduction is small compared to the mass-energy transport. For a 60°F temperature difference over a 2-ft length

$$q_{\text{cond}} = kA\frac{\Delta T}{\Delta x} = (0.5)(25)\frac{60}{2} = 375 \text{ Btu/h} \quad [109.9 \text{ W}] \tag*{[a]}$$

The density of the air at 100°F is

$$\rho_a = \frac{(14.696)(144)}{(53.35)(560)} = 0.07083 \text{ lb}_m/\text{ft}^3 \quad [1.1346 \text{ kg/m}^3] \tag*{[b]}$$

and the mass flow at 1.0 ft/s is

$$\dot{m}_a = \rho Av = (0.07083)(25)(1.0) = 1.7708 \text{ lb}_m/\text{s}$$
$$= 6375 \text{ lb}_m/\text{h} \quad [2891.7 \text{ kg/h}] \tag*{[c]}$$

The corresponding energy transport for a temperature difference of $60°F$ is

$$q = \dot{m} c_{pa} \, \Delta T = (6375)(0.24)(60) = 91{,}800 \text{ Btu/h} \quad [26{,}904 \text{ W}] \qquad [d]$$

and this is much larger than the value in Equation (a).

We now write an energy balance for one of the axial nodes as

Energy transported in $-$ energy transported out $-$ energy lost to surroundings

$=$ rate of energy accumulation of node

or

$$\dot{m} c_{pa} (T^p_{m-1} - T^p_m) - \frac{(T^p_m - T_\infty) P \, \Delta x}{R_\infty} = \rho_r c_r \Delta V_r \frac{(T^{p+1}_m - T^p_m)}{\Delta \tau} \qquad [e]$$

where the air exit temperature from node m is assumed to be the rock temperature of that node (T^p_m). Equation (e) may be solved to give

$$T^{p+1}_m = F \dot{m} c_{pa} T^p_{m-1} + \left[1 - F \left(\dot{m} c_{pa} + \frac{P \, \Delta x}{R_\infty} \right) \right] T^p_m + \frac{F P \, \Delta x}{R_\infty} T_\infty \qquad [f]$$

where

$$F = \frac{\Delta r}{\rho_r c_r \, \Delta V_r}$$

Here P is the perimeter and Δx is the x increment ($P = 4 \times 5 = 20$ ft for this problem). We are thus in a position to calculate the temperatures in the rock bed as time progresses.

The stability requirement is such that the coefficient on the T^p_m terms cannot be negative. Using $\Delta x = 2$ ft, we find that the maximum value of F is 6.4495×10^{-4}, which yields a maximum time increment of 0.54176 h. With a velocity of 3 ft/s the maximum time increment for stability is 0.1922 h. For the calculations we select the following values of $\Delta \tau$ with the resultant calculated values of F:

v	$\Delta \tau$, h	F
1.0	0.2	2.38095×10^{-4}
3.0	0.1	1.190476×10^{-4}

With the appropriate properties and these values inserted into Equation (f) there results

$$T^{p+1}_m = 0.3642943 T^p_{m-1} + 0.630943 T^p_m + 0.1904762 \quad \text{for } v = 1.0 \text{ ft/s} \qquad [g]$$

$$T^{p+1}_m = 0.546430633 T^p_{m-1} + 0.451188 T^p_m + 0.0952381 \quad \text{for } v = 3.0 \text{ ft/s} \qquad [h]$$

The energy storage relative to $40°F$ can then be calculated from

$$E(\tau) = \sum_{m=1}^{5} \rho_r c_r \, \Delta V_r [T_m(\tau) - 40] \qquad [i]$$

as a function of time. The computation procedure is as follows.

1. Initialize all T_m at $40°F$ with T_{m-1} for node 1 at $100°F$ for all time increments.
2. Compute new values of T_m from either Equation (g) or (h), progressing forward in time until a desired stopping point is reached or the temperature attains steady-state conditions.
3. Using computed values of $T_m(\tau)$, evaluate $E(\tau)$ from Equation (i).

Results of the calculations are shown in Figure Example 10-17(b). For $v = 3.0$ ft/s, steady state is reached at about $\tau = 1.5$ h while for $v = 1.0$ ft/s it is reached at about $\tau = 5.5$ h. Note that the steady-state value of E for $v = 1.0$ ft/s is lower than for $v = 3.0$ ft/s because a longer time is involved and more of the energy "leaks out" through the insulation.

This example shows how a rather complex problem can be solved in a straightforward way by using a numerical formulation. The actual calculation may be performed in a very straightforward way using the transient Excel formulation discussed in Appendix D (Section D-5). Displays of the resulting temperature-time profiles for $v = 1$ ft/s are given in Figure Example 10-17(c).

Variable-Properties Analysis of a Duct Heater | EXAMPLE 10-18

A 600-ft-long duct having a diameter of 1 ft serves as a space heater in a warehouse area. Hot air enters the duct at 800°F, and the emissivity of the outside duct surface is 0.6. Determine duct air temperature, wall temperature, and outside heat flux along the duct for flow rates of 0.3, 1.0, and 1.5 lb_m/s. Take into account variations in air properties. The room temperature for both convection and radiation is 70°F.

■ Solution

This is a problem where a numerical solution must be employed. We choose a typical section of the duct with length Δx and perimeter P as shown in Figure Example 10-18A and make the energy balances. We assume that the conduction resistance of the duct wall is negligible. Inside the duct the energy balance is

$$\dot{m}_a c_p T_{m,a} = h_i P \Delta x (T_{m,a} - T_{m,w}) + \dot{m}_a c_p T_{m+1,a} \qquad [a]$$

where h_i is the convection heat-transfer coefficient on the inside that may be calculated from (the flow is turbulent)

$$Nu = \frac{h_i d}{k} = 0.023 \, Re_d^{0.8} \, Pr^{0.3} \qquad [b]$$

with properties evaluated at the bulk temperature of air ($T_{m,a}$). The energy balance for the heat flow through the wall is

$$q_{\text{conv},i} = q_{\text{conv},o} + q_{\text{rad},o}$$

or, by using convection coefficients and radiation terms per unit area,

$$h_i(T_{m,a} - T_{m,w}) = h_c(T_{m,w} - T_\infty) + \sigma\epsilon \left(T_{m,w}^4 - T_\infty^4 \right) \qquad [c]$$

Figure Example 10-18a | Schematic.

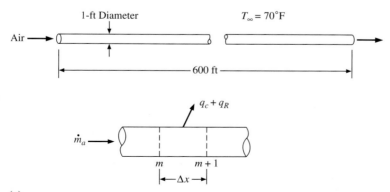

(a)

Figure Example 10-18b | Heat flux.

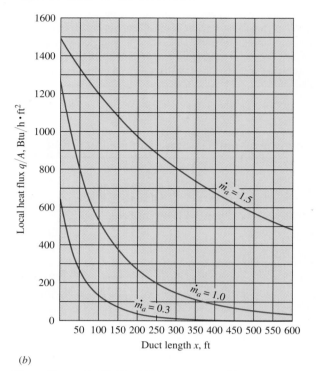

(b)

Figure Example 10-18c | Temperature profiles.

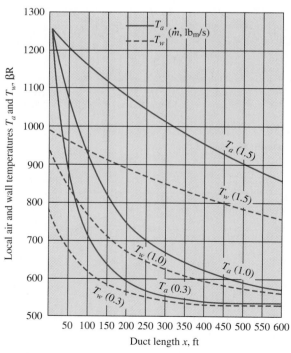

(c)

where the outside convection coefficient can be calculated from the free-convection relation

$$h_c = 0.27 \left(\frac{T_{m,w} - T_\infty}{d} \right)^{1/4} \qquad \text{Btu/h} \cdot \text{ft}^2 \cdot \,^\circ\text{F} \qquad [d]$$

Inserting this relation in Equation (c) gives

$$h_i(T_{m,a} - T_{m,w}) = \frac{0.27}{d^{1/4}}(T_{m,w} - T_\infty)^{5/4} + \sigma\epsilon \left(T_{m,w}^4 - T_\infty^4 \right) \qquad [e]$$

Equation (a) may be solved for $T_{m+1,a}$ to give

$$T_{m+1,a} = \left(1 - \frac{h_i P \Delta x}{\dot{m}_a c_p}\right)_m T_{m,a} + \left(\frac{h_i P \Delta x}{\dot{m}_a c_p}\right)_m T_{m,w} \qquad [f]$$

With these equations at hand, we may now formulate the computational algorithm as follows. Note that all temperatures must be in degrees Rankine because of the radiation term.

1. Select Δx.
2. Starting at $x = 0$, entrance conditions, evaluate h_i from Equation (b) with properties evaluated at $T_{m,a}$. (At entrance $T_{m,a} = 800\,^\circ\text{F} = 1260\,^\circ\text{R}$.)
3. Solve (by iteration) Equation (e) for $T_{m,w}$.
4. Solve for $T_{m+1,a}$ from Equation (f).
5. Repeat for successive increments until the end of the duct ($x = 600$ ft) is reached.
6. The heat lost at each increment is

$$q = P \Delta x h_i (T_{m,a} - T_{m,w})$$

or the heat flux is

$$\frac{q}{A} = h_i(T_{m,a} - T_{m,w}) \qquad [g]$$

7. The results for $T_{m,a}$, $T_{m,w}$, and $(q/A)_m$ may be plotted as in the Figure Examples 10-18(b,c).

For these calculations we have selected $\Delta x = 50$ ft. For the low flow rate (0.3 lb$_m$/s) we note that the air essentially attains the room temperature halfway along the length of the duct, so that little heating is provided past that point. With the 1.0-lb$_m$/s flow rate there is still some heating at the end of the duct, although it is small. The 1.5-lb$_m$/s flow rate contributes substantial heating all along the length of the duct.

As with Example 10-17, the calculation may be performed quite easily using the Excel transient formulation given in Appendix D (Section D-5).

Performance of a Steam Condenser   EXAMPLE 10-19

A shell-and-tube heat exchanger is used to condense atmospheric steam from saturated vapor to saturated liquid at 100°C. Liquid water entering at 35°C is used as the coolant in the tube side of the exchanger. Examine the performance characteristic of the exchanger as a function of the mass flow of cooling water, and establish ranges of operation where the total heat transfer (and thus the total steam condensed) is not a strong function of the mass-flow rate of water. Assume that the value of U and A are constant at 2000 W/m^2 · °C and 1.1 m^2, respectively.

■ **Solution**
For this study, we shall employ the analytical relations of Table 10-3. Because the condensing steam has an effective specific heat that is very large, the water coolant will always be the minimum fluid and $C = C_{min}/C_{max} \approx 0$. Furthermore, for all types of exchangers with $C = 0$,

$$\epsilon = 1 - e^{-N} \qquad [a]$$

Figure Example 10-19a

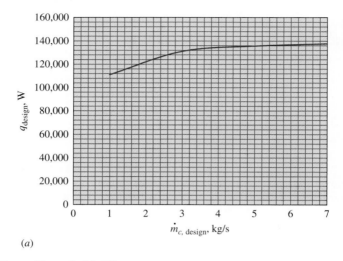

(a)

Figure Example 10-19b

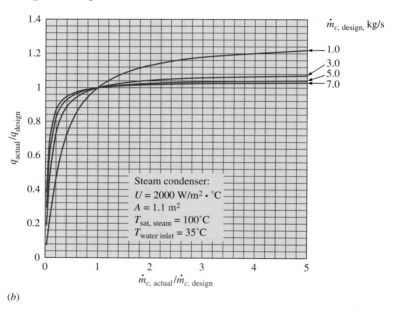

(b)

We also have

$$N = \text{NTU} = \left(\frac{UA}{C_{min}} \right) \qquad [b]$$

$$C_{min} = \dot{m}_c c_c \qquad [c]$$

$$q = C_{min} \Delta T_{\max HX} \times \epsilon = \dot{m}_c c_c (100 - 35) \times \epsilon \qquad [d]$$

Our objective is to study the behavior of the exchanger as a function of the mass-flow rate of cooling water. We thus choose that variable as the primary design parameter. Using $c_c = 4180\,\text{J/kg} \cdot {}^\circ\text{C}$ and selected values of $\dot{m}_c$, the "design" heat transfer can be plotted as shown in Figure Example 10-19a. Note that for cooling water-flow rates above 3 kg/s, the heat transfer rate does not vary more than ± 2.5 percent, and thus is rather insensitive to the relatively larger variations in flow rate.

Now let us select several "design" flow rates and examine the effects of varying the flow rate from the design value on the total heat transfer. Again, Equations (a) through (d) are employed, and the results are displayed in Figure Example 10-19b. These results are presented as the ratio q_{actual}/q_{design} for the ordinate and $\dot{m}_{c,actual}/\dot{m}_{c,design}$ as the abscissa. Plots are presented for four design flow rates of 1, 3, 5, and 7 kg/s. For the three larger flow rates, the curves are very flat (within about 5 percent) from $0.7 < \dot{m}_{c,actual}/\dot{m}_{c,design} < 5$, which encompasses a very wide range of flow rates. The net conclusion is that the total heat transfer rate and steam condensed will not vary much, as long as a threshold cooling-flow rate of 3 kg/s is maintained. In addition, this suggests that any control system that may be deployed in the particular application can probably be a very simple one.

One may question the assumption of constant U in the analysis. The value of the controlling-convection coefficient for the cooling water may be adjusted in the design process by varying the tube diameter (and hence the Reynolds number). Even allowing for some variation, the same characteristic behavior would be observed as shown in the figures. The purpose of this example is to show the utility of the analytical relations of Table 10-3 in anticipating favorable (or unfavorable) operating results in a particular application.

10-9 | HEAT-EXCHANGER DESIGN CONSIDERATIONS

In the process and power industries, or related activities, many heat exchangers are purchased as off-the-shelf items, and a selection is made on the basis of cost and specifications furnished by the various manufacturers. In more specialized applications, such as the aerospace and electronics industries, a particular design is frequently called for. Where a heat exchanger forms a part of an overall machine or device to be manufactured, a standard item may be purchased; or if cost considerations and manufacturing quantities warrant, the heat exchanger may be specially designed for the application. Whether the heat exchanger is selected as an off-the-shelf item or designed especially for the application, the following factors are almost always considered:

1. Heat-transfer requirements
2. Cost
3. Physical size
4. Pressure-drop characteristics

The heat-transfer requirements must be met in the selection or design of any heat exchanger. The way that the requirements are met depends on the relative weights placed on items 2 to 4. By forcing the fluids through the heat exchanger at higher velocities the overall heat-transfer coefficient may be increased, but this higher velocity results in a larger pressure drop through the exchanger and correspondingly larger pumping costs. If the surface area of the exchanger is increased, the overall heat-transfer coefficient, and hence the pressure drop, need not be so large; however, there may be limitations on the physical size that can be accommodated, and a larger physical size results in a higher cost for the heat exchanger. Prudent judgment and a consideration of all these factors will result in the proper design. A practitioner in the field will find the extensive information of Reference 8 to be very useful.

REVIEW QUESTIONS

1. Define the overall heat-transfer coefficient.
2. What is a fouling factor?

3. Why does a "mixed" or "unmixed" fluid arrangement influence heat-exchanger performance?
4. When is the LMTD method most applicable to heat-exchanger calculations?
5. Define effectiveness.
6. What advantage does the effectiveness-NTU method have over the LMTD method?
7. What is meant by the "minimum" fluid?
8. Why is a counterflow exchanger more effective than a parallel-flow exchanger?

LIST OF WORKED EXAMPLES

10-1 Overall heat-transfer coefficient for pipe in air
10-2 Overall heat-transfer coefficient for pipe exposed to steam
10-3 Influence of fouling factor
10-4 Calculation of heat-exchanger size from known temperatures
10-5 Shell-and-tube heat exchanger
10-6 Design of shell-and-tube heat exchanger
10-7 Cross-flow exchanger with one fluid mixed
10-8 Effects of off-design flow rates for exchanger in Example 10-7
10-9 Off-design calculation using ϵ-NTU method
10-10 Off-design calculation of exchanger in Example 10-4
10-11 Cross-flow exchanger with both fluids unmixed
10-12 Comparison of single- or two-exchanger options
10-13 Shell-and-tube exchanger as air heater
10-14 Ammonia condenser
10-15 Cross-flow exchanger as energy conversion device
10-16 Heat-transfer coefficient in compact exchanger
10-17 Transient response of thermal-energy storage system
10-18 Variable-properties analysis of a duct heater
10-19 Performance of a steam condenser

PROBLEMS

10-1 A long steel pipe with a 5-cm ID and 3.2-mm wall thickness passes through a large room maintained at 30°C and atmospheric pressure; 0.6 kg/s of hot water enters one end of the pipe at 82°C. If the pipe is 15 m long, calculate the exit water temperature, considering both free convection and radiation heat loss from the outside of the pipe.

10-2 A counterflow double-pipe heat exchanger operates with hot water flowing inside the inner pipe and a polymer fluid flowing in the annular space between the two pipes. The water-flow rate is 2.0 kg/s and it enters at a temperature of 90°C. The polymer enters at a temperature of 10°C and leaves at a temperature of 50°C while the water leaves the exchanger at a temperature of 60°C. Calculate the value of the overall heat-transfer coefficient expressed in $W/m^2 \cdot °C$, if the area for the heat exchanger is 20 m^2.

10-3 Air at 207 kPa and 200°C enters a 2.5-cm-ID tube at 6 m/s. The tube is constructed of copper with a thickness of 0.8 mm and a length of 3 m. Atmospheric air at 1 atm and 20°C flows normal to the outside of the tube with a free-stream velocity of 12 m/s. Calculate the air temperature at exit from the tube. What would be the effect of reducing the hot-air flow in half?

10-4 Repeat Problem 10-3 for water entering the tube at 1 m/s and 95°C. What would be the effect of reducing the water flow in half?

10-5 Hot water at 90°C flows on the inside of a 2.5-cm-ID steel tube with 0.8-mm wall thickness at a velocity of 4 m/s. Engine oil at 20°C is forced across the tube at a velocity of 7 m/s. Calculate the overall heat-transfer coefficient for this arrangement.

10-6 Hot water at 90°C flows on the inside of a 2.5-cm-ID steel tube with 0.8-mm wall thickness at a velocity of 4 m/s. This tube forms the inside of a double-pipe heat exchanger. The outer pipe has a 3.75-cm ID, and engine oil at 20°C flows in the annular space at a velocity of 7 m/s. Calculate the overall heat-transfer coefficient for this arrangement. The tube length is 6.0 m.

10-7 Air at 2 atm and 200°C flows inside a 1-in schedule 80 steel pipe with $h = 65$ W/m$^2 \cdot$°C. A hot gas with $h = 180$ W/m$^2 \cdot$°C flows across the outside of the pipe at 400°C. Calculate the overall heat-transfer coefficient.

10-8 Hot exhaust gases are used in a finned-tube cross-flow heat exchanger to heat 2.5 kg/s of water from 35 to 85°C. The gases [$c_p = 1.09$ kJ/kg $\cdot$ °C] enter at 200 and leave at 93°C. The overall heat-transfer coefficient is 180 W/m$^2 \cdot$°C. Calculate the area of the heat exchanger using (*a*) the LMTD approach and (*b*) the effectiveness-NTU method.

10-9 Derive Equation (10-12), assuming that the heat exchanger is a counterflow double-pipe arrangement.

10-10 Derive Equation (10-27).

10-11 Water at the rate of 230 kg/h at 35°C is available for use as a coolant in a double-pipe heat exchanger whose total surface area is 1.4 m^2. The water is to be used to cool oil [$c_p = 2.1$ kJ/kg $\cdot$ °C] from an initial temperature of 120°C. Because of other circumstances, an exit water temperature greater than 99°C cannot be allowed. The exit temperature of the oil must not be below 60°C. The overall heat-transfer coefficient is 280 W/m$^2 \cdot$°C. Estimate the maximum flow rate of oil that may be cooled, assuming the flow rate of water is fixed at 230 kg/h.

10-12 The enthalpy of vaporization for water at 120°C is 2202.6 kJ/kg. A finned-tube heat exchanger is used to condense steam from saturated vapor to saturated liquid in the tube side of the exchanger. Air is used on the fin side of the exchanger, producing an overall heat transfer coefficient of 47 W/m$^2 \cdot$°C. The air enters at 30°C and leaves the exchanger at 40°C. What area of exchanger is needed to condense 2500 kg/h of steam?

10-13 Suppose the airflow rate of the exchanger in Problem 10-12 is cut by 40 percent. What decrease in steam condensation rate would result?

10-14 A small shell-and-tube exchanger with one tube pass [$A = 4.64$ m^2 and $U = 280$ W/m$^2 \cdot$°C] is to be used to heat high-pressure water initially at 20°C with hot air initially at 260°C. If the exit water temperature is not to exceed 93°C and the airflow rate is 0.45 kg/s, calculate the water-flow rate.

10-15 A double-pipe heat exchanger having an area of 100 m^2 is used to heat 5 kg/s of water that enters the heat exchanger at 50°C. The heating fluid is oil having a specific heat of 2.1 kJ/kg · °C and a flow rate of 8 kg/s. The oil enters the exchanger at 100°C and the overall heat-transfer coefficient is 120 W/m² · °C. Calculate the exit temperature of the oil and the heat transfer if the exchanger operates in a counterflow mode.

10-16 A counterflow double-pipe heat exchanger is to be used to heat 0.7 kg/s of water from 35 to 90°C with an oil flow of 0.95 kg/s. The oil has a specific heat of 2.1 kJ/kg · °C and enters the heat exchanger at a temperature of 175°C. The overall heat-transfer coefficient is 425 W/m² · °C. Calculate the area of the heat exchanger and the effectiveness.

10-17 Rework Example 6-10, using the LMTD concept. Repeat for an inlet air temperature of 37°C.

10-18 A shell-and-tube heat exchanger operates with two shell passes and four tube passes. The shell fluid is ethylene glycol, which enters at 140°C and leaves at 80°C with a flow rate of 4500 kg/h. Water flows in the tubes, entering at 35°C and leaving at 85°C. The overall heat-transfer coefficient for this arrangement is 850 W/m² · °C. Calculate the flow rate of water required and the area of the heat exchanger.

10-19 The flow rate of glycol to the exchanger in Problem 10-18 is reduced in half with the entrance temperatures of both fluids remaining the same. What is the water exit temperature under these new conditions, and by how much is the heat-transfer rate reduced?

10-20 For the exchanger in Problem 10-8 the water-flow rate is reduced by 30 percent, while the gas flow rate is maintained constant along with the fluid inlet temperatures. Calculate the percentage reduction in heat transfer as a result of this reduced flow rate. Assume that the overall heat-transfer coefficient remains the same.

10-21 Repeat Problem 10-8 for a shell-and-tube exchanger with two tube passes. The gas is the shell fluid.

10-22 Repeat Problem 10-20, using the shell-and-tube exchanger of Problem 10-21.

10-23 It is desired to heat 230 kg/h of water from 35 to 93°C with oil [$c_p = 2.1$ kJ/kg · °C] having an initial temperature of 175°C. The mass flow of oil is also 230 kg/h. Two double-pipe heat exchangers are available:

exchanger 1: $U = 570$ W/m² · °C $A = 0.47$ m²
exchanger 2: $U = 370$ W/m² · °C $A = 0.94$ m²

Which exchanger should be used?

10-24 A small steam condenser is designed to condense 0.76 kg/min of steam at 83 kPa with cooling water at 10°C. The exit water temperature is not to exceed 57°C. The overall heat-transfer coefficient is 3400 W/m² · °C. Calculate the area required for a double-pipe heat exchanger. $T_{sat} = 95.6$°C, $h_{fg} = 2.27 \times 10^6$ J/kg.

10-25 Suppose the inlet water temperature in the exchanger of Problem 10-24 is raised to 30°C. What percentage increase in flow rate would be necessary to maintain the same rate of condensation?

10-26 A counterflow double-pipe heat exchanger is used to heat water from 20 to 40°C by cooling an oil from 90 to 55°C. The exchanger is designed for a total heat transfer of 59 kW with an overall heat-transfer coefficient of 340 W/m² · °C. Calculate the surface area of the exchanger.

10-27 A feedwater heater uses a shell-and-tube exchanger with condensing steam in one shell pass at 120°C. Water enters the tubes at 30°C and makes four passes to produce an overall U value of 2000 W/m² · °C. Calculate the area of the exchanger for 2.5-kg/s mass flow of the water, with a water exit temperature of 100°C.

10-28 A cross-flow heat exchanger is used to cool a hot oil ($c = 1.9\,\text{kJ/kg} \cdot {}^\circ\text{C}$) from 120 to 95°C. The oil flows inside the tubes over which cooling water performs the cooling process. The water enters the exchanger at 20°C and leaves the exchanger at 50°C. The overall heat transfer coefficient is 55 W/m$^2 \cdot$ °C. What size exchanger will be required to cool 3700 kg/h of oil?

10-29 After the exchanger in Problem 10-28 has been sized and purchased, it is discovered that the cooling water is available at 40°C instead of 20°C, but because of other considerations the exit temperature of the water must be maintained at 50°C. Assuming the overall heat transfer coefficient to be constant, what do you recommend?

10-30 Suppose the exchanger in Problem 10-27 has been in service a long time such that a fouling factor of 0.0002 m$^2 \cdot$ °C/W is experienced. What would be the exit water temperature under these conditions?

10-31 An air-to-air heat recovery unit uses a cross-flow exchanger with both fluids unmixed and an airflow rate of 0.5 kg/s on both sides. The hot air enters at 400°C while the cool air enters at 20°C. Calculate the exit temperatures for $U = 40$ W/m$^2 \cdot$ °C and a total exchanger area of 15 m^2.

10-32 In a large air-conditioning application, 1500 m^3/min of air at 1 atm and 10°C are to be heated in a finned-tube heat exchanger with hot water entering the exchanger at 80°C. The overall heat-transfer coefficient is 50 W/m$^2 \cdot$ °C. Calculate the required area for the heat exchanger for an exit air temperature of 37°C and exit water temperature of 48°C.

10-33 A cross-flow finned-tube heat exchanger uses hot water to heat air from 20 to 45°C. The entering water temperature is 75°C and its exit temperature is 45°C. The total heat-transfer rate is to be 35 kW. If the overall heat transfer coefficient is 50 W/m$^2 \cdot$ °C, calculate the area of the heat exchanger.

10-34 Hot oil at 120°C with a flow rate of 95 kg/min is used in a shell-and-tube heat exchanger with one shell pass and two tube passes to heat 55 kg/min of water that enters at 30°C. The area of the exchanger is 14 m^2. Calculate the heat transfer and exit temperature of both fluids if the overall heat-transfer coefficient is 250 W/m$^2 \cdot$ °C.

10-35 A counterflow double-pipe heat exchanger is used to heat liquid ammonia from 10 to 30°C with hot water that enters the exchanger at 60°C. The flow rate of the water is 5.0 kg/s and the overall heat-transfer coefficient is 800 W/m$^2 \cdot$ °C. The area of the heat exchanger is 30 m^2. Calculate the flow rate of ammonia.

10-36 A shell-and-tube heat exchanger has condensing steam at 100°C in the shell side with one shell pass. Two tube passes are used with air in the tubes entering at 10°C. The total surface area of the exchanger is 30 m^2 and the overall heat-transfer coefficient may be taken as 150 W/m$^2 \cdot$ °C. If the effectiveness of the exchanger is 85 percent, what is the total heat-transfer rate?

10-37 Suppose both flow rates in Problem 10-31 were cut in half. What would be the exit temperatures in this case, assuming no change in U? What if the flow rates were doubled?

10-38 Hot water at 90°C is used in the tubes of a finned-tube heat exchanger. Air flows across the fins and enters at 1 atm, 30°C, with a flow rate of 65 kg/min. The overall heat-transfer coefficient is 52 W/m$^2 \cdot$ °C, and the exit air temperature is to be 45°C. Calculate the exit water temperature if the total area is 8.0 m^2.

10-39 5 kg/s of water is to be cooled from 90°C to 70°C in a shell and tube heat exchanger having four shell passes and eight tube passes. The cooling fluid is also water with a flow rate of 5 kg/s that enters at a temperature of 50°C. The overall heat-transfer coefficient is 800 W/m^2 · °C. Calculate the total area of the heat exchanger assuming all four shells are the same size.

10-40 A microprocessor is to be programmed to control the exchanger in Problem 10-38 by varying the water-flow rate to maintain the same exit air temperature for changes in inlet water temperature. Calculate the percentage changes necessary for the water-flow rate for inlet water temperatures of 60, 70, 80, and 100°C. Assume U remains constant.

10-41 High-pressure hot water at 120°C is used to heat an oil from 30 to 40°C. The water leaves the counterflow heat exchanger at a temperature of 90°C. If the total area of the heat exchanger is 5 m^2, calculate the effectiveness of the exchanger. What would be the effectiveness if a parallel-flow exchanger were used with the same area?

10-42 Calculate the number of transfer units for each of the exchangers in Problem 10-41.

10-43 Repeat Problem 10-41 if the hot fluid is a condensing vapor at 120°C.

10-44 Repeat Problem 10-42 if the hot fluid is a condensing vapor at 120°C.

10-45 Hot water enters a counterflow heat exchanger at 99°C. It is used to heat a cool stream of water from 4 to 32°C. The flow rate of the cool stream is 1.3 kg/s, and the flow rate of the hot stream is 2.6 kg/s. The overall heat-transfer coefficient is 830 W/m^2 · °C. What is the area of the heat exchanger? Calculate the effectiveness of the heat exchanger.

10-46 Starting with a basic energy balance, derive an expression for the effectiveness of a heat exchanger in which a condensing vapor is used to heat a cooler fluid. Assume that the hot fluid (condensing vapor) remains at a constant temperature throughout the process.

10-47 Water at 75°C enters a counterflow heat exchanger. It leaves at 30°C. The water is used to heat an oil from 25 to 48°C. What is the effectiveness of the heat exchanger?

10-48 Replot Figures 10-12 and 10-13 as ϵ versus log NTU$_{max}$ over the range 0.1 < NTU$_{max}$ < 100.

10-49 Suppose that the oil in Problem 10-26 is sufficiently dirty for a fouling factor of 0.004 to be necessary in the analysis. What is the surface area under these conditions? How much would the heat transfer be reduced if the exchanger in Problem 10-26 were used with this fouling factor and the same inlet fluid temperatures?

10-50 A shell-and-tube exchanger with one shell pass and two tube passes is used as a water-to-water heat-transfer system with the hot fluid in the shell side. The hot water is cooled from 90 to 70°C, and the cool fluid is heated from 5 to 60°C. Calculate the surface area for a heat transfer of 60 kW and a heat-transfer coefficient of 1100 W/m^2 · °C.

10-51 What is the heat transfer for the exchanger in Problem 10-50 if the flow rate of the hot fluid is reduced in half while the inlet conditions and heat-transfer coefficient remain the same?

10-52 A cross-flow finned-tube heat exchanger uses hot water to heat an appropriate quantity of air from 15 to 25°C. The water enters the heat exchanger at 70°C and leaves at 40°C, and the total heat-transfer rate is to be 29 kW. The overall heat-transfer coefficient is 45 W/m^2 · °C. Calculate the area of the heat exchanger.

10-53 Calculate the heat-transfer rate for the exchanger in Problem 10-52 when the water-flow rate is reduced to one-third that of the design value.

10-54 A gas-turbine regenerator is a heat exchanger that uses the hot exhaust gases from the turbine to preheat the air delivered to the combustion chamber. In an air-standard analysis of gas-turbine cycles, it is assumed that the mass of fuel is small in comparison with the mass of air, and consequently the hot-gas flow through the turbine is essentially the same as the airflow into the combustion chamber. Using this assumption, and also assuming that the specific heat of the hot exhaust gases is the same as that of the incoming air, derive an expression for the effectiveness of a regenerator under both counterflow and parallel-flow conditions.

10-55 Water at 90°C enters a double-pipe heat exchanger and leaves at 55°C. It is used to heat a certain oil from 25 to 50°C. Calculate the effectiveness of the heat exchanger.

10-56 Because of priority requirements the hot fluid flow rate for the exchanger in Problems 10-18 and 10-19 must be reduced by 40 percent. The same water flow must be heated from 35 to 85°C. To accomplish this, a shell-and-tube steam preheater is added, with steam condensing at 150°C and an overall heat-transfer coefficient of 2000 W/m$^2 \cdot$°C. What surface area and steam flow are required for the preheater?

10-57 An engine-oil heater employs ethylene glycol at 100°C entering a tube bank consisting of 50 copper tubes, five rows high with an OD of 2.5 cm and a wall thickness of 0.8 mm. The tubes are 70 cm long with $S_p = S_n = 3.75$ cm in an in-line arrangement. The oil enters the tube bank at 20°C and a velocity of 1 m/s. The glycol enters the tubes with a velocity of 1.5 m/s. Calculate the total heat transfer and the exit oil temperature. Repeat for an inlet glycol velocity of 1.0 m/s.

10-58 An air preheater for a power plant consists of a cross-flow heat exchanger with hot exhaust gases used to heat incoming air at 1 atm and 300 K. The gases enter at 375°C with a flow rate of 5 kg/s. The airflow rate is 5.0 kg/s, and the heat exchanger has $A = 110$ m^2 and $U = 50$ W/m$^2 \cdot$°C. Calculate the heat-transfer rate and exit temperatures for two cases, both fluids unmixed and one fluid mixed. Assume the hot gases have the properties of air.

10-59 A counterflow double-pipe heat exchanger is employed to heat 25 kg/s of water from 20 to 40°C with a hot oil at 200°C. The overall heat-transfer coefficient is 275 W/m$^2 \cdot$°C. Determine effectiveness and NTU for exit oil temperatures of 190, 180, 140, and 80°C.

10-60 A shell-and-tube heat exchanger is designed for condensing steam at 200°C in the shell with one shell pass; 50 kg/s of water are heated from 60 to 90°C. The overall heat-transfer coefficient is 4500 W/m$^2 \cdot$°C. A controller is installed on the steam inlet to vary the temperature by controlling the pressure, and the effect on the outlet water temperature is desired. Calculate the effectiveness and outlet water temperature for steam inlet temperatures of 180, 160, 140, and 120°C. Use the analytical expressions to derive a relation for the outlet water temperature as a function of steam inlet temperature.

10-61 A shell-and-tube heat exchanger with one shell pass and two tube passes is used to heat 5.0 kg/s of water from 30°C to 80°C. The water flows in the tubes. Condensing steam at 1 atm is used in the shell side. Calculate the area of the heat exchanger, if the overall heat-transfer coefficient is 900 W/m$^2 \cdot$°C. Suppose this same exchanger is used with entering water at 30°C, $U = 900$, but with a water-flow rate of 1.3 kg/s. What would be the exit water temperature under these conditions?

10-62 A cross-flow finned-tube heat exchanger has an area of 20 m^2. It operates with water in the tubes entering at 80°C with a flow rate of 0.48 kg/s. Air blows across the tubes with a flow rate of 1.0 kg/s, entering at a temperature of 10°C, and producing an overall heat-transfer coefficient of 50 W/m² · °C. Calculate the exit temperature of the water.

10-63 A double-pipe heat exchanger is used to heat an oil with $c = 2.2$ kJ/kg · °C from 50°C to 100°C. The other fluid having $c = 4.2$ kJ/kg · °C enters the exchanger at 160°C and leaves at 90°C. The overall heat-transfer coefficient is 300 W/m² · °C. Calculate the area and effectiveness of the heat exchanger for a total heat-transfer rate of 600 kW.

10-64 A counterflow double-pipe heat exchanger is used to heat water from 20°C to 40°C with a hot oil that enters the exchanger at 180°C and leaves at 140°C. The flow rate of water is 3.0 kg/s and the overall heat-transfer coefficient is 130 W/m² · °C. Assume the specific heat for oil is 2100 J/kg · °C. Suppose the water-flow rate is cut in half. What new oil flow rate would be necessary to maintain a 40°C outlet water temperature? (The oil flow is *not* cut in half.)

10-65 A home air-conditioning system uses a cross-flow finned-tube heat exchanger to cool 0.8 kg/s of air from 30°C to 7°C. The cooling is accomplished with 0.75 kg/s of water entering at 3°C. Calculate the area of the heat exchanger assuming an overall heat-transfer coefficient of 55 W/m² · °C. If the water-flow rate is cut in half while the same airflow rate is maintained, what percent reduction in heat transfer will occur?

10-66 The same airflow as in Problem 10-65 is to be cooled in a finned-tube exchanger with evaporating Freon in the tubes. It may be assumed that the Freon temperature remains constant at 35°F and that the overall heat-transfer coefficient is 125 W/m² · °C. Calculate the exchanger area required in this case. Also calculate the reduction in heat transfer that would result from cutting the airflow rate by one-third.

10-67 A shell-and-tube heat exchanger with one shell pass and four tube passes is designed to heat 4000 kg/h of engine oil from 40°C to 80°C with the oil in the tube side. On the shell side is condensing steam at 1-atm pressure, and the overall heat-transfer coefficient is 1200 W/m² · °C. Calculate the mass flow of condensed steam if the flow of oil is reduced in half while the inlet temperature and U value are kept the same.

10-68 A heat exchanger with an effectiveness of 75 percent is used to heat 5 kg/s of water from 50°C with condensing steam at 1 atm. Calculate the area for $U = 1200$ W/m² · °C.

10-69 If the flow rate of water for the exchanger in Problem 10-68 is reduced in half, what is the water exit temperature and the overall heat transfer?

10-70 Hot water at 80°C is used to heat air from 7°C to 40°C in a finned-tube cross-flow heat exchanger. The water exit temperature is 52°C. Calculate the effectiveness of this heat exchanger.

10-71 If the mass flow of water in the exchanger in Problem 10-70 is 150 kg/min and the overall heat-transfer coefficient is 50 W/m² · °C, calculate the area of the heat exchanger.

10-72 A shell-and-tube heat exchanger having one shell pass and four tube passes is used to heat 10 kg/s of ethylene glycol from 20 to 40°C on the shell side; 15 kg/s of water entering at 70°C is used in the tubes. The overall heat-transfer coefficient is 40 W/m² · °C. Calculate the area of the heat exchanger.

10-73 The same exchanger as in Problem 10-72 is used with the same inlet temperature conditions but the water flow reduced to 10 kg/s. Because of the reduced water flow rate the overall heat-transfer coefficient drops to 35 W/m² · °C. Calculate the exit glycol temperature.

10-74 A heat-exchanger arrangement having three shell passes and six tube passes is used to heat 4 kg/s of water from 20°C to 60°C in the shell side. Hot water is used in the tubes to accomplish the heating and enters the tubes at 80°C and leaves the tubes at 40°C. Calculate the total area of the heat exchanger if the overall heat-transfer coefficient is 800 W/m² · °C.

10-75 A large condenser is designed to remove 800 MW of energy from condensing steam at 1-atm pressure. To accomplish this task, cooling water enters the condenser at 25°C and leaves at 30°C. The overall heat-transfer coefficient is 2000 W/m² · °C. Calculate the area required for the heat exchanger.

10-76 Suppose the water-flow rate for the exchanger in Problem 10-75 is reduced in half from the design value. What will be the steam condensation rate in kilograms per hour under these conditions if U remains the same?

10-77 A shell-and-tube heat exchanger with one shell pass and two tube passes is used to heat ethylene glycol in the tubes from 25°C to 65°C. The flow rate of glycol is 1.2 kg/s. Water is used in the shell side to supply the heat and enters the exchanger at 95°C and leaves at 55°C. The overall heat-transfer coefficient is 600 W/m² · °C. Calculate the area of the heat exchanger.

10-78 Air at 300 K enters a compact heat exchanger like that shown in Figure 10-19. Inside the tubes, steam is condensing at a constant temperature of 100°C with $h = 9000$ W/m² · °C. If the entering air velocity is 10 m/s, calculate the amount of steam condensed with an array 30 by 30 cm square and 60 cm long.

10-79 Repeat Problem 10-78 for the same airflow stream in configuration D of Figure 10-20.

10-80 If one wishes to condense half as much steam as in Problem 10-78, how much smaller an array can be used while keeping the length at 60 cm?

10-81 A double-pipe heat exchanger is to be designed to cool water from 80 to 60°C with ethylene glycol entering the exchanger at 20°C. The flow rate of glycol is 0.8 kg/s, and the water-flow rate is 0.6 kg/s. Calculate the effectiveness of the heat exchanger. If the overall heat-transfer coefficient is 1000 W/m² · °C, calculate the area required for the heat exchanger.

10-82 A cross-flow heat exchanger uses oil ($c_p = 2.1$ kJ/kg · °C) in the tube bank with an entering temperature of 100°C. The flow rate of oil is 1.2 kg/s. Water flows across the unfinned tubes and is heated from 20 to 50°C with a flow rate of 0.6 kg/s. If the overall heat-transfer coefficient is 250 W/m² · °C, calculate the area required for the heat exchanger.

10-83 Rework Problem 10-82 with the water flowing inside the tubes and the oil flowing across the tubes.

10-84 A shell-and-tube heat exchanger with three shell passes and six tube passes is used to heat 2 kg/s of water from 10 to 70°C in the shell side; 3 kg/s of hot oil ($c_p = 2.1$ kJ/kg · °C) at 120°C is used inside the tubes. If the overall heat-transfer coefficient is 300 W/m² · °C, calculate the area required for the heat exchanger.

10-85 The water-flow rate for the exchanger in Problem 10-84 is reduced to 1.0 kg/s while the temperature of the entering fluid remains the same, as does the value of U. Calculate the exit fluid temperatures under this new condition.

10-86 A shell-and-tube heat exchanger with two shell passes and four tube passes is used to condense Freon 12 in the shell at 37.8°C. Water enters the tubes at 21.1°C and leaves at 26.7°C. Freon is to be condensed at a rate of 0.23 kg/s with an enthalpy of vaporization $h_{fg} = 120$ kJ/kg. If the overall heat-transfer coefficient is 700 W/m$^2 \cdot$ °C, calculate the area of the heat exchanger.

10-87 Calculate the percent reduction in condensation of Freon in the exchanger of Problem 10-86 if the water flow is reduced in half but the inlet temperature and value of U remain the same.

10-88 A shell-and-tube heat exchanger with four shell passes and eight tube passes is used to heat 3 kg/s of water from 10 to 30°C in the shell side by cooling 3 kg/s of water from 80 to 60°C in the tube side. If $U = 1000$ W/m$^2 \cdot$ °C, calculate the area of the heat exchanger.

10-89 For the area of heat exchanger found in Problem 10-88 calculate the percent reduction in heat transfer if the cold-fluid flow rate is reduced to half while keeping the inlet temperature and value of U the same.

10-90 Show that for $C = 0.5$ and 1.0 the effectiveness given in Figure 10-17 can be calculated from an effectiveness read from Figure 10-16 and the equation for n shell passes given in Table 10-3. Note that

$$\text{NTU}(n \text{ shell passes}) = n \times \text{NTU(one shell pass)}$$

10-91 Show that for an n-shell-pass exchanger the effectiveness for each shell pass is given by

$$\epsilon_p = \frac{[(1 - \epsilon C)/(l - \epsilon)]^{1/n} - 1}{[(1 - \epsilon C)/(l - \epsilon)]^{1/n} - C}$$

where ϵ is the effectiveness for the multishell-pass exchanger.

10-92 Hot oil ($c_p = 2.1$ kJ/kg $\cdot$ °C) at a rate of 7.0 kg/s and 100°C is used to heat 3.5 kg/s of water at 20°C in a cross-flow heat exchanger with the oil inside the tubes and the water flowing across the tubes. The effectiveness of the heat exchanger is 60 percent. Calculate the exit temperatures for both fluids and the product UA for the heat exchanger.

10-93 A single-shell-pass heat exchanger with two tube passes is used to condense steam at 100°C (1 atm) on the shell side. Water is used on the tube side and enters the exchanger at 20°C with a flow rate of 1.0 kg/s. The overall heat-transfer coefficient is 2500 W/m$^2 \cdot$ °C, and the surface area of the exchanger is 0.8 m^2. Calculate the exit temperature of the water.

10-94 A shell-and-tube heat exchanger with four shell passes and eight tube passes uses 3.0 kg/s of ethylene glycol in the shell to heat 1.5 kg/s of water from 20 to 50°C. The glycol enters at 80°C, and the overall heat-transfer coefficient is 900 W/m$^2 \cdot$ °C. Determine the area of the heat exchanger.

10-95 After the heat exchanger in Problem 10-94 is sized (i.e., its area determined), it is operated with a glycol flow of only 1.5 kg/s and all other parameters the same. What would the exit water temperature be under these conditions?

10-96 A finned-tube heat exchanger operates with water in the tubes and airflow across the fins. The water inlet temperature is 130°F while the incoming air temperature is 75°F. An overall heat-transfer coefficient of 57 W/m$^2 \cdot$ °C is experienced when the water-flow rate is 0.5 kg/s. The area of the exchanger is 52 m^2 and the airflow is 2.0 kg/s for the given conditions. Calculate the exit water temperature, expressed in °C.

10-97 A finned-tube cross-flow heat exchanger has hot water in the tubes and air blowing across the fins. The overall heat-transfer coefficient is 40 W/m$^2 \cdot$ °C, and the area of the exchanger is 5.0 m^2. The water enters the exchanger at 90°C while the air enters at 20°C. The water-flow rate is 0.2 kg/s and the airflow rate is also 0.2 kg/s. Calculate the heat-transfer rate for the heat exchanger using the effectiveness-NTU method.

10-98 A shell-and-tube heat exchanger with three shell passes and six tube passes is used to heat an oil ($c_p = 2.1$ kJ/kg · °C) in the shell side from 30 to 60°C. In the tube side high-pressure water is cooled from 110 to 90°C. Calculate the effectiveness for each shell pass.

10-99 If the water-flow rate in Problem 10-98 is 3 kg/s and $U = 230$ W/m^2 · °C, calculate the area of the heat exchanger. Using the area, calculate the exit fluid temperatures when the water flow is reduced to 2 kg/s and all other factors remain the same.

10-100 A cross-flow heat exchanger employs water in the tubes with $h = 3000$ W/m^2 · °C and airflow across the tubes with $h = 190$ W/m^2 · °C. If the tube wall is copper having a thickness of 0.8 mm and outside diameter of 25 mm, calculate the overall heat-transfer coefficient based on inside tube area.

10-101 Water flows in each of the tubes of the exchanger of Problem 10-100 at the rate of 0.5 kg/s. Assuming the same overall heat-transfer coefficient applies as calculated in that problem, determine suitable combinations of tube length and inlet air temperature to heat the water from 10°C to 20°C. State assumptions.

10-102 A shell-and-tube heat exchanger employs a liquid in the shell that is heated from 30°C to 55°C by a hot gas in the tubes that is cooled from 100°C to 60°C. Calculate the effectiveness of the heat exchanger.

10-103 A light fuel oil is used in the tube side of a shell-and-tube heat exchanger with two shell passes and four tube passes. Water is heated in the shell side from 10°C to 50°C while the oil is cooled from 90°C to 60°C. The overall heat-transfer coefficient is 53 W/m^2 · °C. The specific heat of the oil is 2.0 kJ/kg · °C. Using both the effectiveness and LMTD methods, calculate the area of the heat exchanger for a total energy transfer of 500 kW. What is the water-flow rate for this heat transfer?

10-104 For the heat exchanger area determined in Problem 10-103, what percentage reduction in water flow is necessary to reduce the total heat transfer rate in half while maintaining the oil flow constant?

10-105 A finned-tube heat exchanger employs condensing steam at 100°C inside the tubes to heat air from 10°C to 50°C as it flows across the fins. A total heat transfer of 44 kW is to be accomplished in the exchanger and the overall heat-transfer coefficient may be taken as 25 W/m^2 · °C. What is the area of the heat exchanger?

10-106 Suppose the value of U for the exchanger in Problem 10-105 varies with the air mass flow rate to the 0.8 power. What percent reduction in mass flow of air would be required to reduce the total heat transfer rate by one-third?

10-107 A steam condenser for a large power plant has steam condensing at 38°C on the shell side of a shell-and-tube exchanger. Water serves as the coolant on the tube side, entering at 20°C and leaving at 27°C. The exchanger involves one shell pass and two tube passes. Using information from Table 10-1, estimate the exchanger area required for a heat-transfer rate of 700 MW. What flow rate of water is required for this heat transfer rate? Be sure to specify the value of U employed in the calculations.

10-108 What condenser area would be required if the exit water temperature in Problem 10-107 is allowed to rise to 34°C for the same heat transfer and U-value?

10-109 Suppose the exchanger of Problem 10-107 is used with the same area and U-value, but the water-flow rate is now changed to allow an exit temperature of 34°C. What percent reduction in the steam condensation rate will occur?

Design-Oriented Problems

10-110 Some of the brine from a large refrigeration system to be used to furnish chilled water for the air-conditioning part of an office building. The brine is available at −15°C, and 105 kW of cooling is required. The chilled water from the conditioned air coolers enters a shell-and-tube heat exchanger at 10°C, and the exchanger is to be designed so that the exit chilled-water temperature is not below 5°C. The overall heat-transfer coefficient for the heat-exchanger is 850 W/m$^2 \cdot$°C. If the chilled water is used on the tube side and two tube passes are employed, plot the heat-exchanger area required as a function of the brine exit temperature.

10-111 Hot engine oil enters a 1-in schedule 40 steel pipe at 80°C with a velocity of 5 m/s. The pipe is submerged horizontally in water at 20°C so that it loses heat by free convection. Calculate the length of pipe necessary to lower the oil temperature to 60°C. Perform some kind of calculation to indicate the effect variable properties have on the results.

10-112 In energy conservation activities, cross-flow heat exchangers with both fluids unmixed are sometimes used under conditions that approximate $C_{min}/C_{max} = C \approx$ 1.0. Using appropriate computer software, calculate and plot NTU = $f(\epsilon)$ for this condition. For the same inlet conditions, what is the effect of doubling the value of the product UA on the overall heat transfer? What is the effect of reducing the value of UA to half, while keeping the same inlet conditions?

10-113 The condenser on a certain automobile air conditioner is designed to remove 60,000 Btu/h from Freon 12 when the automobile is moving at 40 mi/h and the ambient temperature is 95°F. The Freon 12 temperature is 150°F under these conditions, and it may be assumed that the air-temperature rise across the exchanger is 10°F. The overall heat-transfer coefficient for the finned-tube heat exchanger under these conditions is 35 Btu/h $\cdot$ ft$^2 \cdot$°F. If the overall heat-transfer coefficient varies as the seven-tenths power of velocity and air-mass flow varies directly as the velocity, plot the percentage reduction in performance of the condenser as a function of velocity between 10 and 40 mi/h. Assume that the Freon temperature remains constant at 150°F.

10-114 A shell-and-tube heat exchanger is to be designed to heat 7.5 kg/s of water from 85 to 99°C. The heating process is accomplished by condensing steam at 345 kPa. One shell pass is used along with two tube passes, each consisting of thirty 2.5-cm-OD tubes. Assuming a value of U of 2800 W/m$^2 \cdot$°C, calculate the length of tubes required in the heat exchanger.

10-115 Suppose the heat exchanger in Problem 10-114 has been in operation an extended period of time so that the fouling factors in Table 10-2 apply. Calculate the exit water temperature for fouled conditions, assuming the same total flow rate.

10-116 A double-pipe heat exchanger is constructed of copper and operated in a counterflow mode. It is designed to heat 0.76 kg/s of water from 10°C to 79.4°C. The water flows through the inner pipe. The heating is accomplished by condensing steam in the outer pipe at a temperature of 250°F. The water-side heat-transfer coefficient is 1420 W/m$^2 \cdot$°C. Assume a reasonable value for the steam-side coefficient and then calculate the area of the heat exchanger. Estimate what the exit water temperature would be if the water-flow rate were reduced by 60 percent for this exchanger.

10-117 Suppose a fouling factor of 0.0002 $m^2 \cdot °C/W$ is used for the water in Problem 10-36 and 0.0004 $m^2 \cdot °C/W$ for the air. What percent increase in area should be included in the design to take these factors into account for future operation?

10-118 Saturated steam at 100 lb/in^2 abs is to be used to heat carbon dioxide in a cross-flow heat exchanger consisting of four hundred $\frac{1}{4}$-in-OD brass tubes in a square in-line array. The distance between tube centers is $\frac{3}{8}$ in, in both the normal- and parallel-flow directions. The carbon dioxide flows across the tube bank, while the steam is condensed on the inside of the tubes. A flow rate of 1 lb_m/s of CO_2 at 15 lb/in^2 abs and 70°F is to be heated to 200°F. Estimate the length of the tubes to accomplish this heating. Assume that the steam-side heat-transfer coefficient is 1000 Btu/h $\cdot ft^2 \cdot °F$, and neglect the thermal resistance of the tube wall.

10-119 Repeat Problem 10-118 with the CO_2 flowing on the inside of the tubes and the steam condensing on the outside of the tubes. Compare these two designs on the basis of CO_2 pressure drop through the exchanger.

10-120 The refrigerant condenser for an automobile operates with the refrigerant entering as a saturated vapor and leaving as a saturated liquid at 145°F. The cooling air enters the exchanger, which operates in a cross-flow mode, at temperatures varying between 65 and 100°F, depending on weather conditions. Assuming the vehicle moves at constant speed, such that the mass-flow rate of the cooling air remains constant, calculate and plot the ratio q(temperature T)/q(65°F) as a function of the inlet air temperature. State any assumptions that may be necessary, and why they are justified.

10-121 A counterflow double-pipe heat exchanger is currently used to heat 2.5 kg/s of water from 25 to 65°C by cooling an oil [$c_p = 2.1$ kJ/kg $\cdot °C$] from 138 to 93°C. It is desired to 'bleed off' 0.62 kg/s of water at 50°C so that the single exchanger will be replaced by a two-exchanger arrangement that will permit this. The overall heat-transfer coefficient is 450 $W/m^2 \cdot °C$ for the single exchanger and may be taken as this same value for each of the smaller exchangers. The same oil flow is used for the two-exchanger arrangement, except that the flow is split between the two exchangers. Determine the areas of each of the smaller exchangers and the oil flow through each. Assume that the water flows in series through the two exchangers, with the bleed-off taking place between them. Assume the smaller exchangers have the same areas.

10-122 Repeat Problem 10-121, assuming that condensing steam at 138°C is used instead of the hot oil and that the exchangers are of the shell-and-tube type with the water making two passes on the tube side. The overall heat-transfer coefficient may be taken as 1700 $W/m^2 \cdot °C$ for this application.

10-123 A shell-and-tube heat exchanger with four tube passes is used to heat 2.5 kg/s of water from 25 to 70°C. Hot water at 93°C is available for the heating process, and a flow rate of 5 kg/s may be used. The cooler fluid is used on the tube side of the exchanger. The overall heat-transfer coefficient is 800 $W/m^2 \cdot °C$. Assuming that the flow rate of the hot fluid and the overall heat-transfer coefficient remain constant, plot the percentage reduction in heat transfer as a function of the mass flow rate of the cooler fluid.

10-124 Two identical double-pipe heat exchangers are constructed of a 2-in standard schedule 40 pipe placed inside a 3-in standard pipe. The length of the exchangers is 10 ft; 40 gal/min of water initially at 80°F is to be heated by passing through the inner pipes of the exchangers in a series arrangement, and 30 gal/min of water at 120°F and 30 gal/min of water at 200°F are available to accomplish the heating. The two heating streams may be mixed in any way desired before and after they enter the heat exchangers. Determine the flow arrangement for optimum performance (maximum heat transfer) and the total heat transfer under these conditions.

10-125 High-temperature flue gases at 450°C $[c_p = 1.2 \, \text{kJ/kg} \cdot {}^\circ\text{C}]$ are employed in a cross-flow heat exchanger to heat an engine oil from 30 to 80°C. Using the information given in this chapter, obtain an approximate design for the heat exchanger for an oil flow rate of 0.6 kg/s.

10-126 Condensing steam at 100°C with $h = 5000 \, \text{W/m}^2 \cdot {}^\circ\text{C}$ is used inside the tubes of the exchanger in Example 10-16. If the heat exchanger has a frontal area of 0.5 m² and a depth of 40 cm in the airflow direction, calculate the heat-transfer rate and exit air temperature. State assumptions.

10-127 An ammonia condenser is constructed of a 5 by 5 array of horizontal tubes having an outside diameter of 2.5 cm and a wall thickness of 1.0 mm. Water enters the tubes at 20°C and 5 m/s and leaves at 40°C. The ammonia condensing temperature is 60°C. Calculate the length of tubes required. How much ammonia is condensed? Consult thermodynamics tables for the needed properties of ammonia.

10-128 Rework Problem 10-127 for a 10 by 10 array of tubes. If the length of tubes is reduced by half, what reduction in condensate results for the same inlet water temperature? (The exit water temperature is not the same.)

10-129 A compact heat exchanger like that shown in Figure 10-19 is to be designed to cool water from 200°F to 160°F with an airflow that enters the exchanger at 1 atm and 70°F. The inlet airflow velocity is 50 ft/s. The total heat transfer is to be 240,000 Btu/h. Select several alternative designs and investigate each in terms of exchanger size (area and/or volume) and the pressure drop.

10-130 A finned-tube heat exchanger is to be used to remove 30 kW of heat from air at 75°F in a certain air-conditioning application. Two alternatives are to be considered: (1) using cooling water in the tubes, with an entering temperature of 45°F, or (2) using evaporating Freon at 45°F in the tubes. In both cases the air will flow across the fins, and the air convection heat-transfer coefficient should be the controlling factor in the overall heat-transfer coefficient. Assume that the overall heat-transfer coefficient is 55 $\text{W/m}^2 \cdot {}^\circ\text{C}$ when the exit air temperature is 55°C, and that it varies as the mass flow of air to the 0.8 power. Determine the size of the heat exchanger for the conditions given and (1) an exit water temperature of 53°F and (2) a constant Freon temperature of 45°F. Then examine the system behavior under part load conditions where (a) the cooling load is reduced in half, with the mass flow of air and water held the same, along with the inlet air and water temperatures; (b) the cooling load is reduced in half, with the exit air temperature and water-flow rate remaining at the design value, (c) the cooling load is reduced in half, the exit air temperature remains at 55°F, and the water-flow rate is reduced by 25 percent; and (d) the same as in (a) and (b) for the Freon evaporating temperature remaining constant at 45°F.

10-131 In air-conditioning applications the control of inside environmental conditions is sometimes accomplished by varying the flow rate of air across the cooling coils, as examined in Problem 10-130. Assume the objective is to maintain a fixed air temperature at exit from the heat exchanger under varying load conditions. Examine the heat exchanger system of Problem 10-130 further by calculating several values of the air mass flow rate as a function of percent full load conditions, assuming (a) constant inlet water temperature of 45°F and a constant water-flow rate, and (b) constant Freon evaporation temperature of 45°F. Suppose a control system is to be designed to automatically vary airflow rate based on cooling load demand. Assuming a quadratic polynomial variation of air mass flow with load (i.e., $\dot{m} = A + Bq + Cq^2$), devise values of the equation constants to fit your calculation results.

10-132 A steel pipe, 2.5 cm in diameter, is maintained at a surface temperature of 100°C by condensing steam on the inside. Circular steel fins ($k = 43$ W/m · °C) are placed on the outside of the pipe to dissipate heat by free convection to a surrounding room at 20°C with a convection coefficient of 8.0 W/m² · °C for both the bare pipe and fin surfaces. The fins and pipe surfaces are black in color so that they radiate as nearly black surfaces. (The fin-pipe combination will radiate as a cylinder having a diameter equal to the outside diameter of the fins.) Consider several cases of fin outside diameters, fin thickness, and fin spacing to calculate the heat lost to the room per meter of pipe length. State conclusions as appropriate. Assume h is uniform over all the heat-transfer surfaces.

10-133 The same finned-pipe arrangement as in Problem 10-132 has air forced across the surfaces producing a convection coefficient of $h = 20$ W/m² · °C. Perform the same kind of analysis as in the free convection case and state appropriate conclusions. Assume h is uniform over all the heat-transfer surfaces.

10-134 A "run-around" system is used in energy conservation applications as a heat recovery device, as illustrated in Figure P10-134. For the arrangement shown, warm air from the interior of a building is used to transfer 60 kW of heat to a finned-tube heat exchanger with water in the tubes. This water is then pumped to a location 20 m away where the energy is used to preheat outside air entering the occupied space. The outside air enters at 0°C, the conditioned space is at 20°C, and the flow of outside air is about 86 m³/min, which is supplied by suitable fans. The water pipes connecting the two finned-tube heat exchangers are assumed to be perfectly insulated. Using approximate values for U from Table 10-1 or other sources, determine some suitable sizes of heat exchangers that may serve to accomplish the heat transfer objectives in this problem. For this design, assume reasonable values for the temperature rise and fall in the water-flow and estimate the water-flow rates required for each design choice. Comment on your analysis and make recommendations.

Figure P10-134

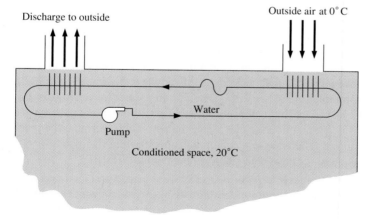

10-135 A shell-and-tube heat exchanger with one shell pass and two tube passes employs condensing steam at 1 atm on the shell side to heat 10 kg/s of engine oil from 50 to 60°C in the tubes. The steam enters as saturated vapor and leaves as saturated liquid with $h_{ag} = 2.26$ MJ/kg. The hot condensate leaving the exchanger is subsequently used to heat water to 60°C in another shell-and-tube exchanger, with the condensate occupying the shell side and the cooler water flowing in the tubes. The cool water enters at 20°C. Using approximate values of U from Table 10-1, or your own calculations, determine the size of the steam condenser and suggest sizes for the exchanger that serves as the water heater. For each exchanger suggested, indicate the amount of water that may be heated. As an alternative, suggest a single choice for the heat exchanger that employs throttling of the water flow. Comment on the suggested arrangements.

10-136 A heat pipe (Section 9-7) operates as a heat recovery device with finned tubes attached to each end. The arrangement is to be used to recover heat from the exhaust of a commercial clothes dryer to preheat incoming air at 10°C. Air from the dryer leaves at 95°C with a volume flow rate of 34 m³/min. Multiple tube heat pipes may be employed with fins to accomplish the heat-transfer objective. Using information from Table 10-1, your own calculations, or other sources, estimate the sizes of the finned-tube exchangers on each end of the heat pipe(s). Assume both exchangers are the same size and the surface of the heat pipe remains constant in temperature. (*Hint:* The combined arrangement will operate very nearly like a single cross-flow exchanger with equal capacities.) Choose reasonable values for exit temperatures of the hot gases and determine the energy cost savings if an electric dryer is used with electricity at $0.085/kWh. Perform the same calculation for a natural gas dryer with energy at $9.00/GJ. Comment on your analysis and make recommendations.

10-137 The cooling radiator for an automobile or diesel power plant has a configuration like that shown in Figure 10-18(*a*), and may be considered as a cross-flow exchanger with both fluids unmixed. The cooling water flows in the oval-tube sections and cooling air flows through the fins. Such an exchanger has heat-transfer characteristics like those shown in Figure 10-19. In a diesel engine power device the amount of energy to be dissipated in the radiator is 1.12 MW, and the cooling water is to leave the radiator at 98°C. Air enters the exchanger at 25°C with a velocity of 15 m/sec that is maintained constant by a suitable fan. Select appropriate exit cooling water temperatures and determine the frontal area and volume of the heat exchanger for each temperature. Select one configuration as a design value.

10-138 In a large cold-storage facility the refrigeration is supplied by an ammonia system. The ammonia condenser operates at 30°C as a shell-and-tube heat exchanger with one shell pass for the condensing ammonia and two tube passes for a water coolant. The total energy that must be removed from the ammonia is 457 kW. The value of h_{fg} for ammonia at 30°C is 1145 kJ/kg. Cooling water is available for the condenser at 16°C. Using approximate values for U from Table 10-1 or other sources and an appropriate assumption for exit water temperature, determine the heat exchanger area required. Then devise a design that specifies the size, length, and number of tubes required for the exchanger. Comment on the design.

10-139 After the exchanger of Problem 10-138 is designed, a part load operation is desired that reduces the ammonia condensing rate by 20 percent while keeping the entering water temperature the same. The reduction will be made by throttling the water flow to some lower flow rate. If the overall heat-transfer coefficient varies with water-flow rate to the 0.8 power, what must the new flow rate be to accomplish the 20 percent condensation reduction?

10-140 In a bread plant the finished loaves of bread leave the oven at about 110°C and are placed on a conveyer belt to cool in room air before being packaged. The cooling process is by a combination of either free convection and radiation or forced convection and radiation. The free-convection process takes longer, and thus a longer transit time is required on the conveyor belt. The final cooled bread temperature is to be 30°C. From an operation-design standpoint, the conveyor belt length must be fixed but the speed of movement may be adjusted to match the cooling transit time requirements. Using appropriate relations for free-convection, radiation, and forced convection, devise transit times for free convection cooling and for two velocities of forced-convection cooling. Assume loaves of bread weighing 0.45 kg. What conveyor belt speeds would you recommend?

10-141 Hot combustion gases leave a heater for a corn-chip fryer at 435°C and are fed to a cross-flow heat exchanger that is used to preheat the air entering the burner. The device is called a *recuperator* and the energy saved is equal to the energy gained by the air. Air may be assumed to enter the exchanger at 30°C, and the hot combustion gases may be assumed to have the same properties as air for purposes of this analysis. To simplify the analysis, the flow rates of hot gases and air are also assumed to be the same. The overall heat-transfer coefficient may be assumed constant at $U = 25$ W/m$^2 \cdot$°C. Selection of a larger heat-transfer area will result in a higher heat-exchanger effectiveness and a larger energy saving. On the other hand, a larger exchanger will involve an increased capital expenditure. For a unit flow rate of hot gases (take as 1 kg/s) investigate different exchanger sizes to determine the influence on net cost savings, where

Cost Savings = (energy savings) × (unit energy cost)
− (heat exchanger cost/unit area)
× (heat exchanger area required for energy savings)

Assume that both the heat exchanger cost per unit area and unit energy costs are independent of the size of the heat exchanger. Assume that the heat exchanger is cross-flow with both gasses unmixed. Comment on the results and assumptions involved in this problem. Would you anticipate that this heat-recovery arrangement would be economical if the energy cost is $6.50/GJ?

10-142 A heat exchanger is to be designed to condense 1100 kg/hr of steam at 150 kPa. For this purpose, cooling water at 50°C is available in sufficient quantity to accomplish the cooling. You are to design a shell-and-tube exchanger with steam in the shell and one tube pass as the condenser. Use information from Chapters 6 and 9 to calculate the convection coefficients for condensation and forced flow through the tubes. Select several alternatives for the tube bank and tube size, determine the length of the heat exchanger, and calculate the pressure drop through the tubes for each selection. Recommend a final design that specifies water-flow rate, tube size, number of tubes, and tube length. Comment on your selection.

10-143 The cooling pipe/tube for a groundwater heat pump is to be buried in a position where the ground temperature is 18°C. Thirty kW of energy is to be dissipated with water entering the pipe at 27°C and leaving at a temperature no greater than 20°C. The thermal conductivity of the soil at this location may be taken as 1.7 W/m · °C. Examine several possibilities for pipe material, diameter, and length to accomplish the cooling objective. Obtain information from commercial sources where necessary. What factors should be considered in selecting a design for this installation? From your analysis, choose a final selection for the design and comment on your selection.

10-144 A finned plate is to be constructed of aluminum and designed to dissipate 125 W to a surrounding room with air at 20°C. The base for the fins is to consist of a flat vertical plate with multiple straight vertical fins having a rectangular profile. The base plate thickness must not exceed 3 mm, and the heat source on the back side of the plate is a power device that must operate at a temperature below 75°C. For the design, you are to determine a suitable set of the following parameters that will accomplish the heat transfer objective: (1) width and height of base plate, (2) number of vertical fins and spacing, (3) length and thickness of fins, and (4) surface treatment (if any) to enhance radiation heat transfer from the fins. For the design choose information on fin performance from Chapter 2, information on free convection from vertical surfaces from Chapter 7, information on radiation heat transfer from Chapter 8, and thermal and radiation properties from Appendix A. Assume that the fins are securely attached to the base plate insofar as thermal contact is concerned. Consider several alternate arrangements for the design and employ Program A of the heat-transfer software in Appendix D to evaluate fin performance, if convenient. Be sure to state all assumptions in the analysis and specify clearly the factors that influenced your final design.

10-145 A shell-and-tube heat exchanger with one shell pass and two tube passes is to be designed to condense 3900 kg/hr of steam at 1 atm. The steam may be assumed to enter the shell side as saturated vapor and leave as saturated liquid. Water is employed as the cooling fluid in the tubes with an inlet temperature of 40°C. Using approximate values for overall heat-transfer coefficients given in Table 10-1, determine a suitable design that specifies (1) the number of tubes in each tube pass, (2) outlet water temperature, (3) tube diameter and length in each tube pass, and (4) the mass flow rate of water. Also, estimate the final inside diameter of the shell that may be required. Consider several alternatives before selecting a final design and discuss the factors that influenced your design selection.

10-146 Suppose the cost of a heat exchanger varies directly with the area. The effectiveness also increases with area (through NTU) but not in a linear fashion. Consider a cross-flow finned-tube exchanger with two gases operating such that $C_{min} = C_{max}$, and $c_{min} = c_{max} = 1.005$ kJ/kg · °C. The exchanger operates as a recuperator to preheat air at 300 K with hot exhaust gases at 600 K. The energy saving is the energy gained by the air and the energy cost saving is that value multiplied by the unit energy cost. Using Table 10-1 as a guide for values of U, devise a relation between unit energy costs and heat-exchanger cost per unit area. Work several cases using realistic unit energy costs to illustrate your analysis. What conclusion can you reach from this study?

10-147 The cost of fouling in a heat exchanger may be expressed by a reduction in heat transfer performance multiplied by some unit energy cost. The problem may be alleviated by cleaning the exchanger or by chemical treatment of the fluid(s) during operation. Using Tables 10-1 and 10-2, devise a heat-exchanger model and carry through an analysis to estimate the cost of fouling. Assume appropriate values for unit energy costs. Can you reach any conclusions from your study?

REFERENCES

1. Siegal, R., and J. R. Howell. *Thermal Radiation Heat Transfer,* 4th ed. New York: Hemisphere Publishing Corp., 1990.
2. "Standards of Tubular Exchanger Manufacturers Association," latest edition.

3. Kays, W. M., and A. L. London. *Compact Heat Exchangers,* 2d ed. New York: McGraw-Hill, 1964.

4. Bowman, R. A., A. E. Mueller, and W. M. Nagle. "Mean Temperature Difference in Design," *Trans. ASME,* vol. 62, p. 283, 1940.

5. Perry, J. H. (ed.). *Chemical Engineers' Handbook,* 4th ed. New York: McGraw-Hill, 1963.

6. American Society of Heating, Refrigeration, and Air Conditioning Engineers Guide, annually.

7. Sparrow, E. M., and R. D. Cess. *Radiation Heat Transfer.* New York: Wadsworth Publishing Co., 1966.

8. Schlunder, E. W. *Heat Exchanger Design Handbook.* New York: Hemisphere Publishing Corp., 1982.

9. Somerscales, E. F. C., and J. G. Knudsen (eds.). *Fouling of Heat Transfer Equipment.* New York: Hemisphere Publishing Corp., 1981.

10. Kraus, A. D., and D. Q. Kern. "The Effectiveness of Heat Exchangers with One Shell Pass and Even Number of Tube Passes," *ASME Paper 65-HT-18,* presented at National Heat Transfer Conference, August 1965.

11. DiGiovanni, M. A., and R. L. Webb. "Uncertainty in Effectiveness–NTU Calculations for Cross-flow Heat Exchangers," *Heat Trans. Engr.,* vol. 10, pp. 61–70, 1989.

12. Bergles, A. E. "Some Perspectives on Enhanced Heat Transfer-Second Generation Heat Transfer Technology," *J. Heat Transfer,* vol. 110, pp. 1082–91, 1988.

13. Bergles, A. E. "Techniques to Augment Heat Transfer," chap. 3. *Handbook of Heat Transfer Applications.* New York: McGraw-Hill, 1985.

14. Bergles, A. E. "Augmentation of Heat Transfer," chap.2.5.11. *Heat Exchanger Design Handbook,* New York: Begell House, 1998.

15. ———. *Heat Exchanger Design Handbook.* New York: Begell House, 1998.

16. Hewitt, G. F., Shires, G. L., and Bott, T. R. *Process Heat Transfer.* Boca Raton: CRC Press, 1994.

17. Webb, R. L. *Principles of Enhanced Heat Transfer.* New York: John Wiley, 1994.

Mass Transfer

11-1 | INTRODUCTION

Mass transfer can result from several different phenomena. There is a mass transfer associated with convection in that mass is transported from one place to another in the flow system. This type of mass transfer occurs on a macroscopic level and is usually treated in the subject of fluid mechanics. When a mixture of gases or liquids is contained such that there exists a concentration gradient of one or more of the constituents across the system, there will be a mass transfer on a microscopic level as the result of diffusion from regions of high concentration to regions of low concentration. In this chapter we are primarily concerned with some of the simple relations that may be used to calculate mass diffusion and their relation to heat transfer. Nevertheless, one must remember that the general subject of mass transfer encompasses both mass diffusion on a molecular scale and the bulk mass transport that may result from a convection process.

Not only may mass diffusion occur on a molecular basis, but accelerated diffusion rates will also occur in turbulent-flow systems as a result of the rapid-eddy mixing processes, just as these mixing processes created increased heat transfer and viscous action in turbulent flow.

Although beyond the scope of our discussion, it is well to mention that mass diffusion may also result from a temperature gradient in a system; this is called *thermal diffusion*. Similarly, a concentration gradient can give rise to a temperature gradient and a consequent heat transfer. These two effects are termed *coupled phenomena* and may be treated by the methods of irreversible thermodynamics. The reader is referred to the monographs by Prigogine [1] and de Groot [2] for a discussion of irreversible thermodynamics and coupled phenomena and their application to diffusion processes.

11-2 | FICK'S LAW OF DIFFUSION

Consider the system shown in Figure 11-1. A thin partition separates the two gases A and B. When the partition is removed, the two gases diffuse through each other until equilibrium is established and the concentration of the gases is uniform throughout the box. The diffusion rate is given by Fick's law of diffusion, which states that the mass flux of a constituent per unit area is proportional to the concentration gradient. Thus

$$\frac{\dot{m}_A}{A} = -D\frac{\partial C_A}{\partial x} \qquad \text{[11-1]}$$

Figure 11-1 | Diffusion of component A into component B.

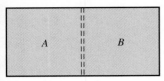

where

D = proportionality constant–diffusion coefficient, m^2/s

$\dot{m}_A$ = mass flux per unit time, kg/s

C_A = mass concentration of component A per unit volume, kg/m^3

An expression similar to Equation (11-1) could also be written for the diffusion of constituent A in either the y or z direction.

Notice the similarity between Equation (11-1) and the Fourier law of heat conduction,

$$\left(\frac{q}{A}\right)_x = -k\frac{\partial T}{\partial x}$$

and the equation for shear stress between fluid layers,

$$\tau = \mu\frac{\partial u}{\partial y}$$

The heat-conduction equation describes the transport of energy, the viscous-shear equation describes the transport of momentum across fluid layers, and the diffusion law describes the transport of mass.

To understand the physical mechanism of diffusion, consider the imaginary plane shown by the dashed line in Figure 11-2. The concentration of component A is greater on the left side of this plane than on the right side. A higher concentration means that there are more molecules per unit volume. If the system is a gas or a liquid, the molecules move about in a random fashion, and the higher the concentration, the more molecules will cross a given plane per unit time. Thus, on the average, more molecules are moving from left to right across the plane than in the opposite direction. This results in a net mass transfer from the region of high concentration to the region of low concentration. The fact that the molecules collide with each other influences the diffusion process strongly. In a mixture of gases there is a decided difference between a collision of like molecules and a collision of unlike molecules. The collision between like molecules does not appreciably alter the basic molecular movement, because the two molecules are identical and it does not make any difference whether one or the other of the two molecules crosses a certain plane. The collision of two unlike molecules, say, molecules A and B, might result in molecule B crossing some particular plane instead of molecule A. The molecules would, in general, have different masses; thus the mass transfer would be influenced by the collision. By using the kinetic theory of gases it is possible to predict analytically the diffusion rates for some systems by taking into account the collision mechanism and molecular weights of the constituent gases.

Figure 11-2 | Sketch illustrating diffusion dependence on concentration profile.

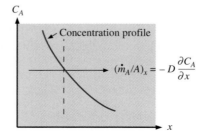

In gases the diffusion rates are clearly dependent on the molecular speed, and consequently we should expect a dependence of the diffusion coefficient on temperature since the temperature indicates the average molecular speed.

11-3 | DIFFUSION IN GASES

Gilliland [4] has proposed a semiempirical equation for the diffusion coefficient in gases:

$$D = 435.7 \frac{T^{3/2}}{p\left(V_A^{1/3} + V_B^{1/3}\right)^2} \sqrt{\frac{1}{M_A} + \frac{1}{M_B}} \qquad \text{[11-2]}$$

where D is in square centimeters per second, T is in degrees Kelvin, p is the total system pressure in pascals, and V_A and V_B are the molecular volumes of constituents A and B as calculated from the atomic volumes in Table 11-1; M_A and M_B are the molecular weights of constituents A and B. Example 11-1 illustrates the use of Equation (11-2) for calculation of diffusion coefficients.

Equation (11-2) offers a convenient expression for calculating the diffusion coefficient for various compounds and mixtures, but it should not be used as a substitute for experimental values of the diffusion coefficient when they are available for a particular system. References 3 and 5 to 9 present more information on calculation of diffusion coefficients. An abbreviated table of diffusion coefficients is given in Appendix A.

Table 11-1 | Atomic volumes.[†]

Air	29.9	In secondary amines	1.20
Bromine	27.0	Oxygen, molecule (O_2)	7.4
Carbon	14.8	Coupled to two other	
Carbon dioxide	34.0	elements:	
Chlorine		In aldehydes and ketones	7.4
Terminal as in R—Cl	21.6	In methyl esters	9.1
Medial as in R—CHCl—R	24.6	In ethyl esters	9.9
Fluorine	8.7	In higher esters and ethers	11.0
Hydrogen, molecule (H_2)	14.3	In acids	12.0
In compounds	3.7	In union with S, P, N	8.3
Iodine	37.0	Phosphorus	27.0
Nitrogen, molecule (N_2)	15.6	Sulfur	25.6
In primary amines	10.5	Water	18.8

[†]For three-membered ring, deduct 6.0. For four-membered ring, deduct 8.5. For five-membered ring, deduct 11.5. For six-membered ring, deduct 15.0. For naphthalene ring, deduct 30.0.

Diffusion Coefficient for CO_2 **EXAMPLE 11-1**

Calculate the diffusion coefficient for CO_2 in air at atmospheric pressure and 25°C using Equation (11-2), and compare this value with that in Table A-8.

■ **Solution**

From Table 11-1

$$V_{CO_2} = 34.0 \qquad M_{CO_2} = 44$$
$$V_{\text{air}} = 29.9 \qquad M_{\text{air}} = 28.9$$

$$D = \frac{(435.7)(298)^{3/2}}{(1.0132 \times 10^5)[(34.0)^{1/3} + (29.9)^{1/3}]^2} \sqrt{\frac{1}{44} + \frac{1}{28.9}}$$

$$= 0.132 \text{ cm}^2/\text{s}$$

From Table A-8

$$D = 0.164 \text{ cm}^2/\text{s} = 0.62 \text{ ft}^2/\text{h}$$

so that the two values are in fair agreement.

We realize from the discussion pertaining to Figure 11-1 that the diffusion process is occurring in two ways at the same time; that is, gas A is diffusing into gas B at the same time that gas B is diffusing into gas A. We thus could refer to the diffusion coefficient for either of these processes.

In working with Fick's law, one may use mass flux per unit area and mass concentration as in Equation (11-1), or the equation may be expressed in terms of molal concentrations and fluxes. There is no general rule to say which type of expression will be most convenient, and the specific problem under consideration will determine the one to be used. For gases, Fick's law may be expressed in terms of partial pressures by making use of the ideal-gas equation of state. (This transformation holds only for gases at low pressures or at a state where the ideal-gas equation of state applies.)

$$p = \rho R T \qquad \text{[11-3]}$$

The density ρ represents the mass concentration to be used in Fick's law. The gas constant R for a particular gas may be expressed in terms of the universal gas constant R_0 and the molecular weight of the gas:

$$R_A = \frac{R_0}{M_A} \qquad \text{[11-4]}$$

where

$$R_0 = 8314 \text{ J/kg} \cdot \text{mol} \cdot \text{K}$$

Then

$$C_A = \rho_A = \frac{p_A M_A}{R_0 T}$$

Consequently, Fick's law of diffusion for component A into component B could be written

$$\frac{\dot{m}_A}{A} = -D_{AB} \frac{M_A}{R_0 T} \frac{dp_A}{dx} \qquad \text{[11-5]}$$

if isothermal diffusion is considered. For the system in Figure 11-1 we could also write for the diffusion of component B into component A

$$\frac{\dot{m}_B}{A} = -D_{BA} \frac{M_B}{R_0 T} \frac{dp_B}{dx} \qquad \text{[11-6]}$$

still considering isothermal conditions. Notice the different subscripts on the diffusion coefficient. Now consider a physical situation called *equimolal counter-diffusion*, as indicated in Figure 11-3; N_A and N_B represent the steady-state molal diffusion rates of components

Figure 11-3 | Sketch illustrating equimolal diffusion.

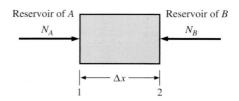

A and *B*, respectively. In this steady-state situation each molecule of *A* is replaced by a molecule of *B*, and vice versa. The molal diffusion rates are given by

$$N_A = \frac{\dot{m}_A}{A} = -D_{AB} \frac{A}{R_0 T} \frac{dp_A}{dx}$$

$$N_B = \frac{\dot{m}_B}{B} = -D_{BA} \frac{A}{R_0 T} \frac{dp_B}{dx}$$

The total pressure of the system remains constant at steady state, so that

$$p = p_A + p_B$$

and

$$\frac{dp_A}{dx} + \frac{dp_B}{dx} = 0$$

or

$$\frac{dp_A}{dx} = \frac{-dp_B}{dx} \qquad\qquad \textbf{[11-7]}$$

Since each molecule of *A* is replacing a molecule of *B*, we may set the molal diffusion rates equal:

$$-N_A = N_B$$

or

$$D_{AB} \frac{A}{R_0 T} \frac{dp_A}{dx} = D_{BA} \frac{A}{R_0 T} \frac{dp_A}{dx}$$

where Equation (11-7) has been used to express the pressure gradient of component *B*. We thus find

$$D_{AB} = D_{BA} = D \qquad\qquad \textbf{[11-8]}$$

The calculation of *D* may be made with Equation (11-2).

We may integrate Equation (11-5) to obtain the mass flux of component *A* as

$$\frac{\dot{m}_A}{A} = \frac{-D M_A}{R_0 T} \frac{p_{A_2} - p_{A_1}}{\Delta x} \qquad\qquad \textbf{[11-9]}$$

corresponding to the nomenclature of Figure 11-3.

Now consider the isothermal evaporation of water from a surface and the subsequent diffusion through a stagnant air layer, as shown in Figure 11-4. The free surface of the water is exposed to air in the tank, as shown. We assume that the system is isothermal and that the total pressure remains constant. We further assume that the system is in steady state. This requires that there be a slight air movement over the top of the tank to remove the water vapor that diffuses to that point. Whatever air movement may be necessary to accomplish this, it is assumed that it does not create turbulence or otherwise alter the concentration profiles in the air in the tank. We further assume that both the air and water vapor behave as ideal gases.

As the water evaporates, it will diffuse upward through the air, and at steady state this upward movement must be balanced by a downward diffusion of air so that the concentration at any *x* position will remain constant. But at the surface of the water there can be no net mass movement of air downward. Consequently, there must be a bulk mass movement

Figure 11-4 | Diffusion of water vapor into air.

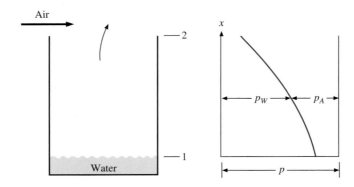

upward with a velocity just large enough to balance the diffusion of air downward. This bulk mass movement then produces an *additional* mass flux of water vapor upward.

The diffusion of air downward is given by

$$\dot{m}_A = \frac{-DAM_A}{R_0 T} \frac{dp_A}{dx}$$ [11-10]

where A denotes the cross-sectional area of the tank. This must be balanced by the bulk mass transfer upward so that

$$-\rho_A A v = -\frac{p_A M_a}{R_0 T} A v$$ [11-11]

where v is the bulk mass velocity upward.

Combining Equations (11-10) and (11-11), we find

$$v = \frac{D}{pA} \frac{dp_A}{dx}$$ [11-12]

The mass diffusion of water vapor upward is

$$\dot{m}_w = -DA \frac{M_w}{R_0 T} \frac{dp_w}{dx}$$ [11-13]

and the bulk transport of water vapor is

$$\rho_w A v = \frac{p_w M_w}{R_0 T} A v$$ [11-14]

The total mass transport is the sum of those given in Equations (11-13) and (11-14). Adding these quantities and making use of Equation (11-12) gives

$$\dot{m}_{w\text{total}} = -\frac{DAM_w}{R_0 T} \frac{dp_w}{dx} + \frac{p_w M_w}{R_0 T} A \frac{D}{p_A} \frac{dp_A}{dx}$$

The partial pressure of the water vapor may be related to the partial pressure of the air by making use of Dalton's law,

$$p_A + p_w = p$$

or

$$\frac{dp_A}{dx} = -\frac{dp_w}{dx}$$

since the total pressure is constant. The total mass flow of water vapor then becomes

$$\dot{m}_{total} = -\frac{DM_wA}{R_0T}\frac{p}{p-p_w}\frac{dp_w}{dx}$$ [11-15]

This relation is called Stefan's law. It may be integrated to give

$$\dot{m}_{w,\,total} = \frac{DpM_wA}{R_0T(x_2-x_1)}\ln\frac{p-p_{w_2}}{p-p_{w_1}} = \frac{DpM_wA}{R_0T(x_2-x_1)}\ln\frac{p_{A_2}}{p_{A_1}}$$ [11-16]

Diffusion of Water in a Tube   EXAMPLE 11-2

Estimate the diffusion rate of water from the bottom of a test tube 10 mm in diameter and 15 cm long into dry atmospheric air at 25°C.

■ **Solution**
We use Equation (11-16) to calculate the mass flux. The partial pressure at the bottom of the test tube is the saturation pressure corresponding to 25°C [77°F], and the vapor pressure may be taken as zero at the top of the test tube since it is diffusing into dry air. Accordingly,

$$p_{A_1} = p - p_{w_1} = 14.696 - 0.4593 = 14.237 \text{ lb/in}^2 \text{ abs} = 9.8155 \times 10^4 \text{ Pa}$$
$$p_{A_2} = p - p_{w_2} = 14.696 - 0 = 14.696 \text{ lb/in}^2 \text{ abs} = 1.0132 \times 10^5 \text{ Pa}$$

From Table A-8

$$D = 0.256 \text{ cm}^2/\text{s} \qquad \dot{m}_w = \frac{DpM_wA}{R_0T(x_2-x_1)}\ln\frac{p_{A_2}}{p_{A_1}}$$

$$\dot{m}_w = \frac{(0.256\times10^{-4})(1.0132\times10^5)(18)(\pi)(5\times10^{-3})^2}{(8314)(298)(0.15)}\ln\frac{1.0132}{0.98155}$$
$$= 3.131\times10^{-10}\text{ kg/s }[0.00113\text{ g/h}]$$

11-4 | DIFFUSION IN LIQUIDS AND SOLIDS

Fick's law of diffusion is also used for problems involving liquid and solid diffusion, and the main difficulty is one of determining the value of the diffusion coefficient for the particular liquid or solid. Unfortunately, only approximate theories are available for predicting diffusion coefficients in these systems. Bird, Stewart, and Lightfoot [9] discuss the calculation of diffusion in liquids, and Jost [6] gives a discussion of the various theories that have been employed to predict values of the diffusion coefficient. The reader is referred to these books for more information on diffusion in liquids and solids.

Diffusion in solids is complex because of the strong influence of the molecular force fields on the process. For these systems Fick's law [Equation (11-1)] is often used, along with an experimentally determined diffusion coefficient, although there is some indication that this relation may not adequately describe the physical processes. The numerical value of the diffusion coefficient for liquids and solids is much smaller than for gases, primarily because of the larger molecular force fields, the increased number of collisions, and the consequent reduction in the freedom of movement of the molecules.

11-5 | THE MASS-TRANSFER COEFFICIENT

We may define a mass-transfer coefficient in a manner similar to that used for defining the heat-transfer coefficient. Thus

$$\dot{m}_A = KA(C_{A_1} - C_{A_2}) \qquad \text{[11-17]}$$

where

$$\dot{m}_A = \text{diffusive mass flux of component A, kg/s}$$
$$K = \text{mass-transfer coefficient, m/s}$$
$$C_{A_1}, C_{A_2} = \text{concentrations through which diffusion occurs, kg/m}^3$$

We recognize as before that the concentrations are, in fact, mass densities of component A in the mixture.

If one considers a steady-state diffusion across a layer of thickness Δx,

$$\dot{m}_A = -\frac{DA(C_{A_2} - C_{A_1})}{\Delta x} = KA(C_{A_2} - C_{A_1})$$

and

$$K = \frac{D}{\Delta x} \qquad \text{[11-18]}$$

For the water-vaporization process described by Equation (11-16)

$$C_{w_1} - C_{w_2} = \frac{M_w}{R_0 T}(p_{w_1} - p_{w_2})$$

so that the mass-transfer coefficient for this situation could be written

$$K = \frac{Dp}{(x_2 - x_1)(p_{w_1} - p_{w_2})} \ln \frac{p - p_{w_2}}{p - p_{w_1}} \qquad \text{[11-19]}$$

Note that the units of the mass-transfer coefficients are in meters per second in SI units.

We have already seen that the phenomenological laws governing heat, mass, and momentum transfer are similar. In Chapter 5 it was shown that the energy and momentum equations of a laminar boundary layer are similar, namely,

$$u\frac{\partial u}{\partial x} + v\frac{\partial u}{\partial y} = \nu\frac{\partial^2 u}{\partial y^2} \qquad \text{[11-20]}$$

$$u\frac{\partial T}{\partial x} + v\frac{\partial T}{\partial y} = \alpha\frac{\partial^2 T}{\partial y^2} \qquad \text{[11-21]}$$

We further observed that the ratio ν/α, the Prandtl number, was the connecting link between the velocity and temperature field and was thus an important parameter in all convection heat-transfer problems. If we considered a laminar boundary layer on a flat plate in which diffusion was occurring as a result of some mass-transfer condition at the surface, we could derive an equation for the concentration of a particular component in the boundary layer. This equation would be

$$u\frac{\partial C_A}{\partial x} + v\frac{\partial C_A}{\partial y} = D\frac{\partial^2 C_A}{\partial^2 y} \qquad \text{[11-22]}$$

where C_A is the concentration of the component that is diffusing through the boundary layer. Note the similarity between Equation (11-22) and Equations (11-20) and (11-21). The concentration and velocity profiles will have the same shape when $\nu = D$ or $\nu/D = 1$.

The dimensionless ratio v/D is called the Schmidt number,

$$\mathrm{Sc} = \frac{v}{D} = \frac{\mu}{\rho D} \qquad\qquad \textbf{[11-23]}$$

and is important in problems where both convection and mass transfer are important. Thus the Schmidt number plays a role similar to that of the Prandtl number in convection heat-transfer problems. Whereas in convection heat-transfer problems we write for the functional dependence of the heat-transfer coefficient for flow over a flat plate

$$\mathrm{Nu}_x = \frac{hx}{k} = f(\mathrm{Re}_x, \mathrm{Pr})$$

in convection mass-transfer problems we should write the functional relation

$$\frac{Kx}{D} = f(\mathrm{Re}_x, \mathrm{Sc})$$

The temperature and concentration profiles will be similar when $\alpha = D$ or $\alpha/D = 1$, and the ratio α/D is called the Lewis number

$$\mathrm{Le} = \frac{\alpha}{D} = \frac{\mathrm{Sc}}{\mathrm{Pr}} \qquad\qquad \textbf{[11-24]}$$

The similarities between the governing equations for heat, mass, and momentum transfer suggest that empirical correlations for the mass-transfer coefficient would be similar to those for the heat-transfer coefficient. This turns out to be the case, and some of the empirical relations for mass-transfer coefficients are presented below. Gilliland [4] presented the equation

$$\frac{Kd}{D} = 0.023 \left(\frac{\rho u_m d}{\mu} \right)^{0.83} \left(\frac{v}{D} \right)^{0.44} \qquad\qquad \textbf{[11-25]}$$

for the vaporization of liquids into air inside circular columns where the liquid wets the surface and the air is forced through the column. The grouping of terms Kx/D or Kd/D is called the Sherwood number

$$\mathrm{Sh} = \frac{Kx}{D} \qquad\qquad \textbf{[11-26]}$$

Note the similarity between Equation (11-25) and the Dittus-Boelter equation (6-4). Equation (11-25) is valid for

$$2000 < \mathrm{Re}_d < 35{,}000 \qquad \text{and} \qquad 0.6 < \mathrm{Sc} < 2.5$$

and is applicable to flow in smooth tubes.

The Reynolds analogy for pipe flow may be extended to mass-transfer problems to express the mass-transfer coefficient in terms of the friction factor. The analogy is written

$$\frac{K}{u_m} \mathrm{Sc}^{2/3} = \frac{f}{8} \qquad\qquad \textbf{[11-27]}$$

that may be compared with the analogy for heat transfer [Equation (6-10)]:

$$\frac{h}{u_m c_p \rho} \mathrm{Pr}^{2/3} = \frac{f}{8} \qquad\qquad \textbf{[11-28]}$$

For flow over smooth flat plates, the Reynolds analogy for mass transfer becomes
Laminar:

$$\frac{K}{u_\infty} \mathrm{Sc}^{2/3} = \frac{C_f}{2} = 0.332 \, \mathrm{Re}_x^{-1/2} \qquad\qquad \textbf{[11-29]}$$

Turbulent:

$$\frac{K}{u_\infty} \text{Sc}^{2/3} = \frac{C_f}{2} = 0.0296 \, \text{Re}_x^{-1/5} \qquad \textbf{[11-30]}$$

Equations (11-29) and (11-30) are analogous to Equations (5-55) and (5-81).

When both heat and mass transfer are occurring simultaneously, the mass- and heat-transfer coefficients may be related by dividing Equation (11-28) by Equation (11-27):

$$\frac{h}{K} = \rho c_p \left(\frac{\text{Sc}}{\text{Pr}}\right)^{2/3} = \rho c_p \left(\frac{\alpha}{D}\right)^{2/3} = \rho c_p \, \text{Le}^{2/3} \qquad \textbf{[11-31]}$$

In addition to using mass density or mass concentration as the driving potential for mass flow (diffusion) and definition of the mass-transfer coefficient, it is also possible to employ molar density and mole fractions as driving forces. Webb [14] suggests a standardized nomenclature applicable to the different quantities.

EXAMPLE 11-3   Wet-Bulb Temperature

Dry air at atmospheric pressure blows across a thermometer that is enclosed in a dampened cover. This is the classical wet-bulb thermometer. The thermometer reads a temperature of 18.3°C. What is the temperature of the dry air?

■ **Solution**

We solve this problem by first noting that the thermometer exchanges no net energy transfer at steady-state conditions and that the heat which must be used to evaporate the water from the cover must come from the air. We therefore make the energy balance

$$hA(T_\infty - T_w) = \dot{m}_w h_{fg}$$

where h is the heat-transfer coefficient and $\dot{m}_w$ is the mass of water evaporated. Now

$$\dot{m}_w = KA(C_w - C_\infty)$$

so that

$$hA(T_\infty - T_w) = KA(C_w - C_\infty)h_{fg}$$

Using Equation (11-31), we have

$$\rho c_p \left(\frac{\alpha}{D}\right)^{2/3} (T_\infty - T_w) = (C_w - C_\infty)h_{fg} \qquad \textbf{[a]}$$

The concentration at the surface C_w is that corresponding to saturation conditions at the temperature measured by the thermometer. From the steam tables at 65°F [18.3°C]

$$p_g = 0.3056 \, \text{lb/in}^2 \, \text{abs} = 2107 \, \text{Pa}$$

and

$$\rho_w = C_w = \frac{p_w}{R_w T_w} = \frac{(2107)(18)}{(8315)(291.3)} = 0.01566 \, \text{kg/m}^3$$

The other properties are

$$\rho_\infty = C_\infty = 0 \text{ (since the free stream is dry air)}$$

$$\rho = \frac{p}{RT} = \frac{1.0132 \times 10^5}{(287)(291.3)} = 1.212 \, \text{kg/m}^3$$

$$c_p = 1.004 \, \text{kJ/kg} \cdot °\text{C} \qquad \text{Le} = \frac{\alpha}{D} = \frac{\text{Sc}}{\text{Pr}} = 0.845$$

$$h_{fg} = 1057 \, \text{Btu/lb}_m = 2.456 \, \text{MJ/kg}$$

Then, from (a)

$$T_\infty - T_w = \frac{(0.01566 - 0)(2.456 \times 10^6)}{(1.212)(1004)(0.845)^{2/3}} = 35.36°C$$

$$T_\infty = 53.69°C$$

The calculation should now be corrected by recalculating the density at the arithmetic-average temperature between wall and free-stream conditions. With this adjustment there results

$$\rho = 1.143 \text{ kg/m}^3 \qquad T_\infty = 55.8°C$$

It is not necessary to correct the ratio Sc/Pr because this parameter does not change appreciably over this temperature range.

| Relative Humidity of Airstream | EXAMPLE 11-4 |

If the airstream in Example 11-3 is at 32.2°C [90°F] while the wet-bulb temperature remains at 18.3°C, calculate the relative humidity of the airstream.

■ **Solution**

From thermodynamics we recall that the relative humidity is defined as the ratio of concentration of vapor to the concentration at saturation conditions for the airstream. We therefore calculate the actual water-vapor concentration in the airstream from

$$\rho c_p \left(\frac{\alpha}{D}\right)^{2/3} (T_\infty - T_w) = (C_w - C_\infty)h_{fg} \qquad [a]$$

and then compare this with the saturation concentration to determine the relative humidity. The properties are taken from Example 11-3:

$$\rho = 1.212 \text{ kg/m}^3 \qquad c_p = 1.004 \text{ kJ/kg} \cdot °C \qquad \frac{\alpha}{D} = \frac{Sc}{Pr} = 0.845$$

$$T_w = 18.3°C \text{ [65°F]} \qquad T_\infty = 32.2°C \text{ [90°F]}$$

$$\rho_w = C_w = 0.01566 \text{ kg/m}^3 \qquad h_{fg} = 2.456 \text{ MJ/kg}$$

We insert the numerical values into Equation (a) and obtain

$$(1.212)(1004)(0.845)^{2/3}(32.2 - 18.3) = (0.01566 - C_\infty)(2.456 \times 10^6)$$

so that

$$\rho_\infty = C_\infty = 0.0095 \text{ kg/m}^3$$

At 32.2°C [90°F] the saturation concentration for the free stream is obtained from the steam tables:

$$\rho_g = C_g[90°F] = 2.136 \times 10^{-3} \text{ lb}_m/\text{ft}^3 = 0.0342 \text{ kg/m}^3$$

The relative humidity is therefore

$$RH = \frac{0.0095}{0.0342} = 27.8\%$$

11-6 | EVAPORATION PROCESSES IN THE ATMOSPHERE

We have already described some evaporation processes and indicated relations between heat and mass transfer. In the atmosphere, the continuous evaporation and condensation of water from the soil, oceans, and lakes influences every form of life and provides many of the day-to-day varieties of climate that govern the environment on earth. These processes are very complicated because in practice they are governed by substantial atmospheric convection currents that are difficult to describe analytically.

Let us first consider the diffusion of water vapor from a horizontal surface into quiescent air as indicated in Figure 11-5. At the surface, the partial pressure of the vapor is p_s. The vapor pressure steadily drops with a rise in elevation z to the "free atmosphere" value of p_∞. The molecular diffusion of the water vapor may be written in the form of Equation (11-13) as

$$\frac{\dot{m}_w}{A} = -D_w \frac{M_w}{R_0 T} \frac{dp_w}{dz} \qquad \text{[11-32]}$$

where A is the surface area under consideration. In hydrologic applications, it is convenient to express this relation in terms of the local atmospheric density and pressure. The total pressure may be expressed as

$$p = \rho \frac{R_0}{M} T \qquad \text{[11-33]}$$

where ρ and M are the density and molecular weight of the moist air, respectively.

Because the molar concentration of water vapor is so small in atmospheric applications, the molecular weight of the moist air is essentially that of dry air, and Equations (11-32) and (11-33) can be combined to give

$$\frac{\dot{m}_w}{A} = -D_w \frac{M_w}{M_a} \frac{\rho}{p} \frac{dp_w}{dz}$$

But, $M_w/M_a = 0.622$, so that

$$\frac{\dot{m}_w}{A} = -0.622 D_w \frac{\rho}{p} \frac{dp_w}{dz} \qquad \text{[11-34]}$$

By using the boundary conditions

$$p_w = p_s \qquad \text{at } z = 0$$
$$p_w = p_\infty \qquad \text{at } z = z_1$$

Eq. (11-34) may be integrated to give

$$\frac{\dot{m}_w}{A} = 0.622 D_w \frac{\rho}{p} \frac{p_s - p_\infty}{z_1} \qquad \text{[11-35]}$$

Figure 11-5 | Diffusion of water vapor from a horizontal surface.

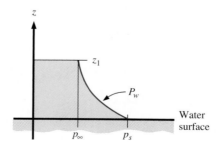

Figure 11-6 | Class A standard pan for measurement of evaporation.

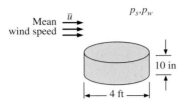

Evaporation processes in the atmosphere are much more complicated than indicated by the simple form of Equation (11-35) for two reasons:

1. The actual diffusion process involves substantial turbulent eddy motion so that the diffusion coefficient D_w may vary significantly with the height z.
2. The air is seldom quiescent and wind currents contribute substantially to the evaporation rate.

As in the many convection problems we have encountered previously, the solution to a complicated problem of this sort is frequently obtained by appealing to carefully controlled measurements in search of an empirical relationship to predict evaporation rates.

In this problem, a "standard pan" is used as shown in Figure 11-6. The mean wind movement is measured 6 in above the pan rim, and water-evaporation rates are measured with the pan placed on the ground (*land pan*) or in a body of water (*floating pan*). For the land pan and with a *convectively stable* atmosphere, the evaporation rate has been correlated experimentally [13] as

$$E_{lp} = (0.37 + 0.0041\bar{u})(p_s - p_w)^{0.88} \qquad \textbf{[11-36]}$$

where

$\quad E_{lp} =$ land-pan evaporation, in/day

$\quad \bar{u} =$ daily wind movement measured 6 in above the pan rim, mi/day

$\quad p_s =$ saturation vapor pressure at dry-bulb air temperature 5 ft above ground surface, in Hg

$\quad p_w =$ actual vapor pressure of air under temperature and humidity conditions 5 ft above ground surface, in Hg

Heat transfer to the pan influences the evaporation rate differently for the ground or water experiments. To convert the pan measurements to those for a natural surface, Equation (11-36) is multiplied by a *pan coefficient* that is 0.7 for the land pan and 0.8 for the floating pan. If the atmosphere is not convectively stable, vertical density gradients can cause substantial deviations from Equation (11-36). These problems are discussed in References 10 to 13.

Water Evaporation Rate ████████ **EXAMPLE 11-5**

A standard land pan is used to measure the evaporation rate in atmospheric air at 100°F and 30 percent relative humidity. The mean wind speed is 10 mi/h. What is the evaporation rate in $lb_m/h \cdot ft^2$ on the land under these conditions?

■ **Solution**

For this calculation we can make use of Equation (11-36). From thermodynamic steam tables

$$p_s = p_g \text{ at } 100°F = 0.9492 \text{ lb/in}^2 \text{ abs} = 1.933 \text{ in Hg} \quad [6.545 \text{ kPa}]$$

$$\text{Relative humidity} = \frac{p_w}{p_s}$$

$$p_w = (0.30)(1.933) = 0.580 \text{ in Hg} \quad [1.964 \text{ kPa}]$$

Also, $\bar{u} = 10 \text{ mi/h} = 240 \text{ mi/day} [386.2 \text{ km/day}]$.

Equation (11-36) now yields, with the application of the 0.7 factor,

$$E_{lp} = 0.7[0.37 + (0.0041)(240)](1.933 - 0.580)^{0.88}$$
$$= 1.237 \text{ in/day} \quad [31.41 \text{ mm/day}]$$

Noting that the standard pan has a diameter of 4 ft, we can use the figure to calculate the mass evaporation rate per unit area as

$$\frac{\dot{m}_w}{A} = \frac{E_{lp}}{12}\rho_w$$
$$= \frac{(1.237)(62.4)}{12} = 6.432 \text{ lb}_m/\text{day} \cdot \text{ft}^2 \quad [31.43 \text{ kg/m}^2 \cdot \text{day}]$$
$$= 0.268 \text{ lb}_m/\text{h} \cdot \text{ft}^2 \quad [1.31 \text{ kg/h} \cdot \text{m}^2]$$

As a matter of interest, we might calculate the molecular-diffusion rate of water vapor from Equation (11-35), taking z_1 as the 5-ft dimension above the standard pan. Since

$$\rho = \frac{p}{RT}$$

Equation (11-35) can be written as

$$\frac{\dot{m}_w}{A} = 0.622 \frac{D_w}{RT} \frac{p_s - p_w}{z_1}$$

From Table A-8,

$$D_w = 0.256 \text{ cm}^2/\text{s} = 0.99 \text{ ft}^2/\text{h}$$

so that

$$\frac{\dot{m}}{A} = \frac{(0.622)(0.99)(0.9492)(1 - 0.3)}{(53.35)(560)(5.0)}$$
$$= 3.94 \times 10^{-4} \text{ lb}_m/\text{h} \cdot \text{ft}^2 \; [19.2 \text{ g/h} \cdot \text{m}^2]$$

This number is negligibly small in comparison with the previous calculation. This means that in the actual evaporation process, turbulent diffusion and convection transport play dominant roles in comparison with molecular diffusion.

REVIEW QUESTIONS

1. How is the diffusion coefficient defined?
2. Define the mass-transfer coefficient.
3. Define the Schmidt and Lewis numbers. What is the physical significance of each?
4. Why cannot atmospheric evaporation rates be calculated with ordinary molecular-diffusion equations?

LIST OF WORKED EXAMPLES

11-1 Diffusion coefficient for CO_2
11-2 Diffusion of water in a tube
11-3 Wet-bulb temperature
11-4 Relative humidity of airstream
11-5 Water evaporation rate

PROBLEMS

11-1 Using physical reasoning, justify the $T^{3/2}$ dependence of the diffusion coefficient as shown by Equation (11-2). *Hint:* Recall that mean molecular velocity is proportional to $T^{1/2}$ and that the density of an ideal gas is inversely proportional to temperature.

11-2 Using Equation (11-2), calculate the diffusion coefficient for benzene in atmospheric air at 25°C.

11-3 Dry air at atmospheric pressure and 25°C blows across a flat plate at a velocity of 1.5 m/s. The plate is 30 cm square and is covered with a film of water that may evaporate into the air. Plot the heat flow from the plate as a function of the plate temperature between $T_w = 15$°C and $T_w = 65$°C.

11-4 The cover on a wet-bulb thermometer is soaked in benzene, and the thermometer is exposed to a stream of dry air. The thermometer indicates a temperature of 26°C. Calculate the free-stream air temperature. The vapor pressure of benzene is 13.3 kPa, and the enthalpy of vaporization is 377 kJ/kg at 26°C.

11-5 Dry air at 25°C and atmospheric pressure flows inside a 5-cm-diameter pipe at a velocity of 3 m/s. The wall is coated with a thin film of water, and the wall temperature is 25°C. Calculate the water-vapor concentration in the air at exit of a 3-m length of the pipe.

11-6 An open pan 20 cm in diameter and 10 cm deep contains water at 25°C and is exposed to atmospheric air at 25°C and 50 percent relative humidity. Calculate the evaporation rate of water in grams per hour.

11-7 A test tube 1.25 cm in diameter and 15 cm deep contains benzene at 26°C and is exposed to dry atmospheric air at 26°C. Using the properties given in Problem 11-4, calculate the evaporation rate of benzene in grams per hour.

11-8 Dry air at 25°C and atmospheric pressure blows over a 30-cm-square surface of ice at a velocity of 1.5 m/s. Estimate the amount of moisture evaporated per hour, assuming that the block of ice is perfectly insulated except for the surface exposed to the airstream.

11-9 The temperature of an airstream is to be measured, but the thermometer available does not have a sufficiently high range. Accordingly, a dampened cover is placed around the thermometer before it is placed in the airstream. The thermometer reads 32°C. Estimate the true air temperature, assuming that it is dry at atmospheric pressure.

11-10 Assume that a human forearm may be approximated by a cylinder 4 in. in diameter and 1 ft long. The arm is exposed to a dry-air environmental temperature of 115°F in a 10-mi/h breeze on the desert and receives a radiant heat flux from the sun of 350 Btu/h · ft^2 of view area (view area for the cylinder $= Ld$). If the arm is perspiring so that it is covered with a thin layer of water, estimate the arm surface temperature.

Neglect internal heat generation of the arm. Assume an emissivity of unity for the water film.

11-11 A 30-cm-square plate is placed on a low-speed wind tunnel; the surface is covered with a thin layer of water. The dry air is at atmospheric pressure and 43°C and blows over the plate at a velocity of 12 m/s. The enclosure walls of the wind tunnel are at 10°C. Calculate the equilibrium temperature of the plate, assuming an emissivity of unity for the water film.

11-12 Calculate the rate of evaporation for the system in Problem 11-11.

11-13 Refine the analysis of Problem 11-10 by assuming a body-heat generation of 1860 W/m^3.

11-14 A small tube 6.4 mm in diameter and 13 cm deep contains water with the top open to atmospheric air at 20°C, 1 atm, and 50 percent relative humidity. Heat is added to the bottom of the tube. Plot the diffusion rate of water as a function of water temperature over the range of 20 to 82°C.

11-15 Dry air at 20°C enters a 1.25-cm-ID tube where the interior surface is coated with liquid water. The mean flow velocity is 3 m/s, and the tube wall is maintained at 20°C. Calculate the diffusion rate of water vapor at the entrance conditions. How much moisture is picked up by the air for a tube 1 m long?

11-16 Dry air at 65°C blows over a 30-cm-square plate at a velocity of 6 m/s. The plate is covered with a smooth porous material, and water is supplied to the material at 25°C. Assuming that the underside of the plate is insulated, estimate the amount of water that must be supplied to maintain the plate temperature at 38°C. Assume that the radiation temperature of the surroundings is 65°C and that the porous surface radiates as a blackbody.

11-17 Dry air at atmospheric pressure blows over an insulated flat plate covered with a thin wicking material that has been soaked in ethyl alcohol. The temperature of the plate is 25°C. Calculate the temperature of the airstream assuming that the concentration of alcohol is negligible in the free stream. Also calculate the mass-transfer rate of alcohol for a 30-cm-square plate if the free-stream velocity is 7 m/s.

11-18 Carrier's equation expresses actual water-vapor partial pressure in terms of wet-bulb and dry-bulb temperatures:

$$p_v = p_{gw} - \frac{(p - p_{gw})(T_{DB} - T_{WB})}{2800 - T_{WB}}$$

where

$$p_v = \text{actual partial pressure, lb/in}^2 \text{ abs}$$
$$p_{gw} = \text{saturation pressure corresponding to wet-bulb temperature, lb/in}^2 \text{ abs}$$
$$p = \text{total mixture pressure, lb/in}^2 \text{ abs}$$
$$T_{DB} = \text{dry-bulb temperature, °F}$$
$$T_{WB} = \text{wet-bulb temperature, °F}$$

Air at 1 atm and 100°F flows across a wet-bulb thermometer, producing a temperature of 70°F. Calculate the relative humidity of the airstream using Carrier's equation and compare with results obtained by methods of this chapter.

11-19 Suppose the wet bulb of Problem 11-18 is exposed to a black-radiation surrounding at 100°F. What radiation equilibrium temperature would it indicate, assuming an emissivity of unity for the wick?

11-20   When employed to measure the evaporation rate into ambient air at 35°C, a standard land-pan apparatus indicates an evaporation rate of 350 g/m² · h when subjected to a crosswind of 250 m/min. Calculate the relative himidity of the airstream.

11-21   A light breeze at 2.2 m/s blows across a standard evaporation pan. The atmospheric conditions are 20°C and 40 percent relative humidity. What is the evaporation rate for a land pan in grams per hour per square meter? What would be the evaporation rate for zero velocity?

11-22   An evaporation rate of 0.3 g/s · m² is experienced for a 4.5-m/s breeze blowing across a standard land pan. What is the relative humidity if the dry-bulb (ambient) air temperature is 40°C?

11-23   The land-pan-evaporation formula may be used to estimate the evaporation rate from swimming pools. Compare the water loss rate for a pool in Pasadena, California, exposed to air at 80°F and 30 percent relative humidity with that for a pool in Phoenix, Arizona, exposed to surroundings at 110°F and 5 percent relative humidity.

11-24   Convert the dimensional equation (11-36) to one in which the units of the variables are $E_{lp} = \text{mm/day}$, $\bar{u} = \text{m/s}$, and p_s and $p_w = \text{Pa}$.

REFERENCES

1. Prigogine, I. *Introduction to Thermodynamics of Irreversible Processes.* Springfield, Ill.: Charles C. Thomas, Publisher, 1955.

2. de Groot, S. R. *Thermodynamics of Irreversible Processes.* Amsterdam: North-Holland Publishing, 1952.

3. Present, R. D. *Kinetic Theory of Gases.* New York: McGraw-Hill, 1958.

4. Gilliland, E. R. "Diffusion Coefficients in Gaseous Systems," *Ind. Eng. Chem.,* vol. 26, p. 681, 1934.

5. Perry, J. H. (Ed.). *Chemical Engineers' Handbook,* 4th ed. New York: McGraw-Hill, 1963.

6. Jost, W. *Diffusion in Solids, Liquids and Gases.* New York: Academic Press, 1952.

7. Reid, R. C., and T. K. Sherwood. *The Properties of Gases and Liquids.* New York: McGraw-Hill, 1958.

8. *Handbook of Chemistry and Physics.* Cleveland, Ohio: Chemical Rubber Publishing, 1960.

9. Bird, R., W. E. Stewart, and E. N. Lightfoot. *Transport Phenomena.* New York: John Wiley & Sons, 1960.

10. Instructions for Climatological Observers, *U.S. Dept. Commerce Weather Bur. Circ. B.,* 10th ed. rev., October 1955.

11. Water-Loss Investigations, vol. 1, *Lake Hefner Studies Tech. Rep., U.S. Geol. Surv. Circ.* 229, 1952.

12. Nordenson, T. J., and D. R. Baker. "Comparative Evaluation of Evaporation Instruments," *J. Geophys. Rec.,* vol. 67, no. 2, p. 671, February 1962.

13. Kohler, M. A., T. J. Nordenson, and W. E. Fox: Evaporation from Pans and Lakes, *U.S. Dept. Commerce Weather Bur. Res. Pap.* 38, May 1955.

14. Webb, R. L. "Standardized Nomenclature for Mass Transfer Processes," *ASHRAE Trans.,* vol. 97, pt. 2, pp. 114–18, 1990.

Summary and Design Information

12-1 | INTRODUCTION

Previous chapters have examined various aspects of conduction, convection, and radiation heat transfer with the caveat always expressed to the reader that most problems involve two or three of the modes. Most conduction problems have some kind of convection boundary condition to contend with, most convection problems must consider delivery of the energy to the convection surface by either or both radiation and conduction, and so on.

When a new practical problem is encountered, it will usually be advantageous to simplify the situation as much as possible and reduce the number and complexity of the modes of heat transfer involved. The reader already has at his or her disposal the information to make such simplifications, but to aid in that process, the summary information presented in this chapter can help speed various considerations by the use of summary charts, which may be employed for preliminary design calculations. In many cases, the information obtained from the charts will be sufficient in itself, while in other cases, it will only point the way to more detailed calculations that may be required. As we have noted several times, the basic uncertainty in many of the convection empirical correlations is of the order of ± 20 to 25 percent, so the numerical values obtained from the charts in this chapter may be quite satisfactory for preliminary design proposes. Obviously, the pertinent equations must be employed in any computer modeling study that may be required later. The equations on which the various charts are based are displayed on the chart or referenced in the accompanying text.

In Sections 3-10, 4-8, 7-14, and 8-18 we have already suggested general ways to proceed when approaching new problems. Some of those considerations will be repeated here (sometimes in a different format) for the sake of compactness and inclusiveness and in recognition of the desirability of some redundancy.

The reader should not fall in the trap of viewing this chapter as an expedient cookbook to solve all problems. As a person becomes more experienced in various aspects of heat transfer calculations and design, much of the information suggested in this chapter will become almost automatic. For the novice or nonspecialist in the field, quite the contrary is true.

12-2 | CONDUCTION PROBLEMS

Chapters 1 through 4 described many types of conduction problems from one-dimensional steady-state solutions to multidimensional transient problems. We now suggest a general approach for examining such problems and affecting a solution. The approach is outlined below and depicted graphically in the block diagram of Figure 12-1.

1. Determine the conduction material and approximate values of its thermal properties—that is, high, low, medium conductivity, dense or lightweight, and so on.

2. Determine the approximate nature of the convection and/or radiation boundary conditions imposed on the conductive solid—that is, natural or forced convection, gas or liquid fluid(s), high or low emissivity, and high or low temperatures for radiation surfaces.

3. Is the problem a steady-state or transient problem? If it is a transient problem, is it possible to use a lumped-capacity approximation to determine the temperature variation with time? Employ Equation (4-6) as a criterion for applicability, that is,

$$\frac{h_{\text{boundary}} \, (V/A_{\text{boundary}})}{k_{\text{solid}}} < 0.1 \qquad \textbf{[4-6]}$$

where h_{boundary} is taken as a total heat-transfer coefficient that includes both convection and radiation at the boundary.

4. Are there significant multidimensional conduction effects or is it possible to describe the system using a one-dimensional model?

5. Is a simple relation available for calculation of temperature distribution in the solid? If not, consider setting up a numerical model for the solution. To aid in that formulation we repeat the nodal equations employed in the finite-difference models of Chapters 3 and 4.

Steady State

$$T_i = \frac{q_i + \sum (T_j/R_{ij})}{\sum (1/R_{ij})}$$

Conduction resistance: $\left(\frac{1}{R_{ij}}\right)_{\text{cond}} = \left(\frac{kA}{\Delta x}\right)_{ij}$

Convection resistance: $\left(\frac{1}{R_{ij}}\right)_{\text{conv}} = h(\Delta A)_{i\text{-fluid}}$

Heat source: $q_i = \dot{q}_i \, \Delta V_i$

Radiation boundary: $q_i = q''_{\text{rad},i} \times (\Delta A)_{i\text{-rad surroundings}}$

Unsteady State

$$T_i^{p+1} = \left(q_i + \frac{\sum T_j^p - T_i^p}{R_{ij}}\right) \frac{\Delta \tau}{C_i} + T_i^p$$

Stability criterion: $\Delta \tau \leq \left[\dfrac{C_i}{\sum \left(\frac{1}{R_{ij}}\right)}\right]_{\text{minimum}}$

Thermal capacity of note $i = C_i = \rho_i c_i \, \Delta V_i$

Figure 12-1 | General conduction procedure.

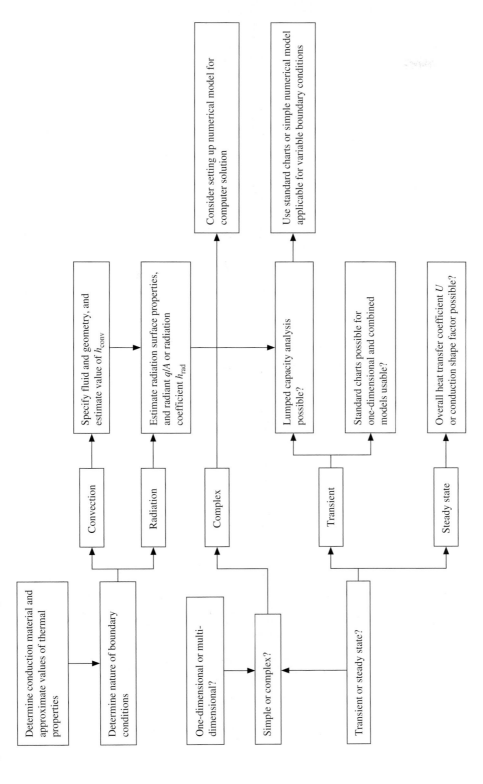

12-3 | CONVECTION HEAT-TRANSFER RELATIONS

We have already given a general procedure for approaching convection heat-transfer problems in Section 7-14 and in Figure 7-15 (also reproduced on the inside back cover). Once again we call attention to the information presented in the textbox of that figure regarding proper areas to be employed in convection problems. The area used in the formula

$$q = hA\,(T_{\text{surface}} - T_{\text{fluid}})$$

is always the heated or cooled surface in contact with the fluid. It is *not* the flow cross-sectional area used to calculate the mass rate of flow in a tube or channel

$$\dot{m} = \rho u_{\text{mean}} A_{\text{cross section}}$$

Gross errors can result if the two areas are interchanged.

The legends on the convection charts are mostly self-explanatory, but we call attention to some features of specific charts. A perusal of the convection charts will reveal that there are several plots given for flow of air at 1 atm and 300 K. In Chapters 5, 6, and 7 we emphasized several times the need to evaluate fluid properties at the proper temperature. In fact, there is a loop in the procedural Figure 7-15 that adds extra emphasis to the need for evaluation of fluid properties at the correct temperature. At this point, we mention that when dealing with convection of an ideal gas, like air, the need for precision in determining proper temperature for property evaluation is not as stringent.

Example 12-3 shows that for ideal gases one has considerable latitude in temperature-property determination without significantly affecting the final calculated value of the convection coefficient h. Thus, the charts, which are presented for a temperature of 300 K, will be found to be applicable, at least for preliminary design purposes, over a rather

Figure 12-2 | Local Nusselt numbers for flow over isothermal flat plates. Properties evaluated at film temperature $T_f = (T_w + T_\infty)/2$: (*a*) linear scale, (*b*) logarithmic scale. See Equations (5-44), (5-81), and (5-82).

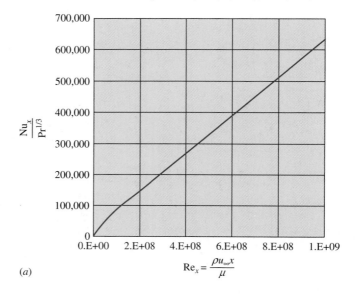

(*a*)

Figure 12-2 | (*Continued*).

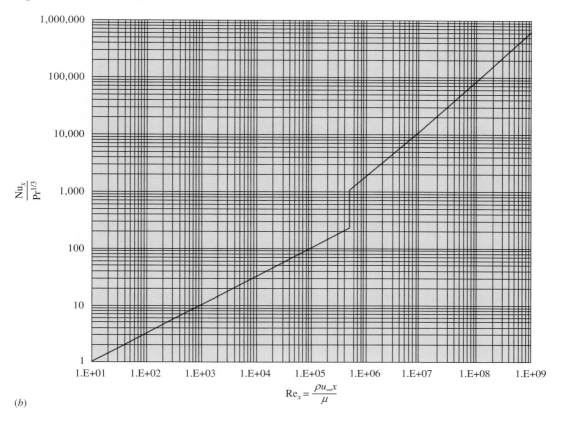

(b)

wide range of temperatures. Note that this statement applies only to flow systems involving ideal gases and would not apply to liquids, particularly viscous oils and petroleum products.

In several cases, convection information has been presented in both linear and logarithmic coordinate systems when it was felt that both systems would be beneficial.

Figures 12-2 and 12-3 present the local and average Nusselt number correlations for flow over flat plates, and the respective equations from which they are derived. Note that fluid properties are evaluated at the film temperature for this system.

Figure 12-4 presents the classic Dittus-Boelter and Gnielinski relations for turbulent flow in smooth tubes. For this system, fluid properties are evaluated at the average bulk temperature. In the absence of more specific correlations, these relations may be expected to apply to noncircular flow cross sections as well, with the hydraulic diameter replacing the physical diameter in the Nusselt and Reynolds numbers.

Figure 12-5 presents the relations for laminar heat transfer in isothermal tubes. For heat transfer with highly viscous fluids and strongly temperature-dependent fluid properties, Chapter 6 should be consulted.

Heat transfer for cross flow over isothermal circular cylinders is plotted in Figure 12-6 over a range of about eight orders of magnitude. Both the Churchill correlation and the simpler one of Equation (6-17) are presented. A similar plot for flow over isothermal spheres is given in Figure 12-7. Note the temperatures used for property determination in these two plots.

Figure 12-3 | Average Nusselt numbers for flow over isothermal flat plates. Properties evaluated at film temperature $T_f = (T_w + T_\infty)/2$. See Equations (5-85) and (5-86).

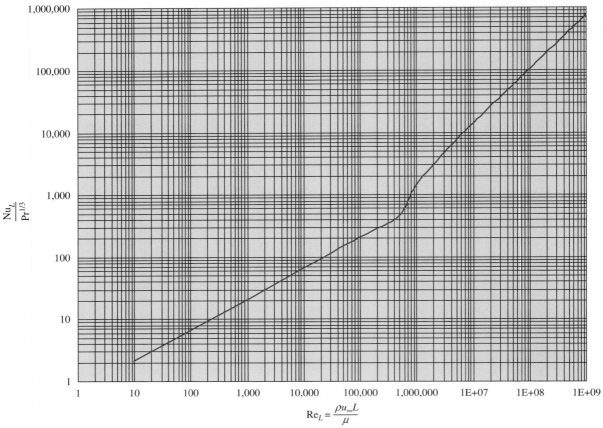

Free-convection heat transfer from isothermal vertical flat plates is given Figure 12-8. As with forced flow, a comparison is shown with the Churchill correlation and the simpler one of Equation (7-25). The correlation range is extremely wide, extending for 13 orders of magnitude. In similar fashion, Figure 12-9 gives the free-convection heat-transfer correlation for isothermal horizontal cylinders. Again, comparisons of the Churchill correlations and the simpler one of Equation (7-25) are shown. These correlations extend over 20 orders of magnitude. Free convection from spheres is shown in Figure 12-10. Note that property evaluation for the free-convection correlations are made at the film temperature.

In some design applications, a quick estimate for free-convection boundary-layer thickness may be needed to determine possible flow interactions. Figures 12-11 and 12-12 give such estimates for the laminar range with air and liquid water at 300 K. Both linear and logarithmic charts are provided. Figures 12-13 and 12-14 provide information that may be used to estimate the maximum flow velocities in free-convection boundary layers. Again, both linear and logarithmic plots have been provided.

Free convection heat transfer across a gap space is important in a number of applications. The effective thermal conductivities for several cases are presented in Figures 12-15, 12-16, and 12-17, and the equations from which the information was derived are noted. Low-density conduction in a vertical air gap is presented in Figure 12-18. Note that this

Figure 12-4 | Average Nusselt numbers for heated, fully developed turbulent flow in smooth tubes. Properties evaluated at average bulk temperature.

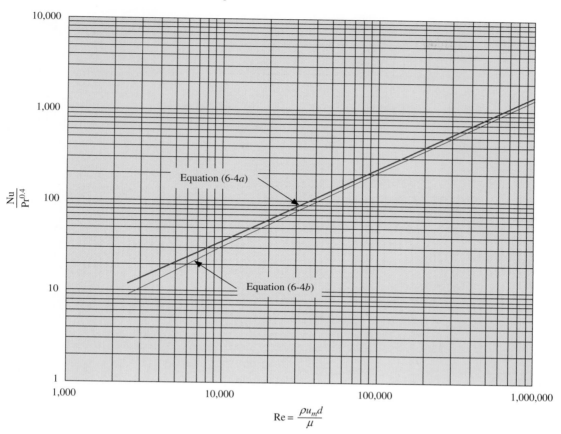

information is given for a temperature of 300 K. The mean free path for the molecules would vary somewhat with temperature, so more precise calculations might be required to take this variation into account.

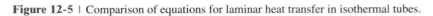

Figure 12-5 | Comparison of equations for laminar heat transfer in isothermal tubes.

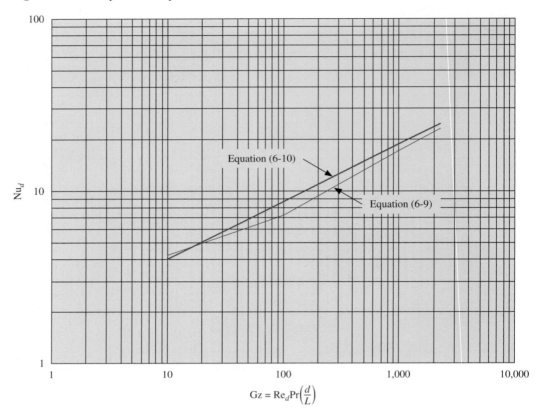

Figure 12-6 | Nusselt numbers for cross flow over isothermal circular cylinders. Properties evaluated at film temperature $T_f = (T_w + T_\infty)/2$.

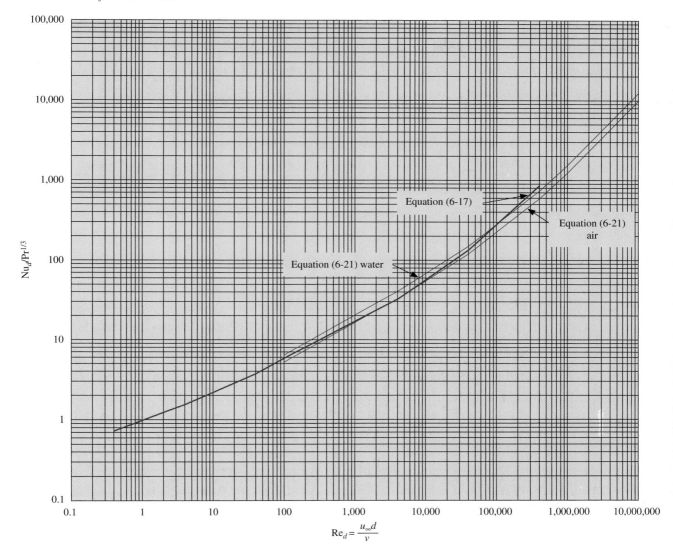

Figure 12-7 | Nusselt numbers for cross flow over isothermal spheres. Properties evaluated at free-stream temperature T_∞, except μ_w.

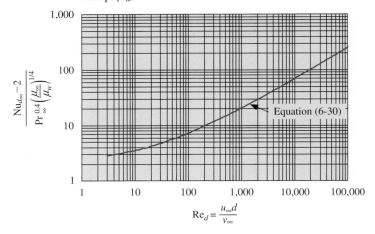

$$\frac{\mathrm{Nu}_{d\infty} - 2}{\mathrm{Pr}_\infty^{0.4}\left(\frac{\mu_\infty}{\mu_w}\right)^{1/4}}$$

Equation (6-30)

$$\mathrm{Re}_d = \frac{u_\infty d}{v_\infty}$$

Figure 12-8 | Average Nusselt numbers for free convection from isothermal vertical flat plates. Properties evaluated at film temperature $T_f = (T_w + T_\infty)/2$.

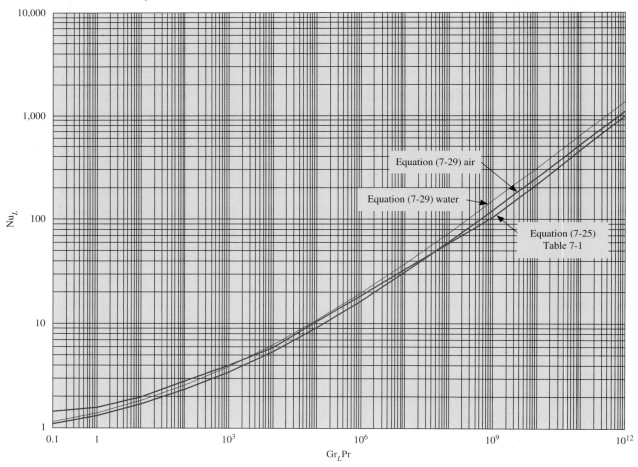

Nu_L

Equation (7-29) air

Equation (7-29) water

Equation (7-25)
Table 7-1

$\mathrm{Gr}_L \mathrm{Pr}$

Figure 12-9 | Average Nusselt numbers for free convection from isothermal horizontal cylinders. Properties evaluated at film temperature $T_f = (T_w + T_\infty)/2$.

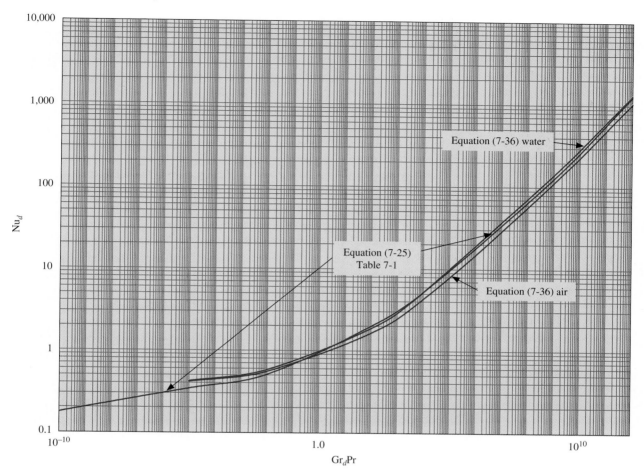

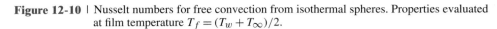

Figure 12-10 | Nusselt numbers for free convection from isothermal spheres. Properties evaluated at film temperature $T_f = (T_w + T_\infty)/2$.

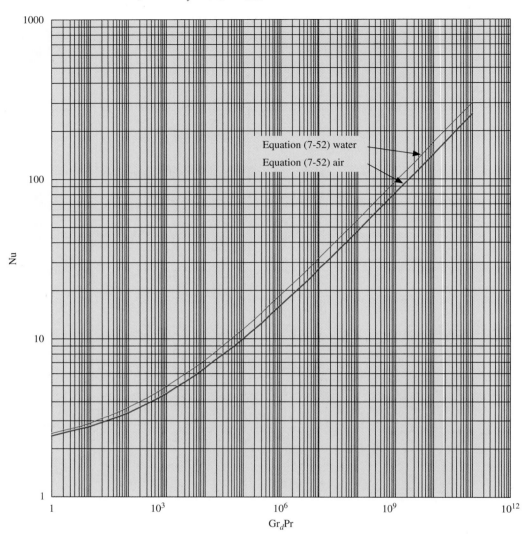

Figure 12-11 | Free-convection boundary-layer thickness on vertical flat plate, log plot, $x =$ distance from leading edge, air at 1 atm pressure, properties evaluated at 300 K, $T_w - T_\infty = 20°C$. For other temperature differences multiply δ by $[20/(T_w - T_\infty)]^{1/4}$ and Gr_x by $(T_w - T_\infty)/20$: (*a*) linear scale, (*b*) log scale. See Equation (7-20*a*).

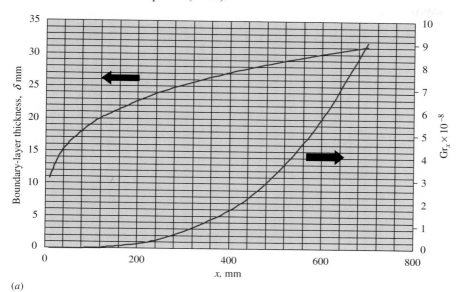

(*a*)

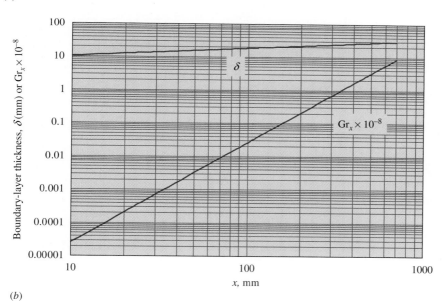

(*b*)

Figure 12-12 | Free-convection boundary-layer thickness on vertical flat plate, x = distance from leading edge, liquid water, properties evaluated at 300 K, $T_w - T_\infty = 20°C$: (*a*) linear plot, (*b*) log plot. See Equation (7-20*a*).

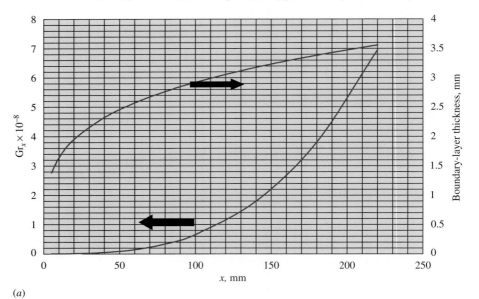

(*a*)

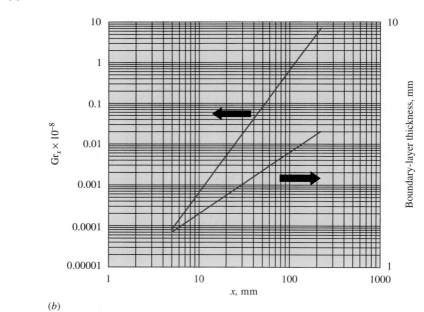

(*b*)

Figure 12-13 | Maximum-flow velocity for free convection of air at 1 atm on vertical flat plate, properties evaluated at 300 K, $T_w - T_\infty = 20°C$, $x =$ distance from leading edge (mm): (a) linear scale, (b) log scale. See Equation (7-20b).

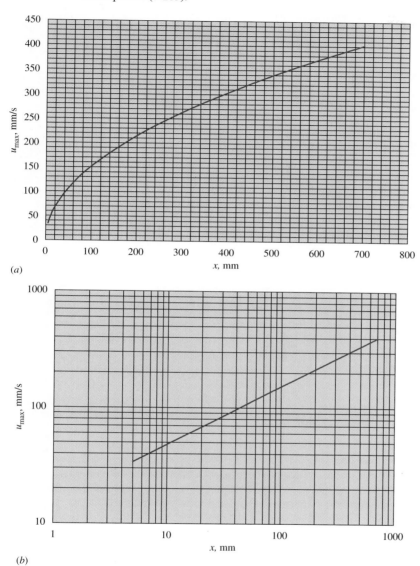

(a)

(b)

Figure 12-14 | Maximum-flow velocity for free convection of liquid water on vertical flat plate, properties evaluated at 300 K, $T_w - T_\infty = 20°C$: (a) linear scale, (b) log scale. See Equation (7-20b).

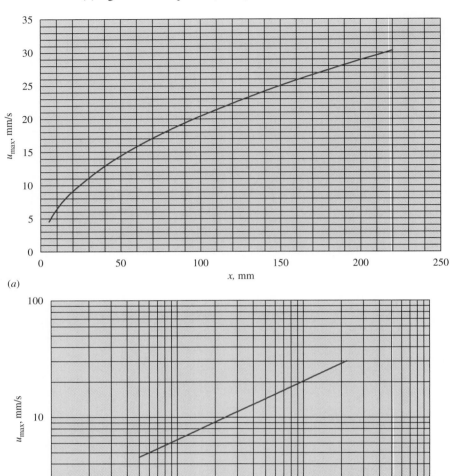

(a)

(b)

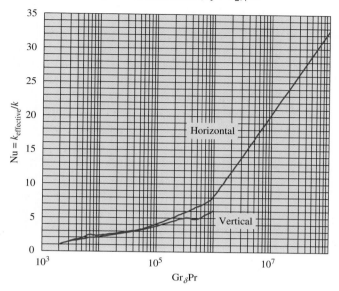

Figure 12-15 | Free convection across vertical and horizontal gaps for liquids. See Equation (7-64) and Table 7-3. Properties evaluated at $(T_1 + T_2)/2$.

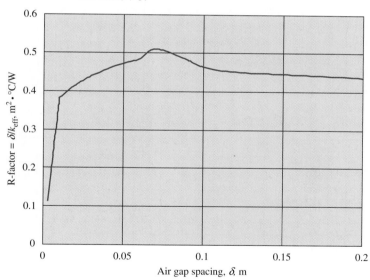

Figure 12-16 | R-factor for free convection across vertical air gap at 1 atm pressure, 300 K, $\Delta T = 10\,\text{K}$, $L = 0.2\,\text{m}$. See Equation (7-64) and Table 7-3.

Figure 12-17 | Free convection across horizontal air gap heated from below at 300 K. See Equation (7-64) and Table 7-3.

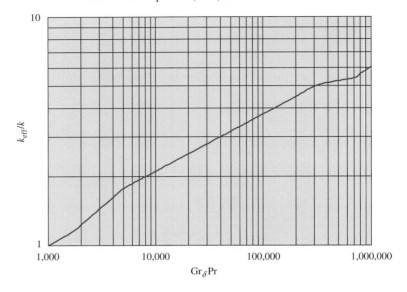

Figure 12-18 | Low-density conduction across vertical air gap at 300 K, $\Delta T = 20\,°C$, q/A in W/m^2: (a) linear scale, (b) logarithmic scale.

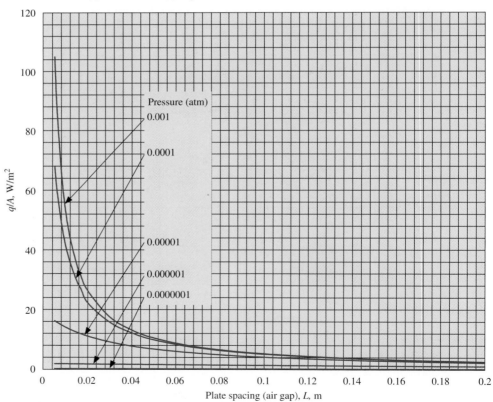

(a)

Figure 12-18 | (*Continued*).

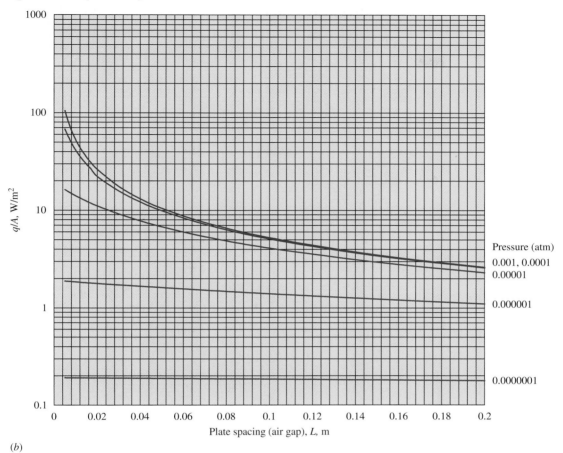

(*b*)

12-4 | RADIATION HEAT TRANSFER

The Planck blackbody radiation spectrum is given in Figure 12-19. Radiant heat transfer between two gray diffuse surfaces is presented in Figure 12-20. Because it is sometimes useful for preliminary design studies, values of the radiation heat-transfer coefficient for a gray body exchanging heat with a large black enclosure at 300 K is presented in Figure 12-21. Otherwise, most of the information regarding radiation is presented in Chapter 8 and in Section 8-18.

A word of caution regarding nongray body behavior is that while many surfaces exhibit near gray body behavior at moderate temperatures (< 700 K), they may have entirely different surface absorption and transmission characteristics when exposed to shorter wavelength thermal radiation, particularly solar radiation. The old phrase "you can't tell by looking" is particularly appropriate. Snow and ice appear quite bright and reflective but are almost black to long-wavelength thermal radiation. Ordinary window glass transmits almost all solar (short-wavelength $\lambda < 2.5\mu$m) radiation but is almost completely opaque for long-wavelength radiation. In most of the problems and examples presented here and in Chapter 8, we have stipulated the values of emissivity or absorptivity to simplify the solution process. When a new problem is encountered, it is always

Figure 12-19 | Planck blackbody spectrum as a function of λT (μm).

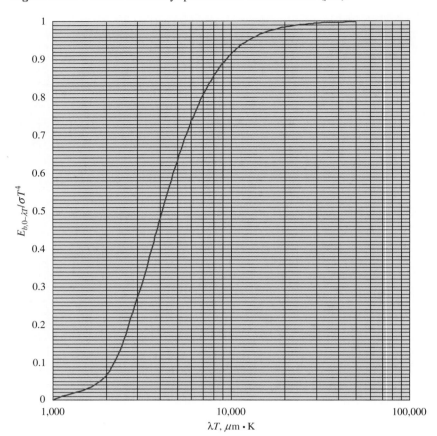

a good idea to check values using information like that in Table A-10 or other source materials.

Figure 12-22 gives an expanded view of the Rohsenow boiling correlation, while Figures 12-23, 12-24, and 12-25 give plots to estimate approximate values of boiling heat-transfer coefficients for water at various pressures. The reader is cautioned to consult reference material for a particular geometric configuration and surface-fluid combination for any final design involving boiling heat transfer.

Figure 12-20 | Radiation heat transfer between two gray, diffuse bodies that see each other and nothing else, $F_{12} = 1.0$.

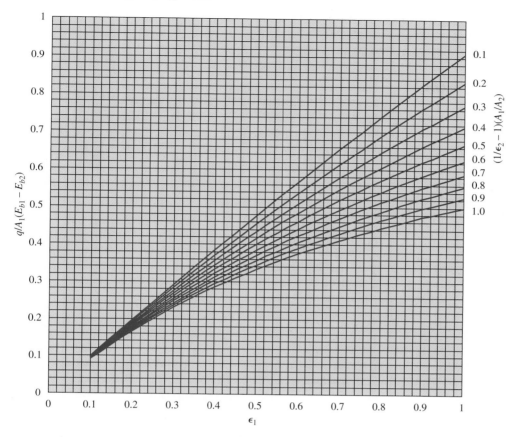

Figure 12-21 | Radiation heat-transfer coefficient for a blackbody in a large enclosure at 300 K. See Equation (8-115). Note:
$$h_{rad}(\text{graybody}) = \epsilon \times h_{rad}(\text{blackbody}).$$

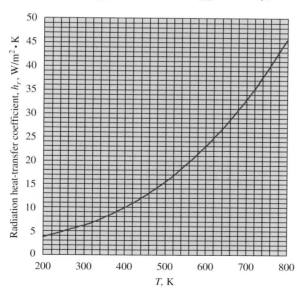

Figure 12-22 | Rohsenow pool boiling correlation for water-platinum. See Equation (9-33) and Table 9-2. Note: $(q/A)_{\text{water-other material}} / (q/A)_{\text{water-platinum}}$
$= \left(0.013 / C_{\text{sf,water-other material}}\right)^3$.

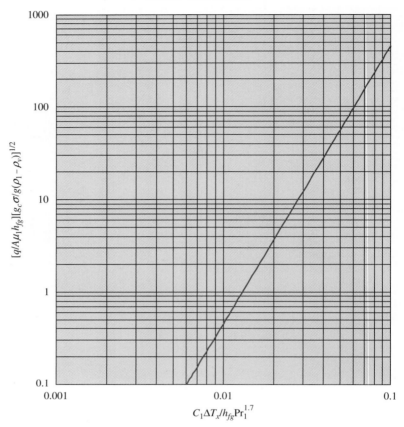

Figure 12-23 | Approximate boiling heat-transfer coefficients for water at low pressures $(0.2 < p < 0.7 \text{ MPa})$. See Equation (9-35).

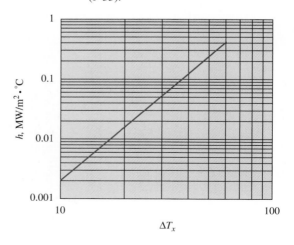

Figure 12-24 | Approximate boiling heat-transfer coefficients for water at higher pressures $(0.7 < p < 14 \text{ MPa})$. See Equation (9-36).

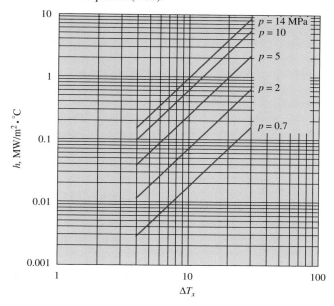

Figure 12-25 | Heat flux and h values for boiling water at 1 atm, horizontal surfaces. See Table 9-3.

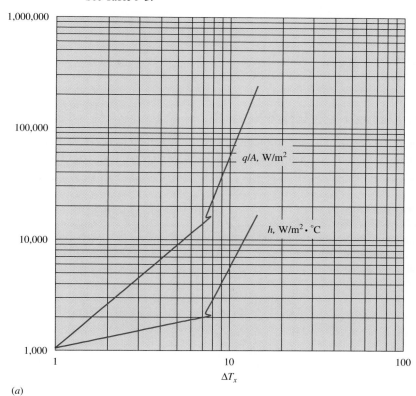

(a)

Figure 12-25 | *(Continued).*

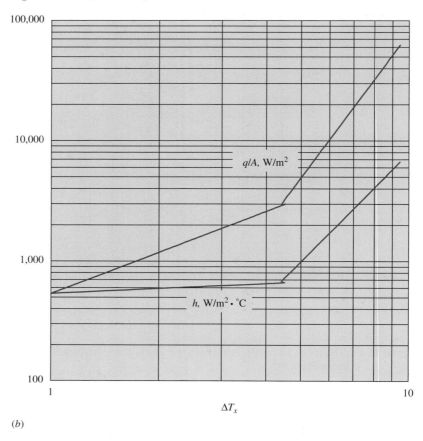

(b)

12-5 | HEAT EXCHANGERS

Most practical heat-exchanger problems involve determinations of convection heat-transfer coefficients for a number of different geometrical configurations. Extensive design information is given in References 3, 8, and 17 of Chapter 10.

| EXAMPLE 12-1 | Cooling of an Aluminum Cube |

An aluminum cube, 5 cm on a side, is heated to a uniform temperature of 500 K and then suddenly allowed to cool in surrounding air at 1 atm and 300 K. The effective radiation temperature of the surroundings is also 300 K. How long will it take the cube to cool to a temperature of 50°C if the surface is coated with a flat black paint?

■ **Solution**

The solid is cooling by a combination of free convection and radiation. Let us assume that the flat black paint produces a surface emissivity very nearly equal to 1.0, that of a blackbody. (See Table A-10.) Furthermore, let us assume that the free convection behaves like a vertical flat plate having a height of 10 cm (see Section 7-6, *Irregular Solids*). The convection coefficients for most free-convection problems are small, so we might anticipate that the solid would behave as a lumped capacity, because aluminum has a relatively high thermal conductivity. We anticipate that the boundary layer will be laminar because the characteristic length is small ($L = 10$ cm). For simplified calculations, the formula of Table 7-2 may be employed. Thus, we perform the calculations:

$$L = 0.1 \text{ m}$$

$$T_{\text{avg solid}} = \frac{500 + 323}{2} = 412 \text{ K}$$

$$\Delta T_{\text{avg}} = 412 - 300 = 112 \text{ K}$$

$$h_{\text{conv, avg}} = (1.42)\left(\frac{112}{0.1}\right)^{0.25} = 8.2 \text{ W/m}^2 \cdot \text{K}$$

For aluminum, from Table A-2,

$$\rho = 2700 \text{ kg/m}^3$$
$$k = 206 \text{ W/m} \cdot \text{K}$$
$$c = 896 \text{ J/kg} \cdot \text{K}$$

$$\Delta V = (0.05)^3 = 0.000125 \text{ m}^3$$
$$A_{\text{surface}} = (6)(0.05)^2 = 0.015 \text{ m}^2$$

For preliminary calculations we may estimate the radiation heat transfer using the concept of the radiation heat-transfer coefficient evaluated at the average temperature of the solid, 412 K. Consulting Figure 12-21 we obtain

$$h_{\text{rad}} = 11 \text{ W/m}^2 \cdot \text{K}$$

We thus assume that the solid behaves as if it is cooling with an average total heat-transfer coefficient of

$$h_{\text{total}} = 8.2 + 11 = 19.2 \text{ W/m}^2 \cdot \text{K}$$

We use this total coefficient to check our anticipation that the solid will cool as a lumped capacity. The criterion of Equation (4-6) is used.

$$\frac{h(V/A)}{k} = \frac{(19.2)(0.000125/0.015)}{206} = 0.00078 < (?) \ 0.1$$

Thus, the criterion is satisfied and we may either use Equation (4-5) or Figure 4-5 to estimate the cooling time. We have

$$\frac{\theta}{\theta_i} = \frac{(323 - 300)}{(500 - 300)} = 0.115$$

Consulting Figure 4-5, we find

$$\text{BiFo} = 2.2 = \frac{(19.2)(0.015/0.000125)\tau}{(2700)(896)}$$

And $\tau = 2310$ s.

In reality, neither the convection coefficient nor the radiation coefficient is constant because they both depend on surface temperature. The convection coefficient has a weak dependence ($\Delta T^{0.25}$), while the radiation-coefficient dependence is much stronger (ΔT^3). To refine the analysis further we would have to take these temperature variations into account. Such an analysis is best performed on a numerical basis using a computer. Example D-6 of the appendix presents such an analysis. When applied to the specific conditions of this example the result is $\tau = 2760$ s.

| **EXAMPLE 12-2** | Cooling of a Finned Block |

We now consider a modification of Example 12-1 to a finned block like that shown in Example 8-9. We will assume that the block has 10 fins instead of the three indicated in the figure for Example 8-9, and that the depth of the fins in the z-direction perpendicular to the page is 50 mm (see Figure Example 12-2). The thickness of the base is 7 mm and the other dimensions shown are given in millimeters. The outside surface of the base is insulated and the finned block is oriented such that the free-convection currents move vertically in the z-direction. As in Example 12-1, the solid material is aluminum and the surface is coated with a flat black paint. As before, the solid is heated to a uniform temperature of 500 K and suddenly exposed to a convection radiation environment at 300 K. We again ask how long it takes for the solid temperature to reach a value of 50°C.

Figure Example 12-2

■ **Solution**

The solution is similar to that obtained in Example 12-1 in that we might still anticipate lumped capacity behavior because of the relatively high value of thermal conductivity for aluminum and relatively low convection coefficients involved with free convection. The free-convection flow system is much more complicated, however, and the effective radiation surface area will no longer be the same as the convection area. The convection area is the total surface area in contact with the fluid (air), which may be calculated as

$$A_{\text{convection}} = (10)(2)(50)(25) + (10)(5)(50) + (9)(10)(50) + (10)(2)(25)(5) + (7)(2)(140)$$
$$+(7)(2)(50) = 35160 \, \text{mm}^2 = 0.03516 \, \text{m}^2$$

On the other hand, the block radiates like a five-sided solid (the base is insulated) having dimensions of $140 \times 32 \times 50$ mm, or an area of

$$A_{\text{radiation}} = (2)[(32)(50) + (32)(140)] + (140)(50) = 19160 \, \text{mm}^2 = 0.01916 \, \text{m}^2$$

or a substantially smaller value than the convection surface area. The solid will therefore cool with an overall or total coefficient of

$$(UA)_{\text{total}} = (hA)_{\text{convection}} + (hA)_{\text{radiation}} \qquad\qquad [a]$$

The value of $h_{\text{radiation}}$ is estimated as in Example 12-1, but obtaining an estimate for $h_{\text{convection}}$ is more problematical.

First, let us examine the free-convection boundary-layer behavior in the vertical channels formed by the fins. Consider a boundary layer that might form on a vertical flat plate having $L = 50$ mm $= 0.05$ m. According to Figure 12-11, $\delta = 11$ mm for a value of $\Delta T = 20°C$ and somewhat larger value for a larger ΔT. Thus, the free-convection boundary layers on each side of the vertical fins will surely interact with one another, and the resultant heat transfer may not be calculated with a simple vertical-flat-plate formula. The two exterior surfaces of the fins, however, would be expected to convect like vertical flat plates. For now, *let us simply recognize the problem and calculate a value for $h_{\text{convection}}$ as if we had a vertical flat plate with L = 50 mm and $\Delta T = 112$ K.* Using the simplified formula of Table 7-2 as before gives

$$h_{\text{convection}} = (1.42)\left(\frac{112}{0.05}\right)^{0.25} = 9.8 \text{ W/m}^2 \cdot \text{K}$$

or a value only 19 percent higher than for the solid of Example 12-1. If the criterion for a lumped-capacity cooling is checked, we would find that it is indeed satisfied. Using the average temperature of 412 K gives the same value of the radiation heat-transfer coefficient as in Example 12-1 or $h_{\text{radiation}} = 11$ W/m$^2 \cdot$ K. The volume of the solid is

$$V = (10)(50)(25) + (140)(7)(50) = 61,500 \text{ mm}^3 = 0.0000615 \text{ m}^3$$

Using the temperatures as before gives $\frac{\theta}{\theta_i} = 0.115$ and BiFo $= 2.2$. Computing the value of $(UA)_{\text{total}}$ from the above in Equation (a),

$$(UA)_{\text{total}} = (9.8)(0.03516) + (11)(0.01916) = 0.5553 \text{ W/K}$$

Using this value for the hA product in the BiFo group,

$$2.2 = \frac{(0.5553/0.0000615)\tau}{(2700)(896)}$$

and $\tau = 589$ s.

When a numerical analysis is performed taking the variation of the convection and radiation coefficients into account, as was done in Example 12-1, the result is $\tau = 660$ s.

■ Comments on Examples 12-1 and 12-2

1. For both examples, the procedure of using heat-transfer coefficients evaluated at some average temperature produces results that agree very closely with a much more detailed procedure, which takes into account variations with temperature throughout the cooling process. While the computer-based routine is easy to execute, if one had to create it and debug it to perform the calculations, considerably more time would be involved. We may therefore be encouraged to apply such simplified averaging techniques when examining complicated processes.

2. The free-convection flow regimes are complicated, as are the radiation processes, but by coupling them with a total heat-transfer coefficient concept, a very quick solution is obtained. Again, we note that it matches very well with a more elaborate computer simulation.

3. Finally, we note once again that a lumped-capacity conduction model may be used to advantage when the major resistance to heat transfer occurs at the boundary, and the internal thermal resistance of the conductor is small—or the solid remains at an essentially uniform temperature throughout, for each time increment.

4. The charts for transient response of a lumped-capacity, and boundary-layer thickness for free convection were used to advantage in the solution procedure.

EXAMPLE 12-3

Temperature for Property Evaluation for Convection with Ideal Gases

The reader will note that part of the protocol for calculation of convection heat-transfer coefficients is a selection of the proper temperature for evaluation of fluid properties. The film temperature is usually selected for flow over external surfaces, while the average bulk temperature is usually elected for flow in tubes and the like. The stronger the dependence of fluid properties on temperature, the more important is the selection of the proper value for calculation of the properties. It turns out that when convection with ideal gases is involved, the selection of temperature for property evaluation is not as critical as it is for liquids.

Figure Example 12-3

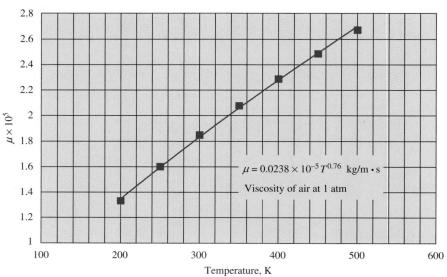

$\mu = 0.0238 \times 10^{-5} T^{0.76}$ kg/m·s

Viscosity of air at 1 atm

(a)

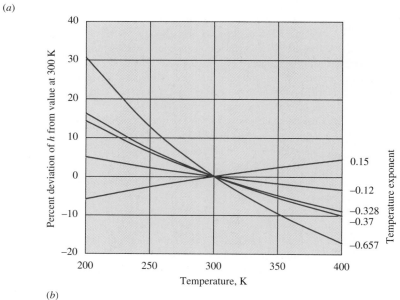

(b)

■ **Solution**

First, recall that an ideal gas is one that obeys the equation of state $p = \rho R T$. Over a pressure range of a few atmospheres a gas like air will have properties like thermal conductivity, k, dynamic viscosity, μ, and specific heat c_p, essentially independent of pressure. Furthermore, we find that the Prandtl number $\text{Pr} = c_p \mu / k$ is very nearly constant with temperature and equal to about 0.7. This implies that μ varies directly as k.

If we plot μ as a function of absolute temperature for air, as obtained from Table A-5, we obtain the curve shown in Figure Example 12-3a. Performing a curve fit to the data indicates that μ varies with the absolute temperature to the 0.76 power. This means that k also varies approximately with T to the 0.76 power. For a constant pressure with an ideal gas we also have

$$\rho \approx \frac{1}{T}$$

$$\beta = \frac{1}{T} \text{ (exactly, for ideal gas only)}$$

$$\mu \approx T^{0.76}$$

$$k \approx T^{0.76}$$

Examining the different correlations for convection we find they take the following forms:

Laminar free convection:

$$h \approx k \times f(\text{body dimensions}, \Delta T) \times \left[\left(\frac{\rho^2 \beta}{\mu^2} \right) \text{Pr} \right]^{1/4}$$

Laminar forced convection over a flat plate:

$$h \approx k \times f(\text{body dimensions, free-stream velocity}) \times \left(\frac{\rho}{\mu} \right)^{1/2} \text{Pr}^{1/3}$$

Turbulent forced convection in smooth tubes:

$$h \approx k \times f(\text{mass flow, tube diameter}) \times \left(\frac{1}{\mu} \right)^{0.8} \text{Pr}^{0.4}$$

Forced convection across tubes $(4{,}000 < Re < 40{,}000)$:

$$h \approx k \times f(\text{body dimension, free-stream velocity}) \times \left(\frac{\rho}{\mu} \right)^{0.618} \text{Pr}^{1/3}$$

Forced convection across tubes $(40{,}000 < Re < 400{,}000)$:

$$h \approx k \times f(\text{body dimensions, free-stream velocity}) \times \left(\frac{\rho}{\mu} \right)^{0.805} \text{Pr}^{1/3}$$

Inserting the temperature dependence of ρ, β, μ, and k in these relations, along with an assumed constant value of Pr gives

$$h \approx T^{-0.37} \quad \text{laminar free convection}$$
$$\approx T^{-0.12} \quad \text{laminar forced convection over a flat plate}$$
$$\approx T^{0.15} \quad \text{turbulent forced convection in a smooth tube}$$
$$\approx T^{-0.328} \quad \text{forced convection over tubes } (4{,}000 < \text{Re} < 40{,}000)$$
$$\approx T^{-0.657} \quad \text{forced convection across tubes } (40{,}000 < \text{Re} < 400{,}000)$$

Plots of the temperature dependence of the convection coefficient h for these different flow circumstances over the temperature range of 200 to 400 K are shown in Figure Example 12-3b. While the variation is large at the outer temperature ranges for some cases, it is surprisingly small for others. For modest variations of ± 25 K from the mean of 300 K, the variation for all cases is small. *The net result of these calculations is that in preliminary design studies of convection heat transfer employing ideal gases, one need not be overly careful about the temperature used for property determination.* For operation near room temperature a value of 300 K will probably be satisfactory. ***Please note that the foregoing analysis and conclusions do not apply to convection problems involving liquids.***

| **EXAMPLE 12-4** | Design Analysis of an Insulating Window |

An insulating window is to be manufactured of two sheets of window glass, each 3 mm thick separated by an air space at approximately atmospheric pressure. For design purposes, the outside of the window is assumed to be exposed to air at −20°C (253 K) and the inside exposed to air at 25°C (298 K) and the convection coefficients for both inside and outside of the window are assumed to be about 5 W/m$^2 \cdot$°C. The effective radiation temperature of the outside surroundings is assumed to be −20°C. Determine a suitable thickness for the air gap between the two glass window plates, and estimate the heat loss in W/m^2.

■ **Solution**

For this problem, we have five thermal resistances in series; the inside and outside convection resistances, the resistance of the air gap, and the two conduction resistances of the glass plates. We will assume that there is also a radiation resistance on the outside of the glass operating in parallel with the convection resistance. The R-values for the elements are (expressed on a unit area basis)

$$R_{glass} = \frac{\text{thickness}}{k}$$

$$R_{conv} = \frac{1}{h_{conv}}$$

$$R_{rad} = \frac{1}{h_{rad}}$$

$$R_{air\ gap} = \frac{\text{thickness}}{k_{eff}}$$

Consulting Figure 12-16, we find that there is little to be gained by increasing the air-gap thickness beyond about $\delta = 2$ cm, producing an R-factor of about 0.4. We therefore set the air-gap thickness of 2 cm as our design value and work from there. From Table A-3, $k_{glass} = 0.78$, so we have

$$R_{glass} = \frac{0.003}{0.78} = 0.0038$$

$$R_{conv} = \frac{1}{5} = 0.2$$

$$R_{air\ gap} = 0.4$$

and R_{rad} is to be determined.

Neglecting the radiation for now, the total R-value is

$$R_{total} = (2)(0.0038) + (2)(0.2) + 0.4 = 0.808$$

The overall temperature difference is $25 - (-20) = 45$°C, so the outside glass temperature is about

$$T = 25 - \left(\frac{0.608}{0.808}\right)(45) = -8.9°C = 261\ K$$

From Table A-10, $\epsilon_{glass} \approx 0.9$ so the radiant loss of a 261 K surface exposed to 253 K surroundings would be about

$$\frac{q}{A} = \sigma\epsilon(261^4 - 253^4) = 27.7\ W/m^2$$

producing a radiation coefficient of about $h_{rad} = 27.7/(261 - 253) = 3.46$. When this coefficient is allowed to operate in parallel with the convection coefficient of 5, the equivalent resistance for the outside surface is about

$$R_{outside} = \frac{1}{5 + 3.46} = 0.118$$

producing a total resistance of about 0.726 for the entire assembly.

The overall heat loss from the insulated window would then be

$$\frac{q}{A} = \frac{25 - (-20)}{0.726} = 61.98 \text{ W/m}^2$$

Assuming the interior radiation temperature is the same as the interior convection temperature, 25°C, a similar approximation for the radiation gain from the inside may be performed. Taking into account just convection, the inside glass temperature would be about

$$T = 25 - (0.2)(0.726)(45) = 12.6°C = 286 \text{ K}$$

The radiation gain from the inside would therefore be about

$$\frac{q}{A} = \sigma\epsilon(298^4 - 286^4) = 61 \text{ W/m}^2$$

producing a radiation coefficient h_{rad} of about $61/(298 - 286) = 5$ W/m² · °C. Combining this value with the interior convection coefficient produces an overall coefficient of about 10 W/m² · °C and an overall surface resistance of 0.1. The overall heat transfer would then be

$$\frac{q}{A} = \frac{25 - (-20)}{0.626} = 71.89 \text{ W/m}^2$$

If no account were taken of radiation on either the inside or outside surface, the overall heat transfer would be

$$\frac{q}{A} = \frac{25 - (-20)}{0.808} = 55.7 \text{ W/m}^2$$

A more precise analysis would take into account a better determination of the temperature difference across the air gap and its effect on k_{eff}. Keep in mind, however, that selection must still be made for the air-gap thickness, and the value chosen ($\delta = 2$ cm), which approximately maximizes the R-factor for the gap is probably quite satisfactory for design purposes. A more precise determination of the free-convection coefficients is probably not worth the effort, because in a practical application, even small air currents can influence the value of h_{conv} considerably.

A quick approach to estimating the effect of radiation could be taken by consulting Figure 12-21 to obtain an estimate for h_{rad}. For the temperature encountered in this example, we find $h_{rad} \approx 5$ to 6. Using the lower limit of 5 gives a total resistance of 0.608, while using $h_{rad} = 6$ gives $R_{total} = 0.586$. The corresponding heat fluxes are 64 and 76.8 W/m², or values within 10 percent of the value already calculated. These values fall easily within the range in uncertainties of the other calculations. It is left as an exercise for the reader to consider the effect of radiation between the two glass plates.

Double-Pipe Heat Exchanger   EXAMPLE 12-5

To illustrate the importance of selection of the correct area(s) to use in design calculations, consider a small double-pipe heat exchanger like that shown in the photographs of Figure 10-3e and f. The cross section through the double-pipe arrangement is shown in Figure Example 12-5, with the following dimensions given:

$$d_{1i} = 3.0 \, \text{mm}$$
$$d_{1o} = 3.2 \, \text{mm}$$
$$d_2 = 6.0 \, \text{mm}$$

Length of tube section perpendicular to page $= L = 1.5$ m

The hot fluid is water that flows inside the inner tube. The cold fluid is ethylene glycol, which flows in the annular region between the two tubes. The mean flow velocity for the water is 2.3 m/s and the mean velocity for the glycol is 2.5 m/s. The mean temperature for the water is 38°C, while the mean temperature of the glycol is 20°C. Determine the overall heat-transfer coefficient of the exchanger, assuming the tubes are constructed of stainless steel with $k = 16$ W/m · °C.

Figure Example 12-5

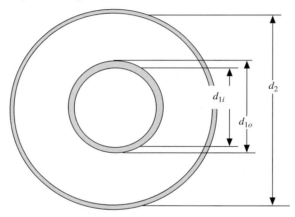

■ Solution

There are several areas that need to be calculated:

$$\text{Flow area for water} = \frac{\pi d_{1i}^2}{4} = \frac{\pi (0.003)^2}{4} = 7.07 \times 10^{-6} \, \text{m}^2$$

$$\text{Flow area for glycol} = \pi \frac{\left(d_2^2 - d_{1o}^2\right)}{4} = 2.023 \times 10^{-5} \, \text{m}^2$$

$$\text{Heat-transfer surface area for water} = \pi d_{1i} L = 0.01414 \, \text{m}^2$$

$$\text{Heat-transfer surface area for glycol} = \pi d_{1o} L = 0.01508 \, \text{m}^2$$

Note that only the outside surface of the inner tube transfers heat to the glycol. The overall annular dimensions, however, influence the hydraulic diameter, which must be used to calculate the Reynolds number, and subsequently the heat-transfer coefficient for the glycol.

The hydraulic diameter for the glycol flowing in the annular section is calculated using Equation (6-14):

$$D_H = \frac{4A_c}{P} = d_2 - d_{1o} = 0.0028 \, \text{m}$$

The fluid properties are obtained from Appendix A, giving
Water at 38°C:

$$c_p = 4174 \, \text{J/kg} \cdot °\text{C}$$
$$\rho = 993 \, \text{kg/m}^3$$

$$\mu = 6.82 \times 10^{-4} \, \text{kg/m} \cdot \text{s}$$
$$k = 0.63 \, \text{W/m} \cdot {}^{\circ}\text{C}$$
$$\text{Pr} = 4.53$$

Glycol at 20°C:

$$c_p = 2382 \, \text{J/kg} \cdot {}^{\circ}\text{C}$$
$$\rho = 1117 \, \text{kg/m}^3$$
$$\mu = 0.0214 \, \text{kg/m} \cdot \text{s}$$
$$k = 0.249 \, \text{W/m} \cdot {}^{\circ}\text{C}$$
$$\text{Pr} = 204$$

The Reynolds numbers are calculated as

$$\text{Water Re} = \frac{\rho d_{1i} u}{\mu} = \frac{(993)(0.003)(2.3)}{6.82 \times 10^{-4}} = 10,046$$

$$\text{Glycol Re} = \frac{\rho D_H u}{\mu} = \frac{(1117)(0.0028)(2.5)}{0.0214} = 365$$

We thus have turbulent flow for the water flowing in the inner tube and laminar flow for the glycol flowing in the outer annulus. In the absence of more specific information regarding the inlet temperatures, we may use Figures 12-4 and 12-5 to estimate the convection heat-transfer coefficients. From Figure 12-4 at Re = 10,000, we have

$$\frac{\text{Nu}}{\text{Pr}^{0.4}} = 36$$

so that

$$h_i = \frac{(36)(4.53)^{0.4}(0.63)}{0.003} = 13,800 \, \text{W/m}^2 \cdot {}^{\circ}\text{C}$$

For laminar flow of the glycol, we need to calculate the Graetz number:

$$\text{Gz} = \text{RePr}\left(\frac{d}{L}\right) = (365)(204)\left(\frac{0.0028}{1.5}\right) = 139$$

and from Figure 12-5 we read the value of the Nusselt number as

$$\text{Nu} = 9.5$$

so that the value of the convection coefficient is calculated as

$$h_o = \frac{(9.5)(0.249)}{0.0028} = 845 \, \text{W/m}^2 \cdot {}^{\circ}\text{C}$$

The respective thermal resistances are now determined as

$$\text{Water side resistance} = \frac{1}{h_i A_i} = \frac{1}{(13,800)(0.01414)} = 0.0051$$

$$\text{Glycol side resistance} = \frac{1}{h_o A_o} = \frac{1}{(845)(0.01508)} = 0.078$$

$$\text{Conduction resistance} = \frac{\ln(d_{1o}/d_{1i})}{2\pi k L} = \frac{\ln(3.2/3.0)}{2\pi(16)(1.5)} = 0.00043$$

The product of overall heat transfer coefficient and area is then

$$UA = \frac{1}{(0.0051 + 0.078 + 0.00043)} = 11.9 \, \text{W/}{}^{\circ}\text{C}$$

If the value of U is based on the inner (water side) tube area,

$$U_i = 845 \, \text{W/m}^2 \cdot {}^\circ\text{C}$$

and if it is based on the outer (glycol side) tube area

$$U_o = 767 \, \text{W/m}^2 \cdot {}^\circ\text{C}$$

The glycol side value of h is clearly the controlling coefficient, and the conduction resistance is negligible within the accuracy of the estimates obtained from the respective figures.

EXAMPLE 12-6	Refrigerator Storage in Desert Climate

A manufacturer of refrigerated transport containers develops a new product for moving frozen foods by rail. It is a refrigerated container about 10 m long and 3 m high and narrow enough to fit on a flatbed railcar. The interior is designed to maintain -10°C conditions when the exterior conditions are quiescent air at 40°C. A reputable testing lab certifies that the container does indeed meet the design specifications. The walls of the container have urethane foam insulation ($k \approx 0.02 \, \text{W/m} \cdot \text{K}$) with a thickness of about 10 cm. Before placing an order for several of the containers, a railroad executive poses the following question to the manufacturer's sales representative: "Your testing lab has certified performance under quiescent air conditions, but what about the performance when the train is moving at a speed of 60 mi/h (27 m/s)? Will it be able to handle the increased heat load of the high-speed hot air?"

■ **Solution**

Most manufacturer's sales representatives would be unable to answer the executive's question on the spot, and would immediately seek help from the appropriate engineering group at the home office. While a heat-transfer "expert" at the home office could probably give a quick answer to the question, it would be prudent to back up the answer with calculations, because a large order is at stake.

There are three main thermal resistances for the container wall: (1) internal convection resistance, (2) conduction resistance through the urethane foam, and (3) convection resistance on the outside of the container. (We ignore any resistance of the metal sheath on the interior and exterior because they are negligible compared to the foam insulation.) The thermal resistance of the foam, on a unit-area basis is

$$R_{\text{foam}} = \frac{\Delta x}{k} = \frac{0.1}{0.02} = 5$$

Whatever the convection coefficient may be on the inside of the container, we assume it does not change when the railcar is in motion or sitting still. When in motion at 60 mi/h, the flow over the outside of the container approximates that of flow over a 10-m-long flat plate. When stopped, the flow approximates that of free-convection on a 3-m-high vertical flat plate. Thus, the crux of the problem is a comparison of the values of the convection coefficients for these two conditions. First, consider the quiescent, free-convection situation. For a 3-m-high plate the value of Gr_x will most certainly be $> 10^9$ (see Figure 12-11), and we may therefore use a simplified relation of Table 7-2 to estimate the value of h. Assuming a value of $\Delta T \approx 10^\circ\text{C}$, we obtain

$$h = 1.31(\Delta T)^{1/3} = (1.31)(10)^{1/3} = 2.8 \, \text{W/m}^2 \cdot {}^\circ\text{C}$$

We next examine the forced-convection situation, with properties of air evaluated at 300 K (see the discussion of Example 12-3 for justification).

$$k = 0.026 \, \text{W/m} \cdot \text{K}$$
$$v = 15.69 \times 10^{-6} \, \text{m}^2/\text{s}$$
$$\text{Pr} = 0.697$$

The Reynolds number for flow over a 10-m-long flat plate at a speed of 27 m/s is

$$\text{Re}_L = \frac{u_\infty L}{v} = \frac{(27)(10)}{15.69 \times 10^{-6}} = 1.7 \times 10^7$$

Consulting Figure 12-3, we obtain

$$\frac{\text{Nu}_L}{\text{Pr}^{1/3}} \approx 20{,}000$$

and the average heat-transfer coefficient is calculated as

$$h = \frac{(0.697)^{1/3}(20{,}000)(0.026)}{10} = 46.1 \, \text{W/m}^2 \cdot \text{K}$$

The respective thermal resistances, on a unit-area basis, are

$$R_{\text{free conv}} = \frac{1}{2.8} = 0.36$$

$$R_{\text{forced conv}} = \frac{1}{46.1} = 0.022$$

Totally neglecting the internal convection resistance gives the total resistances as

$$R_{\text{total, free conv}} = 5.36$$

$$R_{\text{Total, forced conv}} = 5.02$$

The two values differ by only 6 percent. Thus, it makes very little difference whether the railcar is moving or sitting still. The major impediment to heat transfer is the foam insulation, and its high thermal resistance overpowers the convection resistances. The heat-transfer "expert" would have known this from the start. The reason insulation is installed in the first place is to make the interior of the container impervious to the surrounding hot environment.

Cold Draft in a Warm Room   EXAMPLE 12-7

A woman takes her grandchildren to visit a family dwelling located in rural Vermont. The heating is accomplished with an old potbelly woodstove that delivers heat to the surrounding room mainly by radiation. After the stove operates for a while, the room becomes comfortably warm. The woman's favorite chair is located adjacent to an outside window that is about 2 m wide and 1m high. Despite the satisfaction of other persons with the room temperature achieved by the woodstove, she complains of a cold draft that she attributes to air leakage through the old window. On their most recent visit to the cottage, her grandson applied new caulking to seal the old window and eliminate the leakage. That night a cold front moved through the area, and the woman again complained of a "cold draft" when sitting near the window. Her grandson had just completed a college course in *heat transfer* in his engineering studies, and proceeded to devise the following explanation for the cold drafts.

■ Solution

When it is cold outside a free-convection boundary layer forms on the inside of the cold window and air moves *downward* along the window surface. We may estimate the mass flow of the cool air with Equation (7-20c)

$$\dot{m} = \left(\frac{9}{16}\right) \rho u_{\text{max}} \delta \qquad [a]$$

for a unit width of the window. For a window height of 1 m (1000 mm) we consult Figure 12-11a and find that the Grashof number Gr_x will surely exceed the laminar value of 10^9 for $\Delta T = 20°\text{C}$.

We may take the upper limit of Figure 12-11a as a conservative estimate for the boundary-layer thickness or, approximately,

$$\delta \approx 30 \text{ mm}$$

In a similar fashion we may consult Figure 12-13a and obtain an estimate for the maximum flow velocity moving downward on the window as

$$u_{max} \approx 400 \text{ mm/s}$$

Inserting these values in Equation (a), along with the window width of 2 m, gives an estimate for the mass flow of cold air down the window as

$$\dot{m} = \left(\frac{9}{16}\right)(1.2)(0.4)(0.03)(2) = 0.016 \text{ kg/s} \qquad [b]$$

where we have taken the air density as $\rho \approx 1.2$ kg/m^3. This mass flow may not appear very large, but if it is converted to a volume-flow unit employed in typical heating, ventilating, and air-conditioning applications it is equivalent to about 29 cubic feet per minute (CFM), and could easily cause the woman to experience some chill from the draft.

EXAMPLE 12-8   Design of an Evacuated Insulation

An examination of the plots of Figure 12-18 reveals that an air space will conduct less energy when evacuated. This was illustrated also in the calculations of Example 7-12, where it was also shown that radiation can make up a substantial portion of the overall heat transfer. Suppose a special insulating material is to be constructed of a series of evacuated sections like those presented in Figure 12-18, with the inner surfaces of the sections having a special low-emissivity coating to reduce the radiation heat transfer. Because of equipment limitations the lowest pressure that can be economically maintained is 0.00001 atm. We wish to examine the effects of surface emissivity and section thickness on the effective thermal conductivity of a multiple-layer insulating material, neglecting for now the mechanical problems of maintaining the structural integrity of the sandwich construction.

■ **Solution/Analysis**
A review of Example 7-12 and the text material pertaining to the example reveals that the low-density conduction through an evacuated layer is somewhat temperature dependent because the mean free path of the gas molecule is temperature dependent. For this preliminary design analysis we will neglect this temperature dependence and assume that a given thickness section will have a constant thermal resistance independent of the thermal gradient imposed upon it. For unit area, this resistance is the R-value, $\Delta T/(q/A)$. For three thicknesses of Figure 12-18 at a pressure of 0.00001 atm these resistances have the following numerical values:

Δx, m	R-value,$^\circ$ C · m^2/W	$(q/A)_{conduction}$, W/m^2
0.005	1.23	16
0.01	1.42	14
0.02	1.82	11

If we sandwich together several sections of air gaps, the overall thermal-conduction resistance will just be the preceding numbers multiplied by the number of sandwiched sections.

The radiation heat transfer between two large planes having equal emissivities is given in Figure 8-30 as

$$\left(\frac{q}{A}\right)_{radiation} = \frac{E_{b1} - E_{b2}}{(2/\epsilon - 1)} \qquad [a]$$

Figure Example 12-8

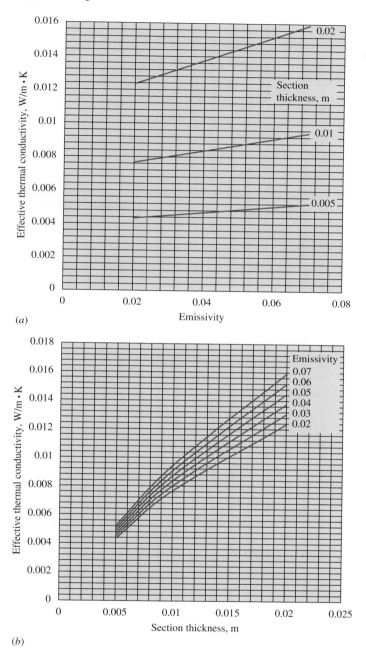

(a)

(b)

For n sections sandwiched together, the denominator in Equation (a) becomes

$$n(2/\epsilon - 1)$$

with the result that the total radiation transfer is also a linear function of the number of sandwich sections. Thus, the total heat transfer is just inversely proportional to the number of sandwich sections. This means, in turn, that an effective thermal conductivity for the assembly will be independent of the number of sandwich sections and can be calculated from

$$k_{\text{effective}} = \left(\frac{q}{A}\right)_{\text{total}} \frac{\Delta x}{\Delta T} \qquad [b]$$

where

$$\left(\frac{q}{A}\right)_{\text{total}} = \left(\frac{q}{A}\right)_{\text{conduction}} + \left(\frac{q}{A}\right)_{\text{radiation}}$$

and $\left(\frac{q}{A}\right)_{\text{radiation}}$ is calculated from Equation (a).

Effective thermal conductivity for the three section thicknesses just given and emissivity values ranging from 0.02 to 0.07 have been calculated and are presented in Figure Example 12-8. For comparison, the thermal conductivity for air at atmospheric pressure is 0.0262 W/m · K, a much larger value. We therefore conclude that evacuated multilayer insulations have the potential to be quite effective. If operation is possible at pressures below the stipulated minimum value of 0.00001 atm, there will be, of course, a further reduction in the effective thermal conductivity.

EXAMPLE 12-9   Radiant Heater

A radiation heat source is available at 800 K and is exposed to a thin-wall tubing carrying air at 300 K. The tubing has a diameter of 25 mm and half is embedded in an insulating material, as shown in Figure Example 12-9a. The emissivity of the tube material is 0.7 and the wall of the tube is sufficiently thin that conduction resistance can be neglected. Determine the total heat transfer to the tube per meter of length as a function of the convection heat-transfer coefficient between the air and the inside of the tube.

■ **Solution**

This problem is a parametric study that requires the calculation of the convection heat transfer for several values of the convection coefficient h. Using the nomenclature of Figure Example 12-9a we have the given numerical information

$$\epsilon = 0.7$$
$$d = 25\,\text{mm} = 0.025\,\text{m}$$
$$A_1 = A_2 = \frac{\pi d}{2} = 0.03927\,\text{m}^2/\text{m length}$$
$$T_s = 800\,\text{K}$$
$$T_a = 300\,\text{K}$$
$$A_d = d = 0.025$$

The energy balance on the exterior tube surface is

$$q_{\text{rad},s-2} = q_{\text{rad},2-1} + q_{\text{conv},2} \qquad [a]$$

and the energy balance on surface 1 (assuming no conduction heat transfer into the insulation) is

$$q_{\text{rad},2-1} = q_{\text{conv},1} \qquad [b]$$

The relations for the different terms in Equations (a) and (b) are

$$q_{\text{rad},s-2} = \sigma \epsilon A_2 \left(T_s^4 - T_2^4\right)$$

Assuming the radiation source is a large enclosure,

$$\Sigma R_{\text{rad}} = \frac{(2)(1-\epsilon)}{\epsilon A_2} + \frac{1}{A_d F_{d2}}; \qquad F_{d2} = 1.0$$
$$q_{\text{conv},2} = h A_2 (T_2 - T_a)$$

Figure Example 12-9

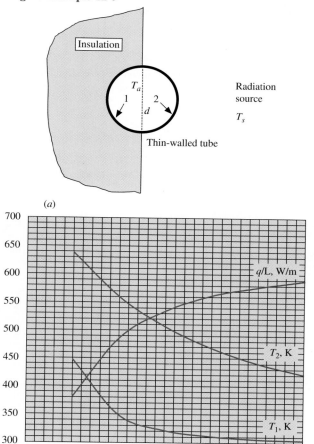

(a)

(b)

$$q_{rad,2-1} = \frac{\sigma\left(T_2^4 - T_1^4\right)}{\sum R_{rad}}$$

$$q_{conv,1} = hA_1\left(T_1 - T_a\right)$$

Inserting the appropriate numerical values in the preceding relations gives

$$(0.7)\sigma(0.03927)\left(800^4 - T_2^4\right) = 0.03h\left(T_2 - 300\right) + \frac{\sigma\left(T_2^4 - T_1^4\right)}{61.827} \qquad [c]$$

$$\frac{\sigma\left(T_2^4 - T_1^4\right)}{61.827} = 0.03927h\left(T_1 - 300\right) \qquad [d]$$

For given values of h, the convection coefficient, Equations (c) and (d) may be solved by iteration for T_1 and T_2, and then subsequently the total convection to the air may be calculated from

$$q_{conv,\ total} = h(0.03927)\left(T_2 + T_1 - 2T_a\right) \qquad [e]$$

These calculations have been made for values of h from 20 to 120 W/m$^2 \cdot$ °F and are plotted in Figure Example 12-9b.

EXAMPLE 12-10 Coolant for a Radiant Heater

Air flows in the tube of Example 12-9 at a mean velocity of 5.5 m/s. Estimate the heat transfer to the air per unit length of tube.

■ **Solution**

We have already solved for the heat flux per unit length of tube in Example 12-9 as a function of the convection heat transfer coefficient h. Evaluating fluid properties at 300 K, we have

$$\rho = 1.178 \, \text{kg/m}^3$$
$$k = 0.02624 \, \text{W/m} \cdot {}^{\circ}\text{C}$$
$$v = 15.69 \times 10^{-6} \, \text{m}^2/\text{s}$$
$$\text{Pr} = 0.71$$

The Reynolds number is therefore

$$\text{Re}_d = \frac{(1.178)(5.5)(0.025)}{15.69 \times 10^{-6}} = 10{,}323$$

which is greater than 2300 so turbulent flow is encountered. Using the Dittus-Boelter equation [Equation (6-4)] or Figure 12-4 gives

$$\frac{Nu}{\text{Pr}^{0.4}} = 37.4$$

and

$$h = \frac{(37.4)(0.02624)}{0.026}(0.71)^{0.4} = 32.9$$

Consulting Figure Example 12-9b we obtain

$$\frac{q}{L} \approx 470 \, \text{W/m}$$

EXAMPLE 12-11 Radiant Electric Stove for Boiling Water

An electric stove uses a radiant heater element to deliver the heating energy to the bottom of a stainless steel pan having a thickness of 6 mm. Inside the pan, water is to be boiled at 1 atm. Determine the radiant heat flux required as a function of the temperature of the radiant element. Assume emissivity ϵ for the element is 0.9 and that for the stainless steel is 0.6 $k = 16 \, \text{W/m} \cdot {}^{\circ}\text{C}$ for the stainless steel.

■ **Solution**

Assuming the radiant element is close enough to the bottom of the pan that the behavior approximates that of parallel infinite planes the radiant heat flux is given by the relation taken from Figure 8-30.

$$\frac{q}{A} = \frac{\sigma\left(T_s^4 - T_{bp}^4\right)}{1/0.9 + 1/0.6 - 1} \qquad [a]$$

This same heat flux must be conducted through the pan bottom so that

$$\frac{q}{A} = \frac{k\left(T_{bp} - T_{ip}\right)}{\text{thickness}} \qquad [b]$$

where T_{bp} and T_{ip} are the temperatures of the bottom and inside of the pan, respectively.

The heat flux is given by the boiling curves of Figure 12-25a or the functional relations of Table 9-3, on which the figure is based.

Figure Example 12-11

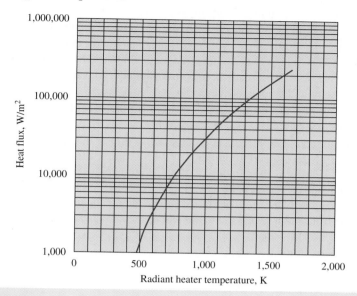

$$\frac{q}{A} = f(T_{bi} - T_{sat})\qquad\qquad\textbf{[c]}$$

Where T_{sat} for water at 1 atm is 100°C = 373 K. Thus, for a particular value of q/A we may solve sequentially for T_{ip} from Equation (c), T_{bp} from Equation (b), and T_s from Equation (a). A chart may then be constructed for q/A as a function of T_s as shown in Figure Example 12-11. Note that only modest heater temperatures are required to maintain a "simmer" condition, while very high temperatures are necessary to establish full nucleate bulk boiling conditions.

LIST OF WORKED EXAMPLES

12-1 Cooling of an aluminum cube

12-2 Cooling of a finned block

12-3 Temperature for property evaluation for convection with ideal gases

12-4 Design analysis of an insulating window

12-5 Double-pipe heat exchanger

12-6 Refrigerator storage in desert climate

12-7 Cold draft in a warm room

12-8 Design of an evacuated insulation

12-9 Radiant heater

12-10 Coolant for radiant heater

12-11 Radiant electric stove for boiling water

PROBLEMS

12-1 Based on information contained in charts in this chapter, specify an approximate height for a vertical flat plate and diameter of a horizontal cylinder for which the simplified equations of Table 7-2 would apply.

12-2 Choose several charts in this chapter and satisfy yourself that they are accurate using whatever information may be available from previous chapters.

12-3 Rework Examples 12-9 and 12-10 for a tube having a diameter of 15 mm.

12-4 Using charts in this chapter, estimate the heat lost by free convection from a vertical flat plate 50 mm high and 25 mm wide maintained at 310 K and exposed to liquid water at 290 K. Also estimate the maximum boundary-layer thickness and maximum flow velocity in the boundary layer.

12-5 Suppose air at 1 atm is mistakenly used as the fluid in Problem 12-4. What percent error would result?

12-6 An airfoil section is placed in a large wind tunnel that produces air speeds of 300 mi/h at free-stream conditions of 290 K and 1 atm pressure. The airfoil is heated to a uniform temperature of 310 K. Estimate the heat lost from both the top and bottom of the airfoil using charts in this chapter. Assume that the airfoil section may be approximated as a flat plate having a length equal to the peripheral distance from the front stagnation point to the rear edge and that this peripheral dimension is measured as 5 m.

12-7 Using charts in this chapter determine the fraction of the solar radiation spectrum that lies between 0.5 and 2.0 μm wavelengths. Assume the overall solar spectrum to be that of a blackbody at 5800 K.

12-8 Rework Example 12-11 for a *vertical* heating surface.

12-9 Compare the convection heat transfer coefficients for air at 300 K flowing in smooth tubes at Reynolds numbers of 100 and 10,000. Choose appropriate values of d and L.

12-10 Common building brick having dimensions of $5 \times 10 \times 20$ cm is heated to a uniform temperature of 400 K and suddenly subjected to air at 1 atm and 300 K. Can the brick be treated as a lumped capacity? If so, determine the time required for the center temperature to reach 350 K.

12-11 Using chart(s) in this chapter determine the heat lost by forced convection from a flat plat maintained at 350 K and subjected to atmospheric air at 300 K and a flow velocity of 50 m/s. The plate length in the direction of flow is 1 m.

12-12 Calculate the heat lost by a circular cylinder maintained at 350 K, having a diameter of 30 cm and subjected to the same flow conditions as in Problem 12-11.

12-13 A thermocouple bead is spherical in shape having a diameter of 2 mm and is subjected to the same flow system as in Problem 12-11. Estimate the heat-transfer coefficient for such a configuration.

12-14 Examine one or more of the formulas in Table 7-2 in the light of the findings of Example 12-3. Discuss.

12-15 A fine wire having a diameter of 0.025 mm is placed in a horizontal position in atmospheric air at 300 K. An electric current is passed through the wire such that the surface temperature reaches a value of 350 K. Estimate the heat lost by the wire per meter of length.

12-16 Suppose the thermocouple bead of Problem 12-13 is maintained at 350 K and exposed to quiescent air at 1 atm and 300 K. Estimate the heat-transfer coefficient that would result in this case.

12-17 A flat plate having a height of 70 mm is maintained at 350 K and exposed to atmospheric air at 300 K. Using charts in this chapter determine the convection heat-transfer coefficient, boundary-layer thickness, and maximum flow velocity in the boundary layer.

12-18 Repeat Problem 12-17 for liquid water at 300 K.

12-19 Two isothermal horizontal plates are maintained at 310 and 290 K, respectively, with the hotter plate below the cooler plate. The spacing between the plates is 1 cm. Estimate the convection heat transfer between the plates per unit area using charts in this chapter, if the fluid is water.

12-20 Repeat Problem 12-19 if the fluid is air at 1 atm.

12-21 A vertical air gap contains air at 300 K and 1 atm. The spacing between the plates is 10 cm and the temperature difference is 20°C. Estimate the percentage reduction in heat transfer that would result from lowering the pressure of the air to 10^{-6} atm. Use charts in this chapter.

12-22 Using charts in this chapter, determine the fraction of blackbody radiation at 300 K that lies below a wavelength of 20 μm.

12-23 A small sphere having diameter of 2 cm and maintained at 475°C is contained in a cube having dimensions of 6 cm on a side and maintained at 300 K. The emissivity of the sphere is 0.5 and the emissivity of the cube is 0.7. Calculate the radiation heat lost by the sphere using charts in this chapter.

12-24 A hot steam pipe at 200°C having a surface emissivity of 0.7 and diameter of 10 cm is placed horizontally in a room of quiescent air at 1 atm and 300 K. Calculate the total heat lost by the pipe per meter of length. State assumptions.

12-25 Apply Figures 12-23 and 12-25 to boiling of water at 1 atm (0.101 MPa). Discuss.

Tables

Table A-1 | The error function.

$\dfrac{x}{2\sqrt{\alpha\tau}}$	$\text{erf}\,\dfrac{x}{2\sqrt{\alpha\tau}}$	$\dfrac{x}{2\sqrt{\alpha\tau}}$	$\text{erf}\,\dfrac{x}{2\sqrt{\alpha\tau}}$	$\dfrac{x}{2\sqrt{\alpha\tau}}$	$\text{erf}\,\dfrac{x}{2\sqrt{\alpha\tau}}$
0.00	0.00000	0.76	0.71754	1.52	0.96841
0.02	0.02256	0.78	0.73001	1.54	0.97059
0.04	0.04511	0.80	0.74210	1.56	0.97263
0.06	0.06762	0.82	0.75381	1.58	0.97455
0.08	0.09008	0.84	0.76514	1.60	0.97636
0.10	0.11246	0.86	0.77610	1.62	0.97804
0.12	0.13476	0.88	0.78669	1.64	0.97962
0.14	0.15695	0.90	0.79691	1.66	0.98110
0.16	0.17901	0.92	0.80677	1.68	0.98249
0.18	0.20094	0.94	0.81627	1.70	0.98379
0.20	0.22270	0.96	0.82542	1.72	0.98500
0.22	0.24430	0.98	0.83423	1.74	0.98613
0.24	0.26570	1.00	0.84270	1.76	0.98719
0.26	0.28690	1.02	0.85084	1.78	0.98817
0.28	0.30788	1.04	0.85865	1.80	0.98909
0.30	0.32863	1.06	0.86614	1.82	0.98994
0.32	0.34913	1.08	0.87333	1.84	0.99074
0.34	0.36936	1.10	0.88020	1.86	0.99147
0.36	0.38933	1.12	0.88079	1.88	0.99216
0.38	0.40901	1.14	0.89308	1.90	0.99279
0.40	0.42839	1.16	0.89910	1.92	0.99338
0.42	0.44749	1.18	0.90484	1.94	0.99392
0.44	0.46622	1.20	0.91031	1.96	0.99443
0.46	0.48466	1.22	0.91553	1.98	0.99489
0.48	0.50275	1.24	0.92050	2.00	0.995322
0.50	0.52050	1.26	0.92524	2.10	0.997020
0.52	0.53790	1.28	0.92973	2.20	0.998137
0.54	0.55494	1.30	0.93401	2.30	0.998857
0.56	0.57162	1.32	0.93806	2.40	0.999311
0.58	0.58792	1.34	0.94191	2.50	0.999593
0.60	0.60386	1.36	0.94556	2.60	0.999764
0.62	0.61941	1.38	0.94902	2.70	0.999866
0.64	0.63459	1.40	0.95228	2.80	0.999925
0.66	0.64938	1.42	0.95538	2.90	0.999959
0.68	0.66278	1.44	0.95830	3.00	0.999978
0.70	0.67780	1.46	0.96105	3.20	0.999994
0.72	0.69143	1.48	0.96365	3.40	0.999998
0.74	0.70468	1.50	0.96610	3.60	1.000000

Table A-2 | Property values for metals.†

Metal	Properties at 20°C ρ kg/m³	c_p kJ/kg·°C	k W/m·°C	$\alpha \times 10^5$ m²/s	Thermal conductivity k, W/m·°C −100°C −148°F	0°C 32°F	100°C 212°F	200°C 392°F	300°C 572°F	400°C 752°F	600°C 1112°F	800°C 1472°F	1000°C 1832°F	1200°C 2192°F
Aluminum:														
Pure	2,707	0.896	204	8.418	215	202	206	215	228	249				
Al-Cu (Duralumin), 94–96% Al, 3–5% Cu, trace Mg	2,787	0.883	164	6.676	126	159	182	194						
Al-Si (Silumin, copper-bearing), 86.5% Al, 1% Cu	2,659	0.867	137	5.933	119	137	144	152	161					
Al-Si (Alusil), 78–80% Al, 20–22% Si	2,627	0.854	161	7.172	144	157	168	175	178					
Al-Mg-Si, 97% Al, 1% Mg, 1% Si, 1% Mn	2,707	0.892	177	7.311		175	189	204						
Lead	11,373	0.130	35	2.343	36.9	35.1	33.4	31.5	29.8					
Iron:														
Pure	7,897	0.452	73	2.034	87	73	67	62	55	48	40	36	35	36
Wrought iron, 0.5% C	7,849	0.46	59	1.626		59	57	52	48	45	36	33	33	33
Steel (C max ≈ 1.5%): Carbon steel C ≈ 0.5%	7,833	0.465	54	1.474		55	52	48	45	42	35	31	29	31
1.0%	7,801	0.473	43	1.172		43	43	42	40	36	33	29	28	29
1.5%	7,753	0.486	36	0.970		36	36	36	35	33	31	28	28	29

Table A-2 | Property values for metals[†] (*Continued*).

Metal	Properties at 20°C ρ kg/m³	c_p kJ/kg·°C	k W/m·°C	$\alpha \times 10^5$ m²/s	Thermal conductivity k, W/m·°C −100°C −148°F	0°C 32°F	100°C 212°F	200°C 392°F	300°C 572°F	400°C 752°F	600°C 1112°F	800°C 1472°F	1000°C 1832°F	1200°C 2192°F
Nickel steel														
Ni ≈ 0%	7,897	0.452	73	2.026										
20%	7,933	0.46	19	0.526										
40%	8,169	0.46	10	0.279										
80%	8,618	0.46	35	0.872										
Invar 36% Ni	8,137	0.46	10.7	0.286										
Chrome steel														
Cr = 0%	7,897	0.452	73	2.026	87	73	67	62	55	48	40	36	35	36
1%	7,865	0.46	61	1.665		62	55	52	47	42	36	33	33	
5%	7,833	0.46	40	1.110		40	38	36	36	33	29	29	29	
20%	7,689	0.46	22	0.635		22	22	22	22	24	24	26		
Cr-Ni (chrome-nickel): 15% Cr, 10% Ni	7,865	0.46	19	0.527										
18% Cr, 8% Ni (V2A)	7,817	0.46	16.3	0.444		16.3	17	17	19	19	22	27	31	
20% Cr, 15% Ni	7,833	0.46	15.1	0.415										
25% Cr, 20% Ni	7,865	0.46	12.8	0.361										
Tungsten steel														
W = 0%	7,897	0.452	73	2.026										
1%	7,913	0.448	66	1.858										
5%	8,073	0.435	54	1.525										
10%	8,314	0.419	48	1.391										
Copper:														
Pure	8,954	0.3831	386	11.234	407	386	379	374	369	363	353			
Aluminum bronze 95% Cu, 5% Al	8,666	0.410	83	2.330										

651

Table A-2 | Property values for metals[†] (*Continued*).

| Metal | Properties at 20°C | | | | Thermal conductivity k, W/m · °C | | | | | | | | | |
	$\rho,$ kg/m³	c_p kJ/kg · °C	k W/m · °C	$\alpha \times 10^5$ m²/s	−100°C −148°F	0°C 32°F	100°C 212°F	200°C 392°F	300°C 572°F	400°C 752°F	600°C 1112°F	800°C 1472°F	1000°C 1832°F	1200°C 2192°F
Bronze 75% Cu, 25% Sn	8,666	0.343	26	0.859										
Red brass 85% Cu, 9% Sn, 6% Zn	8,714	0.385	61	1.804		59	71							
Brass 70% Cu, 30% Zn	8,522	0.385	111	3.412	88		128	144	147	147				
German silver 62% Cu, 15% Ni, 22% Zn	8,618	0.394	24.9	0.733	19.2		31	40	45	48				
Constantan 60% Cu, 40% Ni	8,922	0.410	22.7	0.612	21		22.2	26						
Magnesium: Pure	1,746	1.013	171	9.708	178	171	168	163	157					
Mg-Al (electrolytic) 6–8% Al, 1–2% Zn	1,810	1.00	66	3.605		52	62	74	83					
Molybdenum	10,220	0.251	123	4.790	138	125	118	114	111	109	106	102	99	92
Nickel: Pure (99.9%)	8,906	0.4459	90	2.266	104	93	83	73	64	59				
Ni-Cr 90% Ni, 10% Cr	8,666	0.444	17	0.444		17.1	18.9	20.9	22.8	24.6	22.5			
80% Ni, 20% Cr	8,314	0.444	12.6	0.343		12.3	13.8	15.6	17.1	18.0				
Silver: Purest	10,524	0.2340	419	17.004	419	417	415	412						
Pure (99.9%)	10,525	0.2340	407	16.563	419	410	415	374	362	360				
Tin, pure	7,304	0.2265	64	3.884	74	65.9	59	57						
Tungsten	19,350	0.1344	163	6.271	166	166	151	142	133	126	112	76		
Zinc, pure	7,144	0.3843	112.2	4.106	114	112	109	106	100	93				

[†]Adapted to SI units from E. R. G. Eckert and R. M. Drake, *Heat and Mass Transfer*, 2nd ed. New York: McGraw-Hill, 1959.

Table A-3 | Properties of nonmetals.[†]

Substance	Temperature °C	k W/m · °C	ρ kg/m^3	c kJ/kg · °C	$\alpha \times 10^7$ m^2/s
Structural and heat-resistant materials					
Acoustic tile	30	0.06	290	1.3	1.6
Aluminum oxide,					
sapphire	30	46	3970	0.76	150
Aluminum oxide,					
polycrystalline	30	36	3970	0.76	120
Asphalt	20–55	0.74–0.76			
Bakelite	30	0.23	1200	1.6	1.2
Brick:					
Building brick,					
common	20	0.69	1600	0.84	5.2
Face		1.32	2000		
Carborundum brick	600	18.5			
	1400	11.1			
Chrome brick	200	2.32	3000	0.84	9.2
	550	2.47			9.8
	900	1.99			7.9
Diatomaceous earth,					
molded and fired	200	0.24			
	870	0.31			
Fireclay brick	500	1.04	2000	0.96	5.4
Burnt 2426°F	800	1.07			
	1100	1.09			
Burnt 2642°F	500	1.28	2300	0.96	5.8
	800	1.37			
	1100	1.40			
Missouri	200	1.00	2600	0.96	4.0
	600	1.47			
	1400	1.77			
Magnesite	200	3.81		1.13	
	650	2.77			
	1200	1.90			
Cement, portland		0.29	1500		
Mortar	23	1.16			
Coal,					
anthracite	30	0.26	1300	1.25	1.6
Concrete, cinder	23	0.76			
Stone, 1-2-4 mix	20	1.37	1900–2300	0.88	8.2–6.8
Glass, window	20	0.78 (avg)	2700	0.84	3.4
Corosilicate	30–75	1.09	2200		
Graphite, pyrolytic					
parallel to layers	30	1900	2200	0.71	12,200
perpendicular to					
layers	30	5.6	2200	0.71	36
Gypsum board	30	0.16			
Lexan	30	0.2	1200	1.3	1.3
Nylon	30	0.16	1100	1.6	0.9
Particle board,					
low density	30	0.079	590	1.3	1.0
high density	30	0.17	1000	1.3	1.3
Phenolic	30	0.03	1400	1.6	0.13
Plaster, gypsum	20	0.48	1440	0.84	4.0
Metal lath	20	0.47			
Wood lath	20	0.28			

Table A-3 | Properties of nonmetals[†] (*Continued*).

Substance	Temperature °C	k W/m · °C	ρ kg/m³	c kJ/kg · °C	α × 10⁷ m²/s
					α × 10⁷ m²/s

Substance	Temperature °C	k W/m · °C	ρ kg/m³	c kJ/kg · °C	$\alpha \times 10^7$ m²/s
Structural and heat-resistant materials					
Plexiglass	30	0.2	1200	1.5	1.1
Polyethylene	30	0.33	960	2.1	1.64
Polypropylene	30	0.16	1150	1.9	0.73
Polystrene	30	0.14	1000	1.3	1.1
Polyvinylchloride	30	0.09	1700	1.1	0.48
Rubber, hard	30	0.15	1200	2.0	0.62
Silicon carbide	30	490	3150	0.68	2290
Stone:					
Granite		1.73–3.98	2640	0.82	8–18
Limestone	100–300	1.26–1.33	2500	0.90	5.6–5.9
Marble		2.07–2.94	2500–2700	0.80	10–13.6
Sandstone	40	1.83	2160–2300	0.71	11.2–11.9
Structural concrete					
low density	30	0.21	670		
light weight	30	0.65	1570		
medium weight	30	0.75	1840		
normal weight	30	2.32	2260		
Teflon	30	0.35	2200	1.05	1.5
Titanium dioxide	30	8.4	4150	0.7	29
Wood (across the grain):					
Balsa, 8.8 lb/ft³	30	0.055	140		
Cypress	30	0.097	460		
Fir	23	0.11	420	2.72	0.96
Maple or oak	30	0.166	540	2.4	1.28
Yellow pine	23	0.147	640	2.8	0.82
White pine	30	0.112	430		
Insulating materials					
Asbestos:					
Loosely packed	−45	0.149			
	0	0.154	470–570	0.816	3.3–4
	100	0.161			
Asbestos-cement	20	0.74			
boards					
Sheets	51	0.166			
Felt, 40 laminations/in	38	0.057			
	150	0.069			
	260	0.083			
20 laminations/in	38	0.078			
	150	0.095			
	260	0.112			
Corrugated, 4 plies/in	38	0.087			
	93	0.100			
	150	0.119			
Asbestos cement	—	2.08			
Balsam wood, 2.2 lb/ft³	32	0.04	35		
Cardboard, corrugated	—	0.064			
Celotex	32	0.048			
Cork, regranulated	32	0.045	45–120	1.88	2–5.3
Ground	32	0.043	150		
Corkboard, 10 lb/ft³	30	0.043	160		

Table A-3 | Properties of nonmetals[†] (*Continued*).

Substance	Temperature °C	k W/m · °C	ρ kg/m³	c kJ/kg · °C	α × 10⁷ m²/s
Insulating materials					
Diamond, Type IIa, insulator	30	2300	3500	0.509	12,900
Diatomaceous earth (Sil-o-cel)	0	0.061	320		
Felt, hair	30	0.036	130–200		
Wool	30	0.052	330		
Fiber, insulating board	20	0.048	240		
Glass fiber, duct liner	30	0.038	32	0.84	14.1
Glass fiber, loose blown	30	0.043	16	0.84	32
Glass wool, 1.5 lb/ft³	23	0.038	24	0.7	22.6
Ice	0	2.22	910	1.93	12.6
Insulex, dry	32	0.064			
		0.144			
Kapok	30	0.035			
Magnesia, 85%	38	0.067	270		
	93	0.071			
	150	0.074			
	204	0.080			
Paper (avg.)	30	0.12	900	1.2	1.1
Polyisocyanurate sheet	30	0.023			
Polystyrene, extruded	30	0.028			
Polyurethane foam	30	0.017			
Rock wool, 10 lb/ft³	32	0.040	160		
Loosely packed	150	0.067	64		
	260	0.087			
Sawdust	23	0.059			
Silica aerogel	32	0.024	140		
Styrofoam	32	0.033			
Urethane, cerllular	30	0.025			
Wood shavings	23	0.059			

[†]Adapted to SI units from A. I. Brown and S. M. Marco, *Introduction to Heat Transfer,* 3rd ed. New York: McGraw-Hill, 1958. Other properties from various sources.

APPENDIX A Tables

Table A-4 | Properties of saturated liquids.[†]

$T,°C$	ρ kg/m^3	c_p $kJ/kg \cdot °C$	$v, m^2/s$	k $W/m \cdot °C$	$\alpha, m^2/s$	Pr	β, K^{-1}
			Ammonia, NH_3				
−50	703.69	4.463	0.435×10^{-6}	0.547	1.742×10^{-7}	2.60	
−40	691.68	4.467	0.406	0.547	1.775	2.28	
−30	679.34	4.476	0.387	0.549	1.801	2.15	
−20	666.69	4.509	0.381	0.547	1.819	2.09	
−10	653.55	4.564	0.378	0.543	1.825	2.07	
0	640.10	4.635	0.373	0.540	1.819	2.05	
10	626.16	4.714	0.368	0.531	1.801	2.04	
20	611.75	4.798	0.359	0.521	1.775	2.02	2.45×10^{-3}
30	596.37	4.890	0.349	0.507	1.742	2.01	
40	580.99	4.999	0.340	0.493	1.701	2.00	
50	564.33	5.116	0.330	0.476	1.654	1.99	
			Carbon dioxide, CO_2				
−50	1,156.34	1.84	0.119×10^{-6}	0.0855	0.4021×10^{-7}	2.96	
−40	1,117.77	1.88	0.118	0.1011	0.4810	2.46	
−30	1,076.76	1.97	0.117	0.1116	0.5272	2.22	
−20	1,032.39	2.05	0.115	0.1151	0.5445	2.12	
−10	983.38	2.18	0.113	0.1099	0.5133	2.20	
0	926.99	2.47	0.108	0.1045	0.4578	2.38	
10	860.03	3.14	0.101	0.0971	0.3608	2.80	
20	772.57	5.0	0.091	0.0872	0.2219	4.10	14.00×10^{-3}
30	597.81	36.4	0.080	0.0703	0.0279	28.7	
			Sulfur dioxide, SO_2				
−50	1,560.84	1.3595	0.484×10^{-6}	0.242	1.141×10^{-7}	4.24	
−40	1,536.81	1.3607	0.424	0.235	1.130	3.74	
−30	1,520.64	1.3616	0.371	0.230	1.117	3.31	
−20	1,488.60	1.3624	0.324	0.225	1.107	2.93	
−10	1,463.61	1.3628	0.288	0.218	1.097	2.62	
0	1,438.46	1.3636	0.257	0.211	1.081	2.38	
10	1,412.51	1.3645	0.232	0.204	1.066	2.18	
20	1,386.40	1.3653	0.210	0.199	1.050	2.00	1.94×10^{-3}
30	1,359.33	1.3662	0.190	0.192	1.035	1.83	
40	1,329.22	1.3674	0.173	0.185	1.019	1.70	
50	1,299.10	1.3683	0.162	0.177	0.999	1.61	
			Dichlorodifluoromethane (Freon-12), CCl_2F_2				
−50	1,546.75	0.8750	0.310×10^{-6}	0.067	0.501×10^{-7}	6.2	2.63×10^{-3}
−40	1,518.71	0.8847	0.279	0.069	0.514	5.4	
−30	1,489.56	0.8956	0.253	0.069	0.526	4.8	
−20	1,460.57	0.9073	0.235	0.071	0.539	4.4	
−10	1,429.49	0.9203	0.221	0.073	0.550	4.0	
0	1,397.45	0.9345	0.214×10^{-6}	0.073	0.557×10^{-7}	3.8	
10	1,364.30	0.9496	0.203	0.073	0.560	3.6	
20	1,330.18	0.9659	0.198	0.073	0.560	3.5	
30	1,295.10	0.9835	0.194	0.071	0.560	3.5	
40	1,257.13	1.0019	0.191	0.069	0.555	3.5	
50	1,215.96	1.0216	0.190	0.067	0.545	3.5	

Table A-4 | Properties of saturated liquids[†] (*Continued*).

T, °C	ρ kg/m³	c_p kJ/kg·°C	ν, m²/s	k W/m·°C	α, m²/s	Pr	β, K⁻¹
\multicolumn{8}{c}{Glycerin, $C_3H_5(OH)_3$}							
0	1,276.03	2.261	0.00831	0.282	0.983×10^{-7}	84.7×10^3	
10	1,270.11	2.319	0.00300	0.284	0.965	31.0	
20	1,264.02	2.386	0.00118	0.286	0.947	12.5	0.50×10^{-3}
30	1,258.09	2.445	0.00050	0.286	0.929	5.38	
40	1,252.01	2.512	0.00022	0.286	0.914	2.45	
50	1,244.96	2.583	0.00015	0.287	0.893	1.63	
\multicolumn{8}{c}{Ethylene glycol, $C_2H_4(OH)_2$}							
0	1,130.75	2.294	7.53×10^{-6}	0.242	0.934×10^{-7}	615	
20	1,116.65	2.382	19.18	0.249	0.939	204	0.65×10^{-3}
40	1,101.43	2.474	8.69	0.256	0.939	93	
60	1,087.66	2.562	4.75	0.260	0.932	51	
80	1,077.56	2.650	2.98	0.261	0.921	32.4	
100	1,058.50	2.742	2.03	0.263	0.908	22.4	
\multicolumn{8}{c}{Engine oil (unused)}							
0	899.12	1.796	0.00428	0.147	0.911×10^{-7}	47,100	
20	888.23	1.880	0.00090	0.145	0.872	10,400	0.70×10^{-3}
40	876.05	1.964	0.00024	0.144	0.834	2,870	
60	864.04	2.047	0.839×10^{-4}	0.140	0.800	1,050	
80	852.02	2.131	0.375	0.138	0.769	490	
100	840.01	2.219	0.203	0.137	0.738	276	
120	828.96	2.307	0.124	0.135	0.710	175	
140	816.94	2.395	0.080	0.133	0.686	116	
160	805.89	2.483	0.056	0.132	0.663	84	
\multicolumn{8}{c}{Mercury, Hg}							
0	13,628.22	0.1403	0.124×10^{-6}	8.20	42.99×10^{-7}	0.0288	
20	13,579.04	0.1394	0.114	8.69	46.06	0.0249	1.82×10^{-4}
50	13,505.84	0.1386	0.104	9.40	50.22	0.0207	
100	13,384.58	0.1373	0.0928	10.51	57.16	0.0162	
150	13,264.28	0.1365	0.0853	11.49	63.54	0.0134	
200	13,144.94	0.1570	0.0802	12.34	69.08	0.0116	
250	13,025.60	0.1357	0.0765	13.07	74.06	0.0103	
315.5	12,847	0.134	0.0673	14.02	81.5	0.0083	

[†]Adapted to SI units from E. R. G. Eckert and R. M. Drake, *Heat and Mass Transfer,* 2nd ed. New York: McGraw-Hill, 1959.

Table A-5 | Properties of air at atmospheric pressure.[†]

T, K	ρ kg/m^3	c_p kJ/kg·°C	$\mu \times 10^5$ kg/m·s	$\nu \times 10^6$ m^2/s	k W/m·°C	$\alpha \times 10^4$ m^2/s	Pr
100	3.6010	1.0266	0.6924	1.923	0.009246	0.02501	0.770
150	2.3675	1.0099	1.0283	4.343	0.013735	0.05745	0.753
200	1.7684	1.0061	1.3289	7.490	0.01809	0.10165	0.739
250	1.4128	1.0053	1.5990	11.31	0.02227	0.15675	0.722
300	1.1774	1.0057	1.8462	15.69	0.02624	0.22160	0.708
350	0.9980	1.0090	2.075	20.76	0.03003	0.2983	0.697
400	0.8826	1.0140	2.286	25.90	0.03365	0.3760	0.689
450	0.7833	1.0207	2.484	31.71	0.03707	0.4222	0.683
500	0.7048	1.0295	2.671	37.90	0.04038	0.5564	0.680
550	0.6423	1.0392	2.848	44.34	0.04360	0.6532	0.680
600	0.5879	1.0551	3.018	51.34	0.04659	0.7512	0.680
650	0.5430	1.0635	3.177	58.51	0.04953	0.8578	0.682
700	0.5030	1.0752	3.332	66.25	0.05230	0.9672	0.684
750	0.4709	1.0856	3.481	73.91	0.05509	1.0774	0.686
800	0.4405	1.0978	3.625	82.29	0.05779	1.1951	0.689
850	0.4149	1.1095	3.765	90.75	0.06028	1.3097	0.692
900	0.3925	1.1212	3.899	99.3	0.06279	1.4271	0.696
950	0.3716	1.1321	4.023	108.2	0.06525	1.5510	0.699
1000	0.3524	1.1417	4.152	117.8	0.06752	1.6779	0.702
1100	0.3204	1.160	4.44	138.6	0.0732	1.969	0.704
1200	0.2947	1.179	4.69	159.1	0.0782	2.251	0.707
1300	0.2707	1.197	4.93	182.1	0.0837	2.583	0.705
1400	0.2515	1.214	5.17	205.5	0.0891	2.920	0.705
1500	0.2355	1.230	5.40	229.1	0.0946	3.262	0.705
1600	0.2211	1.248	5.63	254.5	0.100	3.609	0.705
1700	0.2082	1.267	5.85	280.5	0.105	3.977	0.705
1800	0.1970	1.287	6.07	308.1	0.111	4.379	0.704
1900	0.1858	1.309	6.29	338.5	0.117	4.811	0.704
2000	0.1762	1.338	6.50	369.0	0.124	5.260	0.702
2100	0.1682	1.372	6.72	399.6	0.131	5.715	0.700
2200	0.1602	1.419	6.93	432.6	0.139	6.120	0.707
2300	0.1538	1.482	7.14	464.0	0.149	6.540	0.710
2400	0.1458	1.574	7.35	504.0	0.161	7.020	0.718
2500	0.1394	1.688	7.57	543.5	0.175	7.441	0.730

[†]From *Natl. Bur. Stand. (U.S.) Circ.* 564, 1955.

Table A-6 | Properties of gases at atmospheric pressure.[†]

			Values of μ, k, c_p, and Pr are not strongly pressure-dependent for He, H_2, O_2, and N_2 and may be used over a fairly wide range of pressures				
T, K	ρ kg/m^3	c_p kJ/kg·°C	μ, kg/m·s	v, m^2/s	k W/m·°C	α, m^2/s	Pr
Helium							
144	0.3379	5.200	125.5×10^{-7}	37.11×10^{-6}	0.0928	0.5275×10^{-4}	0.70
200	0.2435	5.200	156.6	64.38	0.1177	0.9288	0.694
255	0.1906	5.200	181.7	95.50	0.1357	1.3675	0.70
366	0.13280	5.200	230.5	173.6	0.1691	2.449	0.71
477	0.10204	5.200	275.0	269.3	0.197	3.716	0.72
589	0.08282	5.200	311.3	375.8	0.225	5.215	0.72
700	0.07032	5.200	347.5	494.2	0.251	6.661	0.72
800	0.06023	5.200	381.7	634.1	0.275	8.774	0.72
Hydrogen							
150	0.16371	12.602	5.595×10^{-6}	34.18×10^{-6}	0.0981	0.475×10^{-4}	0.718
200	0.12270	13.540	6.813	55.53	0.1282	0.772	0.719
250	0.09819	14.059	7.919	80.64	0.1561	1.130	0.713
300	0.08185	14.314	8.963	109.5	0.182	1.554	0.706
350	0.07016	14.436	9.954	141.9	0.206	2.031	0.697
400	0.06135	14.491	10.864	177.1	0.228	2.568	0.690
450	0.05462	14.499	11.779	215.6	0.251	3.164	0.682
500	0.04918	14.507	12.636	257.0	0.272	3.817	0.675
550	0.04469	14.532	13.475	301.6	0.292	4.516	0.668
600	0.04085	14.537	14.285	349.7	0.315	5.306	0.664
700	0.03492	14.574	15.89	455.1	0.351	6.903	0.659
800	0.03060	14.675	17.40	569	0.384	8.563	0.664
900	0.02723	14.821	18.78	690	0.412	10.217	0.676
Oxygen							
150	2.6190	0.9178	11.490×10^{-6}	4.387×10^{-6}	0.01367	0.05688×10^{-4}	0.773
200	1.9559	0.9131	14.850	7.593	0.01824	0.10214	0.745
250	1.5618	0.9157	17.87	11.45	0.02259	0.15794	0.725
300	1.3007	0.9203	20.63	15.86	0.02676	0.22353	0.709
350	1.1133	0.9291	23.16	20.80	0.03070	0.2968	0.702
400	0.9755	0.9420	25.54	26.18	0.03461	0.3768	0.695
450	0.8682	0.9567	27.77	31.99	0.03828	0.4609	0.694
500	0.7801	0.9722	29.91	38.34	0.04173	0.5502	0.697
550	0.7096	0.9881	31.97	45.05	0.04517	0.641	0.700
Nitrogen							
200	1.7108	1.0429	12.947×10^{-6}	7.568×10^{-6}	0.01824	0.10224×10^{-4}	0.747
300	1.1421	1.0408	17.84	15.63	0.02620	0.22044	0.713
400	0.8538	1.0459	21.98	25.74	0.03335	0.3734	0.691
500	0.6824	1.0555	25.70	37.66	0.03984	0.5530	0.684
600	0.5687	1.0756	29.11	51.19	0.04580	0.7486	0.686
700	0.4934	1.0969	32.13	65.13	0.05123	0.9466	0.691
800	0.4277	1.1225	34.84	81.46	0.05609	1.1685	0.700
900	0.3796	1.1464	37.49	91.06	0.06070	1.3946	0.711
1000	0.3412	1.1677	40.00	117.2	0.06475	1.6250	0.724
1100	0.3108	1.1857	42.28	136.0	0.06850	1.8571	0.736
1200	0.2851	1.2037	44.50	156.1	0.07184	2.0932	0.748

Table A-6 | Properties of gases at atmospheric pressure[†] (*Continued*).

			Values of μ, k, c_p, and Pr are not strongly pressure-dependent for He, H$_2$, O$_2$, and N$_2$ and may be used over a fairly wide range of pressures				
T, K	ρ kg/m^3	c_p kJ/kg·°C	μ, kg/m·s	v, m^2/s	k W/m·°C	α, m^2/s	Pr
Carbon dioxide							
220	2.4733	0.783	11.105×10^{-6}	4.490×10^{-6}	0.010805	0.05920×10^{-4}	0.818
250	2.1657	0.804	12.590	5.813	0.012884	0.07401	0.793
300	1.7973	0.871	14.958	8.321	0.016572	0.10588	0.770
350	1.5362	0.900	17.205	11.19	0.02047	0.14808	0.755
400	1.3424	0.942	19.32	14.39	0.02461	0.19463	0.738
450	1.1918	0.980	21.34	17.90	0.02897	0.24813	0.721
500	1.0732	1.013	23.26	21.67	0.03352	0.3084	0.702
550	0.9739	1.047	25.08	25.74	0.03821	0.3750	0.685
600	0.8938	1.076	26.83	30.02	0.04311	0.4483	0.668
Ammonia, NH$_3$							
273	0.7929	2.177	9.353×10^{-6}	1.18×10^{-5}	0.0220	0.1308×10^{-4}	0.90
323	0.6487	2.177	11.035	1.70	0.0270	0.1920	0.88
373	0.5590	2.236	12.886	2.30	0.0327	0.2619	0.87
423	0.4934	2.315	14.672	2.97	0.0391	0.3432	0.87
473	0.4405	2.395	16.49	3.74	0.0467	0.4421	0.84
Water vapor							
380	0.5863	2.060	12.71×10^{-6}	2.16×10^{-5}	0.0246	0.2036×10^{-4}	1.060
400	0.5542	2.014	13.44	2.42	0.0261	0.2338	1.040
450	0.4902	1.980	15.25	3.11	0.0299	0.307	1.010
500	0.4405	1.985	17.04	3.86	0.0339	0.387	0.996
550	0.4005	1.997	18.84	4.70	0.0379	0.475	0.991
600	0.3652	2.026	20.67	5.66	0.0422	0.573	0.986
650	0.3380	2.056	22.47	6.64	0.0464	0.666	0.995
700	0.3140	2.085	24.26	7.72	0.0505	0.772	1.000
750	0.2931	2.119	26.04	8.88	0.0549	0.883	1.005
800	0.2739	2.152	27.86	10.20	0.0592	1.001	1.010
850	0.2579	2.186	29.69	11.52	0.0637	1.130	1.019

[†]Adapted to SI units from E. R. G. Eckert and R. M. Drake, *Heat and Mass Transfer,* 2nd ed. New York: McGraw-Hill, 1959.

Table A-7 | Physical properties of some common low-melting-point metals.[†]

Metal	Melting point °C	Normal boiling point °C	Temperature °C	Density, $\rho \times 10^{-3}$ kg/m^3	Viscosity $\mu \times 10^3$ kg/m · s	Heat capacity kJ/kg · °C	Thermal conductivity W/m · °C	Prandtl number
Bismuth	271	1477	316	10.01	1.62	0.144	16.4	0.014
			760	9.47	0.79	0.165	15.6	0.0084
Lead	327	1737	371	10.5	2.40	0.159	16.1	0.024
			704	10.1	1.37	0.155	14.9	0.016
Lithium	179	1317	204	0.51	0.60	4.19	38.1	0.065
			982	0.44	0.42	4.19		
Mercury	−39	357	10	13.6	1.59	0.138	8.1	0.027
			316	12.8	0.86	0.134	14.0	0.0084
Potassium	63.8	760	149	0.81	0.37	0.796	45.0	0.0066
			704	0.67	0.14	0.754	33.1	0.0031
Sodium	97.8	883	204	0.90	0.43	1.34	80.3	0.0072
			704	0.78	0.18	1.26	59.7	0.0038
Sodium-potassium:								
22% Na	19	826	93.3	0.848	0.49	0.946	24.4	0.019
			760	0.69	0.146	0.883		
56% Na	−11	784	93.3	0.89	0.58	1.13	25.6	0.026
			760	0.74	0.16	1.04	28.9	0.058
Lead-bismuth, 44.5% Pb	125	1670	288	10.3	1.76	0.147	10.7	0.024
			649	9.84	1.15			

[†]Adapted to SI units from J. G. Knudsen and D. L. Katz, *Fluid Dynamics and Heat Transfer,* New York: McGraw-Hill, 1958.

Table A-8 | Diffusion coefficients of gases and vapors in air at 25°C and 1 atm.[†]

Substance	D, cm^2/s	$Sc = \dfrac{v}{D}$	Substance	D, cm^2/s	$Sc = \dfrac{v}{D}$
Ammonia	0.28	0.78	Formic Acid	0.159	0.97
Carbon dioxide	0.164	0.94	Acetic acid	0.133	1.16
Hydrogen	0.410	0.22	Aniline	0.073	2.14
Oxygen	0.206	0.75	Benzene	0.088	1.76
Water	0.256	0.60	Toluene	0.084	1.84
Ethyl ether	0.093	1.66	Ethyl benzene	0.077	2.01
Methanol	0.159	0.97	Propyl benzene	0.059	2.62
Ethyl alcohol	0.119	1.30			

[†]From J. H. Perry (ed.), *Chemical Engineers' Handbook,* 4th ed. New York: McGraw-Hill, 1963.

Table A-9 | Properties of water (saturated liquid).[†]

Note: $Gr_x Pr = \left(\dfrac{g\beta\rho^2 c_p}{\mu k}\right) x^3 \Delta T$

°F	°C	c_p kJ/kg · °C	ρ kg/m³	μ kg/m · s	k W/m · °C	Pr	$\dfrac{g\beta\rho^2 c_p}{\mu k}$ 1/m³ · °C
32	0	4.225	999.8	1.79×10^{-3}	0.566	13.25	
40	4.44	4.208	999.8	1.55	0.575	11.35	1.91×10^9
50	10	4.195	999.2	1.31	0.585	9.40	6.34×10^9
60	15.56	4.186	998.6	1.12	0.595	7.88	1.08×10^{10}
70	21.11	4.179	997.4	9.8×10^{-4}	0.604	6.78	1.46×10^{10}
80	26.67	4.179	995.8	8.6	0.614	5.85	1.91×10^{10}
90	32.22	4.174	994.9	7.65	0.623	5.12	2.48×10^{10}
100	37.78	4.174	993.0	6.82	0.630	4.53	3.3×10^{10}
110	43.33	4.174	990.6	6.16	0.637	4.04	4.19×10^{10}
120	48.89	4.174	988.8	5.62	0.644	3.64	4.89×10^{10}
130	54.44	4.179	985.7	5.13	0.649	3.30	5.66×10^{10}
140	60	4.179	983.3	4.71	0.654	3.01	6.48×10^{10}
150	65.55	4.183	980.3	4.3	0.659	2.73	7.62×10^{10}
160	71.11	4.186	977.3	4.01	0.665	2.53	8.84×10^{10}
170	76.67	4.191	973.7	3.72	0.668	2.33	9.85×10^{10}
180	82.22	4.195	970.2	3.47	0.673	2.16	1.09×10^{11}
190	87.78	4.199	966.7	3.27	0.675	2.03	
200	93.33	4.204	963.2	3.06	0.678	1.90	
220	104.4	4.216	955.1	2.67	0.684	1.66	
240	115.6	4.229	946.7	2.44	0.685	1.51	
260	126.7	4.250	937.2	2.19	0.685	1.36	
280	137.8	4.271	928.1	1.98	0.685	1.24	
300	148.9	4.296	918.0	1.86	0.684	1.17	
350	176.7	4.371	890.4	1.57	0.677	1.02	
400	204.4	4.467	859.4	1.36	0.665	1.00	
450	232.2	4.585	825.7	1.20	0.646	0.85	
500	260	4.731	785.2	1.07	0.616	0.83	
550	287.7	5.024	735.5	9.51×10^{-5}			
600	315.6	5.703	678.7	8.68			

[†]Adapted to SI units from A. I. Brown and S. M. Marco, *Introduction to Heat Transfer,* 3rd ed. New York: McGraw-Hill, 1958.

Table A-10 | Normal total emissivity of various surfaces.[†]

Surface	T, °F	Emissivity ε
Metals and their oxides		
Aluminum:		
Highly polished plate, 98.3% pure	440–1070	0.039–0.057
Commercial sheet	212	0.09
Heavily oxidized	299–940	0.20–0.31
Al-surfaced roofing	100	0.216
Brass:		
Highly polished:		
73.2% Cu, 26.7% Zn	476–674	0.028–0.031
62.4% Cu, 36.8% Zn, 0.4% Pb, 0.3% Al	494–710	0.033–0.037
82.9% Cu, 17.0% Zn	530	0.030
Hard-rolled, polished, but direction of polishing visible	70	0.038
Dull plate	120–660	0.22
Chromium (see nickel alloys for Ni-Cr steels), polished	100–2000	0.08–0.36
Copper:		
Polished	242	0.023
	212	0.052
Plate, heated long time, covered with thick oxide layer	77	0.78
Gold, pure, highly polished	440–1160	0.018–0.035
Iron and steel (not including stainless):		
Steel, polished	212	0.066
Iron, polished	800–1880	0.14–0.38
Cast iron, newly turned	72	0.44
turned and heated	1620–1810	0.60–0.70
Mild steel	450–1950	0.20–0.32
Iron and steel (oxidized surfaces):		
Iron plate, pickled, then rusted red	68	0.61
Iron, dark-gray surface	212	0.31
Rough ingot iron	1700–2040	0.87–0.95
Sheet steel with strong, rough oxide layer	75	0.80
Lead:		
Unoxidized, 99.96% pure	240–440	0.057–0.075
Gray oxidized	75	0.28
Oxidized at 300°F	390	0.63
Magnesium, magnesium oxide	530–1520	0.55–0.20
Molybdenum:		
Filament	1340–4700	0.096–0.202
Massive, polished	212	0.071
Monel metal, oxidized at 1110°F	390–1110	0.41–0.46
Nickel:		
Polished	212	0.072
Nickel oxide	1200–2290	0.59–0.86
Nickel alloys:		
Copper nickel, polished	212	0.059
Nichrome wire, bright	120–1830	0.65–0.79
Nichrome wire, oxidized	120–930	0.95–0.98
Platinum, polished plate, pure	440–1160	0.054–0.104
Silver:		
Polished, pure	440–1160	0.020–0.032
Polished	100–700	0.022–0.031

Table A-10 | Normal total emissivity of various surfaces[†] (*Continued*).

Surface	T, °F	Emissivity ϵ
Metals and their oxides		
Stainless steels:		
Polished	212	0.074
Type 301; B	450–1725	0.54–0.63
Tin, bright tinned iron	76	0.043 and 0.064
Tungsten, filament	6000	0.39
Zinc, galvanized sheet iron, fairly bright	82	0.23
Refractories, building materials, paints, and miscellaneous		
Alumina (85–99.5%, Al_2O_3, 0–12% SiO_2, 0–1% Ge_2O_3); effect of mean grain size, microns (μm):		
10 μm		0.30–0.18
50 μm		0.39–0.28
100 μm		0.50–0.40
Asbestos, board	74	0.96
Brick:		
Red, rough, but no gross irregularities	70	0.93
Fireclay	1832	0.75
Carbon:		
T-carbon (Gebrüder Siemens) 0.9% ash, started with emissivity of 0.72 at 260°F but on heating changed to values given	260–1160	0.81–0.79
Filament	1900–2560	0.526
Rough plate	212–608	0.77
Lampblack, rough deposit	212–932	0.84–0.78
Concrete tiles	1832	0.63
Enamel, white fused, on iron	66	0.90
Glass:		
Smooth	72	0.94
Pyrex, lead, and soda	500–1000	0.95–0.85
Paints, lacquers, varnishes:		
Snow-white enamel varnish on rough iron plate	73	0.906
Black shiny lacquer, sprayed on iron	76	0.875
Black shiny shellac on tinned iron sheet	70	0.821
Black matte shellac	170–295	0.91
Black or white lacquer	100–200	0.80–0.95
Flat black lacquer	100–200	0.96–0.98
Aluminum paints and lacquers:		
10% Al, 22% lacquer body, on rough or smooth surface	212	0.52
Other Al paints, varying age and Al content	212	0.27–0.67
Porcelain, glazed	72	0.92
Quartz, rough, fused	70	0.93
Roofing paper	69	0.91
Rubber, hard, glossy plate	74	0.94
Water	32–212	9.95–0.963

[†]Courtesy of H. C. Hottel, from W. H. McAdams, *Heat Transmissions,* 3rd ed. New York: McGraw-Hill, 1954.

Table A-11 | Steel-pipe dimensions.

Nominal pipe size, in	OD, in	Schedule no.	Wall Thickness, in	ID, in	Metal sectional area, in²	Inside cross-sectional area, ft²
$\frac{1}{8}$	0.405	40	0.068	0.269	0.072	0.00040
		80	0.095	0.215	0.093	0.00025
$\frac{1}{4}$	0.540	40	0.088	0.364	0.125	0.00072
		80	0.119	0.302	0.157	0.00050
$\frac{3}{8}$	0.675	40	0.091	0.493	0.167	0.00133
		80	0.126	0.423	0.217	0.00098
$\frac{1}{2}$	0.840	40	0.109	0.622	0.250	0.00211
		80	0.147	0.546	0.320	0.00163
$\frac{3}{4}$	1.050	40	0.113	0.824	0.333	0.00371
		80	0.154	0.742	0.433	0.00300
1	1.315	40	0.133	1.049	0.494	0.00600
		80	0.179	0.957	0.639	0.00499
$1\frac{1}{2}$	1.900	40	0.145	1.610	0.799	0.01414
		80	0.200	1.500	1.068	0.01225
		160	0.281	1.338	1.429	0.00976
2	2.375	40	0.154	2.067	1.075	0.02330
		80	0.218	1.939	1.477	0.02050
3	3.500	40	0.216	3.068	2.228	0.05130
		80	0.300	2.900	3.016	0.04587
4	4.500	40	0.237	4.026	3.173	0.08840
		80	0.337	3.826	4.407	0.7986
5	5.563	40	0.258	5.047	4.304	0.1390
		80	0.375	4.813	6.122	0.1263
		120	0.500	4.563	7.953	0.1136
		160	0.625	4.313	9.696	0.1015
6	6.625	40	0.280	6.065	5.584	0.2006
		80	0.432	5.761	8.405	0.1810
10	10.75	40	0.365	10.020	11.90	0.5475
		80	0.500	9.750	16.10	0.5185

Table A-12 | Conversion factors. (*See also inside cover.*)

Length:
  12 in = 1 ft
  2.54 cm = 1 in
  $1 \, \mu\text{m} = 10^{-6} \, \text{m} = 10^{-4} \, \text{cm}$
Mass:
  $1 \, \text{kg} = 2.205 \, \text{lb}_m$
  $1 \, \text{slug} = 32.16 \, \text{lb}_m$
  $454 \, \text{g} = 1 \, \text{lb}_m$
Force:
  $1 \, \text{dyn} = 2.248 \times 10^{-6} \, \text{lb}_f$
  $1 \, \text{lb}_f = 4.448 \, \text{N}$
  $10^5 \, \text{dyn} = 1 \, \text{N}$

Energy:
  $1 \, \text{ft} \cdot \text{lb}_f = 1.356 \, \text{J}$
  1 kWh = 3413 Btu
  $1 \, \text{hp} \cdot \text{h} = 2545 \, \text{Btu}$
  1 Btu = 252 cal
  $1 \, \text{Btu} = 778 \, \text{ft} \cdot \text{lb}_f$
Pressure:
  $1 \, \text{atm} = 14.696 \, \text{lb}_f/\text{in}^2 = 2116 \, \text{lb}_f/\text{ft}^2$
  $1 \, \text{atm} = 1.01325 \times 10^5 \, \text{Pa}$
  $1 \, \text{in Hg} = 70.73 \, \text{lb}_f/\text{ft}^2$
Viscosity:
  $1 \, \text{centipoise} = 2.42 \, \text{lb}_m/\text{h} \cdot \text{ft}$
  $1 \, \text{lb}_f \cdot \text{s/ft}^2 = 32.16 \, \text{lb}_m/\text{s} \cdot \text{ft}$
Thermal conductivity:
  $1 \, \text{cal/s} \cdot \text{cm} \cdot {}^\circ\text{C} = 242 \, \text{Btu/h} \cdot \text{ft} \cdot {}^\circ\text{F}$
  $1 \, \text{W/cm} \cdot {}^\circ\text{C} = 57.79 \, \text{Btu/h} \cdot \text{ft} \cdot {}^\circ\text{F}$

Useful conversion to SI units

Length:
  1 in = 0.0254 m
  1 ft = 0.3048 m
  1 mi = 1.60934 km
Area:
  $1 \, \text{in}^2 = 645.16 \, \text{mm}^2$
  $1 \, \text{ft}^2 = 0.092903 \, \text{m}^2$
  $1 \, \text{mi}^2 = 2.58999 \, \text{km}^2$
Pressure:
  $1 \, \text{N/m}^2 = 1 \, \text{Pa}$
  $1 \, \text{atm} = 1.01325 \times 10^5 \, \text{Pa}$
  $1 \, \text{lb}_f/\text{in}^2 = 6894.76 \, \text{Pa}$
Energy:
  $1 \, \text{erg} = 10^{-7} \, \text{J}$
  1 Btu = 1055.04 J
  $1 \, \text{ft} \cdot \text{lb}_f = 1.35582 \, \text{J}$
  $1 \, \text{cal} \, (15^\circ\text{C}) = 4.1855 \, \text{J}$
Power:
  1 hp = 745.7 W
  1 Btu/h = 0.293 W
Heat flux:
  $1 \, \text{Btu/h} \cdot \text{ft}^2 = 3.15372 \, \text{W/m}^2$
  $1 \, \text{Btu/h} \cdot \text{ft} = 0.96128 \, \text{W/m}$
Thermal conductivity:
  $1 \, \text{Btu/h} \cdot \text{ft} \cdot {}^\circ\text{F} = 1.7307 \, \text{W/m} \cdot {}^\circ\text{C}$
Heat-transfer coefficient:
  $1 \, \text{Btu/h} \cdot \text{ft}^2 \cdot {}^\circ\text{F} = 5.6782 \, \text{W/m}^2 \cdot {}^\circ\text{C}$

Volume:
  $1 \, \text{in}^3 = 1.63871 \times 10^{-5} \, \text{m}^3$
  $1 \, \text{ft}^3 = 0.0283168 \, \text{m}^3$
  $1 \, \text{gal} = 231 \, \text{in}^3 = 0.0037854 \, \text{m}^3$
Mass:
  $1 \, \text{lb}_m = 0.45359237 \, \text{kg}$
Density:
  $1 \, \text{lb}_m/\text{in}^3 = 2.76799 \times 10^4 \, \text{kg/m}^3$
  $1 \, \text{lb}_m/\text{ft}^3 = 16.0185 \times 10^4 \, \text{kg/m}^3$
Force:
  $1 \, \text{dyn} = 10^{-5} \, \text{N}$
  $1 \, \text{lb}_f = 4.44822 \, \text{N}$

Exact Solutions of Laminar-Boundary-Layer Equations

We wish to obtain a solution to the laminar-boundary-layer momentum and energy equations, assuming constant fluid properties and zero pressure gradient. We have

Continuity:
$$\frac{\partial u}{\partial x} + \frac{\partial v}{\partial y} = 0 \qquad \text{[B-1]}$$

Momentum:
$$u\frac{\partial u}{\partial x} + v\frac{\partial u}{\partial y} = \nu\frac{\partial^2 u}{\partial y^2} \qquad \text{[B-2]}$$

Energy:
$$u\frac{\partial T}{\partial x} + v\frac{\partial T}{\partial y} = \alpha\frac{\partial^2 T}{\partial y^2} \qquad \text{[B-3]}$$

It will be noted that the viscous-dissipation term is omitted from the energy equation for the present. In accordance with the order-of-magnitude analysis of Section 6-1,

$$\delta \sim \sqrt{\frac{\nu x}{u_\infty}} \qquad \text{[B-4]}$$

The assumption is now made that the velocity profiles have similar shapes at various distances from the leading edge of the flat plate. The significant variable is then y/δ, and we assume that the velocity may be expressed as a function of this variable. We then have

$$\frac{u}{u_\infty} = g\left(\frac{y}{\delta}\right)$$

Introducing the order-of-magnitude estimate for δ from Equation (B-4):

$$\frac{u}{u_\infty} = g(\eta) \qquad \text{[B-5]}$$

where

$$\eta = \frac{y}{\sqrt{\nu x/u_\infty}} = y\sqrt{\frac{u_\infty}{\nu x}} \qquad \text{[B-6]}$$

Here, η is called the *similarity variable,* and $g(\eta)$ is the function we seek as a solution. In accordance with the continuity equation, a stream function ψ may be defined so that

$$u = \frac{\partial \psi}{\partial y} \qquad \text{[B-7]}$$

$$v = -\frac{\partial \psi}{\partial x} \qquad \text{[B-8]}$$

Inserting (B-7) in (B-5) gives

$$\psi = \int u_\infty g(\eta)\, dy = \int u_\infty \sqrt{\frac{vx}{u_\infty}}\, g(\eta)\, d\eta \qquad \textbf{[B-9]}$$

or

$$\psi = u_\infty \sqrt{\frac{vx}{u_\infty}}\, f(\eta)$$

where $f(\eta) = \int g(\eta)\, d\eta$.

From (B-8) and (B-9) we obtain

$$v = \frac{1}{2}\sqrt{\frac{vu_\infty}{x}}\left(\eta \frac{df}{d\eta} + f\right) \qquad \textbf{[B-10]}$$

Making similar transformations on the other terms in Equation (B-2), we obtain

$$f\frac{d^2 f}{d\eta^2} + 2\frac{d^3 f}{d\eta^3} = 0 \qquad \textbf{[B-11]}$$

This is an ordinary differential equation, which may be solved numerically for the function $f(\eta)$. The boundary conditions are

Physical coordinates		Similarity coordinates	
$u = 0$	at $y = 0$	$\dfrac{df}{d\eta} = 0$	at $\eta = 0$
$v = 0$	at $y = 0$	$f = 0$	at $\eta = 0$
$\dfrac{\partial u}{\partial y} = 0$	at $y \rightarrow \infty$	$\dfrac{df}{d\eta} = 1.0$	at $\eta \rightarrow \infty$

The first solution to Equation (B-11) was obtained by Blasius.[†] The values of u and v as obtained from this solution are presented in Figure B-1.

Figure B-1 | Velocity profiles in laminar boundary layer. Slope $du/d\eta = 0.332$ at $\eta = 0$.

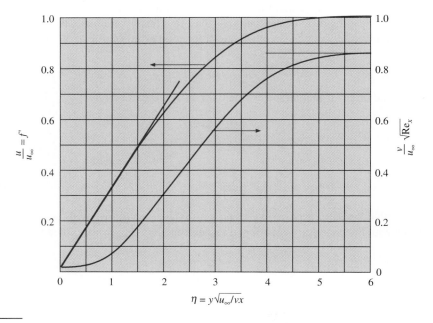

$$\eta = y\sqrt{u_\infty/vx}$$

[†]H. Blasius, *Z. Math. Phys.*, vol. 56, p. 1, 1908.

The energy equation is solved in a similar manner by first defining a dimensionless temperature variable as

$$\theta(\eta) = \frac{T(\eta) - T_w}{T_\infty - T_w} \tag{B-12}$$

where it is also assumed that θ and T may be expressed as functions of the similarity variable η. Equation (B-3) then becomes

$$\frac{d^2\theta}{d\eta^2} + \frac{1}{2} \Pr f \frac{d\theta}{d\eta} = 0 \tag{B-13}$$

with the boundary conditions

$$\theta = \begin{cases} 0 & \text{at } y=0, \ \eta=0 \\ 1.0 & \text{at } y=\infty, \ \eta=\infty \end{cases}$$

Given the function $f(\eta)$, the solution to Equation (B-13) may be obtained as

$$\theta(\eta) = \frac{\int_0^\eta \exp\left(-\frac{\Pr}{2} \int_0^\eta f \, d\eta\right) d\eta}{\int_0^\infty \exp\left(-\frac{\Pr}{2} \int_0^\eta f \, d\eta\right) d\eta} \tag{B-14}$$

This solution is given by Pohlhausen[†] and is shown in Figure B-2. For Prandtl numbers between 0.6 and 15 it was found that the dimensionless temperature gradient at the surface could be represented satisfactorily by

$$\left(\frac{d\theta(\eta)}{d\eta}\right)_{\eta=0} = 0.332 \, \Pr^{1/3} \tag{B-15}$$

The heat-transfer coefficient may subsequently be expressed by

$$\mathrm{Nu}_x = 0.332 \, \mathrm{Re}_x^{1/2} \, \Pr^{1/3} \tag{B-16}$$

in agreement with the results of Chapter 5.

Figure B-2 | Temperature profiles in laminar boundary layer with isothermal wall.

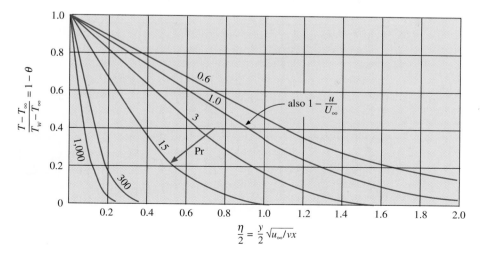

Now let us consider a solution of the complete energy equation, including the viscous-dissipation term. We have

$$u\frac{\partial T}{\partial x} + v\frac{\partial T}{\partial x} = \alpha\frac{\partial T}{\partial y^2} + \frac{\mu}{\rho c_p}\left(\frac{\partial u}{\partial y}\right)^2 \qquad \textbf{[B-17]}$$

The solution to this equation is first obtained for the case of an adiabatic plate. By introducing a new dimensionless temperature profile in terms of the stagnation temperature T_0,

$$\theta(\eta) = \frac{T(\eta) - T_\infty}{T_0 - T_\infty} = \frac{T(\eta) - T_\infty}{u_\infty^2/2c_p}$$

Equation (B-17) becomes

$$\frac{d^2\theta}{d\eta^2} + \frac{1}{2}\operatorname{Pr}\, f\frac{d\theta}{d\eta} + 2\operatorname{Pr}\left(\frac{d^2 f}{d\eta^2}\right)^2 = 0 \qquad \textbf{[B-18]}$$

For the adiabatic-wall case, the boundary conditions are

$$\frac{d\theta}{d\eta} = 0 \qquad \text{at } y = 0, \eta = 0$$

$$\theta = 0 \qquad \text{at } y = \infty, \eta = \infty$$

The solution to Equation (B-18) is given by Pohlhausen as

$$\theta_a(\eta, \operatorname{Pr}) = 2\operatorname{Pr}\int_\eta^\infty \left(\frac{d^2 f}{d\eta^2}\right)^{\operatorname{Pr}}\left[\int_0^\eta \left(\frac{d^2 f}{d\eta^2}\right)^{2-\operatorname{Pr}} d\eta\right] d\eta \qquad \textbf{[B-19]}$$

where the symbol θ_a has been used to indicate the adiabatic-wall solution. A graphical plot of the solution is given in Figure B-3. The recovery factor is given as

$$r = \theta_a(0, \operatorname{Pr})$$

For Prandtl numbers near unity this reduces to the relation given in Equation (5-122)

$$r = \operatorname{Pr}^{1/2} \qquad \textbf{[5-122]}$$

Figure B-3 | Temperature profiles in laminar boundary layer with adiabatic wall.

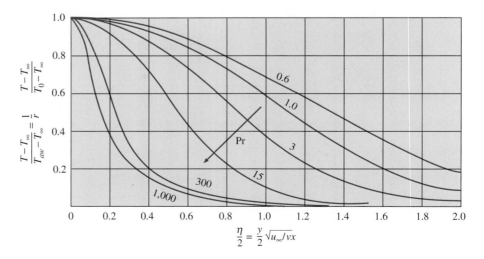

Now consider the case where the wall is maintained at some temperature other than T_{aw}; that is, there is heat transfer either to or from the fluid. The boundary conditions are now expressed as

$$T = \begin{cases} T_w & \text{at } y=0, \ \eta=0 \\ T_\infty & \text{at } y=\infty, \ \eta=\infty \end{cases}$$

We observe that the viscous-heating term in Equation (B-18) contributes a particular solution to the equation. If there were no viscous heating, the adiabatic-wall solution would yield a uniform temperature profile throughout the boundary layer. We now assume that the temperature profile for the combined case of a heated wall and viscous dissipation can be represented by a linear combination of the solutions given in Equations (B-14) and (B-19). This assumption is justified in view of the fact that Equation (B-18) is linear in the dependent variable θ. We then write

$$T - T_\infty = (T_a - T_\infty) + (T_c - T_{aw}) \tag{B-20}$$

where $T_a - T_\infty$ is the temperature distribution from Equation (B-19) and $T_c - T_{aw}$ is the solution from Equation (B-14), with T_{aw} taking the same role as T_∞ in that solution. Equation (B-20) may be written

$$T - T_\infty = \theta_a(T_0 - T_\infty) + \theta(T_{aw} - T_w) + T_w - T_{aw} \tag{B-21}$$

This solution may be tested by inserting it in Equation (B-18). There results

$$(T_0 - T_\infty)\left[\frac{d^2\theta_a}{d\eta^2} + \frac{1}{2}\operatorname{Pr} f\frac{d\theta_a}{d\eta} + 2\operatorname{Pr}\left(\frac{d^2 f}{d\eta^2}\right)^2 \right]$$

$$+ (T_{aw} - T_w)\left(\frac{d^2\theta}{d\eta^2} + \frac{1}{2}\operatorname{Pr} f\frac{d\theta}{d\eta} \right) = 0$$

An inspection of this relation indicates that Equation (B-21) is a valid solution for the actual boundary conditions of the heated wall. The temperature gradient at the wall may thus be expressed as

$$\left.\frac{\partial T}{\partial \eta}\right]_{\eta=0} = (T_0 - T_\infty)\left.\frac{\partial \theta_a}{\partial \eta}\right]_{\eta=0} + (T_{aw} - T_w)\left.\frac{\partial \theta}{\partial \eta}\right]_{\eta=0}$$

The first term is zero, and Equation (B-15) may be used to evaluate the second term. There results

$$\left.\frac{\partial T}{\partial \eta}\right]_{\eta=0} = 0.332(T_{aw} - T_w)\operatorname{Pr}^{1/3} \tag{B-22}$$

This relation immediately suggests the definition of the heat-transfer coefficient for the case where viscous heating is important by the relation

$$\frac{q}{A} = h(T_w - T_{aw}) \tag{B-23}$$

The analysis then proceeds as discussed in Section 5-12.

Analytical Relations for the Heisler Charts

Analytical solutions for the three transient cases covered in the Heisler charts of Chapter 4 are given in Reference 1 of Chapter 4. Heisler (Reference 7 of Chapter 4) was able to show that for $\alpha\tau/L^2$ or $\alpha\tau/r_0^2 > 0.2$, the infinite series solutions for the center temperature ($x = 0$ or $r = 0$) could be approximated within 1 percent with a single term:

$$\frac{\theta_0}{\theta_i} \approx C_B \exp\left(-A_B^2 \frac{\alpha\tau}{s^2}\right) \qquad \text{[C-1]}$$

where $s = L$ for the infinite plate and $s = r_0$ for the cylinder or sphere. C_B and A_B are constants that are functions of the Biot number, hL/k or hr_0/k, and are determined from the following equations.

Infinite Plate

A_B is the solution to

$$A_B \left(\tan A_B\right) = \text{Bi} = \frac{hL}{k} \qquad \text{[C-2]}$$

C_B is obtained from

$$C_B = \frac{4 \sin A_B}{2A_B + \sin 2A_B}$$

The arguments of the trigonometric functions are in radians.

Infinite Cylinder

A_B is the solution to

$$A_B J_1(A_B)/J_0(A_B) = \text{Bi} = \frac{hr_0}{k} \qquad \text{[C-3]}$$

where J_0 and J_1 are Bessel functions of the first kind and are tabulated in a number of references such as References 1 and 3 of Chapter 4. An abbreviated tabulation is given in Table C-1.

C_B for the cylinder is then evaluated from

$$C_B = \frac{2}{A_B} \frac{J_1(A_B)}{J_0^2(A_B) + J_1^2(A_B)} \qquad \text{[C-4]}$$

Table C-1 | Abbreviated tabulation of Bessel functions of the first kind.

u	$J_0(u)$	$J_1(u)$	u	$J_0(u)$	$J_1(u)$
0.0	1.0	0.0	6.0	0.15065	−0.27668
0.2	0.99003	0.09950	6.2	0.20175	−0.23292
0.4	0.96040	0.19603	6.4	0.24331	−0.18164
0.6	0.91200	0.28670	6.6	0.27404	−0.12498
0.8	0.84629	0.36884	6.8	0.29310	−0.06522
1.0	0.76520	0.44005	7.0	0.30008	−0.00468
1.2	0.67113	0.49829	7.2	0.29507	0.05432
1.4	0.56686	0.54195	7.4	0.27860	0.10963
1.6	0.45540	0.56990	7.6	0.25160	0.15921
1.8	0.33999	0.58152	7.8	0.21541	0.20136
2.0	0.22389	0.57672	8.0	0.17165	0.23464
2.2	0.11036	0.55596	8.2	0.12222	0.25800
2.4	0.00251	0.52019	8.4	0.06916	0.27079
2.6	−0.09681	0.47082	8.6	0.01462	0.27275
2.8	−0.18504	0.40970	8.8	−0.03923	0.26407
3.0	−0.26005	0.33906	9.0	−0.09033	0.24531
3.2	−0.32019	0.26134	9.2	−0.13675	0.21741
3.4	−0.36430	0.17923	9.4	−0.17677	0.18163
3.6	−0.39177	0.09547	9.6	−0.20898	0.13952
3.8	−0.40256	0.01282	9.8	−0.23228	0.09284
4.0	−0.39715	−0.06604	10.0	−0.24594	0.04347
4.2	−0.37656	−0.13865	10.2	−0.24962	−0.00662
4.4	−0.34226	−0.20278	10.4	−0.24337	−0.05547
4.6	−0.29614	−0.25655	10.6	−0.22764	−0.10123
4.8	−0.24043	−0.29850	10.8	−0.20320	−0.14217
5.0	−0.17760	−0.32760	11.0	−0.17119	−0.17679
5.2	−0.11029	−0.34322	11.2	−0.13299	−0.20385
5.4	−0.04121	−0.34534	11.4	−0.09021	−0.22245
5.6	0.02697	−0.33433	11.6	−0.04462	−0.23200
5.8	0.09170	−0.31103	11.8	0.00197	−0.23229
			12.0	0.04769	−0.22345

Sphere

A_B is the solution to

$$1 - A_B \cot A_B = \mathrm{Bi} = \frac{hr_0}{k} \qquad \text{[C-5]}$$

and C_B is obtained from

$$C_B = \frac{4(\sin A_B - A_B \cos A_B)}{2A_B - \sin 2A_B} \qquad \text{[C-6]}$$

For convenience, Table C-2 lists the parameters A_B and C_B as a function of Biot number for the three geometries under consideration.

Off-Center Temperatures

Because of the single-term approximation of the series, solutions for the off-center temperatures may also be expressed in the following simple forms.

Infinite plate:

$$\frac{\theta}{\theta_0} = \cos\left(\frac{A_B x}{L}\right) \qquad \text{[C-7]}$$

Table C-2 | Coefficients for Heisler solutions.

$\frac{hL}{k}$ or $\frac{hr_0}{k}$	Infinite plate		Long cylinder		Sphere	
	A_B	C_B	A_B	C_B	A_B	C_B
0.01	0.0998	1.0017	0.1412	1.0025	0.1730	1.0030
0.02	0.1410	1.0033	0.1995	1.0050	0.2445	1.0060
0.04	0.1987	1.0066	0.2814	1.0099	0.3450	1.0120
0.06	0.2425	1.0098	0.3438	1.0148	0.4217	1.0179
0.08	0.2791	1.0130	0.3960	1.0197	0.4860	1.0239
0.1	0.3111	1.0161	0.4417	1.0246	0.5423	1.0298
0.2	0.4328	1.0311	0.6170	1.0483	0.7593	1.0592
0.3	0.5218	1.0451	0.7465	1.0712	0.9208	1.0880
0.4	0.5932	1.0580	0.8516	1.0931	1.0528	1.1164
0.5	0.6533	1.0701	0.9408	1.1143	1.1656	1.1441
0.6	0.7051	1.0814	1.0185	1.1345	1.2644	1.1713
0.7	0.7506	1.0919	1.0873	1.1539	1.3525	1.1978
0.8	0.7910	1.1016	1.1490	1.1724	1.4320	1.2236
0.9	0.8274	1.1107	1.2048	1.1902	1.5044	1.2488
1.0	0.8603	1.1191	1.2558	1.2071	1.5708	1.2732
2.0	1.0769	1.1785	1.5995	1.3384	2.0288	1.4793
3.0	1.1925	1.2102	1.7887	1.4191	2.2889	1.6227
4.0	1.2646	1.2287	1.9081	1.4698	2.4556	1.7202
5.0	1.3138	1.2403	1.9898	1.5029	2.5704	1.7870
6.0	1.3496	1.2479	2.0490	1.5253	2.6537	1.8338
7.0	1.3766	1.2532	2.0937	1.5411	2.7165	1.8674
8.0	1.3978	1.2570	2.1286	1.5526	2.7654	1.8920
9.0	1.4149	1.2598	2.1566	1.5611	2.8044	1.9106
10.0	1.4289	1.2620	2.1795	1.5677	2.8363	1.9249
20.0	1.4961	1.2699	2.2881	1.5919	2.9857	1.9781
30.0	1.5202	1.2717	2.3261	1.5973	3.0372	1.9898
40.0	1.5325	1.2723	2.3455	1.5993	3.0632	1.9942
50.0	1.5400	1.2727	2.3572	1.6002	3.0788	1.9962
100.0	1.5552	1.2731	2.3809	1.6015	3.1102	1.9990

Infinite cylinder:

$$\frac{\theta}{\theta_0} = J_0\left(\frac{A_B r}{r_0}\right) \qquad\qquad \textbf{[C-8]}$$

Sphere:

$$\frac{\theta}{\theta_0} = \frac{r_0}{r A_B}\sin\left(\frac{A_B r}{r_0}\right) \qquad\qquad \textbf{[C-9]}$$

Again, arguments of the trigonometric functions are in radians.

Total Heat Loss

The total heat loss in time τ corresponding to Figures 4-16 to 4-18 can also be obtained as follows.

Infinite plate:

$$\frac{Q}{Q_0} = 1 - \left(\frac{\theta_0}{A_B \theta_i}\right)\sin A_B \qquad\qquad \textbf{[C-10]}$$

Infinite cylinder:

$$\frac{Q}{Q_0} = 1 - \left(\frac{2\theta_0}{A_B \theta_i}\right) J_1(A_B) \qquad \text{[C-11]}$$

Sphere:

$$\frac{Q}{Q_0} = 1 - \left(\frac{3\theta_0}{A_B^3 \theta_i}\right)(\sin A_B - A_B \cos A_B) \qquad \text{[C-12]}$$

In all of the above equations A_B and C_B are the coefficients *for the particular geometry,* tabulated as a function of Biot number in Table C-2.

The analytical solutions obviously afford the user greater accuracy than the charts, but they may be more cumbersome to handle. The method selected for a problem solution may depend on uncertainties in the convection boundary conditions. In some cases a combination may be used, perhaps the analytical method to find θ_0 and the graphs to evaluate θ/θ_0. This might avoid the need for evaluating Bessel functions.

EXAMPLE C-1   Cooling of Small Cylinder

A long, steel cylinder having a diameter of 5 cm has an initial uniform temperature of 250°C and is suddenly exposed to a convection environment at 80°C with $h = 500$ W/m$^2 \cdot$°C. Calculate the temperature at a radius of 1.2 cm after a time of 1 min and the heat lost per unit length during this period. Take the properties of steel as $\rho = 7800$ kg/m^3, $c = 0.48$ kJ/kg $\cdot$ °C, and $k = 35$ W/m $\cdot$ °C.

■ **Solution**
We first calculate the thermal diffusivity as

$$\alpha = \frac{k}{\rho c} = \frac{35}{(7800)(480)} = 9.35 \times 10^{-6} \text{ m}^2/\text{s}$$

and the Biot number as

$$\text{Bi} = \frac{h r_0}{k} = \frac{(500)(0.025)}{35} = 0.357$$

From Table C-2

$$A_B = 0.8064 \qquad C_B = 1.0837$$

From Equation (C-1)

$$\frac{\theta_0}{\theta_i} = (1.0837) \exp\left[\frac{-(0.8064)^2 (9.35 \times 10^{-6})(60)}{(0.025)^2}\right]$$

$$= 0.6045$$

For the off-centerline temperature we must use Equation (C-8) to obtain

$$\frac{\theta}{\theta_0} = J_0\left[\frac{(0.8064)(0.012)}{0.025}\right] = J_0(0.387) = 0.9623$$

Then

$$\frac{\theta}{\theta_i} = (0.6045)(0.9623) = 0.5817$$

and

$$T = (0.5817)(250 - 80) + 80 = 178.9°C$$

To obtain the heat loss we employ Equation (C-11)

$$\frac{Q}{Q_0} = 1 - \left[\frac{(2)(0.6045)}{0.8064} \right] J_1(0.8064)$$

From Table C-1, $J_1(0.8064) = 0.37112$ so that

$$\frac{Q}{Q_0} = 0.4436$$

We have

$$Q_0 = \rho c V \theta_i = (7800)(480)\pi(0.025)^2(250 - 80) = 1.25 \times 10^6 \text{ J}$$

so that

$$Q = (0.4436)(1.25 \times 10^6) = 5.54 \times 10^5 \text{ J/m length}$$

Use of Microsoft Excel for Solution of Heat-Transfer Problems

D-1 | INTRODUCTION

Microsoft Excel is a very versatile spreadsheet software utility. In this appendix we describe some of the features that may be applicable to solution of heat-transfer problems. We assume that the reader has or will acquire at least a perfunctory knowledge of Excel before using this material in order that satisfactory articulation can result. Many sources are available for such articulation including References [1, 2, 3, 4, 5]. Graphical procedures are described in [2, 3] and numerical solutions in [1, 4, 5]. In addition, the Help/Index feature of the software is useful in obtaining on-the-spot operational information. The presentation here is necessarily brief, and makes generous reference to equations, tables, and examples in the body of the text. The exposition is intended to supplement the basic material of the text.

D-2 | EXCEL TEMPLATE FOR SOLUTION OF STEADY-STATE HEAT-TRANSFER PROBLEMS

The grid structure of the Excel worksheet and relative-cell reference feature for copying cell formulas provide a convenient vehicle for steady-state numerical solutions of conduction problems. The nodal equations for $\Delta x = \Delta y$ presented in Table 3-2, in the form suitable for iterative solution, may be listed in general form on a worksheet template, and then copied to appropriate cells on the sheet as required by the geometric configuration of the problem and boundary conditions. The solution of the set of equations is accomplished by a built-in iterative-solution feature of Excel as described below.

An Excel template listing two-dimensional nodal equations with $\Delta x = \Delta y$ is shown in Figure D-1(a) for interior nodes, flat-boundary nodes with convection coefficient h and fluid temperature T_∞, and both interior and exterior corner nodes having a convection exposure. Note that the equation for a right-outside-convection node is the same as for a left-inside-convection surface as illustrated in the nomenclature of Figure D-2. The equations are written in the functional form

$$T_i = f(T'_j\text{s},\ \text{Bi})$$

where the T'_js are the temperatures of the nodes connecting to T_i, and $\text{Bi} = h\Delta x/k$, as noted in Table 3-2. Display of the formulas may be retained on the spreadsheet by clicking

Figure D-1a | Steady-state template.

	A	B	C
1			
2			Generalized nodal formulas for $\Delta x = \Delta y$
3			and $Bi = h\Delta x/k$ $Bi = 0$ for insulated
4			surface
5			
6	Interior node		=(B6+C5+D6+C7)/4
7			
8	Right conv surface, or		=(B8+$B$16*$B$17+(C7+C9)/2)/(2+$B$16)
9	inside left		
10	Left conv surface,		=(D10+$B$16*$B$17+(C9+C11)/2)/(2+$B$16)
11	or inside right		
12	Top conv surface,		=(C13+$B$16*$B$17+(B12+D12)/2)/(2+$B$16)
13	or inside bottom		
14	Bottom conv surface,		=(C13+$B$16*$B$17+(B14+D14)/2)/(2+$B$16)
15	or inside top		
16	Bi =		
17	Conv. Temp =		
18			
19	Top right corner		=($B$16*$B$17+(B19+C20)/2)/(1+$B$16)
20			
21	Top left corner		=($B$16*$B$17+(D21+C22)/2)/(1+$B$16)
22			
23	Bottom right corner		=($B$16*$B$17+(C22+B23)/2)/(1+$B$16)
24			
25	Bottom left corner		=($B$16*$B$17+(C24+D25)/2)/(1+$B$16)
26			
27	Interior top left corner		=($B$16*$B$17+C26+B27+(D27+C28)/2)/(3+$B$16)
28			
29	Interior top right corner		=($B$16*$B$17+D29+C28+(B29+C30)/2)/(3+$B$16)
30			
31	Interior bottom left corner		=($B$16*$B$17+B31+C32+(C30+D31)/2)/(3+$B$16)
32			
33	Interior bottom right corner		=($B$16*$B$17+D33+C34+(B33+C32)/2)/(3+$B$16)
34			

Figure D-1b | Template for composite interfaces.

	A	B	C
34			
35	Top composite surface		=(2*(C36+$B$51*C34)+(1+$B$51)*(B35+D35))/((1+$B$51)*4)
36			
37	Right composite surface		=(2*(B37+$B$51*D37)+(1+$B$51)*(C36+C38))/((1+$B$51)*4)
38			
39	Left composite surface		=(2*($B$51*B39+D39)+(1+$B$51)*(C38+C40))/((1+$B$51)*4)
40			
41	Bottom composite surface		=(2*(C40+$B$51*C42)+(1+$B$51)*(B41+D41))/((1+$B$51)*4)
42			
43	Top right composite corner		=(2*$B$51*(C42+D43)+(1+$B$51)*(B43+C44))/((3*$B$51+1)*2)
44			
45	Top left composite corner		=(2*$B$51*(B45+C44)+(1+$B$51)*(C46+D45))/((3*$B$51+1)*2)
46			
47	Bottom left composite corner		=(2*$B$51*(B47+C48)+(1+$B$51)*(C46+D47))/((3*$B$51+1)*2)
48			
49	Bottom right composite corner		=(2*$B$51*(D49+C50)+(1+$B$51)*(B49+C48))/((3*$B$51+1)*2)
50			
51	[k(ext)]/[k(int)] = K =		

Figure D-2 | Nomenclature for steady-state nodal equations with $\Delta x = \Delta y$.

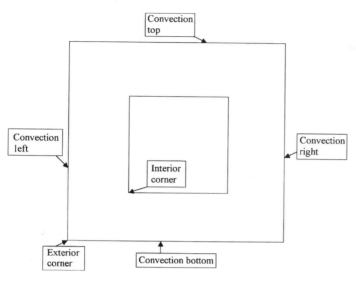

Figure D-3 | Nomenclature for composite boundary.

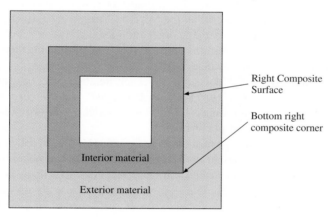

$K = k(\text{exterior material})/k(\text{interior material})$
$K = 1$, no composite
$K = 0.001$ exterior highly insulating, interior highly
          conductive
$K = 1000$ interior highly insulating, exterior highly
          conductive

TOOLS/OPTIONS/VIEW/check formulas. Display of numerical values calculated by the formulas is accomplished by removing the check from the formulas box.

More complex formulas are encountered with an interface surface or corner separating two materials with different thermal conductivities. The nomenclature applied to such composite nodes is given in Figure D-3 and the equations are listed in cells C35:C49 of the template in Figure D-1(b) for $\Delta x = \Delta y$. Note that these formulas involve the parameter $K = k_{\text{exterior}}/k_{\text{interior}}$ and provision is made for its specification in cell B51. This value is called as \$B\$51 in the respective nodal formulas, just as the value of Bi is called as \$B\$16 in the convection surface formulas.

Stability Criteria for Composite Interface

When a transient problem is formulated that involves composite interfaces a maximum time increment must be determined for the calculation. Using the general criterion given by Equation (4-46) the maximum allowable time increments for the composite interfaces may be written for the $\Delta x = \Delta y$ case as

Plane Wall

$$\Delta \tau_{max} = (K + M)(\Delta x)^2 / 4\alpha_{ext}(K + 1)$$

Corner

$$\Delta \tau_{max} = (3K + M)(\Delta x)^2 / 4\alpha_{ext}(3K + 1)$$

with

$$K = k_{ext}/k_{int} \quad \text{and} \quad M = \alpha_{ext}/\alpha_{int}$$

The setup for transient solutions is described on Section D-5 and an example of steady-state calculations for a composite is given in Example D-9.

The procedure for use of the steady-state templates is as follows:

1. On a blank worksheet sketch the nodal system on a part of the worksheet grid away from the cells occupied by the formulas in Figure D-1. Remember that the solution is for $\Delta x = \Delta y$ even though the grid on the worksheet may not be exactly square. Note the type of node (interior, convection, corner, etc.) for each cell or set of cells, namely, all internal nodes will use the formula of C6 on the template, a right exterior convection surface will use the formula at C8, etc.

2. Click TOOLS/OPTIONS/CALCULATION/check iteration box. *Important: Be sure that the box "Workbook Options/Precision Displayed" is NOT checked.* Excel calculates to 15-digit precision unless this box is checked. Severe round-off error can result if full precision is not performed.

3. Copy/Paste the appropriate type formula from Column C into each cell of the worksheet set aside in step 1. Where multiple applications are required as for several interior nodes, or several nodes along a surface, the Drag-Copy feature of Excel may be used to speed the process. Note from Table 3-2 that an insulated surface is accommodated by setting Bi = 0 for that surface. So, an insulated right side surface would use the formula at C8 with zero (0) substituted for $B$16 in the formula.

4. Enter numerical values for specified boundary nodal temperatures in the appropriate cells. These cells do not require a formula entry; only a numerical value.

5. Enter numerical values for Bi and the convection temperature in cells B16 and B17. NOTE: If there are different values of h and T_∞ for different surfaces they must be entered accordingly in the nodal equations *after* the copy/paste operation of step 3.

6. If the formulas have been retained in the display during the above steps, the numerical solution will not yet appear. To display the solution click TOOLS/OPTIONS/ VIEW/remove check from Formulas box. The solution will appear in accordance with number of iterations selected. The default value is 100 iterations, which is often satisfactory. See Excel Help/Index for instructions on modifications of iterations, etc. The number of iterations that will be needed depends on the physical problem and the number of nodes employed in the formulation.

7. A graphical display of the results may be obtained by using the Excel Chart Wizard and a wire-mesh or color-surface chart. Some typical displays are shown in the examples.

In these displays no attempt has been made to finalize the charts with appropriate scale and coordinate labels or cosmetics. They are presented essentially as they initially appear from the Chart Wizard. Addition of proper labeling is important, of course, in final presentation of results.

8. The effects of change(s) in the boundary conditions may be observed by modifying the entry values for Bi, T_∞, or fixed-boundary temperatures. The changes will appear in both the tabulated values of temperatures and the graphical displays. If desired, separate cells may be employed at a convenient location in the worksheet for calculation of Bi from adjustable values of h, Δx, and k.

9. The worksheet templates of Figures D-1(a,b) are available on the text website www.mhhe.com/holman.

D-3 | SOLUTION OF EQUATIONS FOR NONUNIFORM GRID AND/OR NONUNIFORM PROPERTIES

The advantage of the above template and procedure is that writing of the nodal equations is speeded by the relative-cell copy feature of Excel. In addition, the numerical solutions are displayed in a format that matches the configuration of the physical situation. The relative-cell copy feature may only be employed for the case of square grids with $\Delta x = \Delta y$.

Alternatively, the nodal equations can be written separately for individual nodes, and then solved with the same iterative procedure. If the equations are written individually, they may be formulated using the general resistance nomenclature of Equation (3-32), which of course allows for any possible configuration of Δx and Δy, as well as different materials. A value of $Bi = h\Delta x / \Delta k$ no longer has meaning in this case and the value of h is entered for each equation as needed. The procedure is thus:

1. Sketch the geometry of the problem and specify nomenclature for nodal temperatures and values of Δx and Δy.
2. Write the equation for each node in the form of Equation (3-32).
3. Set aside two columns and n rows, with n = number of temperature nodes. Click TOOLS/OPTIONS/VIEW/Formula check and TOOLS/OPTIONS/CALCULATION/ Iteration check.
4. Enter T1 = in 1st column, 1st row; T2 = in 1st column, 2nd row; and so on.
5. Enter the respective nodal temperature formula from step 2 in the second column, referring to adjoining temperatures using their respective cell formula locations (i.e., T1 will be in the 1st row of column 2, T2 will be in the 2nd row of column 2, etc.).
6. Click TOOLS/OPTIONS/VIEW/Uncheck formulas box. Solution will appear. The solution obtained with this procedure is not in a form suitable for easy graphical display.

D-4 | HEAT SOURCES AND RADIATION BOUNDARY CONDITIONS

As noted in Chapter 3, the nodal equations for $\Delta x = \Delta y$ as shown in Table 3-2 may be modified to include heat sources and radiation boundary conditions by adding the following term to the numerator of each equation:

$$q_i = \dot{q}_i \Delta V_i \qquad\qquad \textbf{[D-1]}$$

for heat sources and

$$q_i = q''_{\text{rad},i} \times \Delta A_i \qquad \text{[D-2]}$$

for radiation boundary nodes where $\dot{q}$ = heat generation per unit volume and $q''_{\text{rad},i}$ is determined according to procedures of Chapter 8. In the common case where a surface boundary node having an emissivity of ϵ_i exchanges heat with a large radiation surroundings at T_r, we have for the *net radiant transfer into the surface node:*

$$q_{\text{rad},i} = \sigma \epsilon_i \Delta A_i (T_r^4 - T_i^4) \qquad \text{[D-3]}$$

All temperatures must be in degrees absolute. T_r may or may not be equal to the temperature of a convection environment in contact with the surface node. ΔA_i is the surface area of the node i exposed to the radiation temperature T_r.

   Again, the advantage of the Excel spreadsheet type of solution is that the nodes may be arranged in the same geometric pattern as the physical problem and thus provide the opportunity for a graphical display of temperatures to match. The examples illustrate use of the method.

D-5 | EXCEL PROCEDURE FOR TRANSIENT HEAT TRANSFER

Setup of Worksheet

1. Set aside a block of cells for node calculations with

      Number of temperature nodes = number of columns

      Number of time rows = number of time increments + 3

2. Determine and list material properties in a portion of the worksheet apart from that block set aside for the nodal calculations. Provide cells for listing boundary conditions in this same area. Also provide calculation cells for Bi and Fo for the case where $\Delta x = \Delta y$. Provide a cell for specification of the time increment selection in accordance with the calculation in step 3 below.

3. Determine maximum allowable time increment $\Delta \tau$ for each node in accordance with Equation (4-46). If $\Delta x = \Delta y$, exterior corner nodes will be most restrictive in accordance with the entries in Table 4-2.

In the Block Set Aside for Nodal Temperature Calculations

4. Using a sketch of the node layout for the problem, set nomenclature for node temperatures in the top row of each column and enter T1 =, T2 =, etc.

5. Skip a row.

6. Insert initial values (time = 0) for each nodal temperature in the next row, or enter formulas to calculate these values from other data in the problem, that is to say, a spatial variation that is specified in terms of node location.

7. In the next row, insert (write) formulas for each T^{p+1} in form of Equation (4-47) or respective formula of Table 4-2 for the case of $\Delta x = \Delta y$. In the formulas, refer to values of T^p specified in row of step 6. To retain a display of all formulas during the setup, click TOOLS/OPTIONS/VIEW/Formula check box. Specify the time increment, or Fo and Bi for each nodal formula in terms of the cell locations provided in step 2 above.

8. Drag-copy the formulas for a number of rows equal to the number of time increments desired.

Execution of Program and Graphs of Results

9. Check formulas to see that they are correct!

10. Select a time increment having a value less than the most restrictive condition determined in step 3. Take into consideration the total time desired in the final solution and the number of time increments (rows) available. If needed, drag-copy the formulas for additional rows.

11. Insert all material properties, boundary conditions, and initial conditions in the cells provided in steps above.

12. If formulas are still displayed, click TOOLS/OPTIONS/VIEW/Remove check mark from formula box. Solution will appear.

13. Graphs of the nodal temperatures as a function of the number of time increments (rows) may be obtained by using the Chart Wizard and a *line chart* display, either with or without data markers. If a sufficiently large number of time increments is displayed, the solutions will approach the steady-state values. The number of time increments needed to obtain steady state obviously depends on geometric configuration, material properties, initial conditions, and not insignificantly, the value of $\Delta \tau$ selected.

Graphs of Transient Results with Geometric Configuration

The line charts obtained in step 13 match with the numbers assigned to their respective nodes. The results may also be presented in a format that matches with the geometric configuration of the physical problem by using the following procedure.

14. In an area of the worksheet away from the calculations of step 12 above, or on a new worksheet, set aside a block of cells with configuration to match the physical problem; one cell for each node ($\Delta x = \Delta y$). Be sure to include cells for all fixed-boundary nodal temperatures.

15. For each cell, corresponding to the nomenclature of the problem enter the formula = $column(temperature node), row(time increment). All the row numbers will be the same for the same time increment, even though they appear in different rows in the work sheet.

16. The matrix of numbers obtained in step 15 may be graphed with the Chart Wizard as a wire-mesh, area chart, or 3-D surface chart to display the temperature profile across the geometry for the respective time increment.

17. To obtain the temperature distribution at a subsequent time increment, copy the block of cells obtained in step 15 to another location at row numbers equal to row (step 15) + difference in time increments = row(new block of cells). If the top of the block of cells in step 15 was in row 10 for time increment 2, the top of the block of cells will be in row 13 for time increment 5, and so on.

18. Repeat the procedure of step 17 for as many time increments as desired. Although a single worksheet can accommodate many displays, it may be desirable to copy/paste the solutions on multiple worksheets for printout purposes.

It is also possible to perform the transient calculation by utilizing the recalculation feature of Excel whereby the equations are arranged in two rows for successive time increments and circular references. The number of time increments may be specified for the final result without display of intermediary values. This approach may be preferable for large numbers of time increments, or in those cases where the only result desired is for a specific elapsed time. The method is described in [1]. Use of Visual Basic may also be desirable. The results may still be copied to a spreadsheet block of cells that represents the geometric configuration of the physical problem. This approach, however, does not provide data for construction of the temperature-time line charts described in step 13 above. For complex problems, or those where the results must be displayed in a special format, a combination of methods may be needed.

EXAMPLE D-1 | Temperature Distribution in Two-Dimensional Plate

The two-dimensional plate system described by Equation (3-20) is very easily analyzed numerically using an Excel setup for the nodal equations. We choose $T_1 = 0$ as the constant boundary temperature along the bottom and sides of the plate and $T_2 = 100$ as the constant temperature along the top of the plate. The Excel grid is set up as shown in Figure Example D-1(a) with the constant temperature of 100 inserted in cells E5:Q5. The zero temperatures are inserted in cells E6:E44, Q6:Q44, and F44:P44. All the other nodes are interior nodes and have the same form of nodal equation, i.e., that given in cell C6 of the template. This formula is therefore copied to cell F6 and then drag-copied to the remaining temperature cells. The solution appears as shown in Figure Example D-1(b).

Graphical displays of the results can be presented simply as plan and elevation views of the temperature profiles as shown in Figure Example D-1(c), or as a collection of 3-D charts coupled

Figure Example D-1a | Layout of grid.

Figure Example D-1b

	E	F	G	H	I	J	K	L	M	N	O	P	Q
4													
5	100.00	100.00	100.00	100.00	100.00	100.00	100.00	100.00	100.00	100.00	100.00	100.00	100.00
6	0.00	49.27	68.26	76.71	80.82	82.77	83.35	82.77	80.82	76.71	68.26	49.27	0.00
7	0.00	28.80	47.07	57.76	63.81	66.90	67.85	66.90	63.81	57.76	47.07	28.80	0.00
8	0.00	18.88	33.46	43.46	49.75	53.18	54.27	53.18	49.75	43.46	33.46	18.88	0.00
9	0.00	13.25	24.44	32.86	38.55	41.80	42.85	41.80	38.55	32.86	24.44	13.25	0.00
10	0.00	9.67	18.21	24.98	29.78	32.61	33.55	32.61	29.78	24.98	18.21	9.67	0.00
11	0.00	7.21	13.74	19.08	22.98	25.33	26.12	25.33	22.98	19.08	13.74	7.22	0.00
12	0.00	5.46	10.45	14.62	17.72	19.62	20.26	19.62	17.72	14.62	10.45	5.46	0.00
13	0.00	4.16	7.99	11.22	13.66	15.16	15.67	15.17	13.66	11.22	7.99	4.16	0.00
14	0.00	3.18	6.13	8.63	10.53	11.71	12.11	11.71	10.53	8.63	6.13	3.18	0.00
15	0.00	2.44	4.71	6.64	8.12	9.04	9.35	9.04	8.12	6.64	4.71	2.44	0.00
16	0.00	1.88	3.62	5.11	6.25	6.97	7.21	6.97	6.25	5.11	3.62	1.88	0.00
17	0.00	1.44	2.79	3.94	4.82	5.37	5.56	5.37	4.82	3.94	2.79	1.44	0.00
18	0.00	1.11	2.15	3.04	3.72	4.14	4.29	4.14	3.72	3.04	2.15	1.11	0.00
19	0.00	0.86	1.65	2.34	2.86	3.19	3.31	3.19	2.86	2.34	1.65	0.86	0.00
20	0.00	0.66	1.27	1.80	2.21	2.46	2.55	2.46	2.21	1.80	1.27	0.66	0.00
21	0.00	0.51	0.98	1.39	1.70	1.90	1.96	1.90	1.70	1.39	0.98	0.51	0.00
22	0.00	0.39	0.76	1.07	1.31	1.46	1.51	1.46	1.31	1.07	0.76	0.39	0.00
23	0.00	0.30	0.58	0.82	1.01	1.13	1.17	1.13	1.01	0.82	0.58	0.30	0.00
24	0.00	0.23	0.45	0.64	0.78	0.87	0.90	0.87	0.78	0.64	0.45	0.23	0.00
25	0.00	0.18	0.35	0.49	0.60	0.67	0.69	0.67	0.60	0.49	0.35	0.18	0.00
26	0.00	0.14	0.27	0.38	0.46	0.51	0.53	0.51	0.46	0.38	0.27	0.14	0.00
27	0.00	0.11	0.21	0.29	0.36	0.40	0.41	0.40	0.36	0.29	0.21	0.11	0.00
28	0.00	0.08	0.16	0.22	0.27	0.31	0.32	0.31	0.27	0.22	0.16	0.08	0.00
29	0.00	0.06	0.12	0.17	0.21	0.24	0.24	0.24	0.21	0.17	0.12	0.06	0.00
30	0.00	0.05	0.09	0.13	0.16	0.18	0.19	0.18	0.16	0.13	0.09	0.05	0.00
31	0.00	0.04	0.07	0.10	0.12	0.14	0.14	0.14	0.12	0.10	0.07	0.04	0.00
32	0.00	0.03	0.06	0.08	0.10	0.11	0.11	0.11	0.10	0.08	0.06	0.03	0.00
33	0.00	0.02	0.04	0.06	0.07	0.08	0.09	0.08	0.07	0.06	0.04	0.02	0.00
34	0.00	0.02	0.03	0.05	0.06	0.06	0.07	0.06	0.06	0.05	0.03	0.02	0.00
35	0.00	0.01	0.03	0.04	0.04	0.05	0.05	0.05	0.04	0.04	0.03	0.01	0.00
36	0.00	0.01	0.02	0.03	0.03	0.04	0.04	0.04	0.03	0.03	0.02	0.01	0.00
37	0.00	0.01	0.01	0.02	0.03	0.03	0.03	0.03	0.03	0.02	0.01	0.01	0.00
38	0.00	0.01	0.01	0.02	0.02	0.02	0.02	0.02	0.02	0.02	0.01	0.01	0.00
39	0.00	0.00	0.01	0.01	0.01	0.02	0.02	0.02	0.01	0.01	0.01	0.00	0.00
40	0.00	0.00	0.01	0.01	0.01	0.01	0.01	0.01	0.01	0.01	0.01	0.00	0.00
41	0.00	0.00	0.00	0.01	0.01	0.01	0.01	0.01	0.01	0.01	0.00	0.00	0.00
42	0.00	0.00	0.00	0.00	0.00	0.01	0.01	0.01	0.00	0.00	0.00	0.00	0.00
43	0.00	0.00	0.00	0.00	0.00	0.00	0.00	0.00	0.00	0.00	0.00	0.00	0.00
44	0.00	0.00	0.00	0.00	0.00	0.00	0.00	0.00	0.00	0.00	0.00	0.00	0.00
45													

Figure Example D-1c

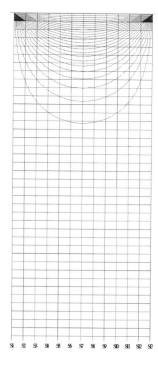

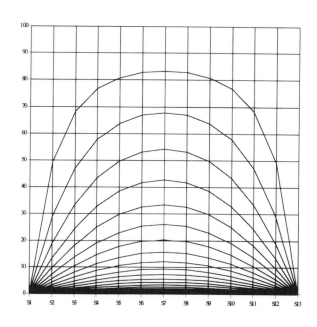

Figure Example D-1d

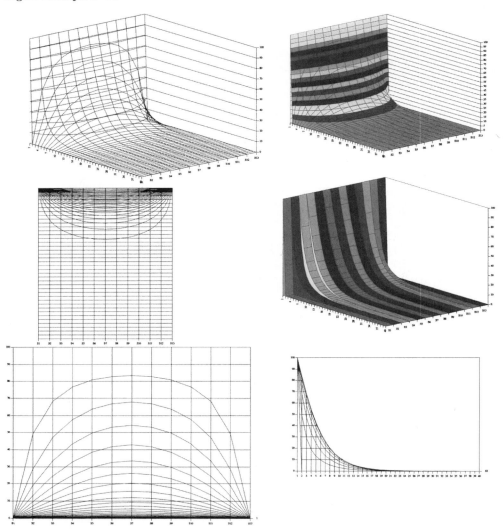

with the plan and elevation views, as shown in Figure Example D-1(d). Color presentations may also be used, but the charts given here are printed in grayscale.

| **EXAMPLE D-2** | Excel Solution and Display of Temperature Distribution in Two-Dimensional Straight Fin |

The thick fin in Figure Example D-2(a) is very long in the direction perpendicular to the page, and its thickness t is sufficiently large compared to the length L to cause a two-dimensional temperature distribution to result when exposed to the convection environment at h, T_∞. A two-dimensional grid is set up as shown in Figure Example D-2(b). The convection environment temperature is 0°C with the base temperature set at 100°C. The nodal equations are written in the format given in Table 4-2 and the respective values of $Bi = h\Delta x/k$. The technique for setting up the problem in Excel is described in Section D-2, and the results of the calculations shown in

Figure Example D-2a,b

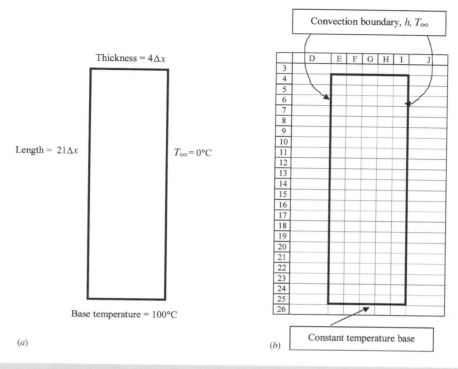

(a)

(b)

Figure Example D-2(c) for different values of the Biot parameter. The 3-D displays illustrate the following behavior of the system:

1. Smaller values of Bi (small convection, large conduction) result in more uniform temperature profiles across the thickness of the fin, i.e., the fin behaves more like a one-dimensional fin as described in Chapter 2.
2. Smaller values of Bi also result in a less rapid decrease in temperature along the length of the fin.
3. Larger values of Bi (high convection, low conduction) cause a rapid drop in temperature along the length of the fin. For Bi sufficiently large, the fin behaves as an "infinite" fin with the tip attaining the environment temperature.

The Excel spreadsheet solutions may also be used to obtain plan and elevation views of the temperature profiles in the fin as illustrated in Figure Example D-2(d) for Bi = 1.0. Color displays may also be provided for presentation purposes, but grayscale is presented here for economy reasons.

Excel Solution of Example 3-5 with and without Radiation Boundary Condition

EXAMPLE D-3

This example is rather straightforward in the Excel format. It involves a set aside of 16 cells, 7 for the fixed boundary temperatures, and 9 for the unknown nodal temperatures. Using the steady-state template and procedure we set aside the block E2:H5 and copy the appropriate formulas as shown in Figure Example D-3(a). Cells $B$16 and $B$17 correspond to $Bi = h\Delta x/k = 1/3$ and $T_\infty = 100°C$. When the iteration procedure is executed the solution appears as shown in

Figure Example D-2c

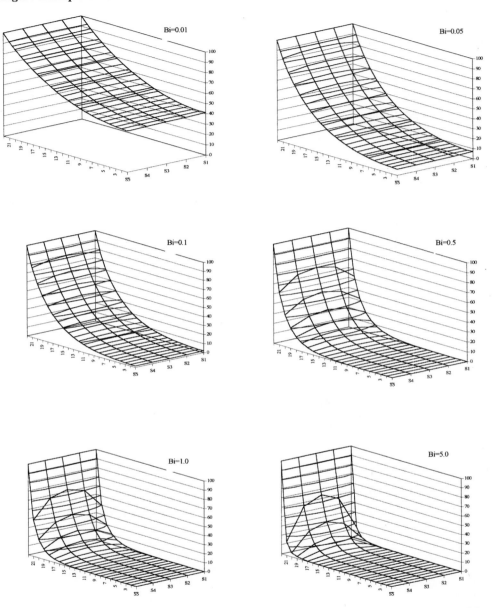

Figure Example D-3(b) along with a wire-mesh chart of the temperatures. The solution agrees with that of Example 3-5 but differs slightly because of the round off of 4 2/3 to 4.67.

Now consider the same example but with the exposed surfaces having $\epsilon = 0.9$ and exchanging radiant energy with a large enclosure having $T_r = 100°C = 373$ K. We may examine the effects of the radiation by including the term

$$q_{\text{rad},i} = \sigma \epsilon_i \Delta A_i \left(T_r^4 - T_i^4 \right) / k \qquad \text{[a]}$$

Figure Example D-2d | Temperature profiles for Bi = 1.0.

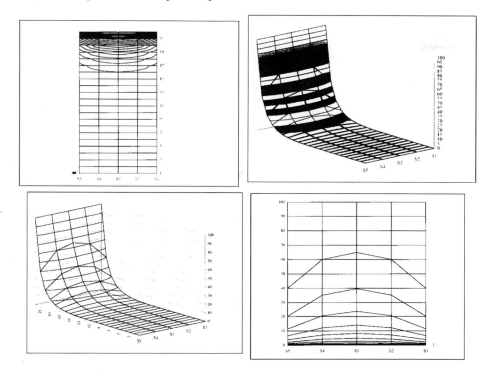

in the numerator of each of the boundary node formulas (nodes 3,6,7,8,9). The surface area for each is $\Delta x = 1/3$ and $T_r = 373$ with T_i written in each formula as

$$T_i = T_i \, (^\circ C) + 273$$

for nodal temperatures expressed in °C. Thus, Equation (a) becomes

$$q_{rad,i} = (5.669 \times 10^{-8})(0.9)(1/3)(393^4 - (T_i + 273)^4)/10 \qquad \textbf{[b]}$$

This term is now entered in the numerator for each boundary node formula as shown in Figure Example D-3(c) for column H. Finally, the iterative solution is executed and the solutions displayed as shown in Figure Example D-3(d). Note that the temperatures are substantially lower than those obtained with a convection boundary alone because of the extra radiant heat loss. The difference is shown graphically in the wire-mesh displays.

Figure Example D-3a | Formulas.

	E	F	G	H
2	500	500	500	500
3	100	=(E3+F2+G3+F4)/4	=(F3+G2+H3+G4)/4	=(G3+$B$16*$B$17+(H2+H4)/2)/(2+$B$16)
4	100	=(E4+F3+G4+F5)/4	=(F4+G3+H4+G5)/4	=(G4+$B$16*$B$17+(H3+H5)/2)/(2+$B$16)
5	100	=(F4+$B$16*$B$17+(E5+G5)/2)/(2+$B$16)	=(G4+$B$16*$B$17+(F5+H5)/2)/(2+$B$16)	=($B$16*$B$17+(H4+G5)/2)/(1+$B$16)

Figure Example D-3b

	E	F	G	H	I	J	K
1							
2	500	500	500	500			
3	100	280.7229	330.4217	309.6386			
4	100	192.4699	231.3253	217.4699			
5	100	157.8313	184.9398	175.9036			
6							

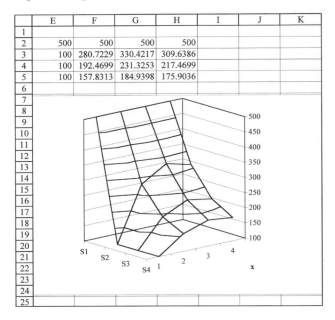

Figure Example D-3c | Boundary node formulas for radiation.

	H
1	
2	500
3	=(G3+$B$16*$B$17+(H2+H4)/2+0.000000001701*(393^4-(H3+273)^4))/(2+$B$16)
4	=(G4+$B$16*$B$17+(H3+H5)/2+0.000000001701*(393^4-(H4+273)^4))/(2+$B$16)
5	=($B$16*$B$17+(H4+G5)/2+0.000000001701*(393^4-(H5+273)^4))/(1+$B$16)

Figure Example D-3d | Convection and radiation
boundary conditions.

	E	F	G	H	I	J	K
1							
2	500.00	500.00	500.00	500.00			
3	100.00	271.55	306.92	252.22			
4	100.00	179.28	203.91	174.24			
5	100.00	141.64	155.20	141.36			
6							

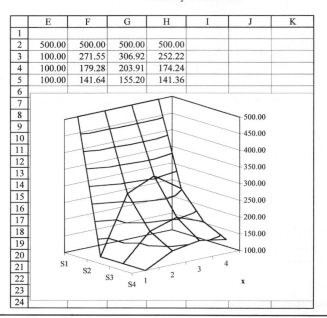

Plate with Boundary Heat Source and Convection   **EXAMPLE D-4**

The two-dimensional plate shown in Figure Example D-4(a) has the bottom three sides maintained at a constant temperature of 100°C while the top surface is exposed to a convection environment at $T_\infty = 0°C$. In addition, a heat source is imposed at the indicated node. The strength of the heat source is $\dot{q} = 400\,k$ where k is the thermal conductivity of the material. The convection boundary is such that Bi = 0.5. We will examine the temperature distribution in the solid for these given conditions, with a setup that allows for changes in specifications.

We employ the steady-state template of Section D-2 and set aside the block of cells E4:M13 for the calculations. Including nodes for the fixed-boundary temperatures requires 90 cells. The fixed-boundary temperatures of 100°C are inserted in cells E4:E13, F13:L13, and M4:M13. Formulas for interior nodes are copied from cell C6 to the block F5:L12. Formulas for a top convection surface are copied from cell C13 to F5:L5. In addition, the value of $\dot{q}/k$ specified in cell $B$15 (in this case 400) is added to the numerator of the formula for cell H4, the cell representing the node with the heat source. A display of a portion of the nodal formulas is shown in Figure Example D-4(b).

When the iterative solution procedure is executed, the results appear as shown in Figure Example D-4(c) along with a wire-mesh chart of the temperature profile. The temperature has an

Figure Example D-4a

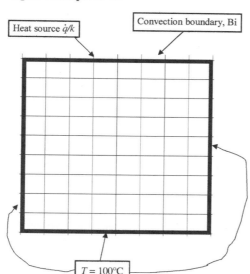

Heat source $\dot{q}/k$

Convection boundary, Bi

$T = 100°C$

Figure Example D-4b | Portion of formulas.

	E	F	G	H
2				
3				
4	100	=(F5+$B$16*$B$17+(E4+G4)/2)/(2+$B$16)	=(G5+$B$16*$B$17+(F4+H4)/2)/(2+$B$16)	=($B$15+H5+$B$16*$B$17+(G4+I4)/2)/(2+$B$16)
5	100	=(E5+F4+G5+F6)/4	=(F5+G4+H5+G6)/4	=(G5+H4+I5+H6)/4
6	100	=(E6+F5+G6+F7)/4	=(F6+G5+H6+G7)/4	=(G6+H5+I6+H7)/4
7	100	=(E7+F6+G7+F8)/4	=(F7+G6+H7+G8)/4	=(G7+H6+I7+H8)/4
8	100	=(E8+F7+G8+F9)/4	=(F8+G7+H8+G9)/4	=(G8+H7+I8+H9)/4
9	100	=(E9+F8+G9+F10)/4	=(F9+G8+H9+G10)/4	=(G9+H8+I9+H10)/4
10	100	=(E10+F9+G10+F11)/4	=(F10+G9+H10+G11)/4	=(G10+H9+I10+H11)/4
11	100	=(E11+F10+G11+F12)/4	=(F11+G10+H11+G12)/4	=(G11+H10+I11+H12)/4
12	100	=(E12+F11+G12+F13)/4	=(F12+G11+H12+G13)/4	=(G12+H11+I12+H13)/4
13	100	100	100	100

Figure Example D-4c

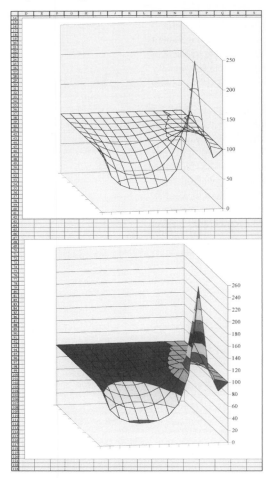

Figure Example D-4d | $\dot{q} = 400k$, $\mathrm{Bi} = 0.5$.

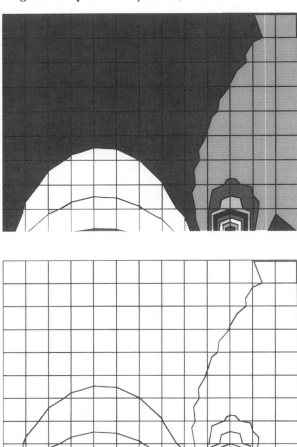

obvious peak at the node containing the heat source, is constant along lower boundaries, and drops off near the convection surface. Plan views of the temperature distribution are also shown in Figure Example D-4(d). The jagged behavior of some of the isotherms results from the relatively coarse grid and the interpolation scheme of the software. Use of a finer grid structure would produce smoother curves.

The effects of different heat-source strengths, convection coefficients, or environment temperatures can be examined by entering new values for the respective quantities in cells B15, B16, and B17.

EXAMPLE D-5

Transient Analysis of Example 3-5 Carried to Steady State

Suppose the two-dimensional solid of Example 3-5 is maintained at an initial uniform temperature of 100°C. The bottom and right surfaces are then suddenly exposed to a convection boundary with $h = 10$ W/m$^2 \cdot$°C and $T_\infty = 100$°C, while the left surface is maintained at 100°C and the top surface is raised to 500°C. The material properties are: $\rho = 8000$ kg/m^3, $c = 290$ J/kg $\cdot$ °C, and $k = 10$ W/m $\cdot$ °C. $\Delta x = \Delta y = 1/3$ m. We will choose an appropriate value of $\Delta\tau$ and compute the transient response using the node system shown in Figure 3-9 for Example 3-5. Two graphical displays will be given for $n = 5$ time increments and for n sufficiently large to achieve steady state. The steady-state solution will be shown to match that obtained in Example 3-5.

First, we calculate the values of Bi and α as

$$Bi = h\Delta x/k = (10)(1/3)/10 = 1/3$$

$$\alpha = k/\rho c = 10/(8000)(290) = 4.31 \times 10^{-6} \text{ m}^2/\text{s}$$

We noted in Chapter 4 that for $\Delta x = \Delta y$ the most restrictive time-increment selection is determined by an exterior corner node so that

$$\Delta\tau_{max} \leq (\Delta x)^2/4\alpha(1 + Bi) = 4833 \text{ sec.}$$

We will choose $\Delta\tau = 4000$ sec for the calculation and determine the Fourier modulus as

$$Fo = \alpha \Delta\tau/(\Delta x)^2 = 0.1862$$

Note that there are 16 nodes in the physical problem but 7 boundary nodal temperatures are specified so only 9 nodes participate in the heating process. Columns A through I are reserved for the nine-node calculation process with row 1 labels as

$$\begin{array}{ccc} A & B & C \\ T1 = & T2 = & T3 = \text{etc.} \end{array}$$

Row 2 is skipped. The initial temperatures for all nodes are entered in row 3 as the given values of 100. The nodal equations are now entered in row 4 using the formulation of Table 4-2. The calculated values of Bi and Fo are inserted in cells M1 and O1 for convenience and then called as $M$1 and $O$1 in the nodal formulas. The resulting equations are shown in Figure Example D-5(a). Next, the equations are drag-copied for 43 rows, which allows for a sufficient number of time increments to achieve steady state. The column display of the transient solution is shown in Figure Example D-5(b).

The nodal temperatures after 5 time increments are displayed in row 8 and steady-state values are achieved for rows greater than about 35. To obtain a graphical display of the solution(s) in a configuration that matches the geometry, we must set aside a block of 16 cells, corresponding to 7 fixed-boundary temperatures plus 9 varying-temperature nodes. We have selected K2:N5 as

Figure Example D-5a | Nodal equations.

	A	B	C
1	T1=	T2=	T3=
2			
3	100	100	100
4	=$O$1*(500+B3+D3+100-4*A3)+A3	=$O$1*(A3+500+C3+E3-4*B3)+B3	=$O$1*(2*$M$1*(100-C3)+2*B3+500+F3-4*C3)+C3
5	=$O$1*(500+B4+D4+100-4*A4)+A4	=$O$1*(A4+500+C4+E4-4*B4)+B4	=$O$1*(2*$M$1*(100-C4)+2*B4+500+F4-4*C4)+C4
6	=$O$1*(500+B5+D5+100-4*A5)+A5	=$O$1*(A5+500+C5+E5-4*B5)+B5	=$O$1*(2*$M$1*(100-C5)+2*B5+500+F5-4*C5)+C5

	D	E	F
1	T4=	T5=	T6=
2			
3	100	100	100
4	=$O$1*(A3+100+E3+G3-4*D3)+D3	=$O$1*(D3+B3+F3+H3-4*E3)+E3	=$O$1*(2*$M$1*(100-F3)+2*E3+C3+I3-4*F3)+F3
5	=$O$1*(A4+100+E4+G4-4*D4)+D4	=$O$1*(D4+B4+F4+H4-4*E4)+E4	=$O$1*(2*$M$1*(100-F4)+2*E4+C4+I4-4*F4)+F4
6	=$O$1*(A5+100+E5+G5-4*D5)+D5	=$O$1*(D5+B5+F5+H5-4*E5)+E5	=$O$1*(2*$M$1*(100-F5)+2*E5+C5+I5-4*F5)+F5

	G	H	I
1	T7=	T8=	T9=
2			
3	100	100	100
4	=$O$1*(2*$M$1*(100-G3)+2*D3+100+H3-4*G3)+G3	=$O$1*(2*$M$1*(100-H3)+2*E3+G3+I3-4*H3)+H3	=2*$O$1*(2*$M$1*(100-I3)+F3+H3-2*I3)+I3
5	=$O$1*(2*$M$1*(100-G4)+2*D4+100+H4-4*G4)+G4	=$O$1*(2*$M$1*(100-H4)+2*E4+G4+I4-4*H4)+H4	=2*$O$1*(2*$M$1*(100-I4)+F4+H4-2*I4)+I4
6	=$O$1*(2*$M$1*(100-G5)+2*D5+100+H5-4*G5)+G5	=$O$1*(2*$M$1*(100-H5)+2*E5+G5+I5-4*H5)+H5	=2*$O$1*(2*$M$1*(100-I5)+F5+H5-2*I5)+I5

Figure Example D-5b | Numerical values.

	A	B	C	D	E	F	G	H	I
1	T1=	T2=	T3=	T4=	T5=	T6=	T7=	T8=	T9=
2									
3	100.00	100.00	100.00	100.00	100.00	100.00	100.00	100.00	100.00
4	174.48	174.48	174.48	100.00	100.00	100.00	100.00	100.00	100.00
5	207.36	221.22	211.98	113.87	113.87	113.87	100.00	100.00	100.00
6	227.03	248.84	236.88	126.11	131.28	127.83	105.16	105.16	105.16
7	239.47	267.43	253.03	137.10	146.70	141.74	111.36	114.25	112.32
8	248.16	280.37	264.66	146.25	160.43	153.65	117.96	123.67	120.94
9	254.49	290.01	273.22	153.99	172.01	164.09	123.98	132.85	128.94
10	259.34	297.40	279.88	160.42	181.86	172.86	129.36	140.98	136.30
11	263.15	303.26	285.13	165.80	190.09	180.29	133.98	148.08	142.65
12	266.21	307.98	289.39	170.27	196.99	186.49	137.91	154.12	148.10
13	268.71	311.83	292.86	174.00	202.74	191.68	141.21	159.23	152.69
14	270.76	314.99	295.71	177.10	207.54	196.00	143.99	163.51	156.56
15	272.45	317.61	298.07	179.69	211.53	199.60	146.30	167.09	159.79
16	273.84	319.77	300.02	181.83	214.86	202.60	148.24	170.08	162.49
17	275.00	321.57	301.64	183.62	217.63	205.10	149.85	172.58	164.74
18	275.97	323.06	302.99	185.11	219.93	207.18	151.19	174.65	166.62
19	276.77	324.30	304.11	186.35	221.84	208.91	152.30	176.38	168.18
20	277.43	325.33	305.04	187.38	223.44	210.35	153.23	177.82	169.47
21	277.99	326.18	305.81	188.23	224.76	211.55	154.01	179.02	170.56
22	278.45	326.90	306.46	188.94	225.87	212.54	154.65	180.01	171.45
23	278.83	327.49	306.99	189.54	226.78	213.37	155.18	180.84	172.20
24	279.15	327.98	307.44	190.03	227.55	214.06	155.63	181.53	172.82
25	279.41	328.39	307.81	190.44	228.18	214.63	156.00	182.10	173.34
26	279.63	328.73	308.11	190.78	228.71	215.11	156.31	182.58	173.77
27	279.82	329.02	308.37	191.07	229.15	215.51	156.56	182.98	174.13
28	279.97	329.25	308.58	191.30	229.52	215.84	156.78	183.31	174.43

Figure Example D-5c,d,e

	K	L	M	N	O
1		Bi=	=1/3	Fo=	=0.1862
2	500	500	500	500	
3	100	=$A8	=$B8	=$C8	
4	100	=$D8	=$E8	=$F8	
5	100	=$G8	=$H8	=$I8	

(c)

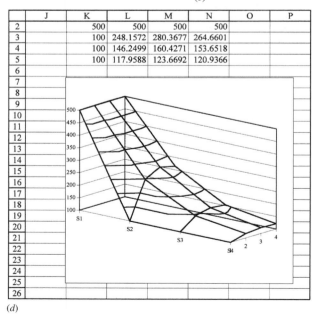

	J	K	L	M	N	O	P
2		500	500	500	500		
3		100	248.1572	280.3677	264.6601		
4		100	146.2499	160.4271	153.6518		
5		100	117.9588	123.6692	120.9366		

(d)

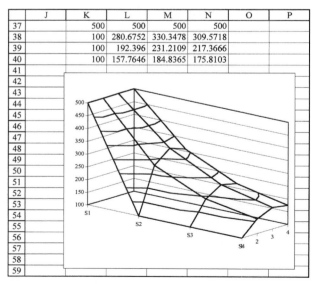

	J	K	L	M	N	O	P
37		500	500	500	500		
38		100	280.6752	330.3478	309.5718		
39		100	192.396	231.2109	217.3666		
40		100	157.7646	184.8365	175.8103		

(e)

such a block as illustrated in Figure Example D-5(c). For 5 time increments (row 8) the formulas are entered as =$A8, =$B8, and so on for nodes 1,2 etc. The fixed boundary temperatures of 100 and 500°C are also entered as constant values for their cell locations. The numerical solutions for $n = 5\Delta\tau$ then appear in the block of 16 cells.

To display the solution at $n = 40\Delta\tau$ (steady state), the block of cells at K2:N5 is selected and copied to another location starting with row 37 (row 2 + increase in number of time increments). In this case we have chosen the block and the temperatures corresponding to 40 time increments as shown in the geometric configuration of the problem. Finally, wire-mesh charts for the two time increments are displayed in Figure Examples D-5(d) and (e). Of course, displays for other time increments could be obtained using the same procedure.

One may choose to present the charts on the same worksheet as the tabulated numerical values, or as charts on separate worksheets, depending on the eventual use to be made of the information. Cosmetic features may also be added for presentation purposes.

Note that the transient solution presented in columns A through I does not display the constant-boundary temperature nodes, while the displays in the geometric configuration format show rather clearly the influence of the hot surface and convection cooling on the temperature distribution in the solid. We note again that inclusion of the constant boundary temperature nodes is an essential part of the displays in Figure Examples D-5(d) and (e).

Comparison with Steady-State Solution of Example 3-5

Note that the steady-state solution of Example 3-5 matches with the transient solution above when carried forward to a sufficiently large number of time increments.

D-6 | FORMULATION FOR HEATING OF LUMPED CAPACITY WITH CONVECTION AND RADIATION

As we have noted several times, heat-transfer problems in practice frequently involve more than one mode of heat transfer. In this section we examine the problem of a lumped-capacity solid receiving (or losing) heat from a convection environment characterized by a surface convection coefficient h_c and fluid temperature T_c, both of which may vary with time or temperature of the solid. The fluid is assumed to be transparent to thermal radiation. The solid-fluid combination is surrounded by a large enclosure having an effective radiation temperature of T_r that exchanges heat with the solid by radiation. The system is illustrated in Figure D-6 where we have shown a finned solid with a hole in its center.

We assume that the solid will radiate with some effective emissivity ϵ from a radiating surface area equal to that of a cylinder enclosing the finned surface and described by the equation shown. We assume that the convection area A_c is the total surface area of the solid exposed to the surrounding fluid with the average convection coefficient h_c.

Assuming that the solid behaves as a lumped capacity, we have the energy balance

Rate of increase of internal energy of solid = heat transfer to solid by convection

+ heat transfer to solid by radiation.

or

$$\rho c\, V\, dT/d\tau = h_c A_c (T_c - T) + \sigma\, \epsilon\, A_r (T_r^4 - T^4) \qquad \textbf{[D-4]}$$

where we have used the relation for radiant exchange in a large enclosure described by Equation (8-43a) or (1-12). Thus, we are assuming that the area of the enclosure is very large compared to A_r. To be general, we allow h_c, T_c, and T_r to vary with time and/or the

Figure D-6

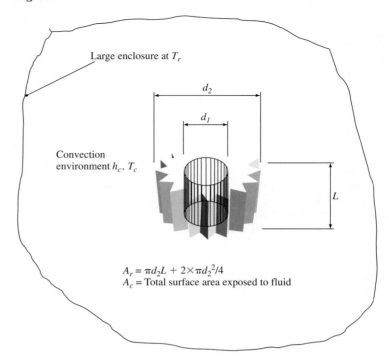

Large enclosure at T_r

Convection environment h_c, T_c

d_2

d_1

L

$A_r = \pi d_2 L + 2 \times \pi d_2{}^2/4$
$A_c = $ Total surface area exposed to fluid

temperature of the solid, and then write Equation (D-4) in finite difference form suitable for numerical solution. Designating the temperatures at the start of a time increment with subscript (p) and the temperatures at the end of time increment $\Delta \tau$ with subscript $(p+1)$, Equation (D-4) becomes

$$T_{p+1} = T_p + \left[h_c A_c (T_{c,p} - T_p) + \sigma \epsilon A_r (T_{r,p}^4 - T_p^4) \right] (\Delta \tau / \rho c V) \qquad \textbf{[D-5]}$$

During a numerical computation it may be desirable to observe the separate temperature effects of radiation and convection by writing Equation (D-5) in the form

$$T_{p+1} = T_p + DT_{\text{conv}} + DT_{\text{rad}} \qquad \textbf{[D-6]}$$

with the obvious designations

$$DT_{\text{conv}} = h_c A_c (T_{c,p} - T_p)(\Delta \tau / \rho c V) \qquad \textbf{[D-7]}$$

and

$$DT_{\text{rad}} = \sigma \epsilon A_r \left(T_{r,p}^4 - T_p^4 \right) (\Delta \tau / \rho c V) \qquad \textbf{[D-8]}$$

All temperatures must be in degrees absolute because of the radiation terms.

When free-convection problems in gases are involved, the convection coefficient frequently takes on a functional form of

$$h_c = B \times (T_c - T)^n \qquad \textbf{[D-9]}$$

where the constant B and exponent n are geometry and fluid dependent. See Table 7-9 for typical values of B and n for free convection in air. For forced-convection situations involving gases (transparent fluids) one would not expect a strong dependence of h_c on temperature of the solid.

When radiation heat transfer is included in the energy balance on the lumped solid, along with the possibility of variable convection coefficients, the time-temperature response may differ considerably from the simple exponential behavior described in Section 4-2. As a result, a description of such a system with a simple time constant of

$$\tau = c\rho V / h A$$

may be in considerable error.

| Cooling of Finned Aluminum Solid | **EXAMPLE D-6** |

A finned aluminum solid like that shown in Figure D-4 has the following physical specifications:

Density $= 2675$ kg/m^3
Specific heat $= 920$ J/kg $\cdot$ °C
Volume $= 3.34 \times 10^{-5}$ m^3
Surface effective emissivity $= 0.85$
Effective radiation area $= 0.01141$ m^2
Total surface area for convection $= 0.02298$ m^2

Figure Example D-6a | Formulas.

	A	B	C
1	**Time, s**	**T, K**	**DTRad**
2	0	533	
3	=A2+$E$2	=B2+C3+D3	=$E$13*(0.00000005668*$E$8*$E$9*($E$12^4-B2^4))
4	=A3+$E$2	=B3+C4+D4	=$E$13*(0.00000005668*$E$8*$E$9*($E$12^4-B3^4))
5	=A4+$E$2	=B4+C5+D5	=$E$13*(0.00000005668*$E$8*$E$9*($E$12^4-B4^4))
6	=A5+$E$2	=B5+C6+D6	=$E$13*(0.00000005668*$E$8*$E$9*($E$12^4-B5^4))
7	=A6+$E$2	=B6+C7+D7	=$E$13*(0.00000005668*$E$8*$E$9*($E$12^4-B6^4))
8	=A7+$E$2	=B7+C8+D8	=$E$13*(0.00000005668*$E$8*$E$9*($E$12^4-B7^4))
9	=A8+$E$2	=B8+C9+D9	=$E$13*(0.00000005668*$E$8*$E$9*($E$12^4-B8^4))

	D	E	F
1	**DTConv**		**Nomenclature**
2		60	Time increment, s
3	=$E$13*($E$14*(ABS($E$7-B2))^$E$15)*$E$10*($E$7-B2)	2675	Density, kg/m^3
4	=$E$13*($E$14*(ABS($E$7-B3))^$E$15)*$E$10*($E$7-B3)	920	Sp. Heat, J/kg-K
5	=$E$13*($E$14*(ABS($E$7-B4))^$E$15)*$E$10*($E$7-B4)	0.0000334	Volume, m^3
6	=$E$13*($E$14*(ABS($E$7-B5))^$E$15)*$E$10*($E$7-B5)	=$E$3*$E$4*$E$5	Ther. Cap. J/K
7	=$E$13*($E$14*(ABS($E$7-B6))^$E$15)*$E$10*($E$7-B6)	293	Conv. Temp., K
8	=$E$13*($E$14*(ABS($E$7-B7))^$E$15)*$E$10*($E$7-B7)	0.85	Emissivity
9	=$E$13*($E$14*(ABS($E$7-B8))^$E$15)*$E$10*($E$7-B8)	0.01141	Rad. Area, m^2
10	=$E$13*($E$14*(ABS($E$7-B9))^$E$15)*$E$10*($E$7-B9)	0.02298	Conv. Area, m^2
11	=$E$13*($E$14*(ABS($E$7-B10))^$E$15)*$E$10*($E$7-B10)	=$E$14*(ABS($E$7-B2))^$E$15	Conv. Coeff., W/m^2-K
12	=$E$13*($E$14*(ABS($E$7-B11))^$E$15)*$E$10*($E$7-B11)	293	Rad. Temp., K
13	=$E$13*($E$14*(ABS($E$7-B12))^$E$15)*$E$10*($E$7-B12)	=$E$2/$E$6	Time parameter
14	=$E$13*($E$14*(ABS($E$7-B13))^$E$15)*$E$10*($E$7-B13)	1.4	Conv. Const.
15	=$E$13*($E$14*(ABS($E$7-B14))^$E$15)*$E$10*($E$7-B14)	0.25	Conv. Exp.
16	=$E$13*($E$14*(ABS($E$7-B15))^$E$15)*$E$10*($E$7-B15)		

Figure Example D-6b | Radiation and convection cooling.

	A	B	C	D	E	F	G	H	I	J	K
1	Time, s	T, K	DTRad	DTConv		Nomenclature					
2	0	533.00			60	Time increment, s					
3	60	481.39	-29.43	-22.18	2675	Density, kg/m^3					
4	120	446.41	-18.59	-16.39	920	Sp. Heat, J/kg-K					
5	180	420.75	-12.98	-12.68	3.34E-05	Volume, m^3					
6	240	401.05	-9.62	-10.09	82.1974	Ther. Cap. J/K					
7	300	385.44	-7.42	-8.18	293	Conv. Temp., K					
8	360	372.81	-5.90	-6.73	0.85	Emissivity					
9	420	362.42	-4.79	-5.60	0.01141	Rad. Area, m^2					
10	480	353.75	-3.97	-4.71	0.02298	Conv. Area, m^2					
11	540	346.44	-3.33	-3.98	5.51037108	Conv. Coeff., W/m^2-K					
12	600	340.22	-2.82	-3.39	293	Rad. Temp., K					
13	660	334.90	-2.42	-2.91	0.729950096	Time parameter					
14	720	330.30	-2.09	-2.50	1.4	Conv. Const.					
15	780	326.32	-1.82	-2.16	0.25	Conv. Exp.					
16	840	322.85	-1.59	-1.88							
17	900	319.81	-1.40	-1.64							
18	960	317.13	-1.24	-1.43							
19	1020	314.78	-1.10	-1.26							
20	1080	312.69	-0.98	-1.10							
21	1140	310.84	-0.88	-0.97							
22	1200	309.19	-0.79	-0.86							
23	1260	307.71	-0.71	-0.76							
24	1320	306.40	-0.64	-0.68							
25	1380	305.22	-0.58	-0.60							
26	1440	304.16	-0.52	-0.54							
27	1500	303.20	-0.48	-0.48							
28	1560	302.34	-0.43	-0.43							
29	1620	301.56	-0.40	-0.38							
30	1680	300.85	-0.36	-0.34							
31	1740	300.22	-0.33	-0.31							
32	1800	299.64	-0.30	-0.28							
33	1860	299.11	-0.28	-0.25							
34	1920	298.63	-0.25	-0.23							
35	1980	298.19	-0.23	-0.20							
36	2040	297.79	-0.22	-0.18							
37	2100	297.43	-0.20	-0.17							
38	2160	297.09	-0.18	-0.15							
39	2220	296.79	-0.17	-0.14							
40	2280	296.51	-0.16	-0.12							
41	2340	296.25	-0.14	-0.11							
42	2400	296.01	-0.13	-0.10							
43	2460	295.80	-0.12	-0.09							
44	2520	295.60	-0.11	-0.08							
45	2580	295.41	-0.11	-0.08							

Radiation-convection cooling

The solid is initially uniform in temperature at $260°C = 533$ K when it is placed in quiescent room air at $20°C$ and allowed to cool. The free-convection heat-transfer coefficient may be assumed to vary according to

$$h = 1.4(T - T_\infty)^{0.25} \text{ W/m}^2 \cdot °C \qquad [a]$$

The effective radiation temperature of the surrounding room walls is also $20°C$.

Determine the temperature-time cooling curve for this situation and compare with what would be experienced if the solid were cooled only by convection with h = value at the start of cooling. What would the time constant be for this latter situation?

■ **Solution**

For this problem, Equations (D-5, D-6, D-7) are entered in the Excel spreadsheet as shown in Figure Example D-6(a). A time increment of 60 sec is selected and Equation (a) inserted to describe the convection coefficient. The formulas are copied for 50 rows or so to produce near steady-state conditions. The value of h at the start of cooling is

$$h_{\text{initial}} = 1.4(533 - 293)^{0.25} = 5.51 \text{ W/m}^2 \cdot °C$$

This value is calculated and indicated in cell E11.

Figure Example D-6c | Convection cooling only.

	A Time, s	B T, K	C DTRad	D DTConv	E	F Nomenclature	G	H	I	J	K
2	0	533.00									
3	60	510.82	0.00	-22.18	60	Time increment, s					
4	120	490.69	0.00	-20.13	2675	Density, kg/m^3					
5	180	472.41	0.00	-18.27	920	Sp. Heat, J/kg-K					
6	240	455.83	0.00	-16.58	3.34E-05	Volume, m^3					
7	300	440.78	0.00	-15.05	82.1974	Ther. Cap. J/K					
8	360	427.12	0.00	-13.66	293	Conv. Temp., K					
9	420	414.73	0.00	-12.40	0	Emissivity					
10	480	403.48	0.00	-11.25	0.01141	Rad. Area, m^2					
11	540	393.26	0.00	-10.21	0.02298	Conv. Area, m^2					
12	600	384.00	0.00	-9.27	5.51	Conv. Coeff., W/m^2-K					
13	660	375.59	0.00	-8.41	293	Rad. Temp., K					
14	720	367.95	0.00	-7.63	0.729950096	Time parameter					
15	780	361.03	0.00	-6.93	5.51	Conv. Const.					
16	840	354.74	0.00	-6.29	0	Conv. Exp.					
17	900	349.03	0.00	-5.71							
18	960	343.85	0.00	-5.18							
19	1020	339.15	0.00	-4.70							
20	1080	334.89	0.00	-4.27							
21	1140	331.02	0.00	-3.87							
22	1200	327.50	0.00	-3.51							
23	1260	324.31	0.00	-3.19							
24	1320	321.42	0.00	-2.89							
25	1380	318.79	0.00	-2.63							
26	1440	316.41	0.00	-2.38							
27	1500	314.25	0.00	-2.16							
28	1560	312.28	0.00	-1.96							
29	1620	310.50	0.00	-1.78							
30	1680	308.88	0.00	-1.62							
31	1740	307.41	0.00	-1.47							
32	1800	306.08	0.00	-1.33							
33	1860	304.87	0.00	-1.21							
34	1920	303.78	0.00	-1.10							
35	1980	302.78	0.00	-1.00							
36	2040	301.88	0.00	-0.90							
37	2100	301.06	0.00	-0.82							
38	2160	300.31	0.00	-0.74							
39	2220	299.64	0.00	-0.68							
40	2280	299.02	0.00	-0.61							
41	2340	298.47	0.00	-0.56							
42	2400	297.96	0.00	-0.51							
43	2460	297.50	0.00	-0.46							
44	2520	297.09	0.00	-0.42							
45	2580	296.71	0.00	-0.38							

Convection cooling, h = const

The resulting cooling curve is plotted in Figure Example D-6(b) along with a few rows of the calculations. Note that the temperature decrements due to radiation and convection are about the same at the start of cooling; thus, to neglect radiation would produce a serious error.

To obtain the cooling curve for $h = 5.51 =$ constant, this value is entered in cell E14 along with an exponent value of zero in cell E15. The resulting cooling curve is shown in Figure Example D-6(c), indicating an obviously longer time for the cooling process than for combined radiation and convection cooling. The time constant is calculated according to Equation (4-5) as

$$\tau = \rho c V/hA = (2675)(920)(0.0000334)/(5.51)(0.02298) = 649 \text{ sec}$$

This is the time to achieve 63.2 percent of the overall temperature drop or a time to reach

$$T = 533 - (0.632)(533 - 293) = 381 \text{ K}$$

For the combined convection-radiation cooling the time to achieve this same temperature is only about 320 sec, a much shorter period. A specification of time constant in terms of the 63.2 percent drop in temperature is obviously an inadequate means for describing the combined convection-cooling problem. A comparison of the two cooling curves is shown in Figure Example D-6(d).

Figure Example D-6d | Comparison of results.

	A	B	C	D	E	F	G	H	I	J	K	L
1												
2	0	533.00	533									
3	60	481.39	510.8177									
4	120	446.41	490.6857									
5	180	420.75	472.4144									
6	240	401.05	455.8318									
7	300	385.44	440.7819									
8	360	372.81	427.123									
9	420	362.42	414.7265									
10	480	353.75	403.4758									
11	540	346.44	393.2649									
12	600	340.22	383.9978									
13	660	334.90	375.5873									
14	720	330.30	367.954									
15	780	326.32	361.0263									
16	840	322.85	354.7389									
17	900	319.81	349.0326									
18	960	317.13	343.8537									
19	1020	314.78	339.1535									
20	1080	312.69	334.8877									
21	1140	310.84	331.0162									
22	1200	309.19	327.5025									
23	1260	307.71	324.3136									
24	1320	306.40	321.4194									
25	1380	305.22	318.7927									
26	1440	304.16	316.4088									
27	1500	303.20	314.2452									
28	1560	302.34	312.2816									
29	1620	301.56	310.4995									
30	1680	300.85	308.8821									
31	1740	300.22	307.4141									
32	1800	299.64	306.0819									

Transient Heating of Electronic Box in an Enclosure

EXAMPLE D-7

A small metal box containing tightly packed electronic equipment has approximate dimensions of a 10-cm cube with mass equal to 0.5 kg and specific heat of 300 J/kg · °C. The surface area of the box is 0.06 m³ and heat is generated in the box at the rate of 200 W. The electronics box is enclosed by an aluminum box having dimensions about 25 cm on a side with a mass of 1.05 kg and specific heat of 890 J/kg · °C. The surface area of the enclosure is 0.375 m² and both the electronics equipment box and aluminum enclosure have surface emissivities of $\epsilon = 0.9$. The space between the two boxes is filled with air at atmospheric pressure with a mass of 0.0176 kg and $c = 718$ J/kg · °C. The system is initially at a uniform temperature of 20°C = 293 K when the electric power input of 200 W is applied. The convection heat-transfer coefficient between all surfaces and the adjoining air is $h = 10$ W/m² · °C, and the entire assembly is enclosed in a large room at 20°C. We will use a numerical technique to determine the time-temperature behavior of the electronics box and its enclosure.

■ **Solution**

The nomenclature for the system is indicated in Figure Example D-7(a). Since $A_3 \gg A_1$, the radiation heat transfer may be calculated with Equation (1-12) as

$$q \text{ rad}_{1\text{-}2} = \sigma \epsilon_1 A_1 \left(T_1^4 - T_3^4 \right)$$

$$q \text{ rad}_{3\text{-}s} = \sigma \epsilon_3 A_3 \left(T_3^4 - T_s^4 \right)$$

Letting

$$C_1 = m_1 c_1; \quad C_2 = C_{\text{air}} = m_{\text{air}} c_{\text{air}}; \quad C_3 = m_3 c_3,$$

Figure Example D-7a | Nomenclature.

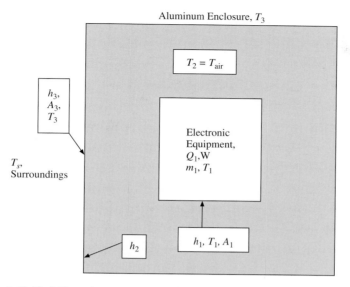

Figure Example D-7b | Formulas.

	C	D
1	T1=	T2=
2		
3	293	293
4	=$H$2*($H$3*0.06*(D3-C3)+0.000000051*0.06*(E3^4-C3^4)+$H$6)/150+C3	=$H$2*($H$3*0.06*(C3-D3)+$H$4*0.375*(E3-D3))/12.58+D3
5	=$H$2*($H$3*0.06*(D4-C4)+0.000000051*0.06*(E4^4-C4^4)+$H$6)/150+C4	=$H$2*($H$3*0.06*(C4-D4)+$H$4*0.375*(E4-D4))/12.58+D4
6	=$H$2*($H$3*0.06*(D5-C5)+0.000000051*0.06*(E5^4-C5^4)+$H$6)/150+C5	=$H$2*($H$3*0.06*(C5-D5)+$H$4*0.375*(E5-D5))/12.58+D5
7	=$H$2*($H$3*0.06*(D6-C6)+0.000000051*0.06*(E6^4-C6^4)+$H$6)/150+C6	=$H$2*($H$3*0.06*(C6-D6)+$H$4*0.375*(E6-D6))/12.58+D6

	E	F	G	H
1	T3=			
2				
3	293		Dt=	5
4	=$H$2*($H$4*0.375*(D3-E3)+$H$5*0.375*(293-E3)+0.000000051*0.06*(C3^4-E3^4)+0.000000051*0.375*(293^4-E3^4))/934+E3		h1=	10
5	=$H$2*($H$4*0.375*(D4-E4)+$H$5*0.375*(293-E4)+0.000000051*0.06*(C4^4-E4^4)+0.000000051*0.375*(293^4-E4^4))/934+E4		h2=	10
6	=$H$2*($H$4*0.375*(D5-E5)+$H$5*0.375*(293-E5)+0.000000051*0.06*(C5^4-E5^4)+0.000000051*0.375*(293^4-E5^4))/934+E5		h3=	10
			Q1=	100

we have

$$C_1 = 150 \text{ J/}^\circ\text{C}; \quad C_2 = 12.63 \text{ J/}^\circ\text{C}; \quad C_3 = 934 \text{ J/}^\circ\text{C}$$

We may now make an energy balance on each of the three nodes, setting the energy rate into each node equal to the time rate of change of internal energy of that node. For node 1, the electronics box, there is also an energy input from the electric power applied, which we designate as Q_1. Thus,

$$C_1 dT_1/d\tau = Q_1 + h_1 A_1 (T_2 - T_1) + \sigma \epsilon_1 A_1 \left(T_3^4 - T_1^4 \right) \qquad \text{[a]}$$

$$C_2 dT_2/d\tau = h_1 A_1 (T_1 - T_2) + h_2 A_3 (T_3 - T_2) \qquad \text{[b]}$$

$$C_3 dT_3/d\tau = h_2 A_3 (T_2 - T_3) + \sigma \epsilon_1 A_1 \left(T_1^4 - T_3^4 \right) + h_3 A_3 (T_s - T_3) + \sigma \epsilon_3 A_3 \left(T_s^4 - T_3^4 \right) \quad \text{[c]}$$

where all temperatures must be in kelvins because of the radiation terms. T_s is designated as the surrounding temperature of the large room. From the data of the problem we have

$$A_1 = 0.06; \quad A_3 = 0.375; \quad h_1 = h_2 = h_3 = 10$$

Figure Example D-7c

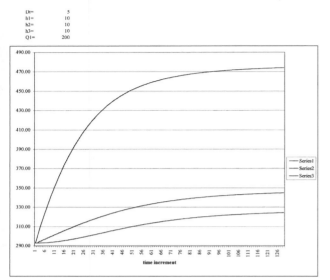

Figure Example D-7d

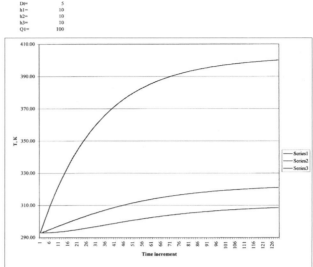

$$\epsilon_1 = \epsilon_3 = 0.9; \quad T_s = 20°C = 293 \text{ K}$$

Equations (a,b,c) may be written in finite difference form by expressing the temperature gradients as

$$d\,T/d\tau = (T^{p+1} - T^p)/\Delta\tau$$

and all the temperatures on the right side of the equation as T^p where, as usual, T^p represents the temperatures at the start of the time increment $\Delta\tau$ and T^{p+1} represents the temperature at the end of the time increment. The resulting formulas have been inserted into an Excel worksheet and are shown in the accompanying Figure Example D-7(b). Dt is used to designate the time increment $\Delta\tau$. The equations for a large number of time increments are then copied to perform the transient calculation. The results for a time increment of 5 sec are shown in Figure Example D-7(c) for the given electric input of 200 W. For comparison, the results for a heat input of 100 W are shown in Figure Example D-7(d), indicating somewhat lower temperatures for each of the three nodes.

Near steady state is reached after about 120 time increments or 10 minutes with the electronic box temperature reaching about 473 K (200°C) for an input of 200 W and only about 400 K (127°C) for the 100 W input. The corresponding steady-state temperatures for the aluminum enclosure are about 52°C and 35°C.

| **EXAMPLE D-8** | Symmetric Formulations |

In some problems a geometric configuration repeats so that advantage can be taken of symmetry in the formulation of nodal equations, thereby reducing the number of nodes needed for solution of the problem. Example 3-8 depicts such a situation. In the present example, we consider a material contained between two walls with protruding ribs as shown in Figure Example D-8(a). The section between the dashed lines repeats, or is a mirror image of itself and may be employed to analyze the total wall problem. The system is set up in the form of 15 by 15 nodes to represent the repeating section. The block A1:O15 is used for the Excel formulation, portions of which are shown in

Figure Example D-8a,b | Schematic and formulas.

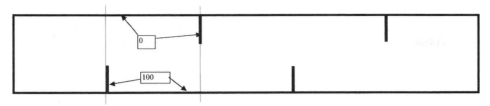

	A	B	C	D
1	0	0	0	0
2	=(A1+2*B3+A3)/4	=(A2+B1+B3+C2)/4	=(B2+C1+C3+D2)/4	=(C2+D1+D3+E2)/4
3	=(A2+2*B4+A4)/4	=(A3+B2+B4+C3)/4	=(B3+C2+C4+D3)/4	=(C3+D2+D4+E3)/4
4	=(A3+2*B5+A5)/4	=(A4+B3+B5+C4)/4	=(B4+C3+C5+D4)/4	=(C4+D3+D5+E4)/4
5	=(A4+2*B6+A6)/4	=(A5+B4+B6+C5)/4	=(B5+C4+C6+D5)/4	=(C5+D4+D6+E5)/4
6	=(A5+2*B7+A7)/4	=(A6+B5+B7+C6)/4	=(B6+C5+C7+D6)/4	=(C6+D5+D7+E6)/4
7	=(A6+2*B8+A8)/4	=(A7+B6+B8+C7)/4	=(B7+C6+C8+D7)/4	=(C7+D6+D8+E7)/4
8	=(A7+2*B9+A9)/4	=(A8+B7+B9+C8)/4	=(B8+C7+C9+D8)/4	=(C8+D7+D9+E8)/4
9	100	=(A9+B8+B10+C9)/4	=(B9+C8+C10+D9)/4	=(C9+D8+D10+E9)/4
10	100	=(A10+B9+B11+C10)/4	=(B10+C9+C11+D10)/4	=(C10+D9+D11+E10)/4
11	100	=(A11+B10+B12+C11)/4	=(B11+C10+C12+D11)/4	=(C11+D10+D12+E11)/4
12	100	=(A12+B11+B13+C12)/4	=(B12+C11+C13+D12)/4	=(C12+D11+D13+E12)/4
13	100	=(A13+B12+B14+C13)/4	=(B13+C12+C14+D13)/4	=(C13+D12+D14+E13)/4
14	100	=(A14+B13+B15+C14)/4	=(B14+C13+C15+D14)/4	=(C14+D13+D15+E14)/4
15	100	100	100	100

	L	M	N	O
1	0	0	0	0
2	=(K2+L1+L3+M2)/4	=(L2+M1+M3+N2)/4	=(M2+N1+N3+O2)/4	0
3	=(K3+L2+L4+M3)/4	=(L3+M2+M4+N3)/4	=(M3+N2+N4+O3)/4	0
4	=(K4+L3+L5+M4)/4	=(L4+M3+M5+N4)/4	=(M4+N3+N5+O4)/4	0
5	=(K5+L4+L6+M5)/4	=(L5+M4+M6+N5)/4	=(M5+N4+N6+O5)/4	0
6	=(K6+L5+L7+M6)/4	=(L6+M5+M7+N6)/4	=(M6+N5+N7+O6)/4	0
7	=(K7+L6+L8+M7)/4	=(L7+M6+M8+N7)/4	=(M7+N6+N8+O7)/4	0
8	=(K8+L7+L9+M8)/4	=(L8+M7+M9+N8)/4	=(M8+N7+N9+O8)/4	=(2*N8+O7+O9)/4
9	=(K9+L8+L10+M9)/4	=(L9+M8+M10+N9)/4	=(M9+N8+N10+O9)/4	=(2*N9+O8+O10)/4
10	=(K10+L9+L11+M10)/4	=(L10+M9+M11+N10)/4	=(M10+N9+N11+O10)/4	=(2*N10+O9+O11)/4
11	=(K11+L10+L12+M11)/4	=(L11+M10+M12+N11)/4	=(M11+N10+N12+O11)/4	=(2*N11+O10+O12)/4
12	=(K12+L11+L13+M12)/4	=(L12+M11+M13+N12)/4	=(M12+N11+N13+O12)/4	=(2*N12+O11+O13)/4
13	=(K13+L12+L14+M13)/4	=(L13+M12+M14+N13)/4	=(M13+N12+N14+O13)/4	=(2*N13+O12+O14)/4
14	=(K14+L13+L15+M14)/4	=(L14+M13+M15+N14)/4	=(M14+N13+N15+O14)/4	=(2*N14+O13+O15)/4
15	100	100	100	100

Figure Example D-8(b). The 100° rib protrudes up from the 100° surface with fixed values entered for nodes A9:A15. Similarly, the 0° rib protrudes down from the 0° surface so constant values of zero are entered in cells O1:O7.

All of the nodal formulas are of the interior node variety, but the ones in cells A2:A8 and O8:O14 involve mirror-image temperatures such that the temperature to the left of node A2 would be the same as that at B3. Similarly, the temperature to the right of node O8 would be the same as N8. Thus, 2*B3 and 2*N8, are entered in the formulas and so on. Once one characteristic formula is entered, the formulas for similar cells can be drag-copied. The other interior node formulas can be drag copied in the usual fashion, and some of the entries are shown in Figure Example D-8(b).

When the iterative solution is effected, the results appear as shown in Figure Example D-8(c) along with a 3-D wire-mesh chart and plan and elevation views of the temperature profiles. Grayscale depictions of the solutions are shown in Figure Example D-8(d).

For comparison, the temperature profiles for a wall and single protruding rib at 100°C surrounded by three walls at 0°C is shown in Figure Example D-8(e).

Figure Example D-8c

	A	B	C	D	E	F	G	H	I	J	K	L	M	N	O
1	0	0	0	0	0	0	0	0	0	0	0	0	0	0	0
2	29.59636	20.39049	15.7993	13.09533	11.28135	9.935189	8.848008	7.898257	7.002796	6.096361	5.123574	4.038889	2.813939	1.45158	0
3	46.05282	36.16582	29.71092	25.30025	22.09445	19.61097	17.55817	15.74183	14.0162	12.25875	10.35878	8.217839	5.765153	2.99233	0
4	57.59702	48.50836	41.57767	36.29961	32.18458	28.8554	26.03123	23.49407	21.06085	18.56316	15.83452	12.70821	9.036301	4.752507	0
5	66.94972	58.69217	51.79098	46.13514	41.48799	37.59399	34.21643	31.14158	28.16923	25.09783	21.70739	17.74376	12.91906	6.981295	0
6	75.16313	67.5188	60.75807	54.96104	50.03727	45.81515	42.09793	38.68563	35.37577	31.95076	28.15276	23.63984	17.91453	10.25348	0
7	82.77968	75.46109	68.76059	62.91272	57.88389	53.53036	49.67344	46.1262	42.69647	39.1758	35.31229	30.74771	24.84531	16.11794	0
8	90.38572	82.7846	75.90966	70.04441	65.05419	60.74787	56.93819	53.44819	50.10707	46.74273	43.17206	39.19269	34.60056	29.37269	24.76861
9	100	89.38141	82.0483	76.30019	71.5396	67.46769	63.8822	60.62022	57.53986	54.51503	51.43967	48.24968	44.99089	42.00312	40.32875
10	100	92.69232	86.60129	81.56767	77.33545	73.70014	70.5017	67.60964	64.91615	62.33695	59.82104	57.37469	55.10949	53.31951	52.53983
11	100	94.78621	90.09637	86.03307	82.5336	79.49489	76.81395	74.3996	72.17728	70.09476	68.13205	66.31783	64.75218	63.62495	63.19123
12	100	96.35587	92.96447	89.9341	87.27038	84.93118	82.8589	80.99679	79.29789	77.73207	76.2939	75.01176	73.95585	73.23631	72.97493
13	100	97.67259	95.47126	93.46807	91.6822	90.10004	88.69315	87.43024	86.28486	85.24121	84.29921	83.47899	82.82268	82.38907	82.23568
14	100	98.86309	97.77972	96.78447	95.89002	95.09331	94.3831	93.74581	93.16979	92.64839	92.18243	91.78201	91.46653	91.26132	91.18958
15	100	100	100	100	100	100	100	100	100	100	100	100	100	100	100

Figure Example D-8d

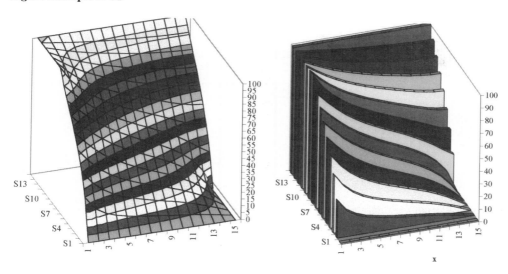

Figure Example D-8e | Single rib.

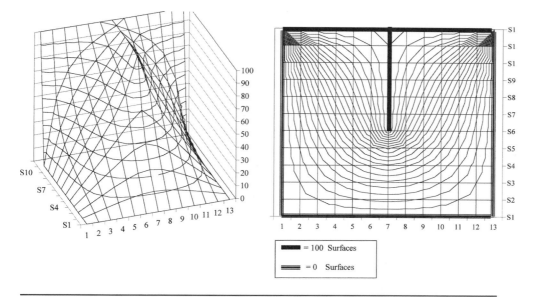

= 100 Surfaces

= 0 Surfaces

Solid with Composite Materials   EXAMPLE D-9

As a final example of use of the Excel templates, consider the composite solid shown in Figure Example D-9(a). The exterior material is 10 by 24 cm with an interior material insert 4 by 10 cm centered vertically and horizontally. We set up the problem with the top surface specified as $T = 0$, while the indicated cells F12, G12, and H12 may be used to provide entries for the temperatures of the other 3 surfaces. K is specified in the template with cell B51.

There are no convection boundaries so we have only interior nodes, which may be written using cell C6 and drag-copying, and the interface nodes between the two materials. The cell locations for the worksheet are indicated in Figure Example D-9(a). The setup provides for any number of alternatives through assignment of values to the boundary temperature. We choose the set:

Figure Example D-9a | Nomenclature, composite example.

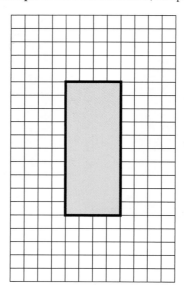

Composite Example. Cells E28:Q48 for exterior material
Cells I33:M43 for interior material; centered
Left surface temperature = F12
Right surface temperature = G12
Bottom surface temperature = H12
Top surface temperature = 0
$k_{\text{exterior}}/k_{\text{interior}} = K$ = cell B51

Left surface temperature $= 100$

Right surface temperature $= 300$

Bottom surface temperature $= 0$

Top surface temperature $= 0$.

The solutions for three extremes in thermal conductivity ratios are presented:

1. $K = 1.0$. Both materials have the same properties.
2. $K = 0.0001$. Interior material good conductor, exterior material good insulator. Sample materials: copper and fiberglass.
3. $K = 10000$. Interior material good insulator, exterior material good conductor.

There are too many nodes to display all of the numerical solutions so we have given 3-D wire-mesh charts of the temperature distributions to illustrate the effects. Case 1 is shown in Figure Example D-9(b) and a smooth variation of the temperature surface is exhibited. This solution is independent of the numerical value of the thermal conductivity of the material.

The solution for case 2 is given in Figure Example D-9(c) and indicates what one would expect; strong temperature gradients through the exterior insulating material, and very small temperature gradients in the interior material. Note the "flat top" of the high conductivity interior material that results from the small temperature gradients.

The solution for case 3 is displayed in Figure Example D-9(d) and illustrates just the opposite of case 2; strong temperature gradients for the interior material and relatively shallow gradients for the exterior material.

Plan and elevation views have been included for all three cases to further illustrate the temperature-gradient effects.

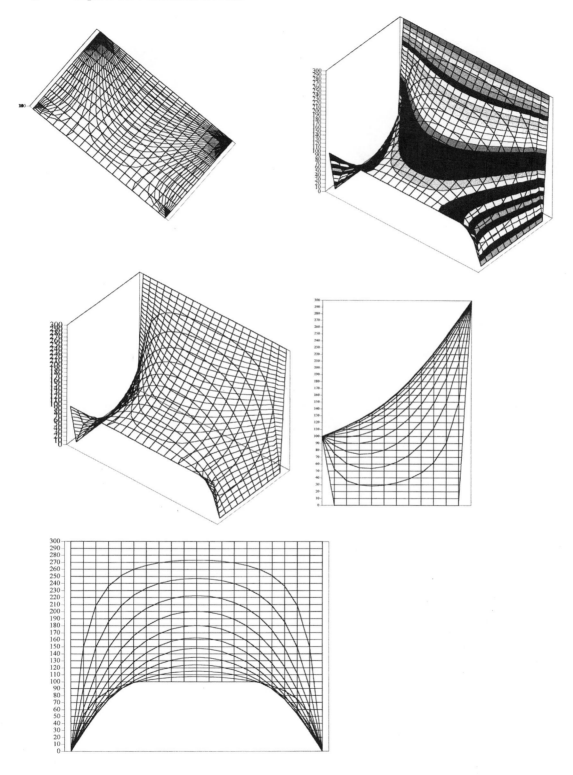

Figure Example D-9c | Results for $K = 0.0001$.

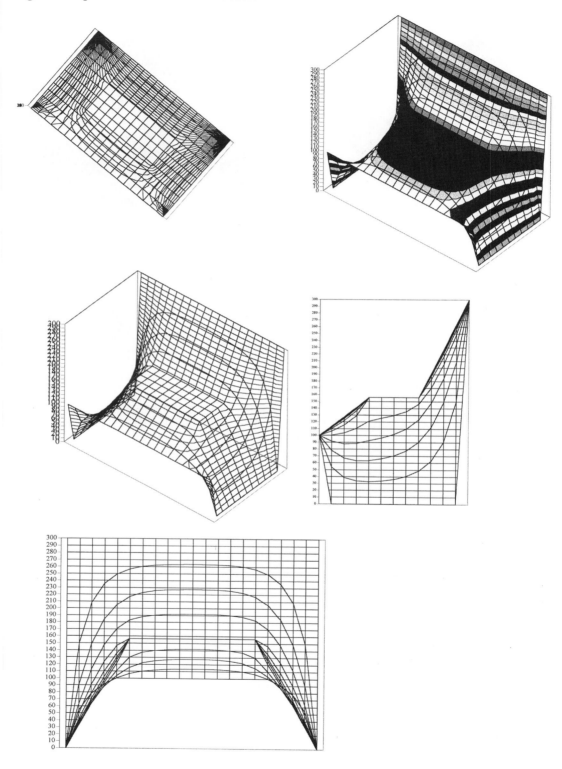

Figure Example D-9d | Results for $K = 10,000$.

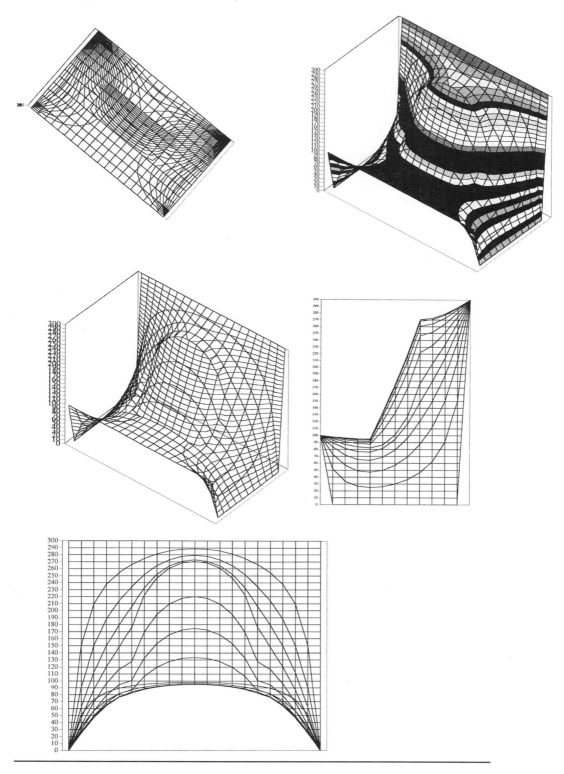

LIST OF WORKED EXAMPLES

D-1 Temperature distribution in two-dimensional plate
D-2 Excel solution and display of temperature distribution in two-dimensional straight fin
D-3 Excel solution of Example 3-5 with and without radiation boundary condition
D-4 Plate with boundary heat source and convection
D-5 Transient analysis of Example 3-5 carried to steady state
D-6 Cooling of finned aluminum solid
D-7 Transient heating of electronic box in an enclosure
D-8 Symmetric formulations
D-9 Solid with composite materials

REFERENCES

1. Orvis, W. J. *Excel for Scientists and Engineers*, 2nd ed. San Francisco: SYBEX, 1996.

2. Holman, J. P. *What Every Engineer Should Know About EXCEL*. Boca Raton, FL: CRC Press, 2006.

3. Gottfried, B. *Spreadsheet Tools for Engineers-Excel 97 Version*. New York: McGraw-Hill, 1998.

4. Liengme, B. V. *A Guide to Microsoft Excel for Scientists and Engineers*. New York: JohnWiley, 1997.

5. Chapra, S. C., and R. P. Canale. *Numerical Methods for Engineers*, 3rd ed. New York: McGraw-Hill, 1998.

INDEX